KU-797-980

LADY MARGARET HALL LIBRARY
OXFORD

302210206G

Soil Mechanics in Engineering Practice

Soil Mechanics in Engineering Practice

Karl Terzaghi

Late Professor of the Practice of Civil Engineering
Harvard University
Lecturer and Research Consultant in Civil Engineering
University of Illinois

Ralph B. Peck

Professor of Foundation Engineering
University of Illinois

Second Edition

JOHN WILEY & SONS, New York • Chichester • Brisbane • Toronto

ISBN 0 471 85273 2
Library of Congress Catalog Card Number: 67-17356
Printed in the United States of America

30 29 28 27 26 25 24 23 22

PREFACE TO SECOND EDITION

Much of the work of revising this book had been accomplished before Karl Terzaghi's death on October 25, 1963. The scope of the changes had been agreed upon in considerable detail, and Dr. Terzaghi himself had prepared the drafts of the sections for which he took primary responsibility. The first drafts of most of the other major revisions had also undergone his scrutiny. The writer is grateful that these pages reflect as much of Dr. Terzaghi's contribution as they do, but he regrets and must accept the responsibility for the inevitable shortcomings that would not have escaped Dr. Terzaghi's attention in his final perusal and criticism of the manuscript.

In particular, Dr. Terzaghi prepared substantial revisions of the discussions of stability of natural slopes, and added extensively to the articles on dams and their foundations. Since dams occupied an increasingly larger proportion of his activities during his later years, these additions may be regarded as the essence of his experience and thought on the subject.

The text has been supplemented by references and lists of selected reading that may serve as a guide to the literature. A new chapter on performance observations has been added to aid the engineer in the use of the observational method which is at the very heart of successful application of soil mechanics.

The enormous growth of the literature on soil mechanics in the last two decades has vastly increased the problem of selecting the information to be included. In making the selection, the authors have been guided by the title of the book.

Mrs. Josephine B. Hegenbart has gone far beyond the usual duties of typing in the preparation of the manuscript. Her unending assistance is greatly appreciated.

Ralph B. Peck

Urbana, Illinois
January, 1967

PREFACE TO FIRST EDITION

Soil mechanics originated several decades ago under the pressure of necessity. As the practical problems involving soils broadened in scope, the inadequacy of the scientific tools available for coping with them became increasingly apparent. Efforts to remedy the situation started almost simultaneously in the United States and in Europe, and within a short period they produced an impressive array of useful information.

The initial successes in this field of applied science were so encouraging that a new branch of structural analysis appeared to be in the making. As a consequence, the extent and profundity of the theoretical investigations increased rapidly, and experimental methods were developed to a high degree of refinement. Without the results of these painstaking investigations a rational approach to the problems of earthwork engineering could not have been attempted.

Unfortunately, the research activities in soil mechanics had one undesirable psychological effect. They diverted the attention of many investigators and teachers from the manifold limitations imposed by nature on the application of mathematics to problems in earthwork engineering. As a consequence, more and more emphasis has been placed on refinements in sampling and testing and on those very few problems that can be solved with accuracy. Yet, accurate solutions can be obtained only if the soil strata are practically homogenous and continuous in horizontal directions. Furthermore, since the investigations leading to accurate solutions involve highly specialized methods of sampling and testing, they are justified only in exceptional cases. On the overwhelming majority of jobs no more than an approximate forecast is needed, and if such a forecast cannot be made by simple means it cannot be made at all. If it is not possible to make an approximate forecast, the behavior of the soil must be observed during construction, and the design may subsequently have to be modified in accordance with the findings. These facts cannot be ignored without defying the purpose of soil mechanics. They govern the treatment of the subject in this book.

Part A deals with the physical properties of soils and Part B with the theories of soil mechanics. These two parts are very short, but they contain all that engineering students and the average engineer need to know about soil mechanics proper at the present time. The heart of the book is Part C.

Part C deals with the art of getting satisfactory results in earthwork and foundation engineering at a reasonable cost, in spite of the complexity of the structure of natural soil strata and in spite of the inevitable gaps in our knowledge of the soil conditions. To achieve this goal the engineer must take advantage of all the methods and resources at his disposal—experience, theory, and soil testing included. Yet all these resources are of no avail unless they are used with careful discrimination, because almost every practical problem in this field contains at least some features without precedent.

Every discussion of practical problems in Part C starts with a critical survey of conventional methods and proceeds step by step to whatever improvements have been realized with the assistance of the results of research in soil mechanics. Therefore, the experienced engineer is advised to start reading the book at the beginning of this part. He should use Parts A and B only for reference, to get information about concepts with which he is not yet familiar. Otherwise he would be obliged to digest a considerable amount of material before he would be in a position to realize its function in his field of interest.

The details of the methods for coping with the practical problems covered by Part C may change as experience increases, and some of them may become obsolete in a few years because they are no more than temporary expedients. Yet the merits of the semiempirical approach advocated in Part C are believed to be independent of time. At the end of each article of Part C the reader will find a list of references. In their choice priority was given to those publications that are likely to foster the urge and capacity for careful and intelligent field observations. In connection with these references it should be emphasized that some of the discussions and closures may contain more important information than the articles themselves.

Since the field of soil engineering is too broad to be covered adequately in a single volume, various important topics such as highway, airport, and tunnel engineering had to be excluded. Brief references concerning these fields have been assembled in an appendix.

In its early stages, the manuscript was critically studied by Professor C. P. Siess, whose comments were especially helpful. The authors also appreciate the suggestions of the several practicing engineers who read various portions of the text. In particular, they are indebted to Mr. A. E. Cummings, Mr. O. K. Peck, and Mr. F. E. Schmidt for criticisms of Part C, to Dr. R. E. Grim for review of Article 4, and to Dr. Ruth D. Terzaghi for assistance in the preparation of Article 63.

Tables and figures taken in whole or in part from other sources are acknowledged where they occur in the text. The drawings are the work of Professor Elmer F. Heater. For his co-operative interest and skilful work the authors are indeed grateful.

KARL TERZAGHI *and* RALPH B. PECK

CONTENTS

Part II. Theoretical Soil Mechanics

Part III. Problems of Design and Construction

SYMBOLS

The symbols used in this text conform generally to those in the first edition.

In the United States at the present time, it is customary to express the results of laboratory tests in terms of metric units, whereas the English system is used in the field and the design office. In conformity with this practice, the soil constants and test results contained in Part I are given in terms of metric units. In Parts II and III, which deal with theories and practical applications, only the English system is used. Fortunately, the values of quantities which enter most frequently into the computations involved in soil mechanics can be converted without mental effort from one system to the other by means of the closely approximate relation

$$1\ \text{kg/cm}^2 \approx 1\ \text{ton/ft}^2 \approx 1\ \text{atmosphere} = 34\ \text{ft of water} = 15\ \text{lb/in.}^2$$

In this relation, the ton is the short ton of 2000 lb. Other conversion factors which may be required are

$$\begin{aligned} 1\ \text{lb} &= 454\ \text{gm} \\ 1\ \text{kg} &= 2.2\ \text{lb} \\ 1\ \text{ft} &= 30.5\ \text{cm} \end{aligned}$$

In the following list, the dimensions of the quantities are given in the metric (cm-gm-sec) system. If the numerical value of a quantity is given in this system, it can readily be expressed in the English (ft-lb-sec) system by means of the conversion factors just given. For example, we may express the value,

$$E = 120{,}000\ \text{gm/cm}^2$$

in the English system by introducing into the preceding equation,

$$1\ \text{gm} = \frac{1}{454}\ \text{lb} \quad \text{and} \quad 1\ \text{cm} = \frac{1}{30.5}\ \text{ft}$$

whence

$$\begin{aligned} E &= 120{,}000\ \frac{1}{454}\ \text{lb} \times \frac{30.5^2}{\text{ft}^2} = 120{,}000\ (2.05\ \text{lb/ft}^2) \\ &= 245{,}000\ \text{lb/ft}^2 \end{aligned}$$

If no dimension is added to a symbol, the symbol indicates a pure number.

A (cm^2) = area
$\bar{\mathbf{A}}$ = pore-pressure coefficient = $u_d/\Delta p$
A_p (cm^2) = base area of pile or pier
A_r = area ratio of sampling spoon
a_v (cm^2/gm) = coefficient of compressibility
B (cm) = width
$\mathbf{B}$ = pore-pressure coefficient = u_a/p_3
C (any dimension) = constant
C (gm) = resultant cohesion
C_a (gm) = total adhesion
C_c = compression index for soil in field; creep ratio
C_c' = compression index of remolded soil
C_s = swelling index
C_w = weighted creep ratio (failure by piping)
c (gm/cm^2) = cohesion
c (in.) = constant in Engineering News formula
c_1 (gm/cm^2) = cohesion intercept for overconsolidated clay
c_a (gm/cm^2) = adhesion between soil and pile, pier, wall, or sheeting
c_v (cm^2/sec) = coefficient of consolidation
D (cm) = grain size; depth; diameter; spacing between centers of piles
D_{10} (cm) = effective grain size
D_f (cm) = depth of foundation
D_r = relative density of cohesionless soil
d (cm) = diameter of pile; distance
E (gm/cm^2) = modulus of elasticity
E (volt) = difference in electric potential
E = efficiency of cutoff
E (gm/cm) = normal force on side of slice (stability analysis)
E_i (gm/cm^2) = initial tangent modulus
e = void ratio
e (coulomb/cm^2) = electric charge per unit of area
e_0 = void ratio in loosest state; void ratio under effective overburden pressure p_0
e_{min} = void ratio in densest state
e_w = volume of water per unit volume of solid matter (for saturated soil $e_w = e$)
e_c = critical void ratio
F (gm) = reaction; resultant force
F = factor of safety
f_s (gm/cm^2) = sum of friction and adhesion between soil and pile or pier
f = coefficient of friction between soil and base of structure
f_0 (1/sec) = natural frequency (vibrations)
f_1 (1/sec) = frequency of impulse (vibrations)
G_a = air-space ratio (drainage)
H (cm) = thickness of stratum except when used in connection with consolidating layer. In this event, H = thickness of half-closed layer or half-thickness of open layer
H (cm) = height of fall of hammer (pile driving)
H_c (cm) = critical height of slope
ΔH (cm) = position head (hydraulics)
h (cm) = hydraulic head
h_w (cm) = piezometric head
Δh (cm) = potential drop (hydraulics)

h_c (cm) = height of capillary rise; critical head for failure by piping
h_{cc} (cm) = height of complete saturation of drained soil
h_{cr} (cm) = critical head for failure by piping according to computation based on line of creep method
h_r = relative vapor pressure
h_{ra} = relative humidity
I_l = liquidity index
I_w = plasticity index
i = hydraulic gradient
i_c = critical hydraulic gradient
i_e (volts/cm) = potential gradient, electro-osmosis
i_p (gm/cm^3) = pressure gradient
K = ratio between intensities of horizontal and vertical pressures at a given point in a mass of soil
K_0 = coefficient of earth pressure at rest (value of K for initial state of elastic equilibrium)
K_A = coefficient of active earth pressure
K_P = coefficient of passive earth pressure
K (cm^2) = permeability
K_s (gm/cm^3) = coefficient of subgrade reaction
k (cm/sec) = coefficient of permeability
k_I (cm/sec) = coefficient of permeability in direction parallel to bedding planes
k_{II} (cm/sec) = coefficient of permeability in direction perpendicular to bedding planes
k_r (cm/sec) = coefficient of permeability of remolded clay
k_h, k_v (gm/cm^3) = coefficients for computing pressure of backfill against retaining wall
k_h (cm/sec) = coefficient of permeability in horizontal direction
k_v (cm/sec) = coefficient of permeability in vertical direction
k_e (cm/sec) = coefficient of electro-osmotic permeability
L (cm) = length of line of creep; length
L_w = liquid limit
l (cm) = length
M_c (gm cm) = moment of cohesive forces
m = reduction factor (earth pressures against bracing in open cuts)
m_v (cm^2/gm) = coefficient of volume compressibility
m_α = coefficient (stability analysis) Eq. 35.11
N = dimensionless factor (N_c, $N\gamma$, and N_q = bearing capacity factors; N_s = stability factor in theory of stability of slopes); number of blows on sampling spoon during performance of standard penetration test
$N\phi$ = flow value = $\tan^2 (45° + \phi/2)$
N_d = number of equipotential drops (flow net)
N_f = number of flow channels (flow net)
n = porosity; number of piles in group
n_a = ratio between distance from bottom of lateral support to point of application of earth pressure, and total height of lateral support
n_d = depth factor (stability of slopes)
n_g = ratio between maximum acceleration produced by earthquake and acceleration of gravity
P = per cent of grains smaller than given size
P (gm or gm/cm) = resultant pressure, normal force

P_A (gm/cm) = active earth pressure if arching is absent (retaining walls; active Rankine state)
P_a (gm/cm) = active earth pressure if arching is present (bracing in open cuts)
P_g (gm) = resultant force of gravity on particle
P_P (gm/cm) = passive earth pressure. May be subdivided into P_P' which depends on unit weight of the soil, and P_P'' which depends on cohesion and surcharge. P_P'' may be further subdivided into P_c and P_q, respectively.
P_s (gm) = resultant of forces having seat on surface of particle
P_w (gm/cm) = resultant water pressure
P_w = plastic limit
ΔP_A (gm/cm) = part of active earth pressure due to line load q'
p (gm/cm^2) = pressure or normal stress; subgrade reaction
$p_{1,2,3}$ (gm/cm^2) = major, intermediate, and minor principal stresses
$\bar{p}$ (gm/cm^2) = effective pressure (bar may be omitted); effective overburden pressure, where used in expression $c/\bar{p}$
p_A (gm/cm^2) = intensity of active earth pressure
p_a (gm/cm^2) = pressure due to atmosphere
p_c (gm/cm^2) = confining pressure; all-around pressure; initial consolidation pressure
p_h (gm/cm^2) = horizontal pressure against vertical plane
p_v (gm/cm^2) = vertical pressure against horizontal plane
p_k (gm/cm^2) = capillary pressure
p_q (gm/cm^2) = increase in pressure on retaining wall due to surcharge q per unit of area
p_q' (gm/cm) = increase in pressure on retaining wall due to surcharge q' per unit of length parallel to crest
p_s (gm/cm^3) = seepage pressure
p_u (gm/cm^2) = pressure corresponding to point b, Figure 13.5*a*.
p_0 (gm/cm^2) = initial pressure; present overburden pressure
p_0' (gm/cm^2) = maximum consolidation pressure on soil in field
Δp (gm/cm^2) = change in pressure; consolidation stress; supplementary axial pressure (triaxial test)
Δp_b (gm/cm^2) = bond strength
Δp_f (gm/cm^2) = stress difference at failure
Δp_u (gm/cm^2) = ultimate value of stress difference
Q (cm^3 or cm^2) = total discharge per unit of time
Q (gm) = concentrated load; strut load
Q_a (gm) = allowable load on pile
Q_d (gm) = ultimate static resistance of pile
Q_d (gm or gm/cm) = critical load on footing or pier resting on dense or stiff soil. May be subdivided into Q' due to weight of soil and Q'' due to cohesion and surcharge. Bearing capacity of circular footing is denoted by Q_{dr} and of square footing by Q_{ds}
Q_d' (gm or gm/cm) = critical load on footing or pier resting on loose or soft soil
Q_{dy} (gm) = dynamic resistance to penetration of pile
Q_f (gm) = skin friction (total)
Q_g (gm) = ultimate bearing capacity of pile group
Q_s (gm) = frictional resistance of pile or pier
Q_t (gm) = excess load on footing or raft, consisting of net dead load Q_{dn} and live load Q_l; load on pile, consisting of Q exerted by building, and $Q' + Q''$ due to negative skin friction

q (gm/cm^2) = uniformly distributed load; surcharge per unit of area
q' (gm/cm) = uniformly distributed line load
q_a (gm/cm^2) = allowable soil pressure
q_d (gm/cm^2) = ultimate bearing capacity for dense or stiff soil. Value for loose or soft soil denoted by q_d'. Bearing capacity of circular footing denoted by q_{dr}, and of square footing by q_{ds}.
q_p (gm/cm^2) = bearing capacity of soil beneath base of pile or pier; cone penetration resistance
q_u (gm/cm^2) = unconfined compressive strength
R = ratio of size of filter material to size of material to be protected
R (cm) = radius of influence of well; radius of curvature of deformed slope
r (cm) = radius
r_f (cm) = radius of friction circle (stability of slopes)
r_0 (cm) = radius of logarithmic spiral
S (gm/cm) = total sliding resistance between base of dam and subsoil
S (cm) = settlement; penetration of pile under hammer blow
S_e (cm) = temporary elastic compression of pile under hammer blow
S_r = degree of saturation
S_t = degree of sensitivity
S_w = shrinkage limit
s (gm/cm^2) = shearing resistance
s_r (gm/cm^2) = residual shear strength
T (gm/cm) = shear force on side of slice (stability analysis)
T (degrees centigrade) = temperature
T_s (gm/cm) = surface tension of liquid
T_v = time factor (theory of consolidation)
t (sec) = time
t (gm/cm^2) = shearing stress
U (gm/cm) = total neutral pressure on base of dam; total excess hydrostatic pressure
U = degree of consolidation; uniformity coefficient = D_{60}/D_{10}
u (gm/cm^2) = excess hydrostatic pressure
u_a (gm/cm^2) = porewater pressure caused by all-around pressure p_3; increment in all-around cell pressure
u_d (gm/cm^2) = porewater pressure caused by stress difference Δp in undrained triaxial test
u_f (gm/cm^2) = porewater pressure at failure in consolidated-undrained triaxial test
u_g (gm/cm^2) = pressure in air or vapor phase of soil
u_w (gm/cm^2) = neutral stress, porewater pressure
V (cm^3) = total volume
V_v (cm^3) = total volume of voids
v (cm/sec) = discharge velocity
v_s (cm/sec) = seepage velocity
W (gm or gm/cm) = weight
W_H (gm) = weight of ram of pile driver
W_P (gm) = weight of pile
W_s (gm) = effective weight of soil replaced by footing or basement
w = water content in per cent of dry weight
z (cm) = depth
z_c (cm) = depth of tension cracks

α = angle
α = reduction factor on strength of clay adjacent to shaft of pier
β (degrees) = slope angle
γ (gm/cm^3) = unit weight (soil, water and air)
γ' (gm/cm^3) = submerged unit weight
γ_d (gm/cm^3) = unit weight of soil if water is entirely replaced by air
γ_w (gm/cm^3) = unit weight of water
γ_s (gm/cm^3) = unit weight of solid constituents
Δ = increment
Δ (gm cm) = energy lost in pile driving
δ (degrees) = angle of wall friction; angle between resultant stress on plane and normal to plane
ϵ = base of Naperian logarithms; unit strain
η (gm/cm^2 sec) = viscosity
θ (degrees) = angle; central angle
μ = Poisson's ratio; micron
Φ = velocity potential (flow net)
ϕ (degrees) = angle of internal friction; angle of shearing resistance in Revised Coulomb equation (16.5)
ϕ_{cu} (degrees) = consolidated-undrained angle of shearing resistance
ϕ_f (degrees) = friction angle between particles at their points of contact
ϕ_r (degrees) = angle of residual shearing resistance
ϕ_1 (degrees) = angle of shearing resistance for overconsolidated clay
χ = coefficient relating pore pressure in gas and water phases of soil
$\log a$ = Naperian (natural) logarithm of a
$\log_{10} a$ = logarithm of a to the base 10
$\overline{ab}$ = distance ab measured along a straight line
$\widehat{ab}$ = distance ab measured along an arc
$\approx$ means approximately equal
15.3 indicates Eq. 3 in Article 15. The article number appears at the top of each right-hand page

INTRODUCTION

Soil Mechanics in Engineering Practice is divided into the following three parts:

I. Physical Properties of Soils.
II. Theoretical Soil Mechanics.
III. Problems of Design and Construction.

Part I deals with the physical and mechanical properties of homogeneous specimens of undisturbed and remolded soils. It discusses those properties which serve as convenient criteria for distinguishing between different soils and provides instructions for describing soils adequately. It also deals with those soil properties that have a direct bearing on the behavior of soil masses during and after construction operations.

Part II provides the reader with an elementary knowledge of the theories required for solving problems involving the stability or bearing capacity of soils or the interaction between soil and water. All these theories are based on radically simplifying assumptions regarding the mechanical and hydraulic properties of the soils. Nevertheless, when properly applied, the results obtained by means of these approximate procedures are accurate enough for most practical purposes.

Part III deals with the application of our present knowledge of soil behavior and of the theories of soil mechanics to design and construction in the field of foundation and earthwork engineering.

The physical properties of soils could be discussed quite properly in a general study of the engineering properties of materials, and the theories of soil mechanics constitute a part of the general subject of theoretical mechanics. However, design and construction in the field of foundation and earthwork engineering, which constitutes the third and largest part of this book, is an independent subject in its own right, because it involves methods of reasoning and procedure that have no counterpart in other fields of structural engineering. In all other fields, the engineer is concerned with the effect of forces on structures made of manufactured products such as steel and concrete or carefully selected natural materials such as timber or stone. Since the properties of these materials can be determined reliably, the problems associated with design can almost always be solved by the direct application of theory or the results of model tests.

On the other hand, every statement and conclusion pertaining to soils in the field involves many uncertainties. In extreme cases the concepts on which a design is based are no more than crude working hypotheses that may be far from the truth. In such cases the risk of partial or total failure can be eliminated only by using what may be called the observational procedure. This procedure consists of making appropriate observations soon enough during construction to detect any signs of departure of the real conditions from those assumed by the designer and of modifying either the design or the method of construction in accordance with the findings.

These considerations determine the subject matter and method of presentation of Part III. Instead of starting with instructions for applying theoretical principles to design, Part III deals first of all with the technique for securing information about the soil conditions at the chosen site by boring, sounding, sampling, and testing. In spite of the great amount of time and labor involved in such exploratory work, the results commonly leave much room for interpretation.

Subsequent chapters of Part III contain a discussion of the general principles of the design of structures such as retaining walls, earth dams, and foundations. The behavior of all such structures depends chiefly on the physical soil properties and the subsoil conditions. Because our knowledge of subsoil conditions is always incomplete, uncertainties inevitably enter into the fundamental design assumptions. These uncertainties require and receive continuous attention in the text. Similar discussions are not required in textbooks pertaining to other fields of structural design, because the reliability of the fundamental assumptions concerning the properties of the other common construction materials can almost always be taken for granted.

Soil Mechanics in Engineering Practice

PART I

Physical Properties of Soils

The subject matter of Part I is divided into three chapters. The first deals with the procedures commonly used to discriminate between different soils or between different states of the same soil. The second deals with the hydraulic and mechanical properties of soils and with the experimental methods used to determine numerical values representative of these properties. The third chapter deals with the physical processes involved in the drainage of soils.

Chapter 1

INDEX PROPERTIES OF SOILS

ART. 1 PRACTICAL IMPORTANCE OF INDEX PROPERTIES

In foundation and earthwork engineering, more than in any other field of civil engineering, success depends on practical experience. The design of ordinary soil-supporting or soil-supported structures is necessarily based on simple empirical rules, but these rules can be used safely only by the engineer who has a background of experience. Large projects involving unusual features may call for extensive application of scientific methods to design, but the program for the required investigations cannot be laid out wisely, nor can the results be interpreted intelligently, unless the engineer in charge of design possesses a large amount of experience.

Since personal experience is necessarily somewhat limited, the engineer is compelled to rely at least to some extent on the records of the experiences of others. If these records contain adequate descriptions of the soil conditions, they constitute a storehouse of valuable information. Otherwise, they may actually be misleading. In the field of structural engineering, an account of the failure of a beam would be of little value unless it contained, in addition to other essential data, a statement as to whether the beam was made of steel or of cast iron. In all the older records of foundation experience, the nature of the soils is indicated merely by such general terms as "fine sand" or "soft clay." Yet, the differences between the mechanical properties of two fine sands from different localities can be greater and more significant than those between cast iron and steel. As a consequence, one of the foremost aims in recent attempts to reduce the hazards in dealing with soils has been to find methods for discriminating between the different kinds of soil in a given category. The properties on which the distinctions are based are known as *index properties*, and the tests required to determine the index properties are *classification tests*.

The nature of any given soil can be altered by appropriate manipulation. Vibrations, for example, can transform a loose sand into a dense one. Hence, the behavior of a soil in the field depends not only on the significant properties of the individual constituents of the soil mass, but also on those properties which are due to the arrangement of the particles within the mass. Accordingly, it is convenient to divide index properties into two classes: *soil grain properties* and *soil aggregate properties.* The principal soil grain properties are the size and shape of the grains and, in clay soils, the mineralogical character of the smallest grains. The most significant aggregate property of cohesionless soils is the relative density, whereas that of cohesive soils is the consistency.

The discussion of the soil grain and aggregate properties will be preceded by a description of the principal types of soil, and it will be followed by a condensed review of the minimum requirements for adequate soil descriptions to be incorporated in the records of field observations.

ART. 2 PRINCIPAL TYPES OF SOILS

The materials that constitute the earth's crust are rather arbitrarily divided by the civil engineer into the two categories, *soil* and *rock.* Soil is a natural aggregate of mineral grains that can be separated by such gentle mechanical means as agitation in water. Rock, on the other hand, is a natural aggregate of minerals connected by strong and permanent cohesive forces. Since the terms "strong" and "permanent" are subject to different interpretations, the boundary between soil and rock is necessarily an arbitrary one. As a matter of fact, there are many natural aggregates of mineral particles that are difficult to classify either as soil or as rock. In this text, however, the term soil will be applied only to materials that unquestionably satisfy the preceding definition.

Although the terminology described in the preceding paragraph is generally understood by civil engineers, it is not in universal use. To the geologist, for example, the term rock implies all the material which constitutes the earth's crust, regardless of the degree to which the mineral particles are bound together, whereas the term soil is applied only to that portion of the earth's crust which is capable of supporting vegetation. Therefore, if the civil engineer makes use of information prepared by workers in other fields, he must be certain that he understands the sense in which the terms soil and rock are used.

On the basis of the origin of their constituents, soils can be divided into two large groups, those which consist chiefly of the results of chemical and physical rock weathering, and those which are chiefly of organic origin. If the products of rock weathering are still located at the place where they originated, they constitute a *residual soil.* Otherwise they constitute a *transported soil,* regardless of the agent which performed the transportation.

Residual soils which have developed in semiarid or temperate climates are usually stiff and stable, and do not extend to great depth. However, particularly in warm humid climates where the time of exposure has been long, residual soils may extend to depths of hundreds of feet. They may be strong and stable, but they may also consist of highly compressible materials surrounding blocks of less weathered rock (Article 49). Under these circumstances they may give rise to difficulties with foundations and other types of construction. Many deposits of transported soils are soft and loose to a depth of several hundred feet and may also lead to serious problems.

Soils of organic origin are formed chiefly *in situ,* either by the growth and subsequent decay of plants such as peat mosses, or by the accumulation of fragments of the inorganic skeletons or shells of organisms. Hence a soil of organic origin can be either organic or inorganic. The term *organic soil* ordinarily refers to a transported soil consisting of the products of rock weathering with a more or less conspicuous admixture of decayed vegetable matter.

The soil conditions at the site of a proposed structure are commonly explored by means of test borings or test shafts. The foreman on the job examines samples of the soil as they are obtained. He classifies them in accordance with local usage and prepares a boring log or shaft record containing the name of each soil and the elevation of its boundaries. The name of the soil is modified by adjectives indicating the stiffness, color, and other attributes. At a later date the record may be supplemented by an abstract of the results of tests made on the samples in the laboratory.

The following list of soil types includes the names commonly used by experienced foremen and practical engineers for field classification.

Sand and *gravel* are cohesionless aggregates of rounded subangular or angular fragments of more or less unaltered rocks or minerals. Particles with a size up to $\frac{1}{8}$ in. are referred to as sand, and those with a size from $\frac{1}{8}$ in. to 6 or 8 in. as gravel. Fragments with a diameter of more than 8 in. are known as *boulders.*

Hardpan is a soil that offers an exceptionally great resistance to the penetration of drilling tools. Most hardpans are extremely dense well-graded somewhat cohesive aggregates of mineral particles.

Inorganic silt is a fine-grained soil with little or no plasticity. The least plastic varieties generally consist of more or less equidimensional grains of quartz and are sometimes called *rock flour*, whereas the most plastic types contain an appreciable percentage of flake-shaped particles and are referred to as *plastic silt*. Because of its smooth texture, inorganic silt is often mistaken for clay, but it may be readily distinguished from clay without laboratory testing. If shaken in the palm of the hand, a pat of saturated inorganic silt expels enough water to make its surface appear glossy. If the pat is bent between the fingers, its surface again becomes dull. This procedure is known as the *shaking test*. After the pat has dried, it is brittle, and dust can be detached by rubbing it with the finger. Silt is relatively impervious, but if it is in a loose state it may rise into a drill hole or shaft like a thick viscous fluid. The most unstable soils of this category are known locally under different names, such as bull's liver.

Organic silt is a fine-grained more or less plastic soil with an admixture of finely divided particles of organic matter. Shells and visible fragments of partly decayed vegetable matter may also be present. The soil ranges in color from light to very dark gray, and it is likely to contain a considerable quantity of H_2S, CO_2, and various other gaseous products of the decay of organic matter which give it a characteristic odor. The permeability of organic silt is very low and its compressibility very high.

Clay is an aggregate of microscopic and submicroscopic particles derived from the chemical decomposition of rock constituents. It is plastic within a moderate to wide range of water content. Dry specimens are very hard, and no powder can be detached by rubbing the surface of dried pats with the fingers. The permeability of clay is extremely low. The term *gumbo* is applied, particularly in the western United States, to clays which are distinguished in the plastic state by a soapy or waxy appearance and by great toughness. At higher water contents they are conspicuously sticky.

Organic clay is a clay that owes some of its significant physical properties to the presence of finely divided organic matter. When saturated, organic clay is likely to be very compressible, but when dry its strength is very high. It is usually dark gray or black in color, and it may have a conspicuous odor.

Peat is a somewhat fibrous aggregate of macroscopic and microscopic fragments of decayed vegetable matter. Its color ranges between light brown and black. Peat is so compressible that it is almost always unsuitable for supporting foundations. Various techniques have been developed for carrying earth embankments across peat deposits without the risk of breaking into the ground, but the settlement of these embankments is likely to be large and to continue at a decreasing rate for many years.

If a soil is made up of a combination of two different soil types, the predominant ingredient is expressed as a noun, and the less prominent ingredient as a modifying adjective. For example, silty sand indicates a soil which is predominantly sand but contains a small amount of silt. A sandy clay is a soil which exhibits the properties of a clay but contains an appreciable amount of sand.

The aggregate properties of sand and gravel are described qualitatively by the terms *loose, medium,* and *dense,* whereas those of clays are described by *hard, stiff, medium,* and *soft.* These terms are usually evaluated by the boring foreman on the basis of several factors, including the relative ease or difficulty of advancing the drilling and sampling tools and the consistency of the samples. However, since this method of evaluation may lead to a very erroneous conception of the general character of the soil deposit, the qualitative descriptions should be supplemented by quantitative information whenever the mechanical properties are likely to have an important influence on design. The quantitative information is commonly obtained by means of laboratory tests on relatively undisturbed samples (Article 7), or by suitable field tests (Article 44).

A record of the color of the different strata encountered in adjacent borings reduces the risk of errors in correlating the boring logs. Color may also be an indication of a real difference in the character of the soil. For example, if the top layer of a submerged clay stratum is yellowish or brown and stiffer than the underlying clay, it was probably exposed temporarily to desiccation combined with weathering. Terms such as mottled, marbled, spotted, or speckled are used when different colors occur in the same stratum of soil. Dark or drab colors are commonly associated with organic soils.

Under certain geological conditions soils form which are characterized by one or more striking or unusual features such as a root-hole structure or a conspicuous and regular stratification. Because of these features, such soils can easily be recognized in the field and, as a consequence, they have been given special names by which they are commonly known. The following paragraphs contain definitions and descriptions of some of these materials.

Till is an unstratified glacial deposit of clay, silt, sand, gravel, and boulders. It covers part of the rock surface in those regions which were glaciated during the ice age.

Tuff is a fine-grained water- or wind-laid aggregate of very small mineral or rock fragments ejected from volcanoes during explosions.

Loess is a uniform cohesive wind-blown sediment, commonly light brown in color. The size of most of the particles ranges between the narrow limits of 0.01 and 0.05 mm. The cohesion is due to the presence of a binder that may be predominantly calcareous or clayey. Because of the universal presence of continuous vertical root holes, the permeability in horizontal directions is much less than in the vertical direction; moreover, the material has the ability to stand on nearly vertical slopes. True loess deposits have never been saturated. On saturation the bond between particles is weakened and the surface of the deposit may settle.

Modified loess is a loess that has lost its typical characteristics by secondary processes, including temporary immersion; erosion and subsequent deposition; chemical changes involving the destruction of the bond between the particles; or chemical decomposition of the more perishable constituents such as feldspar. Thorough chemical decomposition produces *loess loam,* characterized by greater plasticity than other forms of modified loess.

Diatomaceous earth (kieselguhr) is a deposit of fine, generally white, siliceous powder, composed chiefly or wholly of the remains of diatoms. The term diatom applies to a group of microscopic unicellular marine or fresh-water algae characterized by silicified cell walls.

Lake marl or *boglime* is a white fine-grained powdery calcareous deposit precipitated by plants in ponds. It is commonly associated with beds of peat.

Marl is a rather loosely used term for various fairly stiff or very stiff marine calcareous clays of greenish color.

Adobe is a term applied in the southwestern part of the United States and other semiarid regions to a great variety of light-colored soils ranging from sandy silts to very plastic clays.

Caliche refers to layers of soil in which the grains are cemented together by carbonates such as lime. These layers commonly occur at a depth of several feet below the surface, and their thickness may range between a few inches and several feet. A semiarid climate appears to be necessary for their formation.

Varved clay consists of alternating layers of medium gray inorganic silt and darker silty clay. The thickness of the layers rarely exceeds one-half inch, but occasionally very much thicker varves are encountered. The constituents were transported into freshwater lakes by melt water at the close of the ice age. Varved clays are likely to combine the undesirable properties of both silts and soft clays.

Bentonite is a clay with a high content of montmorillonite (Article 4). Most bentonites were formed by chemical alteration of volcanic ash. In contact with water, dried bentonite swells more than other dried clays, and saturated bentonite shrinks more on drying. Bentonite deposits occur in practically every state west of the Mississippi; in Tennessee, Kentucky, and Alabama; and to a minor extent in several other states. They are also common in Mexico.

Each term used in the field classification of soils includes a rather great variety of different materials. Furthermore, the choice of terms relating to stiffness and density depends to a considerable extent on the person who examines the soil. Because of these facts, the field classification of soils is always more or less uncertain and inaccurate. More specific information can be obtained only by physical tests that furnish numerical values representative of the properties of the soil.

The methods of soil exploration, including boring and sampling, and the procedures for determining average numerical values for the soil properties are a part of the design and construction program. They are discussed in Chapter 7, Part III.

ART. 3 SIZE AND SHAPE OF SOIL PARTICLES

The size of the particles that constitute soils may vary from that of boulders to that of large molecules.

Grains larger than approximately 0.06 mm can be inspected with the naked eye or by means of a hand lens. They constitute the *very coarse* and *coarse* fractions of the soils.

Grains ranging in size from about 0.06 mm to 2μ ($1\mu = 1$ micron $= 0.001$ mm) can be examined only under the microscope. They represent the *fine fraction.*

Grains smaller than 2μ constitute the *very fine fraction.* Grains having a size between 2μ and about 0.1μ can be differentiated under the microscope, but their shape cannot be discerned. The shape of grains smaller than about 1μ can be determined by means of an electron microscope. Their molecular structure can be investigated by means of X-ray analysis.

The process of separating a soil aggregate into fractions, each consisting of grains within a different size range, is known as *mechanical analysis.* By means of mechanical analysis, it has been found that most natural soils contain grains representative of two or more soil fractions. The general character of mixed-grained soils is determined almost entirely by the character of the smallest soil constituents. In this respect soils are somewhat similar to concrete. The properties of concrete are determined primarily by the cement, whereas the aggregate, which constitutes most of the concrete, is inert. The "aggregate," or the inert portion of a mixed-grained soil, comprises about 80 or 90% of the total dry weight. The decisive or active portion constitutes the remainder.

Very coarse fractions, for example gravel, consist of rock fragments each composed of one or more minerals. The fragments may be angular, subangular, rounded, or flat. They may be fresh, or they may show signs of considerable weathering. They may be resistant or crumbly.

Coarse fractions, exemplified by sand, are made up of grains usually composed chiefly of quartz. The individual grains may be angular, subangular, or rounded. Some sands contain a fairly high percentage of mica flakes that make them very elastic or springy.

In the fine and very fine fractions, any one grain usually consists of only one mineral. The particles may be angular, flake-shaped, or, rarely, tubular. Rounded particles, however, are conspicuously absent. Exceptionally, the fine fraction contains a high percentage of porous fossils, such as diatoms or Radiolaria, that produce abnormal mechanical properties. In general, the percentage of flaky constituents in a given soil increases with decreasing grain size of the soil fraction.

If the size of most of the grains in an aggregate of soil particles is within the limits given for any one of the soil fractions, the aggregate is called a *uniform soil.* Uniform very coarse or coarse soils are common, but uniform very fine or colloidal soils are very seldom encountered. All clays contain fine, very fine, and colloidal constituents, and some clays contain even coarse particles. The finest grain-size fractions of clays consist principally of flake-shaped particles.

The widespread prevalence of flake-shaped particles in the very fine fractions of natural soils is a consequence of the geological processes of soil formation. Most soils originate in the chemical weathering of rocks. The rocks themselves consist partly of chemically very stable and partly of less stable minerals. Chemical weathering transforms the less stable minerals into a friable mass of very small particles of secondary minerals that commonly have a scale-like or flaky crystal form, whereas the stable minerals remain practically unaltered. Thus the process of chemical weathering reduces the rock to an aggregate consisting of fragments of unaltered or almost unaltered minerals embedded in a matrix composed chiefly of discrete scaly particles. During subsequent transportation by running water the aggregate is broken up, and the constituents are subjected to impact and grinding. The purely mechanical process of grinding does not break up the hard equidimensional grains of unaltered minerals into fragments smaller than about 10μ (0.01 mm). On the other hand, the friable cohesiveness. Nevertheless, even the finest fractions do not exhibit plasticity—the ability to be rolled out into threads within a certain range of water content.

The grains of biotite, in contrast to those of quartz, are characteristically platy. For a thin plate-shaped particle, the ratio of volume to surface area and consequently the ratio P_g/P_s are relatively much smaller than those for equidimensional particles, and the influence of grain size on the porosity and other physical properties of the aggregate is still more striking. In addition to acquiring cohesiveness with decreasing grain size, the saturated aggregate also acquires a considerable degree of plasticity.

The important differences between the behavior of quartz and biotite particles have their origin in the difference in crystal structure of the two minerals. The crystal structure of quartz leads to a bulky habit, whereas that of biotite leads to a platy habit. It has been determined that the platy habit exhibited by some minerals is invariably associated with a sheeted crystal structure. It has further been found that the very fine fractions of different minerals with a sheeted crystal structure also exhibit somewhat different properties, because the electrical characteristics of the surfaces of the sheets depend on the particular crystal structure associated with each mineral.

Practically all the minerals with a sheeted structure which are encountered in the very fine soil fraction belong to a group known as the *clay minerals*. Most of the minerals of this group can be referred to one of three subgroups known as the *kaolinites*, the *illites*, and the *montmorillonites*. Each of these is characterized by an arrangement of atoms leading to a negative electrical charge on the flat surfaces of the crystals.

A single particle of clay may consist of many sheets piled one another. Because each sheet has a definite thickness but is not limited in dimensions at right angles to its thickness, clay particles likely to be plate-shaped or to exhibit flat, terraced surfaces (Fig.). The flat surfaces carry residual negative electrical charges, but broken edges of the plates or the edges of the terraces may carry either positive or negative charges, depending upon the environment. In problems of interest to the civil engineer, clay particles are always in contact with water. The interactions among the clay particles, the water, and the various materials dissolved in the water are primarily responsible for the properties of the soil consisting of the particles.

Pure water consists primarily of molecules of H_2O, but a few of molecules always dissociate into hydrogen ions H^+ and hydroxyl flake-shaped particles of secondary minerals, although small, are readily ground and broken into still sma Hence, the very fine fractions of natural soils consist flake-shaped particles of secondary minerals.

ART. 4 PROPERTIES OF VERY FINE SOIL FR

If we crush and grind a specimen of any mineral, fractions of different grain sizes, and then saturate th find that the finest fraction exhibits properties absen ones. Moreover, we observe that these properties de extent on the nature of the mineral.

The influence of the size of the particles and of t mineral can be illustrated by comparing certain p different fractions of quartz, and by comparing cert corresponding grain-size fractions of quartz and of of the fractions of quartz, which consists of fairly or bulky grains, is shaken with distilled water and it will be observed that successively finer fraction having successively higher porosities. In the fir smallest particles remain in suspension for many w hand, if a drop of a solution containing an elect the suspension, sedimentation starts almost instant more, the porosity of the sediment is far in ex loosest sediment precipitated from distilled water. indicate that each of the particles is acted upo force of gravity P_g, which tends to pull the par by other forces, of which the resultant is designa seat at the surface of the particle where they ca movement of adjacent particles. The forces P_s an electrical nature.

With decreasing diameter D of the nearly eq grains, the force P_g acting on a particle decrease whereas the surface force P_s decreases in propor ratio P_g/P_s decreases directly with the diamete cube of quartz with a volume of 1 cm^3 were s ones with sides of 1μ, then the ratio P_g/P_s factor 10^{-4}. For very small cubes the gravity fo negligible in comparison to the surface for would assume a major influence on the prop Thus, although the coarse fraction of quartz i with decreasing grain size quartz acquires a

Fig. 4.1. Electron photomicrograph of terrace-shaped particles of kaolinite.

ions OH^-. If impurities are present, such as acids or bases, the impurities also dissociate into positively charged cations and negatively charged anions. Salt, for example, breaks up into Na^+ and Cl^-. Since the plane surfaces of the clay minerals carry negative electrical charges, the cations including the H^+ furnished by the water itself are attracted toward the surfaces of the particles. Such a cation is said to be *adsorbed*. The various clay minerals differ widely in their ability to adsorb cations; the approximate cation exchange capacity (expressed in terms of the total number of positive charges adsorbed per 100 gm) of different clay minerals with approximately equal particle size is shown in Table 4.1.

Table 4.1

Mineral	No. of positive charges adsorbed per 100 gm ($\times 10^{20}$)
Montmorillonite	360–500
Illite	120–240
Kaolinite	20–90

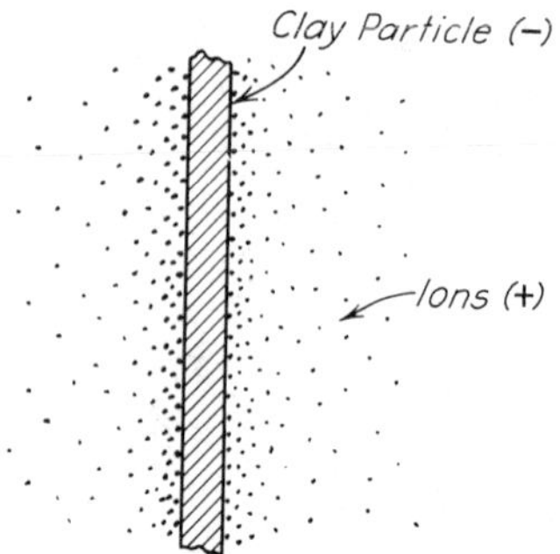

Fig. 4.2. Diagrammatic representation of distribution of cations adjacent to clay particle with negative surface charge.

The adsorbed ions are not permanently attached to the clay mineral. If, for example, a clay containing adsorbed ions of Na^+ is washed by a solution of KCl, most of the Na^+ ions are replaced by K^+ ions. This process is known as *cation exchange.*

The water adjacent to the negatively charged faces of the mineral particles may itself undergo alteration. The water molecules may become organized into a pattern determined by the location and nature of the adsorbed cations and to some extent by the spacing of the crystal lattice of the clay mineral. The water is then said to be adsorbed and to have a structure. The thickness of the adsorbed water varies considerably with the type of clay mineral and the character of any cations that may be present. The properties of the adsorbed water have not yet been adequately investigated, but they may exert an important influence on the mechanical properties of the clay mass. The adsorbed ions together with the adsorbed water constitute the *adsorption complex.*

The cations adsorbed by a clay mineral particle are in ceaseless motion due to thermal agitation. They distribute themselves statistically near the surface in an assemblage having the greatest density of ions near the surface and a decreasing density with increasing distance, as indicated in Fig. 4.2. They constitute a positively charged zone or layer which, together with the negatively charged surface of the particle itself, is known as the *electric double layer*. The electric double layers surrounding two adjacent approximately parallel particles exert repulsions on each other, of intensities that depend greatly on the nature and concentration of the ions in the water. In addition to the repulsions associated with the double layers, several other fields of force surround the charged particles. These include attractive as well as repulsive forces. Although the nature of the various fields of force is fairly well understood, the factors affecting the magnitude of the forces have not been fully analyzed. Nevertheless, present information permits reasonable but grossly simplified interpretations of many of the observed phenomena, and serves to illustrate their complexity.

One consequence of the forces associated with the surface of clay particles is the *structure* that may develop during sedimentation. If clay particles are introduced into distilled water, the negative charge on each particle causes repulsion of any other particles that chance to approach it. No particle adheres to another, the force of gravity upon any particle remains negligibly small, and the particles settle out very slowly or remain in suspension exhibiting Brownian movement. In natural waters which contain a sufficient concentration of electrolytes, such as the waters in limestone regions, the surfaces of at least some of the particles attract and adsorb ions of opposite sign. Such particles can then be attracted to others and may accumulate in flocs which may become large enough to settle to the bottom under the influence of gravity. Under some circumstances, especially if the broken ends of the plates forming the particles carry positive charges, the particles in the flocs may possess an *edge-to-face* structure (Fig. 4.3*a*); in others the flocs may consist of particles in an essentially *parallel structure* (Fig. 4.3*b*). Sediments formed exclusively of clay minerals are, therefore, likely to consist of aggregations of flocs made up of clay particles. The flocs may have a loose structure among themselves; within each floc the particles may all have an edge-to-face structure, a parallel structure, or some intermediate structure. Moreover, most sediments also contain coarser particles which significantly alter the arrangement (Article 18).

If the pressure acting on a sediment increases because of the addition of overlying sediments or the application of an external load, the water content of the sediment decreases, the particles are forced closer together, and the soil is said to *consolidate*. Most of the energy that must be expended to consolidate the sediment is consumed in producing structural breakdown of the flocs and in the work done

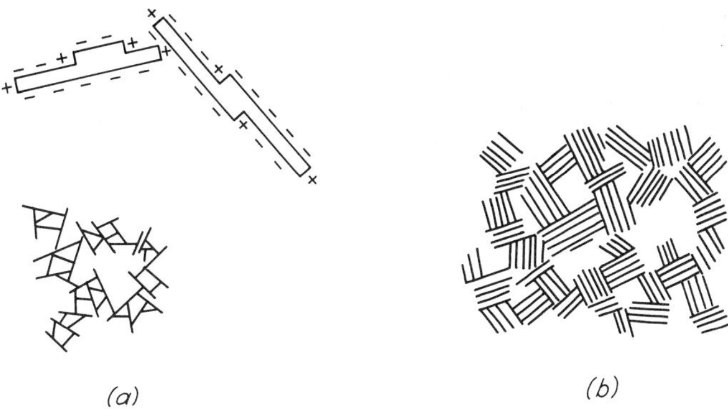

Fig. 4.3. **(*a*) Edge-to-face arrangement of terrace-shaped or plate-shaped clay particles, and combination into flocs. (*b*) Flocs of clay particles in parallel arrangement.**

against the repulsive forces between the particles. Part of the energy is accounted for by elastic deformation of the particles.

If at any stage the pressure is removed while the soil remains in contact with free water, the water content and the volume increase. This phenomenon is known as *swelling*. Some of the energy recovered as a consequence of the swelling represents the work done by the repulsive forces in separating the particles. Part of the energy is elastic restitution.

The causes of consolidation and swelling are likely to be different for different grain-size fractions. If the pressure on a coarse-grained sand-mica mixture is altered, for example, much of the consolidation or swelling is caused by elastic deformation or restitution of the grains. In the very fine soil fraction, however, the phenomena associated with the electrical charges may predominate.

If a single floc of clay particles with parallel orientation, such as one of those shown in Fig. 4.3*b*, could be subjected to shearing deformation, the resistance to shear along surfaces located midway between the particles would be extremely small, provided the surfaces of the particles were planes. In reality the particles are not plane but possess a terraced configuration which involves some interference and resistance to shear. If a sediment consisting of many flocs, each with parallel but different orientation, were subjected to shear, considerably greater resistance per floc would develop because of the interference between flocs. If a sediment consisted of flocs having an edge-to-face structure (Fig. 4.3*a*), substantial interference between particles would occur. Moreover, resistance would also be offered by the attractions at the contacts between the edges and faces of the particles. The various interferences and attractions constitute the shearing resistance of the sediment.

If a natural sediment is thoroughly kneaded or remolded, the flocs are largely disrupted and many of the clay particles become oriented in nearly parallel arrays. As a consequence, the shearing resistance may be substantially decreased. The clay is, therefore, said to be *sensitive* to disturbance.

Certain marine clays found in the Scandinavian countries and in eastern Canada are characterized by extraordinarily high sensitivity and are, therefore, often referred to as *quick clays*. After a disturbance like a landslide, these clays assume the character of a viscous fluid and commonly flow to a great distance (Article 49). Their high sensitivity is attributed to a reduction of the concentration of sodium ions in the porewater as a consequence of leaching. This theory is supported by field data as well as by the results of laboratory experiments.

When the quick clays were deposited, the voids were occupied by sea water with a substantial salt content; the concentration may have been as high as 35gm per liter. Chemical analyses of the porewater in a number of Scandinavian quick clays have shown that they now contain little or no salt, whereas in the same locality the porewater of other similar marine clays of only moderate sensitivity contains a considerable concentration of salt. In general, among the Scandinavian marine clays that have been analyzed, the lower sensitivities are associated with the higher salt contents (Skempton and Northey 1952).

If sodium chloride is added to a remolded sample of the quick clay and the clay is then allowed to stand, the sensitivity does not increase significantly. However, if the salt content of the remolded clay is then removed by leaching, the clay again becomes highly sensitive (Rosenqvist 1946). Aging without the addition of salt is not associated with a notable increase in sensitivity.

After a sample of a very fine soil fraction is thoroughly remolded, the positions of the particles with respect to each other are not necessarily associated with equilibrium of the various attractive and repulsive forces. Hence, the particles may tend to rotate and assume more stable configurations, at unaltered volume. The shearing strength may correspondingly increase. The soil thus exhibits *thixotropy*.

A somewhat similar phenomenon, known as *syneresis*, causes the porosity of the top layer of many fresh sediments to decrease slowly at a decreasing rate until the layer is reduced to a small fraction of its original volume. The gradual contraction cannot be accounted for by gravity. In some clays, it produces a network of hair cracks.

As a result of the manifold practical implications of the intricate physico-chemical processes and the high demand for clays with specified physical properties for industrial purposes, much research has been carried out during the last decades in clay mineralogy and in the interactions between clay particles and the surrounding media. Numerous investigations have also been made of the relationships between the physico-chemical processes and the engineering properties of clay soils. Nevertheless, in connection with most practical problems in earthwork engineering, benefits resulting from such research are still very limited because of the great number of factors responsible for the significant properties of clay. The combined influence of all the physico-chemical interactions is reflected in the index properties (Article 1), which are expedient and inexpensive to determine. A similar situation prevails in concrete technology. The processes by which Portland cement acquires its strength are also very intricate and still incompletely known. Nevertheless, concrete design is a fairly old and well-established branch of structural engineering. The assumptions on which it is based have been derived from purely mechanical laboratory tests on concrete specimens; certain properties of the cement, such as the increase of its strength with time, are disregarded. Nevertheless, the theories and procedures for design based on these simplifying assumptions are accurate enough for most practical purposes.

Selected Reading

The principal steps in the development of present ideas concerning the influence of structure and physico-chemical processes on the properties of fine-grained soils are contained in the following references, in chronological sequence:

Atterberg, A. (1911). "On the investigation of the physical properties of soils and the plasticity of clays," (German) *Int. Mitt. Bodenkunde,* **1,** p. 10.

Terzaghi, K. (1925). "Structure and volume of voids of soils," Pages 10–13, in *Erdbaumechanik auf Bodenphysikalisher Grundlage,* translated by A. Casagrande in *From theory to practice in soil mechanics,* New York, John Wiley and Sons, (1960), pp. 146–148.

Casagrande, A. (1932*b*). "The structure of clay and its importance in foundation engineering," *J. Boston Soc. Civil Engrs.,* **19,** No. 4, p. 168.

Terzaghi, K. (1941*a*). "Undisturbed clay samples and undisturbed clays," *J. Boston Soc. Civil Engrs.,* **28,** No. 3, pp. 211–231.

Skempton, A. W. and R. D. Northey (1952). "The sensitivity of clays," *Géot.,* **3,** pp. 30–53.

Rosenqvist, I. Th. (1953). "Considerations on the sensitivity of Norwegian quick clays," *Géot.,* **3,** pp. 195–200.

Grim, R. E. (1953). *Clay mineralogy.* New York, McGraw-Hill, 384 pp.

Bjerrum, L. (1954). "Geotechnical properties of Norwegian marine clays," *Géot.,* **4,** pp. 49–69.

Bolt, G. H. (1956). "Physico-chemical analysis of the compressiblity of pure clays," *Géot.,* **6,** pp. 86–93.

Lambe, T. W. (1960). "Structure of compacted clay," *Trans. ASCE,* **125,** pp. 682–705.

Mitchell, J. K. (1961). "Fundamental aspects of thixotropy in soils," *Trans. ASCE,* **126,** Part 1, pp. 1586–1620.

ART. 5 MECHANICAL ANALYSIS OF SOILS

Methods of Mechanical Analysis

The purpose of mechanical analysis is to determine the size of the grains which constitute a soil and the percentage of the total weight represented by the grains in various size ranges. The most direct method for separating a soil into grain-size fractions is the use of sieves. However, since the openings of the finest mesh readily available have a width of 0.07 mm, the use of sieves is restricted to analysis of clean sands. If a soil contains grains smaller than 0.07 mm, it may be separated into two parts by washing with water. As the water becomes turbid, it is drawn off. The coarser portion of the soil remains in the container and can be subjected to a sieve analysis. The soil particles in the turbid liquid, which are too fine to be collected on sieves, can be subjected to wet mechanical analysis or elutriation.

The methods for performing wet mechanical analysis are based on Stokes's law, which determines the velocity at which a spherical particle of given diameter settles in a quiet liquid. In the method commonly used for engineering purposes, 20 to 40 gm of clay soil or 50 to 100 gm of sandy soil are mixed with one liter of water, agitated, and poured into a container. The density of the suspension is measured at various times by means of a hydrometer of special design. At any given time, the size of the largest particles remaining in suspension at the level of the hydrometer can be computed by means of Stokes's law, whereas the weight of the particles finer than that size can be computed from the density of the suspension at the same level. The performance of a test requires several days.

By means of wet mechanical analysis soil fractions can be separated down to a size of about 0.5μ. Still finer fractions can be obtained by means of a centrifuge, but the results of such refined methods are of interest only in connection with scientific research.

Agitation in water transforms many clays into suspensions, not of individual particles, but of flocs. In order to break up the flocs into individual grains, or to disperse the soil, a deflocculating agent must be added to the water. The most common errors in the results of wet mechanical analysis are caused by inadequate dispersion.

The results of wet mechanical analysis are not strictly comparable to those of sieve analysis, because soil grains are never exactly spherical, and the smallest ones are commonly of a flaky shape. In a sieve analysis the width of the flake is measured, whereas the dimension determined by means of elutriation methods is the diameter of a sphere which sinks at the same rate as the flake. This diameter may be much smaller than the width of the actual flake.

The most convenient representation of the results of a mechanical analysis is the semilogarithmic grain-size curve shown in Fig. 5.1. The abscissas of this curve represent the logarithm of the grain size. The ordinates represent the percentage P, by weight, of grains smaller

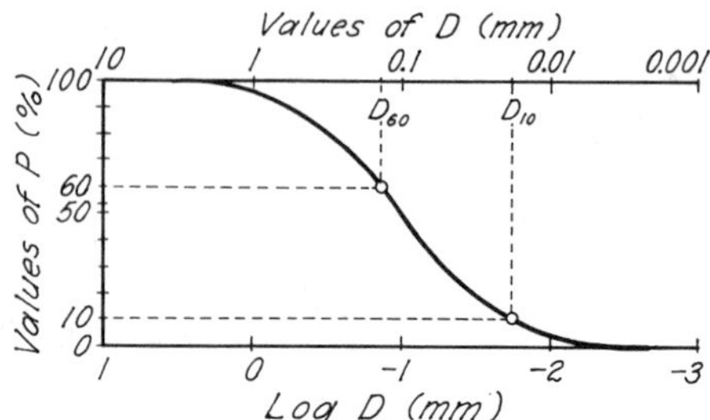

Fig. 5.1. Semilogarithmic plot of results of mechanical analysis.

than the size denoted by the abscissa. The more uniform the grain size, the steeper is the slope of the curve; a vertical line represents a perfectly uniform powder. The most important advantage of a semilogarithmic plot is that the grain-size curves of soils having equal uniformity are identical in shape, regardless of the average grain size. In addition, the horizontal distance between two curves of the same shape is equal to the logarithm of the ratio of the average grain sizes of the corresponding soils.

Figure 5.2 shows several typical grain-size curves. Curve *a* is a common type. It closely resembles the normal frequency curve that represents one of the fundamental laws of statistics. Since grain size is a statistical phenomenon, attempts have been made to utilize the terms and concepts of statistics to describe the results of mechanical analyses. Such refinements, however, are usually not warranted in connection with soil mechanics for engineering purposes.

If a sample has the grain-size distribution shown in Fig. 5.2*a*, the uniformity of the fraction having grains larger than D_{50} (corresponding to $P = 50\%$) is approximately equal to that of the fraction having grains smaller than D_{50}. If the distribution resembles that shown in *b*, the coarser half of the sample is relatively uniform, whereas the size of the grains in the finer half varies over a wide range. Conversely, the distribution represented in *c* corresponds to a sample in which the coarser grains are of widely different sizes and the finer ones are more uniform. The curves represented in *d* and *e* are said to be composite.

The grain-size curves of immature residual soils are usually similar to that shown in Fig. 5.2*b*. With increasing age of the soil, the average grain size decreases because of weathering, and the curves become more nearly straight (Fig. 5.2*a*). The grain-size curves of mature soils resemble that shown in Fig. 5.2*c*. Distributions represented by *b* and *c* are also common among soils of glacial or fluvioglacial origin. Absence of a medium grain size in a sedimentary soil, as exemplified by the curve in Fig. 5.2*d*, appears to be common among sand–gravel mixtures that were deposited by swiftly flowing rivers carrying a large load of sediment. Gravels of this type are said to be poorly, or gap, graded. A curve such as that in Fig. 5.2*d* may also be obtained if the materials from two different layers are mixed before the mechanical analysis is performed.

A conspicuous break in the continuity of the grain-size curve may also indicate the simultaneous deposition of the soil by two different agents. For instance, one fraction might be washed into a glacial lake by a river and another fraction dropped from melting ice floats. Thus, a knowledge of the shape of grain-size curves may assist in determining the geological origin of a soil and thereby reduce the risk of error in the interpretation of the data obtained from test borings.

Abbreviated Representation of Grain-Size Characteristics

In order to represent the essential results of the mechanical analysis of a great number of soils, it may be convenient to express the grain-

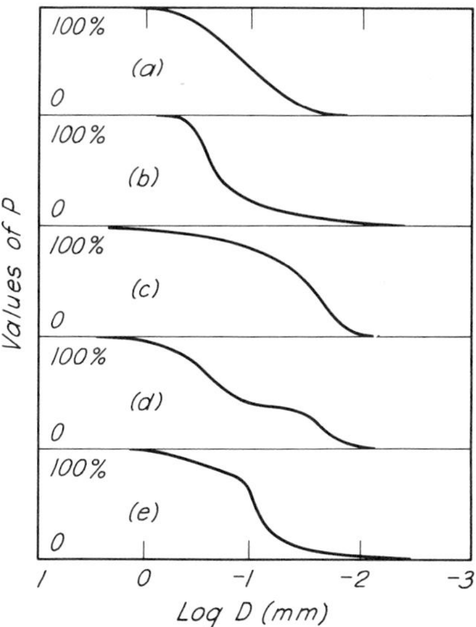

Fig. 5.2. Typical grain-size curves. (*a*) Normal frequency curve. (*b* and *c*) Curves for soils having coarser and finer fractions of different uniformity. (*d* and *e*) Composite curves.

size characteristics of each soil either by numerical values indicative of some characteristic grain size and of the degree of uniformity or else by names or symbols that indicate the dominant soil fraction. The most common procedure based on numerical values is known as Allen Hazen's method. On the basis of a great number of tests with filter sands, Hazen (1892) found that the permeability of these sands in a loose state depends on two quantities that he called the effective size and the uniformity coefficient. The *effective size* is the diameter D_{10} which corresponds to $P = 10\%$ on the grain-size diagram. In other words, 10% of the particles are finer and 90% coarser than the effective size. The *uniformity coefficient* U is equal to D_{60}/D_{10}, wherein D_{60} is the grain size corresponding to $P = 60\%$.

Hazen's findings led other investigators to the more or less arbitrary assumption that the quantities D_{10} and U are also suitable for expressing the grain-size characteristics of mixed-grained natural soils. With increasing knowledge concerning fine-grained soils, it has become evident that the character of such soils depends chiefly on the finest 20% and that it might be preferable to select D_{20} and D_{70} as the significant quantities. However, the advantage is not sufficiently important to justify a departure from well-established procedure. The use of symbols to indicate the grain-size characteristics of a soil is described in Article 8.

Selected Reading

Techniques for performing mechanical analyses and other classification tests are described in Lambe, T. W. (1951): *Soil testing for engineers,* New York, John Wiley and Sons, 165 pp.

ART. 6 SOIL AGGREGATE

Introduction

The term *aggregate* refers to the soil itself, in contrast to its constituent parts. Qualitatively, soil aggregates may differ in texture, structure, and consistency. Quantitatively, they may differ in porosity, relative density, water and gas content, and also in consistency. The qualitative information is obtained in the field by visual inspection. It constitutes the basis for preparing the boring logs or other records which describe the succession of strata in the underground. The quantitative information is obtained by means of laboratory and field tests. Without such information the description of any soil is inadequate.

Texture, Structure, and Consistency

The term *texture* refers to the degree of fineness and uniformity of a soil. It is described by such expressions as *floury, smooth, gritty,* or *sharp,* depending on the sensation produced by rubbing the soil between the fingers.

Fine-grained soils may be stable even if each particle does not touch several of its neighbors. If no coarse particles are present, the soil may have a *dispersed structure* in which all the particles are oriented parallel to each other, or a *cardhouse* or *flocculent structure* in which many of the particles have an edge-to-face attitude (Article 4). If the soil consists of loosely arranged bundles of particles, irrespective of the arrangement of the particles within the bundles, it is said to have a *honeycombed structure.*

Almost invariably, very fine-grained natural soils contain coarser particles. The coarse grains disrupt the structures described in the preceding paragraph, so that such structures are rarely found in nature. In some instances the coarser grains may form a skeleton with its interstices partly filled by a relatively loose aggregation of the finest soil constituents. This arrangement of the particles is called a *skeleton structure.* It may be partly responsible for the remarkable instability of many slightly cohesive soils with a grain size between about 0.05 and 0.005 mm (Article 17). In soft clays the inherent instability of the structure is masked by cohesion.

A few rather exceptional soils, including some marls, consist of relatively large composite grains. These grains constitute an aggregate with a single-grained or a honeycombed structure. However, the grains themselves are clusters of densely packed silt or clay particles. Such soils are said to have a *cluster structure.* Cluster structure has been encountered in both residual and sedimentary clay deposits. The geological processes responsible for it are not yet known and may be very different for different soils. However, the influence of cluster structure on the engineering properties of the soil is always beneficial. Although soils with this structure are very compressible, their swelling due to removal of load is imperceptible, and remolding at unaltered water content reduces their permeability to a small fraction of that of the same soil *in situ* (Terzaghi 1958*b*, FitzHugh et al. 1947).

Every sediment contains at least a small percentage of scale-like or disk-shaped particles. While these particles settle out of a suspension, their flat sides tend to maintain a horizontal position. Therefore, in the sediment, many of these particles are oriented more or less parallel to horizontal planes. Subsequent increase of the overburden pressure further accentuates this orientation. A sediment containing oriented particles is said to possess *transverse isotropy.*

Since a visual inspection of the structure of fine- or very fine-grained soils is not practicable, the structure of such soils must be judged on the basis of their porosity and various other properties.

Stiff clays may contain tubular root holes extending to a depth of tens of feet below the surface, or they may be divided by hair cracks into prismatic or irregular fragments that fall apart as soon as the confining pressure is removed. Such hair cracks are known as *joints.* Striations produced by movements along the walls of the joints are known as *slickensides.* The origin, nature, and practical implications of such defects of soil strata are discussed in Part III, Chapter 7.

The term *consistency* refers to the degree of adhesion between the soil particles and to the resistance offered against forces that tend to deform or rupture the soil aggregate. The consistency is described by such terms as *hard, stiff, brittle, friable, sticky, plastic,* and *soft.* The more nearly a soil approaches the characteristics of a clay, the greater is the variety of states of consistency in which it may be found. The degree of plasticity is sometimes expressed by the terms *fat* and *lean.* A lean clay is one that is only slightly plastic because it contains a large proportion of silt or sand. Further information concerning the consistency of clays is given in Article 7.

Porosity, Water Content, and Unit Weight

The *porosity* n is the ratio of the volume of voids to the total volume of the soil aggregate. The term *volume of voids* refers to that portion of the volume of the soil not occupied by mineral grains. If the porosity is expressed as a percentage, it is referred to as the *percentage of voids.*

The *void ratio* e is the ratio of the volume of voids to the volume of the solid substance. If

$$V = \text{total volume}$$

$$V_v = \text{total volume of voids}$$

then

$$n = \frac{V_v}{V} \tag{6.1a}$$

and

$$e = \frac{V_v}{V - V_v} \tag{6.1b}$$

The relation between void ratio and porosity is expressed by the equations,

$$e = \frac{n}{1 - n} \tag{6.2a}$$

and

$$n = \frac{e}{1 + e} \tag{6.2b}$$

The porosity of a stable mass of equal cohesionless spheres depends on the manner in which the spheres are arranged. In the densest possible arrangement, n is equal to 26%, and in the loosest state to 47%. Natural sands are found with porosities varying from about 25 to 50%. The porosity of a natural sand deposit depends on the shape of the grains, the uniformity of grain size, and the conditions of sedimentation.

The effect of the shape of the grains on the porosity of the aggregate can be demonstrated by mixing various percentages of mica with a uniform angular sand. If the percentage of mica, by weight, is equal successively to 0, 5, 10, 20, and 40, the porosities of the resultant mixtures when loosely dumped into a vessel are, respectively, about 47, 60, 70, 77, and 84% (Gilboy 1928). The porosity of soft natural clays, which contain an appreciable percentage of flat particles, usually ranges between 30 and 60%. It can even exceed 90%.

Because of the great influence of the shape of the grains and of the degree of uniformity on the porosity, the porosity itself does not indicate whether a soil is loose or dense. This information can be obtained only by comparing the porosity of the given soil with that of the same soil in its loosest and densest possible states. The looseness or denseness of sandy soil can be expressed numerically by the *relative density* D_r, defined by the equation

$$D_r = \frac{e_0 - e}{e_0 - e_{min}} \tag{6.3}$$

in which e_0 = void ratio of the soil in its loosest stable state
e_{min} = void ratio in the densest state which can be obtained in the laboratory
e = void ratio of the soil in the field

To bring a medium or coarse sand into its loosest state, corresponding to the void ratio e_0, the sand is first dried and then poured from a small height into a vessel. Fine and very fine sands may in some instances be brought into the loosest state by mixing a sample with enough water to transform it into a thick suspension that is then allowed to settle; the value of e_0 is equal to the final void ratio of the sediment. In other instances the loosest state can be established by carefully depositing the sand in a slightly moist state, such that the capillary forces produce a honeycomb structure, and by then permitting a very slow upward flow of water that causes the unstable structure to collapse. The densest state of clean sands can be established by prolonged vibrations, under a small vertical load, at a frequency of 20 to 30 per second.

The relative density of sand has a well-defined meaning because its value is practically independent of the static pressure to which the sand is subjected. It depends primarily on the procedure used in placing and compacting the sand. On the other hand, the degree of density of clays and other cohesive soils depends chiefly on the loads that these soils have carried and, in some instances, on the rate at which the loads were applied. The degree of density of these soils is most clearly reflected by the liquidity index I_l (Article 7).

The *water content* w of a soil is defined as the ratio of the weight of water to the dry weight of the aggregate. It is usually expressed as a percentage. In sands located above the water table, part of the voids may be occupied by air. If e_w represents the volume occupied by water per unit volume of solid matter, the ratio,

$$S_r(\%) = \frac{100e_w}{e} \tag{6.4}$$

represents the *degree of saturation*. The degree of saturation of sands is commonly expressed by such words as dry or moist. Table 6.1 gives a list of such descriptive terms and of the corresponding degrees of saturation. The nomenclature represented in Table 6.1 applies only to sands or very sandy soils. A clay in the state of desiccation repre-

Table 6.1
Degree of Saturation of Sand in Various States

Condition of sand	Degree of saturation (%)
Dry	0
Humid	1–25
Damp	26–50
Moist	51–75
Wet	76–99
Saturated	100

*Table 6.2**
Unit Weight of Most Important Soil Constituents

	gm/cm³		gm/cm³
Gypsum	2.32	Dolomite	2.87
Montmorillonite	2.6–2.8	Aragonite	2.94
Orthoclase	2.56	Biotite	3.0–3.1
Kaolinite	2.6	Augite	3.2–3.4
Illite†	2.6	Hornblende	3.2–3.5
Chlorite	2.6–3.0	Limonite	3.8
Quartz	2.66	Hematite, hydrous	4.3±
Talc	2.7	Magnetite	5.17
Calcite	2.72	Hematite	5.2
Muscovite	2.8–2.9		

* From Larsen and Berman (1934).
† Theoretical values computed on the basis of the atomic weights of the constituents of the space lattice (according to R. E. Grim).

sented by $S_r = 90\%$ might be so hard that it would be called dry instead of wet.

Coarse sands located above the water table are usually humid. Fine or silty sands are moist, wet, or saturated. Clays are almost always completely or nearly saturated, except in the layer of surface soil that is subject to seasonal variations of temperature and moisture. If a clay contains gas, the gas is present in bubbles scattered throughout the material. The bubbles may be composed of air that entered the deposit during sedimentation, or of gas produced at a later date by chemical processes such as the decomposition of organic material. The gas may be under pressure great enough to cause the clay to swell energetically at constant water content if the confining pressure is decreased. The determination of the gas content of a clay is extremely difficult. If it can be accomplished at all, it requires special equipment and is not a routine test.

The *unit weight* of the soil aggregate is defined as the weight of the aggregate (soil plus water) per unit of volume. It depends on the unit weight of the solid constituents, the porosity of the aggregate, and the degree of saturation. It may be computed as follows: Let

γ_s = average unit weight of solid constituents

γ_w = unit weight of water

n = porosity (expressed as a ratio)

The unit weight of dry soil ($S_r = 0\%$) is

$$\gamma_d = (1 - n)\gamma_s \tag{6.5}$$

and of saturated soil ($S_r = 100\%$) is

$$\gamma = (1 - n)\gamma_s + n\gamma_w = \gamma_s - n(\gamma_s - \gamma_w) \tag{6.6}$$

The unit weight of the principal solid constituents of soils is given in Table 6.2. For sand grains the average unit weight is usually about 2.65 gm/cm^3. For clay particles the unit weight varies from 2.5 to 2.9 with a statistical average of approximately 2.7.

Given in Table 6.3 are the porosity and the saturated unit weight of typical soils. For sandy soils the weight of dry soil has also been included. The weights have been computed on the assumption that

Table 6.3
Porosity, Void Ratio, and Unit Weight of Typical Soils in Natural State

Description	Porosity, n (%)	Void ratio, e	Water content, w (%)	Unit weight grams/cm^3 γ_d	Unit weight grams/cm^3 γ	Unit weight lb/ft^3 γ_d	Unit weight lb/ft^3 γ
1. Uniform sand, loose	46	0.85	32	1.43	1.89	90	118
2. Uniform sand, dense	34	0.51	19	1.75	2.09	109	130
3. Mixed-grained sand, loose	40	0.67	25	1.59	1.99	99	124
4. Mixed-grained sand, dense	30	0.43	16	1.86	2.16	116	135
5. Glacial till, very mixed-grained	20	0.25	9	2.12	2.32	132	145
6. Soft glacial clay	55	1.2	45	–	1.77	–	110
7. Stiff glacial clay	37	0.6	22	–	2.07	–	129
8. Soft slightly organic clay	66	1.9	70	–	1.58	–	98
9. Soft very organic clay	75	3.0	110	–	1.43	–	89
10. Soft bentonite	84	5.2	194	–	1.27	–	80

w = water content when saturated, in per cent of dry weight.
γ_d = unit weight in dry state.
γ = unit weight in saturated state.

the value of γ_s is 2.65 gm/cm^3 for sandy soils and 2.70 gm/cm^3 for clays. The tabulated values should be considered only as approximations. Before final computations are made on a given job, the actual unit weight of the soil should always be determined.

Problems

1. A sample of saturated clay weighed 1526 gm in its natural state, and 1053 gm after drying. Determine the natural water content. If the unit weight of the solid constituents was 2.70 gm/cm^3, what was the void ratio? the porosity? the weight per cubic foot?

Ans. $w = 45.0\%$; $e = 1.22$; $n = 0.55$; $\gamma = 111$ lb/ft^3.

2. A sample of hardpan had a weight of 129.1 gm and a volume of 56.4 cm^3 in its natural state. Its dry weight was 121.5 gm. The unit weight of the solid constituents was found to be 2.70 gm/cm^3. Compute the water content, the void ratio, and degree of saturation.

Ans. $w = 6.3\%$; $e = 0.25$; $S_r = 0.67$.

3. The unit weight of a sand backfill was determined by field measurements to be 109 lb/ft^3. The water content at the time of the test was 8.6%, and the unit weight of the solid constituents was 2.60 gm/cm^3. In the laboratory the void ratios in the loosest and densest states were found to be 0.642 and 0.462, respectively. What were the void ratio and the relative density of the fill?

Ans. $e = 0.616$; $D_r = 0.14$.

4. A dry quartz sand weighs 96 lb/ft^3. What is its unit weight when saturated?

Ans. $\gamma = 122$ lb/ft^3.

5. A sample of silty clay was found, by immersion in mercury, to have a volume of 14.88 cu cm. Its weight at the natural water content was 28.81 gm and after oven drying was 24.83 gm. The unit weight of solid constituents was 2.70 gm/cm^3. Calculate the void ratio and the degree of saturation of the sample.

Ans. $e = 0.617$; $S_r = 0.701$.

6. Given the values of porosity n for the soils in Table 6.3, check the values of water content w and unit weight γ (lb/ft^3). For soils 1–5, $\gamma_s = 2.65$ gm/cm^3; for soils 6–10, $\gamma_s = 2.70$ gm/cm^3.

ART. 7 CONSISTENCY AND SENSITIVITY OF CLAYS

Consistency and Sensitivity of Undisturbed Soils

The consistency of clays and other cohesive soils is usually described as *soft, medium, stiff,* or *hard.* The most direct quantitative measure of consistency is the load per unit of area at which unconfined prismatic or cylindrical samples of the soil fail in a simple compression test. This quantity is known as the *unconfined compressive strength* of the soil. Values of the compressive strength corresponding to the various degrees of consistency are given in Table 7.1.

Table 7.1
Consistency of Clay in Terms of Unconfined Compressive Strength

Consistency	Unconfined compressive strength, q_u (kg/cm^2)
Very soft	Less than 0.25
Soft	0.25–0.5
Medium	0.5–1.0
Stiff	1.0–2.0
Very stiff	2.0–4.0
Hard	Over 4.0

Clays share with many other colloidal substances the property that kneading or working at unaltered water content makes the material softer. The process of kneading or working is commonly referred to as *remolding,* and clays that have been subjected to the process are called *remolded clays.* The softening effect is probably due to two different causes: destruction of the orderly arrangement of the molecules in the adsorbed layers, and injury to the structure that the clay acquired during the process of sedimentation. That part of the loss of strength caused by the disturbance of the adsorbed layers may be gradually regained, at unaltered water content, after the working has ceased. The remainder, probably caused by permanent alteration of the structure, is irrecoverable unless the water content of the clay is reduced. The ratio between these two parts of the loss of strength is very different for different clays.

The term *sensitivity* indicates the effect of remolding on the consistency of a clay, regardless of the physical nature of the causes of the change. The degree of sensitivity is different for different clays, and it may also be different for the same clay at different water contents. If a clay is very sensitive, a slide may turn it into a mass of lubricated chunks capable of flowing on a gently sloping base, whereas a similar slide in a clay with low sensitivity merely produces a conspicuous local deformation. The change in consistency produced by the disturbance of a sensitive clay is always associated with a change of the permeability.

The degree of sensitivity S_t of a clay is expressed by the ratio between the unconfined compressive strength of an undisturbed specimen and the strength of the same specimen at the same water content but in a remolded state. That is,

$$S_t = \frac{\text{unconfined compressive strength undisturbed}}{\text{unconfined compressive strength remolded}} \tag{7.1}$$

The values of S_t for most clays range between 2 and about 4. For sensitive clays they range from 4 to 8. However, extrasensitive clays are encountered with values of S_t between 8 and 16, and in some localities clays with even higher sensitivities are found; these are known as *quick clays*. High degrees of sensitivity may be due to a well-developed honeycomb or skeleton structure, or to leaching of soft glacial clays deposited in salt water and subsequently uplifted (Article 4). The quick clays of Scandinavia and of the St. Lawrence valley are of this category. On the other hand, the extra-sensitive clays of Mexico City were derived from the decomposition of volcanic ash.

The remolded strengths of some saturated clays may be so low that an unconfined specimen cannot stand without excessive deformation under its own weight. Under these conditions the degree of sensitivity S_t may be evaluated by comparing the undisturbed and remolded shearing strength determined by such other procedures as the vane shear tests (Article 44).

Consistency of Remolded Soils

After a cohesive soil has been remolded, its consistency can be changed at will by increasing or decreasing the water content. Thus, for instance, if the water content of a clay slurry is gradually reduced by slow desiccation, the clay passes from a liquid state through a plastic state, and finally into a solid state. The water contents at which different clays pass from one of these states into another are very different. Therefore, the water contents at these transitions can be used for identification and comparison of different clays. However, the transition from one state to another does not occur abruptly as soon as some critical water content is reached. It occurs gradually over a fairly large range in the value of the water content. For this reason every attempt to establish criteria for the boundaries between

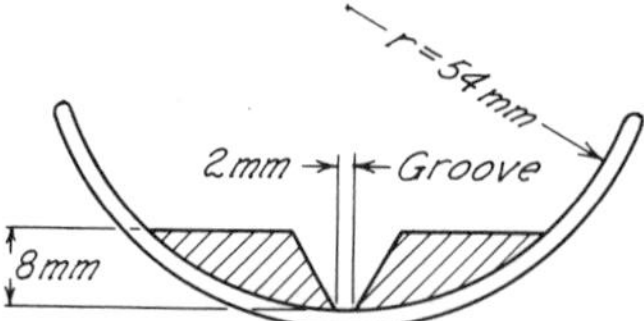

Fig. 7.1. Cross section through soil pat for liquid-limit test (after A. Casagrande 1932*a*).

the limits of consistency involves some arbitrary elements. The method that has proved most suitable for engineering purposes was taken over from agronomy. It is known as Atterberg's method, and the water contents that correspond to the boundaries between the states of consistency are called the *Atterberg limits* (Atterberg 1911).

The *liquid limit* L_w is the water content in per cent of the dry weight at which two sections of a pat of soil having the dimensions shown in Fig. 7.1 barely touch each other but do not flow together when subjected in a cup to the impact of sharp blows from below. The personal equation has an important influence on the test results. In order to eliminate this factor, a standardized mechanical device is used (A. Casagrande 1932*a*).

The *plastic limit* P_w or lower limit of the plastic state is the water content at which the soil begins to crumble when rolled out into thin threads.

The record of the results of the plastic-limit test should also contain a statement as to whether the threads, immediately before crumbling, were very tough like those of a gumbo, moderately tough like those of an average glacial clay, or weak and spongy like those of an organic or of a micaceous inorganic soil.

The *shrinkage limit* S_w or lower limit of volume change is the water content below which further loss of water by evaporation does not result in a reduction of volume. As soon as the soil passes below the shrinkage limit, it becomes slightly lighter in color.

The range of water content within which a soil possesses plasticity is known as the *plastic range*, and the numerical difference between the liquid limit and the plastic limit is the *plasticity index* I_w. As the water content w of a cohesive soil approaches the lower limit P_w of the plastic range, the stiffness and degree of compaction of the soil increase. The ratio,

$$I_l = \frac{w - P_w}{L_w - P_w} = \frac{w - P_w}{I_w} \tag{7.2}$$

is called the *liquidity index* of the soil. If the water content of a natural soil stratum is greater than the liquid limit (liquidity index greater than 1.0), remolding transforms the soil into a thick viscous slurry. If the natural water content is less than the plastic limit (liquidity index negative), the soil cannot be remolded. The unconfined compressive strength of undisturbed clays with a liquidity index near unity commonly ranges between 0.3 and 1.0 kg/cm². If the liquidity index is near zero, the compressive strength generally lies between 1 and 5 kg/cm².

In addition to the Atterberg limits, a knowledge of the *dry strength* is useful in the identification and comparison of cohesive soils. The strength of air-dry specimens of clay ranges from about 2 to more that 200 kg/cm², and a trained experimenter can distinguish among the degrees of *very low, low, medium, high,* and *very high,* merely by pressing an angular fragment of the soil between his fingers. The strength is called medium if the fragment can be reduced to powder only with great effort. Fragments with very high strength cannot be injured at all, whereas those of very low strength disintegrate completely on gentle pressure. The fragments should be obtained by molding a cylindrical specimen about one inch high and one inch in diameter from a paste at a water content close to the plastic limit. After the cylinder has dried at room temperature, it is broken into smaller pieces, and fragments for the examination are selected from the interior of the specimen.

Plasticity Chart

It has been observed (A. Casagrande 1932*a*) that many properties of clays and silts, such as their dry strength, their compressibility, their reaction to the shaking test, and their consistency near the plastic limit, can be correlated with the Atterberg limits by means of the *plasticity chart* (Fig. 7.2). In this chart, the ordinates represent the plasticity index I_w and the abscissas the corresponding liquid limit L_w.

The chart is divided into six regions, three above line A and three below. The group to which a given soil belongs is determined by the name of the region that contains the point representing the values of I_w and L_w for the soil. All points representing inorganic clays lie above line A, and all points for inorganic silts lie below it. Therefore, if a soil is known to be inorganic, its group affiliation can be ascertained on the basis of the values of I_w and L_w alone. However, points representing organic clays are usually located within the same region as those representing inorganic silts of high compressibility, and points representing organic silts in the region assigned to inorganic silts of medium compressibility. Usually, the organic soils can be distinguished from the inorganic by their characteristic odor and their dark-gray or black color. In doubtful cases the liquid limit should be determined for an oven-dry specimen as well as a fresh one. If drying decreases the value of the liquid limit by 30% or more, the soil may usually be classified as organic, although in some instances other constituents, such as the clay mineral halloysite, similarly lower the liquid limit. Finally, if an inorganic and an organic soil are represented in Fig. 7.2 by approximately the same point, the dry strength of the organic soil is considerably greater than that of the inorganic soil. Experience has shown that the points which represent different samples from the same soil stratum define a straight line that is roughly parallel to line A. As the liquid limit of soils represented by such a line increases, the plasticity and the compressibility of the soils also increase. The dry strength of inorganic soils represented by points on lines located above A increases from medium for samples with a liquid limit below 30 to very high for samples with a liquid limit of 100. On the other hand, if the line representative of inorganic samples from a given stratum is located at a considerable distance below A, the dry strength of samples with a liquid limit less than 50 is very low, and that of samples with a liquid limit close to 100 is only medium. In accordance with these relationships, the dry strength of inorganic soils from different localities but with equal liquid limits increases in a general way with increasing plasticity index. Figure 7.3 shows the plasticity characteristics of several well-defined types of clay.

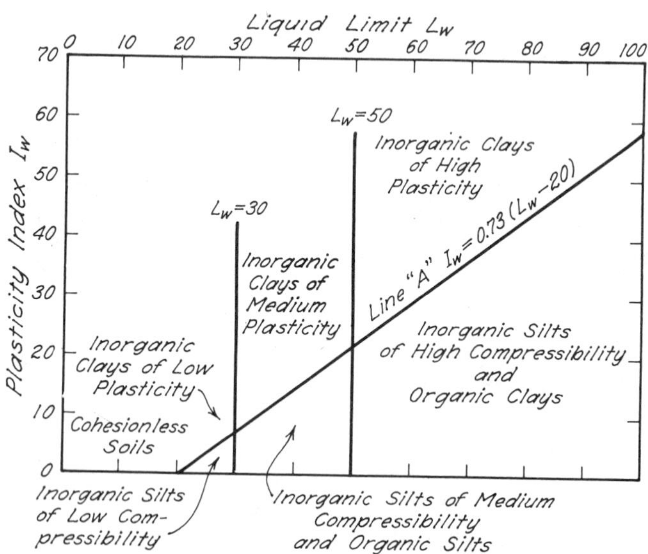

Fig. 7.2. Plasticity chart (after A. Casagrande 1932*a*).

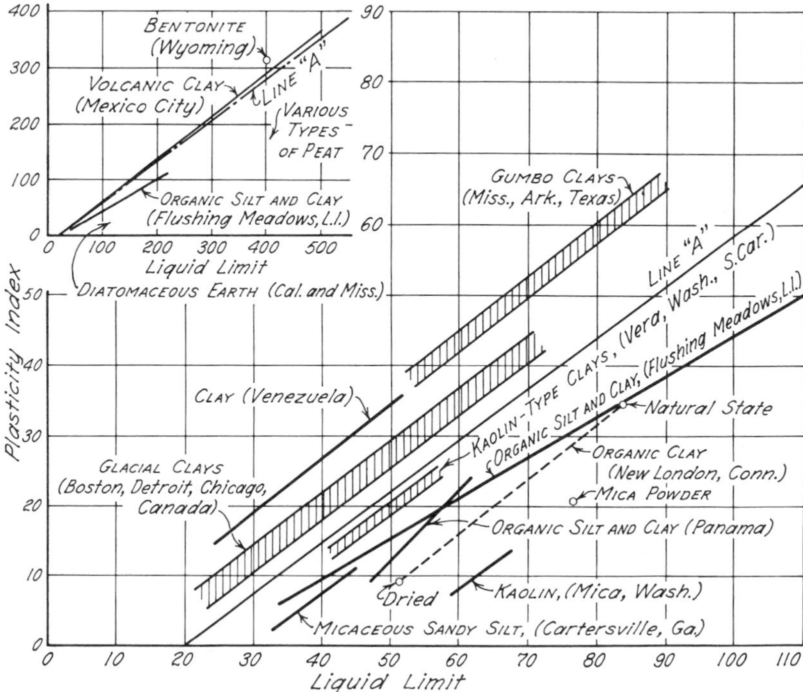

Fig. 7.3. Relation between liquid limit and plasticity index for typical soils (after A. Casagrande 1932*a*).

The samples required for Atterberg-limit tests need not be undisturbed, and the technique of making the tests is simple. Yet, even at the present still incipient state of our knowledge, a great amount of useful information can be derived from the test results. Therefore, the investigation of statistical relations between the Atterberg limits and the other physical properties of cohesive soils constitutes one of the most promising fields for research in soil physics. Every well established statistical relation of this type broadens the scope of conclusions that can be drawn from the results of limit tests. Two useful relations of this kind are shown in Figs. 13.6 and 14.3.

Selected Reading

The classic study of the Atterberg limits and their uses for engineering purposes is Casagrande, A. (1932*a*): "Research on the Atterberg limits of soils," *Public Roads*, **13**, pp. 121–136.

ART. 8 SOIL CLASSIFICATION

Practical Significance of Soil Classification

Ever since the physical properties of soils became a matter of interest, frequent attempts have been made to correlate the results of simple classification tests with the soil constants needed for solving practical problems. Most of the early correlations were related to the grain-size characteristics. The results of the endeavors to base systems of classification exclusively on grain size, however, have been consistently disappointing. Attempts to compute the coefficient of permeability of soils on the basis of the results of mechanical analysis have failed because the permeability depends to a large extent on a shape of the grains, which can be very different for soils with identical grain-size characteristics. Furthermore, it is usually cheaper to perform a permeability test than a mechanical analysis, and the results are more reliable. It has also been claimed that the internal friction of compacted well-graded sands is greater than that of compacted uniform sands. Field experience suggests that this may be the case. However, since the angle of internal friction of a sand (see Article 17) depends not only on the grain-size characteristics but also on the shape of the grains and the roughness of their surface, the internal friction of two compacted sands with identical grain-size characteristics can be very different. As a matter of fact, no well-defined relation between grain-size characteristics and the angle of internal friction has yet been observed. Attempts to correlate the grain-size characteristics of fine-grained soils such as silt or clay with the internal friction have been even less successful. The reason is illustrated by Fig. 8.1.

In Fig. 8.1 the heavy uppermost curve represents what is known as the *grain-size frequency curve* for a glacial clay from southeastern Canada. On the horizontal axis are plotted the logarithms of the

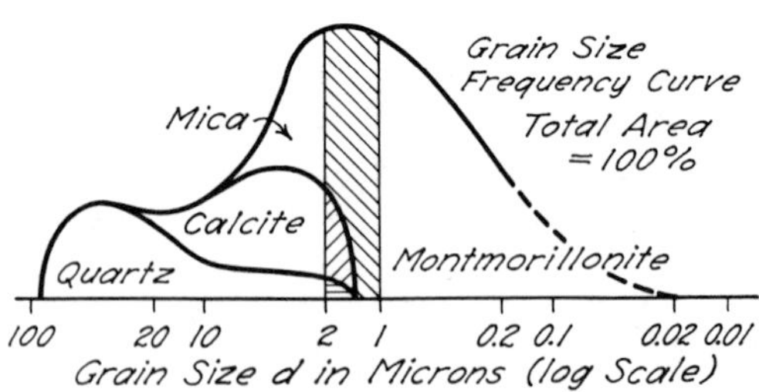

Fig. 8.1. Grain size and mineralogical composition of a glacial marine clay (courtesy R. E. Grim).

grain size. The area of the strip located above an arbitrary grain-size range, for instance 2μ to 1μ, represents the quantity of soil particles within this range, in per cent of the total weight of the dried clay. According to the diagram the macroscopic fraction (>0.06 mm), like that of most other clays, consists chiefly of quartz. The microscopic fraction (0.06 to 0.002 mm) consists partly of quartz and calcite and partly of mica flakes. The mica content of this fraction is very different for different clays, and it has a decisive influence on the compressibility and other properties of the clay. The colloidal fraction (<0.002 mm) consists almost exclusively of montmorillonite, whereas that of other clays may consist chiefly of clay minerals of the kaolin or illite groups. The physical properties of the clay depend to a large extent on the type of clay mineral that dominates the colloidal fraction. They also depend to a large extent on the substances that are present in the adsorbed layers (Article 4). Hence, two clays with identical grain-size curves can be extremely different in every other respect.

Because of these conditions, well-defined statistical relations between grain-size characteristics and significant soil properties such as the angle of internal friction have been encountered only within relatively small regions where all the soils of the same category, such as all the clays or all the sands, have a similar geological origin. In such regions the grain-size characteristics can be used as a basis for judging the significant properties of the soils. This is commonly and successfully done. However, none of the procedures that grow out of experience in such regions can safely be used outside the boundaries of the region where they originated.

Since the properties of fine-grained soils can be correlated in a general way with the plasticity of the materials, classification systems for such soils are preferably based on the Atterberg limits rather than on grain size. Classification of mixed-grained soils containing both coarse and fine fractions should be based not only on the grain-size characteristics of the coarse fractions but also on the plasticity of the fine and very fine fractions.

Classification Based on Grain Size

In spite of their shortcomings, soil classifications based on grain-size characteristics are widely used, especially for preliminary or general descriptions. It is customary, in connection with such classifications, to assign the names of soils, such as "silt" or "clay," to different grain-size fractions. The most widely accepted conventions of this type are shown in graphical form in Fig. 8.2. From an engineering point of view, the MIT classification is preferable to the others (Glossop and Skempton 1945). In many instances, records concerning soils and their behavior contain no more than the results of a mechanical analysis of the coarse-grained fraction and the percentage of the total that passes the 200-mesh sieve. The latter includes all the soil particles smaller than 0.074 mm. A grain size of 0.074 mm is slightly greater than the value 0.06 mm which, in the MIT classification, represents the boundary between fine sand and coarse silt.

However, any system of classification based on grain size alone is likely to be misleading, because the physical properties of the finest soil fractions depend on many factors other than grain size (see Article 4). For example, according to any one of the commonly used conventions represented in Fig. 8.2, a soil consisting of quartz grains of colloidal size should be called a clay, whereas in reality it does not possess even a remote resemblance to clay. Hence, if the words "silt" or "clay" are used to express grain size, they should be combined with the word "size," as in the expression "clay-size particle." Since the grain-size classifications are not yet standardized, the descriptive adjectives must be supplemented by numerical values that indicate the grain-size range represented by the adjectives.

With few exceptions natural soils consist of a mixture of two or more different grain-size fractions. Hence, on the basis of its grain-size composition a natural soil can be designated by the names of its

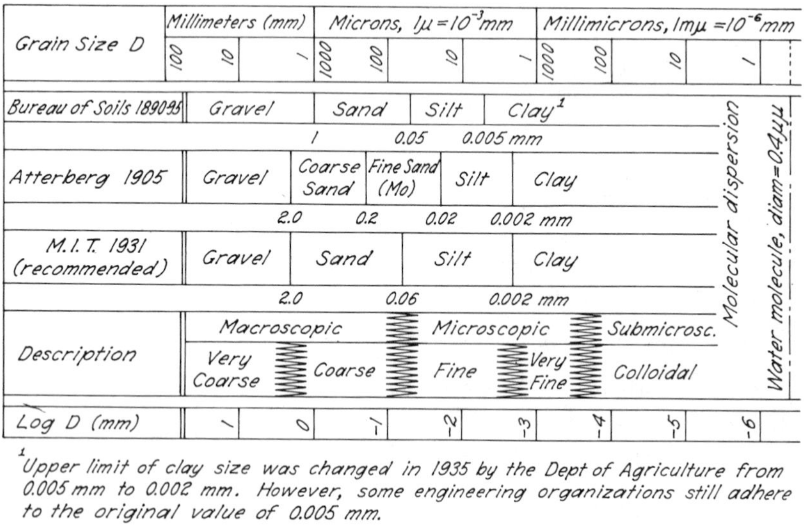

Grain Size D	Millimeters (mm): 100, 10, 1	Microns, 1μ = 10^{-3} mm: 1000, 100, 10, 1	Millimicrons, 1mμ = 10^{-6} mm: 1000, 100, 10, 1
Bureau of Soils 1890-95	Gravel / Sand / Silt / Clay[1] (boundaries: 1, 0.05, 0.005 mm)		
Atterberg 1905	Gravel / Coarse Sand / Fine Sand (Mo) / Silt / Clay (boundaries: 2.0, 0.2, 0.02, 0.002 mm)		
M.I.T. 1931 (recommended)	Gravel / Sand / Silt / Clay (boundaries: 2.0, 0.06, 0.002 mm)		
Description	Macroscopic / Microscopic / Submicrosc.; Very Coarse / Coarse / Fine / Very Fine / Colloidal		
Log D (mm)	1, 0, -1, -2, -3, -4, -5, -6		

Molecular dispersion — Water molecule, diam=0.4μμ

[1] Upper limit of clay size was changed in 1935 by the Dept of Agriculture from 0.005 mm to 0.002 mm. However, some engineering organizations still adhere to the original value of 0.005 mm.

Fig. 8.2. Soil classification based on grain size.

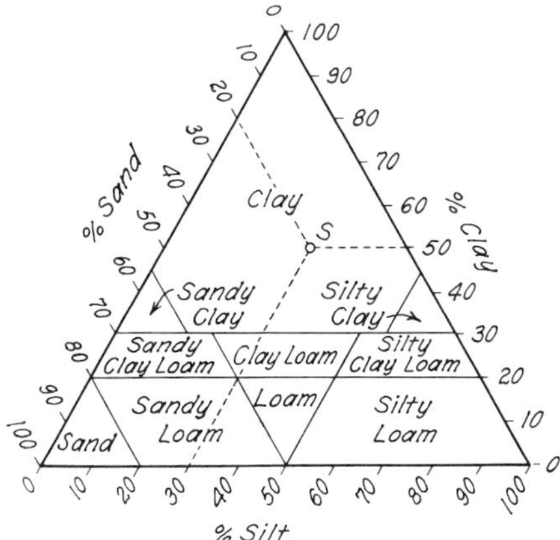

Fig. 8.3. Soil classification chart developed by Bureau of Public Roads.

principal components, such as "silty clay," or "sandy silt." Or it may be assigned some symbol that identifies it with one of several standard mixtures of grain-size fractions.

The designation of soils by the names of their principal constituents is facilitated by the use of diagrams such as that adopted by the Bureau of Public Roads, Fig. 8.3 (Rose 1924). In this diagram, each of the three coordinate axes pertains to one of the grain-size fractions, designated as sand, silt, or clay. The chart is divided into regions to which the names of soil types are assigned. The three co-ordinates of a point represent the percentages of the three fractions present in a given soil and determine the type to which the soil belongs. For example, a mixed-grained soil composed of 20% sand, 30% silt, and 50% clay, represented by point S, is classified as a clay.

The identification of a given soil by comparison with standard mixtures can be accomplished rapidly by means of master plots on transparent paper. In these plots each of various standard mixtures of grain-size fractions is represented by a grain-size curve that bears an identifying symbol. To classify a real soil, the transparent master plot is placed over a sheet on which the grain-size curve for the soil has been drawn. The soil is given the symbol of the standard curve that most nearly resembles that of the real soil.

Unified Soil Classification System

The unsatisfactory nature of most systems of soil classification, as well as their great number and diversity, led to a critical review of the problem (A. Casagrande 1948) and the proposal of the Unified

Soil Classification System, adopted in 1952 by the U.S. Corps of Engineers and Bureau of Reclamation, and subsequently by many other organizations (USBR 1963).

According to this system, all soils are divided into three major groups: coarse-grained, fine-grained, and highly organic (peaty). The peaty soils are readily identified by the characteristics listed in Article 2. The boundary between coarse-grained and fine-grained soils is taken to be the 200-mesh sieve (0.074 mm). In the field the distinction is based on whether the individual particles can be seen with the unaided eye. If more than 50% of the soil by weight is judged to consist of grains that can be distinguished separately, the soil is considered to be coarse-grained.

The coarse-grained soils are divided into gravelly (G) or sandy (S) soils, depending on whether more or less than 50% of the visible grains are larger than the No. 4 sieve (3/16 in.). They are each divided further into four groups:

- W: *well graded* (uniformity coefficient $U > 4$); *fairly clean* (<5% finer than 0.074 mm).
- P: *poorly graded* (gap-graded, or $U < 4$ for gravels or 6 for sands); *fairly clean* (<5% finer than 0.074 mm).
- C: *dirty* (>12% finer than 0.074 mm); *plastic* (clayey) *fines* ($I_w > 7$, also plots above A-line in plasticity chart).
- M: *dirty* (>12% finer than 0.074 mm); *nonplastic or silty fines* ($I_w < 4$, or plots below A-line in plasticity chart).

The soils are represented by symbols such as GW or SP. Borderline materials are represented by a double symbol, as GW-GP.

The fine-grained soils are divided into three groups: inorganic silts (M), inorganic clays (C), and organic silts and clays (O). The soils are further divided into those having liquid limits lower than 50% (L), or higher (H).

The distinction between the inorganic clays C and the inorganic silts M and organic soils O is made on the basis of a modified plasticity chart (Fig. 8.4). Soils CH and CL are represented by points above the A-line, whereas soils OH, OL, and MH correspond to positions below. Soils ML, except for a few clayey fine sands, are also represented by points below the A-line. The organic soils O are distinguished from the inorganic soils M and C by their characteristic odor and dark color or, in doubtful instances, by the influence of oven-drying on the liquid limit (Article 7). In the field, the fine-grained soils can be differentiated by their dry strength, their reaction

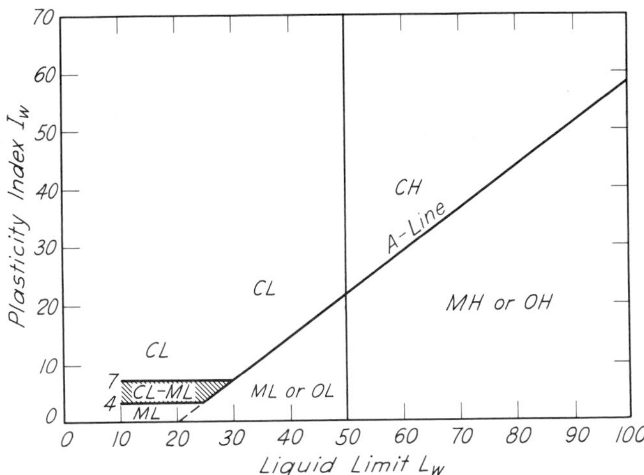

Fig. 8.4. Modified Plasticity Chart for use with Unified Soil Classification System. Soils represented by points within shaded area are considered borderline and are given dual symbols (after USBR 1963).

to the shaking test, or their toughness near the plastic limit (Article 7). The pertinent characteristics are indicated in Table 8.1. Borderline materials are represented by a double symbol, as *CL-ML*.

The Unified Soil Classification System permits reliable classification on the basis of relatively few and inexpensive laboratory tests. With experience it also provides a practicable basis for visual or field classification. Like all procedures based on grain size or the properties of remolded materials, it cannot take into consideration the characteristics of the intact materials as found in nature. Hence, it can serve

Table 8.1
Classification of Fine-Grained Soils Unified Soil Classification System

Group	Dry strength	Reaction to shaking test	Toughness at plastic limit
ML	None to Very Low	Rapid to Slow	None
CL	Medium to High	None to Very Slow	Medium
OL	Very Low to Medium	Slow	Slight
MH	Very Low to Medium	Slow to None	Slight to Medium
CH	High to Very High	None	High
OH	Medium to High	None to Very Slow	Slight to Medium

only as a starting point for the description of the engineering properties of soil masses or soil deposits.

Selected Reading

The definitive discussion of soil classification, as well as the basis for the Unified Soil Classification System, are contained in Casagrande, A. (1948): "Classification and identification of soils," *Trans. ASCE,* **113,** pp. 901–992.

ART. 9 MINIMUM REQUIREMENTS FOR ADEQUATE SOIL DESCRIPTION

In Article 8 suitable procedures were described for dividing soils into several large groups on the basis of their grain-size characteristics and their plasticity. If the engineer knows the group to which a given soil belongs, he also knows generally the more outstanding physical characteristics of the soil. However, each group includes soils with a great variety of properties and, furthermore, every soil can occur in the field in very different states. In order to distinguish among the individual members of each group and the different states of each member, two different procedures can be used. Either the principal groups can be further subdivided, or else the group name can be supplemented by numerical values representing the pertinent index properties.

The first of these two procedures is suitable for classifying the soils within geographically limited districts, because within such districts the number of different types and states of soil is likely to be fairly limited. As a consequence, the method is used extensively and to advantage by local construction organizations such as state highway departments. However, attempts to use a similar procedure for establishing a universal system of soil classification have little prospects for success, because the required terminology would inevitably be so complex that it would lead to ultimate confusion.

The second procedure, on the other hand, can be used profitably under any circumstances, provided the engineer chooses those index properties which are indicative of the essential physical characteristics of the soil. The properties required for adequate description of the various types of soil are summarized in Table 9.1. The soils listed in this table have been described in Article 2, which contains all the information required for at least a tentative classification of the soil. After the type has been recognized, the engineer turns to Table 9.1 and performs all the soil tests prescribed for this type. The test results represent the criteria for distinguishing between different soils of the same type.

With the exception of till, hardpan, and peat, all the soils listed in Table 9.1 consist either exclusively of coarse grains such as sand and gravel, or exclusively of fine grains having the size of silt or clay particles. Soils that consist of a mixture of these ingredients are regarded as composite. To describe a composite soil, it is first necessary to determine the natural void ratio e, the natural water content w, and the grain-size distribution. The soil is then divided into two parts, one of which contains all the grains larger than about 0.07 mm (width of the openings in the 200-mesh sieve), and the other of which contains the remainder. The coarse fraction is submitted to the classification tests prescribed for sand and gravel, and the remainder to those for silts and clays.

If the soils encountered on a given job are submitted to tests other than those listed in Table 9.1, the significant results of these tests should be included in the record. Since soil strata are seldom homogeneous, even an apparently homogeneous soil stratum cannot be considered adequately described unless the index properties of several samples from the stratum have been determined. The record should also contain a brief statement of whatever can be learned about the geological history of the stratum.

Today most large construction organizations, such as the Corps of Engineers of the United States Army, the United States Bureau of Reclamation, and many state highway departments, maintain soil laboratories in which the classification tests are made as a matter of routine. However, the results of these tests are of such practical importance that they should also be made by every engineer who deals with soils. The performance of the tests increases the engineer's familiarity with the various properties of the soils with which he deals, and the test results greatly increase the value of his field records.

After an engineer has personally tested several dozen samples of soil from one locality, he is likely to discover that he can estimate the index properties of most of the soils from that locality without any tests. He will also acquire the ability to discriminate between different soils or different states of the same soil which previously he had considered identical.

Every engineer should develop the habit of expressing his opinion on the plasticity and grain-size characteristics of the soils he encounters by numerical values rather than by adjectives. The grading of a sand should be expressed by the estimated value of the uniformity

Table 9.1
Data Required for Soil Identification

Type of Soil	General Information						Results of Classification Tests											
							Intact Samples[1]						Disturbed Samples					
	Color	Odor[2]	Texture[3]	Dilatancy[4]	Grain properties[5]	Dry strength[6]	Natural void ratio e[7]	Natural water content w	Unit weight, natural state	Unit weight, oven-dried γ_d	Unconf. compressive strength q_u	Sensitivity S_t[8]	Max. void ratio e_{max}[9]	Min. void ratio e_{min}[10]	Liquid limit L_w[11]	Plastic limit P_w[12]	Mechanical analysis[13]	Carbonate content[14]
Hardpan[15]	x	–	x	–	x	–	–	–	x	–	–	–	–	–	–	–	–	–
Sand, gravel	x	–	–	–	x	–	x	–	–	–	–	–	x	x	–	–	x	–
Inorganic silt	x	–	x	x	–	x	–	x	x	x	x	x	–	–	x	x	x	x
Organic silt	x	x	x	x	–	x	–	x	x	x	x	x	–	–	x	x	x	x
Clay	x	–	x	–	–	x	–	x	x	x	x	x	–	–	x	x	–	x
Organic clay	x	x	x	–	–	x	–	x	x	x	x	x	–	–	x	x	–	x
Peat	x	x	x	–	x	–	–	–	x	x	–	–	–	–	–	–	–	–
Till	x	–	–	–	x	–	x	x	x	–	–	–	x	x	–	–	x	–
Tuff, fine-grained	x	–	x	–	–	x	–	–	x	x	x	x	–	–	x	x	x	–
Loess[16]	x	–	x	x	–	x	x	x	x	x	x	–	x	x	x	x	x	x
Modified loess	x	–	x	x	–	x	x	x	x	x	x	–	x	x	x	x	x	x
Adobe	x	–	x	x	–	x	x	x	x	x	x	–	x	x	x	x	x	x
Marl	x	–	x	x	–	x	–	x	x	x	x	x	–	–	x	x	x	x
Lake marl	x	–	x	x	–	x	–	x	x	x	x	x	–	–	x	x	x	x
Gumbo	x	–	x	–	–	x	–	x	x	x	x	x	–	–	x	x	x	x

[1] If no undisturbed or tube samples were obtained, use the spoon samples (Article 44).

[2] If the odor is faint, heat the sample slightly. This intensifies the odor.

[3] Describe appearance of fresh fracture of intact sample (granular, dull, smooth, glossy). Then rub small quantity of soil between the fingers, and describe sensation (floury, smooth, gritty, sharp). If large specimens break up readily into smaller fragments describe appearance of walls of cracks (dull, slickensided) and average spacing of cracks.

[4] Perform shaking test, page 6. Describe results (conspicuous, weak, none) depending on intensity of phenomena observed.

[5] Describe shape (angular, subangular, subrounded, rounded, well rounded)

coefficient, $U = D_{60}/D_{10}$ (Article 5) and not by the words "well graded" or "poorly graded." The degree of plasticity should be indicated by the estimated value of the plasticity index I_w (Article 7) and not by the words "trace of plasticity" or "highly plastic." This habit is so important that it should be encouraged from the beginning by the instructor in the classroom. The use of the numerical system prevents misunderstandings and is an incentive to check from time to time the degree of accuracy of the estimates. Without occasional check tests the progressive deterioration of the ability to estimate may pass unnoticed.

---

and mineralogical characteristics of macroscopic soil particles only. Mineralogical characteristics include types of rocks and minerals represented among the grains so far as they can be discerned by inspection under the hand lens. Describe rock fragments (fresh, slightly weathered, or thoroughly decomposed; hard or friable). If a sand contains mica flakes, indicate mica content (slightly, moderately, or very micaceous). In connection with peat, the term grain properties refers to the type and state of preservation of the predominant visible remnants of plants such as fibers, twigs, or leaves.

[6] Crush dry fragment between fingers and indicate hardness (very low, low, medium, high, very high).

[7] If no undisturbed samples were obtained, substitute results of standard penetration test (Article 44) or equivalent.

[8] Applies only to clay and fine silt at a water content above the plastic limit.

[9] Prepare sample as described on page 25.

[10] Determine as described on page 26 for sands or gravels, or for other materials by means of Proctor method, page 445.

[11] If soil may be organic, determine L_w first in fresh state and then after drying in oven at 105°C.

[12] In addition to numerical value of P_w, state whether threads were tough, firm, medium, or weak.

[13] Present results either in form of semilogarithmic graph, or else by numerical values of D_{10} and $U = D_{60}/D_{10}$ (Article 5) accompanied by adjectives indicating the type of grain-size grading (see Fig. 5.2.)

[14] Calcium carbonate content can be detected by moistening the dry material with HCl. Describe result of test (strong, weak, or no effervescence).

[15] Add to data on texture a description of general appearance, structure, and degree of cohesiveness of chunks in fresh state and after soaking in water.

[16] Add to data on texture a description of the macroscopic features of the loess, such as diameter and spacing of root holes.

Chapter 2

HYDRAULIC AND MECHANICAL PROPERTIES OF SOILS

ART. 10 SIGNIFICANCE OF HYDRAULIC AND MECHANICAL PROPERTIES OF SOILS

In the preceding chapter, we have dealt with the index properties of soils. Since these properties reflect the general character of a given soil, they serve to indicate the extent to which soils from different localities may or may not be similar. In addition, they constitute the basis for recording construction experience and for utilizing this experience on subsequent jobs.

It has been emphasized that foundation and earthwork engineering is based chiefly on experience. However, it must also be emphasized that civil engineering in general did not emerge from a state of relative stagnation until the accumulated stock of experience became fertilized by applied science. The function of science was to disclose the relations between events and their causes.

In order to establish these relations in the realm of foundation and earthwork engineering, it has been necessary to investigate the physical properties of the different types of soils, just as it was necessary in structural engineering to investigate the properties of steel and concrete. A given steel or concrete is adequately described for most practical purposes if its strength and modulus of elasticity are known. On the other hand, practical problems involving soils may require the consideration of a variety of soil properties. Foremost among these are the permeability, the compressibility, the resistance against flow and shear, and the stress-deformation relationships. In the following articles, these properties are discussed in detail.

ART. 11 PERMEABILITY OF SOILS

Introduction

A material is said to be permeable if it contains continuous voids. Since such voids are contained in all soils including the stiffest clays, and in practically all nonmetallic construction materials including sound granite and neat cement, all these materials are permeable. Furthermore, the flow of water through all of them obeys approximately the same laws. Hence the difference between the flow of water through clean sand and through sound granite is merely one of degree.

The permeability of soils has a decisive effect on the cost and the difficulty of many construction operations, such as the excavation of open cuts in water-bearing sand, or on the rate at which a soft clay stratum consolidates under the influence of the weight of a superimposed fill. Even the permeability of dense concrete or rock may have important practical implications, because water exerts a pressure on the porous material through which it percolates. This pressure, which is known as *seepage pressure*, can be very high. The erroneous but widespread conception that stiff clay and dense concrete are impermeable is due to the fact that the entire quantity of water that percolates through such materials toward an exposed surface is likely to evaporate, even in a very humid atmosphere. As a consequence, the surface appears to be dry. However, since the mechanical effects of seepage are entirely independent of the rate of percolation, the absence of visible discharge does not indicate the absence of seepage pressures. Striking manifestations of this fact may be observed while an excavation is being made in very fine rock flour. The permeability of this material is very low. Yet, a slight change in the pressure conditions in the porewater may suffice to transform a large quantity of the material into a semiliquid.

Definitions and Darcy's Law

As water percolates through a permeable material, the individual water particles move along paths which deviate erratically but only slightly from smooth curves known as *flow lines.* If adjacent flow lines are straight and parallel, the flow is said to be *linear*.

The hydraulic principles involved in linear flow are illustrated by Fig. 11.1. In this figure, the points a and b represent the extremities of a flow line. At each extremity a standpipe, known as a piezometric tube, has been installed to indicate the level to which the water rises at these points. The water level in the tube at b is designated as the *piezometric level* at b, and the vertical distance from this level to point b is the *piezometric head* at b. The vertical distance between a and b represents the *position head* ΔH. If the water in the hydraulic system stands at the same elevation in the piezometric tubes at a and b, the system is in a state of rest, regardless of the magnitude of the position head. Flow can occur only if the piezometric levels at a and b differ by a distance h

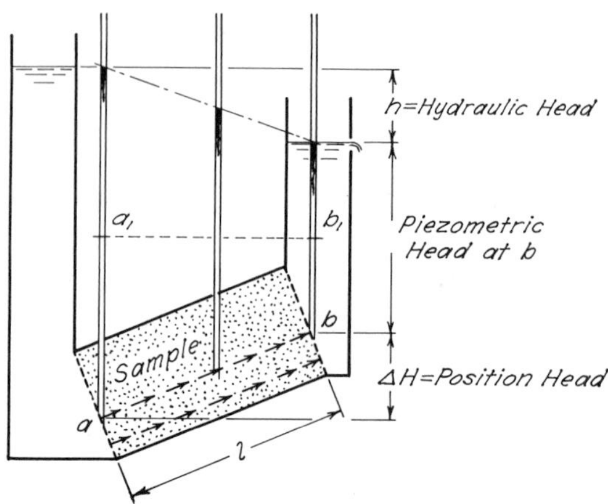

Fig. 11.1. Diagram illustrating meaning of hydraulic head and piezometric head associated with linear flow of water through soil sample.

known as the *hydraulic head* at a with respect to b. The distance h is also referred to as the *difference in piezometric level* between a and b. It should be observed that the difference in piezometric level is equal to the difference in the piezometric heads at a and b only if the position head ΔH is zero.

In Fig. 11.1, a_1 and b_1 represent any two points at the same elevation in the piezometric tubes rising from a and b respectively. Since the unit weight of the water is γ_w (grams per cubic centimeter), the hydrostatic pressure at a_1 exceeds that at b_1 by the amount $\gamma_w h$. The difference $\gamma_w h$ is the pressure that drives the water through the soil between a and b. It is referred to as the *excess hydrostatic pressure* at a with respect to b. The ratio,

$$i_p = \gamma_w \frac{h}{l} = \frac{u}{l} \tag{11.1}$$

in which u is the excess hydrostatic pressure, represents the pressure gradient (grams per cubic centimeter) from a to b. The ratio,

$$i = \frac{i_p}{\gamma_w} = \frac{1}{\gamma_w}\frac{u}{l} = \frac{h}{l} \tag{11.2}$$

is known as the *hydraulic gradient*. It is a pure number.

The *discharge velocity* v is defined as the quantity of water that percolates in a unit time across a unit area of a section oriented at right angles to the flow lines. In a statistically isotropic porous material the porosity of a plane section is equal to the volume porosity n. Hence, the average velocity v_s at which the water percolates through the voids of the material is equal to the discharge velocity divided by the porosity. The value v_s represents the *seepage velocity*. If the term velocity is used without qualification in connection with permeability, it always indicates the discharge and not the seepage velocity.

If water percolates through fine saturated sand or other fine-grained completely saturated soils without affecting the structure of the soil, the discharge velocity is almost exactly determined by the equation,

$$v = \frac{K}{\eta} i_p \tag{11.3}$$

in which η (gram-seconds per square centimeter) is the viscosity of the water, and K is an empirical constant referred to as the *permeability*. The viscosity of water decreases with increasing temperature, as shown in Fig. 11.2. The value K (square centimeters) is a constant for any permeable material with given porosity characteristics, and it is independent of the physical properties of the percolating liquid. From Eqs. 11.2 and 11.3 we obtain for the discharge velocity the expression,

$$v = \frac{K}{\eta} \gamma_w i \tag{11.4}$$

Seepage problems encountered in civil engineering deal almost exclusively with the flow of ground water at moderate depths below

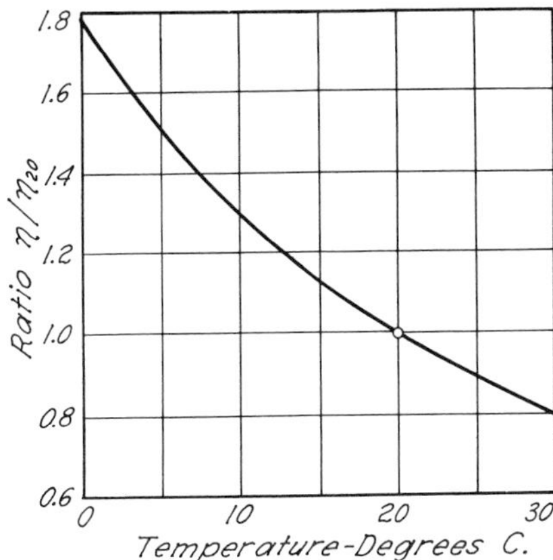

Fig. 11.2. Relation between temperature and viscosity of water.

the surface and with leakage out of reservoirs. The temperature of the percolating water varies so little that the unit weight γ_w is practically constant and, in addition, the viscosity η varies within fairly narrow limits. Therefore, it is customary to substitute in Eq. 11.4

$$k = K\frac{\gamma_w}{\eta} \tag{11.5}$$

whence

$$v = ki \tag{11.6}$$

In civil engineering, the value k is commonly called the *coefficient of permeability*. Equation 11.6 is known as *Darcy's law* (Darcy 1856).

It should be emphasized that the permeability characteristics of a porous material are expressed by K (square centimeters) and not by k (centimeters per second). The coefficient K is independent of the properties of the liquid, whereas k depends not only on the properties of the porous material, but also on the properties of the liquid. The use of k in this book, or in civil engineering in general, is justified only by convenience.

The channels through which the water particles travel in a mass of soil have a variable and irregular cross section. As a consequence, the real velocity of flow is extremely variable. However, the average rate of flow through such channels is governed by the same laws that determine the rate of flow through straight capillary tubes having a uniform cross section. If the cross section of the tube is circular, the velocity of flow increases, according to Poiseuille's law, with the square of the diameter of the tube. Since the average diameter of the voids in soil at a given porosity increases practically in proportion to the grain size D, it is possible to express k on the basis of Poiseuille's law as

$$k = \text{constant} \times D^2$$

From his experiments with loose filter sands of high uniformity (uniformity coefficient not greater than about 2), Allen Hazen obtained the empirical equation,

$$k(\text{cm/sec}) = C_1 D_{10}^2 \tag{11.7}$$

in which D_{10} is the effective size in centimeters (see Article 5), and C_1 (1/cm sec) varies from about 100 to 150. It should be noted that Eq. 11.7 is applicable only to fairly uniform sands in a loose state.

Relation between Void Ratio and Permeability

When a soil is compressed or vibrated, the volume occupied by its solid constituents remains practically unchanged, but the volume of the

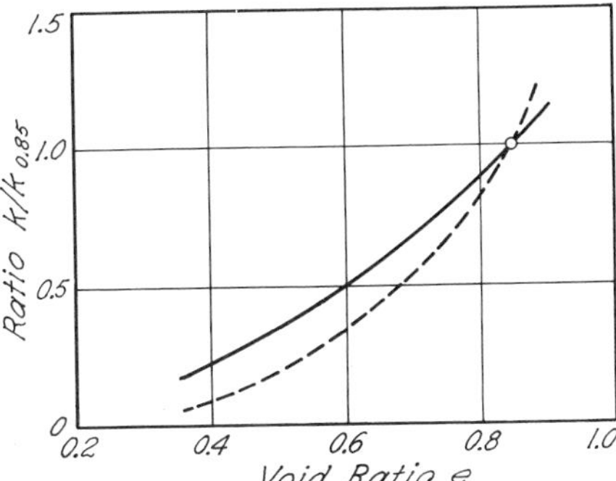

Fig. 11.3. Relation between void ratio and permeability of mixed-grained sand (full line) and soil with flaky constituents (dash line).

voids decreases. As a consequence, the permeability of the soil also decreases. The influence of the void ratio on the permeability is illustrated by Fig. 11.3. In this figure the abscissas represent the void ratio. The ordinates represent the ratio $k/k_{0.85}$ between the coefficient of permeability k of the soil at any given void ratio e and that of the same soil at a void ratio of 0.85. The plain curve shows the relation between e and $k/k_{0.85}$ for fine or medium clean sands with bulky grains. This relation can be expressed fairly accurately by various simple equations, such as A. Casagrande's unpublished equation

$$k = 1.4e^2k_{0.85} \tag{11.8}$$

In connection with foundation problems, clean sands are seldom encountered. If a sand contains a high percentage of scale-like particles such as mica flakes, the relation between e and $k/k_{0.85}$ resembles that indicated by the dash curve below the plain curve in Fig. 11.3. Fine-grained soils always contain flaky constituents, but, since the proportion is different for different soils, the corresponding $e - k/k_{0.85}$ curves are different.

In a soil that contains air bubbles, the size of the bubbles decreases with increasing water pressure. As a consequence, the coefficient of permeability of such a soil also increases with increasing water pressure. In clays with root holes or open cracks, percolation is almost inevitably associated with internal scour. The detached particles gradually clog the narrowest parts of the water passages whereupon the coefficient of permeability decreases to a small fraction of its initial value. Hence, Darcy's law is not valid unless the volume and shape of the water passages are independent of pressure and time.

Permeability Tests

The principal types of apparatus for determining the coefficient of permeability of soil samples are illustrated in Fig. 11.4. The constant-head permeameter (*a* and *b*) is suitable for very permeable soils, and the falling-head permeameter (*c*) for less permeable ones. In order to perform a test with any of these types of apparatus, a hydraulic gradient is established within the sample, and water flows through the soil.

In the constant-head permeameter (Fig. 11.4*a*) the hydraulic head h is kept constant, and the discharge is measured. In the falling-head permeameter (Fig. 11.4*c*) the water flows out of a narrow tube P with cross-sectional area A_1, through the sample which has a cross-sectional area A_2, into a stationary vessel V. The coefficient of permeability k is computed on the basis of the observed rate at which the water level descends in the tube P, while the water level in the vessel V remains unchanged.

The most important sources of experimental error in a permeability test are the formation of a filter skin of fine material on the surface of the sample and the segregation of air in the form of bubbles within the soil. Both of these phenomena reduce the average permeability of the

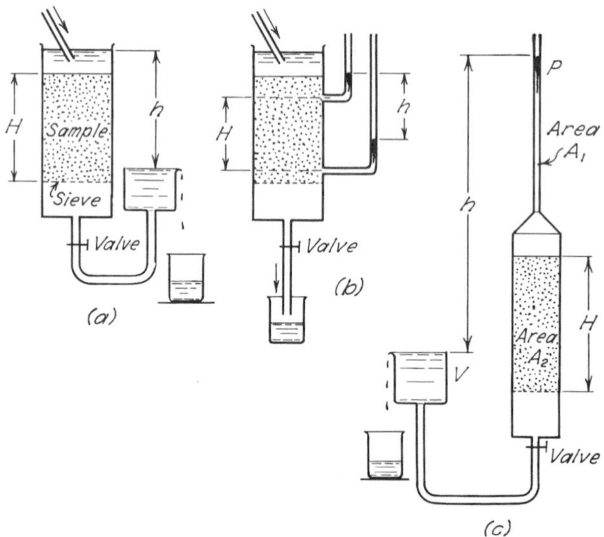

Fig. 11.4. (*a* and *b*) Constant-head permeameters. (*c*) Falling-head permeameter.

specimen. The error due to the formation of a filter skin can be eliminated by measuring the loss of head between two points located in the interior of the sample, as shown in Fig. 11.4*b*.

The value of the coefficient of permeability determined by means of a permeability test depends on the temperature at which the test is performed, because k (Eq. 11.5) is a function of the unit weight of the water γ_w and of the viscosity η. Both of these quantities vary with temperature. However, since the variation of γ_w is negligible in comparison with that of η, we may compute the value of k for any temperature T by means of the equation,

$$k = \frac{\eta_1}{\eta} k_1 \tag{11.9}$$

In this expression, derived from Eq. 11.5, k_1 is the coefficient of permeability corresponding to the test temperature, and η_1 is the corresponding viscosity. It is customary to express values of k at a standard temperature of 20°C. In Fig. 11.2, the ordinates represent the ratio between the values of η corresponding to the temperatures given by the abscissas, and the value η_{20} corresponding to $T = 20°C$.

Equation 11.9 was derived on the assumption that the coefficient of viscosity of the water is independent of the porosity and that it changes with temperature in accordance with the law represented by the curve in Fig. 11.2. In clays temperature seems to have a greater influence on viscosity than it has in coarser soils. Furthermore, the average viscosity of the porewater of clay appears to increase with decreasing pore space. At a given porosity, the average viscosity seems to decrease temporarily after remolding, even if the temperature is held constant. These facts exclude Eq. 11.9 from application to clays and other very fine-grained soils, but they do not invalidate Darcy's law (Eq. 11.6).

If a clay is remolded at unaltered water content, its coefficient of permeability is likely to decrease from the original value k to a smaller value k_r. For most inorganic clays, the ratio k/k_r is not greater than about two. However, for organic clays and for marls with a cluster structure, it may be as great as 30.

For coarse-grained soils with approximately equidimensional grains, such as quartz sand, the relation between the void ratio e and the coefficient of permeability k can be expressed with satisfactory accuracy by a single equation, such as Eq. 11.8, or by a single curve, such as the plain curve in Fig. 11.3. Hence, it is sufficient to determine the value of k corresponding to one arbitrary value of e. The values of k for other values of e can be derived from the test result by means of Eq.

11.8 or Fig. 11.3. On the other hand, the value of k for micaceous sands and for practically all the fine-grained soils encountered in the field depends to a large extent on the percentage of flaky constituents and on various other factors that are independent of the void ratio. For this reason, it has already been mentioned that the dash curve in Fig. 11.3 serves merely to illustrate the general character of the relation for such soils and cannot be used as a basis for computation. Hence, if a soil is micaceous or if it contains fine or very fine constituents, it is necessary to establish the relation between void ratio and permeability by performing permeability tests on at least three specimens with widely different void ratios.

Table 11.1 contains information regarding the range of the coefficient of permeability for various soils and the most appropriate methods for performing the permeability tests on these soils.

Permeability of Stratified Masses of Soil

Natural transported soils commonly consist of layers which have different permeability. In order to determine the average coefficient of permeability of such deposits, representative samples are secured from each of the layers and are tested. Once the values of k are known for the individual strata, the averages can be computed by using the following method. Let

$k_1, k_2 \cdots k_n$ = coefficients of permeability of the individual strata
$H_1, H_2 \cdots H_n$ = thicknesses of corresponding strata
$H = H_1 + H_2 + \cdots H_n$ = total thickness
k_I = average coefficient of permeability parallel to bedding planes (usually horizontal)
k_{II} = average coefficient of permeability perpendicular to bedding planes (usually vertical)

If the flow is parallel to the bedding planes, the average discharge velocity v is

$$v = k_I i = \frac{1}{H}[v_1H_1 + v_2H_2 + \cdots v_nH_n]$$

Furthermore, since the hydraulic gradient must be the same in every layer,

$$k_I i = \frac{1}{H}[k_1iH_1 + k_2iH_2 + \cdots k_niH_n]$$

whence

$$k_I = \frac{1}{H}[k_1H_1 + k_2H_2 + \cdots k_nH_n] \qquad (11.10)$$

Table 11.1
Permeability and Drainage Characteristics of Soils*

Coefficient of Permeability k in cm per sec (log scale): 10^{2}, 10^{1}, 1.0, 10^{-1}, 10^{-2}, 10^{-3}, 10^{-4}, 10^{-5}, 10^{-6}, 10^{-7}, 10^{-8}, 10^{-9}

	Range of k (cm per sec)	
Drainage	10^{2} to 10^{-4}	Good
	10^{-4} to 10^{-6}	Poor
	10^{-6} to 10^{-9}	Practically Impervious
Soil types	10^{2} to 1.0	Clean gravel
	1.0 to 10^{-3}	Clean sands, clean sand and gravel mixtures
	10^{-3} to 10^{-7}	Very fine sands, organic and inorganic silts, mixtures of sand silt and clay, glacial till, stratified clay deposits, etc.
	10^{-2} to 10^{-7}	"Impervious" soils modified by effects of vegetation and weathering
	10^{-7} to 10^{-9}	"Impervious" soils, e.g., homogeneous clays below zone of weathering
Direct determination of k	10^{2} to 10^{-4}	Direct testing of soil in its original position—pumping tests. Reliable if properly conducted. Considerable experience required
	10^{2} to 10^{-3}	Constant-head permeameter. Little experience required
Indirect determination of k	1.0 to 10^{-3}	Falling-head permeameter. Reliable. Little experience required
	10^{-3} to 10^{-6}	Falling-head permeameter. Unreliable. Much experience required
	10^{-6} to 10^{-9}	Falling-head permeameter. Fairly reliable. Considerable experience necessary
	10^{2} to 10^{-3}	Computation from grain-size distribution. Applicable only to clean cohesionless sands and gravels
	10^{-7} to 10^{-9}	Computation based on results of consolidation tests. Reliable. Considerable experience required

* After Casagrande and Fadum (1940).

For flow at right angles to the bedding planes, the hydraulic gradient across the individual layers is denoted by $i_1, i_2 \cdots i_n$. The hydraulic gradient across the series of layers is h/H, where h equals the total loss in head. The principle of continuity of flow requires that the velocity be the same in each layer. Therefore

$$v = \frac{h}{H} k_{II} = k_1 i_1 = k_2 i_2 = \cdots k_n i_n$$

Also,

$$h = H_1 i_1 + H_2 i_2 + \cdots H_n i_n$$

Combining these equations, we obtain

$$k_{II} = \frac{H}{\dfrac{H_1}{k_1} + \dfrac{H_2}{k_2} + \cdots \dfrac{H_n}{k_n}} \tag{11.11}$$

It can be demonstrated theoretically that for every stratified mass k_{II} must be less than k_I.

Scour and Scour Prevention at Boundaries

The engineer is often compelled to divert percolating water out of the soil into wells or ditches, or toward conduits located beneath foundations. This procedure is known as *drainage* (see Article 21). Wells usually consist of perforated pipes, and conduits of perforated pipes or pipe lines with open joints. The space between the natural soil and the pipes is filled with a coarse-grained material known as a *filler*. If the voids of the filler are very much larger than the finest grains of the adjoining natural soil, the finest soil particles are likely to be washed into the interstices of the filler where they accumulate and gradually obstruct the flow. On the other hand, if the voids in the filler are almost as small as those in the natural soil, the filler may be washed into the conduits and carried away. Both conditions are equally undesirable. If they are to be prevented, the filler must consist of a material with a grain size that meets certain requirements. Such a material is known as a *filter*.

The essential requirements for filter materials have been determined by experiment (Terzaghi 1922, USBR 1947). They are based primarily on the grain-size distribution of the filter materials and of the materials to be protected. They are summarized in Table 11.2.

If a filter extends across a boundary between coarse and fine soils, different materials should be used for covering the areas on either side of the boundary.

Table 11.2
Requirements for Filter Materials (after USBR 1963)

Character of Filter Materials	Ratio R_{50}	Ratio R_{15}
Uniform grain-size distribution (U = 3 to 4)	5 to 10	–
Well graded to poorly graded (non-uniform); subrounded grains	12 to 58	12 to 40
Well graded to poorly graded (non-uniform); angular particles	9 to 30	6 to 18

$$R_{50} = \frac{D_{50} \text{ of filter material}}{D_{50} \text{ of material to be protected}} \qquad R_{15} = \frac{D_{15} \text{ of filter material}}{D_{15} \text{ of material to be protected}}$$

Notes: If the material to be protected ranges from gravel (over 10% larger than No. 4 sieve) to silt (over 10% passing No. 200), limits should be based on fraction passing No. 4. Maximum size of filter material should not exceed 3 in. Filters should contain not over 5% passing No. 200. Grain-size curves (semi-logarithmic plot) of filter and of material to be protected should be approximately parallel in finer range of sizes.

Since it is desirable to reduce the loss of head due to percolation through the filter to the smallest value compatible with the grain-size requirements, large filters are usually made up of several layers. Each of these layers satisfies the conditions illustrated by Table 11.2 with respect to the preceding layer. Such composite filters are said to be *graded.*

The emergence of water from the ground at the boundary between a coarse and a fine soil may cause scour of the finer material, provided the velocity of the discharging water is great enough. Scour usually begins with the formation of small springs at different points along the boundary, from which channels are eroded in a backward direction toward the area where the water enters the soil. Hence, the process is known as *backward erosion.* It is one of the most dangerous menaces to dams, and it has been responsible for some of the most catastrophic dam failures (Article 63). Since the erosion cannot occur unless a large amount of soil is gradually washed out of the ground, it can be prevented effectively by constructing a filter over the area where springs may develop.

Problems

1. A sample of coarse sand, 15 cm high and 5.5 cm in diameter, was tested in a constant-head permeameter. Water percolated through the soil under a hydrostatic head of 40 cm for a period of 6.0 sec. The discharge water was collected and found to weigh 400 gm. What was the coefficient of permeability at the void ratio and temperature of the test?

Ans. $k = 1.05$ cm/sec.

2. A bed of sand consists of three horizontal layers of equal thickness. The value of k for the upper and lower layers is 1×10^{-4} cm/sec, and of the middle layer is 1×10^{-2} cm/sec. What is the ratio of the average permeability of the bed in the horizontal direction to that in the vertical direction?

Ans. 23 to 1.

3. A sample of mixed-grained sand with rounded particles has a void ratio of 0.62 and a coefficient of permeability of 2.5×10^{-2} cm/sec. Estimate the value of k for the same material at a void ratio of 0.73.

Ans. $k = 3.5 \times 10^{-2}$ cm/sec.

ART. 12 EFFECTIVE AND NEUTRAL STRESSES AND CRITICAL HYDRAULIC GRADIENT

Effective and Neutral Stresses

Figure 12.1*a* shows a cross section through a thin layer of soil that covers the bottom of a container. If a load p per unit of area is applied to the surface of the sample, for example by covering it with lead shot, the void ratio of the soil decreases from e_0 to e_1. The pressure p also produces a change in all of the other mechanical properties of the soil such as its shearing resistance. For this reason, it is known as an *effective pressure*. It is given the symbol $\bar{p}$.

If, instead, the vessel is filled with water to such a height h_w that $h_w = p/\gamma_w$, the normal stress on a horizontal section through the sample is also increased by p. Nevertheless, the increase in pressure due to

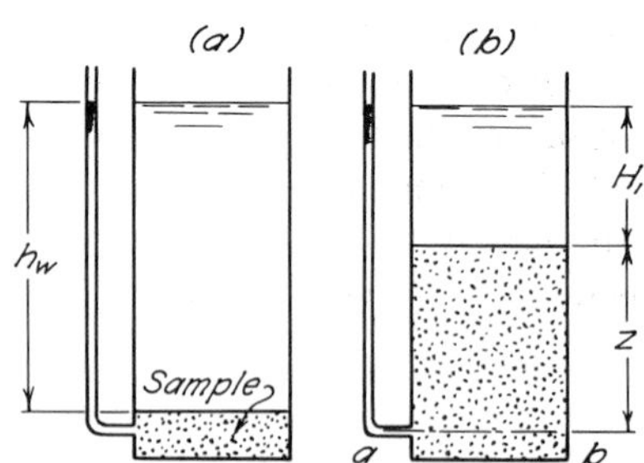

Fig. 12.1. Apparatus for demonstrating difference between effective and neutral stresses.

the weight of the water does not have a measurable influence on the void ratio or on any other mechanical property of the soil such as the shearing resistance. Therefore, the pressure produced by the water load is called a *neutral pressure*. It is said to be zero if it is equal to atmospheric pressure. Hence, the neutral pressure is equal to the piezometric head h_w times the unit weight of water γ_w, or

$$u_w = \gamma_w h_w \tag{12.1}$$

The total normal stress p at any point on a section through a saturated soil consists, therefore, of two parts. One part, u_w, acts in the water and in the solid in every direction with equal intensity; this part is known as the *neutral stress* or the *porewater pressure*. The remaining part $\bar{p} = p - u_w$ represents an excess over the neutral stress u_w and has its seat exclusively in the solid phase of the soil. This fraction of the total stress is called the *effective stress*.

A change in the neutral stress produces practically no volume change and has practically no influence on the stress conditions for failure, whereas all the measurable effects of a change in stress, such as compression, distortion, and a change in shearing resistance, are due exclusively to changes in the effective stress $\bar{p}$. Hence, every investigation of the stability or settlement of a saturated body of soil requires the knowledge of both the total and neutral stresses, and the equation

$$p = \bar{p} + u_w \tag{12.2}$$

is one of the most important in soil mechanics (Terzaghi 1936*b*).

The lower part of the container shown in Fig. 12.1*b* is filled with saturated soil having a unit weight γ. Water stands to a height H_1 above the surface of the soil. After equilibrium is established, the piezometric head h_w at depth z is $H_1 + z$, the neutral stress is

$$u_w = (H_1 + z)\gamma_w \tag{12.3}$$

and the total normal stress is

$$p = H_1\gamma_w + z\gamma \tag{12.4}$$

Hence the effective stress at depth z is

$$\bar{p} = p - u_w = H_1\gamma_w + z\gamma - (H_1 + z)\gamma_w = z(\gamma - \gamma_w) = z\gamma' \tag{12.5}$$

in which

$$\gamma' = \gamma - \gamma_w \tag{12.6}$$

The quantity γ' is called the *submerged unit weight* of the soil. It is equal to the difference between the unit weight γ of the saturated soil and the unit weight of water γ_w.

Critical Hydraulic Gradient

In the derivation of Eq. 12.5, the water in the voids of the soil is assumed to be in a state of rest. If, instead, water is flowing through the voids, Eq. 12.5 must be replaced by an expression that contains the hydraulic gradient i. This can be demonstrated by means of the apparatus shown in Fig. 12.2*a*. The cylindrical vessel A contains a layer of dense sand resting on a screen. The thickness of the layer is H, and the rim of the vessel is located at a distance H_1 above the top of the sand. The space below the screen communicates through a tube with vessel B. The water level is maintained at the elevation of the upper rim of each vessel. Hence, whatever the position of the water level in B, the total normal stress p on a horizontal section at depth z below the surface of the sand is always equal to p (Eq. 12.4). The corresponding effective normal stress $\bar{p}$ is equal to

$$\bar{p} = p - u_w$$

Hence, if the neutral stress in the water decreases or increases by Δu_w, the effective stress increases or decreases by the same amount, or

$$\Delta\bar{p} = -\Delta u_w \tag{12.7}$$

As long as the water level in both vessels is at the same elevation, the effective pressure at depth z is equal to $\bar{p} = z\gamma'$ (Eq. 12.5). If the vessel B is lowered through a distance h, water percolates through the sand in a downward direction under a hydraulic gradient $i = h/H$. The neutral stress at depth H is reduced by $h\gamma_w = iH\gamma_w$, and that at

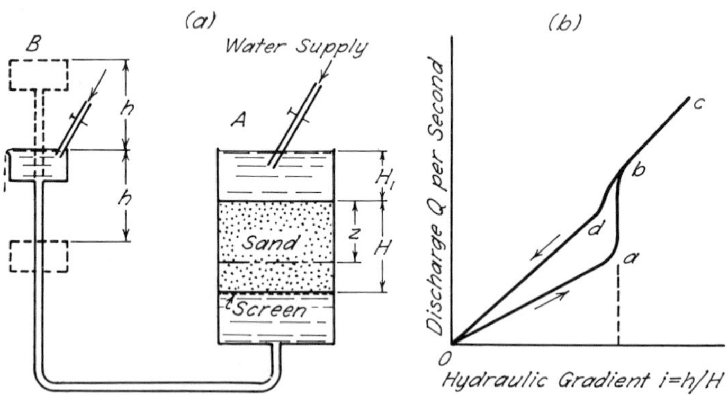

Fig. 12.2. (*a*) Apparatus for illustrating hydraulic conditions associated with boiling of sand. (*b*) Relation between upward hydraulic gradient and discharge through sand in apparatus shown in (*a*).

any other depth z is proportionately reduced by the amount $\Delta u_w = iz\gamma_w$. The effective stress is increased by the same amount.

On the other hand, if the vessel B is lifted through a distance h, the neutral stress at depth z increases by $\Delta u_w = iz\gamma_w$, and the effective stress decreases to

$$\bar{p} = z\gamma' - iz\gamma_w \tag{12.8}$$

The increase Δu_w of the neutral stress is caused exclusively by the transition of the porewater from a stationary state into a state of flow. The corresponding change Δu_w of the effective pressure in the sand is referred to as the *seepage pressure*. It is produced by the friction between the percolating water and the walls of the voids and can be described as a "drag." If the water percolates in a downward direction, the current drags the soil particles down and thereby increases the effective pressure in the sand. On the other hand, if the water flows in an upward direction, the friction between the water and the walls tends to lift the soil grains. As soon as the hydraulic gradient i in Eq. 12.8 becomes equal to

$$i_c = \frac{\gamma'}{\gamma_w} \tag{12.9}$$

the effective stress becomes equal to zero at any depth in the layer of sand. In other words, the average seepage pressure becomes equal to the submerged weight of the sand. The value i_c represents the *critical hydraulic gradient*.

Figure 12.2*b* illustrates the mechanical effect of the upward flow of water on the properties of the sand. In this diagram, the abscissas represent the hydraulic gradient, and the ordinates the corresponding discharge Q per unit of time. The curve *Oabc* represents the relation between discharge and hydraulic gradient as the hydraulic gradient is steadily increased. As long as i is less than i_c, the discharge increases in accordance with Darcy's law (Eq. 11.6) in direct proportion to i, and the value of k remains constant. This fact indicates that the mutual position of the sand grains remains practically unaltered. However, at the instant when i becomes equal to i_c, the discharge increases suddenly, involving a corresponding increase of the coefficient of permeability (Terzaghi 1929*a*). If a weight previously rested on the surface of the sand, it now sinks down as if the sand were a liquid. During a further increase of i the discharge again increases in direct proportion to i, and the coefficient of permeability retains the value that it acquired immediately after the hydraulic gradient exceeded the critical value. The decrease of discharge due to lowering the hydraulic gradient from a value greater than i_c is indicated by the line *cbdO*. As soon as i becomes roughly equal to i_c, the permeability decreases and then remains constant throughout any further decrease of i. Since the line *bdO* is located above the line *Oab*, the corresponding coefficient of permeability is greater than the original value. This fact suggests that the event represented by the step *ab* in the line *Oab* causes a permanent decrease of the density of the sand.

The process represented by the step *ab* is accompanied by a violent and visible agitation of the sand particles. Hence, it is commonly referred to as a *boiling* of the sand. Sand starts to boil in any open excavation if the ground water rises toward the bottom of the excavation at a hydraulic gradient greater than the critical value i_c. It has often been claimed that the boiling occurs only in certain types of sand known as quicksands. Therefore, it may be well to emphasize that it takes place in every sand and even in gravel, as soon as the hydraulic gradient becomes equal to i_c. The term true quicksand should be reserved for the members of a small group of very fine and very loose sands capable of becoming "quick" even if the hydraulic gradient of the seepage water is less than the critical value, and even if there is no perceptible external provocation. The little that is known about the characteristics of true quicksands is discussed in Article 17.

The boiling of ordinary sand can be prevented by constructing a loaded filter above the area in which the seepage emerges from the ground. A properly designed filter has almost no effect on the neutral stress in the soil. Hence, its entire weight serves to increase the effective pressure and to keep the sand particles in their original positions.

Problems

1. A sand is composed of solid constituents having a unit weight of 2.60 gm/cm^3. The void ratio is 0.572. Compute the unit weight of the sand when dry and when saturated, and compare with the effective unit weight when submerged.

Ans. $\gamma_d = 103.2$; $\gamma = 125.9$; $\gamma' = 63.5$ lb/ft^3.

2. The water table in a deep deposit of very fine sand is 4 ft below the ground surface. Above the water table, the sand is saturated by capillary water. The unit weight of the saturated sand is 127 lb/ft^3. What is the effective vertical pressure on a horizontal plane at a depth of 12 ft below the ground surface?

Ans. 1025 lb/ft^2.

3. A submerged stratum of clay has a thickness of 50 ft. The average water content of samples taken from the stratum is 54%, and the unit weight of the solid constituents is 2.78 gm/cm^3. What is the effective vertical pressure, due to the weight of the clay, at the base of the stratum?

Ans. 2220 lb/ft^2.

4. The unit weight of the particles of a sand is 2.66 gm/cm^3, the porosity in the loose state is 45%, and in the dense state is 37%. What is the critical hydraulic gradient for these two states?

Ans. 0.91; 1.05.

5. A large open excavation was made in a stratum of stiff clay with a saturated unit weight of 110 lb/ft^3. When the depth of the excavation reached 25 ft, the bottom rose, gradually cracked, and was flooded from below by a mixture of sand and water. Subsequent borings showed that the clay was underlain by a bed of sand with its surface at a depth of 37 ft. Compute the elevation to which the water would have risen from the sand into a drill hole before the excavation was started.

Ans. 21.2 ft above top of sand.

Selected Reading

The history and significance of the concept of effective stress are discussed by A. W. Skempton in "Terzaghi's discovery of effective stress," included in *From theory to practice in soil mechanics,* New York, John Wiley and Sons, 1960, pp. 42–53.

ART. 13 COMPRESSIBILITY OF CONFINED LAYERS OF SOIL

Introduction

If a stratum of soft clay is located directly beneath the footings of a building, the footings are likely to settle excessively and perhaps even to break into the soil. However, since unfavorable soil conditions of this kind are readily recognized, designers generally foresee the possible dangers and avoid difficulty by establishing the footings on piers or piles that pass through the soft stratum to a firm layer below.

On the other hand, if a thin layer of soft clay is buried beneath a thick layer of sand, the consequences of the presence of the clay layer are not so obvious. Before the advent of soil mechanics most engineers believed that the settlement of a footing depended only on the nature of the soil located immediately below the footing. Hence, if the soft clay was located more than 10 or 15 ft below the base of the footings, its presence was usually ignored. Even today, many engineers fail to take the clay into account although, because of its gradual consolidation under the weight of the building, excessive and unequal settlement of the building is likely to develop (Article 54).

As a result of the relative frequency of unexpected settlements of this kind, the compressibility of confined clay strata has received increasing attention during the last decades and methods have been developed for computing or estimating the amount and the distribution of the settlement. If the computed settlements are found to exceed a tolerable amount, the foundation is redesigned.

Adhesion and friction at the boundaries of confined clay strata appreciably restrain the strata from stretching in horizontal directions. Hence, the information required for computing the settlement due to the compression of confined clay strata can be derived from compression tests on laterally confined specimens.

Method of Testing

A confined compression test is made by placing the sample in a ring as shown in Fig. 13.1. The load is applied to the top of the sample through a rigid slab, and the compression is measured by means of a dial indicator. If the soil is saturated, the sample is placed between two porous disks that permit the escape of water during compression.

The results of the test are presented graphically. The void ratio e is plotted to a natural scale in the vertical direction. If the intensity of pressure p is plotted to a natural scale in the horizontal direction, the resulting curve is designated as an e–p curve. If the pressure is plotted to a logarithmic scale, the result is called an e–$\log p$ curve. Since each method of plotting has advantages, diagrams of both types will be used, and shown.

A distinction must be made between soils in their natural state and soils in which the original structure has been destroyed by remolding (see Article 7). The constituents of remolded soils are brought into their final positions by a process of kneading that involves slippage along the points of contact, whereas those of a sedimentary deposit are laid down grain by grain. These two processes may lead to very different structural patterns. Furthermore, in the ground, the constituents of most natural soils have not changed their relative positions for hundreds or even thousands of years, whereas those of a remolded soil or of a mineral powder obtained by a process of crushing or grinding reached their final positions only a few hours or days before the test. A

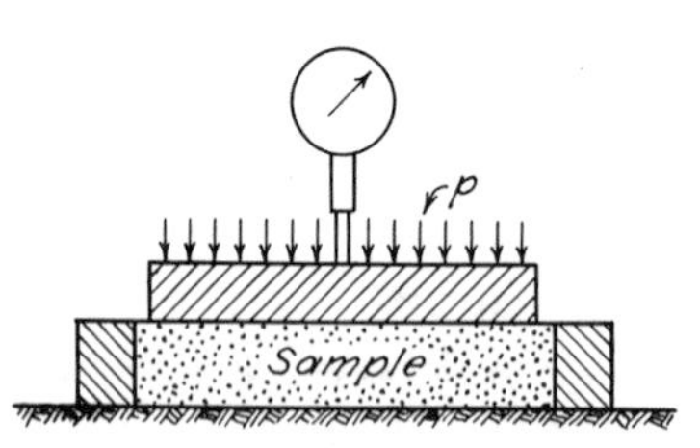

Fig. 13.1. Apparatus for performing laterally confined compression test on soil sample.

point contact of long duration may create molecular intergranular bonds which are wholly absent in a remolded soil. Therefore, the relations between void ratio and pressure for remolded and undisturbed soils are likely to be different. They are discussed under separate subheadings.

Compressibility of Crushed Minerals and Remolded Soils

Typical *e–p* curves for various crushed minerals and remolded soils are shown in Fig. 13.2*a*, and the corresponding *e*–log *p* curves in Fig. 13.2*b*. The effect of the shape of the grains on the compressibility of the grain aggregate is demonstrated by curves *a*, *b*, and *d*, in Fig. 13.2*a*. Curve *a* corresponds to a mixture of 80% sand and 20% mica; curve *b* to 90% sand and 10% mica; and curve *d* to 100% sand. Each sample was initially compacted by rodding and vibrating (Gilboy 1928). These curves demonstrate that the compressibility increases greatly with increasing percentages of scale-shaped particles. Furthermore, Fig. 13.2*a* shows that the average slope of the curve *d* for dense sand is considerably flatter than that of curve *c* for the same sand in a loose state and that the void ratio of a loose sand, even under very great pressure, is greater than that of the same sand in a dense state subjected to no pressure.

Figure 13.2*a* also shows that the curve *e* corresponding to a remolded sample of a soft clay is very similar to that for a mixture of 90% sand

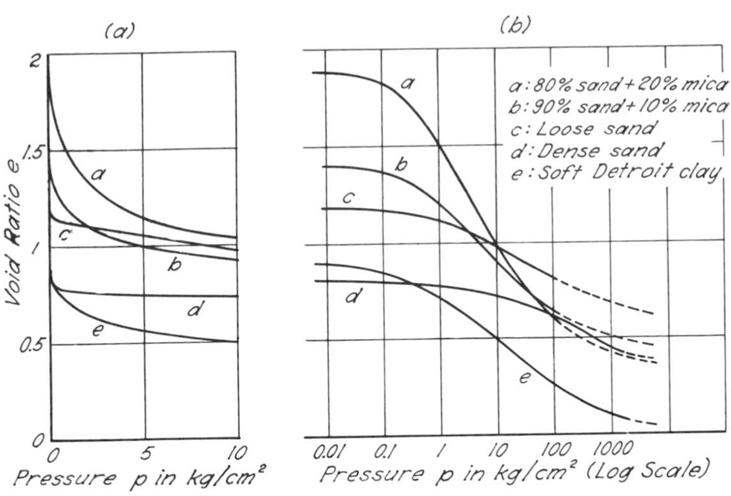

Fig. 13.2. **(*a*) Typical *e-p* curves. (*b*) Corresponding *e*-log *p* curves representing results of compression tests on laterally confined laboratory soil aggregates.**

and 10% mica, but that the void ratio of the clay at any given pressure is much smaller than the corresponding void ratio of the sand-mica mixture.

All of the *e*–log *p* curves shown in Fig. 13.2*b* have certain characteristics in common. Each curve starts with a horizontal tangent and probably ends with a tangent that is nearly horizontal. The middle sloping part of each curve is fairly straight. For sands the middle part is straight from a pressure of about 10 to about 100 kg/cm^2. At this pressure the grains begin to crush, and the slope increases. The slope then remains fairly constant up to about 1000 kg/cm^2, whereupon it begins to decrease (Hendron 1963). The slope of the middle part of the curves for soft remolded clays decreases so slightly throughout the range from about 1 to 2000 kg/cm^2 that the curve can be regarded as straight within this entire range (Akagi 1960). The middle sections of the curves for sand-mica mixtures are practically straight within the range from 1 to 10 kg/cm^2. The slope then decreases as the curves approach a nearly horizontal tangent.

Two other phenomena are of special interest in connection with the compressibility of soils in general. They are the time rate at which the compression takes place and the volume change caused by temporary removal of load.

The time effects associated with the compression of sand are illustrated by Fig. 13.3. In this figure, the curve K_l represents the decrease of the void ratio of a loose sand due to a pressure that increases at a constant and fairly rapid rate. If the process of loading is interrupted, the void ratio decreases at constant load, as indicated by the vertical step in the *e*–*p* curve, and by the corresponding *e–time* curve. If, after an intermission, the process of loading is resumed at the original rate, curve K_l merges smoothly into the curve that would have been obtained by loading the sand at a constant rate without intermission. The decrease of the void ratio at constant load is due to a lag in the adjustment of the position of the grains to the increasing pressure.

Similar time effects due to the same cause are also observed when a sample of saturated remolded clay is tested. However, they are combined with the much more important lag due to the low permeability of the clay. Because of the time lag, an *e*–*p* curve has no definite physical meaning unless every point corresponds to a stage at which the void ratio has become practically constant at a constant load.

Figure 13.3 also shows the change in void ratio due to temporary removal of the load. The removal of the load is represented by the *decompression curve bc*, and the subsequent reapplication of load by the *recompression curve cd*. For clays *bc* represents the *swelling curve*. The

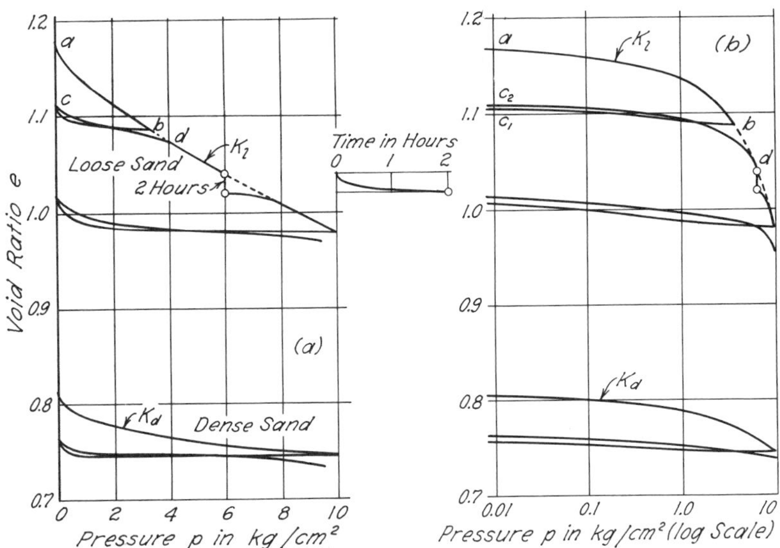

Fig. 13.3. **(*a* and *b*) Relation between *e* and *p* corresponding to results of compression tests on laterally confined sand.**

area between the decompression and recompression curves is a *hysteresis loop*. Hysteresis loops for different soils differ only in slope and in width. In arithmetic plots they are concave upward, whereas in semilogarithmic plots they are concave downward. Figure 13.4 shows a hysteresis loop for a dense mixture consisting of 90% sand and 10% mica. Hysteresis loops for remolded clays are very similar.

Undisturbed Sand

In nature all sands are more or less stratified. The compressibility of a stratified deposit in the direction of the bedding planes is somewhat smaller than that in the direction at right angles to them. In addition, most natural sands contain at least traces of cementing material, and above the water table they also contain some soil moisture. Both ingredients produce cohesion. Furthermore, some sands in the natural state have a relative density greater than that which can be obtained by any artificial means other than vibrations. Other sands in a natural state have a very unstable structure that can be approximated in the laboratory only by preparing extremely loose specimens by special procedures (Article 17). These facts suggest that the structure of sands in their natural state may be slightly different from that of the same

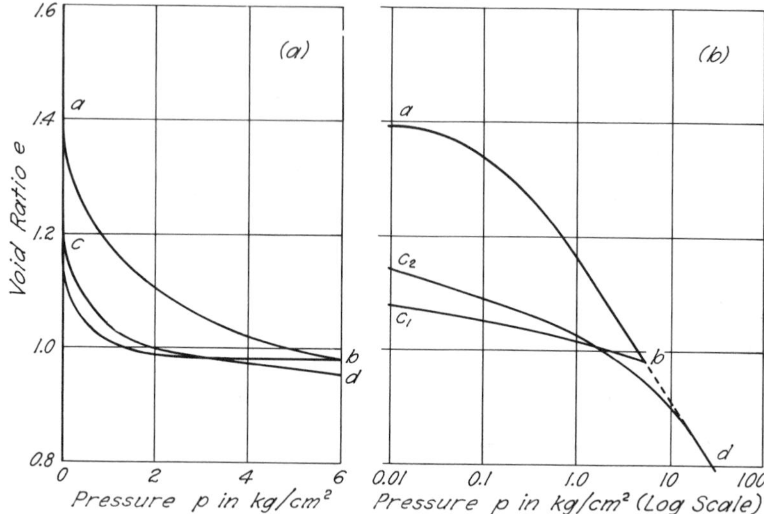

Fig. 13.4. Relation between e and p for laterally confined dense sample consisting of 90% sand and 10% mica (Gilboy 1928). A similar relation is obtained for remolded clays.

sands in samples made in the laboratory. However, if the void ratios of the sands are the same in both states, the compressibilities are also likely to be approximately equal.

Undisturbed Normally Loaded Insensitive Clays

The following discussion is limited to clays that have never been subjected to a pressure greater than that which corresponds to the present overburden. Such clays are referred to as *normally loaded*. Experience indicates that the natural water content w of normally loaded clays is commonly close to the liquid limit L_w. If w is considerably lower than L_w, the sensitivity (Article 7) of the clay is likely to be exceptionally low. On the other hand, if w is considerably greater than L_w, the clay is likely to have a high sensitivity. In any event, normally loaded clays are always soft to a considerable depth below the surface.

In order to obtain information about the compressibility of a confined stratum of an insensitive normally loaded clay located at depth D below the surface, we may test an undisturbed sample taken at that depth from a test shaft or boring. In Fig. 13.5 the coordinates of point a represent the natural void ratio e_0 of the sample and the effective overburden pressure p_0 on the soil at depth D. The pressure p_0 is equal to the sum of the submerged weight of the soil located between depth D and the water table and the full weight of the soil and soil moisture located above the water level. It is expressed in weight per unit of area.

During the process of sampling, the overburden pressure on the clay that constitutes the sample is reduced to a very small value, although the water content remains almost unchanged. In Fig. 13.5 this process is represented by the dash line ae_0. If the pressure on the sample is again increased by loading the sample in the consolidation device, the void ratio of ordinary clays with medium or low sensitivity decreases with increasing load as indicated by the line K_u. The curved portion of K_u represents a recompression curve such as curve c_2d in Fig. 13.4*b*. It merges into a straight line. The upward continuation of the straight part of K_u corresponds to the tangent db to the curve c_2d in Fig. 13.4*b*. It intersects the horizontal line through a, Fig. 13.5, at point b. Experience shows that for normally loaded clays point b is always located on the left side of point a.

If we transform the clay sample into a thick paste by mixing it with water and then gradually consolidate the paste under an increasing

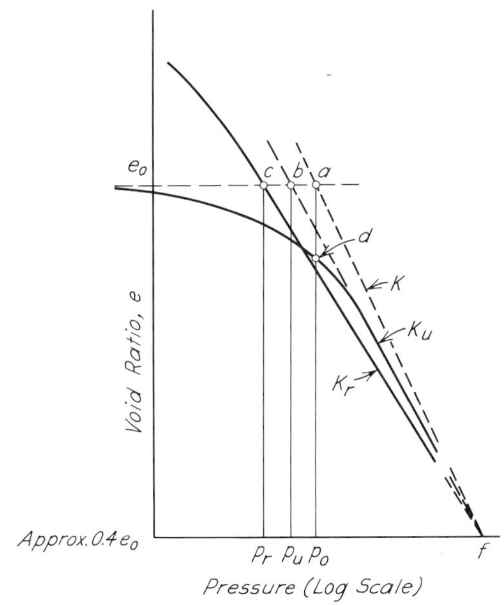

Fig. 13.5. Relations between e and p for clay of ordinary sensitivity corresponding to K_r remolded and K_u undisturbed states in the laboratory, and K natural state in the field.

pressure, we obtain the e–log p line K_r (Fig. 13.5). Below point c this line is almost straight. Its slope is somewhat smaller than that of the straight part of K_u, and its downward continuation intersects the straight projection of K_u at point f, at a void ratio equal to approximately $0.4e_0$ (Schmertmann 1953).

The consolidation line K that represents the relation between e and log p in the field must pass through a. Yet, neither of the two laboratory lines K_u and K_r passes through this point. Hence, it is obvious that the line K can be determined only by some process of extrapolation from the results of the laboratory tests. If the two lines K_u and K_r are straight and intersect at about $e = 0.4e_0$, it seems reasonable to assume that the e–log p curve for the soil in the field is also a straight line that passes through a and, if continued downward, also intersects the line $e = 0.4e_0$ at point f. The line so obtained is referred to as the *field consolidation line*. If no undisturbed samples are available, point f can be determined with sufficient accuracy from the e–log p curve for a remolded sample, K_r in Fig. 13.5, provided the load on the sample is increased to at least 20 kg/cm^2.

The value of the ratio p_u/p_0 between the pressures represented by the abscissas of b and a (Fig. 13.5) indicates the degree to which the structure of the sample has been disturbed. It ranges from about 0.3 to 0.7, with an average value of 0.5. A considerable scattering from the average value is characteristic even for samples taken with the same tool from the same boring. Hence, it seems that the value p_u/p_0 depends to a considerable extent on accidental factors, such as variations in the sensitivity of the clay and whether the test specimen was taken from the soil in the lower, middle, or upper part of the sampling spoon.

The field consolidation line K in Fig. 13.5 represents the basis for the computation of the settlement of structures located above confined strata of normally loaded clay. The weight of a fill or of a structure increases the pressure on the clay from the overburden pressure p_0 to the value $p_0 + \Delta p$. The corresponding void ratio decreases from e_0 to e. Hence, for the range in pressure from p_0 to $p_0 + \Delta p$, we may write

$$e_0 - e = \Delta e = a_v \Delta p$$

The value,

$$a_v(\text{cm}^2/\text{gm}) = \frac{e_0 - e}{\Delta p(\text{gm}/\text{cm}^2)} \qquad (13.1)$$

represents the *coefficient of compressibility* for the range p_0 to $p_0 + \Delta p$. For a given difference in pressure, the value of the coefficient of compressibility decreases as the pressure increases. The decrease in porosity Δn per unit of the original volume of the soil, corresponding to a decrease in void ratio Δe, may be obtained by means of Eq. 6.2. The resulting expression is

$$\Delta n = \frac{\Delta e}{1 + e_0}$$

in which e_0 is the initial void ratio. Therefore

$$\Delta n = \frac{a_v}{1 + e_0} \Delta p = m_v \Delta p \tag{13.2}$$

in which

$$m_v(\text{cm}^2/\text{gm}) = \frac{a_v(\text{cm}^2/\text{gm})}{1 + e_0} \tag{13.3}$$

is known as the *coefficient of volume compressibility*. It represents the compression of the clay, per unit of original thickness, due to a unit increase of the pressure. If H is the thickness of a bed of clay under a pressure p, an increase of the pressure from p to $p + \Delta p$ reduces the thickness of the stratum by

$$S = H \cdot \Delta p \cdot m_v \tag{13.4}$$

The field consolidation line K for ordinary clays appears in a semilogarithmic diagram as a straight line, as shown in Fig. 13.5. This line can be represented by the equation,

$$e = e_0 - C_c \log_{10} \frac{p_0 + \Delta p}{p_0} \tag{13.5}$$

in which C_c (dimensionless) is the *compression index*. It is equal to the tangent of the slope angle of the straight part of line K. In contrast to a_v and m_v, which decrease rapidly with increasing values of the pressure p_0, the value C_c is a constant, and Eq. 13.5 which contains this constant is valid within a fairly large range of pressure.

In a semilogarithmic plot, the decompression curve, such as bc_1 in Fig. 13.4*b*, is also fairly straight over a wide range of pressure. If the pressure is reduced from p to $p - \Delta p$, the corresponding decompression curve can be expressed by the equation,

$$e = e_1 + C_s \log_{10} \frac{p_0 + \Delta p}{p_0} \tag{13.5a}$$

in which C_s (dimensionless) is the *swelling index*. It is a measure of the volume increase due to the removal of pressure.

By combining Eq. 13.5 with Eq. 13.1 and 13.3, we obtain

$$a_v = \frac{C_c}{\Delta p} \log_{10} \frac{p_0 + \Delta p}{p_0} \tag{13.6}$$

and

$$m_v = \frac{C_c}{\Delta p(1 + e_0)} \log_{10} \frac{p_0 + \Delta p}{p_0} \tag{13.7}$$

Substituting the value of m_v into Eq. 13.4, we find that the compression S of a confined stratum of normally loaded ordinary clay is

$$S = H \frac{C_c}{1 + e_0} \log_{10} \frac{p_0 + \Delta p}{p_0} \tag{13.8}$$

If a clay is remolded, its e–log p curve changes from K (Fig. 13.5) to K_r. Since the line K_r is straight over a large range of pressure, it can be represented by the equation,

$$e = e_0 - C_c' \log_{10} \frac{p_0 + \Delta p}{p_0} \tag{13.9}$$

which is an analogue to Eq. 13.5. The symbol C_c', which represents the compression index for the clay in a remolded state, is equal to the tangent of the slope angle of the straight part of the line K_r. The values of C_c' for different clays increase consistently with increasing liquid limit as shown in Fig. 13.6. The abscissas of the points shown in the diagram represent the liquid limit L_w, and the ordinates the corresponding values of C_c' for different clays. The samples were selected at random. They came from different parts of the world, and the assortment includes both ordinary and extrasensitive clays. All the points are located close to a straight line with the equation,

$$C_c' = 0.007(L_w - 10\%) \tag{13.10}$$

in which L_w is the liquid limit in per cent of the dry weight of the clay. The scattering of the real values of C_c' from those determined by equation 13.10 is about $\pm 30\%$ (Skempton 1944).

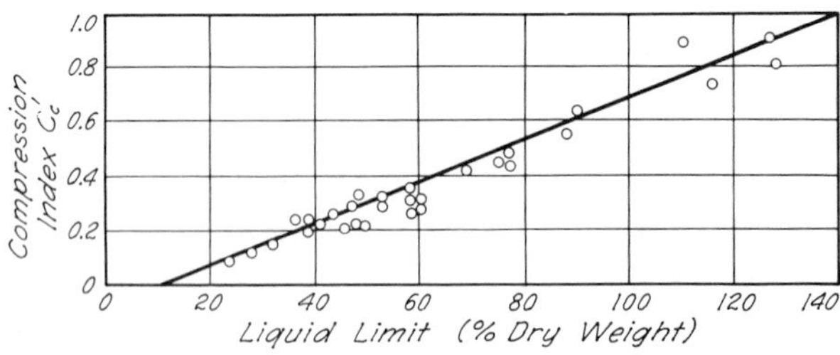

Fig. 13.6. Relation between liquid limit and compression index for remolded clays (after Skempton 1944 and others).

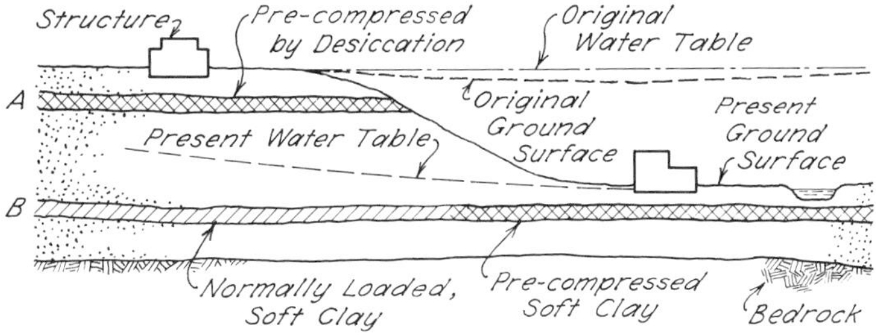

Fig. 13.7. Diagram illustrating two geological processes leading to precompression of clays.

For an ordinary clay of medium or low sensitivity, both the e–log p lines K_r and K are straight over a wide range of pressure, and the value of C_c corresponding to the field consolidation line K appears to be roughly equal to 1.30 C_c' (Eq. 13.10). That is,

$$C_c \sim 1.30C_c' = 0.009(L_w - 10\%) \qquad (13.11)$$

If the value of C_c for a given layer of clay is known, the compression of the layer due to a surcharge Δp can be computed by means of Eq. 13.8. For normally loaded clays with low or moderate sensitivity the value of C_c can be estimated roughly by means of Eq. 13.11. Hence, the order of magnitude of the settlement of a structure located above a stratum of such clay can be determined without making any tests other than liquid-limit tests.

Undisturbed Precompressed Clay

A clay is said to be precompressed if it has ever been subjected to a pressure in excess of its present overburden pressure. The temporary excess pressure may have been caused by the weight of soil strata that were later eroded, by the weight of ice that later melted, or by desiccation due to temporary exposure. If the excess pressure Δp_0 was smaller than about 4 kg/cm^2, the clay may still be soft. If Δp_0 was much greater, however, the clay is stiff.

Two of the processes which lead to the precompression of clays are illustrated in Fig. 13.7. All of the strata located above bedrock were deposited in a lake at a time when the water level was located above the level of the present high ground. When parts of the strata were removed by erosion, the water content of the clay in the right-hand portion of stratum B increased slightly, whereas that of the left-hand portion decreased considerably because of the lowering of the water table. Nevertheless, with respect to the present overburden, the clay on the right-hand side is a precompressed soft clay, and that on the left-hand side is a normally loaded soft clay.

While the water table descended from its original to its final position below the floor of the eroded valley, the sand strata above and below the upper clay layer A became drained. As a consequence, the layer A gradually dried out. In Article 21 it is shown that such a process of desiccation constitutes the mechanical equivalent of consolidation under load. Therefore, layer A is said to be *precompressed by desiccation.*

If a bed of clay is formed by sedimentation in an open body of water subject to seasonal or cyclic variations in water level, the highest portions of the surface of the sediment may emerge from time to time. Beneath these areas dry crusts are formed by desiccation. After the surface is flooded again, the crusts are buried under freshly deposited sediments, but their water content remains abnormally low. Hence, they constitute layers or lenses of precompressed clay located between layers of normally loaded clay.

If a layer of stiff clay is located above a layer of soft clay of the same type, it is likely that the upper layer has been precompressed by desiccation. Furthermore, if the upper layer was exposed to the atmosphere for a long time, it is also likely to be discolored by oxidation. For example, in the Chicago area a thick layer of soft normally loaded clay of grayish color is covered by a layer of stiff precompressed yellow and gray clay between 2 and 6 ft thick. Precompressed layers of glacial clay located between normally loaded layers of soft clay of the same type have been encountered in southern Sweden. In some instances, stiff crusts may have formed without emergence by a process of subaqueous weathering or cation exchange (Moum and Rosenqvist 1957).

The influence of precompression on the relation between void ratio and pressure is shown in Fig. 13.8. Both diagrams are plotted to a natural scale. Fig. 13.8a represents the relation between e and p for the normally loaded part of the clay stratum B in Fig. 13.7, and Fig. 13.8b shows the corresponding relation for the precompressed part of the same stratum. In both diagrams point a' represents the state of the clay before erosion started. At that time the water table was located above stratum A, and the effective overburden pressure for the entire stratum B was equal to p_0' per unit of area. Since erosion was associated with a lowering of the water table at almost constant total overburden pressure, the effective overburden pressure on the left-hand part of stratum B increased from p_0' to p_0, and the point that represents the state of the clay (Fig. 13.8a) moved from a' to a.

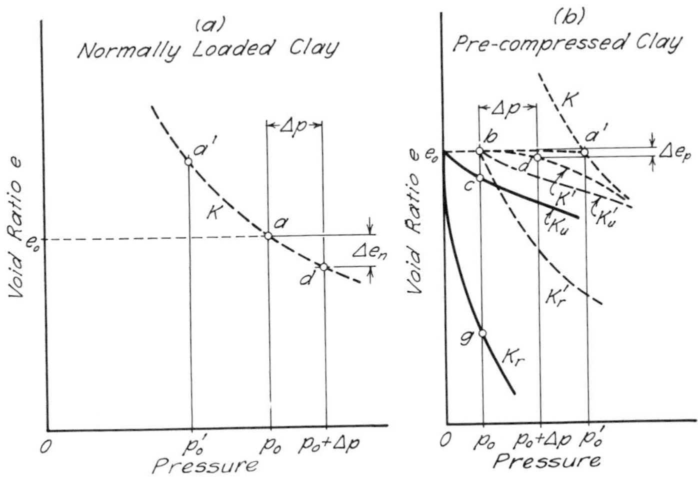

Fig. 13.8. **(*a*) Field relation between *e* and *p* for normally loaded clay. (*b*) Relations between *e* and *p* for similar clay in a precompressed state.**

In the right-hand part of stratum B the lowering of the water table took place simultaneously with the removal of most of the overburden. Hence, the effective pressure on the right-hand part of the stratum decreased from p_0' to p_0, and the clay passed from state a' (Fig. 13.8b) into state b. The transition was associated with a slight increase of the void ratio.

An increase of Δp of the effective pressure on the normally loaded part of stratum B, caused by such an operation as the construction of a large heavy building on the high ground, reduces the void ratio of the clay located beneath the building by Δe_n (Fig. 13.8a) and the clay passes from state a into state d. An increase of the effective pressure on the precompressed right-hand part of stratum B by the same amount Δp reduces the void ratio of the clay by Δe_p (Fig. 13.8b) and the clay passes from state b into state d.

If disturbed samples were taken from both parts of stratum B, they would probably create the impression that the precompressed clay is softer than the normally loaded clay, because the water content of the precompressed part of the stratum at the time of sampling would be appreciably greater than that of the normally loaded part. Nevertheless, if Δp is smaller than about one half of $p_0' - p_0$, the compression Δe_p of the precompressed stratum will be much smaller than the compression Δe_n of the normally loaded stratum. This is due to the fact that the point which represents the state of the normally loaded clay in the ground advances from a to d (Fig. 13.8a) on a curve representing the decrease of void ratio due to a steadily increasing pressure, whereas the corresponding point for the precompressed clay moves on a recompression curve from b to d (Fig. 13.8b). As shown in Figs. 13.3 and 13.4, the slope of a recompression curve is very much smaller than that of a direct compression curve.

Some conception of the amount of compression that the precompressed part of stratum B would experience under the weight of the building can be obtained from the results of consolidation tests on representative samples of this part. However, because of the precompression, the e–p curve for the soil in the field is likely to differ to a considerable extent from that obtained by means of laboratory tests. The magnitude of the difference depends on the degree of disturbance of the samples.

If the sample is badly disturbed, the laboratory relation between e and p resembles the steep curve K_r in Fig. 13.8b. By adding the distance bg to the ordinates of this curve we obtain the curve K_r' which passes through point b representing the state of the clay in the ground. However, experience shows that the curve K_r' has no resemblance to the field consolidation line bd.

If the consolidation test is made on an undisturbed sample carefully carved out of the ground in a shaft, the curve K_u is obtained. By adding the distance cb to the ordinates of this curve we obtain the curve K_u' which passes through b. Although the slope of K_u' is much smaller than that of K_r', it has been found that, if Δp is smaller than about one half of $p_0' - p_0$, the compression of the clay computed on the basis of K_u' is still two to five times greater than the actual compression of the clay in the field. Hence, extrapolation from test results to field conditions is very uncertain, irrespective of the care with which sampling operations are carried out.

Computation of the relation between e and p for a clay with a given liquid limit on the basis of Eq. 13.11 leads to a curve through b which is steeper than K_r'. The ordinates of this curve with reference to a horizontal line through b are equal to at least twice the ordinates of K_u', which in turn are two to five times greater than those of the field e–p curve K'. Hence, the use of Eq. 13.11 for estimating the compressibility of a precompressed clay leads to values between four and ten or more times greater than the correct ones. Since the same equation furnishes reasonably accurate values when applied to normally loaded clays, it is obvious that the load history of a clay is of outstanding practical importance.

Under the conditions illustrated by Fig. 13.7, the maximum consolidation pressure p_0' can be estimated rather accurately on the basis of geological evidence. The geology and physiography of the site leave no doubt that the original ground surface was located at or above the level of the present high ground and that the water table was fairly close to the original ground surface. However, if the geological evidence is not unmistakable, or if the precompression was caused by the weight of an ice sheet which melted without leaving any evidence of its thickness, a geological estimate of the maximum consolidation pressure is very uncertain. In such instances, the only remaining procedure for gaining at least a general conception of the value of p_0' is to make an estimate based on the results of laboratory tests.

Several methods have been proposed for determining the value of the maximum consolidation pressure from the results of laboratory tests. The one most commonly used is illustrated by Fig. 13.9 (A. Casagrande 1936*b*). This figure shows the *e*-log *p* curve for an undisturbed clay sample. Through point *c*, at which the radius of curvature is a minimum, a horizontal line is drawn. The bisector of the angle α between this line and the tangent to K_u at *c* intersects the upward continuation of the straight lower part of K_u at point *d*. The abscissa of *d* is assumed to be equal to p_0'.

The method illustrated by Fig. 13.9 is based on the observed effect of cyclic loading on the void ratio of undisturbed clay samples. It has been found to give good agreement with the existing overburden effective pressures in deposits known to be normally loaded, provided the tests were made on undisturbed samples of the highest quality. In the few instances in which the maximum consolidation pressure of a preloaded clay has been reliably determined by geologic evidence or other independent means, agreement between the actual maximum consolidation pressure and that determined by means of the graphical procedure has also been quite satisfactory provided the samples for the consolidation tests were undisturbed.

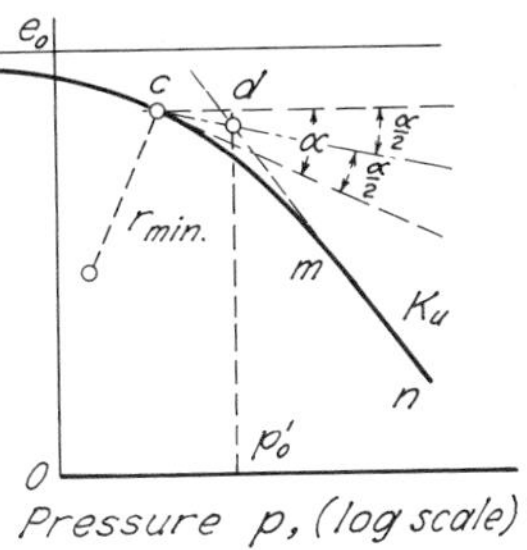

Fig. 13.9. Diagram illustrating commonly used graphical construction for estimating value of maximum consolidation pressure (after A. Casagrande 1936*b*).

If a clay has been heavily preconsolidated, it may not be practicable to increase the pressure on an undisturbed sample in a consolidation test until the steep, fairly straight portion of the *e*-log *p* curve above the maximum consolidation pressure is well

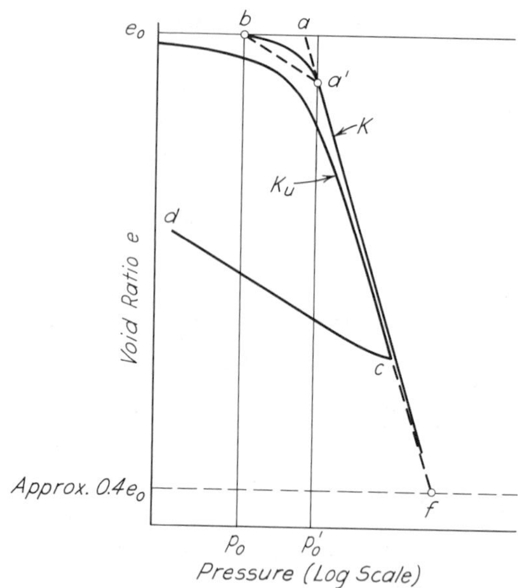

Fig. 13.10. Graphical construction for approximating field relation between *e* and *p* for precompressed clay (after Schmertmann 1953).

defined. However, if the magnitude of the precompression permits the determination of this part of the curve, a somewhat more reliable estimate of the field *e*-log *p* curve can be made by means of a graphical procedure (Schmertmann 1953). The construction requires unloading the sample in increments after the maximum pressure has been reached, in order to obtain a laboratory rebound curve. The laboratory curve is represented by K_u (Fig. 13.10). Point *b* represents the void ratio e_0 and effective overburden pressure p_0 of the clay as it existed in the ground before sampling. The field *e*-log *p* curve must pass through this point. The vertical line p_0' corresponds to the maximum consolidation pressure as determined by the graphical construction (Fig. 13.9). The portion of the field *e*-log *p* curve between p_0 and p_0' is a recompression curve. Since in the laboratory there is little difference in slope between rebound and recompression curves, it is assumed that the field curve between p_0 and p_0' is parallel to the laboratory rebound curve. Accordingly, a line is drawn from *b* parallel to *cd*; its intersection with the vertical at p_0' is denoted by *a'*. The field curve for pressures above p_0' is approximated by the straight line *a'f*, where *f* is the intersection of the downward extension of the steep straight portion of K_u and the horizontal line

$e = 0.4e_0$. Between b and a' a smooth curve is sketched in as indicated in Fig. 13.10.

For practical purposes it is often sufficient to know whether a clay is heavily precompressed. This decision can usually be made without recourse to the graphical construction (Fig. 13.9). If a clay is normally loaded, the points b (Fig. 13.5) are invariably located to the left of the points a. Hence, if several undisturbed samples from a clay stratum have been tested, and if all of the points b obtained from the test results are so located, the value of p_0' is certainly not much greater than the present overburden pressure, and the effect of the precompression on the settlement can be disregarded. On the other hand, if the precompression pressure was considerably greater than the present overburden pressure, at least some of the points b are located to the right of a. In this event the settlement of the structure to be erected above the clay will be small compared to that predicted on the basis of the test results, because the relation between the laboratory and the field consolidation curves for such a clay resembles that between the curves K_u' and K' in Fig. 13.8b.

If part of a normally loaded clay stratum has been precompressed by desiccation, the water content of the precompressed layers is relatively low. Hence, the location and thickness of these layers can be inferred from the water-content profile. In the settlement computation, the precompressed layers can often be considered incompressible.

Undisturbed Extrasensitive Clays

For undisturbed samples of ordinary clays, the e–log p curve K_u (Fig. 13.5) is roughly parabolic. However, the corresponding curve for extrasensitive clays has the shape indicated by K_u in Fig. 13.11. It remains nearly horizontal until the pressure on the sample approaches or somewhat exceeds the effective overburden pressure p_0, whereupon it turns downward quite abruptly. As the intensity of pressure is increased, the slope of the curve again decreases appreciably until finally the curve passes into an inclined straight line K_t. The upward projection of the tangent to the steep portion of K_u at its point of inflection c intersects the horizontal line e_0 at b'.

If the deposit of extrasensitive clay were normally loaded and if the sample were perfectly undisturbed, the point b' might be expected to coincide with point a having the coordinates (p_0, e_0). If the sample were slightly disturbed, b' should be to the left of a. Under these conditions, the construction of a building that contributes even a very slight increase to the pressure p_0 should be followed by a dramatic settlement of the building. In reality it has generally been found possible to

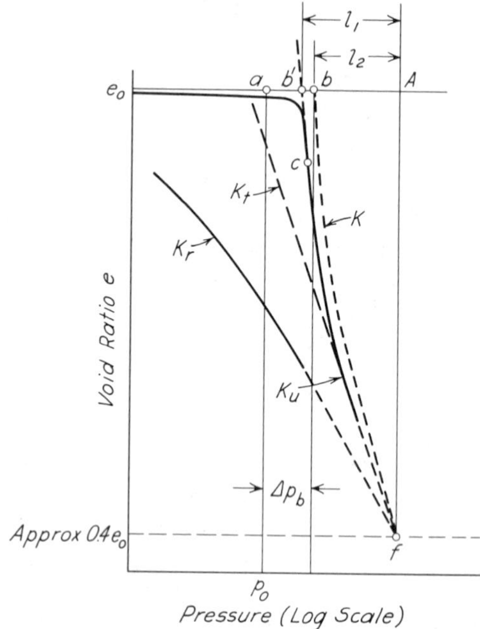

Fig. 13.11. Relations between *e* and *p* for extrasensitive clay corresponding to K_r remolded and K_u undisturbed states in the laboratory, and *K* natural state in the field.

increase the stress from p_0 to a somewhat greater value $p_0 + \Delta p_b$ without a disproportionate increase in settlement, but for values of Δp exceeding Δp_b the behavior corresponds to an e–log p curve at least as steep as the very steep upper part of K_u (Fig. 13.11). The ability of an extrasensitive clay to withstand without greatly increased settlement a pressure exceeding the existing overburden pressure may in some instances be a consequence of a slight degree of overconsolidation, as discussed under the previous subheading. On the other hand, it may be a consequence of the growth of bonds between the clay particles (Article 4); hence, the stress Δp_b is sometimes referred to as a *bond strength* (Terzaghi 1941*a*).

If the bond strength can be estimated, the field curve K can be approximated in the following manner. The straight lower part of K_u is extended downward to point f on the horizontal line $e = 0.4e_0$. Point b is located on the line $e = e_0$ at a value of p equal to $p_0 + \Delta p_b$. Finally a vertical line is traced through f which intersects the horizontal line $e = e_0$ at A. The curve K is constructed such that for any value of e the ratio between the horizontal distance from K to fA and the distance from K_u to fA is equal to

$$\frac{l_2}{l_1} = \frac{bA}{b'A}$$

In some localities such as Mexico City the bond strength can be estimated quite reliably on the basis of field experience. If no reliable estimate can be made, it is preferable to consider points b and b' to coincide.

The curve K_u can be obtained only by testing an undisturbed sample. If the sample is badly disturbed, or remolded and mixed with enough water to transform the clay into a thick paste, the e–log p curve K_r for the remolded material resembles in every respect the e–log p curve K_r (Fig. 13.5) for ordinary clays. It is practically straight over a wide range of pressure, and its slope is slightly less than that of the tangent K_t to the lower part of the curve K_u in Fig. 13.11. In other words, the disturbance of the structure of the clay obliterates the properties which cause the sharp bend in the curve K_u below point b in Fig. 13.11. Therefore, the information required for constructing the field consolidation line for extrasensitive clays can be obtained only from consolidation tests on undisturbed samples. Fortunately, by means of thin-walled piston samplers (Article 44) very good undisturbed samples can often be obtained of extrasensitive clays because the soil at the cutting edge of the sampler is so completely remolded that it offers virtually no resistance to penetration and forms a thin protective sheath of nearly frictionless soil over the undistorted core as the sampling tube is lowered.

If a clay is extrasensitive, the slope of the upper part of the field consolidation curve K may be several times that of K_r for the soil in a remolded state. For such clays, the approximate method of calculating the compression of a layer based on Eq. 13.11 furnishes merely a lower limiting value for the compression of the clay. The real compression may be several times greater. Fortunately clays of this type are rather rare. They include the clays of Mexico City which are of volcanic origin, certain types of marine clays in southeastern Canada and in the Scandinavian countries, and various highly organic clays. If a clay has a liquid limit greater than 100%, if its natural water content at a depth of more than 20 or 30 ft below the surface is greater than the liquid limit, or if it contains a high percentage of organic material, it is likely to have the consolidation characteristics illustrated in Fig. 13.11. The sensitivity S_t (Eq. 7.1) of these clays is greater than about 4, whereas that of ordinary clays is less. If the sensitivity of a clay is greater than 8, it is fairly certain that the clay will have the consolidation characteristics illustrated by Fig. 13.11.

Summary of Methods for Evaluating the Compressibility of Natural Soil Strata

If the soil beneath a structure contains layers of sand or stiff clay alternating with layers of soft clay, the compressibility of the sand and stiff-clay strata can be disregarded.

The compressibility of layers of clay depends primarily on two factors: the liquid limit of the clay, and the magnitude of the greatest pressure that has acted on the clay since its deposition. If this pressure has never exceeded the present effective overburden pressure, the layer is said to be normally loaded. Otherwise, it is said to be precompressed.

The compressibility of a normally loaded layer of clay with a known liquid limit can be estimated roughly by means of the empirical Eq. 13.11, provided the clay has no unusual properties. However, if the clay has a liquid limit above 100, if its natural water content at a depth of 20 or 30 ft is greater than the liquid limit, or if it contains a high percentage of organic material, the compressibility of the layer may be several times as great as that computed by means of Eq. 13.11. Hence, if a building is to be constructed above a layer of such an exceptional clay, it is advisable to determine the compressibility of the clay by means of consolidation tests on undisturbed samples.

The compressibility of a precompressed clay depends not only on the liquid limit of the clay but also on the ratio $\Delta p/(p_0' - p_0)$, in which Δp is the pressure added by the structure to the present overburden pressure p_0, and p_0' is the maximum pressure that has ever acted on the clay. If this ratio is less than 50%, the compressibility of the clay is likely to be from 10 to 25% of that of a similar clay in a normally loaded state. With increasing values of the ratio the effect of the precompression on the compressibility of the clay decreases. For values greater than 100% the influence of the precompression on the settlement of the structure can be disregarded.

The precompression of a clay can be due to the weight of soil strata that were removed by erosion, to the weight of ice that melted away, or to desiccation. If the precompression is due to a load that has been removed, the excess pressure that acted on the soil was the same at every point along a vertical line below the ground surface. However, if it was due to desiccation, the excess pressure probably decreased in a downward direction from the former surface of evaporation, and the total depth of the precompressed layer may not exceed a few feet.

The compressibility of heavily precompressed beds of clay is usually irrelevant, unless the engineer is required to construct above a thick bed of stiff clay an unusually large and heavy structure that would be damaged even by moderate differential settlement. If the problem warrants a settlement computation, consolidation tests must be made on undisturbed samples, preferably taken from test shafts. The sources and the importance of the errors involved in settlement computations based on the results of tests on such samples have been discussed on page 78.

Problems

1. A stratum of clay with an average liquid limit of 45% is 25 ft thick. Its surface is located at a depth of 35 ft below the present ground surface. The natural water content of the clay is 40%, and the unit weight of the solid clay particles is 2.78 gm/cm^3. Between the ground surface and the clay, the subsoil consists of fine sand. The water table is located at a depth of 15 ft below the ground surface. The average submerged unit weight of the sand is 65 lb/ft^3, and the unit weight of the moist sand located above the water table is 110 lb/ft^3. From geological evidence, it is known that the clay is normally loaded. The weight of the building that will be constructed on the sand above the clay increases the present overburden pressure on the clay by 1.2 tons/ft^2. Estimate the average settlement of the building.

Ans. 10 in.

2. The clay stratum *B* shown in Fig. 13.7 has a thickness of 25 ft. Its surface is located at a depth of 30 ft below the average water level in the river and 35 ft below the present valley floor. The surface of the high ground adjoining the valley is located 150 ft above the present valley floor, and the original water table was 5 ft above this surface. The clay is covered with sand having the same unit weight as that in the preceding problem. Compute the maximum consolidation pressure for the right-hand part of the stratum.

Ans. 4.76 tons/ft^2 in excess of the present overburden pressure.

3. The building shown on the valley floor in Fig. 13.7 increases the average pressure on the clay stratum by 1.2 tons/ft^2. The average liquid limit of the clay is 45%. The other data concerning the thickness of the stratum and the location of the site are the same as those in problem 2. The average natural water content of the clay is 35%, and the unit weight of the solid clay particles is 2.78 gm/cm^3. Estimate the upper and lower limits for the settlement of the building.

Ans. Not more than 25% of 11.6 in., or 2.9 in., and probably not less than 10% of 11.6 in., or 1.2 in.

Selected Reading

A general discussion of the engineering properties of sediments, with emphasis on their compressibility, is found in Terzaghi, K. (1955*a*): "Influence of geological factors on the engineering properties of sediments," *Economic Geology, Fiftieth Anniversary Volume,* pp. 557–618. The paper includes a carefully selected list of references. Although written to acquaint geologists with the engineering aspects of the properties of sediments, the article is of much interest to the engineer.

ART. 14 CONSOLIDATION OF CLAY LAYERS

In the preceding article, it was mentioned that the compression of clay due to an increase in load proceeds very slowly. The source of a small part of the delay is the gradual adjustment of the position of the grains to the increase in pressure. This source is common to both sand and clay. However, in clays the major part of the delay is due to the very low permeability of the material. As a consequence, a long time is required to drain out the excess water. The gradual decrease of the water content at constant load is known as *consolidation.*

The mechanics of the delaying effect of low permeability on the compression of an elastic layer under constant load can be demonstrated by means of the device shown in Fig. 14.1. It consists of a cylindrical vessel that contains a series of pistons separated by springs. The space between the pistons is filled with water, and the pistons are perforated. When a pressure p per unit of area is applied to the surface of the uppermost piston, the height of the springs is at the first instant unchanged, because sufficient time has not yet elapsed for the escape of any water from between the pistons. Since the springs cannot carry load until their height decreases, the load p per unit of area must at first be carried entirely by an excess hydrostatic pressure $h_1\gamma_w = p$ in the water. At this stage, the water in each of the piezometric tubes stands at the height h_1.

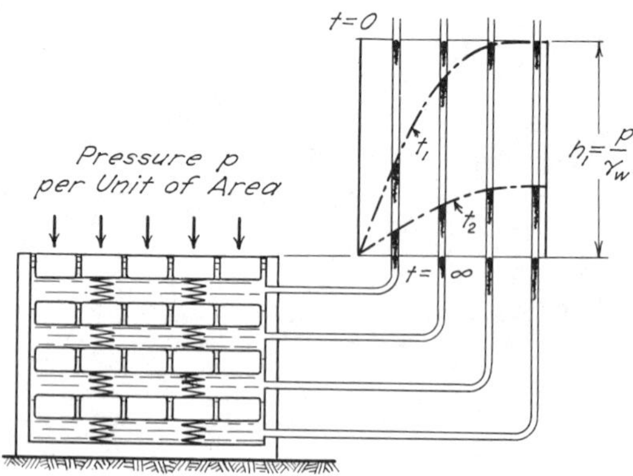

Fig. 14.1. Device for demonstrating mechanics of process of consolidation.

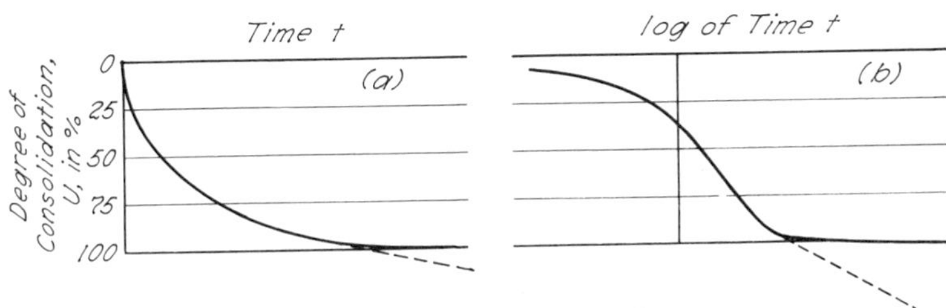

Fig. 14.2. Time-consolidation curves. Solid lines represent relation for mechanical device illustrated in Fig. 14.1. Dash lines represent relation for clay sample with similar consolidation characteristics.

After a short time t_1 has elapsed, some water will have escaped from the upper compartment, but the lowest compartment will still be practically full. The decrease in volume of the upper compartment is accompanied by a compression of the upper set of springs. Therefore, the upper springs begin to carry a portion of the pressure p, whereupon the water pressure in the upper compartments decreases. In the lower compartments conditions are still unaltered. At this stage the water levels in the piezometric tubes are located on a curve t_1 that merges into a horizontal line at elevation h_1. The corresponding compression, or decrease in thickness of the set of pistons, is S_1. Any curve, such as t_1, which connects the water levels in the piezometric tubes at a given instant, is known as an *isochrone*. At a later stage the water levels in the tubes are located on the curve t_2. Finally, after a very long time the excess hydrostatic pressure becomes very small, and the corresponding final compression is $S = S_\infty$. For a clay the final compression is determined by the initial thickness of the layer and by Eq. 13.4. The ratio

$$U(\%) = \frac{S}{S_\infty} \tag{14.1}$$

represents the *degree of consolidation* at the time t.

The rate of consolidation of the system of pistons and springs can be computed on the basis of the principles of hydraulics. The relation between the degree of consolidation of such a system and the elapsed time is indicated by the solid curves in Figs. 14.2*a* and 14.2*b*.

The rate of consolidation of a clay sample may be investigated in the laboratory by means of the confined compression test described in Article 13. Up to a degree of consolidation of about 80%, the shape of the experimental time–consolidation curves is very similar to that of the curves for spring–piston systems. However, instead of approaching horizontal asymptotes, the curves for clay continue on a gentle slope, as indicated in Fig. 14.2 by dash lines. In the semi-logarithmic plot (Fig. 14.2*b*) the dash lines can be either straight or slightly curved. The average slope of the dash portion is very different for different clays. For organic clays, the initial slope can be almost as great as that of the preceding portion of the plain curve. The progressive consolidation represented by the plain curves is known as *primary consolidation*, whereas that represented by the vertical distance between the full and dash curves is known as the *secondary time effect*.

The secondary time effect is probably a consequence of the fact that the compression of a layer of clay is associated with slippage between grains. Since the bond between grains consists of layers of adsorbed water with a very high viscosity (Article 4), the resistance of these layers to deformation by shear would delay the compression even if the time lag resulting from the low permeability of the clay were negligible. In the system of pistons and springs to which the primary consolidation corresponds, the lag in compression is due only to the resistance against rapid escape of the excess water.

On inorganic soils the rate of settlement of structures due to secondary time effects ranges between almost zero and about one inch per year. Although the secondary time effects can be observed and measured during the performance of consolidation tests, the results of various attempts to predict the settlement of full-sized structures due to secondary time effects on the basis of the results of laboratory tests have not yet been consistently satisfactory.

The results of consolidation tests performed on clay samples have disclosed several simple relationships. For a given clay the time required to reach a given degree of consolidation increases in proportion to the square of the thickness of the layer. For equally thick layers of different clays the time required to reach a given degree of consolidation increases in direct proportion to m_v/k, where m_v is the coefficient of volume compressibility (Eq. 13.3) and k is the coefficient of permeability. The ratio,

$$c_v(\text{cm}^2/\text{sec}) = \frac{k}{m_v}\frac{1}{\gamma_w} \tag{14.2}$$

is known as the *coefficient of consolidation*. With decreasing void ratio, both k and m_v decrease rapidly, but the ratio k/m_v is fairly constant over a considerable range of pressure. The values of c_v for different clays decrease in a general way with the liquid limit, as shown by the diagram (Fig. 14.3). In this figure, the abscissas represent values of the liquid limit, and the ordinates the corresponding values of the coefficient of consolidation of undisturbed samples of clays under normal pressures between 1 and 4 kg/cm². The figure shows that the coefficient

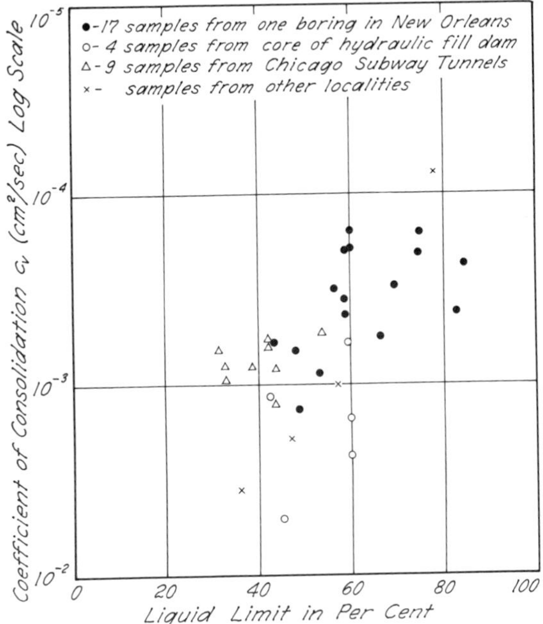

Fig. 14.3. Relation between liquid limit and coefficient of consolidation for undisturbed samples of clay.

of consolidation for clays with a given liquid limit varies within a wide range.

If the pressure in a natural clay stratum is relieved, for instance by the excavation of a shaft or tunnel, the corresponding volume expansion of the clay may not begin for a week or more after the excavation is completed. In a few instances it has even been observed that the consolidation of such strata under the influence of superimposed loads did not start for a few weeks after the load was applied. These delays in the reaction of clay to a change in stress, like the secondary time effect and the influence on c_v of the magnitude of the load increment, cannot be explained by means of the simple mechanical concept on which the theory of consolidation is based. Their characteristics and conditions for occurrence can be investigated only by observation.

In spite of the radical simplifications involved, the theory of consolidation serves a useful purpose, because it permits at least a rough estimate of the rate of settlement due to consolidation, on the basis of the results of laboratory tests. Therefore, the theory is presented briefly in Part II, Article 25.

Problems

1. The results of a consolidation test on a sample of clay having a thickness of 0.75 in. indicate that half the ultimate compression occurs in the first 5 min. Under similar drainage conditions, how long will be required for a building on a 12-ft layer of the same clay to experience half its final settlement? (Neglect the secondary time effect.)

Ans. $t = 128$ days.

2. The void ratio of clay A decreased from 0.572 to 0.505 under a change in pressure from 1.2 to 1.8 kg/cm^2. The void ratio of clay B decreased from 0.612 to 0.597 under the same increment of pressure. The thickness of sample A was 1.5 times that of B. Nevertheless, the time required for 50% consolidation was three times longer for sample B than for sample A. What is the ratio of the coefficient of permeability of A to that of B?

Ans. 31 to 1.

3. The subsoil of a building consists of a thick deposit of sand that contains, at about mid-thickness, one layer of soft clay 10 ft thick. A laboratory sample of the clay, drained at both top and bottom, reaches 80% consolidation in 1 hr. The sample is 1 in. thick. How much time will elapse before the degree of consolidation of the clay stratum becomes equal to 80%?

Ans. $t = 600$ days.

ART. 15 STRESS AND STRAIN IN SOILS

Practical Considerations

The relations between stress and strain in soils determine the settlement of soil-supported foundations. They also determine the change in earth pressure due to small movements of retaining walls or other earth supports.

If the settlement of a foundation is due chiefly to the consolidation of strata of soft soil located between layers of relatively incompressible material, it can be computed or estimated as explained in Article 13. However, this simple procedure is valid only if the horizontal deformation of the compressible layers is negligible in comparison to the vertical deformation. Under all other conditions, local application of load causes a yield of the soil mass in every direction. The stress-strain properties that determine the yield are too complex to be expressed in the form of quantitative relations that can be utilized in settlement computations. Reliable estimates of settlement can be made only on the basis of experience gained from observations concerning the settlement of other buildings supported by similar soils. However, since the chances are slight that a proposed foundation would have the same dimensions as an existing one, even an estimate based on experience requires a knowledge of the influence on the settlement of the size of the loaded area, the depth of foundation, and other factors. These influences are governed to a very large extent by the general relations between stress and strain for soils.

The stress-strain relationships for soils are much more complex than those for manufactured construction materials such as steel. Whereas the stress-strain relationships for steel can be described adequately for many engineering applications by two numerical values expressing the modulus of elasticity and Poisson's ratio, the corresponding values for soils are functions of stress, strain, time, and various other factors. Furthermore, the experimental determination of these values for soils is much more difficult. The investigations are usually carried out by means of triaxial compression tests.

Description of Triaxial Apparatus

In a triaxial test, a cylindrical specimen of soil is subjected to an equal all-around pressure, known as the *cell pressure*, in addition to an axial pressure that may be varied independently of the cell pressure.

The essential features of the triaxial apparatus are shown diagrammatically in Fig. 15.1. The cylindrical surface of the sample is covered by a rubber membrane sealed to a pedestal at the bottom and to a cap at the top. The assemblage is contained in a chamber into which water may be admitted under any desired pressure; this pressure acts laterally on the cylindrical surface of the sample through the rubber membrane and vertically through the top cap. The additional axial load is applied by means of a piston passing through the top of the chamber.

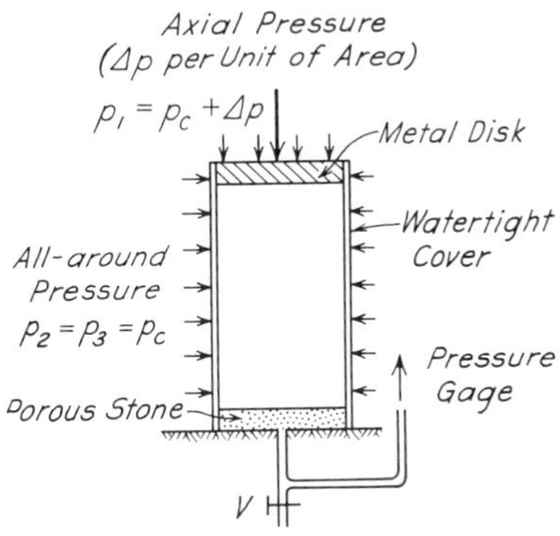

Fig. 15.1. Diagram illustrating principal features of triaxial-test apparatus.

A porous disk is placed against the bottom of the sample and is connected to the outside of the chamber by tubing. By means of the connection the pressure in the water contained in the pores of the sample can be measured if drainage is not allowed. Alternatively, if flow is permitted through the connection, the quantity of water passing into or out of the sample during the test can be measured. As the loads are altered, the vertical deformation of the specimen is measured by a dial gage.

A test is usually conducted in two steps: the application of the cell pressure, followed by the additional axial load.

Behavior under Initial All-Around Compression

The sample shown in Fig. 15.2*a* is assumed to be completely consolidated under a cell pressure p_c known as the *initial consolidation pressure*. The sample is then subjected instantaneously to a cell pressure p_3.

If the drainage lines of the apparatus are open so the sample can drain freely, water may be expelled from the soil and the volume of the sample may decrease by a process of consolidation. If the sample is initially saturated, the decrease in volume per unit of volume, $\Delta V/V$, will occur as indicated in Fig. 15.2*b* in which time is plotted to an arithmetic scale, or as indicated in Fig. 15.2*c* wherein time is plotted to a logarithmic scale. The rate of volume change takes place in accordance with the laws of consolidation for the particular boundary conditions associated with the test arrangement. It is commonly observed that the secondary consolidation is relatively small for an equal all-around consolidation pressure, possibly because there are no externally applied shearing stresses associated with this type of loading.

If the sample is only partly saturated, a portion of the volume change takes place almost instantaneously as the air is compressed. The remainder of the volume change is associated with the expulsion of water, air, or both.

If, on the other hand, the drainage lines are closed prior to the application of the cell pressure p_3, no drainage is possible. If the sample is saturated, the porewater pressure is found to be equal to the applied cell pressure p_3, as shown in Fig. 15.2*d*, and the volume change is zero (Fig. 15.2*e*). If, on the contrary, the sample is partly saturated, volume change occurs on account of compression of the air. The corresponding pressure in the porewater is less than that for the same soil in a saturated condition. However, for greater values of cell pressure p_3, the air is compressed to a greater extent, and a larger fraction of the free air is dissolved in the porewater. At some value of p_3 the free air is completely dissolved whereupon the sample becomes saturated. At this stage the slope of the diagram (Fig. 15.2*d*) representing the relation

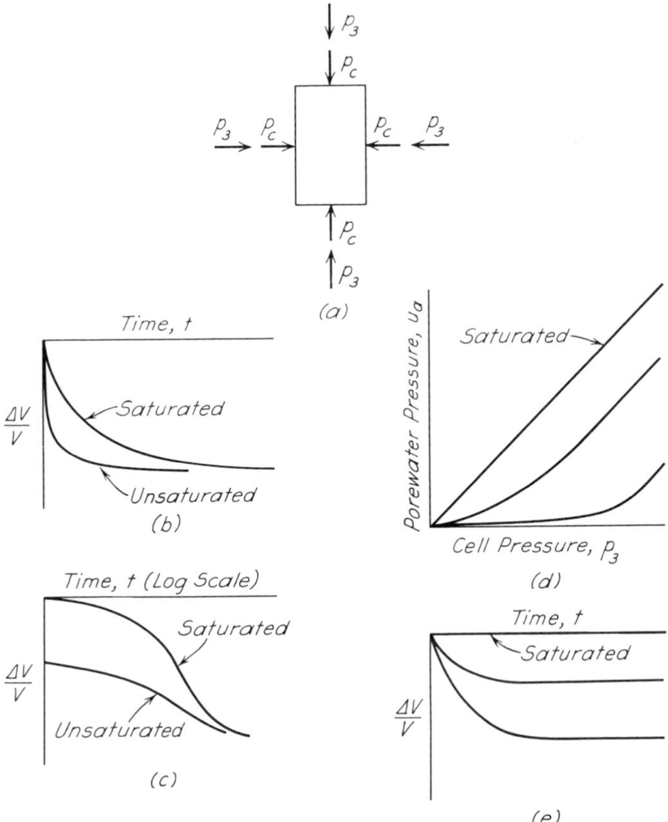

Fig. 15.2. Behavior of triaxial specimen under initial all-around compression. (*a*) Principal stresses acting on specimen. Decrease in volume if drainage is permitted, plotted against time on (*b*) arithmetic and (*c*) logarithmic scale. (*d*) Porewater pressure as function of cell pressure if drainage is not permitted. (*e*) Decrease in volume plotted against time if drainage is not permitted.

between porewater pressure and cell pressure becomes equal to that for a saturated material. The ratio of the porewater pressure u_a, caused by the all-around pressure p_3, and the pressure p_3 is the *pore-pressure coefficient* **B** (Skempton 1954); that is,

$$\mathbf{B} = \frac{u_a}{p_3} \tag{15.1}$$

It is apparent that the value of the pore-pressure coefficient **B** for a soil initially saturated before application of p_3 is equal to 1.0. For partially saturated soils the value of **B** is less than unity.

If a sample of normally consolidated clay of low sensitivity were transformed in the triaxial apparatus into an extra-sensitive clay before application of the cell pressure, the cell pressure would cause the collapse of the metastable structure whereupon the pore-pressure coefficient **B** would rise to a value much greater than unity. Such a transformation could, for instance, be produced by leaching the salt out of a specimen of marine clay (Article 4) after it had been consolidated under the all-around pressure p_c.

Drained and Undrained Conditions

The stress-strain characteristics of soils, like their pressure-volume relationships, depend greatly upon whether the water content can adjust itself to the state of stress (A. Casagrande 1934). Two extreme conditions may be recognized: *drained conditions,* under which the changes in stress are applied so slowly with respect to the ability of the soil to drain that no excess pore pressures develop, and *undrained conditions,* under which the stresses are changed so rapidly with respect to the ability of the soil to drain that no dissipation of pore pressure takes place. These extreme conditions are rarely fully realized in the field. They can be produced in the laboratory, however, and because they represent limiting conditions they are valuable guides to understanding the behavior of soil masses.

Drained Test with Increasing Stress Difference

Tests in which full pore-pressure dissipation is permitted are known as *drained tests.* A sample at the start of a drained test is permitted first to consolidate or swell freely under a uniform all-around cell pressure p_3 (Fig. 15.3*a*) until the cell pressure is carried entirely by an effective stress $\bar{p}_3$ within the sample. The strains associated with the stress $\bar{p}_3$ represent a volume change and, for an isotropic material, are equal in all directions. They are not plotted in Fig. 15.3.

As soon as consolidation is complete under $\bar{p}_3$, the final stage of the test is begun. The axial stress is increased by small increments or at a rate slow enough that no appreciable pore pressures are developed within the specimen. For loose sands or normally loaded clays of low sensitivity the relation between the axial strain and the vertical stress difference Δp is shown by the plain curve in Fig. 15.3*b*. The corresponding changes in volume are shown by the plain curve in Fig. 15.3*c*; the volume decreases continuously with increasing values of Δp and approaches a limiting value. The relations for a clay of high sensitivity are shown by the dash curves.

If a similar test is performed on a sample of dense sand or of highly overconsolidated clay, the stress-strain curve corresponding to an increase in axial pressure Δp has the shape indicated in Fig. 15.3*d*. The curve representing the volume change (Fig. 15.3*e*) is strikingly different from that for loose sand or normally consolidated clay (Fig. 15.3*c*). The volume decreases somewhat during the earliest stages of loading but

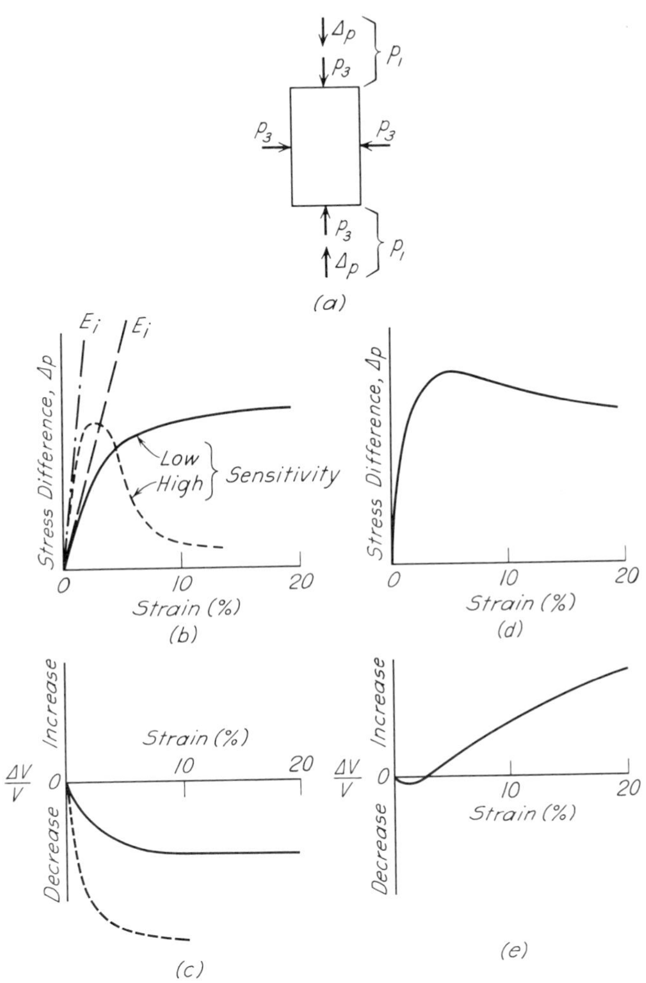

Fig. 15.3. Behavior of saturated triaxial specimen under drained conditions as stress difference Δp is increased. (*a*) Principal stresses acting on specimen. (*b* and *c*) Stress difference and volume change as functions of strain for loose sand or normally loaded clay. (*d* and *e*) Stress difference and volume change as functions of strain for dense sand or highly overconsolidated clay.

with increased strain the volume of the sample increases and at large values of strain the volume of the sample is greater than the initial volume in spite of the fact that the sample has shortened in the vertical direction. The tendency of the volume to increase under increasing stress difference is known as *dilatancy*.

If the relative density of a sand has a particular value intermediate between the loose and dense states, the sand will experience very small volume changes under drained conditions. If the void ratio at large strains is identical with that before the application of the stress difference, the sand is said to be at the critical void ratio (A. Casagrande 1936*a*). The significance of the critical void ratio will be discussed in Article 17.

The reason for the difference in the volume-change characteristics of the materials may be visualized readily in connection with sands. If the sand is in a loose state, a distortion of the sample tends to cause the grains to slide with respect to each other and to assume a more closely packed position. On the other hand, if the grains are initially closely packed, a distortion of the sample cannot occur, unless the individual grains are themselves broken, without an increase in distance between the centers of the sand particles. For clays the phenomena associated with volume change are somewhat more complex. Nevertheless, the structure of a soft clay is such that the grains can readily be oriented into a more compact position, whereas the grains of a highly compressed clay are tightly packed like those of a dense sand. Therefore, distortion is associated with an increase of volume.

The slope of the tangent (stress/strain) to the curves b and d in Fig. 15.3 at the point of origin of each curve is designated as the *initial tangent modulus* E_i of the specimen. For small stress differences, Δp, the stress-strain relations for the soil closely approximate those for a perfectly elastic, homogeneous material with a modulus of elasticity E_i. For all soils the value of E_i increases with increasing consolidation pressure $\bar{p}_c$ according to the relation

$$E_i = C\bar{p}_c \qquad (15.2)$$

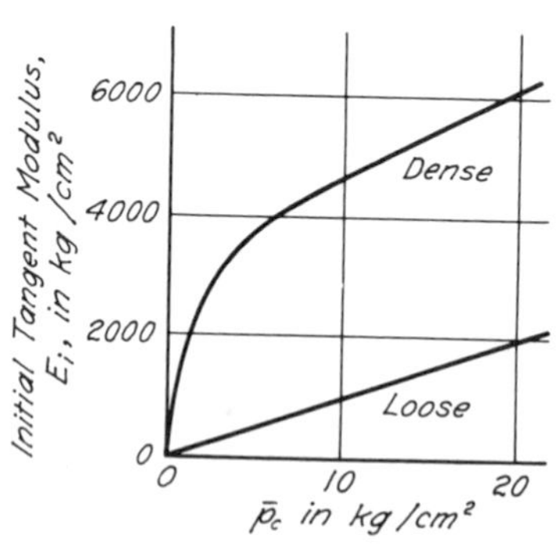

Fig. 15.4. Relation between initial tangent modulus and all-around pressure for sand (after Scheidig 1931).

For sand this relation is shown in Fig. 15.4. It can be seen that for loose sand C is practically independent of $\bar{p}_c$ and roughly equal to 100, whereas for dense sands the value of C is high for low values of $\bar{p}_c$ and decreases with increasing values of $\bar{p}_c$. For clays the corresponding relationship for the drained state is not yet known.

Consolidated-Undrained Tests with Increasing Stress Difference

A test in which no dissipation of pore pressure is permitted after the sample is initially brought into hydraulic equilibrium under the all-around consolidation pressure p_3 is known as a *consolidated-undrained test*. If such a test is carried out on an initially saturated sample of loose sand or soft insensitive clay, the results are similar to those indicated by the plain curves in Fig. 15.5*b* to *d*. The dash curves refer to clays of high sensitivity. After the sample is initially allowed to come to equilibrium under the cell pressure p_3, the drainage connections are closed. If then the axial stress $p_1 = \Delta p + p_3$ is increased steadily or in increments, the relationship between the stress difference Δp and the strain is as shown in Fig. 15.5*b*.

Moreover, as the strain increases, the pore pressure u_d associated with the stress difference Δp increases as shown in Fig. 15.5*c*. The ratio between the pore pressure u_d produced by the stress difference, and the stress difference itself, is the *pore-pressure coefficient* $\bar{\mathbf{A}}$ (Skempton 1954); that is,

$$\bar{\mathbf{A}} = \frac{u_d}{\Delta p} \tag{15.3}$$

The relation between $\bar{\mathbf{A}}$ and strain may be derived from the curves in Fig. 15.5*b* and 15.5*c* and is shown in 15.5*d*. For most loose sands and insensitive normally loaded clays, the value of $\bar{\mathbf{A}}$ at low strains is less than unity but with increasing strains it increases approximately to unity and maintains this value throughout most of the test. However, in extremely loose sands or extra-sensitive clays, the application of the stress difference may tend to cause a collapse of the metastable structure of the material. The dash lines (Fig. 15.5*b* to *d*) are then obtained, and the values of $\bar{\mathbf{A}}$ may exceed unity (Article 18).

If consolidated-undrained tests are carried out on a dense sand or on a highly overconsolidated clay, the results are similar to those represented in Fig. 15.5*e* to *g*. The relation between the stress difference Δp and the strain is shown in Fig. 15.5*e*. The pore pressure at small strains is likely to increase but at higher strains it tends to decrease and to become negative with respect to atmospheric pressure (Fig. 15.5*f*). The decrease in pore pressure is associated with the dilatancy of the soil. Since volume increase cannot take place because the drainage of

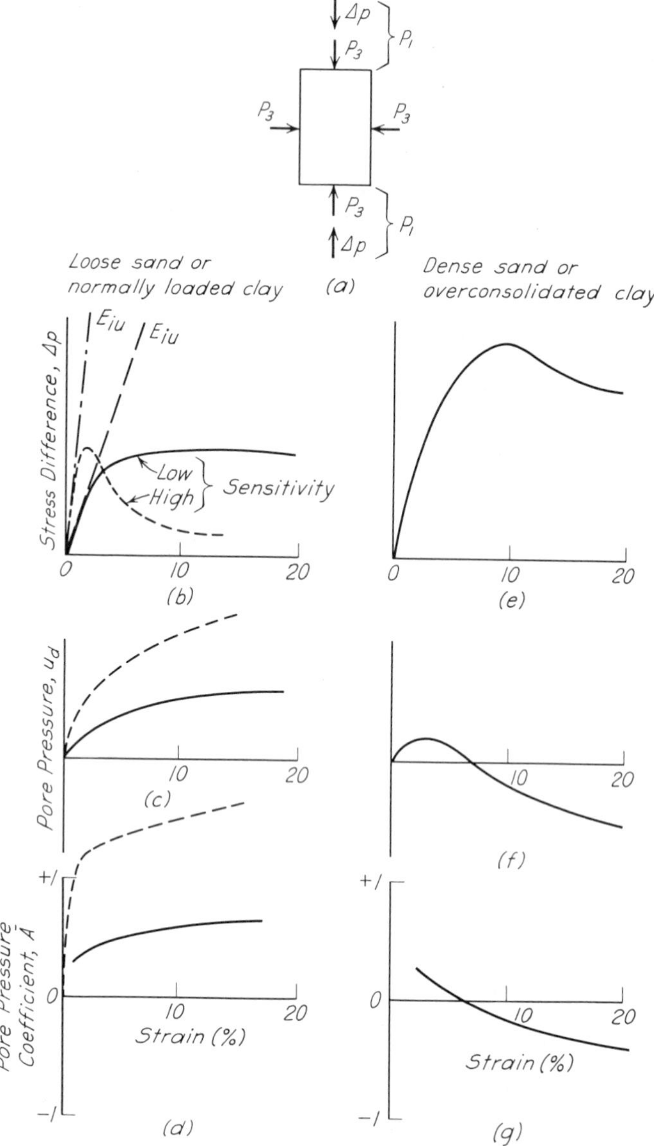

Fig. 15.5. Behavior of saturated triaxial specimen in consolidated-undrained test as stress difference Δp is increased. (*a*) Principal stresses acting on specimen. (*b* to *d*) Stress difference, pore pressure, and pore-pressure coefficient $\bar{A}$ as functions of strain for loose sand or normally loaded clay. (*e* to *g*) Stress difference, pore pressure, and pore-pressure coefficient $\bar{A}$ as functions of strain for dense sand or highly overconsolidated clay.

the sample is prevented, a deficiency in stress is developed in the water in the effort to draw additional water into the specimen.

The corresponding pore-pressure coefficient $\bar{\mathbf{A}}$ has a positive value at low strains but decreases with increasing strain and may become negative (Fig. 15.5*g*). In this respect the behavior of a dense or overconsolidated material differs radically from that of a loose or normally loaded material. At the critical void ratio, a sample of sand tested under undrained conditions experiences very small or negligible changes in pore pressure.

The slope of the tangent to the plain or the dash curve in Fig. 15.5*b* at the point of its origin represents the initial tangent modulus E_{iu} for the soil in a consolidated-undrained state. During a consolidated-undrained test on loose sand or normally loaded clay the porewater pressure remains positive throughout the test and, if plotted to the same scales, the plain stress-strain curve in Fig. 15.5*b* is flatter than the corresponding plain curve in Fig. 15.3*b*. For dense sand or overconsolidated clay (Fig. 15.5*e*) the curve is steeper than the corresponding curve in Fig. 15.3*d*. As a consequence, in Fig. 15.4, representing the relation between confining pressure and the values of E_i, the curve showing the corresponding relation for E_{iu} for loose sand would be located below the E_i-line for loose sand, whereas that for dense sand would be above the E_i-line for dense sand.

If a specimen remains unsaturated at the end of its initial consolidation under the all-around pressure p_3, closure of the drainage connections prior to application of the external stress Δp does not prevent volume change, because of the compressibility of the air within the sample. As the stress difference is increased, the pore pressure increases not only within the water contained in the pores but also within the air. The relation between the pressure in the air and in the water is complex, and the separate measurement of pore-air and porewater pressures is not yet a routine procedure. The stress-strain relations depend to a considerable extent on the initial degree of saturation. They also are significantly influenced, for compacted soils, by the method of compaction (Seed et al. 1960).

Unconfined Compression Test

If a sample of saturated clay that has been completely consolidated under an all-around pressure p_3 is removed from the triaxial chamber, the pressure p_3 is replaced by a capillary pressure p_k of equal intensity (Eq. 21.4); as a consequence, both the water content and the effective stresses in the clay remain practically unaltered. Hence, if the specimen is subjected to a compression test in its unconfined state, the re-

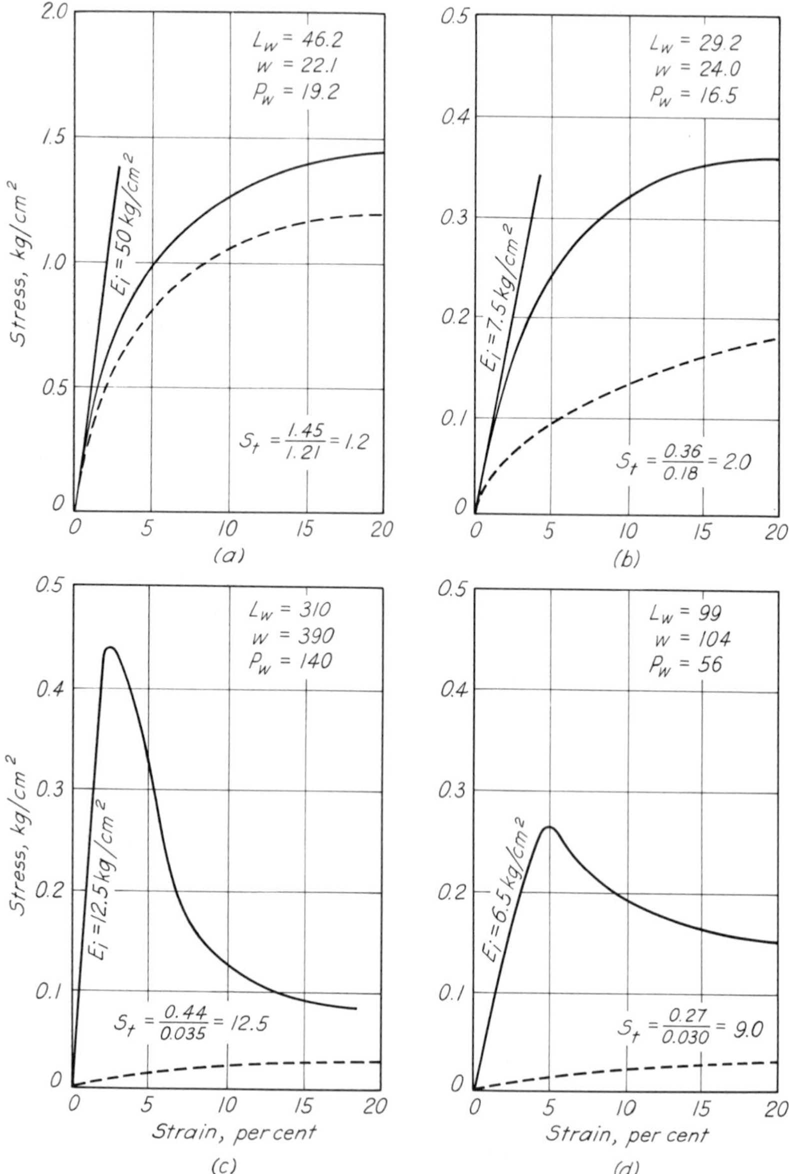

Fig. 15.6. Stress-strain curves in unconfined compression for undisturbed (full lines) and remolded (dash lines) samples of four typical clays.

sults of the test are practically identical with those of a normal consolidated-undrained compression test performed on the same material.

If the clay has been normally consolidated in the ground, the horizontal consolidation pressure $\bar{p}_h$ is always somewhat smaller than the overburden pressure $\bar{p}_v$. The ratio $\bar{p}_h/\bar{p}_v$ appears to range between about 0.6 for lean and 0.8 for highly plastic clays. Therefore, if a perfectly undisturbed sample is recovered from the ground the initial effective stresses around the periphery of the sample are replaced by an all-around capillary pressure with an intensity of about

$$p_3 = \tfrac{1}{3}(\bar{p}_v + 2\bar{p}_h) = (0.7 \text{ to } 0.9)\bar{p}_v \tag{15.4}$$

Hence, the results of an unconfined compression test on a perfectly undisturbed sample are approximately the same as those of a consolidated-undrained test performed on the same clay under a confining pressure p_3 (Eq. 15.4). This relationship makes it possible to get information concerning the stress-strain characteristics of a clay under consolidated-undrained conditions without the need for triaxial apparatus. Typical test results are shown in Fig. 15.6. The plain curves shown in Fig. 15.6*a* and *b* correspond to the plain curves in Fig. 15.5*b* to *d*, whereas the plain curves in Fig. 15.6*c* and *d* correspond to the dash curves in Fig. 15.5*b* to *d*.

If the tests represented by the plain curves in Fig. 15.6 are repeated on the same specimens after remolding at unaltered water content, the dash curves are obtained. The difference between the ordinates of the plain and the corresponding dash curves indicates the degree of sensitivity of the clay as defined in Article 7.

It can be seen that the slope of the stress-strain curve for an undisturbed sample of a clay of low sensitivity decreases steadily with increasing strain like that of the plain curve in Fig. 15.5*b*, whereas for highly sensitive clays the slope remains almost constant until the point of failure is reached (dash curve in Fig. 15.5*b*). As a consequence, highly sensitive clays perform in an undisturbed state like brittle materials, but upon remolding they acquire the consistency of very viscous liquids.

Stress-Strain Relationships under Varying Stress

In engineering practice the loads on soils located beneath most structures vary periodically between lower and upper limiting values as, for example, those corresponding to dead load and dead plus live loads. Laboratory tests and experience both have shown that the reduction and subsequent reapplication of a stress on a soil of any type is associated with an increase of strain, as indicated in Fig. 15.7 for a

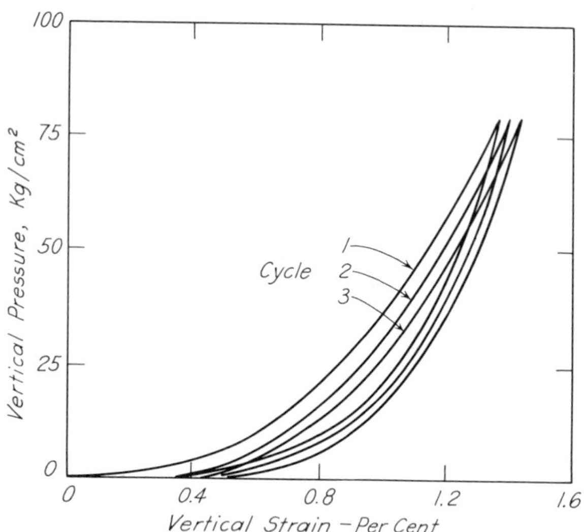

Fig. 15.7. Relation between stress and vertical strain for confined moderately dense coarse uniform sand subject to repeatedly applied vertical load (after Hendron 1963).

confined specimen of a fairly dense sand. However, the magnitude of the increase decreases with increasing number of stress cycles. Hence, in estimating the ultimate settlement of structures carrying widely varying loads, such as storage bins or crane runways, the consequences of the variations in load must be considered.

Selected Reading

Andresen, A. and N. E. Simons (1960). "Norwegian triaxial equipment and technique," *Proc. ASCE Research Conf. on Shear Strength of Cohesive Soils*, pp. 695–709. Discussion of refined equipment and procedures for triaxial tests.

Bishop, A. W. and D. J. Henkel (1962). *The measurement of soil properties in the triaxial test.* 2nd ed., London, Edward Arnold, 228 pp. Comprehensive discussion of apparatus, techniques, and typical results.

ART. 16 CONDITIONS OF FAILURE FOR SOILS

Mohr's Rupture Diagram and Coulomb's Equation

Soils, like most solid materials, fail either in tension or in shear. Tensile stresses may cause the opening of cracks that, under some circumstances of practical importance, are undesirable or detrimental. In the majority of engineering problems, however, only the resistance to failure by shear requires consideration.

Shear failure starts at a point in a mass of soil when, on some surface passing through the point, a critical combination of shearing and normal stresses is reached. Various types of equipment have been developed to determine and investigate these critical combinations. At present the most widely used is the triaxial apparatus described in Article 15. Because only principal stresses can be applied to the boundaries of the specimen in this equipment, the state of stress on any other than principal planes must be determined indirectly.

According to the principles of mechanics, the normal stress and the shearing stress on a plane inclined at angle α to the plane of the major principal stress and perpendicular to the plane of the intermediate principal stress (Fig. 16.1*a*) are determined by the following equations

$$p = \tfrac{1}{2}(p_1 + p_3) + \tfrac{1}{2}(p_1 - p_3)\cos 2\alpha \tag{16.1}$$

$$t = \tfrac{1}{2}(p_1 - p_3)\sin 2\alpha \tag{16.2}$$

These equations represent points on a circle in a rectangular system of coordinates (Fig. 16.1*b*) in which the horizontal axis is that of normal stresses and the vertical axis is that of shearing stresses. Similar expressions may be written for the normal and shearing stresses on planes on which the intermediate principal stress acts. The corresponding components of stress are represented by points on the dash circles plotted on the same axes in Fig. 16.1*b*. Since, in the usual triaxial test, the major principal stress acts in a vertical direction and the cell pressure represents both the intermediate and minor principal stresses which are equal, we are generally concerned only with the outer circle associated with the major and minor principal stresses p_1 and p_3. This is known as the *circle of stress*.

Every point, such as D, on the circle of stress represents the normal stress and shearing stress on a particular plane inclined at an angle α to the direction of the plane of the major principal stress. From the geometry of the figure it can be shown that the central angle $AO'D$ is equal to 2α.

If the principal stresses p_1 and p_3 correspond to a state of failure in the specimen, then at least one point on the circle of stress must represent a combination of normal and shearing stresses that led to failure on some plane through the specimen. Moreover, if the coordinates of that point were known, the inclination of the plane upon which failure took place could be determined from a knowledge of the angle α.

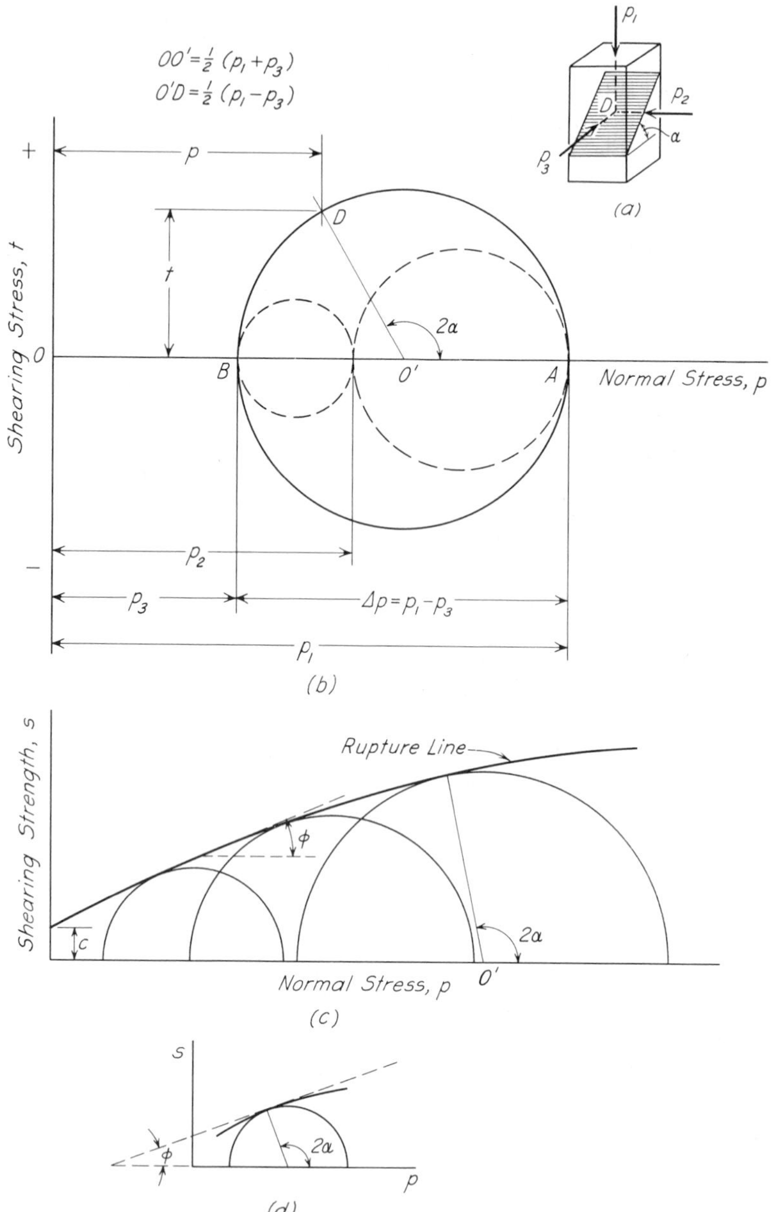

Fig. 16.1. Diagram illustrating Mohr's circle of stress and rupture diagrams. (*a*) Principal stresses and inclined plane on which normal and shearing stresses p and t act. (*b*) Circle of stress. (*c*) Rupture line from series of failure circles. (*d*) Relation between angles α and ϕ.

If a series of tests is performed and the circle of stress corresponding to failure is plotted for each of the tests, at least one point on each circle must represent the normal and shearing stresses associated with failure. As the number of tests increases indefinitely, and if the material is homogeneous and isotropic, it is apparent that the envelope of the failure circles (Fig. 16.1*c*) represents the locus of points associated with failure of the specimens. The envelope is known as the *rupture line* for the given material under the specific conditions of the series of tests.

From the geometry of Fig. 16.1*d* it may be seen that for any failure circle

$$2\alpha = 90° + \phi$$

Therefore the angle between the plane on which failure occurs and the plane of the major principal stress is

$$\alpha = 45° + \frac{\phi}{2} \tag{16.3}$$

In general the rupture line for a series of tests on a soil under a given set of conditions is curved. However, it may often be approximated by a straight line with the equation

$$s = c + p \tan \phi \tag{16.4}$$

This expression is known as *Coulomb's equation*. In this equation the symbol t, representing shearing stress, is replaced by s, known as the *shearing resistance* or *shearing strength*, because points on the rupture line refer specifically to states of stress associated with failure.

Evaluation of c and ϕ

Equations 16.3 and 16.4 are valid only if $\tan \phi$ has the same value for every plane section passing through a given point in the stressed material. If the voids of an isotropic soil are occupied only by air under atmospheric pressure, this condition is satisfied. On the other hand, if they are filled with a liquid under stress u_w, one part $\bar{p}$ of the pressure p (Eq. 16.4) is carried by the solid constituents which exhibit a definite value of the parameter $\tan \phi$, whereas the balance $p - \bar{p}$ is carried by the liquid for which $\tan \phi = 0$. The ratio $\bar{p}/u_w$ is different for different sections through the same point. Therefore the equations and the preceding physical interpretation of Mohr's rupture line are valid only on the condition that p in Eqs. 16.1 to 16.4 is replaced by the effective stress $\bar{p} = p - u_w$, whence

$$s = c + (p - u_w) \tan \phi = c + \bar{p} \tan \phi \tag{16.5}$$

This equation will be called the *Revised Coulomb Equation* (Terzaghi 1938*a*).

If the abscissas in Mohr's diagram represent effective pressures $\bar{p}$ and if the rupture line is straight, the inclination of the rupture line is usually designated the *angle of shearing resistance* ϕ of the material, and the intercept at $\bar{p} = 0$ is commonly called the *cohesion*. Materials for which these conditions are strictly satisfied are called *ideal plastic materials*. Their shear characteristics are defined by the two parameters c and ϕ.

The value ϕ in Eq. 16.5 is assumed to be a property of the material. In reality the part $\bar{p}$ tan ϕ of the shearing resistance represents the combined result of two very different components. One component is $\bar{p}$ tan ϕ_f, wherein ϕ_f, the friction angle between the particles at their points of contact, depends on the composition of the particles and the liquid which occupies the voids (Horn and Deere 1963). Practically no strain is required to mobilize this part of the shearing resistance. The second, much more important, component depends on the shape of the grains and the degree of interlock between the particles located on both sides of the surface of sliding. Its magnitude depends on the relative density or the liquidity index of the material. Mobilization of this component is associated with a rotational displacement of the particles with reference to each other and, therefore, requires a considerable strain. Furthermore, once a surface of sliding has developed, the subsequent sliding is associated with an increasingly less intimate degree of interlock between the grains located on the two sides of the surface of sliding than existed at the instant of failure. In cohesive soils, the failure is also commonly associated with a decrease of the cohesion. Therefore, in all soils other than loose cohesionless sands the slip is associated with a permanent decrease of the shearing resistance along the surface of failure. This fact is responsible for the treacherous character of slopes on which previous slides have occurred.

In soil mechanics the mathematical solution of practically all stability problems is preceded by the experimental determination of the values c and ϕ and by the subsequent replacement of the real soil by an ideal plastic material to which the shear characteristics c and ϕ are assigned. This replacement involves the assumption that both c and ϕ are independent of strain. Therefore, the soils should not fail until the shearing stress at every point along a continuous potential surface of sliding reaches the value s defined by Eq. 16.5. Failures of this type are called *simultaneous*. The stress-strain curve for a triaxial test of an ideal plastic material exhibiting simultaneous failure, therefore, resembles one of those shown in Fig. 16.2*a*. Failure is said to occur when the stress

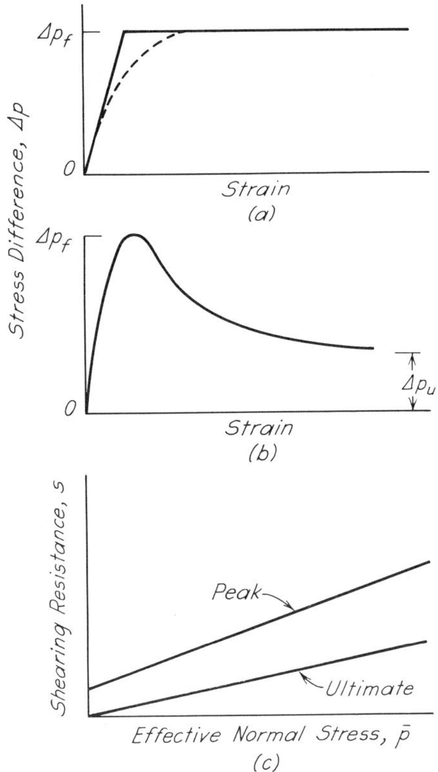

Fig. 16.2. (*a*) Stress-strain curves for ideal plastic material exhibiting simultaneous failure. (*b*) Stress-strain curve for real soil, illustrating peak and ultimate strengths. (*c*) Typical rupture lines for peak and ultimate strengths of same soil.

difference reaches the value Δp_f, and there is no ambiguity about the position of the rupture line plotted on the basis of effective stresses.

In contrast, the stress-strain curve for a real soil is likely to exhibit a peak corresponding to a small strain (Fig. 16.2*b*) whereupon the value of Δp that the specimen can support decreases from its maximum value Δp_f and approaches, at large strains, a smaller value Δp_u designated the *ultimate value*. The position of the rupture line then depends upon which value of Δp is considered to represent failure. Peak values of Δp correspond to the upper rupture line (Fig. 16.2*c*). On the other hand, if the lower limit of the shearing strength is of interest, a rupture line based on ultimate values may be constructed. However, whenever a material is characterized by a stress-strain curve exhibiting a peak, the conditions for simultaneous failure are likely to be violated because, even in a homogeneous material, the strains along a potential surface of sliding are not likely to be uniform. As a consequence, the soil along part of the surface of sliding may be exerting its peak strength while that along the remainder may be exerting a smaller value. Under extreme conditions as, for example, if a clay is extrasensitive, a small shear strain suffices to reduce the shear strength s to a small fraction of its peak value (Fig. 15.6*c*). Hence, failure of a body of extrasensitive clay starts at the point where the shearing stress becomes equal to s (Eq. 16.5) and from this point is likely to spread over the balance of the potential surface of failure. Failures of this type are said to be *progressive.* They invalidate the results of computations based on the conventional assumption of simultaneous failure.

Because of these differences between real and ideal soils, stability computations based on test results and on Eq. 16.5 are strictly valid only for the ideal plastic material that was substituted for the real soil. The practical consequences of the observed differences between real soils and their ideal substitutes must be compensated by adequate factors of safety. The importance of the differences depends on the type of soil and, for a given soil, on its load history. The only real soil which performs almost exactly like its equivalent ideal plastic material is clean cohesionless sand at a void ratio close to the critical value.

With respect to their shearing characteristics, real soils are commonly divided into two categories: cohesionless soils such as gravels, sands, and nonplastic silts, and cohesive soils such as clays and plastic silts. Cohesionless soils will be discussed in Article 17, and cohesive soils in Article 18.

Selected Reading

Henkel, D. J. (1960). "The shear strength of saturated remolded clays," *Proc. ASCE Research Conf. on Shear Strength of Cohesive Soils,* pp. 533–554. Summary of triaxial tests in terms of fundamental stress relationships.

Newmark, N. M. (1960). "Failure hypotheses for soils," *Proc. ASCE Research Conf. on Shear Strength of Cohesive Soils,* pp. 17–32. General discussion of failure hypotheses for ideal materials and their possible applications to soils. See also "Discussions," pp. 987–995.

Schmertmann, J. H. and J. O. Osterberg (1960). "An experimental study of the development of cohesion and friction with axial strain in saturated cohesive soils," *Proc. ASCE Research Conf. on Shear Strength of Cohesive Soils,* pp. 643–694.

Bishop, A. W. (1966). "The strength of soils as engineering materials," *Géot.,* **16,** pp. 91–128.

ART. 17 SHEARING RESISTANCE OF COHESIONLESS SOILS

Sands and Inorganic Silts

The shear characteristics of sands and inorganic silts can, unless the soil is exceptionally loose, be represented reliably by Eq. 17.1.

$$s = (p - u_w) \tan \phi = \bar{p} \tan \phi \tag{17.1}$$

Deposits of natural sands and silts may be encountered in any state between loose and dense. Depending principally upon the relative density, the value of ϕ may range between fairly wide limits; the grain-size distribution and the shape of the grains also have an influence on ϕ. Representative values for ϕ under effective pressures $\bar{p}$ less than about 5 kg/cm^2 are given in Table 17.1.

Table 17.1
Representative Values of ϕ for Sands and Silts

	Degrees	
Material	*Loose*	*Dense*
Sand, round grains, uniform	27.5	34
Sand, angular grains, well graded	33	45
Sandy gravels	35	50
Silty sand	27–33	30–34
Inorganic silt	27–30	30–35

Since most of the shearing strength is caused by interlocking of the grains, the values of ϕ are not appreciably different whether the soil is wet or dry.

As the pressure $\bar{p}$ is increased from about 5 to about 50 kg/cm^2, the values of ϕ decrease gradually by about 10°. The decrease is associated with an increase in the percentage of grains that are crushed as the state of failure is approached.

Figures 15.3*c* and *e* pointed out the tendency of a loose sand to decrease in volume, and of a dense sand to dilate, during shear. The permeability of very fine saturated sand and silt is so low that the rapid application of a shearing stress is associated with a temporary increase of pore pressure u_w (Eq. 17.1) if the soil is loose, or with a temporary decrease of u_w if the soil is dense. The strength of the soil is correspondingly temporarily decreased or increased. Hence, for example, if piles are being driven into one of these materials in a loose saturated state, the piles encounter only a slight resistance which is almost independent of depth, whereas in the same material in the dense state the piles may meet refusal.

If the sand or silt is at the critical void ratio (Article 15) the pore pressure u_w and, consequently, the shearing resistance, remain practically constant. Therefore, in order to avoid a reduction of shearing strength when a shearing stress is applied, it is commonly considered advisable to compact fills of sand or silt to a void ratio below the critical value. Inasmuch as the critical void ratio decreases somewhat with increasing confining pressure, greater compaction may be required to accomplish this purpose below heavily loaded foundations or beneath high fills than under lighter loads.

Spontaneous Liquefaction and True Quicksands

In a few localities fine sands have been encountered which are so loose that a slight disturbance such as a mild shock causes an important decrease in volume at unaltered value of p (Eq. 17.1). If this decrease takes place below the water table, it is preceded by a temporary increase in u_w to a value almost equal to p, whereupon $\bar{p} = p - u_w$ becomes almost zero and the sand flows like a viscous fluid. This phenomenon is known as *spontaneous liquefaction*. It has occurred both in loose sand fills and in natural sand deposits. Examples of slope failures on natural sand deposits (*true quicksands*) will be given in Article 49.

Experience indicates that spontaneous liquefaction most commonly occurs in fine silty sands. This fact, combined with the observed performance of true quicksands, suggests that the aggregate formed by the sand grains possesses a *metastable structure;* that is, the structure is stable only because of the existence of some supplementary stabilizing influence. A clean sand deposited under water is stable, although it may be loose, because the grains roll down into stable positions. In a sand capable of spontaneous liquefaction, some agent must interfere with this process.

If an artificial deposit of damp sand is placed above the water table, the interfering agent consists of films of moisture; the apparent cohesion produced by the films is sufficient to prevent the grains from rolling into stable positions. This process and its consequences have been reproduced in the laboratory (Geuze 1948, Bjerrum et al. 1961). The experiments also indicated that the relative density of true quicksands is very much lower than that corresponding to the critical void ratio.

Although clean sand deposited under water has a stable structure even if loose, sand deposited simultaneously with silt may develop a metastable structure. The depressions between the grains of sand on the surface of the sediment are partly filled with loose silt which prevents the sand grains from reaching stable conditions. Subsequent consolidation under static pressure, with no lateral strain, is resisted by friction at the points of contact between the grains of sand. However, if slip at the points of contact occurs, for instance on account of a shock with an intensity exceeding a certain threshold value, the metastable structure breaks down and liquefaction takes place. The resulting failure appears to be progressive, starting at one point and proceeding by a chain reaction.

A metastable structure in a natural sand deposit is very difficult to detect, because the structure collapses during sampling and subsequent transportation. Yet, if a layer of true quicksand is located beneath the base of a structure or of an earth dam, it is a potential source of danger. Experience suggests that true quicksands may occur in layers or large lenses between layers of loose or moderately dense sands. Such occurrences are probably the result of seasonal variations in the silt content of the turbid water which transported the sand to the site of deposition. Hence, if a dam is to be built above a thick layer of loose sand, the sand should be compacted as described in Article 50 because it may contain zones of true quicksand.

Liquefaction under Reversals of Stress or Strain

In Article 15 it was pointed out that each reduction and reapplication of stress to a soil is accompanied by an increase in strain, although the magnitude of the increase becomes smaller for each cycle. If the soil is saturated and drainage is prevented, each reduction and reapplication of stress or strain is accompanied by a change in pore pressure, although the magnitude of the change decreases for each cycle.

If a specimen of loose saturated sand is consolidated in a triaxial device under an all-around pressure p_c, and if then at the same cell pressure the axial stress is caused to alternate between $p_c + \Delta p$ and $p_c - \Delta p$ under conditions of no drainage, each alternation produces an increment Δu of pore pressure within the specimen (Fig. 17.1). After a certain number of alternations the value of u_w becomes equal to the effective stress p_c that prevailed before the cyclic loading began, whereupon the specimen loses its strength and is no longer able to maintain its shape. The sudden loss of strength and rigidity corresponds to liquefaction of the sand.

If the test is repeated with the same sand in a dense state, the values of u_w build up in a similar fashion except that the increments of Δu are much smaller per cycle, and the number of cycles to induce liquefaction is greatly increased. An increase in the consolidation pressure p_c, other

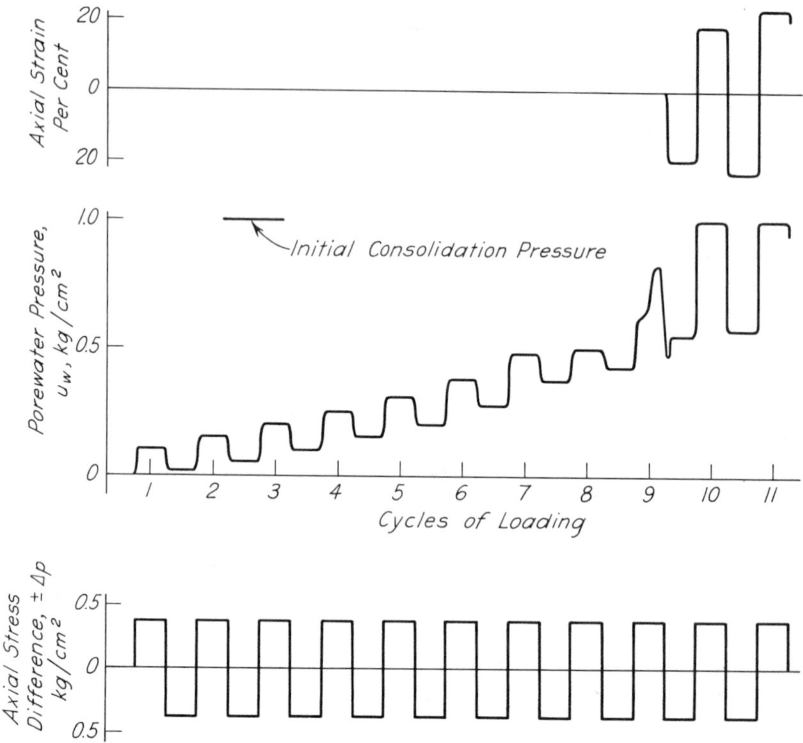

Fig. 17.1. Results of undrained triaxial test on loose saturated sand wherein axial stress alternates between 1 ± 0.39 kg/cm² while cell pressure is maintained at equivalent of 1 kg/cm² (after Seed and Lee 1966).

variables being equal, increases the number of cycles required to induce liquefaction, whereas an increase in Δp has the opposite effect (Seed and Lee 1966). Similar behavior occurs if alternations of strain rather than of stress are imposed on the specimen.

Bodies of loose, relatively fine, uniform sand below the water table are susceptible to liquefaction during an earthquake, especially if the duration of the quake is long enough for the occurrence of a large number of oscillations involving repeated reversals of shearing strains of large amplitude. After a violent earthquake has been in progress for a sufficient time, a loose zone at moderate depth may liquefy, whereupon the excess water will rise to the surface, often in more or less equally spaced sand boils. The sand above the liquefied zone is subjected to an upward hydraulic gradient and also loses its shearing resistance; consequently, footings supported on the sand may sink into the ground (IISEE 1965). Dense sands are less likely to experience liquefaction under these circumstances because the duration of most violent earthquakes is not long enough to encompass the required number of repetitions.

Very loose fills or natural deposits of saturated sand may liquefy, even if they do not possess a metastable structure, under the small provocation of mild vibrations or of a few repeated shocks. One such fill constituted the upstream portion of a dike, with a vertical clay core, located between a concrete storage dam and the right bank of a river valley. The slope had a height of about 50 ft; its inclination decreased from 2:1 at the crest to 4:1 at the toe. The sand, of which 80% of the grains ranged between 0.3 and 1.5 mm, was dumped in irregular layers above the water table, in a moist state without compaction. The slope was stable during the first filling of the reservoir and during a subsequent drawdown of 5 ft; it remained stable while the contractors started by blasting to demolish the upstream cofferdam, about 500 ft upstream. However, the intensity of the charges was gradually increased. About 8 or 10 minutes after the last charge was exploded, the slope started to fail at the junction between the dike and the concrete dam. Within approximately 20 seconds the movement spread over the full length of the slope to a distance of about 250 ft from the starting point. The sand spread out on the floor of the reservoir like a thick blanket and left the major part of the upstream face of the clay core without support.

Selected Reading

Rutledge, P. C. (1947). *Review of the cooperative triaxial shear research program of the Corps of Engineers.* Waterways Experiment Station. Chapter IV, "Detailed results for cohesionless soils," contains considerable data on properties of sands and gravels.

Chen, L. S. (1948). "An investigation of stress-strain and strength characteristics of cohesionless soils by triaxial compression tests," *Proc. 2nd Int. Conf. Soil Mech.*, Rotterdam, **5**, pp. 35–43.

Penman, A. D. M. (1953). "Shear characteristics of a saturated silt, measured in triaxial compression," *Géot.*, **3**, pp. 312–328.

Holtz, W. G. and H. J. Gibbs (1956*b*). "Triaxial shear tests on pervious gravelly soils," *ASCE J. Soil Mech.*, **82,** No. SM1, Paper No. 867, 9 pp.

Wu, T. H. (1957). "Relative density and shear strength of sands," *ASCE J. Soil Mech.*, **83,** No. SM1, Paper No. 1161, 23 pp.

Bjerrum, L., S. Kringstad and O. Kummeneje (1961). "The shear strength of a fine sand," *Proc. 5th Int. Conf. Soil Mech., Paris,* **1,** pp. 29–37.

ART. 18 SHEARING RESISTANCE OF COHESIVE SOILS

Normally Loaded Undisturbed Clays of Low to Moderate Sensitivity

The results of drained triaxial tests on normally loaded cohesive soils can be expressed with satisfactory accuracy by Coulomb's equation in which $c = 0$. Thus

$$s = \bar{p} \tan \phi \tag{18.1}$$

The values of ϕ for such materials, whether in a remolded or an undisturbed state, are related to the plasticity index. Approximate values may be estimated with the aid of Fig. 18.1, although the scattering from the curve for most clays may be on the order of 5° (Bjerrum and Simons 1960). However, the exceptionally high value $\phi = 47°$ was obtained (Lo 1962) for clay with a liquid limit of 426% from Mexico City. Hence, it is apparent that the statistical relation represented by Fig. 18.1 is not of general validity and should be used with caution.

Under conditions usually encountered in the field, the low permeability of clays greatly retards the drainage; as a consequence the pore pressures u_w associated with the forces tending to shear the clay may not dissipate readily. Since the pore pressures associated with shear are positive (Fig. 15.5*c*), the strength indicated by Eq. 18.1 may not be developed for a very long time; the time required for dissipation is governed by the consolidation characteristics and dimensions of the cohesive body (Articles 14 and 25).

The conditions associated with complete lack of drainage may be approximated in consolidated-undrained triaxial tests (Article 15). The results of such a test, in which $\bar{p}_1$ and $\bar{p}_3$ are the effective principal

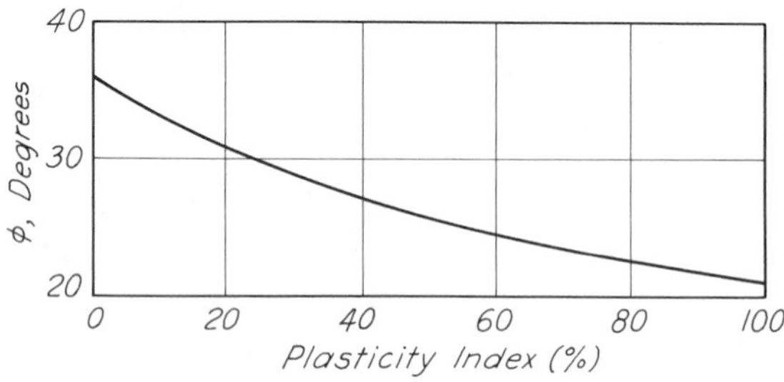

Fig. 18.1. Relation between ϕ and plasticity index for clays of moderate to low sensitivity under drained conditions.

stresses at failure, are represented by the rupture circle E, Fig. 18.2*a*; this circle is tangent to the rupture line defined by Coulomb's equation

$$s = \bar{p} \tan \phi \tag{18.1}$$

At the time of failure, positive pore pressures u_f act in all directions in the sample (see Fig. 18.2*a*). Hence, the total principal stresses at failure are

$$p_1 = \bar{p}_1 + u_f \tag{18.2}$$

and

$$p_3 = \bar{p}_3 + u_f \tag{18.3}$$

The rupture circle in terms of total stresses is then circle A; it has the same diameter as E but is displaced to the right a distance $\bar{\mathbf{A}}_f \Delta p_f$ equal to the pore pressure u_f induced in the sample at failure.

If several tests are carried out under undrained conditions on the same clay initially consolidated under different cell pressures p_3, then, in terms of total stresses, the envelope to the rupture circles is also approximately a straight line passing through the origin (dash line in Fig. 18.2*a*), with the equation

$$s = p \tan \phi_{cu} \tag{18.4}$$

where ϕ_{cu}, known as the *consolidated-undrained angle of shearing resistance*, is appreciably smaller than ϕ. The relation between ϕ and ϕ_{cu} is determined by the value of the pore pressure induced by the stress difference $p_1 - p_3$ at failure; for normally loaded clays of low to moderate sensitivity this value is approximately equal to the stress difference itself.

It should be noted that the failure circle for a given test has the same diameter whether it is plotted in terms of effective stresses or total stresses. The pore pressure acts with equal intensity in all directions; hence the increment of pore pressure is the same for both the major and minor principal stresses. This conclusion leads to an extremely useful concept, known as the $\phi = 0$ condition. In Fig. 18.2*b* the solid circle E is the effective stress circle shown in Fig. 18.2*a*. The total stress circle A corresponds to the consolidated-undrained test in which the pore pressure at the start of the test was zero and the pore pressure at the end of the test was u_f. If, however, after the initial consolidation under the cell pressure p_3, the cell pressure had been increased by an amount u_a without allowing drainage, the initial pore pressure in the sample would have been u_a and the pore pressure at failure $u_a + u_f$. The corresponding failure circle would have been B (Fig. 18.2*b*). The effective stress circle would, nevertheless, still be E.

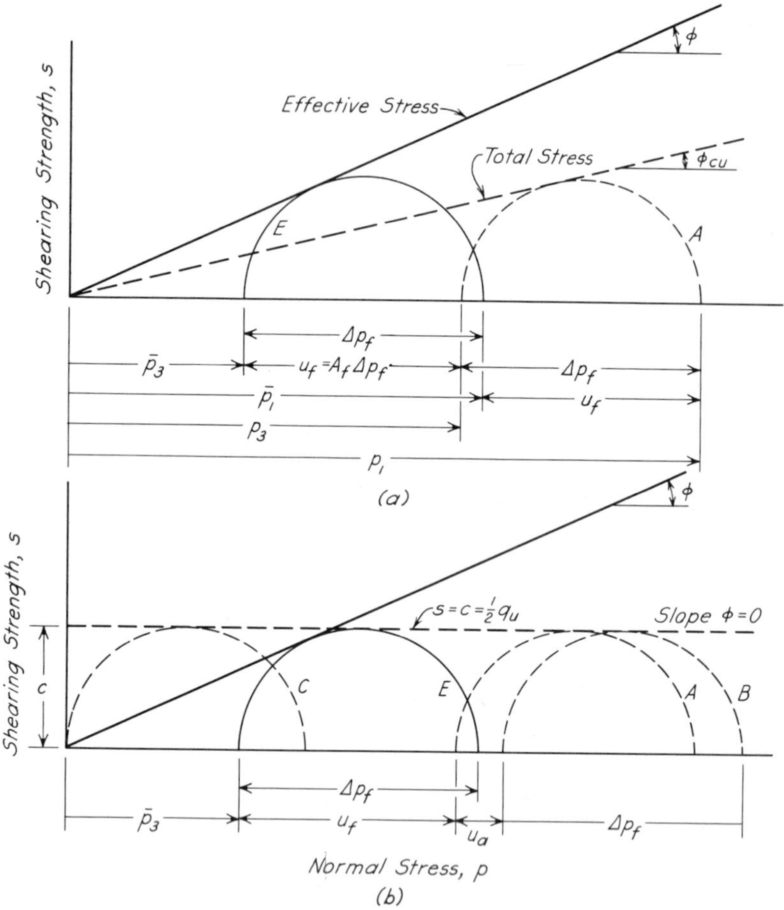

Fig. 18.2. (*a*) Results of consolidated-undrained triaxial tests on normally loaded clay of moderate sensitivity. (*b*) Diagram illustrating $\phi = 0$ condition.

Since any change u_a in cell pressure could have been chosen, it follows that if several samples are consolidated under the same cell pressure $\bar{p}_3$ and then are tested under undrained conditions at different cell pressures, the rupture line *with respect to total stresses* is horizontal. The line may be regarded as a special case of Coulomb's equation in which $s = c$ and $\phi = 0$. Hence, these particular circumstances are known as the *$\phi = 0$ condition* (Skempton 1948). Inasmuch as an unconfined compression test is merely a triaxial test in which the total minor principal stress p_3 is zero (circle C in Fig. 18.2*b*), the shearing strength under $\phi = 0$ conditions may be evaluated on the basis of unconfined compression tests as

$$s = c = \tfrac{1}{2}q_u \tag{18.5}$$

In connection with soils of such low permeabilities as those possessed by most clays and some silts, there are many practical problems in which we can assume that the water content of the soil does not change for an appreciable time after the application of a stress. That is, undrained conditions prevail. Moreover, if a sample is extracted at the same water content and is tested without allowing change in water content, either in unconfined compression or with a cell pressure $p_3 + u_a$, the strength of the soil with respect to total stresses will be approximately (within the limitations imposed by Eq. 15.4) the value c as determined readily from Eq. 18.5. Hence, as a consequence of the $\phi = 0$ concept, the unconfined compression test assumes unusual practical importance.

Moreover, when undrained conditions can be expected to prevail in deposits of saturated clay in the field, other expedient types of tests can often be used advantageously for evaluating c. Foremost among these are several varieties of *vane shear tests* as shown in Fig. 44.17. (The equipment for performing vane shear tests in the field is described in Article 44). Similar vanes of smaller size are often used in the laboratory, especially for investigating the strength of samples of very weak or remolded clays. Among the most convenient modifications (Fig. 18.3) is the portable *torvane* (Sibley and Yamane 1965). The vanes are pressed to their full depth into the clay, whereupon a torque is applied through a calibrated spring until the clay fails along the cylindrical surface circumscribing the vanes and, simultaneously, along the circular surface constituting the base of the cylinder. The value of c is read directly from the indicator on the calibrated spring. By means of such a device, rapid and detailed surveys of c can be carried out (see Fig. 45.5).

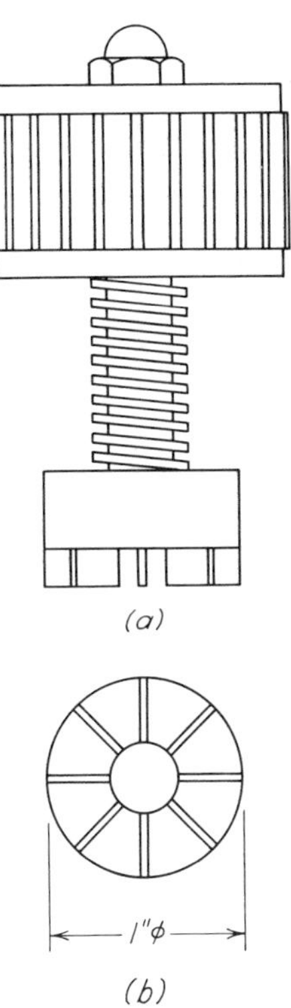

Fig. 18.3. Torvane for determining shear strength of materials for which $s = c$. (*a*) Side view. (*b*) Bottom view of vanes.

Several examples of the use of the $\phi = 0$ concept will be developed in Part III.

If a normally loaded clay is consolidated under an all-around pressure p_3 and then failed under undrained conditions, the failure circle with respect to total stresses is represented by A in Fig. 18.2a. The shearing strength under $\phi = 0$ conditions is measured by the radius c of this circle. By geometry (Fig. 18.4a)

$$\frac{c}{p_3 + c} = \frac{b}{c}$$

whence

$$\frac{c}{p_3} = \frac{\sin \phi_{cu}}{1 - \sin \phi_{cu}}$$

which, for a given clay, is a constant. This relation has suggested (Skempton 1957) that a similar constant ratio should exist between the undrained shear strength of normally loaded natural deposits, as determined by means of unconfined compression or vane tests, and the effective overburden pressure at the depths corresponding to the strength tests. It has been found that this ratio, designated as $c/\bar{p}$, is indeed constant for a given normally loaded deposit, provided the plasticity index is approximately the same throughout the deposit. Moreover, it has also been found that the field $c/\bar{p}$ values for various deposits or fairly homogeneous portions of deposits are correlated closely with the plasticity index, as shown in Fig. 18.4b. Like all statistical relations, Fig. 18.4b entails the possibility that exceptions may appear, but so far the relation has been found applicable over a wide range of types of sedimented clays.

The $c/\bar{p}$ ratio, estimated by means of Fig. 18.4b, makes possible a rough evaluation of the undrained shear strength of normally loaded deposits on the basis of the results of Atterberg limit tests. Conversely, if the undrained strength has been determined by independent tests, comparison with values based on Fig. 18.4b may indicate whether the clay is normally loaded or precompressed.

Extrasensitive and Quick Clays

Most natural clay deposits consist of more or less well graded mixtures of particles of sizes intermediate between those of fine sand and clay, and are relatively insensitive. However, clays consisting primarily of clay-size particles in an edge-to-face structure, or possessing a flocculent structure (Article 4) are likely to experience appreciable loss of strength upon remolding and may exhibit at least moderate sensi-

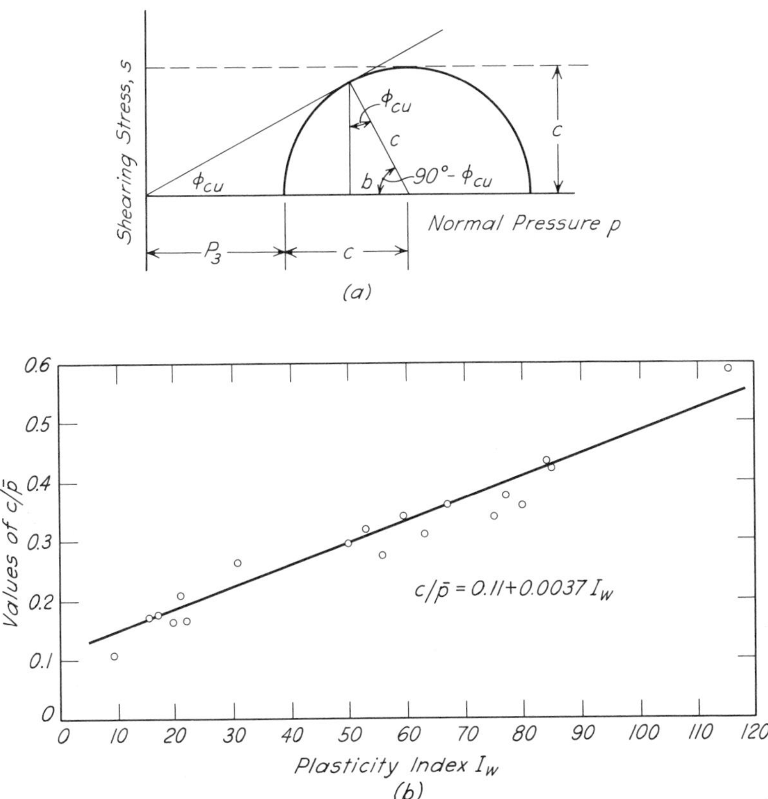

Fig. 18.4. (*a*) Mohr rupture diagram for calculating relation between c and $\bar{p}_3$ for consolidated-undrained test. (*b*) Statistical relation between $c/\bar{p}$ ratio and plasticity index (after Skempton 1957).

tivity. Some natural clay deposits, moreover, consist of a mixture of particles of fairly uniform fine sand and clay. While sedimentation proceeds, the simultaneous deposition of the flaky constituents of the finest fraction and of the equidimensional sand grains interferes with the rolling of the sand grains into stable arrangements. Therefore, if the sand grains touch each other, their configuration may be as metastable as that of true quicksands. However, the interstices between the sand grains are occupied by the clay-size materials which acquire, as a result of such physico-chemical processes as thixotropy and syneresis, appreciable strength as sedimentation proceeds. As a consequence, although the clay is sensitive, it does not exhibit the properties of true quicksands. In many respects, the states of transition from loose sand to true quicksands have their counterparts in the states between clays with low and very high sensitivity.

The failure of extrasensitive clays, like that of true quicksands, appears to be progressive. However, instead of turning throughout into a viscous liquid, extrasensitive clays break up into relatively solid chunks floating in a viscous liquid that can travel on valley floors to distances of several miles at a rate up to 10 miles per hour. One eye-witness, who had the misfortune to be standing on top of one of the chunks in such a slide, graphically described the character of the material in the following words (Terzaghi 1950):

". . . . after reaching the bottom I was thrown about in such a manner that at one time I found myself facing upstream toward what had been the top of the gully . . . . The appearance of the stream was that of a huge, rapidly tumbling, and moving mass of moist clayey earth. . . . At no time was it smooth looking, evenly flowing or very liquid. Although I rode in and on the mass for some time my clothes afterwards did not show any serious signs of moisture or mudstains . . . as I was carried further down the gully away from the immediate effect of the rapid succession of collapsing slices near its head . . . it became possible to make short scrambling dashes across its surface toward the solid ground at the side without sinking much over the ankles."

Quick clays are normally consolidated marine clays that differ from other extrasensitive clays inasmuch as they have acquired their present degree of sensitivity in two steps: the first during deposition, and the second, far more important, by leaching after being lifted above sea level as described in Article 4. In an undisturbed state such clays are as brittle as other extrasensitive clays. A slope failure on such clays commonly starts at the foot of the slope and proceeds by progressive failure in an uphill direction, even on very gentle slopes. Examples of quick-clay flows are discussed in Article 49.

Intact Overconsolidated Clays

The shear-strength characteristics of an overconsolidated clay under drained conditions are illustrated by Fig. 18.5*a*. The rupture line corresponding to the peak strengths of normally loaded samples is represented by the straight line *Od*. We may, however, consolidate a number of identical samples under the same cell pressure $\bar{p}_3$. If one such sample is tested under drained conditions by increasing the vertical pressure, the stress on the failure plane at failure is represented by point *a* on the

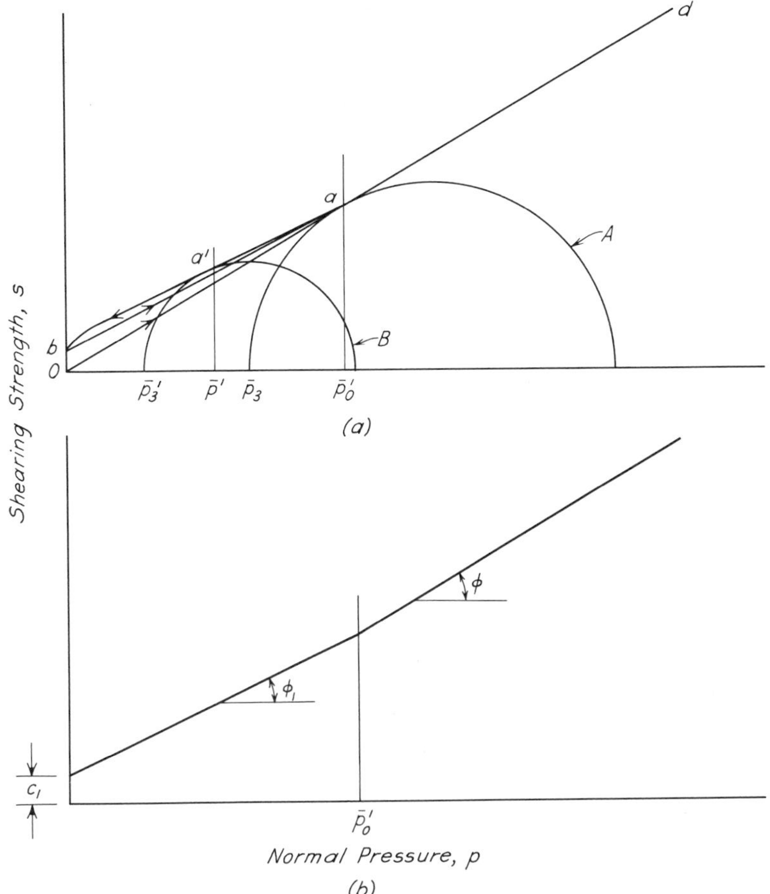

Fig. 18.5. (*a*) **Rupture diagram for clay under drained conditions, preconsolidated to $\bar{p}_0'$.** (*b*) **Simplified rupture lines for same clay.**

circle of stress A. The normal stress on the failure plane is $\bar{p}_0'$. Circle A corresponds to a normally loaded sample.

If one of the samples previously consolidated to $\bar{p}_3$ is allowed to swell under a cell pressure $\bar{p}_3'$ and is then tested under drained conditions, the strength of the sample (circle B) exceeds that of a normally loaded sample tested under the same conditions. The failure envelope $aa'b$ for such samples lies above the line Oa representative of the normally loaded material. The curve $aa'b$ corresponds to the rebound curve bc_1 in the e-log p diagram (Fig. 13.4). If several samples are first consolidated to the stress $\bar{p}_3$, then allowed to swell under zero pressure, and finally are consolidated under various pressures before the performance of drained tests, it is found that the rupture line resembles the lower line *ba* for pressures less than $\bar{p}_0'$, but for greater pressures is nearly identical to the rupture line *Od* for the normally loaded clay. The lower line *ba* corresponds to the reloading portion of the *e*-log *p* curve (Fig. 13.4).

As a rough approximation, the rebound and reloading branches *aa'b* and *ba* of the rupture line may be replaced (Fig. 18.5*b*) up to the pressure $\bar{p}_0'$ by the straight line

$$s = c_1 + \bar{p} \tan \phi_1 \tag{18.6}$$

in which, for a given clay, ϕ_1 is found to be nearly constant whereas c_1, known as the *cohesion intercept*, is found to depend on $\bar{p}_0'$. For pressures greater than $\bar{p}_0'$, the expression

$$s = \bar{p} \tan \phi \tag{18.7}$$

is applicable.

Since for most clays the value of c_1 is very small and ϕ_1 is only slightly smaller than ϕ, a small error on the safe side is made if Eq. 18.7 is considered applicable for all values of $\bar{p}$. Hence, the strength of an intact, moderately overconsolidated clay under drained conditions does not differ significantly from that of normally loaded clays.

In contrast, under undrained conditions the strength of a preloaded clay may be smaller or larger than the drained strength, depending on the value of the overconsolidation ratio. If the overconsolidation ratio lies in the range between 1.0 and about 4 to 8, the volume of the clay tends to decrease during shear and the undrained strength, like that of a normally loaded clay, is less than the drained strength. On the other hand, for values of overconsolidation ratio greater than about 4 to 8, the volume tends to increase, the value of u_w correspondingly decreases, and the undrained strength exceeds the drained value. For high overconsolidation ratios the excess may be very large. However, the strong negative pore pressures associated with high overconsolidation ratios tend to draw water into the soil and cause it to swell, whereupon the strength is reduced. For this reason the undrained strength often cannot be depended upon. Moreover, in most practical problems the attempt to apply the $\phi = 0$ concept for an overconsolidated clay would lead to results on the unsafe side, whereas for a normally loaded clay the tendency toward consolidation would lead to errors in the conservative direction. Hence, except for overconsolidation ratios as low as possibly 2 to 4, the $\phi = 0$ concept should not be used for preloaded clays.

Heavily overconsolidated clays and clay shales are likely to exhibit high peak strengths even when tested under fully drained conditions because of the strong bonds that have developed between the particles

(Article 49). However, after a surface of sliding forms and extensive slip occurs, the bonds are destroyed and the particles along the surface of sliding assume an orientation favorable to a low resistance to shear along the surface. The ultimate shearing resistance after very large displacements under fully drained conditions is known as the *residual strength* (Skempton 1964). It cannot be investigated in conventional triaxial tests because the amount of slip in such tests is restricted; special direct shear or torsional shear devices are required (Haefeli 1950). The residual shear strength may be expressed as

$$s_r = \bar{p} \tan \phi_r \tag{18.8}$$

where ϕ_r varies from about 30° for clays having low plasticity indices and a small clay-size fraction to as little as 5° to 12° for some highly plastic clays with a large percentage of clay-size particles (<0.002 mm). Because of the nearly complete destruction of the structure of the natural clay along the surface of sliding, it is likely that the values of ϕ_r are the same irrespective of the stress-history of the clay, and can be determined with sufficient accuracy on remolded specimens (Skempton 1964).

Fissured Overconsolidated Clays

The continuity of heavily overconsolidated clays is commonly disrupted by a network of hair cracks. If the average pressure in such clays has been reduced, either by excavation or by geological processes such as erosion, the shearing resistance decreases at constant shearing stress; it may ultimately become as small as 0.2 ton/ft^2 irrespective of its original value. Therefore, the failure of slopes in open cuts underlain by such materials may occur many years after the cut is made.

The mechanics of the process of softening is explained in Article 49. At any given time the shearing resistance of the clay increases rapidly with increasing depth below the surface. After a slide occurs the material underlying the newly exposed surface begins to soften and the process continues until another slide occurs. Hence, the side slopes of valleys located in such clays are subject to intermittent superficial landslides from the time the valleys originate; the process does not stop until the slope angle becomes compatible with the softest consistency the clay can acquire. Thus the slopes become gentler. In some regions, such as the valley of the Saskatchewan River south of Saskatoon in Canada, slides still occur without provocation on slopes rising at 1 vertical on 15 horizontal. The problem of determining the shear characteristics of such clays for design purposes has not yet been solved (Peterson et al. 1960).

Shear Characteristics of Cohesive Fills

For reasons explained in Article 50, cohesive fills are commonly placed and compacted at a water content close to the plastic limit. The processes of excavation, transportation and compaction completely destroy the original structure of the soil. Therefore, the finished product has the shear characteristics of a remolded, moderately overconsolidated clay. The values of ϕ under drained conditions depend primarily on the plasticity index; they may be estimated by means of Fig. 18.1. For most practical purposes the value of c may be considered zero.

If in the field the clay becomes saturated, its strength depends, in accordance with the Revised Coulomb equation, on the pore pressure. The investigations under this condition are identical with those in connection with undisturbed preloaded clays. If the degree of compaction of the clay is such that the clay would tend to consolidate under the load to which it may be subjected, and if the rate of pore-pressure dissipation is slow with respect to the rate of loading, then the $\phi = 0$ concept may be utilized. If the clay would tend to swell under its loading or as a consequence of shear, results of analyses based on the $\phi = 0$ condition would be unconservative.

If the fill remains unsaturated, the Revised Coulomb equation

$$s = c + \bar{p} \tan \phi \tag{16.5}$$

remains approximately applicable, but the pore pressure has different values in the air and water phases of the voids. If u_g denotes the pressure in the air or vapor phase and u_w that in the liquid phase, Eq. 16.5 may be rewritten (Bishop, Alpan et. al. 1960; Skempton 1961*a*) as

$$s = c + [p - u_g - \chi(u_w - u_g)] \tan \phi \tag{16.6}$$

where the factor χ depends on the character of the soil and on the degree of saturation; for saturated soils $\chi = 1.0$, and for perfectly dry soils $\chi = 0$. Inasmuch as the techniques for measuring u_g and u_w or for evaluating χ are complex and still suited only for research, it is current practice in connection with practical problems to investigate the strength of partly saturated soils by means of triaxial tests in which only total stresses are measured, and in which the laboratory test conditions are made to duplicate, as closely as possible, those anticipated in the field. The results of four series of undrained tests on samples of an inorganic clay (*CL*), shown in Fig. 18.6 (Casagrande and Hirschfeld 1960), may be regarded as typical. All the samples were compacted initially to the same dry density. In each series the initial degree of saturation S_r (Eq. 6.4) was constant, but the initial value of S_r was

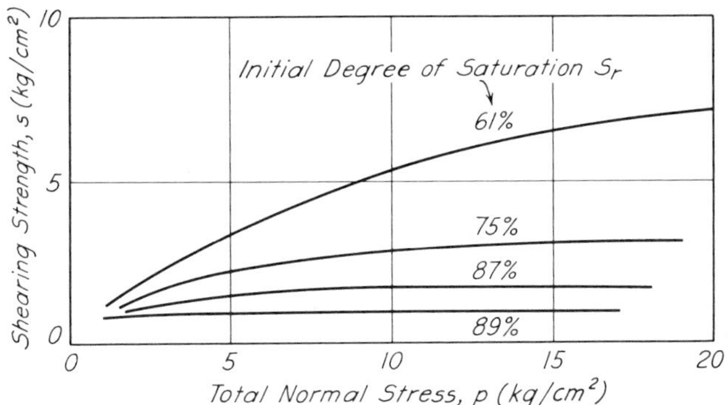

Fig. 18.6. Rupture lines for undrained tests on a lean clay, in terms of total stresses, at various initial degrees of saturation.

different for each series. Drainage was prevented in all tests during the application of the all-around cell pressure as well as during the application of the subsequent stress difference. It is seen that the strength of the samples in the series having the lowest initial degree of saturation ($S_r = 61\%$) exceeds that, at corresponding values of normal stress, of samples having higher initial degrees of saturation, and that the rupture line is markedly curved. As the initial degree of saturation increases, the corresponding rupture lines occupy successively lower positions in Fig. 18.6.

With increasing pressure on an undrained partially saturated sample the volume of the air decreases in accordance with Boyle's law. Furthermore, under increasing pressure, the solubility of air in water increases. Hence, in any series of tests on samples initially at the same degree of saturation, the degree of saturation increases with increase of total pressure on the sample or total normal stress on the plane of failure. If at some pressure all the air is dissolved in the water, the sample becomes saturated and the rupture line with respect to total stresses becomes horizontal ($\phi = 0$ condition). Hence, all the rupture lines (Fig. 18.6) approach horizontal asymptotes, but the $\phi = 0$ condition is reached at lower pressures for the series of tests at the higher initial degrees of saturation.

A compacted fill is ordinarily placed in a partially saturated condition. The strength at the time of compaction depends, for a given compaction procedure, upon the placement moisture content. This is illustrated by the results of unconsolidated-undrained tests on a silty clay (Fig. 18.7*a*). Ultimately, however, if the fill becomes nearly or completely saturated, the strength may differ significantly from that at

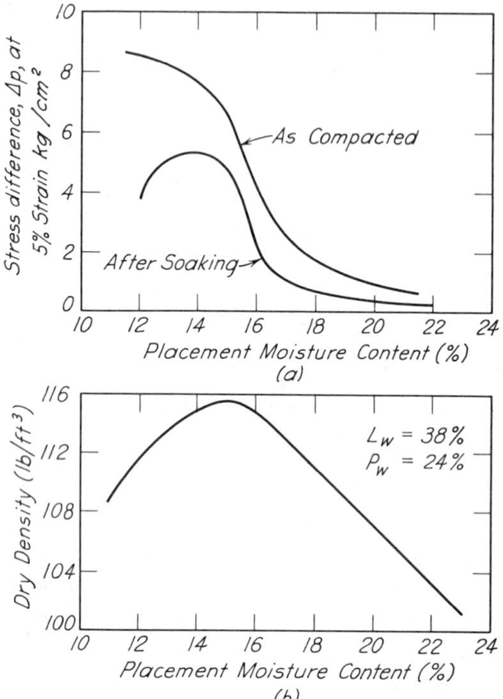

Fig. 18.7. **(*a*) Strength of compacted lean clay, as compacted and after soaking, as function of placement moisture content. (*b*) Moisture-density curve for same material.**

placement, as shown in the figure. The relationships exemplified by Fig. 18.7 differ greatly for different soils, and for the same soils subjected to different compaction procedures (Seed et al. 1960). They also depend on whether the moisture change occurs with or without volume change of the soil.

Because of the complex nature of the phenomena associated with the shear strength of partly saturated soils, considerable experience is needed to select the appropriate test procedures and interpret the results.

Creep

If the shearing stress acting on a sample of clay is less than a value known as the *creep strength*, the clay deforms during and within a short time after application of the shearing stress and thereafter experiences no progressive deformation. On the other hand, if the creep strength is exceeded, the clay deforms continuously under constant shearing stress. Investigations of the rate of creep require special equipment, such as ring or torsion shear devices, in which the area of the surface of failure does not decrease as the shearing strain increases. The results of one such investigation, on an overconsolidated highly plastic remolded clay under fully drained conditions (Hvorslev 1937, 1960), are shown in Fig. 18.8*a*. In the investigation the relation between time and shearing strain was determined for each increment of shearing stress. The deformations during the first 100 hours after the application of each increment, which include the immediate response to the change in stress, are not plotted; only the subsequent deformations are shown. It is apparent that the rate of creep increased with increasing values of the shearing stress. Failure, as evidenced by continuous rotation at a constant rate, occurred at a shear stress of 0.5 kg/cm².

If the area of the surface of failure decreases with increasing deformation, the rate of deformation under a given shear stress is likely to accelerate after approaching a nearly constant value, whereupon failure occurs suddenly. This phenomenon is illustrated by Fig. 18.8*b*, which represents the results of unconfined compression tests carried out under

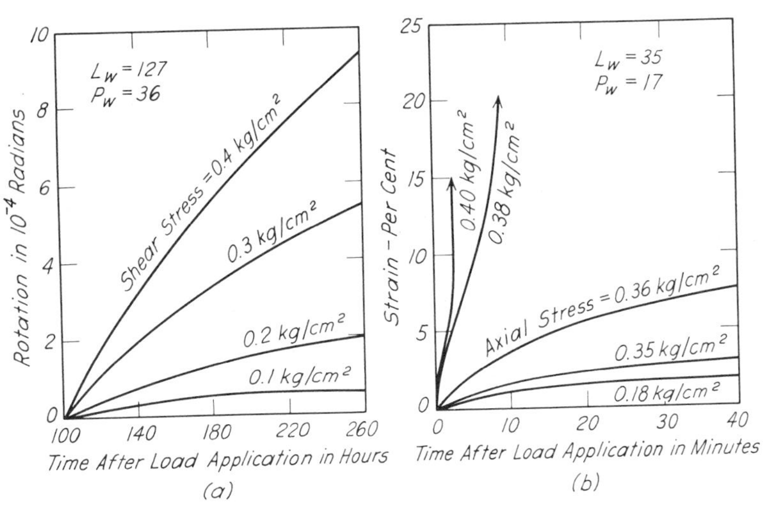

Fig. 18.8. (*a*) **Relation between angular deformation and time for drained specimen of remolded overconsolidated plastic clay in torsion shear test (after Hvorslev 1937).** (*b*) **Relation between strain and time for identical specimens of undisturbed Chicago clay of low plasticity tested under undrained conditions in unconfined compression.**

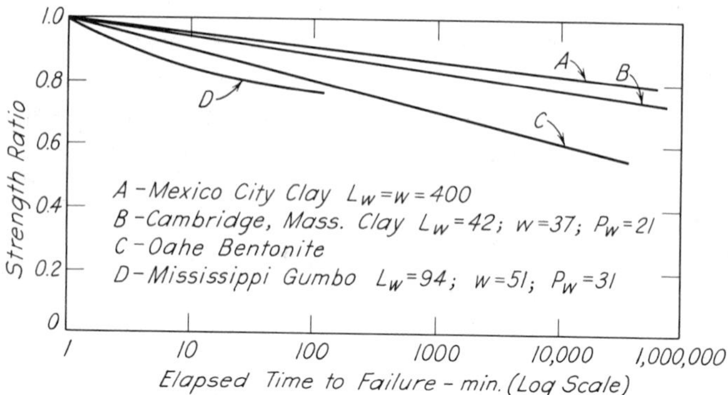

Fig. 18.9. Undrained strength of clay soils reaching failure at various times compared to strength when tested to failure in one minute (after Casagrande and Wilson 1951).

undrained conditions on identical undisturbed samples of a clay of low plasticity from Chicago.

The ratio between creep strength and peak shearing resistance has not been extensively investigated. For some insensitive clays it appears to be as small as 0.3, whereas for brittle clays it may be on the order of 0.8. Stresses in excess of the creep strength have been an important cause of progressive lateral movement of such structures as retaining walls and abutments.

Rate of Loading

In all the conventional shear tests, failure is usually reached within a few hours or days. In some soils the value of s decreases with decreasing rate of load application as shown for undrained tests in Fig. 18.9 (Casagrande and Wilson 1951). These findings lead to the suspicion that the strength of such clays may also decrease with increasing time at constant stress. This possibility requires consideration in the selection of allowable values of shear strength in connection with stability problems.

Problems

1. The results of a series of drained triaxial tests on a lean clay are expressed with sufficient accuracy by the equation $s = \bar{p} \tan 31°$. A consolidated-undrained test on the same material is carried out by first consolidating a specimen under an all-around pressure of 2 tons/ft^2 and then by increasing the axial load without drainage until failure occurs. The sample fails at an axial

stress of 1.80 tons/ft^2 in excess of the cell pressure. What is the value of the pore-pressure coefficient $\bar{\mathbf{A}}_f$ at failure? What is the value of ϕ_{cu}?

Ans. 0.64; 18.1°.

2. The unconfined compressive strength of a sample of clay is found to be 2 tons/ft^2. The clay has a plasticity index of 40; its drained angle of internal friction should, according to Fig. 18.1, be about 27.7°. On the assumption that this value of ϕ is valid, what is the magnitude of the porewater pressure in the unconfined specimen at failure?

Ans. 1.15 tons/ft^2 negative.

3. A normally loaded deposit of undisturbed clay has a plasticity index of 65% and a saturated unit weight of 114.3 lb/ft^3. The clay extends to a depth of about 50 ft below the ground surface; the groundwater level coincides with the ground surface. A very good undisturbed sample is taken from a depth of 30 ft. What unconfined compressive strength is it likely to exhibit?

Ans. About 0.55 ton/ft^2.

4. Two consolidated-undrained triaxial tests are made on remolded samples of the same clay. One sample is consolidated under a cell pressure of 1.70 kg/cm^2. It fails under an added axial stress of 1.24 kg/cm^2. The porewater pressure is measured throughout the test and at failure is found to have a positive value of 1.07 kg/cm^2. The other sample is consolidated under a cell pressure of 4.27 kg/cm^2. The added axial stress at failure is 3.12 kg/cm^2, and the porewater pressure at failure is 2.70 kg/cm^2. What are the values of ϕ_{cu} and ϕ for the samples?

Ans. 15.5°, 30°.

Selected Reading

The current state of knowledge concerning the shear strength of cohesive soils was reviewed at a Research Conference on Shear Strength of Cohesive Soils sponsored by the American Society of Civil Engineers at Boulder, Colorado, in June 1960. The proceedings of that conference contain many useful papers among which the following are especially pertinent to Article 18:

Bishop, A. W., I. Alpan, G. E. Blight, and I. B. Donald. *Factors controlling the strength of partly saturated cohesive soils*, p. 503.

Bishop, A. W. and L. Bjerrum. *The relevance of the triaxial test to the solution of stability problems*, p. 437.

Bjerrum, L. and N. E. Simons. *Comparison of shear strength characteristics of normally consolidated clays*, p. 711.

Casagrande, A. and R. C. Hirschfeld. *Stress-deformation and strength characteristics of a clay compacted to a constant dry unit weight*, p. 359.

Hvorslev, M. J. *Physical components of the shear strength of saturated clays*, p. 169.

Peterson, R., J. L. Jaspar, P. J. Rivard, and N. L. Iverson. *Limitations of laboratory shear strength in evaluating stability of highly plastic clays*, p. 765.

Seed, H. B., J. K. Mitchell, and C. K. Chan. *The strength of compacted cohesive soils*, p. 877.

Simons, N. E. *Comprehensive investigations of the shear strength of an undisturbed Drammen clay*, p. 727.

Simons, N. E. *The effect of overconsolidation on the shear strength characteristics of an undisturbed Oslo clay*, p. 747.

ART. 19 EFFECT OF VIBRATIONS ON SOILS

It is a matter of common experience that vibrations due to pile driving, traffic, or the operation of machinery usually increase the density of a sand and cause its surface to subside. Damage to buildings may be caused by the subsidence and is often the subject of lawsuits against the parties responsible for the vibrations. On the other hand, vibrations are also one of the most economical means for compacting embankments of sand or natural layers of loose sand prior to the construction of foundations (Article 50). Hence, the effect of vibrations on soils may be harmful or beneficial, but it always deserves attention.

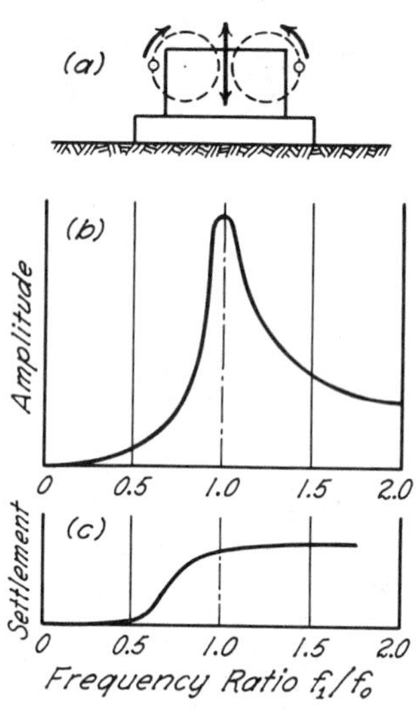

Fig. 19.1. (*a*) Principle of soil vibrator. (*b*) Relation between frequency and amplitude of vibrations. (*c*) Relation between frequency and settlement of vibrator base (after Hertwig et al. 1933).

In order to investigate the factors that influence the compacting effect of vibration, apparatus shown diagramatically in Fig. 19.1*a* (Hertwig et al. 1933) has been used. It consists of a bearing plate and two equal eccentric weights which rotate in opposite directions. The total force exerted on the ground by the base plate of the apparatus consists of a static force equal to the weight of the equipment, plus a pulsating force with a maximum value equal to the centrifugal force of the two eccentric weights. The number of revolutions of the eccentric weights per unit of time is the *frequency*, usually expressed in cycles per second. The greatest vertical distance through which the base moves from its equilibrium position is called the *amplitude* of vibration of the base. At a certain frequency the amplitude is a maximum (see Fig. 19.1*b*). This frequency is approximately equal to the natural frequency f_0 of the vibrator and the vibrating portion of the supporting soil.

The term *natural frequency* indicates the frequency of the vibrations that ensue if a body with well-defined boundaries is acted on by a single impulse. If the impulse is periodic, the amplitude of the resulting *forced vibrations* increases as the frequency f_1 of the impulse approaches the natural frequency of the body. At a frequency close to the natural frequency, the amplitude is a maximum. This state is called *resonance*. In Fig. 19.1*b* it is represented by a peak.

Table 19.1 contains values of the resonant frequency of a vibrator such as that shown in Fig. 19.1 operating on different soils and soft rocks (Lorenz 1934). The vibrator had a weight of 6000 lb and a contact area of 10.7 ft^2. The values were obtained by steadily increasing the frequency of the impulse up to and beyond the occurrence of resonance.

Table 19.1
Resonant Frequency of Vibrator on Various Types of Soil

Supporting Soil or Rock	Frequency, cycles per second
Loose fill	19.1
Dense artificial cinder fill	21.3
Fairly dense medium sand	24.1
Very dense mixed-grained sand	26.7
Dense pea gravel	28.1
Soft limestone	30.0
Sandstone	34.0

The resonant frequency depends not only on the properties of the supporting soil but also to a certain extent on the weight and dimensions of the vibrator. These variables have been investigated by the U.S. Corps of Engineers in two series of tests, one on a cohesive silty clay and the other on cohesionless sand. The weight of the vibrator and its base varied from 13,000 to 65,000 lbs, the diameters of the loaded areas from 5 to 16 ft, and the contact areas from 21 to 200 ft^2. Several modes of vibration were induced separately (WES 1963). The results have substantially extended the range of the pertinent variables, but do not differ fundamentally from those illustrated in Fig. 19.1.

If a particular vibrator is used on different soils, the resonant frequency increases with increasing density and decreasing compressibility of the soil. By taking advantage of this fact, extensive use has been made of such equipment for determining the degree of compaction of artificial fills and for comparing the effectiveness of different methods of compaction.

If a vibrator operates on sand, the sand beneath the bearing plate becomes compacted. At constant frequency of the impulses, the size of the zone of compaction increases at a rate that decreases with time. The ultimate size of the zone depends on the intensity of the periodic impulses exerted by the vibrator and on the initial density of the sand. Beyond the boundaries of this zone the density of the sand remains practically unchanged.

Since the vibrator rests on the surface of the zone of compaction, the process of compaction is associated with a settlement of the vibrator. If the frequency of the impulse is gradually increased, the corresponding settlement of the vibrator increases as shown in Fig. 19.1*c*. As the resonant frequency is approached, the settlement increases rapidly and becomes many times greater than the settlement produced by a static load of the same magnitude as the pulsating force. The range of frequencies within which the increase of settlement is greatest is called the *critical range*. It seems to extend from $\frac{1}{2}$ to $1\frac{1}{2}$ times the resonant frequency.

If the frequency of a vibrating engine supported on sand is within the critical range for the sand, the resulting settlement is very much greater than that which would be caused by the equivalent static forces. The frequency of vibrations caused by the slight but inevitable eccentricity of the rotating parts of steam turbines happens to be within the critical range for sand (Article 60). Therefore, foundations for steam turbines on strata of loose sand settle excessively unless the sand is artificially compacted before the turbine foundations are constructed. Whatever the subsoil conditions may be, it is advisable to make special provisions to reduce the amplitude of the forced vibrations.

The effect of vibration on clays is far less conspicuous than on sand because the cohesive bond between clay particles interferes with intergranular slippage. Nevertheless, even a soft clay consolidates to a moderate extent when it is continually subjected to intense vibrations having a frequency close to the natural frequency of the clay.

In reality, vibrating engines oscillate not only vertically but in several other modes each of which may be characterized by a different resonant frequency. The resulting motions are very complex and cannot usually be predicted reliably, although for simple cases the resonant frequencies can be approximated (Barkan 1962, Lysmer and Richart 1966).

Similar phenomena of resonance may be induced if a vibrator is mounted at the top of a pile. The principle has found application in pile driving. In this instance the vibrator is operated at the natural frequency of longitudinal vibrations in the pile itself, whereupon the pile may penetrate readily into the ground (ASCE 1961).

Chapter 3

DRAINAGE OF SOILS

ART. 20 WATER TABLE, SOIL MOISTURE, AND CAPILLARY PHENOMENA

Definitions

The terms *water level, water table,* and *phreatic surface* designate the locus of the levels to which water rises in observation wells in free communication with the voids of the soil in situ. The water table can also be defined as the surface at which the neutral stress u_w (Article 12) in the soil is equal to zero.

If the water contained in a soil were subject to no force other than gravity, the soil above the water table would be perfectly dry. In reality, every soil in the field is completely saturated for a certain distance above the water table and is partly saturated above this level. The water that occupies the voids of the soil located above the water table constitutes *soil moisture.*

If the lower part of a mass of dry soil comes into contact with water, the water rises in the voids to a certain height above the free-water surface. The upward flow into the voids of the soil is attributed to the *surface tension* of the water. The seat of the surface tension is located at the boundary between air and water. Within the boundary zone the water is in a state of tension comparable to that in a stretched rubber membrane attached to the walls of the voids of the soil. However, in contrast to the tension in a stretched membrane, the surface tension in the boundary film of water is entirely unaffected by either the contraction or stretching of the film. The concepts regarding the molecular interactions that produce surface tension are still in a controversial state. Nevertheless, the existence of a tensile stress in the surface film was established beyond any doubt more than a century ago, and the intensity of this stress has since been determined by very different methods with consistent results.

Rise of Water in Capillary Tubes

The phenomenon of capillary rise can be demonstrated by immersing the lower end of a very small-diameter glass tube into water.

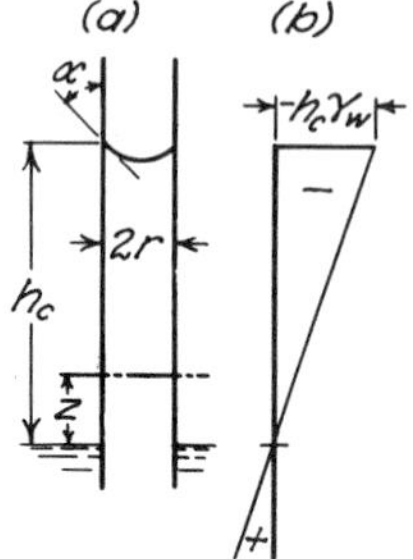

Fig. 20.1. (*a*) **Rise of water in capillary tube.** (*b*) **State of stress of water in capillary tube.**

Such a tube is known as a capillary tube. As soon as the lower end of the tube comes into contact with the water, the attraction between the glass and the water molecules combined with the surface tension of the water pulls the water up into the tube to a height h_c above the water level (Fig. 20.1*a*). The height h_c is known as the *height of capillary rise*. The upper surface of the water assumes the shape of a cup, called the *meniscus*, that joins the walls of the tube at an angle α known as the *contact angle*. The value of α depends on the material that constitutes the wall and on the type of impurities that cover it. For glass tubes with chemically clean or wetted walls α is equal to 0°, and the water rises in such tubes to the greatest height compatible with the diameter of the tube and the surface tension of the water. If the walls are not chemically clean, α is likely to have some value intermediate between 0° and 90°, and the corresponding height of capillary rise is smaller than h_c for $\alpha = 0°$. Finally, if the walls are covered with a thin film of grease, α is greater than 90°, and the meniscus is located below the free-water level. This phenomenon is ascribed to a repulsion between the molecules of the water and the grease.

If T_s denotes the surface tension in grams per centimeter, and γ_w the unit weight of water, equilibrium requires that

$$h_c \pi r^2 \gamma_w = 2\pi r T_s \cos \alpha$$

whence

$$h_c = \frac{2T_s}{r\gamma_w} \cos \alpha \tag{20.1}$$

The value of T_s decreases slightly with increasing temperature. At room temperature it is about 0.075 gm/cm, and γ_w is equal to 1 gm/cm³. Therefore,

$$h_c(\text{cm}) = \frac{0.15}{r(\text{cm})} \cos \alpha \tag{20.2}$$

Above the free-water level the hydrostatic pressure in the water u_w is negative. At an elevation z it is

$$u_w = -z\gamma_w \tag{20.3}$$

Capillary Rise of Water in Soils

In contrast to capillary tubes, the continuous voids in soils and most other porous materials have a variable width. They communicate with each other in every direction and constitute an intricate network of voids. If such a network is invaded by water from below, the lower part of the network becomes completely saturated. In the upper part, however, the water occupies only the narrowest voids, and the wider ones remain filled with air. In the laboratory, the rise of water into the voids of a dry sand due to surface tension can be demonstrated by the test arrangement shown in Fig. 20.2*a*. The sand is poured into an upright glass tube with a screen across the bottom. The bottom of the tube is then placed just below a free-water surface, whereupon the water rises into the sand. That part of the sand in which the voids become partly or completely occupied by liquid assumes a dark color, whereas the remainder is light. To a height h_{cc} above the water level the sand is completely saturated. Between h_{cc} and h_c it is partially saturated, as shown in Fig. 20.2*b*. The height h_c is called the *height of capillary rise*. Figure 20.2*c* shows the time rate at which the upper surface of the moistened zone approaches its equilibrium position at elevation h_c.

As the effective grain size decreases, the size of the voids also decreases, and the height of capillary rise increases. The height h_c (cm) is approximately equal to

$$h_c = \frac{C}{eD_{10}} \tag{20.4}$$

in which e is the void ratio, D_{10} (centimeters) is Allen Hazen's effective size (Article 5), and C (square centimeters) is an empirical con-

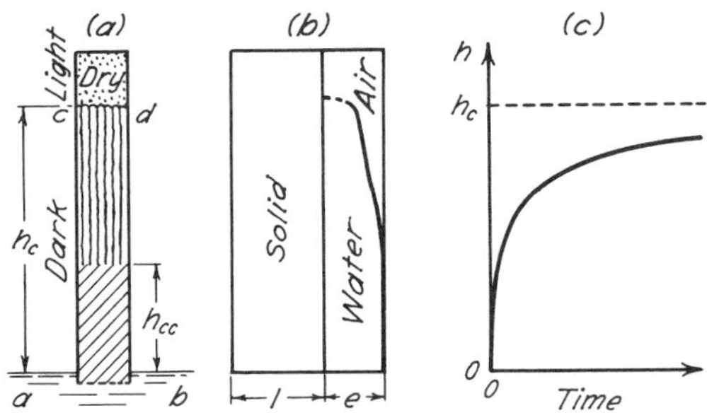

Fig. 20.2. Capillary rise of water into dry sand.

stant that depends on the shape of the grains and on the surface impurities. It ranges between 0.1 and 0.5 cm^2. However, since the decrease in permeability associated with a decrease in effective size reduces the rate of capillary rise, the height to which water will rise within a specified time, such as 24 hr, is a maximum at some intermediate grain size. In Fig. 20.3 the abscissas represent the logarithm of the grain size of a uniform quartz powder in a fairly dense state, and the ordinates represent the height to which the water rises in 24 hr. The height of rise is a maximum for a grain size of about 0.02 mm. For a 48-hr rise the optimum grain size would be slightly smaller.

Capillary Siphoning

Capillary forces are able to raise water against the force of gravity not only into capillary tubes or the voids in columns of dry soil, but also into narrow open channels or V-shaped grooves. This fact can be demonstrated by the apparatus shown in Fig. 20.4. If the highest point of the groove is located below the level to which the surface tension can lift the water, the capillary forces will pull the water into the descending part of the groove and will slowly empty the vessel. This process is known as *capillary siphoning.* The same process may also occur in the voids of a soil in the field. For example, water may flow over the crest of an impermeable core in a dam or dike, as shown in Fig. 20.5, in spite of the fact that the elevation of the free-water surface is below that of the crest of the core. Capillary siphoning over the crest of cores was found to cause a loss of 450 gal/min for a length of 12 miles of the canal between Berlin and Stettin in Germany. The impermeable core of the dikes extended one foot above the water table. When the height of the core was increased 16 in., the loss was reduced to 100 gal/min.

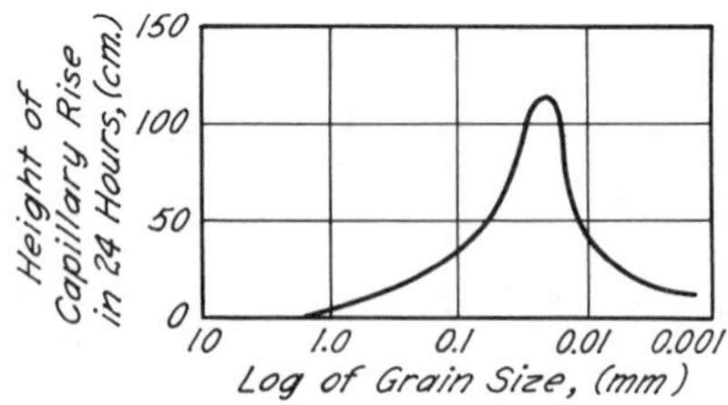

Fig. 20.3. Relation between grain size of uniform quartz powder and height of capillary rise for 24-hr period (after Atterberg 1908).

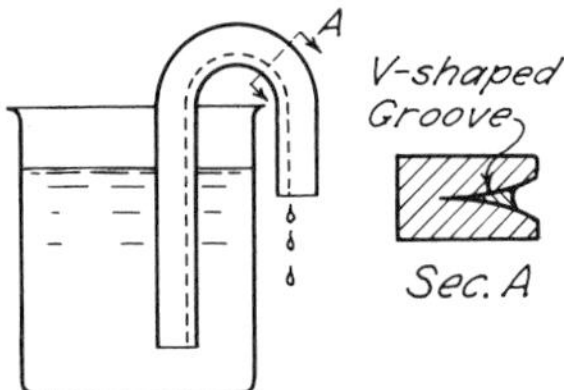

Fig. 20.4. Capillary flow through V-shaped groove.

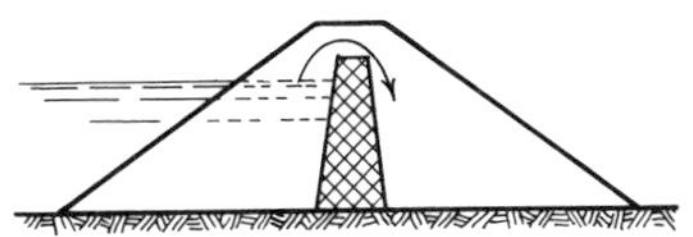

Fig. 20.5. Capillary flow over impermeable core in earth fill.

Discontinuous Soil Moisture

Between the heights h_{cc} and h_c (Fig. 20.2*a*) some of the void space is occupied by continuous channels of air. The remainder is occupied by threads of water. Since these threads are also continuous, the stress in the water up to an elevation h_c is governed by Eq. 20.3. However, if a sand is only moist, the water particles do not communicate with each other, and Eq. 20.3 is not applicable.

The water contained in a moist sand is known as *contact moisture* because each drop of water surrounds a point of contact between two soil grains, as shown in Fig. 20.6. The surface tension at the boundary between the water and the air in the adjoining voids pulls the soil grains together with a force P known as the *contact pressure.* The frictional resistance produced by the contact pressure has the same effect as if the sand had a certain amount of cohesion (see Figs. 21.3*a* and *b*). As soon as the sand is immersed, the surface tension is eliminated, the contact pressure becomes equal to zero, and the sand disintegrates.

The mechanical effect of the cohesion due to contact moisture depends on the relative density of the sand. If the sand is dense, the cohesion increases its shearing resistance to such an extent that vertical slopes with a height of several feet can stand without lateral support. On the other hand, if a damp sand is loosely deposited, for instance by dumping, the cohesion prevents the soil particles from settling into stable positions and reduces the bearing capacity of the sand almost to zero. The volume of such a sand may exceed that of the same sand in a loose dry state by 20 to 30%. This phenomenon is known as *bulking*. Since

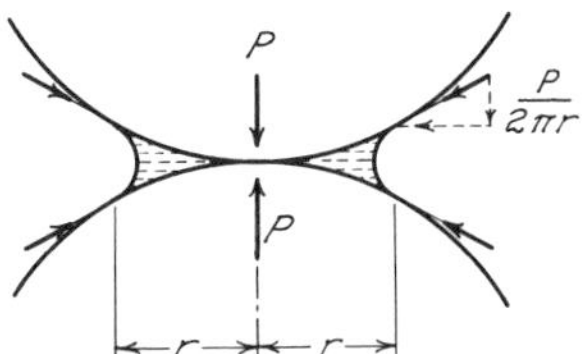

Fig. 20.6. Forces produced by contact moisture.

the forces that maintain the grains in their unstable positions are extremely small, bulking occurs only within a few feet below the surface of the sand. Watering reduces the porosity of the sand to that of the same sand in a dry or saturated loose state because it eliminates the surface tension of the water.

Common Misconceptions

Since the physical causes of the capillary movement of water through soils are not so obvious as those of percolation under the influence of gravity, several misconceptions have found their way into publications. It has, for instance, been maintained that water cannot rise in a capillary tube higher than in the suction tube of a pump (about 30 ft). The height to which water can be lifted by suction depends on the atmospheric pressure, and it is independent of the diameter of the suction tube. On the other hand, the height to which water can be lifted by capillarity is independent of atmospheric pressure, and it increases with decreasing diameter of the tube. Thus it is evident that the two types of rise have nothing in common. In a vacuum water cannot be lifted in a suction tube to any height, whereas the height of capillary rise is the same as under atmospheric pressure.

It has also been maintained that most of the water contained in fine sand does not participate in the flow of seepage because it is held in the sand by molecular attraction. This opinion is incompatible with the well-established fact that the thickness of the layers of water that are bound to the solid by molecular attraction does not exceed about 0.1μ. Beyond these layers the water is normal and able to flow as freely as in a pipe line. Since the quantity of water located within a distance of 0.1μ from the surface of the grains in a saturated sand is negligible compared to the total, practically the entire water content consists of normal water which participates in seepage flow.

Problems

1. The effective size of a very fine sand is 0.05 mm, and its void ratio is 0.6. What is the height h_c of capillary rise for this sand?

Ans. Between 33 and 165 cm.

2. The unconfined compressive strength of a dense fine moist sand is 0.2 kg/cm^2, and its angle of internal friction is 40°. What is the intensity of the all-around pressure p_c that would be required to produce the same effect on the strength of the sand as the cohesion produced by the contact moisture?

Ans. 0.056 kg/cm^2.

ART. 21 PROCESSES OF DRAINAGE

Purpose and Types of Drainage

In engineering practice drainage is used wherever it is desirable to eliminate seepage pressure, to reduce the danger of frost damage, or to increase the shearing resistance of the soil by reducing the neutral stresses (see Articles 12 and 17). It consists of lowering the water table below the base of the mass of soil which requires protection or reinforcement.

In order to lower the water table to a given elevation, it is necessary to establish below this level a system of collectors located in wells, galleries, or ditches. The water flows by gravity out of the soil into the collectors, and it is removed from the collectors by pumping or other appropriate means. Since the hydraulic gradient at the walls of every collector is very high, the finer soil particles are gradually washed out of the ground into the collectors unless the walls are surrounded by filters which consist of wire mesh or of screened sand or gravel. The openings of filter screens should be roughly equal to the 60% grain size D_{60} of the adjoining natural soil. Sand or gravel filters should satisfy the grain-size requirements specified at the end of Article 11.

Drainage wells are commonly lined with steel tubes called *well casings*, which are perforated where they are in contact with water-bearing strata. If the casing has a diameter less than about $2\frac{1}{2}$ in., it is called a *well point*. The water is pumped out of a group of well points through a *header pipe* that interconnects the upper ends of all the casings. If the diameter of a drainage well is 12 in. or more, the water is commonly pumped out through a suction tube with a very much smaller diameter, and the space between the tube and the walls of the hole is filled with coarse sand or gravel. Such wells are known as *filter wells*. The annular filter acts as a substitute for the perforated well casing. Collectors in ditches or galleries usually consist of open-jointed pipe lines embedded in sand or gravel that satisfies the grain-size requirements for a filter.

In sand part of the water that flows out of the voids into the collectors is replaced by air (*drainage by air invasion*). However, very fine-grained soils remain in a saturated state, and the volume of the voids of the soil decreases by an amount equal to that of the expelled water (*drainage by consolidation*).

The drainage of any type of soil can also be accomplished by evaporation from a surface exposed to the atmosphere. This process is called *drainage by desiccation.* Depending on the type of soil, it can take place by air invasion, by consolidation, or by air invasion preceded by consolidation.

Very fine-grained soils can also be drained by passing an electric current through them. This process is known as *drainage by electro-osmosis.* If the uppermost part of a saturated mass of very fine-grained soil is exposed to a temperature below the freezing point, water is drawn out of the lower part and accumulates in the upper part, where it participates in the formation of ice layers. The seepage pressure of the percolating water consolidates the layer of soil located beneath the zone of freezing. Therefore, it may be said that this layer is subject to *drainage by frost action.* On the other hand, the average water content of the soil contained within the zone of freezing increases. The following paragraphs contain a description of the different processes of drainage.

Drainage by Gravity

The smallest value to which the water content of a soil can be reduced by gravity drainage is known as the *water-holding capacity* of the soil. In order to obtain numerical values for comparing the water-holding capacity of different soils, various laboratory procedures are used. In some, known as *gravity methods,* the water drains out of the sample under the influence of gravity alone. In others, known as *suction methods,* the force of gravity is augmented by applying a vacuum at the bottom of the sample or air pressure at the top. In a third type, known as *centrifuge methods,* the gravity forces are replaced by inertia forces of much greater intensity.

If the void ratio of a soil after drainage, the unit weight of the solid soil particles, and the water-holding capacity are known, the degree of saturation S_r (Article 6) and the air-space ratio G_a of the drained soil can be computed. The *air-space ratio* is defined by the equation,

$$G_a = \frac{\text{air space}}{\text{total void space}} = 1 - \frac{S_r(\%)}{100} \tag{21.1}$$

Curves A and B in Fig. 21.1 represent the relation between the air-space ratio and the effective grain size for different soil fractions that were drained by using two different methods. The data for plotting curve A were obtained by submitting saturated samples to drainage by suction. A vacuum was applied for 2 hr at the bottom of samples 4 in. high. Curve B represents the results of tests made

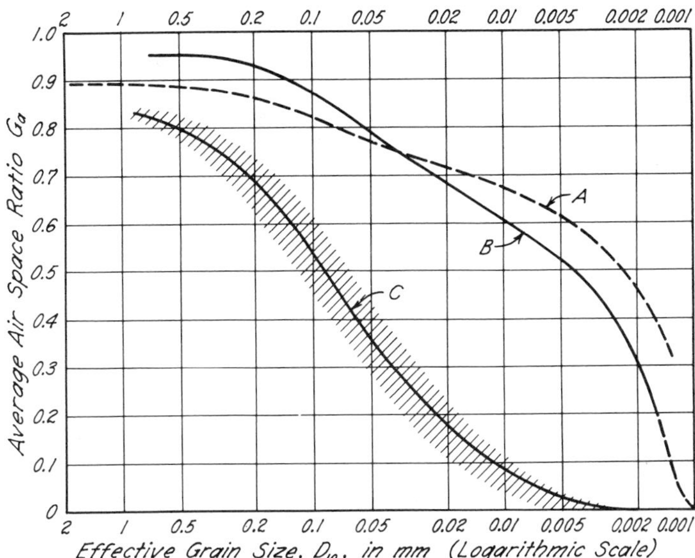

Fig. 21.1. Relation between grain size and degree of saturation after drainage. Curve *A* (after Zunker 1930) obtained by suction method; curve *B* (after Lebedeff 1928) by centrifuge method; curve *C* by field measurements.

by the centrifuge method, in which the samples were subjected for 2 min to a force 18,000 times that of gravity (Lebedeff 1928).

In the laboratory the drainage of sand under the influence of gravity continues for years at a decreasing rate, even if the sand is fairly coarse. Figure 21.2*a* represents two samples of sand $2\frac{1}{2}$ years after the start of drainage. In both samples the air-space ratio increased quite rapidly with increasing elevation above the water table, as shown in Fig. 21.2*b*. Furthermore, even at the end of $2\frac{1}{2}$ years, the average air-space ratio of both samples was still increasing (King 1899).

In the field every process of drainage by gravity is accompanied periodically by an inflow of water from rain or melting snow and ice. The effect of the recharge on the average moisture content of a drained soil in the field depends not only on the amount of recharge and evaporation, but also to a considerable extent on details of stratification. Furthermore, experience indicates that the air-space ratio for a drained soil in the field is practically independent of the elevation above the water table, whereas in drained laboratory specimens it increases consistently in an upward direction, as shown in Fig. 21.2*b*. Hence, there is no definite relation between the water-holding

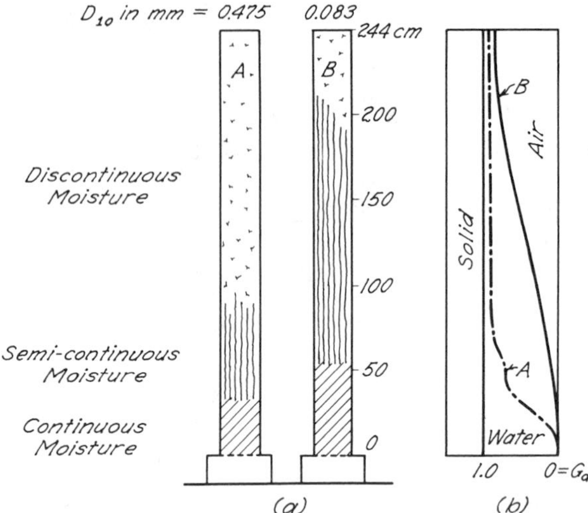

Fig. 21.2. Soil moisture in two different samples of sand after drainage for $2\frac{1}{2}$ years in laboratory (after King 1899).

capacity of a soil after drainage in the laboratory and that of the same soil after drainage in the field. This can be seen by comparing the laboratory curves A and B in Fig. 21.1 with curve C. The shaded area surrounding curve C represents the relation between air-space ratio and effective grain size of various soils after gravity drainage in the field under climatic conditions similar to those in the east-central part of the United States. Observations in regions with different climatic conditions may lead to very different curves. Yet no field curves can be expected to have more than the general trend in common with the laboratory curves.

Fortunately, in connection with engineering operations, the quantity of water drained out of the ground is seldom of importance. Of much greater significance are the mechanical effects of the drainage and the time required to produce them.

Rate and Effect of Drainage by Gravity

As stated before, drainage can be accomplished by pumping from filter wells, by tapping the water-bearing strata with galleries, or by diverting the water into drainage ditches. Whatever the method of drainage may be, the time required for draining the soil is always a factor of outstanding importance.

Theoretical methods for estimating the rate of drainage by air invasion are still rather unsatisfactory. Hence, in order to estimate the time required for draining a sand stratum, the engineer must rely primarily on experience. The drainage of a stratum of clean coarse sand by pumping from filter wells spaced not in excess of 40 ft can usually be accomplished within a few days (very rapid drainage). On the other hand, the same operation in a very fine sand may require several months (slow drainage). The methods available for draining soils and the conditions for their successful application are discussed in Article 47. The settlements associated with lowering the water table are the subject of Article 59.

Desiccation of Soils

If a specimen of soft clay is exposed to the air, water is drawn from the interior of the specimen toward the surface where it evaporates. During this process, the clay becomes stiffer and finally very hard. The state at which evaporation ceases depends on the relative humidity of the surrounding air. According to the laws of physics, water evaporates at every air–water boundary unless the relative humidity of the air is at least equal to a certain value that is a function of the tension in the water. The *relative humidity* h_{ra} is defined as the ratio between the weight of water vapor actually contained in the air at a given temperature and the greatest quantity of vapor that can be contained in the air at the same temperature. In humid climates the relative humidity commonly ranges between 0.15 and 0.95 and exceptionally reaches 0.99. If the relative humidity of the air above a free-water surface is smaller than 1.0, water evaporates until the relative humidity of the superincumbent air is equal to 1.0, or until the water has completely evaporated. If the water is in a state of tension, it ceases to evaporate at a lower value of relative humidity. This lower value h_r is designated as the *relative vapor pressure* of the water. Within a temperature range of 10° to 30°C, and a relative vapor-pressure range of 0.7 to 1.0, the relationship between the neutral stress u_w in the water and the relative vapor pressure h_r of the water can be expressed approximately by the equation

$$u_w(\text{kg/cm}^2) = -1500(1 - h_r) \tag{21.2}$$

For example, if $h_r = 0.90$, $u_w = -150$ kg/cm². Hence, if the neutral stress in an exposed specimen of clay is equal to -150 kg/cm², the water content of the clay does not remain constant unless the relative humidity of the surrounding air is equal to 0.90. If the relative vapor pressure is smaller, the clay continues to lose water by evaporation, whereas, if it is greater, water condenses on the surface of the clay and causes the clay to swell until the tension in the water drops to the value determined by Eq. 21.2. This fact can be used as a basis for computing the tension in the water contained in fine-grained porous materials such as clays.

If the water evaporates at the ends of a capillary tube having a radius r (centimeters), the curvature of the menisci and the stress u_w in the water increase until u_w equals $-h_c\gamma_w$. Substitution of h_c from Eq. 20.2 leads to

$$u_{w\ \max}(\mathrm{gm/cm^2}) = -\frac{0.15\ (\mathrm{gm/cm})}{r(\mathrm{cm})}\cos\alpha \qquad (21.3)$$

Further evaporation causes the water to withdraw into the interior of the tube at a constant neutral stress. A similar process takes place in the porewater of desiccating soils. When a soil dries, the value u_w first increases until it assumes the greatest value compatible with the size of the voids at the surface of the soil. Further evaporation causes air to penetrate the specimen, whereupon the color of the soil changes from dark to light. At the beginning of this second stage, the water content of the specimen is equal to the shrinkage limit (Article 7). However, during this stage, the neutral stress u_w can increase further because the water withdraws into the narrowest corners and grooves. Evaporation does not cease until the relative vapor pressure h_r (Eq. 21.2) becomes equal to the relative humidity h_{ra}.

The water that remains in the dried soil constitutes the contact moisture mentioned in Article 20. After desiccation at room temperature, the water content of soils ranges from almost zero for clean sand to 6 or 7% for typical clays. The corresponding air-space ratio ranges from 1.0 to about 0.8. At this state a perfectly clean sand is cohesionless, where a clay is very hard.

If an oven-dried soil specimen is cooled in contact with the atmosphere, its water content increases. The water taken up by the soil particles from the surrounding atmosphere is called the *hygroscopic moisture.* The amount of hygroscopic moisture varies for a given specimen with the temperature and relative humidity of the air. In a general way it increases with decreasing grain size. For sands it is negligible. For silty soils it is very small, yet sufficient to induce bulking. In clay it may amount to more than 5% of the dry weight.

When an air-dry clay specimen is heated to a temperature somewhat above the boiling point of water, its water content decreases slightly. At this stage some of the physical properties of the clay undergo changes that appear to be permanent. The changes are disclosed by permanent changes in the Atterberg limits. Further increase of the temperature to several hundred degrees centigrade above the boiling point leads to actual fusion between the grains at their points of contact. This process produces a strong and permanent bond between the grains and gives the clay the characteristics of a solid body. The transformation of sand–clay mixtures into brick occurs in a similar manner.

The rate at which water evaporates from the surface of clay specimens under given conditions of exposure decreases with decreasing water content. At the liquid limit the rate of evaporation is approximately equal to that from a free surface. At such a surface the rate of evaporation depends on the temperature, the relative humidity, and the wind velocity. In the United States the area of lowest evaporation from large free-water surfaces is the Great Lakes region, where the rate of evaporation ranges from 15 to 20 in. per year. To the west and south of the Great Lakes region it gradually increases. It amounts to about 70 in. in southwest Texas and southeast New Mexico. In the central parts of the Imperial Valley, California, values up to 90 in. per year have been recorded. Even if a clay sample coated with paraffin is stored in a humid room, it gradually shrinks away from its shell. This shrinkage indicates the escape of water vapor through invisible but continuous voids in the paraffin.

As the water content of a desiccating clay decreases, the rate of evaporation also decreases because the tension in the porewater increases. According to Eq. 21.2, the increase in tension involves a decrease in the relative vapor pressure. Such a decrease has the same retarding effect on the rate of evaporation at constant relative humidity as an increase of the relative humidity has on the rate of evaporation from a free-water surface.

Below the shrinkage limit the rate of evaporation is further retarded, because the relative humidity of the air in the voids is always higher than in the adjoining open air. As soon as the relative vapor pressure in the pore space becomes equal to the relative humidity of the surrounding air, further evaporation ceases. If the relative humidity then increases, the water content of the clay increases slightly.

Effect of Desiccation on Strength of Soils

While a soil is drying, tension develops in the porewater. This tension increases with decreasing water content, whereas the total normal stress on a given section through the soil remains practically unaltered. Since the total normal stress is equal to the sum of the neutral and effective stresses, the increasing tension in the porewater involves an equivalent increase of the effective pressure. As desiccation increases the tension in the porewater from zero to $-u_w$, the surface tension simultaneously produces an effective all-around pressure

$$p_k = -u_w \tag{21.4}$$

This pressure is known as *capillary pressure*. It increases the shearing resistance of the soil along any section by

$$\Delta s = p_k \tan \phi \tag{21.5}$$

wherein ϕ represents the angle of internal friction for sand or the consolidated-undrained value of the angle of shearing resistance for clays.

At the shrinkage limit, air invades the voids of the specimen and, as a consequence, the soil moisture ceases to be continuous. The tension in the water that remains in the clay produces contact pressures, as illustrated in Fig. 20.6, and the contact pressures in turn produce shearing resistance. However, because of the discontinuity of the porewater, the relation between Δs and u_w is no longer governed by Eqs. 21.3 and 21.5.

Because of capillary pressure even perfectly cohesionless materials such as fine clean sands may temporarily acquire the characteristics of cohesive materials. As a consequence, samples of the material possess a definite compressive strength when unconfined. Since the cohesion of such soils disappears completely after immersion, it is referred to as *apparent cohesion*.

The water content at which the unconfined compressive strength q_u of a desiccating soil sample is a maximum depends chiefly on the grain size. This statement is illustrated by Fig. 21.3, which shows the effect of a decrease of the water content, due to desiccation, on the compressive strength of three different soils. The water content of each soil at the shrinkage limit is denoted by S_w. For values of w smaller than S_w the degree of saturation (Eq. 6.4) is approximately equal to $100w/S_w$.

For a perfectly clean fine sand moistened with distilled water (Fig. 21.3*a*), q_u is a maximum for a degree of saturation of about 80%. Further desiccation ultimately reduces q_u to zero. However, if the interstices are filled with tap water, the impurities are precipitated during evaporation and form a very thin but continuous layer that

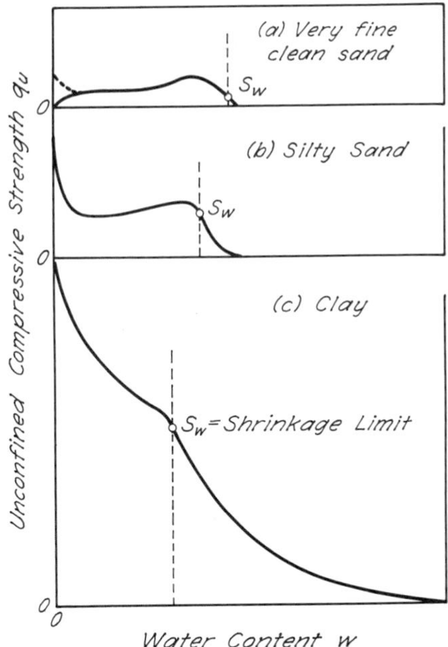

Fig. 21.3. Unconfined compressive strength of various soils at water contents above and at various stages of desiccation below the shrinkage limit (*b* and *c* after Atterberg 1916).

adheres to the grains and interconnects them at their points of contact. Thus, during the last stage of desiccation, the sand acquires a slight cohesion, as indicated in Fig. 21.3*a* by the dash line.

The relation between w and q_u for a fine silty sand is shown in Fig. 21.3*b*. As the water content approaches the shrinkage limit, the strength increases. At the shrinkage limit, air invades the voids of the specimen, and the strength decreases gradually until the degree of saturation becomes approximately equal to 10%. Thereafter it increases and becomes greater than at the shrinkage limit (Atterberg 1916).

The strength of clays (Fig. 21.3*c*) below the shrinkage limit increases at an increasing rate as the dry state is approached.

Desiccation under Field Conditions

In nature desiccation takes place whenever the surface of the soil is not permanently flooded. The apparent cohesion of very fine-grained silty sands due to periodic desiccation may be quite large. Since rain water does not expel more than a small part of the air contained in the voids, the cohesion survives even wet spells of long duration. As a consequence, such soils have often been mistaken, particularly in semiarid and arid regions, for soft rocks. However, if the surface of the soil is flooded, the cohesion gradually disappears, and the soil may start to slide.

The desiccation of a soft clay layer proceeds very slowly from the exposed surface in a downward direction and leads to the formation of a crust that becomes thicker with age. As explained in Article 13, if such a crust is buried beneath clay sediments and is permanently flooded, it constitutes a stiff precompressed layer located between soft normally loaded strata. Thick layers of soft clay can be consolidated by circulating hot dry air through systems of ventilation tunnels, but such a procedure is rarely economical.

In semiarid regions such as western Texas the desiccation of clays in the dry season proceeds to a depth as great as 20 ft (Simpson 1934). Within this depth the clay is broken up by shrinkage cracks. During the rainy season water enters the cracks, and the clay swells. The swelling causes an important rise of the ground surface. Beneath areas covered by buildings the loss of water due to evaporation is very much smaller than the corresponding loss beneath the adjacent areas. Hence, the water content of the clay located beneath the covered areas increases for many years at a decreasing rate and causes a heave of the central part of the areas with reference to their outer boundaries. The amount of heave is practically independent of the weight of the buildings, but its effect on the buildings is very similar to that of unequal settlement. Under unfavorable climatic and soil conditions, the heave can, in the course of time, become greater than a foot.

If the basement floor of a centrally heated building rests on clay, the porewater of the clay may evaporate through the voids of the concrete, whereupon the clay shrinks away from the concrete and deprives the floor of its support. This undesirable development can be prevented by covering the surface of the clay with a bituminous layer before placing the concrete.

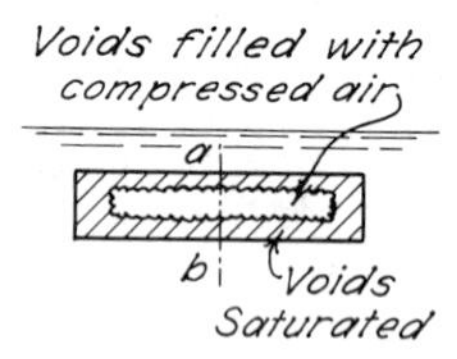

Fig. 21.4. Diagram illustrating process of slaking of dried clay.

Slaking

When a dried specimen of clay (Fig. 21.4) is rapidly immersed in water, the outer portions of the specimen become saturated, and air is trapped in the inner portions. The pressure in the air produces a tension in the solid skeleton that is likely to cause failure in tension along some surface such as *ab*. This process is known as *slaking*. It is responsible for the breaking up and ultimate sloughing of unprotected clay slopes.

Drainage by Electro-Osmosis

If two electrodes are driven into a saturated soil and an electric current is made to flow from one to the other, the water contained in the soil migrates from the positive electrode (anode) toward the negative one (cathode). If the cathode consists of a well point, the water seeps into it, whence it can be removed by pumping.

The movement of the water is due to the negative charge carried by the surface of the soil particles (see Article 4). Accordingly, positively charged ions in the water are attracted toward the soil particles, and the film of water adjacent to the soil is positively charged because of the preponderance of positive ions. Although there is no sharp boundary between the positively charged and the neutral water, for present purposes the seat of the electric charges may be regarded as a well-defined layer (Fig. 21.5*a*) known as the *electric double layer*. The positive ions, which are concentrated in the water near the soil particles, are attracted by the negative electrode and repelled by the positive one. Hence, the positive layer together with the enclosed column of neutral water migrates toward the cathode. The flow of water produced by the electric current is known as an *electro-osmotic phenomenon*.

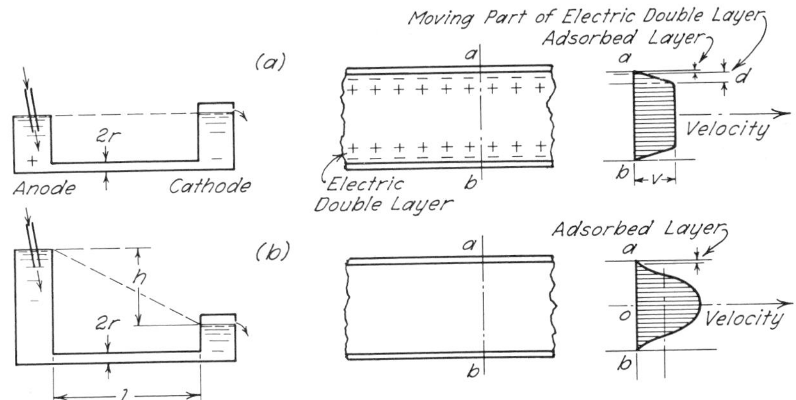

Fig. 21.5. Diagrams illustrating difference between flow through capillaries and soils produced by a hydraulic head (*b*) and by an electric current (*a*).

It should be noted that the velocity of flow is constant over the cross section of the entire column of water surrounded by the double layer, whereas the velocity of gravity flow through a capillary increases from the walls toward the center of the tube as shown in Fig. 21.5*b*.

The velocity v (centimeters per second) at which water flows by electro-osmosis through a cylindrical tube is given approximately by the equation,

$$v = \frac{1.02 \times 10^{-4} deE}{\eta l} \tag{21.6}$$

where e (coulombs/cm²) = electric charge per unit of area of the walls of the tube

E (volts) = difference in electric potential of the two ends of the tube

d (cm) = thickness of the electric double layer

η (gm sec/cm²) = viscosity of the water

l (cm) = length of the tube

For a given material constituting the tubes and for a relatively small range in temperature, e, d, and η are approximately constant and Eq. 21.6 may be rewritten as

$$v = k_e i_e \tag{21.7}$$

where k_e is designated the *coefficient of electro-osmotic permeability*, and i_e is the potential gradient E/l (volts/cm). The coefficient of electro-osmotic permeability is expressed as the velocity of flow (cm/sec) under a potential gradient of 1 volt/cm. Although the width of capillaries in a bundle of capillary tubes is constant, whereas that of the voids in soils varies from point to point, Eq. 21.7 represents at least crudely the velocity of electro-osmotic flow through soils. It is analogous to Eq. 11.6 which represents the flow of water under the influence of a hydraulic gradient.

In contrast to the hydraulic coefficient of permeability k, which varies over an extremely wide range depending on the size of the voids in the soil and, consequently, on the grain size, the coefficient of electro-osmotic permeability is almost independent of grain size. For most soils it lies within the range 0.4 to 0.6×10^{-4} cm/sec. Hence, for such fine-grained soils as silts, which cannot be drained effectively by gravity, electro-osmosis may prove particularly advantageous (L. Casagrande 1949, 1962).

As soon as the electrical potential is applied to the soil, water starts to flow toward the cathode. Seepage pressures (Article 23) are immediately set up which, if directed away from the exposed face of an excavation, may greatly increase the stability of the excavation. Consequently, stabilization of cut slopes in saturated silty soils is one of the more common applications of electro-osmosis (Article 47).

The application of an electrical potential to a compressible fine-grained soil such as clay leads to expulsion of water at the cathodes and, therefore, to consolidation of the clay. The consolidation is not only associated with an increase of strength, but also with the formation of cracks and fissures, especially near the anodes. The anodes are generally corroded away as metallic ions are transported into the soil; the deposition of these ions, as well as the replacement of low-valent ions by those carrying higher charges, lead to permanent changes in the Atterberg limits and other physical characteristics of the soils. The physico-chemical phenomena involved are complex and are not yet well understood.

Frost Heave and Frost-Heave Prevention

If the water contained in the voids of a saturated clean sand or gravel freezes, the structure of the soil remains unchanged. The process of freezing merely increases the volume of each void by 9% because of the expansion of the water contained in the void. On the other hand, if a saturated fine-grained soil freezes, the process involves the formation of lenses of clear ice oriented roughly parallel to the surface exposed to low temperature. The thickness of the individual ice lenses may increase to several inches, and the soil subject to freezing assumes the character of a stratified material consisting of alternate lenses of soil and clear ice (Taber 1930).

Although the molecular mechanics involved in the formation of the ice lenses and the intensity of the forces involved have been the subject of intensive investigation, a quantitative understanding of the phenomena has not yet been achieved (Yong and Warkentin 1966). Nevertheless, the conditions for the formation of the lenses and the means for preventing it are well established (A. Casagrande 1931, Beskow 1935).

Ice lenses develop only in fine-grained soils. However, the critical grain size marking the boundary between soils that are subject to ice-lens formation and those that are not depends on the uniformity of the soil. In perfectly uniform soils ice lenses do not develop unless the grains are smaller than 0.01 mm. Fairly uniform soils must contain at least 10% of grains smaller than 0.02 mm. The formation of ice lenses in mixed-grained soils requires, as a rule, that grains with a

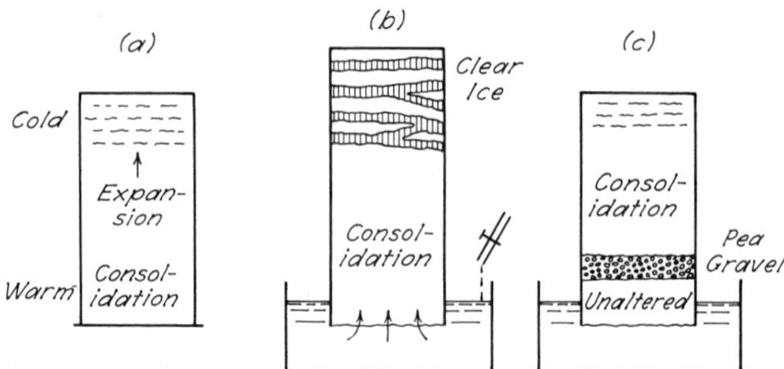

Fig. 21.6. Diagram illustrating frost action in soils. (*a*) Closed system. (*b*) Open system. (*c*) Method of transforming open into closed system by means of a layer of coarse sand that intercepts capillary flow toward zone of freezing.

size less than 0.02 mm constitute at least 3% of the total aggregate. In soils with less than 1% of grains smaller than 0.02 mm ice lenses are not formed under any conditions which may be encountered in the field.

Figure 21.6 represents three cylindrical specimens of a fine saturated silt. Specimen *a* is surrounded by air, whereas the lower ends of specimens *b* and *c* are immersed in water. The temperature of the upper end of each specimen is kept below the freezing point. In *a* the water that enters the ice lenses is drawn out of the lower part of the specimen. As a consequence, the lower part consolidates in the same manner as if the water were pulled toward a surface of evaporation at the upper end. The growth of the ice lenses probably continues until the water content of the lower part is reduced to the shrinkage limit. Since all the water entering the ice lenses comes from within the specimen, the sample is referred to as a *closed system*. The volume increase associated with the freezing of a closed system does not exceed the volume increase of the water contained in the system. It ranges between about 3 and 5% of the total volume.

In *b* the water required for the initial growth of the ice lenses is also drawn out of the specimen, whereupon the lower part of the sample consolidates. However, as the consolidation progresses, more and more water is drawn from the pool of free water located below the specimen. Finally, both the rate of flow toward the zone of freezing and the water content of the unfrozen zone through which the water percolates become constant. Such a sample constitutes an *open system*. The total thickness of the ice lenses contained in such a system can, at least theoretically, increase indefinitely.

The open system represented by sample *b* can be transformed into a closed system by inserting a layer of coarse-grained material between the zone of freezing temperature and the water table, as shown by *c*. Since the water cannot rise by capillarity through the coarse layer, the upper part of the sample represented in *c* constitutes a closed system.

In engineering practice open systems are encountered wherever the vertical distance between the water table and the frost line is smaller than the height of capillary rise of the soil. Since the water that migrates out of the groundwater reservoir is continually replenished, the ice lenses grow continually during the frost periods, and the ground surface located above the zone of freezing rises. This phenomenon is commonly known as *frost heave*. Even in regions with moderate winter climate, such as New England, frost heaves up to 6 in. are by no means uncommon. Since the thickness of the ice lenses reflects very conspicuously the variations of the permeability of the underlying soil, the frost heave is usually nonuniform. Consequently, highway pavements located above the heaving zone are likely to be broken up. Subsequent thaws transform the soil containing the ice lenses into a zone of supersaturated material of a mushy consistency. This condition is likely to be even more detrimental to pavements than the preceding heave.

The tendency of the ice lenses to develop and grow increases rapidly with decreasing grain size. On the other hand, the rate at which the water flows in an open system toward the zone of freezing decreases with decreasing grain size. Hence it is reasonable to expect that the worst frost-heave conditions would be encountered in soils having an intermediate grain size. As a matter of fact, experience has shown that the greatest difficulties with frost heave occur in fine silts and sand–silt mixtures, somewhat finer than the soils for which the capillary rise in a 24-hr period is a maximum (see Fig. 20.3). In a body of soil with given grain-size characteristics, constituting a closed system, the intensity of the growth of ice lenses increases with increasing compressibility of the soil.

Frost action in humid regions with severe winters constitutes a counterpart to the annual volume changes due to desiccation in semiarid regions with hot summers, such as central Texas. It not only damages roads but also displaces retaining walls (see Article 46) and lifts shallow foundations. However, by inserting a layer of gravel between the highest water table and the frost line, the body of soil subject to freezing can be transformed from an open into a closed system, and frost heave can usually be kept within tolerable limits.

Problems

1. The water content of a sample of drained soil is 16.0%, the porosity 42.0%, and the unit weight of the solid soil particles is 2.70 gm/cm^3. Compute the air-space ratio.

Ans. 0.40.

2. An undisturbed sample of very soft clay is kept without any protection in a humid room. It is noticed that the clay becomes stiffer until its unconfined compressive strength ultimately becomes equal to 10 kg/cm^2. The consolidated-undrained value of the angle of shearing resistance of the clay is 20°. Compute the relative humidity of the air.

Ans. 0.9936.

Selected Reading

Casagrande, A. (1931). "Discussion: A new theory of frost heaving," *Proc. Hwy. Res. Board,* **11,** pp. 168–172.

Beskow, G. (1935). "Tjälbildningen och Tjällyftningen med Särskild Hänsyn till Vägar och Järnvägar" (Soil freezing and frost heaving with special application to roads and railroads). *Sveriges Geologiska Undersokning, Stockholm,* Series Cv, No. 375, 242 pp.

Osterberg, J. O. (1940). "A survey of the frost-heaving problem," *Civil Eng.,* **10,** pp. 100–102. Contains a condensed bibliography on the subject.

Physics of the Earth—Part IX, "Hydrology" (1942). Edited by O. E. Meinzer, New York, McGraw-Hill. 1st ed., pp. 331–384. Review of present knowledge concerning soil moisture.

Yong, R. N. and B. P. Warkentin (1966). "Soil freezing and permafrost," Chapter 12 in *Introduction to soil behavior,* New York, MacMillan, pp. 391–428.

PART II

Theoretical Soil Mechanics

Theoretical soil mechanics deals principally with the interaction between soil and water, Chapter 4; with the limiting conditions for the equilibrium of soil masses, Chapter 5; and with the deformations produced by external forces, Chapter 6. The soil constants that appear in the final equations are either estimated on the basis of experience or else obtained by taking the average of values furnished by laboratory tests on what are believed to be representative samples. Hence, none of the theories should be regarded as more than a means for making rough estimates. Some of them, such as those which deal with the settlement of footings on unstratified masses of soil, are intended to serve merely as a guide to judgment in the process of establishing semiempirical rules based on construction experience.

Because of the unavoidable uncertainties involved in the fundamental assumptions of the theories and in the numerical values of the soil constants, simplicity is of much greater importance than accuracy. If a theory is simple, one can readily judge the practical consequences of various conceivable deviations from the assumptions and can act accordingly. If a theory is complicated, it serves no practical purpose until the results are condensed into graphs or tables that permit rapid evaluation of the final equations on the basis of several different assumptions. In this book only the simple theories are set forth in detail. In those exceptional cases in which theoretical refinements are justified the reader should consult the extensive literature on this subject.

Chapter 4

HYDRAULICS OF SOILS

ART. 22 SCOPE OF HYDRAULIC PROBLEMS

The interaction between soil and percolating water enters into several groups of problems in earthwork engineering. One group involves the estimate of the quantity of water that will enter a pit during construction, or the quantity of stored water that will be lost by percolation through a dam or its subsoil (Article 23). A second group deals with the influence of the permeability on the rate at which the excess water drains from loaded clay strata (Article 25). A third group of problems deals with the effect of the seepage pressure on the stability of slopes and foundations. Since the problems of this group also involve consideration of the equilibrium of masses of soil, discussion of hydraulic problems in this category will be deferred to Chapter 5, "Plastic Equilibrium in Soils."

The theoretical solution of each of these problems is based on the assumption that the mass of soil through which the water percolates is homogeneous or that it is composed of a few homogeneous strata with well-defined boundaries. Similar assumptions will be made in the derivation of the theories dealing with earth pressure, stability, and settlement. However, the practical implications of the assumptions are fundamentally different in the hydraulic problems.

Earth pressure, stability, and settlement depend merely on the average values of the soil properties involved. Therefore, even a considerable scattering of the values from the average is of little practical consequence. On the other hand, in connection with hydraulic problems, apparently insignificant geological details may have a decisive influence on both the amount of seepage and the distribution of the seepage pressures throughout the soil. The following example illustrates this point.

If a thick deposit of sand contains a few thin layers of dense fine silt or stiff clay, the presence of these layers has practically no effect on the lateral pressure exerted by the sand against the bracing of an open cut above the water table, on the ultimate bearing capacity of the sand, or on the settlement of a structure resting on the sand. Hence, in connection with these problems the presence of such layers can safely be ignored, and it makes no difference whether or not the boring foreman noticed them.

On the other hand, in connection with any practical problem involving the flow of water through the sand, for instance from a pond on the upstream side of a row of sheet piles to the downstream side, the presence or absence of thin layers of relatively impermeable soil is of decisive importance. If one of the layers is continuous and located above the lower edge of the sheet piles, it intercepts the flow almost completely. If the layers are discontinuous, it is impossible to estimate their influence on the amount and direction of the seepage without knowing the degree of their continuity. Yet, this degree cannot be determined by any practicable means. As a matter of fact, the test borings may not even disclose the presence of the layers at all.

Every natural soil stratum and every man-made earth fill contain undetected or undetectable inclusions of material with exceptionally high or low permeability, and the location of the horizontal boundaries of these inclusions can only be a matter of conjecture. Therefore, the difference between reality and the results of any investigation involving the flow of water through soil can be very important, irrespective of the thoroughness and care with which the subsoil is explored. Yet, if no investigation is made at all, the engineer is entirely at the mercy of chance. Therefore, sound engineering calls for the following procedure in dealing with hydraulic problems. The design should be based on the results of a conscientious hydraulic investigation. However, during the entire period of construction and, if necessary, for several years afterwards, all the field observations should be made that are required for finding out whether and to what extent the real hydraulic conditions in the subsoil differ from the assumed ones. If the observations show that the real conditions are less favorable than the designer anticipated, the design must be modified in accordance with the findings. By means of this procedure, which is illustrated by several examples in Part III, many dam failures could have been avoided.

ART. 23 SEEPAGE COMPUTATIONS

Fundamental Relationships

In the following analysis, it is assumed that the flow of water through the soil follows Darcy's law (Eq. 11.6) and that the soil consists of relatively incompressible material such as sand, silty sand, or rock flour.

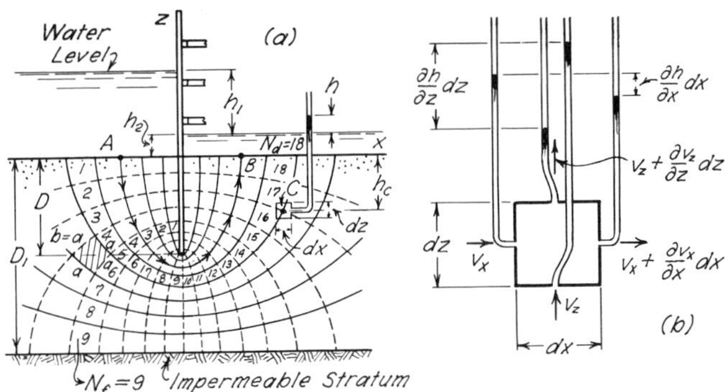

Fig. 23.1. **(*a*) Flow of water around lower edge of single row of sheet piles in homogeneous sand. (*b*) Hydrostatic pressure conditions at four faces of element of sand shown in (*a*).**

In order to compute the rate of flow of water through such soils, it is necessary to determine the intensity and distribution of the neutral stresses, commonly known as the *porewater pressures.* These stresses can be determined by constructing a graph called the *flow net,* which represents the flow of water through an incompressible soil (Forchheimer 1917). To illustrate the method, we shall compute the quantity of water which escapes out of a pond by percolation through the subsoil of the single sheet-pile cofferdam shown in Fig. 23.1*a*. The row of sheet piles is assumed to be impermeable. The piles are driven to a depth D into a homogeneous sand stratum having a depth D_1. The sand rests on a horizontal impermeable base. The hydraulic head h_1 (see Article 11) is kept constant. Water entering the sand at the upstream surface travels along curves known as *flow lines.* Curve AB, marked by arrows, is one such flow line.

Figure 23.1*b* shows a prismatic element of the permeable layer drawn to a larger scale. The lengths of the sides of this element in the plane of the paper are dx and dz. The length perpendicular to the paper is dy. Let

v_x = component of discharge velocity in horizontal direction
i_x = $\partial h/\partial x$, hydraulic gradient in horizontal direction
v_z, and i_z = $\partial h/\partial z$, corresponding values for vertical direction
h = hydraulic head at place occupied by element

The total quantity of water that enters the element per unit of time is

$$v_x\,dz\,dy + v_z\,dx\,dy$$

The quantity that leaves it is

$$v_x\,dz\,dy + \frac{\partial v_x}{\partial x}\,dx\,dz\,dy + v_z\,dx\,dy + \frac{\partial v_z}{\partial z}\,dz\,dx\,dy$$

If the liquid is perfectly incompressible, and the volume of voids occupied by the water is constant, the quantity of water which enters the element is equal to that which leaves it. Therefore,

$$\left(v_x\,dz\,dy + \frac{\partial v_x}{\partial x}\,dx\,dz\,dy + v_z\,dx\,dy + \frac{\partial v_z}{\partial z}\,dz\,dx\,dy\right) - (v_x\,dz\,dy + v_z\,dx\,dy) = 0$$

or

$$\frac{\partial v_x}{\partial x} + \frac{\partial v_z}{\partial z} = 0 \tag{23.1}$$

Equation 23.1 is known as the *continuity condition* for flow parallel to the XZ plane. Since both water and soil are at least slightly compressible, the flow of water through soil does not strictly satisfy the continuity condition. However, in connection with practical seepage problems this fact can usually, although not always, be disregarded.

By combining Eq. 23.1 with Eq. 11.6, we obtain

$$v_x = ki_x = k\frac{\partial h}{\partial x} \quad \text{and} \quad v_z = ki_z = k\frac{\partial h}{\partial z}$$

It may be seen from these equations that the velocities v_x and v_z can be regarded as the partial derivatives with respect to x and z of a quantity,

$$\Phi = kh$$

known as a *velocity potential*. Substituting the values,

$$v_x = \frac{\partial \Phi}{\partial x} \quad \text{and} \quad v_z = \frac{\partial \Phi}{\partial z}$$

into Eq. 23.1, we obtain

$$\frac{\partial^2 \Phi}{\partial x^2} + \frac{\partial^2 \Phi}{\partial z^2} = 0 \tag{23.2}$$

This expression, known as *Laplace's equation*, governs the steady flow of incompressible fluid through an incompressible porous material when the flow can be considered two-dimensional. Graphically, the equation can be represented by two sets of curves that intersect at right angles. The curves of one set are called *flow lines*, whereas the curves of the other set are known as *equipotential lines*. At all points along an equipotential line the water would rise in a piezometric tube to a certain elevation known as the *piezometric level* (Article 11) corresponding to the given equipotential line. The water particles travel along the flow lines, in a direction at right angles to the equipotential lines.

In the problem illustrated by Fig. 23.1*a* the upstream surface of the sand is one of the equipotential lines, and the downstream surface is another. On the other hand, the surface of the impermeable base of the sand is a flow line. These are called the *hydraulic boundary conditions* of the problem. By solving Eq. 23.2 in accordance with these boundary conditions, we obtain the data required for constructing the flow net shown in Fig. 23.1*a*. Every strip located between two adjacent flow lines, as shown in the figure, is called a *flow channel*, and every section of a flow channel located between two equipotential lines is known as a *field*. It is convenient to construct the equipotential lines such that the difference between the piezometric levels for any two adjacent equipotential lines is a constant. This difference is called the *potential drop* Δh. If h_1 is the total hydraulic head and N_d is the number of potential drops ($N_d = 18$ in Fig. 23.1*a*), the potential drop is equal to

$$\Delta h = \frac{h_1}{N_d} \tag{23.3}$$

Once the flow net has been constructed, the porewater pressure at any point located within the flow net, such as point C in Fig. 23.1*a*, can be determined readily on the basis of the following reasoning. If there were no flow, that is, if the downstream ground surface were perfectly impervious, the piezometric head at point C would be equal to the sum of the hydraulic head h_1 and the position head $h_2 + h_c$. However, as a consequence of the flow that does occur, there is a drop in head between the upstream ground surface and point C. Since C is located on the right-hand boundary of the 16th equipotential drop and $N_d = 18$, this drop in head is equal to $16h_1/18$. Therefore, the pressure in the water at point C is

$$u_w = (h_1 + h_2 + h_c - \tfrac{16}{18}h_1)\gamma_w$$

That part

$$(h_1 - \tfrac{16}{18}h_1)\gamma_w = h\gamma_w$$

due only to the flow of the water is known as the *excess hydrostatic pressure*.

Computation of Seepage and of Seepage Pressure

In order to derive the equations necessary to compute the quantity of seepage, we shall consider the field indicated by the shaded area in

Fig. 23.1*a*. The length of its side in the direction of the flow lines is a. The hydraulic gradient across the field is

$$i = \frac{\Delta h}{a}$$

and the discharge velocity is

$$v = ki = k\frac{\Delta h}{a} = \frac{k}{a}\frac{h_1}{N_d}$$

If the width of the field measured at right angles to the flow lines is taken equal to an arbitrary value b, the quantity of water that flows through the field, per unit of width of the sheet piles, is

$$\Delta Q = bv = k\frac{b}{a}\frac{h_1}{N_d}$$

In order to simplify the computation of seepage, flow nets are constructed such that $b = a$, or, in other words, such that every field is square. On this assumption we obtain

$$\Delta Q = k\frac{ah_1}{aN_d} = k\frac{h_1}{N_d} \tag{23.4}$$

If N_f is the total number of flow channels ($N_f = 9$ in Fig. 23.1*a*), the seepage Q per unit of width of sheet piles and per unit of time is

$$Q = N_f\,\Delta Q = kh_1\frac{N_f}{N_d} \tag{23.5}$$

By means of this equation the seepage can be computed readily, after the flow net has been constructed.

The total excess hydrostatic pressure on the upstream side of the cubical element with side a is

$$a^2 \times 15\Delta h\gamma_w$$

and on the downstream side is

$$a^2 \times 14\Delta h\gamma_w$$

The difference between these two pressures,

$$p_s = a^2\Delta h\gamma_w = a^3\frac{\Delta h}{a}\gamma_w$$

is transferred by the water onto the soil grains. Since $\Delta h/a$ is equal to the hydraulic gradient i, and a^3 is the volume of the element, the water exerts a force against the soil equal to

$$p_s = i\gamma_w \tag{23.6}$$

per unit of volume. This force is known as the *seepage pressure*. It has the dimension of a unit weight, and at any point its line of action is tangent to the flow line.

Construction of Flow Net

The data required for plotting a flow net can be obtained by solving Eq. 23.2, but a mathematical solution is not practicable unless the boundary conditions are very simple. The boundary conditions corresponding to most hydraulic structures do not satisfy this condition. Although flow nets for such structures can be obtained by various experimental methods, by far the most convenient and least expensive of the procedures is to construct the flow net graphically by trial and error. The steps in performing the graphical construction are illustrated in Fig. 23.2. In this figure *a* represents a vertical section through an overflow dam with a sheet-pile cutoff wall.

Before starting the construction of the flow net, we must examine the hydraulic boundary conditions of the problem and ascertain their effect on the shape of the flow lines. The upstream and downstream ground surfaces in Fig. 23.2*a* represent equipotential lines. The base of the dam and the sides of the cutoff wall represent the uppermost flow line,

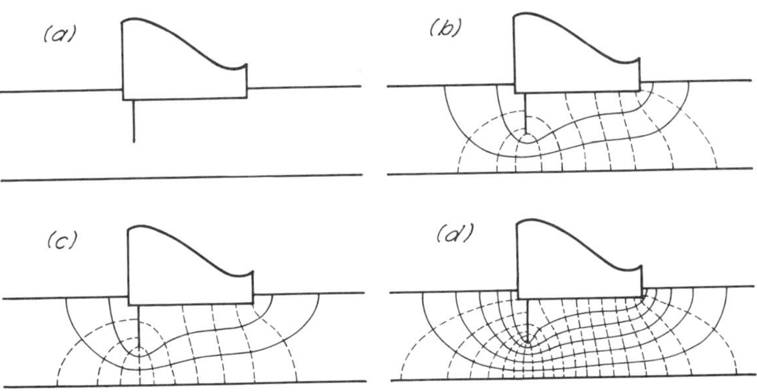

Fig. 23.2. Steps in constructing a flow net. (*a*) Cross section through pervious stratum. (*b*) Result of first attempt to construct flow net. (*c*) Result of adjusting flow net constructed in (*b*). (*d*) Final flow net.

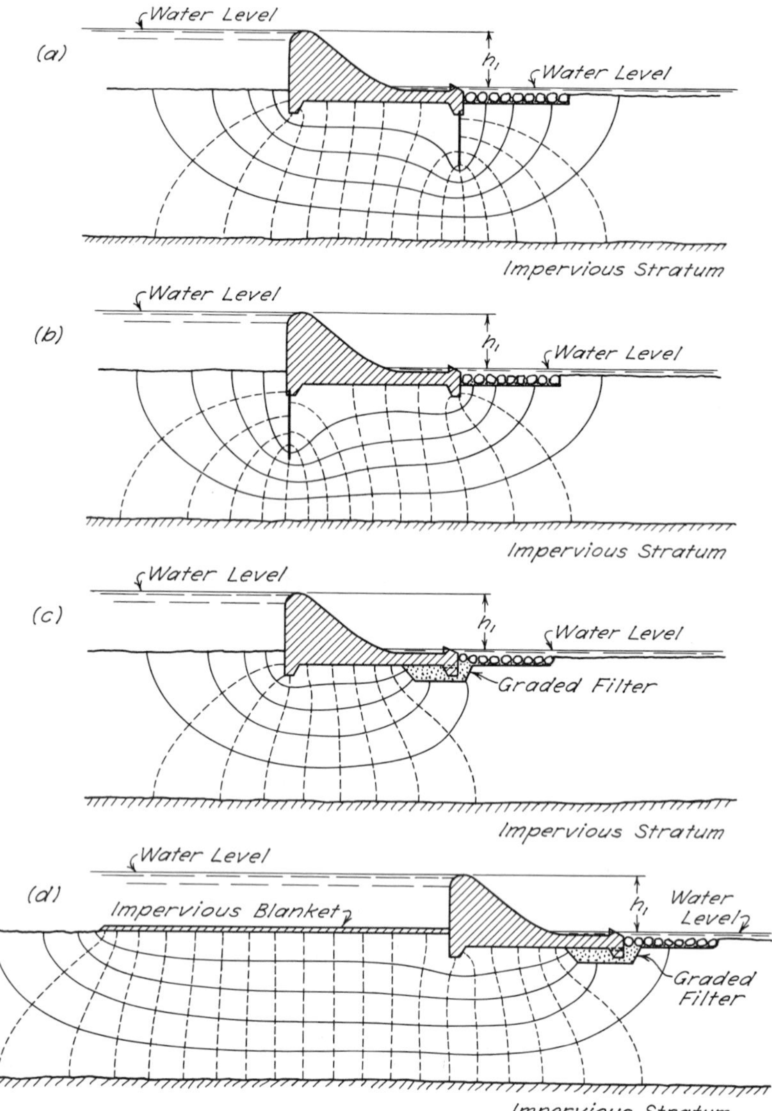

Fig. 23.3. Seepage through homogeneous sand beneath base of concrete dam (after A. Casagrande 1935*a*).

and the base of the pervious stratum represents the lowest flow line. The other flow lines lie between these two, and their shapes must represent a gradual transition from one to the other. Furthermore, all the flow lines must be vertical where they meet the upstream and downstream ground surfaces. The first step in constructing the flow net is to draw several smooth curves representing flow lines (plain curves in Fig. 23.2*b*) that satisfy these requirements. Then several equipotential lines, which should intersect the flow lines at right angles, are drawn so that the fields are at least roughly square. In this manner a first rough approximation to the flow net is obtained.

The next step is to examine the trial flow net carefully in order to detect the most conspicuous defects. In the trial flow net shown in Fig. 23.2*b*, the flow lines and the equipotential lines do intersect at approximately right angles, but several of the fields are not yet square. Therefore, a new flow net is drawn in which the fields are more nearly square. The process of adjustment is continued until all of the fields are roughly square. The flow net at this stage is represented by Fig. 23.2*c*.

Finally, the fields in Fig. 23.2*c* are subdivided, and the flow net is adjusted until each small field is square. The result is shown in Fig. 23.2*d*. Each field in Fig. 23.2*c* has been subdivided into four small fields, and minor inaccuracies have been eliminated.

For all practical purposes the flow net is satisfactory as soon as all of the fields are roughly square. Even an apparently inaccurate flow net gives remarkably reliable results. Figures 23.3 and 23.4 may serve as a guide for constructing flow nets that satisfy various hydraulic boundary conditions. The flow net in Fig. 23.4*a* contains one line that represents a free-water surface located entirely within the pervious medium. Along this surface, the vertical distance between each adjacent pair of equipotential lines is a constant and is equal to Δh.

Every flow net is constructed on the assumption that the soil within a given stratum through which the water percolates is uniformly permeable. In a natural soil stratum, the permeability varies from point to point, especially along lines at right angles to the boundaries of the stratum. Therefore, the difference between even a very roughly sketched flow net and an accurate one is commonly small compared to the difference between the flow pattern in the real soil and that indicated by the accurate flow net. Because of this universal condition, refinements in the construction of flow nets or elaborate model studies are entirely unwarranted.

The use of models based on the analogy between the flow of water in a pervious medium and the flow of electricity in a conductor affords a convenient means for constructing a flow net such as Fig. 23.4*a* that

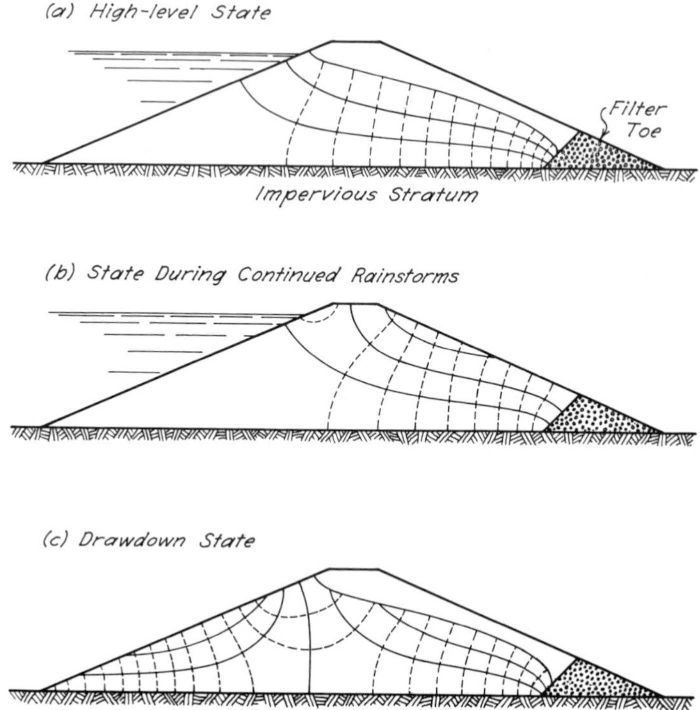

Fig. 23.4. Seepage through imaginary homogeneous dam consisting of very fine clean sand.

contains a free-water surface. However, the assembling of the necessary equipment is not warranted unless many flow nets of this type have to be drawn.

Seepage through Soils with Transverse Isotropy

The flow nets shown in Figs. 23.1 to 23.4 have been constructed on the assumption that the soil is hydraulically isotropic. In nature every mass of soil is more or less stratified. Therefore, as stated in Article 11, the average permeability k_I in a direction parallel to the planes of stratification is always greater than the average permeability k_{II} at right angles to these planes. To construct a flow net for such a stratified mass of soil, we substitute for the real soil a homogeneous material having horizontal and vertical permeabilities equal, respectively, to k_I and k_{II}. A medium with such properties is said to possess *transverse isotropy.*

In order to prepare a flow net for a homogeneous medium with transverse isotropy, we proceed as follows: A drawing is made showing a vertical section through the permeable layer parallel to the direction of flow. The horizontal scale of the drawing is reduced by multiplying all horizontal directions by $\sqrt{k_{II}/k_I}$. For this transformed section we construct the flow net as if the medium were isotropic. The horizontal dimensions of this flow net are then increased by multiplying them by $\sqrt{k_I/k_{II}}$. The quantity of seepage is obtained by substituting the quantity,

$$k = \sqrt{k_I k_{II}}$$

into Eq. 23.5. The expression for the quantity of seepage per unit width of the medium is then

$$Q = h_1 \frac{N_f}{N_d} \sqrt{k_I k_{II}} \tag{23.7}$$

The procedure is illustrated by Fig. 23.5.

The preceding method (Samsioe 1931) has been developed on a purely mathematical basis without any simplifying assumptions. Therefore, the results are as reliable as Darcy's law and the values of k_I and k_{II} that enter the computation.

The average value of k_I for almost all natural soil strata is considerably greater than k_{II}. However, the ratio k_I/k_{II} ranges from about two or three to several hundred, and there is no way to determine the value accurately for a given deposit. Therefore, it is advisable to sketch two flow nets, one on the basis of the greatest probable value for k_I/k_{II}, and the other on the basis of the least probable one. In selecting these values, consideration should be given to the fact that k_I/k_{II} cannot be less than unity, nor greater than the ratio between the coefficients of permeability

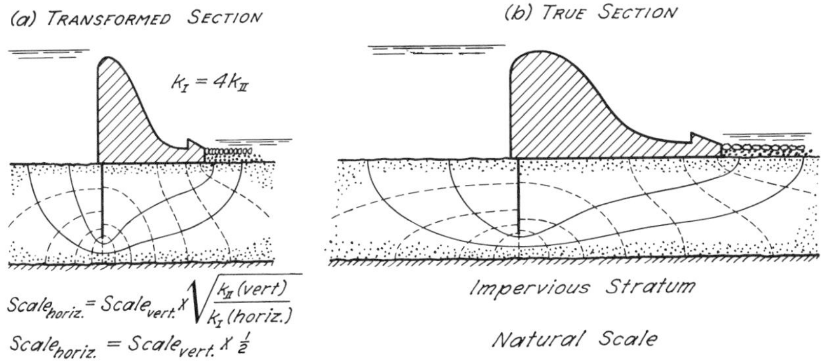

Fig. 23.5. Construction of flow net if coefficients of permeability of sand stratum are different in horizontal and vertical directions. (*a*) Transformed section. (*b*) True section.

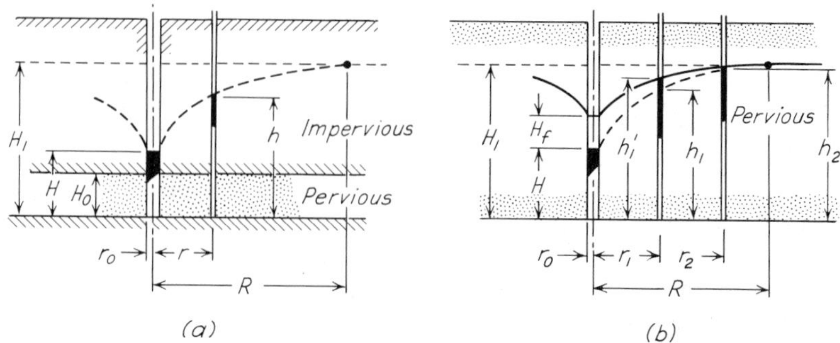

Fig. 23.6. Diagram illustrating flow of water toward well during pumping test. (*a*) If piezometric level lies above pervious layer. (*b*) If free-water surface lies within pervious layer.

of the most and least permeable layers. For design purposes, that flow net should be retained which represents the most unfavorable conditions, or else provisions should be made to ascertain during construction whether the difference between the real and the anticipated seepage conditions is on the side of safety.

Seepage Toward Single Well

Figure 23.6*a* is a vertical section through a well, with radius r_0, extending to the bottom of a pervious horizontal layer located between impervious deposits. The layer has a thickness H_0 and a uniform coefficient of permeability k. By pumping at a constant rate Q from the well until a steady state of flow is achieved, the height of the water in the well with respect to the bottom of the pervious layer is lowered from H_1 to H, and that in observation wells at distance r is lowered from H_1 to h. It is assumed that the water flows toward the well in horizontal, radial directions. The total flow across the boundary of any cylindrical section of radius r is then, according to Eq. 11.6,

$$Q = kiA = k\frac{dh}{dr}2\pi r H_0$$

Whence, by integration

$$Q\int_{r_1}^{r_2}\frac{dr}{r} = 2\pi H_0\,k\int_{h_1}^{h_2}dh$$

$$Q = \frac{2\pi H_0\,k(h_2 - h_1)}{\log_\epsilon \dfrac{r_2}{r_1}} \tag{23.8}$$

Or, if the well is being pumped to evaluate k,

$$k = \frac{Q}{2\pi H_0(h_2 - h_1)} \log_\epsilon \frac{r_2}{r_1} \tag{23.9}$$

The permeability can be determined most accurately by measuring h_1 and h_2 at corresponding radii r_1 and r_2 (Article 44). However, a rough estimate can be made by making use of the conditions that $h_1 = H$ at $r_1 = r_0$, and that at a large value of $r_2 = R$, h_2 approaches H_1. The dimension R, known as the *radius of influence* of the well, represents the distance beyond which the water table remains essentially horizontal. It does not need to be known with accuracy because as R/r_0 increases its logarithm increases much more slowly. Hence, if at least the order of magnitude of R is known, k can be approximated without the assistance of observation wells.

If, on the other hand, the well penetrates to the bottom of an open pervious layer (Fig. 23.6*b*) the water table at the boundary of the well cannot be drawn down to the water level within the well itself because a considerable quantity of flow enters the well through the exposed free surface H_f. The discharge from such a well was first evaluated (Dupuit 1863) on the simplifying assumptions that $H_f = 0$ (dash curve in Fig. 23.6*b*), and that at any radius r the hydraulic gradient causing horizontal flow toward the well is equal to the slope of the assumed drawdown curve at the radius r. On these assumptions

$$Q = ki\,A = k\frac{dh}{dr}2\pi rh$$

whence

$$Q = \frac{\pi k(h_2^2 - h_1^2)}{\log_\epsilon \dfrac{r_2}{r_1}} \tag{23.10}$$

or

$$k = \frac{Q}{\pi(h_2^2 - h_1^2)} \log_\epsilon \frac{r_2}{r_1} \tag{23.11}$$

For the boundary conditions $h_1 = H$ at $r_1 = r_0$ and $h_2 = H_1$ at $r_2 = R$,

$$Q = \frac{\pi k(H_1^2 - H^2)}{\log_\epsilon \dfrac{R}{r_0}} \tag{23.12}$$

Both theory (Boreli 1955) and experiments (Babbitt and Caldwell 1948) have demonstrated that Eq. 23.12 leads to reliable values of Q even if H is reduced to zero. On the other hand, the difference between the ordinates h_1 and h_1' of the Dupuit drawdown curve and that determined by taking proper account of the presence of the discharge surface H_f becomes significant at distances from the well less than about 1.0 to 1.5 H_1 and increases rapidly as the well is approached or as H decreases.

Problems

1. The sand beneath the dams shown in Fig. 23.3 has a permeability in every direction of 4.2×10^{-3} cm/sec. The head h_1 is 25 ft. Compute the seepage loss in cubic feet per second, per lineal foot along the axis of each dam.

Ans. (*a*) 1.15×10^{-3}; (*b*) 1.15×10^{-3}; (*c*) 1.91×10^{-3}; (*d*) 0.86×10^{-3} ft^3/sec.

2. Estimate the hydrostatic uplift pressure in excess of that at tailwater level, at a point midway between the upstream and downstream faces of the concrete base of the dams of problem 1.

Ans. (*a*) 15; (*b*) 9; (*c*) 6; (*d*) 2.5 ft of head.

3. The subsoil of the dam shown in Fig. 23.3*b* contains a horizontal layer of silt, 1 in. thick, that intersects the row of sheet piles a short distance above the bottom of the piles. There are no means for detecting the presence of such a layer by any practicable method of soil exploration. The coefficient of permeability of the sand is 4.2×10^{-3} cm/sec, whereas that of the silt is 2.1×10^{-6} cm/sec. The total thickness of the sand stratum upstream from the dam is 55 ft, and the lower edge of the sheet piles is located 25 ft above the base of the sand. (*a*) Describe how the influence of the silt layer on the seepage loss could be evaluated on the assumption that the silt layer is continuous over a large area. (*b*) Describe the effect on the seepage loss of gaps in the silt layer. (*c*) How can the degree of continuity of the silt layer be determined in advance?

Ans. (*a*) The silt layer has the same effect as increasing the thickness of the sand layer from 55 to 221 ft, and the penetration of the sheet piles from 30 to 196 ft. Therefore, the seepage loss could be evaluated by sketching a flow net for these fictitious soil conditions. Since the gap beneath the sheet piles in the fictitious profile is small compared to the depth of sheet-pile penetration, the loss of water computed on the basis of this flow net would be only a small fraction of that through the sand without a silt layer. (*b*) Depending on the size and location of the gaps in the layer, a discontinuous silt layer may have any effect varying from almost nothing to that of a continuous layer. (*c*) It cannot.

4. Compute the seepage loss per foot of length of the dam shown in Fig. 63.6*b*, assuming $k = 1 \times 10^{-3}$ cm/sec. Estimate the uplift pressure on the base of the dam at the back of the high masonry section.

Ans. 1.1×10^{-3} ft^3/sec per lin ft; 64 ft of head.

5. The average coefficient of permeability of the sand beneath the dam shown in Fig. 23.5 is 16×10^{-4} cm/sec in the horizontal direction and 4×10^{-4} cm/sec in the vertical direction. What is the seepage loss per lineal foot of dam, when the head is 30 ft?

Ans. 2×10^{-4} ft^3/sec.

6. Construct the flow net for the dam shown in Fig. 23.5*b*, if the value of k is equal to 36×10^{-4} cm/sec in the horizontal direction and 4×10^{-4} cm/sec in the vertical direction. The base width of the dam is 83 ft, the thickness of the pervious layer is 38 ft, and the length of the sheet piles is 29 ft. The head is 30 ft. What is the seepage loss per lineal foot of dam? Compare this value with the seepage loss beneath the same dam if k is equal to 12×10^{-4} cm/sec in every direction.

Ans. 3.9×10^{-4}; 2.5×10^{-4} ft^3/sec.

7. What is the approximate intensity of the horizontal hydrostatic excess pressure against the left-hand side of the sheet-pile wall in Fig. 63.6*a* at the lowest point of the wall?

Ans. 2620 lb/ft^2.

Selected Reading

Casagrande, A. (1935*b*). "Seepage through dams," *J. New England Water Works Assn.*, **51,** No. 2, pp. 131–172. Reprinted in *Contributions to soil mechanics 1925–1940,* Boston Soc. of Civil Engrs., 1940, and as *Harvard Univ. Soil Mech. Series No. 5.* A classic presentation of the flow-net method and its applications.

The following treatises deal with advanced aspects of seepage computations:

Muskat, M. (1937). *The flow of homogeneous fluids through porous media,* New York, McGraw-Hill, 63 pp. Reprinted by J. W. Edwards, Ann Arbor, 1946.

Polubarinova-Kochina, P. Ya. (1962). *Theory of ground water movement.* Translated from the Russian by J. M. R. de Wiest, Princeton Univ. Press, 613 pp.

Harr, M. E. (1962). *Groundwater and seepage.* New York, McGraw-Hill, 315 pp.

An excellent presentation of the fundamentals of seepage, with applications, is contained in Cedergren, H. R. (1967): *Seepage, drainage, and flow nets,* New York, John Wiley and Sons, 489 pp.

ART. 24 MECHANICS OF PIPING

Definition of Piping

Many dams on soil foundations have failed by the apparently sudden formation of a pipe-shaped discharge channel or tunnel located between the soil and the foundation. As the stored water rushed out of the reservoir into the outlet passage, the width and depth of the passage increased rapidly until the structure, deprived of its foundation, collapsed and broke into fragments that were carried away by the torrent. An event of this type is known as a *failure by piping.*

Failures by piping can be caused by two different processes. They may be due to scour or subsurface erosion that starts at springs near the downstream toe and proceeds upstream along the base of the structure or some bedding plane (Article 63). Failure occurs as soon as the upstream or intake end of the eroded hole approaches the bottom of the reservoir. The mechanics of this type of piping defy theoretical approach. However, piping failures have also been initiated by the sudden rise of a large body of soil adjoining the downstream toe of the structure. A failure of this kind occurs only if the seepage pressure of the water that percolates upward through the soil beneath the toe becomes greater than the effective weight of the soil. Failures of the first category will be referred to as *failures by subsurface erosion,* and those of the second as *failures by heave.* The following paragraphs deal exclusively with failures by heave.

The magnitude and distribution of the excess hydrostatic pressure are determined by the flow net. In Article 23 it has been emphasized that the theoretical flow net is never identical with the one that represents the flow of water through the real soil strata. Indeed, the two flow nets may have no resemblance whatsoever. Therefore, the results of theoretical investigations into the mechanical effects of the flow of seepage serve merely as a guide for judgment and as a basis for planning appropriate installations for surveillance during and after construction.

Mechanics of Piping due to Heave

The mechanics of failure by piping due to heave are illustrated by Fig. 24.1*a* which represents a vertical section through one side of a single-wall sheet-pile cofferdam. To a depth h_1 below the water level, the soil outside the cofferdam consists of coarse gravel, whereas the gravel within the cofferdam has been removed by dredging. The gravel rests on a bed of uniform sand. The loss of head in the gravel is so small that it can be disregarded. We wish to compute the factor of safety F with respect to piping, after the water level on the inside has been pumped down to the surface of the sand.

Before making this computation, we shall consider the hydrostatic conditions at the instant of failure. As soon as the water level within the cofferdam is lowered by pumping, water begins to flow downward through the sand on the left side of the sheet piles and upward on the right. The excess hydrostatic pressure on a horizontal section such as Ox (Fig. 24.1b) reduces the effective pressure on that section. As soon as the average effective pressure on and above a portion of Ox near the sheet piles becomes equal to zero, the water that flows through the sand can straighten and widen the flow channels without meeting any resistance. This process greatly increases the permeability of the sand adjoining the sheet piles, as explained in Article 12, and it diverts an additional part of the seepage toward this zone. The surface of the sand then rises (see Fig. 24.1a). Finally, the sand starts to boil, and a mixture of water and sand rushes from the upstream side of the sheet piles, through the space below the lower edge of the sheet piles, and toward the zone where the boiling started.

By model tests (Terzaghi 1922) it has been found that the rise of the sand occurs within a distance of about $D/2$ from the sheet piles. The failure, therefore, starts within a prism of sand having a depth D and a width $D/2$. At the instant of failure the effective vertical pressure on any horizontal section through the prism is approximately equal to zero. At the same time the effective lateral pressure on the sides of the prism is also approximately zero. Therefore, piping occurs as soon as the excess hydrostatic pressure on the base of the prism becomes equal to the effective weight of the overlying sand.

In order to compute the excess hydrostatic pressure a flow net must be constructed. After this has been done (Fig. 24.1a) the intensity of this pressure can be determined readily at every point on the base of the prism at depth D by means of the procedure described in Article 23. In Fig. 24.1b these values are represented by the ordinates of curve C with reference to a horizontal axis through O. Within the distance $D/2$

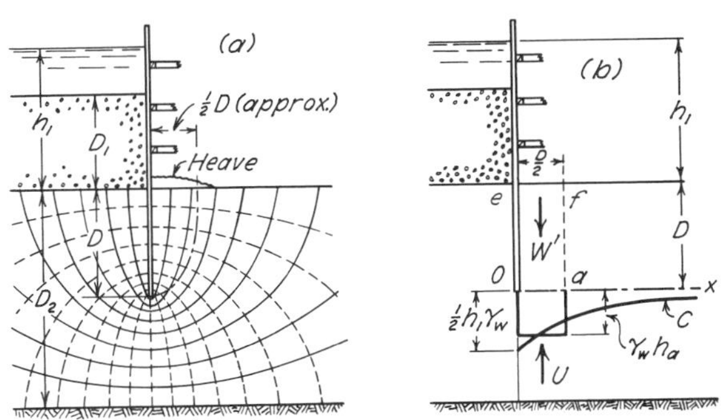

Fig. 24.1. Use of flow net to determine factor of safety of row of sheet piles in sand with respect to piping. (a) Flow net. (b) Forces acting on sand within zone of potential heave.

from the sheet piles the average excess hydrostatic pressure on the base of the prism has the value $\gamma_w h_a$, and the total excess hydrostatic pressure on the base is $U = \frac{1}{2}D\gamma_w h_a$. Failure by piping occurs as soon as U becomes equal to the effective weight of the sand which, in turn, is equal to the submerged weight $W' = \frac{1}{2}D^2\gamma'$. Therefore, the factor of safety with respect to piping is

$$F = \frac{W'}{U} = \frac{D\gamma'}{h_a\gamma_w} \tag{24.1}$$

In a similar manner, we may compute the factor of safety for a dam with a sheet-pile cutoff.

Uplift Compensation by Loaded Filters

If the factor of safety against failure by piping is too small, it may be increased by establishing on top of the prism *Oafe* (Fig. 24.1*b*) an inverted filter which has a weight W. The presence of the filter does not alter the excess hydrostatic pressure U, but it increases the effective weight of the prism from W' to $W' + W$. Hence, it increases the factor of safety with respect to piping from F (Eq. 24.1) to

$$F' = \frac{W + W'}{U} \tag{24.2}$$

The stabilizing effect of loaded inverted filters has been demonstrated repeatedly by experiment and by experience with filter-protected structures. In order to be effective, the filters must be coarse enough to permit the free outflow of the seepage water, but fine enough to prevent the escape of soil particles through their voids. The design of filters to satisfy both requirements is discussed in Article 11.

Problems

1. In Fig. 24.1 the head h_1 is 25 ft. The penetration of the sheet piles into the sand layer is 19 ft. If the saturated unit weight of the sand is 113 lb/ft^3, what is the weight of an inverted filter required to increase the factor of safety with respect to piping to 2.5?

Ans. 340 lb/ft^2.

2. The sand layer mentioned in Problem 1 contains a seam of clay too thin to be detected by the boring crew, but thick enough to constitute a relatively impermeable membrane. The numerical data regarding the head and the depth of sheet piles are identical with those given in Problem 1. The clay seam is located a few feet above the lower edge of the sheet piles. Its left-hand boundary is located a few feet upstream from the sheet piles, and on the downstream side it is continuous. On the downstream side the sand stratum carries an inverted filter weighing 340 lb/ft^2 which provides a factor of safety of 2.5 on the assumption that the sand contains no obstacle against flow. (*a*) To what value does the clay seam reduce the factor of safety? (*b*) What procedure could be used to detect the danger?

Ans. (*a*) 0.83. The sand at the downstream side of the sheet piles would blow up as soon as the head reached 21 ft. (*b*) Install a single observation well on the downstream side of the sheet piles, with its lower end a few feet below the level of the bottom of the sheet-pile wall.

ART. 25 THEORY OF CONSOLIDATION

Process of Consolidation

If the load on a layer of highly compressible porous saturated soil such as clay is increased, the layer is compressed, and excess water drains out of it. This constitutes a process of consolidation (Article 14). During the process the quantity of water that enters a thin horizontal slice of the soil is smaller than the quantity that leaves it. Therefore, the continuity condition expressed by Eq. 23.1, on which the theory of flow nets and seepage is based, is no longer applicable.

The added load or pressure per unit of area that produces consolidation is known as the *consolidation pressure* or *consolidation stress*. At the instant of its application, the consolidation pressure is carried almost entirely by the water in the voids of the soil (see Article 14). Therefore, at the beginning of a process of consolidation, there is an initial excess pressure in the water almost exactly equal to the consolidation stress. As time goes on, the excess water pressure decreases, and the corresponding average effective pressure in the layer increases. At any point in the consolidating layer, the value u of the excess hydrostatic pressure at a given time may be determined by Eq. 12.1, written in the form,

$$u = \gamma_w h \tag{25.1}$$

in which h is the hydraulic head with respect to the ground-water level above the consolidating layer. After a very great time the excess hydrostatic pressure u becomes equal to zero, and the entire consolidation pressure becomes an effective stress transmitted from grain to grain. If the consolidation pressure at any point is denoted by Δp, equilibrium requires that

$$\Delta p = \Delta \bar{p} + u \tag{25.2}$$

where $\Delta \bar{p}$ represents that portion of the consolidation stress which, at a given time, is transmitted from grain to grain, and u is the corresponding excess hydrostatic pressure.

Graphical Representation of Progress of Consolidation

Since Δp in Eq. 25.2 is a constant, the progress of consolidation at a given point can be visualized by observing the variation of u at that point or, according to Eq. 25.1, by observing the variation in h by means of an imaginary standpipe rising from that point.

Figure 25.1 illustrates the consolidation of a compressible layer located between two layers of sand. Because of the construction of a large building or the placement of a fill on the ground surface, the compressible layer is subjected to a consolidation stress Δp. It is assumed that the layer can drain freely at both its upper and lower surfaces and that within the layer the water flows only in a vertical direction. Furthermore, it is assumed that Δp does not vary from top to bottom of the layer.

The progress of consolidation within the layer can be studied by observing the position of the water level in a series of standpipes. The lower ends of the pipes are located on a vertical line through the layer as shown in Fig. 25.1. Since the excess hydrostatic pressure is independent of the position of the water table, the water table is assumed to be identical with the top surface of the consolidating layer. If the standpipes are arranged in such a manner that the horizontal distances 1–2′, 1–3′, etc., are equal to the corresponding vertical distances 1–2, 1–3, etc., as shown in the figure, the curve that represents the locus of the water levels in the standpipes at a given time represents the *isochrone* (see Article 14). The hydraulic gradient i at any depth d below a is equal to

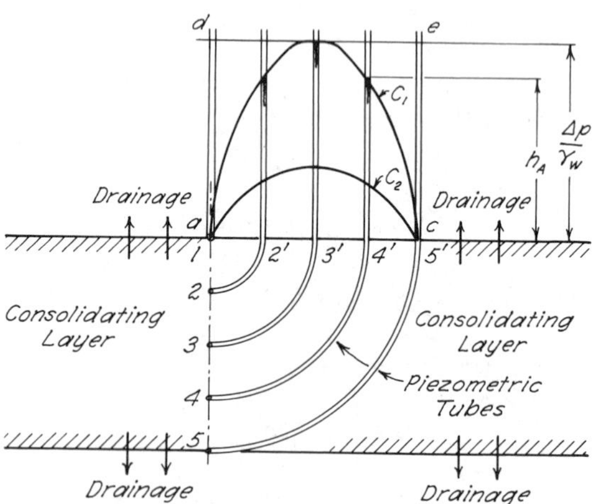

Fig. 25.1. Diagram illustrating consolidation of compressible layer of clay.

the slope of the isochrone at a horizontal distance d from a. Furthermore, if the slope at any point of the isochrone is upward toward the right, the flow is upward at the corresponding point in the layer.

The distribution of the initial excess hydraulic head over vertical sections through the clay layer is represented by the horizontal line de located at an elevation $\Delta p/\gamma_w$ above the free-water surface. This line is the initial isochrone. According to Article 14, the consolidation of a layer of clay proceeds from the drainage surface or surfaces toward the interior. Hence, at an early stage of consolidation the piezometric levels for the central part of the layer are still unchanged while those for the outer parts have already dropped as shown by the isochrone C_1. In an advanced stage, represented by C_2, all of the levels have dropped, but the elevations decrease from the central part toward zero at the drainage surfaces. Finally, after a very long time, all the excess hydrostatic pressure has disappeared, and the final isochrone is represented by the horizontal line ac.

Figure 25.2 shows the isochrones for different processes of consolidation. If the consolidating layer is free to drain through both its upper and lower surfaces, the layer is called an *open layer*, and its thickness is denoted by $2H$. If the water can escape through only one surface, the layer is called *half-closed*. The thickness of half-closed layers is denoted by H. In Fig. 25.2, the layers labeled a, b, c, and e are open, whereas the layers d and f are half-closed.

Figure 25.2a is a simplified replica of Fig. 25.1. The piezometric tubes are not shown. The diagram represents the consolidation of an open layer of clay under the influence of a consolidation stress that is uniform from top to bottom of the layer.

If the consolidating layer is fairly thick with respect to the width of the loaded area, the consolidation pressure due to the weight of a structure or a fill decreases with depth in a manner similar to that indicated by the curve C_a (Fig. 40.3). Under the simplifying assumption that the decrease of the pressure with depth is linear, the initial isochrone may be represented by the line de in Fig. 25.2b, and the consolidation pressures at the top and bottom of the layer are Δp_t and Δp_b, respectively.

If the consolidating layer is very thick compared to the width of the loaded area, the pressure Δp_b is likely to be very small compared to Δp_t. Under this condition it can be assumed with sufficient accuracy that $\Delta p_b = 0$. The corresponding isochrones are shown in Fig. 25.2c for an open layer, and in Fig. 25.2d for a half-closed layer. It should be noticed that the consolidation of the half-closed layer in Fig. 25.2d is associated with a temporary swelling of the clay in the lower part of the layer.

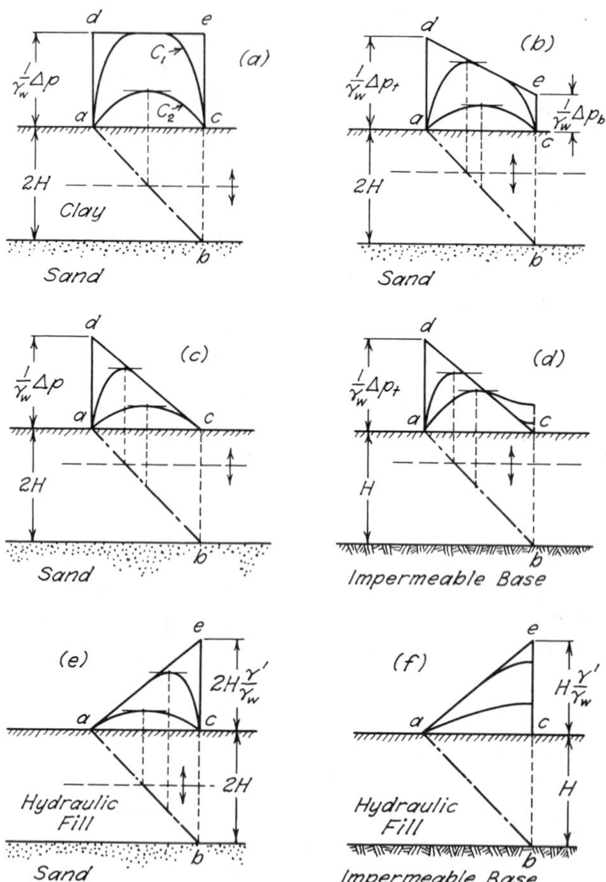

Fig. 25.2. Isochrones representing progress of consolidation of a layer of ideal clay for different types of drainage and different distributions of consolidation pressure in the vertical direction (after Terzaghi and Frölich 1936).

Figures 25.2*e* and *f* illustrate the consolidation of hydraulically placed layers acted on by no force other than their own weight. The water table is assumed to be located at the top surface of the layers, and the consolidation that occurs during construction is disregarded. The fill shown in Fig. 25.2*e* rests on a stratum of sand (open layer), whereas that in Fig. 25.2*f* rests on an impermeable stratum (half-closed layer). At a time $t = 0$, the entire submerged weight of the soil in either layer (γ' per unit of volume) is carried by the water, and the consolidation pressure increases from zero at the surface to $H\gamma'$ at the base. Therefore, the final result of the consolidation is the same for both layers. However, the difference in the shape of the isochrones for intermediate stages of consolidation indicates that the rate at which the final stage is approached is very different for the two layers.

Computation of Rate of Consolidation

In order to compute the rate of consolidation and the degree of consolidation $U\%$ (Eq. 14.1) for the processes illustrated in Fig. 25.2, we make the following simplifying assumptions:

1. The coefficient of permeability k (Eq. 11.6) is the same at every point in the consolidating layer and for every stage of consolidation.
2. The coefficient of volume compressibility m_v (Eq. 13.3) is the same at every point in the layer and for every stage of consolidation.
3. The excess water drains out only along vertical lines.
4. The time lag of the compression is caused exclusively by the low permeability of the material. Thus the secondary time effect discussed in Article 14 is disregarded.

Figure 25.3*a* represents a vertical section through a thin horizontal slice of a consolidating layer. The thickness of the slice is dz. Water flows

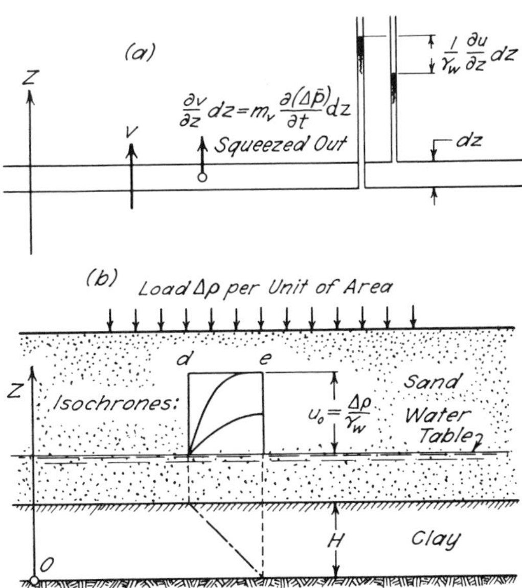

Fig. 25.3. (*a*) Vertical section through thin horizontal slice of consolidating layer showing hydraulic pressure conditions at boundaries of slice. (*b*) Section through consolidating layer, showing hydraulic boundary conditions.

through the layer at a rate v. The unbalanced hydrostatic pressure is $(\partial u/\partial z)dz$. Darcy's law (Article 11) requires that

$$v = ki = -k\frac{\partial h}{\partial z} = -k\frac{1}{\gamma_w}\frac{\partial u}{\partial z} \tag{25.3}$$

If the layer were incompressible, the quantity of water flowing out of the layer would equal that which flows in, and we could write

$$\frac{\partial v}{\partial z} = 0 \tag{25.4}$$

This condition is identical with the continuity condition expressed by Eq. 23.1. However, in a consolidating compressible layer with a thickness equal to unity, the quantity of water that leaves the layer per unit of time exceeds that which enters it by an amount equal to the corresponding volume decrease of the layer. Hence, by making use of Eq. 13.2, we can write

$$\frac{\partial v}{\partial z} = m_v \frac{\partial(\Delta\bar{p})}{\partial t}$$

Since Δp is a constant, Eq. 25.2 leads to

$$\frac{\partial(\Delta\bar{p})}{\partial t} = -\frac{\partial u}{\partial t}$$

whence

$$\frac{\partial v}{\partial z} = -m_v\frac{\partial u}{\partial t}$$

By combining this equation with Eq. 25.3, we obtain

$$\frac{\partial v}{\partial z} = -m_v\frac{\partial u}{\partial t} = -\frac{k}{\gamma_w}\frac{\partial^2 u}{\partial z^2}$$

or

$$\frac{\partial u}{\partial t} = \frac{k}{\gamma_w m_v}\frac{\partial^2 u}{\partial z^2} \tag{25.5}$$

Equation 25.5 is the differential equation of every process of consolidation that involves linear drainage. It can be simplified by substituting

$$c_v(\text{cm}^2/\text{sec}) = \frac{k(\text{cm/sec})}{\gamma_w(\text{gm/cm}^3)m_v(\text{cm}^2/\text{gm})} \tag{25.6}$$

The coefficient c_v represents the coefficient of consolidation (Eq. 14.2). Hence,

$$\frac{\partial u}{\partial t} = c_v \frac{\partial^2 u}{\partial z^2} \tag{25.7}$$

The solution of this equation must satisfy the hydraulic boundary conditions. These conditions depend on the loading and drainage conditions as shown in the diagrams in Fig. 25.2. The boundary conditions that determine the consolidation of a half-closed layer and a uniform pressure distribution may serve as an example. According to Fig. 25.3*b*, the boundary conditions are as follows:

1. At $t = 0$ and at any distance z from the impervious surface, the excess hydrostatic pressure is equal to Δp.
2. At any time t at the drainage surface $z = H$, the excess hydrostatic pressure is zero.
3. At any time t at the impervious surface $z = 0$, the hydraulic gradient is zero (that is $\partial u/\partial z = 0$).
4. After a very great time, at any value of z, the excess hydrostatic pressure is zero.

Equation 25.7 combined with the boundary conditions determines the degree of consolidation $U\%$ for a given time t. The equation for $U\%$ is

$$U\% = f(T_v) \tag{25.8}$$

In this expression,

$$T_v = \frac{c_v}{H^2} t \tag{25.9}$$

is a pure number called the *time factor*. Since the soil constants and the thickness of the compressible layer enter Eq. 25.8 only in the combination represented by the dimensionless time factor T_v, the value $U\% = f(T_v)$ is the same for every layer that consolidates under specified conditions of loading and drainage. It has been determined for every condition of practical importance by means of the differential Eq. 25.7. The results have been presented in the form of graphs or tables. By means of these graphs and tables, all the problems likely to be met in practice can be solved without any computation other than the evaluation of Eq. 25.9. Figure 25.4 represents the solutions of the problems illustrated in Fig. 25.2. The following instructions serve as a guide for using the graphs.

For every open layer (thickness $2H$) the relationship between $U\%$ and T_v is determined by the curve C_1, regardless of the slope of the zero isochrone *de*. Therefore, the curve C_1 represents the solution for all the consolidation problems represented by Fig. 25.2*a*, *b*, *c*, and *e*. If the zero isochrone is horizontal, indicating a uniform distribution of the consolidation pressure throughout the consolidating layer, curve C_1 also represents the process of consolidation for a half-closed layer with thickness H. The following example illustrates the procedure for using the graph (Fig. 25.4*a*).

The coefficient of consolidation of an open layer with thickness $2H$ is c_v. We wish to determine the time t at which the degree of consolidation of the layer due to the weight of a superimposed building becomes equal to 60%. From Eq. 25.9 we obtain

$$t = T_v \frac{H^2}{c_v}$$

According to curve C_1 in Fig. 25.4*a*, a degree of consolidation of 60% corresponds to the time factor 0.28, whence

$$t = 0.28 \frac{H^2}{c_v} \qquad (25.10)$$

regardless of the slope of the zero isochrone. If the zero isochrone for a half-closed layer of clay with thickness H is horizontal, the degree of consolidation of this layer after time t (Eq. 25.10) will also be equal to 60%.

If the consolidation pressure for a half-closed layer decreases from some value Δp_t at the top to zero at the bottom, as shown in Fig. 25.2*d*, the relation between U and T_v is given by the curve C_2. If it increases from zero at the top to Δp_b at the bottom, as in Fig. 25.2*f*, curve C_3 furnishes the required information. For intermediate types of vertical distribution of consolidation pressure, sufficiently accurate results can be obtained by interpolation. Figure 25.4*b* shows the curves C_1 to C_3 plotted to a semilogarithmic scale. Small values of U can be obtained somewhat more accurately from the semilogarithmic curves. The semilogarithmic plot of C_1 corresponds to the solid curve in Fig. 14.2*b*.

Because of the simplifying assumptions listed at the outset of the preceding analysis, the computation of the rate of settlement has the character of a crude estimate. The most important discrepancy between theory and reality has been referred to as the secondary time effect (Article 14). According to the theory of consolidation, the time–settlement curve should approach a horizontal asymptote whereas in reality it merges into an inclined tangent as shown in Fig. 14.2*a*. At present the secondary settlement cannot be predicted reliably on the basis of test results. Experience shows that the rate of the secondary settlement of

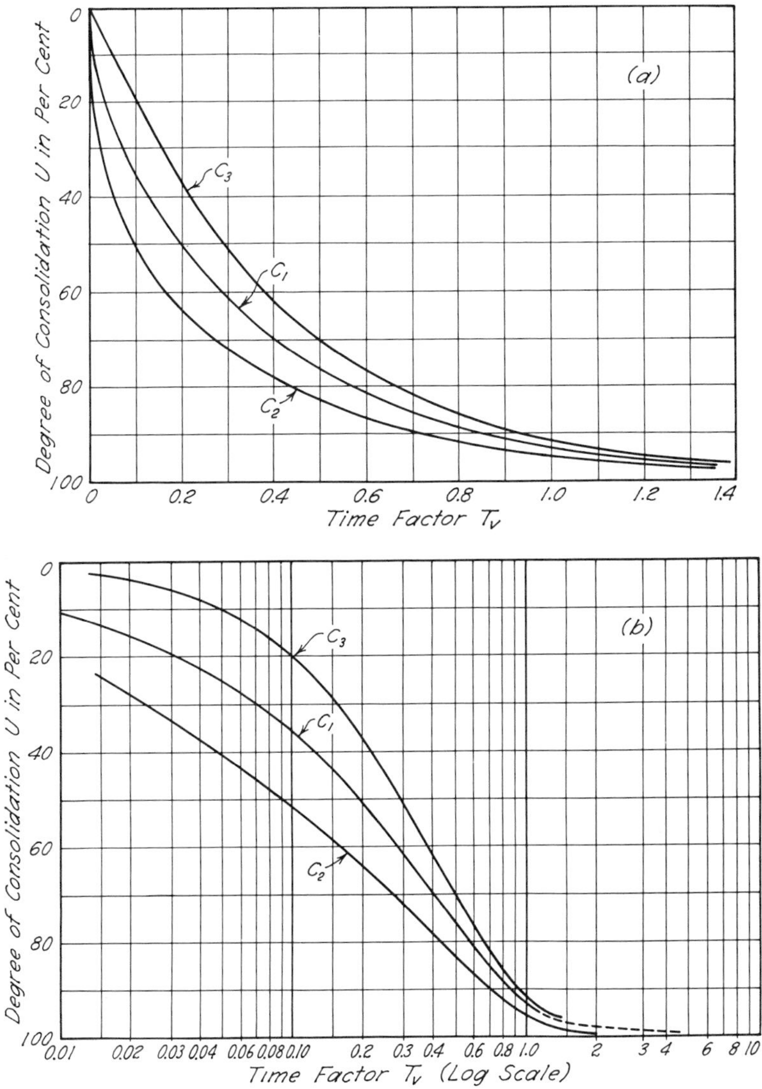

Fig. 25.4. Relation between time factor and degree of consolidation. In (*a*) the time factor is plotted to an arithmetic and in (*b*) to a logarithmic scale. The curves C_1, C_2, C_3, correspond to different conditions of loading and drainage, represented by *a*, *d*, and *f*, respectively, in Fig. 25.2 (after Terzaghi and Fröhlich 1936).

buildings resting on normally loaded clay ranges, during the first decades after construction, between $\frac{1}{8}$ and $\frac{1}{2}$ in per year. Exceptional rates as high as one inch per year have been observed.

It is obvious that the results of a settlement computation are not even approximately correct unless the assumed hydraulic boundary conditions are in accordance with the drainage conditions in the field. Every continuous sand or silt seam located within a bed of clay acts like a drainage layer and accelerates consolidation of the clay, whereas lenses of sand and silt have no effect. If the test boring records indicate that a bed of clay contains partings of sand or silt, the engineer is commonly unable to find out whether or not these partings are continuous. In such instances the theory of consolidation can be used only for determining an upper and a lower limiting value for the rate of settlement. The real rate remains unknown until it is observed.

Furthermore, in reality water escapes from the clay beneath a loaded foundation not only in vertical directions, but also by flow in horizontal or inclined directions. Problems of three-dimensional consolidation have been solved for relatively simple boundary and stress conditions (Biot 1941, Gibson and McNamee 1963). For more complex conditions, solutions can be obtained by numerical procedures (Abbott 1960, Gibson and Lumb 1953).

Problems

1. Representative samples were obtained from a layer of clay 20 ft thick, located between two layers of sand. By means of consolidation tests, it was found that the average value of c_v for these samples was 4.92×10^{-4} cm²/sec. By constructing a building above the layer, the average vertical pressure in the layer was everywhere increased and the building began to settle. Within how many days did half the ultimate settlement occur?

Ans. 438 days.

2. If the clay layer in problem 1 contained a thin drainage layer located 5 ft below its upper surface, how many days would be required to attain half the ultimate settlement?

Ans. 127 days.

3. A layer of clay 30 ft thick rests on an impermeable rock base. The consolidation stress along a given vertical line is assumed to vary uniformly from a maximum at the top of the layer to zero at the rock surface. The value for c_v for the clay is 9.5×10^{-5} cm²/sec. How many years will elapse after the construction of a building until the settlement becomes equal to 30% of the final value? Solve the same problem on the assumption that the clay rests on a pervious sand bed instead of rock.

Ans. 6.5; 4.9 years.

Selected Reading

Solutions for the consolidation of masses of soil having various boundary conditions may be found in the following references.

Terzaghi, K. and O. K. Fröhlich (1936). *Theorie der Setzung von Tonschichten* (Theory of settlement of the clay layers). Leipzig, Deutike, 166 pp.

Gray, H. (1945). "Simultaneous consolidation of contiguous layers of unlike compressible soils," *Trans. ASCE,* **110,** pp. 1327–1344.

Barron, R. A. (1948). "Consolidation of fine-grained soils by drain wells," *Trans. ASCE,* **113,** pp. 718–742.

Gibson, R. E. and P. Lumb (1953). "Numerical solution of some problems in the consolidation of clay," *Proc. Inst. Civil Engrs.,* London, Part 1, **2,** pp. 182–198.

Carslaw, H. S. and J. C. Jaeger (1959). *Conduction of heat in solids,* Oxford, Clarendon Press, 2nd ed., 510 pp.

Abbott, M. B. (1960). "One-dimensional consolidation of multi-layered soils," *Géot.,* **10,** pp. 151–165.

Gibson, R. E. and J. McNamee (1963). "A three-dimensional problem of the consolidation of a semi-infinite clay stratum," *Quart. J. Mech. and Appl. Math.,* **16,** Part 1, pp. 115–127.

Chapter 5

PLASTIC EQUILIBRIUM IN SOILS

ART. 26 FUNDAMENTAL ASSUMPTIONS

This chapter deals with the earth pressure against lateral supports such as retaining walls or the bracing in open cuts, with the resistance of the earth against lateral displacement, with the bearing capacity of footings, and with the stability of slopes. Problems of this kind merely require the determination of the factor of safety of the lateral support or the slope with respect to failure. The solution is obtained by comparing the magnitude of two sets of forces: those that tend to produce a failure, and those that tend to prevent it. Such an investigation is called a *stability computation*. In order to make a stability computation, it is necessary to determine the position of the potential surface of sliding and to compute or to estimate the resistance against sliding along this surface.

The sliding resistance s per unit of area depends not only on the type of soil but also on the effective normal stress $p - u_w$ on the surface of sliding and a number of other factors. These have been discussed in Articles 16 through 18. The selection of the appropriate value of s for a particular problem requires experience and judgment. Nevertheless, a reasonable approximation can usually be expressed in one of the following forms:

$s = (p - u_w) \tan \phi$, representing cohesionless sands (17.1)

$s = c + (p - u_w) \tan \phi$, crudely representing overconsolidated clays and unsaturated clay fills (16.5)

$s = \frac{1}{2}q_u = c$, representing soils, especially saturated clays, under undrained ($\phi = 0$) conditions (18.5)

Stiff clays commonly contain a network of hair cracks that make the conditions for their stability dependent to a large extent on the degree and duration of exposure to atmospheric agencies (Article 43). Therefore, such clays are beyond the scope of theoretical treatment.

Each of the stability problems will be solved first for a dry ($u_w = 0$) cohesionless sand to which Eq. 17.1 is applicable and then for a cohesive material to which Eq. 16.5 applies. After the reader is able to solve problems on the basis of these two equations, he can readily solve similar problems dealing with partly or completely submerged sand or with saturated clay under undrained conditions.

In a partly submerged mass of sand in which the water is at rest, the neutral stress u_w at any depth z below the water table is

$$u_w = \gamma_w z$$

This stress reduces the effective unit weight of that part of the sand below water level from γ to the submerged unit weight γ' (Eq. 12.6). Hence a stability calculation dealing with a partly submerged sand can be made on the assumption that the sand is dry, provided that the unit weight γ of the soil below water level is replaced by γ'. The pressure exerted by a partly submerged mass of sand against a lateral support is equal to the sum of the sand pressure, computed on the basis just mentioned, and the full water pressure. However, if the water percolates through the voids of the soil instead of being stagnant, this procedure is not applicable because the seepage pressure of the percolating water must be taken into account. Problems dealing with seepage pressure are discussed in Articles 35 and 36.

By substituting $\phi = 0$, Eq. 16.5 can be reduced to Eq. 18.5. Hence, theoretical expressions derived on the basis of Eq. 16.5 can be used in connection with many problems of practical importance dealing with saturated soft to medium clays. Nevertheless, it should be emphasized that calculations based on such expressions are strictly applicable only if the water content of the clay remains constant, and that changes in the stability with time must be taken into consideration. The time effects can be predicted in some instances by estimating the changes in pore pressures, but usually they can be taken into account only in a general way on the basis of a knowledge of the physical properties of the soil. Some of the effects of time have been discussed in Article 18. Others are indicated in Part III.

The condition for failure represented by equation 16.5 corresponds to Mohr's rupture diagram in which the failure envelope is a straight line (Fig. 26.1). Consequently a definite relation exists at failure between the major and minor principal stresses p_1 and p_3 respectively. By geometry

$$p_1 + d = OA + AB = OA(1 + \sin\phi)$$

$$p_3 + d = OA - AB = OA(1 - \sin\phi)$$

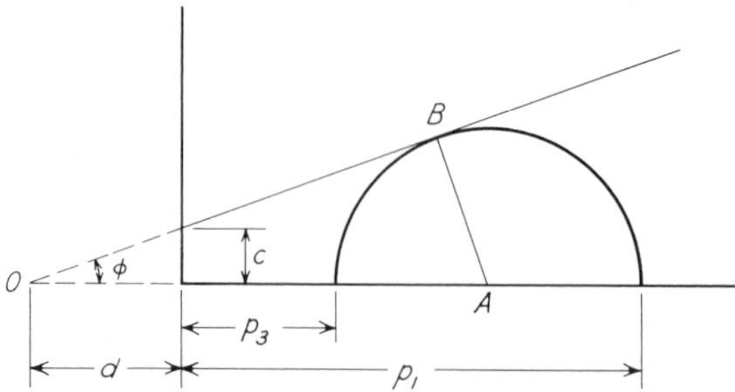

Fig. 26.1. Mohr rupture diagram for condition in which failure envelope is straight line.

whence

$$p_1 = p_3 \frac{1 + \sin \phi}{1 - \sin \phi} + d \left(\frac{1 + \sin \phi}{1 - \sin \phi} - 1 \right)$$

But

$$d = c \frac{\cos \phi}{\sin \phi} = c \frac{\sqrt{1 - \sin^2 \phi}}{\sin \phi}$$

Therefore

$$p_1 = p_3 \frac{1 + \sin \phi}{1 - \sin \phi} + 2c \sqrt{\frac{1 + \sin \phi}{1 - \sin \phi}}$$

$$p_1 = p_3 \tan^2 \left(45° + \frac{\phi}{2} \right) + 2c \tan \left(45° + \frac{\phi}{2} \right)$$

Or, if

$$N_\phi = \tan^2 \left(45° + \frac{\phi}{2} \right) \tag{26.1}$$

$$p_1 = p_3 N_\phi + 2c \sqrt{N_\phi} \tag{26.2}$$

The quantity N_ϕ is known as the *flow value*. If $c = 0$,

$$p_1 = p_3 N_\phi \tag{26.3}$$

and if $\phi = 0$

$$p_1 = p_3 + 2c \tag{26.4}$$

Problems

1. A sample of dense dry sand is subjected to a triaxial test. The angle of internal friction is believed to be about 37°. If the minor principal stress is 2 kg/cm², at what value of the major principal stress is the sample likely to fail?

Ans. 8.0 kg/cm².

2. Solve problem 1 on the assumption that the sand has a slight cohesion, equal to 0.10 kg/cm².

Ans. 8.4 kg/cm².

3. The shearing resistance of a soil is determined by the equation, $s = c + p \tan \phi$. Two triaxial tests are performed on the material. In the first test the all-around pressure is 2 kg/cm², and failure occurs at an added axial unit stress of 6 kg/cm². In the second test the all-around pressure is 3.5 kg/cm², and failure occurs at an added stress of 10.5 kg/cm². What values of c and ϕ correspond to the test results?

Ans. 0; 37°.

ART. 27 STATES OF PLASTIC EQUILIBRIUM

Fundamental Concepts

A body of soil is in a *state of plastic equilibrium* if every part of it is on the verge of failure. Rankine (1857) investigated the stress conditions corresponding to those states of plastic equilibrium which can be developed simultaneously throughout a semi-infinite mass of soil acted on by no force other than gravity. States of plastic equilibrium identical with those which Rankine considered are referred to as *Rankine states of plastic equilibrium.* A discussion of the Rankine states in a semi-infinite mass serves primarily as an introduction to the more complicated states of plastic equilibrium encountered in connection with practical problems.

The Rankine states are illustrated by Fig. 27.1. In this figure, AB represents the horizontal surface of a semi-infinite mass of cohesionless sand with a unit weight γ, and E represents an element of the sand with a depth z and a cross-sectional area equal to unity. Since the element is symmetrical with reference to a vertical plane, the normal stress on the base

$$p_v = \gamma z \tag{27.1}$$

is a principal stress. As a consequence, the normal stresses p_h on the vertical sides of the element at depth z are also principal stresses.

According to Eq. 26.3, the ratio between the major and minor principal stresses in a cohesionless material cannot exceed the value

$$\frac{p_1}{p_3} = N_\phi = \tan^2\left(45° + \frac{\phi}{2}\right)$$

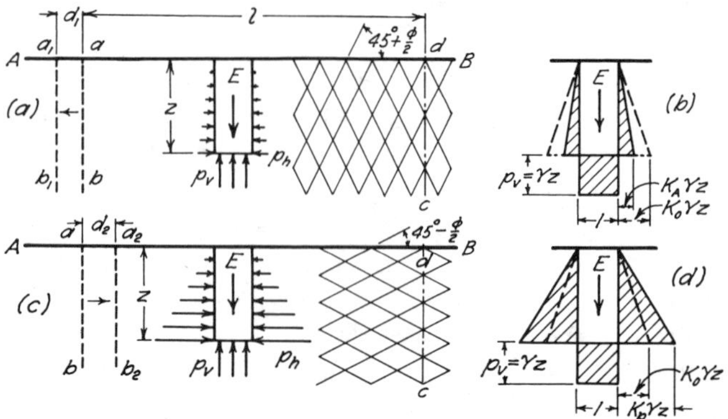

Fig. 27.1. (*a* and *b*) **Diagrams illustrating active Rankine state in semi-infinite mass of sand.** (*c* and *d*) **Corresponding diagrams for passive Rankine state.**

Since the vertical principal stress p_v in the mass of sand shown in Fig. 27.1*a* can be either the major or the minor principal stress, the ratio $K = p_h/p_v$ can assume any value between the limits,

$$K_A = \frac{p_h}{p_v} = \frac{1}{N_\phi} = \tan^2\left(45° - \frac{\phi}{2}\right) \tag{27.2}$$

and

$$K_P = \frac{p_h}{p_v} = N_\phi = \tan^2\left(45° + \frac{\phi}{2}\right) \tag{27.3}$$

After a mass of sand has been deposited by either a natural or an artificial process, K has a value K_0 intermediate between K_A and K_P, and

$$p_h = K_0 p_v \tag{27.4}$$

wherein K_0 is an empirical coefficient known as the *coefficient of earth pressure at rest.* Its value depends on the relative density of the sand and the process by which the deposit was formed. If this process did not involve artificial compaction by tamping, the value of K_0 ranges from about 0.40 for dense sand to 0.50 for loose sand. Tamping in layers may increase the value to about 0.8.

In order to change the value of K for a mass of sand from K_0 to some other value, it is necessary to give the entire mass an opportunity either to stretch or to be compressed in a horizontal direction. Since the weight of sand above any horizontal section remains unchanged, the vertical pressure p_v is unaltered. The horizontal pressure $p_h = Kp_v$, however, decreases if the mass stretches and increases if it compresses.

As the mass stretches, any two vertical sections such as *ab* and *cd* move apart, and the value of K decreases until it becomes equal to K_A (Eq. 27.2). The sand is then in what is known as the *active Rankine state*. In this state the intensity of the horizontal pressure at any depth z is equal to

$$p_h = K_A p_v = K_A \gamma z = \gamma z \frac{1}{N_\phi} \tag{27.5}$$

in which K_A is called the *coefficient of active earth pressure*. The distribution of pressure over the sides and base of an element such as E is shown in Fig. 27.1*b*. Further stretching of the mass has no effect on p_h (Eq. 27.5), but sliding occurs along two sets of plane surfaces as indicated on the right-hand side of Fig. 27.1*a*. According to Eq. 16.3, such surfaces of sliding intersect the direction of the minor principal stress at the angle $45° + \phi/2$. Since the minor principal stresses in the active Rankine state are horizontal, the shear planes rise at an angle of $45° + \phi/2$ with the horizontal. The pattern formed by the traces of the shear planes on a vertical section parallel to the direction of stretching is known as the *shear pattern*.

A horizontal compression of the entire mass of sand causes *ab* to move toward *cd*, as shown in Fig. 27.1*c*. As a consequence, the ratio $K = p_h/p_v$ increases. As soon as K becomes equal to K_P (Eq. 27.3) the sand is said to be in the *passive Rankine state*. At any depth z the horizontal pressure is

$$p_h = K_P p_v = K_P \gamma z = \gamma z N_\phi \tag{27.6}$$

in which K_P is the *coefficient of passive earth pressure*. Since the minor principal stress in the passive Rankine state is vertical, the surfaces of sliding rise at an angle of $45° - \phi/2$ with the horizontal, as shown in Fig. 27.1*c*.

The active and the passive Rankine states constitute the two limiting states for the equilibrium of the sand. Every intermediate state, including the state of rest, is referred to as a *state of elastic equilibrium*.

Local States of Plastic Equilibrium

The Rankine states illustrated by Fig. 27.1 were produced by uniformly stretching or compressing every part of a semi-infinite mass of sand. They are known as *general states of plastic equilibrium*. However, in a stratum of real sand, no general state of equilibrium can be produced except by a geological process such as the horizontal compression by tectonic forces of the entire rock foundation of the sand strata. Local events, such as the yielding of a retaining wall, cannot produce a radical change in the state of stress in the sand except in the immediate vicinity of the source of the disturbance. The rest of the sand remains in a state of elastic equilibrium.

Local states of plastic equilibrium can be produced by very different processes of deformation. The resulting states of stress in the plastic zone and the shape of the zone itself depend to a large extent on the type of deformation and on the degree of roughness of the surface of contact between the soil and its support. These factors constitute the *deformation* and the *boundary conditions*. The practical consequences of these conditions are illustrated by Figs. 27.2 and 27.3.

Figure 27.2*a* is a vertical section through a prismatic box having a length l equal to the distance between the vertical sections *ab* and *cd* in Fig. 27.1. If sand is deposited in the box by the same process that was responsible for the formation of the semi-infinite mass represented in Fig. 27.1, the states of stress in both masses are identical. They represent states of elastic equilibrium.

When the state of the semi-infinite mass of sand (Fig. 27.1*a*) was changed from that of rest to the active Rankine state, the vertical section *ab* moved through the distance d_1. In order to change the state of the entire mass of sand contained in the box (Fig. 27.2*a*) into the active Rankine state, the wall *ab* must be moved through the same distance. This constitutes the deformation condition. While the wall *ab* (Fig. 27.2*a*) moves out, the height of the mass of sand decreases, and its length increases. These movements involve displacements between the sand and

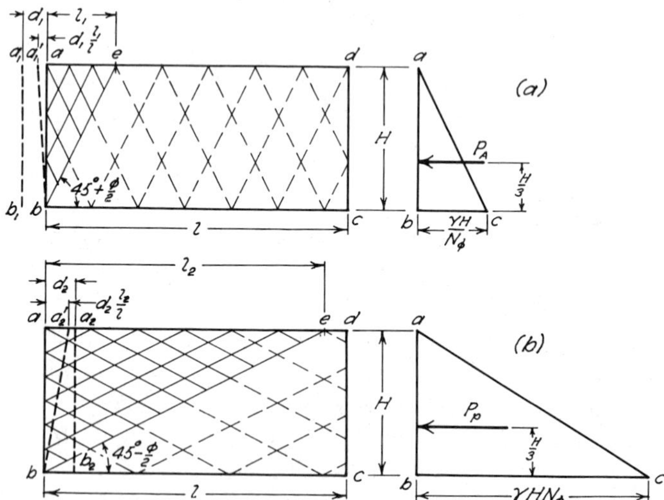

Fig. 27.2. (*a*) Diagrams illustrating local active Rankine state in sand contained in rectangular box. (*b*) Corresponding diagrams for local passive Rankine state.

all of the surfaces of the box with which it is in contact. If the contact surfaces are rough, shearing stresses will develop along vertical and horizontal planes. Since in the active Rankine state the shearing stresses on these planes are zero, this state cannot materialize unless the sides and bottom of the box are perfectly smooth. This requirement constitutes the boundary condition for the transition of the sand in the box to the active Rankine state. If this condition is satisfied, the sand passes into an active Rankine state as soon as the wall *ab* reaches the position a_1b_1. At this stage, the unit stretch of the soil is d_1/l. Any further movement of the wall causes slippage along the two sets of surfaces of sliding indicated by dash lines in Fig. 27.2*a*, but the stress conditions remain unchanged.

If the wall *ab* is perfectly smooth but the bottom of the box is rough, the sand located between the wall *ab* and the potential surface of sliding *be* is free to deform in exactly the same manner as it does in a box with a smooth bottom, but the state of stress in the balance of the sand cannot change materially because the friction along the bottom prevents the required deformation. Hence, an outward movement of the wall *ab* produces an active Rankine state only within the wedge-shaped zone *abe*. Since the width of the wedge increases from zero at the bottom to l_1 at the top, the unit stretch d_1/l required to establish the active Rankine state in the wedge is attained as soon as the left-hand boundary of the wedge moves from *ab* to $a_1'b$ (Fig. 27.2*a*). This is the deformation condition for the development of an active Rankine state within the wedge. As soon as the wall *ab* passes beyond this position, the wedge slides downward and outward along a plane surface of sliding *be* which rises at an angle of $45° + \phi/2$ with the horizontal.

If the wall *ab* is pushed toward the sand, and if both the walls and the bottom of the box are perfectly smooth, the entire mass of sand is transformed into the passive Rankine state (Fig. 27.2*b*) as soon as the wall moves beyond a distance d_2 from its original position. The planes of sliding rise at an angle of $45° - \phi/2$ with the horizontal. If the wall *ab* is perfectly smooth but the bottom of the box is rough, the passive Rankine state develops only within the wedge-shaped zone *abe*. The transition from the elastic to the plastic state does not occur until *ab* moves into or beyond the position $a_2'b$.

If the end of the box is free to move outward at the bottom but is restrained at the top, as indicated in Fig. 27.3, the sand fails by shear along some surface of sliding as soon as the tilt becomes perceptible, because the deformations compatible with an elastic state of equilibrium are very small. However, even at the state of failure, the sand between the wall and the surface of sliding does not pass into the active Rankine

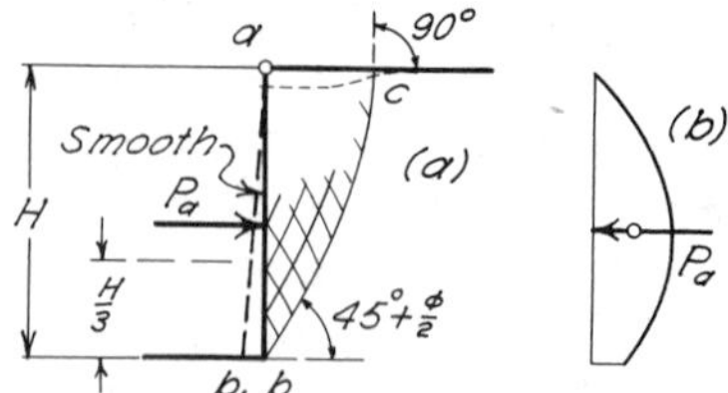

Fig. 27.3. Failure of sand behind smooth vertical wall when deformation condition for active Rankine state is not satisfied. (*a*) Section through back of wall. (*b*) Stress against back of wall.

state because the upper part of the wall cannot move, and, as a consequence, the deformation condition for the active Rankine state within the sliding wedge is not satisfied.

Theoretical and experimental investigations regarding the type of failure caused by a tilt of the lateral support about its upper edge have led to the conclusion that the surface of sliding starts at *b* (Fig. 27.3*a*) at an angle of $45° + \phi/2$ with the horizontal and that it becomes steeper until it intersects the ground surface at a right angle. The upper part of the sliding wedge remains in a state of elastic equilibrium until the lower part of the wedge has passed completely into a state of plastic equilibrium. The distribution of pressure against the lateral support is roughly parabolic (Fig. 27.3*b*) instead of triangular (Fig. 27.1*b*).

Similar investigations regarding the effect of pushing the bottom of the support toward the soil (Fig. 27.4*a*) have shown that the surface of sliding rises from *b* at an angle $45° - \phi/2$ with the horizontal and that it also intersects the ground surface at a right angle. The corresponding distribution of pressure is shown in Fig. 27.4*b*.

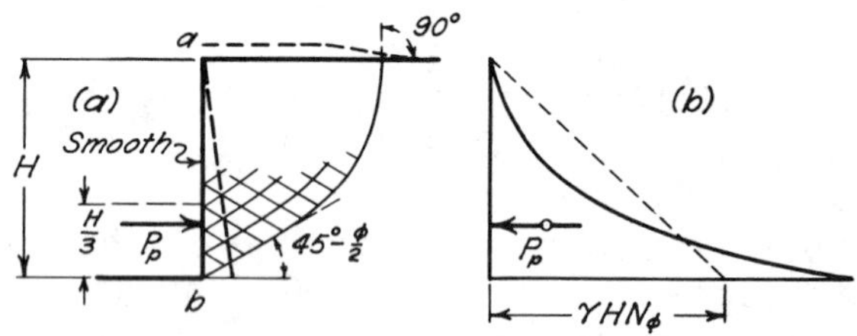

Fig. 27.4. Failure of sand behind smooth vertical wall when deformation condition for passive Rankine state is not satisfied. (*a*) Section through back of wall. (*b*) Stress on back of wall.

Selected Reading

A general discussion of the application of the theory of plasticity to states of limiting equilibrium, including problems of earth pressure, stability of slopes, and bearing capacity, is contained in Sokolovski, V. V. (1960): *Statics of soil media.* Translated from the Russian by D. H. Jones and A. N. Schofield, London, Butterworths, 237 pp.

Mathematical methods for solving problems with mixed boundary conditions are developed in Hansen, B. (1965): *A theory of plasticity for ideal frictionless materials,* Copenhagen, Teknisk Forlag, 471 pp.

ART. 28 RANKINE'S EARTH-PRESSURE THEORY

Earth Pressure against Retaining Walls

Retaining walls serve the same function as the vertical sides of the box shown in Fig. 27.2. The soil adjoining the wall is known as the *backfill.* It is always deposited after the wall is built. While the backfill is being placed, the wall yields somewhat under the pressure. The ultimate value of the pressure depends not only on the nature of the soil and the height of the wall but also on the amount of yield. If the position of the wall is fixed, the earth pressure is likely to retain forever a value close to the earth pressure at rest (Article 27). However, as soon as a wall yields far enough, it automatically satisfies the deformation condition for the transition of the adjoining mass of soil from the state of rest into an active state of plastic equilibrium. Hence, the factor of safety of a retaining wall capable of yielding must be adequate with respect to the active earth pressure, but does not need to be investigated for greater values of earth pressure.

Although the back of every real retaining wall is rough, approximate values of the earth pressure can be obtained on the assumption that it is smooth. In the following paragraphs, this assumption is made. Methods for obtaining more accurate values will be described in subsequent articles.

Active Earth Pressure of Cohesionless Soil Against Smooth Vertical Walls

If the surface of a sand backfill is horizontal, and if the back of the retaining wall is vertical and perfectly smooth, the magnitude and the distribution of pressure against the back of the wall are identical with those of the active pressure against the fictitious plane *ab* in Fig. 27.1*a*. Therefore, the earth pressure can be computed on the basis of the equa-

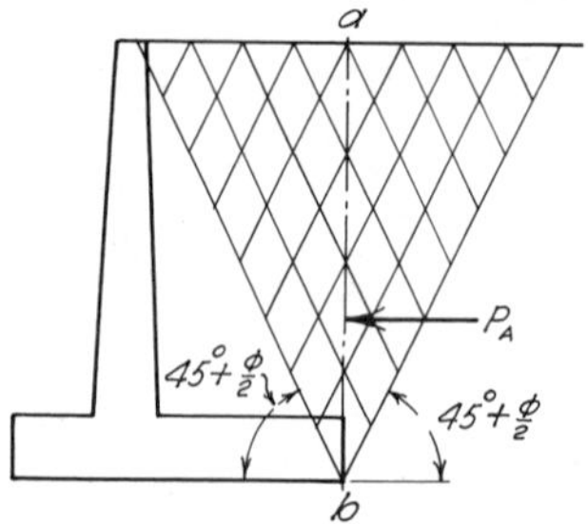

Fig. 28.1. Failure of sand behind cantilever retaining wall; deformation condition for active Rankine state is almost satisfied.

tions already derived. In reality, there are no perfectly smooth surfaces. However, the equations based on this assumption are so simple that they are quite commonly used for evaluating the earth pressure against real retaining walls and other structures acted on by earth pressure. It is shown subsequently that the roughness of the back of a wall commonly reduces the active and increases the passive earth pressure. Hence, as a rule, the error associated with the assumption is on the safe side.

Furthermore, in one case of considerable practical importance, the assumption of a smooth vertical wall is almost strictly correct. This case is illustrated by Fig. 28.1 which represents a cantilever wall. If such a wall yields under the influence of the earth pressure, the sand fails by shear along two planes rising from the heel of the wall at angles of $45° + \phi/2$ with the horizontal. Within the wedge-shaped zone located between these two planes, the sand is in the active Rankine state, and no shearing stresses act along the vertical plane *ab* through the heel. Hence, the earth pressure against this plane is identical with that against a smooth vertical wall.

If the sand backfill is perfectly dry, the active pressure against a smooth vertical wall at any depth z is

$$p_h = \gamma z \frac{1}{N_\phi} \tag{27.5}$$

It increases in simple proportion to the depth, as indicated by the pressure triangle *abc* (Fig. 27.2*a*). The total pressure against the wall is

$$P_A = \int_0^H p_h\, dz = \frac{1}{2}\gamma H^2 \frac{1}{N_\phi} \tag{28.1}$$

The point of application of P_A is located at a height $H/3$ above b.

If the wall is pushed into the position $a_2'b$ in Fig. 27.2*b* the pressure p_h against the wall assumes a value corresponding to the passive Rankine state,

$$p_h = \gamma z N_\phi \tag{27.6}$$

and the total pressure against the wall becomes equal to

$$P_P = \int_0^H p_h \, dz = \frac{1}{2} \gamma H^2 N_\phi \tag{28.2}$$

Active Earth Pressure of Partly Submerged Sand Supporting a Uniform Surcharge

In Fig. 28.2*a* the line *ab* represents the smooth vertical back of a wall with height H. The effective unit weight of the sand when dry is γ_d and when submerged is γ' (see Article 12); the unit weight of water is γ_w. The surface of the horizontal backfill carries a uniformly distributed surcharge q per unit of area. Within the backfill the water table is located at depth H_1 below the crest of the wall. The angle of internal friction of both the dry and submerged sand is assumed to be ϕ.

As the wall yields from position *ab* into position $a_1'b$, the pressure against its back decreases from the value of the earth pressure at rest to that of the active Rankine pressure. In Article 26, it was shown that the entire effect of the porewater pressure on the effective stresses in the sand can be taken into account by assigning to the submerged part of the sand the submerged unit weight γ' (Eq. 12.6). Within the depth H_1 the pressure on the wall due to the weight of the adjoining sand is represented by the triangle *ace* in Fig. 28.2*b*. At any depth z' below the water table the effective vertical pressure on a horizontal section through the sand is

$$p_v = H_1\gamma + z'\gamma'$$

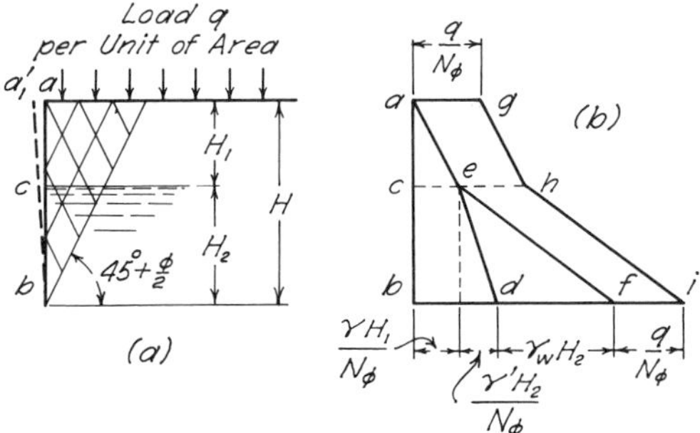

Fig. 28.2. Active earth pressure of partly submerged sand supporting a uniform surcharge. (*a*) Section through back of supporting structure. (*b*) Pressure against back of structure.

For the corresponding horizontal active Rankine pressure we obtain by means of Eq. 27.5

$$p_h = \frac{p_v}{N_\phi} = (H_1\gamma + z'\gamma')\frac{1}{N_\phi} \tag{28.3}$$

The total effective horizontal pressure below the water level is represented by the area *bced* in Fig. 28.2*b*. To this pressure must be added the total water pressure,

$$P_w = \tfrac{1}{2}\gamma_w H_2{}^2 \tag{28.4}$$

which acts against the lower part *cb* of the wall. In Fig. 28.2*b*, the water pressure is represented by the triangle *def*.

If the fill carries a uniformly distributed surcharge q per unit of area, the effective vertical stress p_v increases at any depth by q, and the corresponding horizontal active Rankine pressure increases by

$$\Delta p_h = \frac{q}{N_\phi} \tag{28.5}$$

In Fig. 28.2*b* the pressure produced by the surcharge q is represented by the area *aefihg*.

Active Earth Pressure of Cohesive Soils against Smooth Vertical Surfaces

In Fig. 28.3*a* the line *ab* represents the smooth vertical back of a wall in contact with a cohesive soil having a unit weight γ. The shearing resistance of the soil is defined by the equation,

$$s = c + p \tan \phi$$

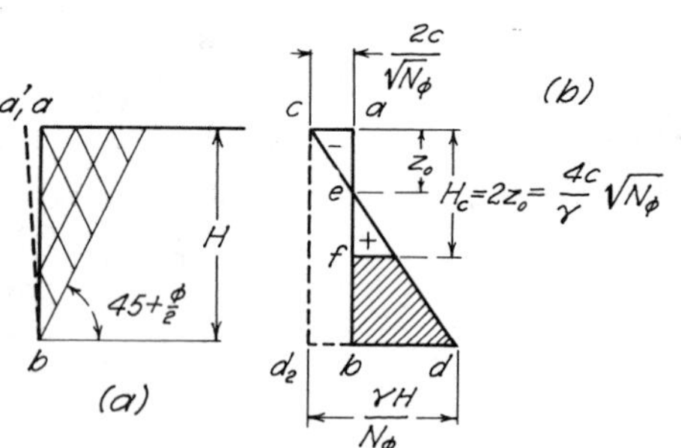

Fig. 28.3. Failure of clay behind smooth vertical wall when deformation condition for active earth pressure is satisfied. (*a*) Section through back of wall. (b) Pressure against back of wall.

which, according to Article 26, applies at least crudely to dry or moist cohesive soil located above the water table. The relation between the extreme values of the principal stresses in such soils is determined by the expression,

$$p_1 = p_3 N_\phi + 2c\sqrt{N_\phi} \tag{26.2}$$

wherein p_1 and p_3 are, respectively, the major and minor principal stresses, and

$$N_\phi = \tan^2\left(45° + \frac{\phi}{2}\right) \tag{26.1}$$

is the flow value. In Article 16 it is also shown that the surfaces of sliding intersect the direction of the minor principal stress at an angle $45° + \phi/2$, regardless of the value of c (Eq. 26.2).

Since the back of the wall is smooth, the vertical principal stress at depth z below the horizontal surface of the backfill is $p_v = \gamma z$. Before the support ab moves, it is acted on by the earth pressure at rest. In this state the horizontal stress p_h is the minor principal stress. An outward movement of the support into or beyond the position $a_1'b$ reduces p_h to the value corresponding to the active Rankine pressure. Substituting $p_v = p_1 = \gamma z$ and $p_h = p_3$ into Eq. 26.2, we obtain

$$p_h = \gamma z \frac{1}{N_\phi} - 2c \frac{1}{\sqrt{N_\phi}} \tag{28.6}$$

This stress at any depth z is represented by the horizontal distance between the lines ab and cd in Fig. 28.3*b*. At depth,

$$z_0 = \frac{2c}{\gamma}\sqrt{N_\phi} \tag{28.7}$$

the stress p_h is equal to zero. At a depth less than z_0, the pressure against the wall is negative, provided that a crack does not open up between the wall and the uppermost part of the soil. The total earth pressure against the wall is

$$P_A = \int_0^H p_h\, dz = \frac{1}{2}\gamma H^2 \frac{1}{N_\phi} - 2c \frac{H}{\sqrt{N_\phi}} \tag{28.8}$$

If the wall has a height,

$$H = H_c = \frac{4c}{\gamma}\sqrt{N_\phi} = 2z_0 \tag{28.9}$$

the total earth pressure P_A is equal to zero. Hence, if the height of a vertical bank is smaller than H_c, the bank should be able to stand without lateral support. However, the pressure against the wall increases from $-2c/\sqrt{N_\phi}$ at the crest to $+2c/\sqrt{N_\phi}$ at depth H_c, whereas on the vertical face of an unsupported bank the normal stress is zero at every point. Because of this difference the greatest depth to which a cut can be excavated without lateral support of its vertical sides is slightly smaller than H_c (see Article 35).

For the $\phi = 0$ condition (Article 18), $N_\phi = 1$. Therefore

$$P_A = \tfrac{1}{2}\gamma H^2 - 2cH \tag{28.10}$$

and

$$H_c = \frac{4c}{\gamma} \tag{28.11}$$

Since the soil does not necessarily adhere to the wall, it is commonly assumed that the active earth pressure of cohesive soils against retaining walls is equal to the pressure represented in Fig. 28.3*b* by the triangular area *bde*, equal to area cdd_2 − area $cebd_2$. Therefore,

$$P_A = \frac{1}{2}\gamma H^2 \frac{1}{N_\phi} - 2cH \frac{1}{\sqrt{N_\phi}} + \frac{2c^2}{\gamma} \tag{28.12}$$

For the $\phi = 0$ condition,

$$P_A = \frac{1}{2}\gamma H^2 - 2cH + \frac{2c^2}{\gamma} \tag{28.13}$$

Passive Earth Pressure of Cohesive Soils in Contact with Smooth Vertical Surfaces

If the face *ab* of the wall or block that supports the soil and its uniform surcharge *q* is pushed toward the backfill as indicated in Fig. 28.4*a*, the horizontal principal stress p_h increases and becomes greater than p_v. As soon as *ab* arrives at or beyond the position $a_2'b$, which represents the

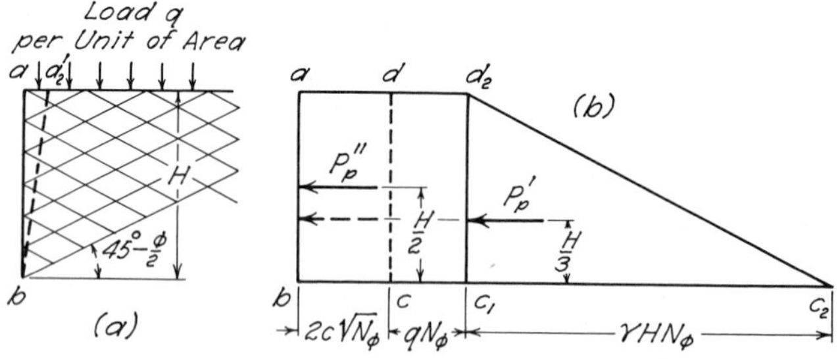

Fig. 28.4. Failure of clay behind smooth vertical wall when deformation condition for passive earth pressure is satisfied. (*a*) Section through back of wall. (*b*) Stress on back of wall.

deformation condition for the passive Rankine state, the stress conditions for failure (Eq. 26.2) are satisfied. Since p_h represents the major principal stress, we may substitute $p_h = p_1$ and $p_v = p_3 = \gamma z + q$ into Eq. 26.2 and obtain

$$p_h = \gamma z N_\phi + 2c\sqrt{N_\phi} + qN_\phi \tag{28.14}$$

The stress p_h can be resolved into two parts. One part

$$p_h' = \gamma z N_\phi$$

increases like a hydrostatic pressure in simple proportion to depth. In Fig. 28.4*b* the stresses p_h' are represented by the width of the triangle $c_1c_2d_2$ with the area

$$P_P' = \tfrac{1}{2}\gamma H^2 N_\phi \tag{28.15}$$

The point of application of P_P' is located at an elevation $H/3$ above b. The quantity P_P' represents the total passive earth pressure of a cohesionless material with an angle of internal friction ϕ and a unit weight γ.

The second part of p_h is

$$p_h'' = 2c\sqrt{N_\phi} + qN_\phi$$

This part is independent of the depth. It is represented by the width of the rectangle abc_1d_2 in Fig. 28.4*b*. The total pressure is equal to the area of the rectangle. Hence,

$$P_P'' = H(2c\sqrt{N_\phi} + qN_\phi) \tag{28.16}$$

The point of application of P_P'' is at mid-height of the surface ab. Since Eq. 28.16 does not contain the unit weight γ, the value P_P'' can be computed on the assumption that the backfill is weightless. From Eqs. 28.15 and 28.16, we find that the total passive earth pressure is

$$P_P = P_P' + P_P'' = \tfrac{1}{2}\gamma H^2 N_\phi + H(2c\sqrt{N_\phi} + qN_\phi) \tag{28.17}$$

According to the preceding discussion, P_P can be computed by means of two independent operations. First, P_P' is computed on the assumption that the cohesion and the surcharge are zero ($c = 0$, $q = 0$). The point of application of P_P' is located at the lower third-point of H. Secondly, P_P'' is computed on the assumption that the unit weight of the backfill is zero ($\gamma = 0$). The point of application of P_P'' is at the midpoint of H. In the following articles this simple procedure is used repeatedly for determining the point of application of the passive earth pressure of cohesive soils. The subdivision of P_P into the two parts P_P' and P_P'' is strictly correct only when the back of the wall is vertical and perfectly smooth. For all other conditions, the procedure is approximate.

Problems

1. A wall with a smooth vertical back 10 ft high retains a mass of dry cohesionless sand that has a horizontal surface. The sand weighs 113 lb/ft^3 and has an angle of internal friction of 36°. What is the approximate total pressure against the wall, if the wall is prevented from yielding? if the wall can yield far enough to satisfy the deformation condition for the active Rankine state?

Ans. 2260 to 2830 lb/lin ft; 1470 lb/lin ft.

2. The water level behind the wall described in problem 1 rises to an elevation 4 ft below the crest. The submerged unit weight of the sand is 66 lb/ft^3. If the deformation condition for the active Rankine state is satisfied, what is the total pressure that the earth and water exert against the wall? At what height above the base does the resultant of the earth and water pressures act?

Ans. 2380 lb/lin ft; 2.83 ft.

3. What is the total lateral pressure against the yielding wall in problem 1, if the sand mass supports a uniformly distributed load of 400 lb/ft^2? At what height above the base of the wall is the center of pressure?

Ans. 2510 lb/lin ft; 4.02 ft.

4. The space between two retaining walls with smooth backs is filled with sand weighing 113 lb/ft^3. The foundations of the walls are interconnected by a reinforced-concrete floor, and the crests of the walls by heavy steel tie rods. The walls are 15 ft high and 50 ft apart. The surface of the sand is used for storing pig iron weighing 300 lb/ft^2. If the coefficient of the earth pressure at rest is $K_0 = 0.50$, what is the total pressure against the walls before and after the application of the surcharge?

Ans. 6360 lb/lin ft; 8610 lb/lin ft.

5. The same wall as in problem 1 supports a purely cohesive soil having a cohesion $c = 200$ lb/ft^2 and a unit weight of 110 lb/ft^3. The value of ϕ is zero. What is the total active Rankine pressure against the wall? At what distance above the base is the center of pressure? At what depth is the intensity of pressure zero?

Ans. 1500 lb/lin ft; −1.11 ft; 3.64 ft.

6. A vertical bank was formed during the excavation of a plastic clay having a unit weight of 120 lb/ft^3. When the depth of excavation reached 18 ft, the bank failed. On the assumption that $\phi = 0°$, what was the approximate value of the cohesion of the clay?

Ans. 540 lb/ft^2.

7. A smooth vertical wall 20 ft high is pushed against a mass of soil having a horizontal surface and a shearing resistance given by Coulomb's equation in which $c = 400$ lb/ft^2 and $\phi = 15°$. The unit weight of the soil is 120 lb/ft^3. Its surface carries a uniform load of 200 lb/ft^2. What is the total passive Rankine pressure? What is the distance from the base of the wall to the center of pressure? Determine the intensity of lateral pressure at the base of the wall.

Ans. 68,400 lb/lin ft; 8.01 ft; 5460 lb/ft^2.

ART. 29 INFLUENCE OF WALL FRICTION ON THE SHAPE OF THE SURFACE OF SLIDING

The back of the wall in Fig. 29.1*a* is assumed to be rough. Otherwise it is identical with that shown in Fig. 27.2*a*. The backfill consists of clean sand. If the wall moves outward, the sliding wedge subsides, and the sand moves downward along the back of the wall. The downward movement of the sand with reference to the wall develops frictional forces that cause the resultant active earth pressure to be inclined at an angle δ to the normal to the wall. This angle is known as the *angle of wall friction*. It is considered positive when the resultant reaction is oriented such that its tangential component acts in an upward direction (Fig. 29.1*a*). Advanced theoretical analyses (Ohde 1938) as well as experiments have shown that the corresponding surface of sliding *bc* consists of a curved lower portion and a straight upper part. Within the section *adc* of the sliding wedge the shear pattern is identical with the

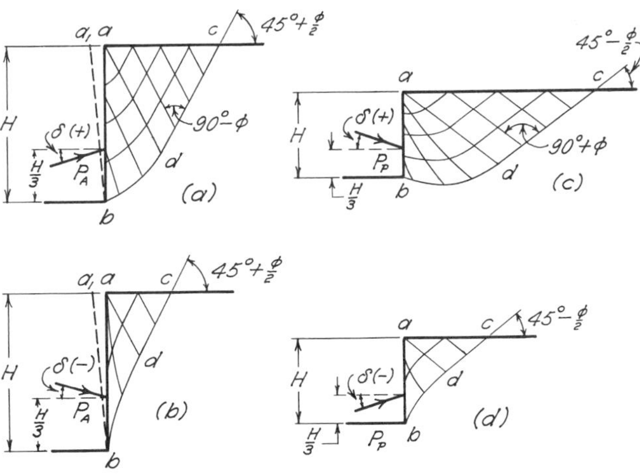

Fig. 29.1. Shear patterns associated with failure of sand behind rough vertical wall.

active Rankine pattern (Fig. 27.2*a*). Within the area *adb* the shear pattern consists of two sets of curved lines.

If the wall is forced down with reference to the backfill, for instance by the action of a heavy load on its crest, the value of δ becomes negative, and the curvature of the lower part of the surface of sliding is reversed, as shown in Fig. 29.1*b*.

If the wall is pushed toward the fill, the movement is resisted by the passive earth pressure. If the weight of the wall is greater than the friction between the sand and the wall, the sand rises with reference to the wall, and the reaction to the resultant passive earth pressure acts at an angle δ with the normal to the back of the wall. The tangential component of this force tends to prevent the rise of the sand. Under this condition the angle δ (Fig. 29.1*c*) is considered positive. The straight portion of the surface of sliding rises at an angle $45° - \phi/2$ with the horizontal. Within the isosceles triangle *adc* the shear pattern is identical with that shown in Fig. 27.2*b*, and the material is in the passive Rankine state. Within the area *adb* both sets of lines which constitute the shear pattern are curved.

If the weight of the wall is smaller than the friction between the sand and the wall, the angle between the normal to the back of the wall and the reaction to the resultant passive pressure is smaller than δ. Finally, if the wall is acted on by an upward force equal to the sum of the weight of the wall and the friction between the sand and the wall, the resultant passive earth pressure is oriented as shown in Fig. 29.1*d*, and the angle of wall friction is considered negative. The curvature of the curved portion of the surface of sliding is reversed.

The deformation conditions for the plastic states represented by the shear patterns in Figs. 29.1*a* and *b* require a certain minimum lengthening of every horizontal element of the wedge. The deformation conditions for the plastic states represented in Figs. 29.1*c* and *d* require a certain minimum shortening of every horizontal element. These requirements are the equivalent of those for producing the active or passive Rankine states in the backfill of a perfectly smooth wall, as illustrated by Figs. 27.2*a* and *b*.

ART. 30 COULOMB'S THEORY OF ACTIVE EARTH PRESSURE AGAINST RETAINING WALLS

Introduction

Since the back of every real retaining wall is more or less rough, the boundary conditions for the validity of Rankine's theory are seldom satisfied, and earth-pressure computations based on this theory usually involve an appreciable error. Most of this error can be avoided by using Coulomb's theory (Coulomb 1776). Coulomb's method can be adapted to any boundary condition, but, in exchange, it involves a simplifying assumption regarding the shape of the surface of sliding. However, the error due to this assumption is commonly small compared to that associated with the use of Rankine's theory. When the boundary conditions for the validity of Rankine's theory are satisfied, the two theories lead to identical results.

Both Coulomb's and Rankine's theories are based on the assumptions that the wall is free to move into or beyond the position a_1b (Fig. 29.1*a*) and that the water contained in the voids of the soil does not exert any appreciable seepage pressure. It is also quite obviously assumed that the soil constants that appear in the equations have definite values that can be determined.

Coulomb's Theory

The surface of sliding in the backfill of a real retaining wall is slightly curved, as shown in Figs. 29.1*a* and *b*. In order to simplify the computations, Coulomb assumed it to be plane. The error due to disregarding the curvature is, however, quite small.

The forces that act on the sliding wedge are shown in Fig. 30.1*a* in which the straight line bc_1 is arbitrarily assumed to represent the surface of sliding. The wedge abc_1 is in equilibrium under the weight W_1, the reaction to the resultant earth pressure P_1, and the reaction F_1. The reaction F_1 is inclined at the angle ϕ to the normal to bc_1 because the frictional resistance is assumed to be fully developed along the surface of sliding. If the retaining wall rests on a firm foundation, the force P_1 is inclined to the normal to the back of the wall at the angle of wall friction $+\delta$, as indicated by the solid arrow in the figure. On the other hand, if it is likely that the wall may settle more than the backfill, the force P_1 will be inclined at the angle $-\delta$, as shown by the dash arrow. Since the magnitude of W_1 is known and the directions of all three forces are also known, the earth pressure P_1 can be scaled from the polygon of forces (Fig. 30.1*b*). Inasmuch as bc_1 is not necessarily the real surface of sliding, similar constructions are made to determine the earth pressures P_2, P_3, etc., for other arbitrarily selected surfaces bc_2, bc_3, etc. (not shown). The greatest value of the earth pressure obtained in this manner is equal to the active earth pressure P_A.

Culmann's Graphical Construction

An expedient method was devised by Culmann (1875) for performing the graphical constructions described in the preceding paragraph. It is illustrated in Fig. 30.1*c*. The first step in Culmann's procedure is to trace the line bS which passes through the bottom edge b of the back of the wall and rises at the angle ϕ above the horizontal base of the backfill. This line is known as the *slope line* because it represents the natural slope of the backfill material. The next step is to trace the *earth-pressure line* bL, which is located below the slope line and which intersects it at the angle θ. The angle θ is equal to the angle between the vertical and the direction of the resultant earth pressure P_A, as shown in Fig. 30.1. It depends on the angle of wall friction δ and the inclination α of the back of the wall.

In order to determine the earth pressure P_1 exerted by a wedge located above an arbitrary plane surface of sliding bc_1, it is first necessary to compute the weight W_1 of this wedge. This weight is laid off along bS at any convenient scale of forces. Thus point d_1 is obtained. The line d_1e_1 is then traced parallel to bL. Since the triangle e_1d_1b in Fig. 30.1*c* is similar to the force polygon (Fig. 30.1*b*), the distance d_1e_1 is equal to the

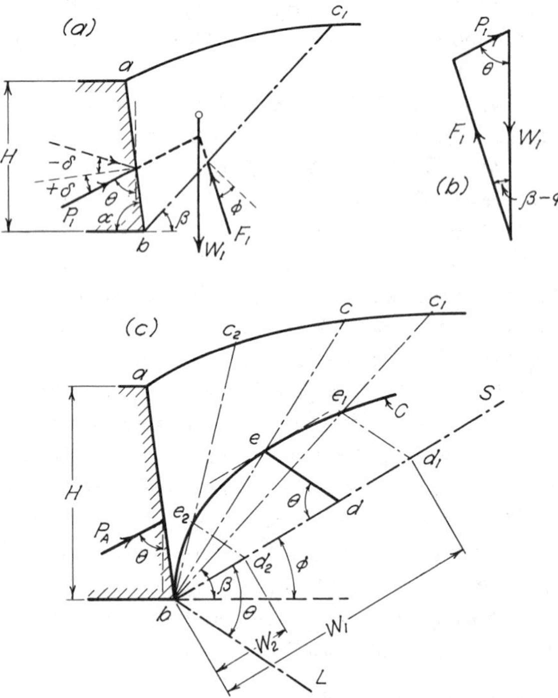

Fig. 30.1. (*a* and *b*) **Diagrams illustrating assumptions upon which Coulomb's theory of active earth pressure is based.** (*c*) **Culmann's graphical method for determining earth pressure of sand.**

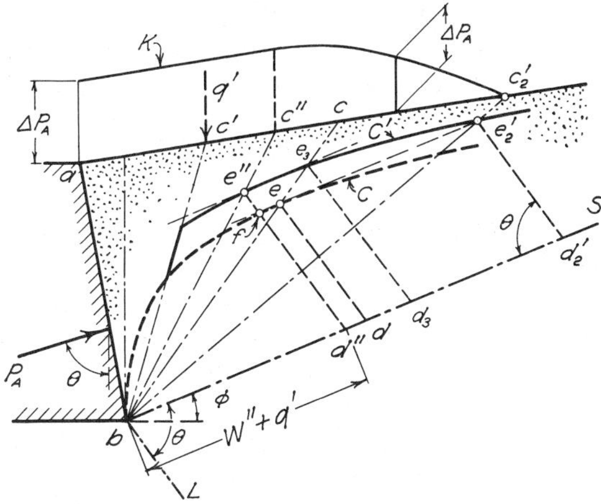

Fig. 30.2. Culmann's graphical method for determining earth pressure exerted by sand backfill that carries a line load.

earth pressure corresponding to the surface of sliding bc_1. To find the active earth pressure P_A, the construction is repeated for different planes bc_2, etc. The points e_1, e_2, etc., are connected by a curve C, known as the *Culmann line*, and a tangent to C is traced parallel to bS. The distance ed represents P_A, and the real surface of sliding passes through point e.

Earth Pressure Due to Line Load

Figure 30.2 is a section through a wall that supports a mass of sand with an inclined surface. Along a line parallel to the crest of the wall, at a distance ac' from the crest, the surface of the backfill carries a load q' per unit of length of the line. The procedure for determining the active earth pressure against the wall is essentially the same as that illustrated in Fig. 30.1c. However, if the right-hand boundary of a wedge intersects the ground surface to the right of c', the distance to be laid off on the slope line bS is proportional to the weight of the sand that constitutes the wedge, plus the line load q' (Fig. 30.2).

If the ground surface carried no surcharge, the Culmann line C (dash curve) in Fig. 30.2 would correspond to the line C in Fig. 30.1c. If the surface carries a line load q' at some point c', the Culmann line consists of two sections. The section to the left of plane bc' is identical with C, because the wedges bounded by planes to the left of bc' carry no surcharge. On the right side of bc' the Culmann line for the loaded backfill is located above C, as indicated by the solid curve C' in Fig. 30.2, because every wedge bounded by a plane to the right of bc' is acted on by the weight q'. Therefore the complete Culmann line consists of the curve C to the left of bc', and the curve C' to the right. It has a discontinuity at the plane bc' which passes through the point of application of the line load.

If the load is located on the left side of c_2', the value of the active earth pressure exerted by the loaded backfill corresponds to the greatest distance between the Culmann line C' and the slope line bS measured in a direction parallel to the earth pressure line bL. If the line load acts at any position on the surface of the fill between points a and c'', the greatest distance is $d''e''$. Therefore, the slip occurs along the plane bc'' which passes through e''. The quantity $d''e''$–de represents that part ΔP_A of the active earth pressure due to the line load q'.

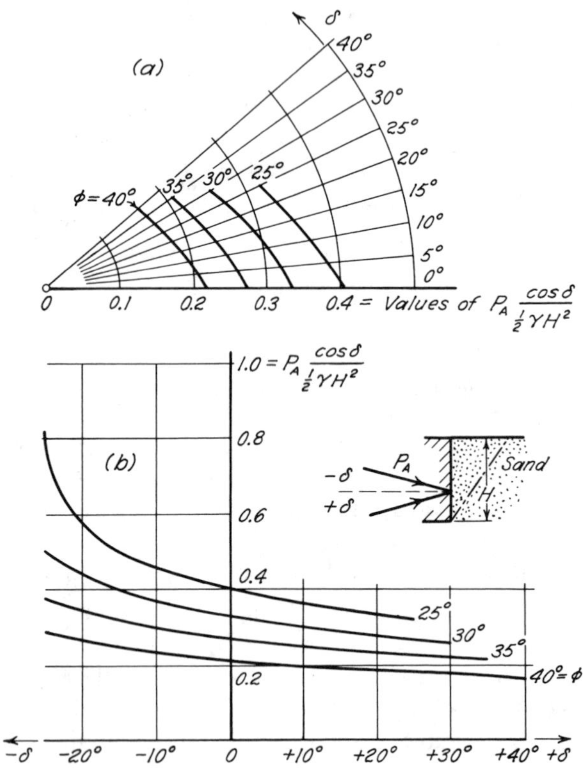

Fig. 30.3. Two types of charts that furnish coefficients for computation of active earth pressure (*a* after Syffert 1929).

The ordinates of the curve K (Fig. 30.2) with reference to the ground surface represent the values of ΔP_A corresponding to the various positions c' at which q' may be located. Between a and c'', K is straight and parallel to the surface of the backfill because ΔP_A is independent of the position of q' between these two points.

If q' moves to the right beyond c'' to such a position as c, the Culmann line consists of the dash curve C to the left of bc and the solid curve C' to the right. The maximum value P_A of the earth pressure is represented by the line e_3d_3. The failure plane passes through the point e_3 and intersects the ground surface at the line of application of q'. As the line of application of q' moves to the right, the value of ΔP_A decreases steadily as indicated by the ordinates of curve K (Fig. 30.2) until at c_2' it becomes equal to zero.

Finally, if the line of action of q' is at c_2', the value of the earth pressure $e_2'd_2'$, determined by means of curve C', is equal to the value ed that represents the active earth pressure when there is no surcharge. If q' moves to the right of c_2', the earth pressure determined by means of C' becomes smaller than ed. Hence, if the line load acts on the right side of c_2', it no longer has any effect on the active earth pressure, and the surface of sliding has the same position bc as it does for a backfill that carries no surcharge. The greater the line load q', the farther c_2' moves to the right. Therefore, the distance within which q' has an influence on the earth pressure depends on the magnitude of q'.

Culmann's method is chiefly used if the wall has an inclined or broken back, and if the backfill has an irregular surface or carries a surcharge. If a vertical wall supports a cohesionless backfill with a horizontal surface, it is more expedient to obtain the value of P_A from charts prepared for this purpose. Figure 30.3 consists of two different charts of this kind.

Problems

1. A vertical retaining wall 20 ft high supports a cohesionless fill that weighs 115 lb/ft^3. The upper surface of the fill rises from the crest of the wall at an angle of 20° with the horizontal. The angle of internal friction is 28°, and the angle of wall friction is 20°. By Culmann's method compute the total active earth pressure against the wall.

Ans. 10,600 lb/lin ft.

2. The stem of a cantilever retaining wall is 36 ft high. It retains a storage pile of cohesionless iron ore. The wall has a cross section symmetrical about its vertical center line. At the top its width is 6 ft, and at the base of the stem is 12 ft. From a point on the back of the wall 4 ft below the crest, the surface of the ore pile rises at an angle of 35° with the horizontal to a maximum height of 65 ft above the base of the stem. The remainder of the pile is level. If ϕ and δ are each equal to 36° and γ is 160 lb/ft³, what is the total lateral pressure of the ore above the base of the stem? If the entire lateral force against the stem is resisted by steel tie rods 3 in. square, stressed to 27,000 lb/in.², what spacing of the rods is required?

Ans. 48,800 lb/lin ft; 5 ft.

3. A vertical wall 18 ft high supports a cohesionless fill weighing 105 lb/ft³. The surface of the fill is horizontal. The values of ϕ and δ are 31° and 20°, respectively. The fill supports two line loads of 2000 lb/lin ft, parallel to the crest of the wall, at distances of 8 and 13 ft, respectively. Compute the value of the total active earth pressure against the wall. Determine the horizontal distance from the back of the wall to the point at which the surface of sliding intersects the surface of the fill.

Ans. 6310 lb/lin ft; 13 ft.

4. A retaining wall 15 ft high with a vertical back is just adequate to support a level fill of sand having a unit weight of 115 lb/ft³ and a value of ϕ equal to 32°. The value of δ is 20°. A vertical load of 5000 lb/lin ft is to be added along a line parallel to the crest of the wall. What is the minimum horizontal distance at which the load can be located from the back of the wall without increasing the earth pressure against the wall?

Ans. 16.2 ft.

5. If the fill in problem 3 carries no surcharge, what is the magnitude of the active earth pressure? Check the graphical computation by means of the charts, Fig. 30.3.

Ans. 4870 lb/lin ft.

Selected Reading

Details of calculation of earth pressure by the graphical trial wedge method, identical in principle to Culmann's method, are given for a variety of conditions and for materials possessing both c and ϕ by Huntington, W. C. (1957): *Earth pressures and retaining walls,* New York, John Wiley and Sons, 534 pp.

ART. 31 POINT OF APPLICATION OF EARTH PRESSURE

The procedure described in Article 30 makes it possible to determine the magnitude of the total earth pressure provided its direction is known. However, it does not furnish any information regarding the point of application of the pressure. In order to get this information, Coulomb assumed that every point on the back of a wall represents the foot of a potential surface of sliding. For example, the point d on the curved line ab in Fig. 31.1a represents the lower extremity of a potential surface of sliding de. The earth pressure P_A on ad can be computed by means of Culmann's procedure as described in Article 30. If the depth to the foot of the potential surface of sliding is increased from z to $z + dz$, the earth pressure is increased by

$$dP_A = p_A\, dz$$

where p_A is the average intensity of pressure over the increment of depth dz. Therefore,

$$p_A = \frac{dP_A}{dz} \tag{31.1}$$

By means of this equation the distribution of the earth pressure on the back of the wall can be determined. When the distribution is known, the point of application of the resultant pressure can be located by means of a suitable analytical or graphical method. At any point the line of action of the pressure p_A makes an angle δ with the normal to the back of the wall.

In practice, this method is rather cumbersome. Therefore, simplified methods are used that give approximately the same results. For instance, in Fig. 31.1a, the point of application O_1 is located approximately at the point of intersection of the back of the wall and a line OO_1, which

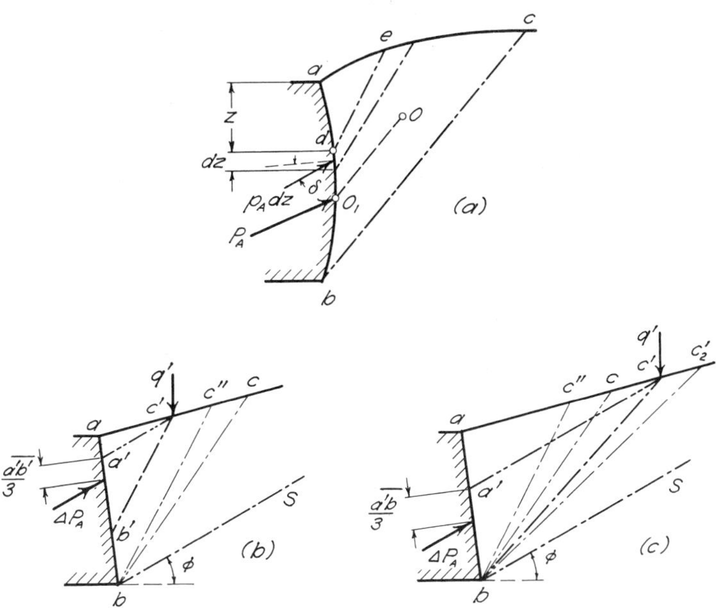

Fig. 31.1. Diagrams illustrating simplified procedure for determining point of application of active earth pressure.

is parallel to the surface of sliding bc and which passes through the center of gravity O of the sliding wedge abc.

Figures 31.1*b* and *c* illustrate a simplified method for estimating the position of the point of application of the additional pressure ΔP_A produced by a line load q'. The lines bc, bc'', etc., correspond to the lines bc, bc'', etc., in Fig. 30.2. If q' acts between a and c'' (Fig. 31.1*b*), $b'c'$ is traced parallel to the surface of sliding bc'', and $a'c'$ is traced parallel to bS, the slope line (see Fig. 30.2). The force ΔP_A acts at the upper third-point of $a'b'$. If q' acts between c'' and c_2', $a'c'$ is traced parallel to bS, and ΔP_A acts at the upper third-point of $a'b$, as shown in Fig. 31.1*c*.

All these procedures are based on Coulomb's assumption that every point on the back of a wall represents the foot of a potential surface of sliding. This assumption is justified in connection with retaining walls, because most such walls can hardly fail without yielding in a manner that satisfies the deformation condition for the plastic state. Coulomb, however, did not specify this deformation condition. Consequently, the theory was commonly used for computing the active earth pressure against lateral supports that did not satisfy the deformation condition, such as the bracing in open cuts (Article 37). Since it was found that the results of the computations did not agree with reality, many experienced engineers concluded that the theory as such was unreliable. Therefore, it should be emphasized that Coulomb's theory is as satisfactory as any theory in structural engineering, provided the deformation condition for its validity is satisfied.

Problems

1. At what distance above the base of the stem of the retaining wall of problem 2, Article 30, does the resultant earth pressure act?

Ans. 10.8 ft.

2. Locate the center of pressure of the added earth pressure due to each of the two line loads in problem 3, Article 30, on the assumption that the influence of each of the two loads can be considered separately.

Ans. 10 ft; 6.75 ft from the bottom of the wall.

ART. 32 PASSIVE EARTH PRESSURE AGAINST ROUGH CONTACT FACES

Definition

In the broadest sense, the term passive earth pressure indicates the resistance of a mass of soil against displacement by lateral pressure. The object that exerts the lateral pressure may consist of the foundation of a retaining wall, the outer face of the buried part of a sheet pile bulkhead, or a block of masonry such as the abutment of a loaded arch. It may also consist of a mass of soil that exerts a horizontal pressure because it supports a vertical load. The soil beneath a loaded footing acts in this manner. Since the stability of almost any lateral earth support and the bearing capacity of every shallow foundation depend to some extent on the passive earth pressure, the problem of computing this pressure is of outstanding practical importance.

The surface of contact between the soil and the object that exerts the lateral pressure is called the *contact face*. Coulomb computed the passive earth pressure against rough contact faces on the simplifying assumption that the surface of sliding is plane (Fig. 32.1*a* and *b*). The error due to this assumption is always on the unsafe side. If the angle of wall friction δ is small, the surface of sliding is really almost plane and the error is tolerable. However, if δ is large, the error is excessive, and Coulomb's method should not be used.

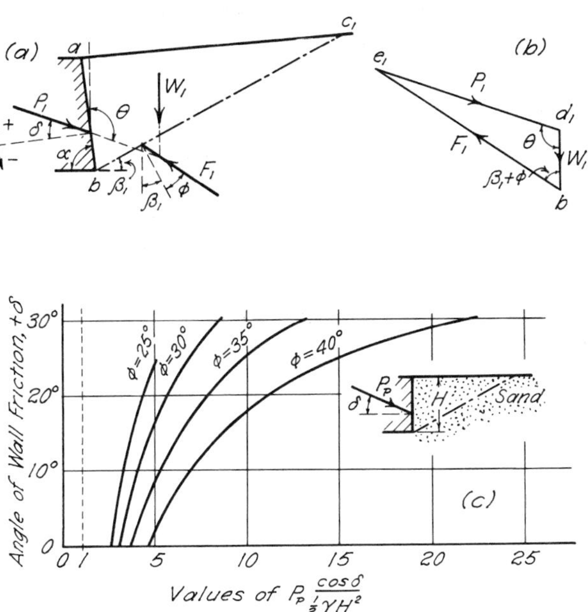

Fig. 32.1. **(*a* and *b*) Diagrams illustrating assumptions on which Coulomb's theory of passive earth pressure is based. (*c*) Chart that furnishes coefficients for computation of passive earth pressure.**

Coulomb's Theory of the Passive Earth Pressure of Sand

The Coulomb value of the passive earth pressure can be determined graphically by Culmann's method. The procedure is identical with that described in Article 30 except that the slope line bS (Fig. 30.1c) must be drawn at an angle ϕ below the horizontal instead of above.

Figure 32.1c shows the influence of the angle of wall friction δ on the Coulomb value of the passive earth pressure. According to this chart, the earth pressure increases rapidly with increasing values of the angle of wall friction. However, if δ is greater than about $\phi/3$, the surface of sliding is strongly curved (Fig. 29.1c). As a consequence, the error due to Coulomb's assumption of a plane surface increases rapidly. For $\delta = \phi$ it may be as great as 30%. Hence, for values of δ greater than $\phi/3$, the curvature of the surface of sliding must be taken into consideration.

Passive Earth Pressure of Cohesive Soils

In order to illustrate the methods for determining the passive earth pressure without assuming a plane surface of sliding, the problem illustrated by Fig. 32.2 will be solved. In this figure ab is a section through a contact face that is pushed toward a mass of ideal cohesive soil. The shearing resistance of the soil is determined by Eq. 16.4

$$s = c + p \tan \phi$$

The surface of the soil is horizontal. The angle of wall friction is denoted by δ, and the total adhesion between the soil and the contact face by C_a. The real surface of sliding is bde. It consists of a curved part bd and a straight part de. According to Article 29, the soil within the isosceles triangle ade is in the passive Rankine state. Therefore, the shearing

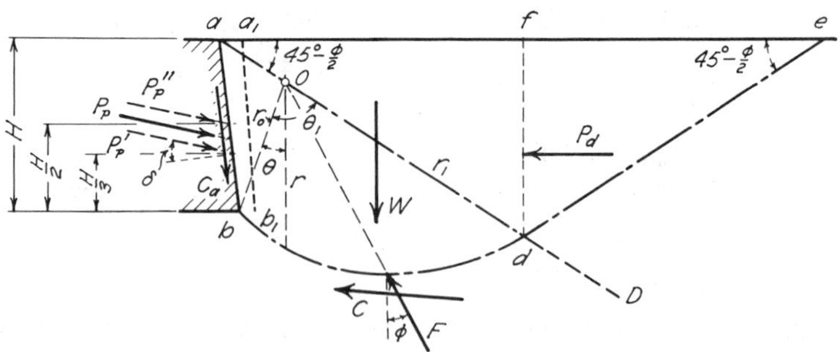

Fig. 32.2. Diagram illustrating assumptions on which theory of passive earth pressure against rough contact faces is based.

stresses on the vertical section df are zero, and the pressure P_d on this section is horizontal. It can be computed by means of Eq. 28.17. The body of soil $abdf$ is acted on by the following forces: its weight W; the pressure P_d; the resultant C of the cohesion along bd; the adhesion C_a along ab; the resultant F of the normal and frictional stresses along bd; and the resultant P_P of the normal and frictional components of the passive earth pressure.

Since the point of application of P_P is not known, we make use of the approximations discussed at the end of Article 28 and replace P_P by the two forces, P_P' and P_P''. Each of these forces acts at an angle δ with the normal to the contact face. One force P_P' maintains equilibrium with the weight of the mass $abdf$ and the friction due to the weight. The other, P_P'', maintains equilibrium with the cohesion on the surface of sliding and the friction due to forces other than the weight. The force P_P' acts at the lower third-point of ab, whereas P_P'' acts at the midpoint. Since the point of application and the direction of each of these forces are known, we may compute each force individually. The resultant of these two forces represents the total passive earth pressure P_P.

The methods for determining the real shape of the surface of sliding are so involved that they are unsuitable for practical purposes. However, sufficiently accurate results can be obtained on the simplifying assumption that the curved portion bd of the real surface of sliding is either an arc of a circle or a logarithmic spiral that has the equation,

$$r = r_0 \epsilon^{\theta \tan \phi} \tag{32.1}$$

In the following paragraphs, the curved part of the surface of sliding is assumed to be a logarithmic spiral. Since the spiral is tangent at d to the straight part de of the surface of sliding, the center O of the spiral must be located on the line aD (Fig. 32.2) which is inclined at $45° - \phi/2$ to the horizontal. According to Eq. 32.1, every radius of the spiral makes an angle ϕ with the normal to the spiral at the point where it intersects the curve. Since ϕ is the angle of internal friction, the resultant dF of the normal stress and the frictional resistance on any element of the surface of sliding also makes an angle ϕ with the normal to the element, and its direction coincides with that of the radius that subtends the element. Since every radius of the spiral passes through point O, the resultant F of the normal and frictional forces on bd also passes through the center O. This fact is utilized in the following calculations.

In order to compute P_P' (the value of P_P if $c = 0$), we arbitrarily select a surface of sliding bd_1e_1 (Fig. 32.3*a*) consisting of the logarithmic spiral bd_1 with its center at O_1, and the straight line d_1e_1 which makes an angle of $45° - \phi/2$ with the horizontal. The lateral pressure required to produce a slip on this surface is designated as P_1'. We then evaluate the force P_{d_1}', which acts at the lower third-point of f_1d_1, by means of the equation,

$$P_{d_1}' = \tfrac{1}{2}\gamma H_{d_1}^2 N_\phi$$

Finally, we take moments of the forces P_1', P_{d_1}', W_1, and F_1' about O_1. Since the moment of F_1' about O_1 is zero,

$$P_1'l_1 = W_1l_2 + P_{d_1}'l_3$$

from which

$$P_1' = \frac{1}{l_1}[W_1l_2 + P_{d_1}'l_3] \tag{32.2}$$

The value of P_1' is plotted to scale above f_1. It is represented by the point C_1'. Similar computations are performed for other arbitrarily selected surfaces of sliding, and a curve P' is drawn through the points C_1', etc. If the soil has no cohesion ($c = 0$), the second component P_P'' of the passive earth pressure P_P is equal to zero, and the value of P_P is represented by the minimum ordinate of the curve P', at point C'. The surface of sliding passes through the point d which is located on aD vertically below C'.

If the soil possesses cohesion, we must also compute P_P'' (the value of P_P if $\gamma = 0$). In order to calculate the value P_1'' which corresponds to the arbitrary surface of sliding bd_1e_1, we must consider the forces involved in the computation (see Fig. 32.3*b*). The value of P_{d_1}'' is obtained by making $q = 0$ and $H = H_{d_1}$ in Eq. 28.16. Hence,

$$P_{d_1}'' = 2cH_{d_1}\sqrt{N_\phi}$$

The point of application of this force is at the mid-height of d_1f_1. The influence of the cohesion along the curve bd_1 may be evaluated by considering an element having a length ds (Fig. 32.3*c*). The cohesion along the length ds is equal to $c\,ds$. The moment of $c\,ds$ about O_1 is

$$dM_c = rc\,ds\cos\phi = rc\frac{r\,d\theta}{\cos\phi}\cos\phi = cr^2\,d\theta$$

and the moment of the total cohesion along bd_1 is

$$M_{c_1} = \int_0^{\theta_1} dM_c = \frac{c}{2\tan\phi}(r_1^2 - r_0^2) \tag{32.3}$$

The force F_1'' passes through O_1. By taking moments about this point, we obtain

$$P_1''l_1 = M_{c_1} + P_{d_1}''l_3 - C_al_4$$

from which

$$P_1'' = \frac{1}{l_1}[M_{c_1} + P_{d_1}''l_3 - C_a l_4] \tag{32.4}$$

In Fig. 32.3*a* the value P_1'' is plotted to scale at C_1 above point C_1'. Since P_1' and P_1'' represent the forces required to overcome the two parts of the resistance against sliding along the same surface bd_1e_1, the

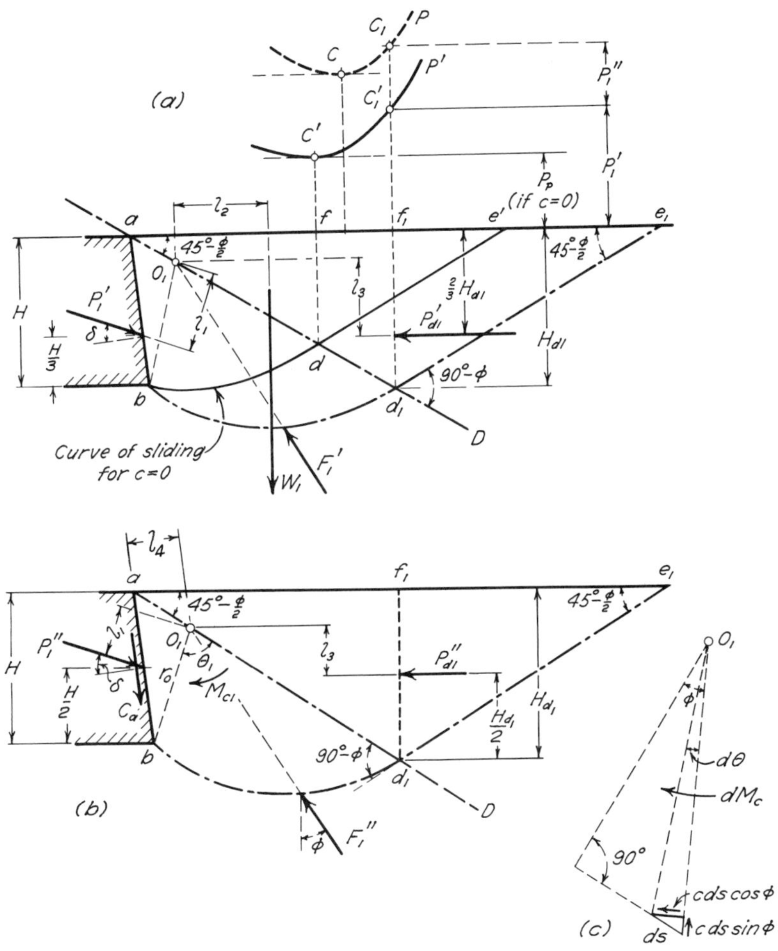

Fig. 32.3. Logarithmic spiral method for determining passive earth pressure. (*a*) Forces entering into computation of component due to weight of soil, neglecting cohesion. (*b*) Forces entering into computation of component due to friction and cohesion, neglecting weight of soil. (*c*) Diagram illustrating computation of moment due to cohesion.

ordinate of point C_1 represents the total force required to produce a slip along this surface. Similarly, values of P'' are obtained for other arbitrary surfaces of sliding, and a curve P is drawn through the points C_1, etc. The passive earth pressure P_P is represented by the minimum ordinate of the curve P, and the surface of sliding passes through a point on aD directly below the point C at which curve P is closest to ae_1. The total force against the contact face is equal to the resultant of P_P and the adhesion force C_a.

The shape of the curved part of the real surface of sliding is intermediate between that of an arc of a circle and that of a spiral. Since the difference between the shapes of these two curves is small, the error due to replacing the real curve by either a circle or a logarithmic spiral is negligible. As a matter of fact, comparisons between the approximate and the exact methods have shown that values of the passive earth pressure computed by means of the approximate methods are at least as accurate as values of the active earth pressure computed by Coulomb's method in which it is assumed that the real slightly curved surface of sliding is a plane.

The preceding investigations are based on the assumption that the mass of soil adjoining the contact face is pushed into a position located entirely beyond a_1b_1 (Fig. 32.2). If the upper part of the contact face does not advance so far as a_1b_1, the surface of sliding is curved throughout its full length, and only the lowest part of the sliding mass passes into the passive Rankine state. If the lower part of the face stops short of a_1b_1, the soil adjoining this part does not pass into a state of plastic equilibrium at all. In these instances, the total passive earth pressure and its distribution over the contact face depend on the type of restriction imposed on the movement of the contact face.

Problems

1. Construct a logarithmic spiral for $\phi = 36°$. The value of r_0 should be taken as 1 in., with values of θ ranging from $-30°$ to $270°$.

2. Compute by the logarithmic-spiral method the total passive earth pressure against a vertical face in contact with a sand fill having a level surface. The contact face is 20 ft high, and the angle of wall friction is $+20°$. The fill has a unit weight of 112 lb/ft^3 and an angle of internal friction of 36°. In order to facilitate use of the spiral constructed in problem 1, the graphical solution should be laid out on tracing paper. Use the scale 1 in. = 10 ft.

Ans. 175,000 lb/lin ft.

3. Compute the value of passive earth pressure in problem 2, assuming a plane surface of sliding.

Ans. 200,000 lb/lin ft.

4. Compute the passive earth pressure against the contact surface of problem 2 if, in addition to frictional resistance, the soil possesses a cohesion of 500 lb/ft^2. The adhesion between the soil and the contact face is also 500 lb/ft^2. Locate the resultant earth pressure P_P.

Ans. 255,000 lb/lin ft; 8 ft above base.

Selected Reading

Tables of active and passive earth pressures, for materials possessing friction, cohesion, or both, and for all angles of wall friction, are contained in Caquot, A. and J. Kerisel (1948): *Tables for the calculation of passive pressure, active pressure and bearing capacity of foundations.* Translated from the French by Maurice A. Bec., Paris, Gauthier-Villars, 120 pp.

ART. 33 BEARING CAPACITY OF SHALLOW FOOTINGS

Fundamental Assumptions

When a load is applied on a limited portion of the surface of a soil, the surface settles. The relation between the settlement and the average load per unit of area may be represented by a *settlement curve* (Fig. 33.1). If the soil is fairly dense or stiff, the settlement curve is similar to curve C_1. The abscissa q_d of the vertical tangent to the curve represents the *bearing capacity* of the soil. If the soil is loose or fairly soft, the settlement curve may be similar to C_2, and the bearing capacity is not always well defined. The bearing capacity of such soils is commonly assumed to be equal to the abscissa q_d' of the point at which the settlement curve becomes steep and straight.

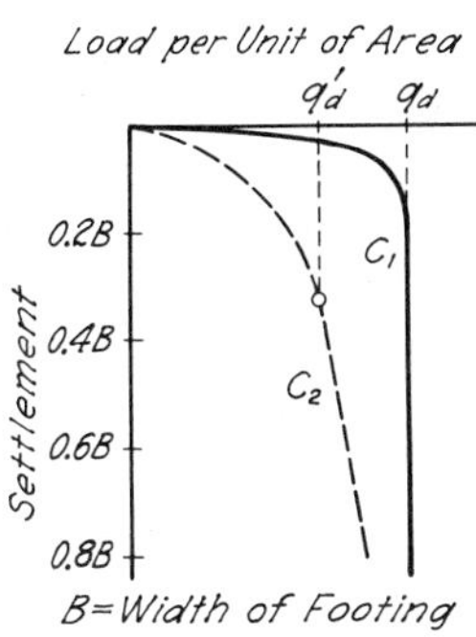

Fig. 33.1. Relation between intensity of load and settlement of a footing on C_1 dense or stiff and C_2 loose or soft soil.

In practice, loads are transmitted to the soil by means of footings, as shown in Fig. 33.2. The footings may be *continuous*, having a long rectangular shape, or they may be *spread footings*, which are usually square or circular. The *critical load* is the load per unit of length of a continuous footing or the total load on a spread footing at which the soil support fails. The distance from the level of the ground surface to the base of the footing is known as the *depth of foundation* D_f. A footing that has a width B equal to or greater than D_f is considered a *shallow footing*. In computations dealing

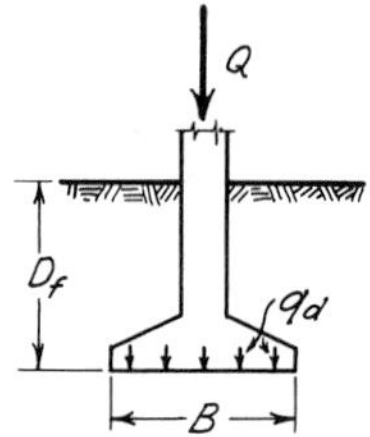

Fig. 33.2 Section through continuous shallow footing.

with shallow footings, the weight of the soil above the base level of the foundation may be replaced by a uniform surcharge,

$$q = \gamma D_f \tag{33.1}$$

This substitution simplifies the computations. The error involved is small and on the safe side.

States of Plastic Equilibrium Beneath Shallow Continuous Footings

The results of mathematical investigations concerning the state of plastic equilibrium beneath continuous footings are not fully satisfactory. No general solution has been found that rigorously satisfies Eq. 16.5 and also takes into account the weight of the soil, the influence of the depth of surcharge D_f, and the real distribution of vertical and horizontal forces on the base of the footing. Furthermore, the existing theories commonly assume that the volume of the soil does not change as the critical load is approached. Yet these shortcomings are not of serious practical importance because the accuracy of even approximate solutions is limited by our ability to evaluate the appropriate physical properties of the soil that enter into the equations rather than by the defects in the theories themselves.

The following general conclusions can be drawn from the theoretical studies. If the base of a continuous footing rests on the surface of a weightless soil possessing cohesion and friction, the loaded soil fails, as shown in Fig. 33.3*a*, by plastic flow along the composite surface $fede_1f_1$. This region can be divided into five zones, one marked *I* and two pairs of zones marked *II* and *III*. Because of friction and adhesion between the soil and the base of the footing, zone *I* remains in an elastic state. It acts as if it were part of the footing and penetrates the soil like a wedge. Its boundaries rise at an angle of $45° + \phi/2$ with the horizontal. In zones *II* and *III* shear patterns develop. Those in zone *III* are identical with that for the passive Rankine state (Article 27); the boundaries for the passive Rankine zone rise at $45° - \phi/2$ with the horizontal. The zones *II* located between *I* and *III* are known as *zones of radial shear*, because the lines that constitute one set in the shear pattern in these zones radiate from the outer edge of the base of the footing. These lines are straight. The lines of the other set are logarithmic spirals with their centers located at the outer edge of the base of the footing. The bearing capacity per unit of area is found to be (Prandtl 1921)

$$q_d = N_c\, c \tag{33.2}$$

where c is the cohesion and N_c, known as a *bearing capacity factor*, depends only on ϕ. It is equal to

$$N_c = \cot \phi \left[\epsilon^{\pi \tan \phi} \tan^2 \left(45° + \frac{\phi}{2} \right) - 1 \right] \tag{33.3}$$

If the surface of the ground is acted upon by a uniformly distributed surcharge q the shear pattern remains the same and the bearing capacity is increased by an amount $N_q\, q$ (Reissner 1924) where

$$N_q = \epsilon^{\pi \tan \phi} \tan^2 \left(45° + \frac{\phi}{2} \right) \tag{33.4}$$

whence

$$N_c = \cot \phi (N_q - 1) \tag{33.5}$$

If $\phi = 0°$ the spirals become arcs of circles and the corresponding values of N_c and N_q become $(2 + \pi)$ and 1.0 respectively. Hence, for a footing at the ground surface

$$q_d = (2 + \pi)c = 5.14c = 2.57q_u \tag{33.6}$$

where q_u is the unconfined compressive strength. Moreover, for $\phi = 0°$, the shear pattern and Eq. 33.6 remain valid even if the weight of the soil is not zero.

The right side of Fig. 33.3*a* shows the deformations of the soil located within the zones of plastic flow. The soil in zones *III* is compressed laterally. Its surface rises and terminates at the side of the footing in a sharp edge that conveys the impression that the soil has been punched.

If the soil is cohesionless but possesses friction and weight, the shear pattern is represented by Fig. 33.3*b*. The boundaries of the elastic zone *I* are curved. The two branches intersect at d at an angle of $90° - \phi$ and the boundaries de and de_1 of the zones *II* merge smoothly at d into the boundaries of zone *I*. In zone *II* the radial lines are curved. In zone *III* the shear pattern again corresponds exactly to that of the passive Rankine state. A rigorous general solution for the bearing capacity under these conditions has not been obtained but solutions for particular cases are available (Lundgren and Mortensen 1953).

Approximate Methods for Computing the Bearing Capacity of Continuous Footings

Real soils possess weight and in general exhibit both cohesion and friction. Moreover, the bases of most footings are located at least a short distance below the surface of the surrounding ground. Rigorous

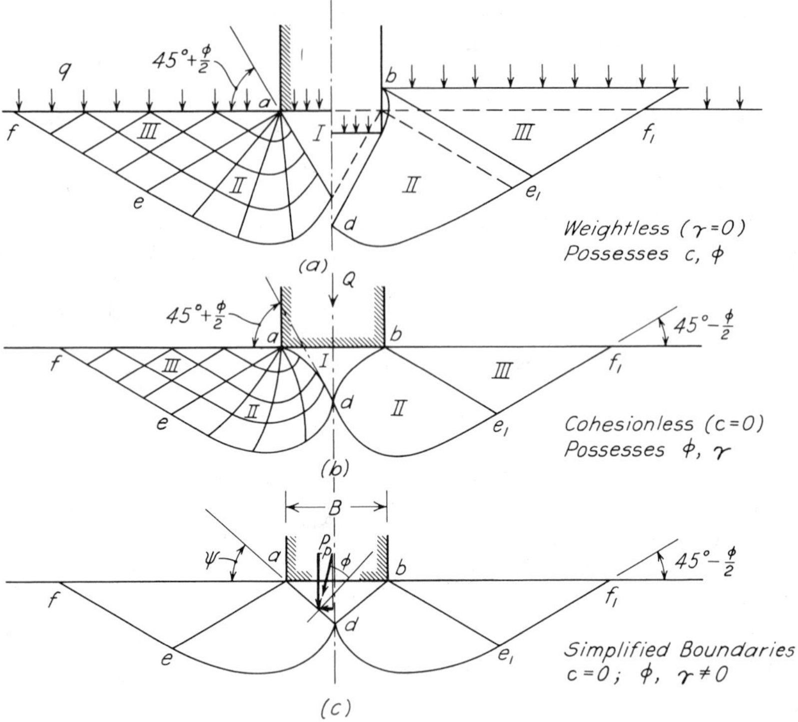

Fig. 33.3. Boundaries of zone of plastic equilibrium after failure of soil beneath continuous footing.

methods are not available for computing the bearing capacity under these circumstances, but for practical purposes no other than approximate methods are needed.

When the bearing capacity of a real footing is exceeded, the soil fails along a surface of rupture similar to those indicated by $fede_1f_1$ (Fig. 33.3). The actual surface is not likely, however, to coincide with either of those in Fig. 33.3 corresponding to ideal materials. In the approximate methods it is assumed that the bearing capacity consists in general of the sum of three components, computed separately, representing respectively the contributions of (1) the cohesion and friction of a weightless material carrying no surcharge, (2) the friction of a weightless material upon addition of a surcharge q on the ground surface, and (3) the friction of a material possessing weight and carrying no surcharge. Each component is computed on the assumption that the surface of sliding corresponds to the conditions for that particular component.

Since the surfaces differ from each other and from the surface for the real material, the result is an approximation. The error is small and on the side of safety.

The approximate value of the bearing capacity is given by the equation

$$q_d = cN_c + \gamma D_f N_q + \tfrac{1}{2}\gamma B N_\gamma \tag{33.7}$$

in which N_c and N_q are the bearing capacity factors with respect to cohesion and surcharge, respectively. They can be evaluated by means of Eqs. 33.5 and 33.4. The surcharge is represented by the weight per unit area γD_f of the soil surrounding the footing. The bearing capacity factor N_γ accounts for the influence of the weight of the soil. All the bearing capacity factors are dimensionless quantities depending only on ϕ.

Since a theoretical solution is not available for evaluating N_γ an approximate procedure is used. In this procedure the curved boundaries *ad* and *bd* of the elastic zone *abd* (Fig. 33.3*b*) are replaced by straight lines (Fig. 33.3*c*) rising at angles ψ to the horizontal. The unit weight of the soil is γ. At the instant of failure, the pressure on each of the surfaces *ad* and *bd* is equal to the passive earth pressure P_P. Since slip occurs along these faces, the resultant earth pressure acts at an angle ϕ to the normal on each face. If the weight of the soil within *adb* is disregarded, the equilibrium of the footing in the vertical direction requires that

$$Q = 2P_P \cos(\psi - \phi)$$

The average vertical pressure, corresponding to the average bearing capacity, is then

$$q_\gamma = \frac{Q}{B} = \frac{2P_P}{B} \cos(\psi - \phi) \tag{33.8}$$

The problem, therefore, is reduced to determining the passive earth pressure P_P (Article 32). The point of application of P_P is located at the lower third-point of *ad*. By introducing the symbol

$$N_\gamma = \frac{4P_P}{\gamma B^2} \cos(\psi - \phi) \tag{33.9}$$

into Eq. 33.8, we obtain

$$q_\gamma = \tfrac{1}{2}\gamma B N_\gamma \tag{33.10}$$

the third term in Eq. 33.7. Since the bearing capacity factor N_γ is dimensionless and depends only on ϕ, values can be computed once for all by the methods explained in Article 32. However, the inclination ψ

is not known. Hence, the computations must be repeated for a given value of ϕ with various inclinations ψ until the minimum value of N_γ is found. The results are conservative but agree well with those calculated for particular cases by the more advanced procedures (Meyerhof 1955). Meyerhof's values are plotted in the chart (Fig. 33.4) together with the values of N_c and N_q obtained from Eqs. 33.5 and 33.4. The use of the chart greatly facilitates the computation of the bearing capacity.

The soil does not fail as shown in Fig. 33.3*c* unless it is fairly dense or stiff, so that its settlement curve resembles C_1 in Fig. 33.1. Otherwise the footing sinks into the ground before the state of plastic equilibrium spreads beyond e and e_1 (Fig. 33.3) and the corresponding settlement curve has no well-defined break (curve C_2 in Fig. 33.1). An approximate value for the bearing capacity q_d of continuous footings on such soils can be obtained by assuming that the cohesion and friction of the soil are equal to two-thirds of the corresponding values in Coulomb's equation, or that

$$c' = \tfrac{2}{3}c \tag{33.11a}$$

and

$$\tan\phi' = \tfrac{2}{3}\tan\phi \tag{33.11b}$$

If the angle of shearing resistance is ϕ' instead of ϕ, the bearing-capacity factors assume values N_c', N_q', and N_γ'. These values are

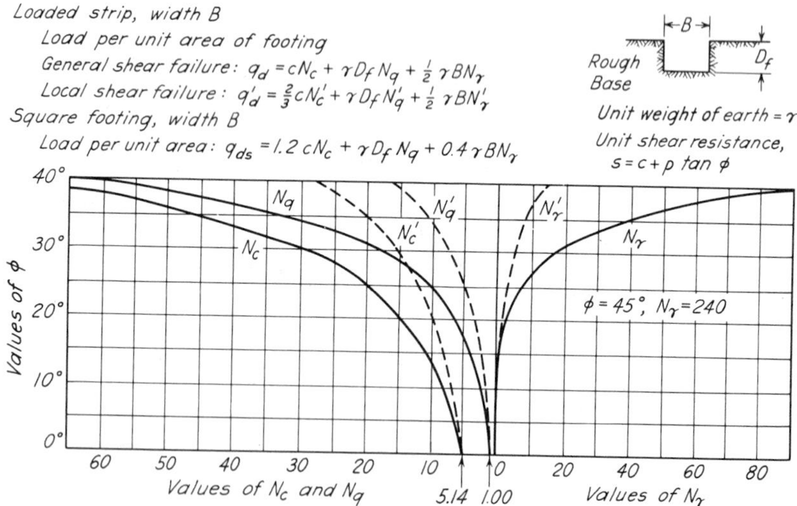

Fig. 33.4. Chart showing relation between ϕ and bearing capacity factors (values of N_γ after Meyerhof 1955).

given by the dash curves in Fig. 33.4. The bearing capacity is then obtained from the equation

$$q_d' = (\tfrac{2}{3}cN_c' + \gamma D_f N_q' + \tfrac{1}{2}\gamma B N_\gamma') \tag{33.12}$$

Experience has shown that even uniformly loaded foundations always fail by tilting. This fact, however, does not invalidate the reasoning in the preceding paragraphs. It merely demonstrates that there are no perfectly uniform subgrades. With increasing load the settlement above the weakest part of the subgrade increases more rapidly than that above the rest. Because of the tilt, the center of gravity of the structure shifts toward the weak part and increases the pressure on that part, whereas the pressure on the stronger parts decreases. These factors almost exclude the possibility of a failure without tilting.

Bearing Capacity of Footings of Finite Length

All the preceding discussions refer to continuous footings. For computing the bearing capacity of square or circular bases, only a few special cases have been solved rigorously; the solutions require the use of numerical procedures. On the basis of the results and of experiments a semiempirical equation has been developed for the bearing capacity q_{dr} per unit of area of a circular footing with radius r resting on a fairly dense or stiff soil

$$q_{dr} = 1.2cN_c + \gamma D_f N_q + 0.6\gamma r N_\gamma \tag{33.13}$$

The corresponding value for square footings, $B \times B$, on dense or stiff soil, is

$$q_{ds} = 1.2cN_c + \gamma D_f N_q + 0.4\gamma B N_\gamma \tag{33.14}$$

The values of N are given by the ordinates of the solid curves in Fig. 33.4.

If $\phi = 0$ conditions prevail and the soil possesses cohesion, the bearing capacity at the ground surface is

$$q_{dr} = q_{ds} = 6.2c = 3.1q_u \tag{33.15}$$

which is considerably greater than the value $q_d = 5.14c$ (Eq. 33.6). On the other hand, if $c = 0$ and $D_f = 0$, the bearing capacity q_{dr} per unit of area is considerably smaller than q_d for a continuous footing with a width equal to the diameter of the circular footing.

If $\phi = 0$ and $c > 0$, the increase in bearing capacity per unit of area produced by the surcharge γD_f is exactly compensated by the weight of soil removed for construction of the footing. Therefore it is convenient to deal with the *net bearing capacity*

$$q_{d\,\text{net}} = q_d - \gamma D_f \tag{33.16}$$

In reality, because the strength of the clay above footing level is not actually zero, the net bearing capacity increases slightly with increasing values of D_f. For values of D_f/B not exceeding 2.5, Skempton (1951) has proposed the following simple expression for the net bearing capacity of a rectangular footing with width B and length L

$$q_{d\ \mathrm{net}} = 5c\left(1 + 0.2\frac{D_f}{B}\right)\left(1 + 0.2\frac{B}{L}\right) \tag{33.17}$$

It is apparent that the value of N_c has been rounded conservatively from 5.14.

If the supporting soil is fairly loose or soft, the values of N must be replaced by the values N', determined from the dash curves in Fig. 33.4, and the value of c must be replaced by c' (Eq. 33.11a).

Problems

1. Compute the bearing capacity per unit of area of a continuous footing 8 ft wide, supported on a soil for which $c = 400$ lb/ft^2, $\phi = 17°$, and $\gamma = 120$ lb/ft^3. The load-settlement curve resembles C_1 in Fig. 33.1, and the relation between normal stress and shearing resistance is $s = c + p \tan \phi$. The depth of foundation is 6 ft.

Ans. 9,200 lb/ft^2.

2. Compute the bearing capacity per unit of area of a footing 10 ft square on dense sand ($\phi = 37°$), if the depth of foundation is, respectively, 0, 2, 5, 10, and 15 ft. The unit weight of the soil is 126 lb/ft^3.

Ans. 26,000; 36,000; 53,000; 79,000; 106,000 lb/ft^2.

3. A load test was made on a bearing plate 1 ft square on the surface of a cohesionless deposit of sand having a unit weight of 110 lb/ft^3. The load-settlement curve approached a vertical tangent at a load of 4000 lb. What was the value of ϕ for the sand?

Ans. 39.5°.

4. A load test was made on a plate 1 ft square on a dense cohesionless sand having a unit weight of 115 lb/ft^3. The bearing plate was enclosed in a box surrounded by a surcharge 2 ft deep. Failure occurred at a load of 12,000 lb. What would be the failure load per unit of area of the base of a footing 5 ft square located with its base at the same depth in the same material?

Ans. 22,000 lb/ft^2.

5. A structure was built on a mat foundation 100 ft square. The mat rested at the ground surface on a stratum of uniform soft clay which extended to a depth of 150 ft. If failure occurred at a uniformly distributed load of 4500 lb/ft^2, what was the average value of c for the clay? Because of the great depth of the zone of plastic equilibrium, the consolidation of the clay prior to failure can be disregarded.

Ans. 730 lb/ft².

Selected Reading

Meyerhof, G. G. (1951). "The ultimate bearing capacity of foundations," *Géot.*, **2**, pp. 301–332. Approximate theoretical solutions for shallow and deep foundations, supplemented by model tests.

Skempton, A. W. (1951). "The bearing capacity of clays," *Proc. British Bldg. Research Congress*, **1**, pp. 180–189. Discussion for $\phi = 0$ condition; influence of compressibility on bearing capacity.

Meyerhof, G. G. (1955). "Influence of roughness of base and ground-water conditions on the ultimate bearing capacity of foundations," *Géot.*, **5**, pp. 227–242. Review of 1951 paper in light of subsequent developments.

Sokolovski, V. V. (1960). *Statics of soil media.* London, Butterworths, 237 pp. General discussion of theory of critical equilibrium, with solutions for several problems of practical importance.

Hansen, J. Brinch (1961). "A general formula for bearing capacity" *Ingeniøren*, **5**, pp. 38–46; also *Bull. 11, Danish Geot Inst.* Brief summary of present state of theoretical developments.

ART. 34 BEARING CAPACITY OF PIERS AND PILES

Definitions

A pier is a slender prismatic or cylindrical body of masonry that transfers a load through a poor stratum onto a better one. A pile is essentially a very slender pier that transfers a load either through its lower end onto a firm stratum or else through side friction onto the surrounding soil. The relation between the load on a pier or pile and the corresponding settlement is very similar to that for footings. The load-settlement curve approaches either a vertical or an inclined tangent, as shown in Fig. 33.1. The definition of the bearing capacity of piers and piles is identical with that of footings (Article 33).

Bearing Capacity of Cylindrical Piers

Part of the load carried by a pier is transmitted directly to the soil immediately beneath its base, and part is transferred to the surrounding soil by friction and adhesion between the sides of the pier and the soil. At failure the load on a pier with depth D_f may be expressed as

$$Q_d = Q_p + Q_s = q_p A_p + 2\pi r f_s D_f \qquad (34.1)$$

in which q_p is the bearing capacity per unit of area of the soil beneath the base, A_p is the base area and r the radius of the pier, and f_s is the average value at failure of the combined friction and adhesion per unit

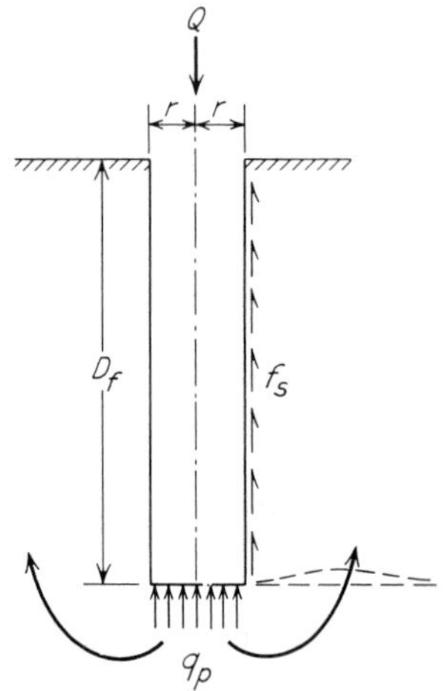

Fig. 34.1. Section through cylindrical pier.

of contact area between the sides of the pier and the soil. It is commonly called the *skin friction.*

Failure of the soil beneath the base cannot occur without the displacement of at least part of the mass in an outward or an outward and upward direction as indicated by the curved arrows in Fig. 34.1. If the soil within the depth D_f is appreciably more compressible than that beneath the base, the displacements produce negligible shearing stresses within the depth D_f. Consequently, the influence of the surrounding soil is identical to that of a surcharge having an intensity γD_f. Under these circumstances the bearing-capacity factors may be taken from Fig. 33.4 and q_p may be considered equal to q_{dr} or q_{ds} (Eqs. 33.13 or 33.14). On the other hand, if the soil is homogeneous the shearing stresses set up in the soil above base level as a consequence of the displacements have two significant effects: they may alter the shear pattern so that the bearing-capacity factors (Fig. 33.4) are no longer applicable, and they may alter the intensity of vertical pressure in the soil near the base of the pier. The latter effect appears to be the more important. On account of it, the term $\gamma D_f N_q$ (Eqs. 33.13 and 33.14) should be replaced by the expression $p_v N_q$, in which p_v is the actual intensity of the effective vertical pressure immediately adjacent to the pier at foundation level, at the time the failure of the pier occurs (Vesic 1963).

In reality the state of stress near the base of a deeply buried pier at failure is very complex and is not yet understood. Large-scale experiments in homogeneous sands (Vesic 1963, Kerisel 1964) have indicated that, for values of $D_f/2r$ greater than about 5, the base resistance Q_p no longer increases with depth in accordance with forecasts based on the term $\gamma D_f N_q$, and that, for $D_f/2r$ greater than about 15, the base resistance remains roughly constant irrespective of the depth D_f. These findings suggest that for values of $D_f/2r$ greater than about 15, the pressure p_v remains practically constant with increasing depth, and depends only on ϕ. In homogeneous clays under $\phi = 0$ conditions the net base resistance per unit of area becomes practically constant for values of $D_f/2r$ greater than about 4 and may be taken as $9c$ (Skempton 1951).

The second term on the right-hand side of Eq. 34.1 contains the skin friction f_s. The value of f_s is usually considered to be the sum of two terms

$$f_s = c_a + p_h \tan \delta \tag{34.2}$$

in which c_a is the adhesion per unit of area between pier and soil, p_h is the average horizontal pressure on the vertical surface of the pier at failure, and δ is the angle of friction between pier and soil. The values of c_a and δ can in some instances be determined approximately by means of laboratory tests. However, both quantities depend, among other factors, on the method of installation. Moreover, the stress conditions at the contact surface are unknown and as complex as those associated with the vertical stress p_v. Therefore, f_s is commonly and preferably estimated on the basis of empirical data derived from field observations (Article 57).

Bearing Capacity of Piles

Since piles are essentially very slender cylindrical piers, their bearing capacity can also be estimated by Eq. 34.1. The quantity Q_p is called the *point resistance*. If Q_p is large compared to Q_f, the pile is said to be *point-bearing*. On the other hand, if Q_p is relatively insignificant, the pile is known as a *friction pile*.

The principal difference between piles and piers lies in the method of installation. The construction of piers is preceded by excavation, whereas the driving of piles, which usually consist either of solid bodies or of shells closed at the lower end, involves a displacement of the soil. Occasionally the driving is facilitated by removing part of the soil located in the path of the pile by means of a water jet or a pre-excavator, but the volume of the soil so removed is usually small compared to the total volume of the piles.

If the piles are driven through compressible material to a firm base, a lower limiting value for the point resistance Q_p can be obtained by means of Eqs. 33.13 and 33.14 for circular and square cross-sections respectively. If the piles are surrounded and underlain by a homogeneous cohesionless material, Eqs. 33.13 and 33.14 may overestimate the point resistance for the reasons discussed under the preceding subheading. The bearing capacity of friction piles depends on the skin friction f_s (Eq. 34.1). The evaluation of the skin friction on the basis of laboratory tests is even more unreliable for piles than it is for piers, because the computation of the stresses produced by the partial or total displacement of the soil during the process of pile driving is beyond the power of analysis. Therefore, the bearing capacity Q_d of a friction pile can be determined only by load tests on piles in the field or else, less accurately, on the basis of empirical values for f_s. Values of f_s corresponding to the principal types of soil are given in Article 56. In those cities where friction piles are extensively used, empirical values for f_s, derived from local experience, are likely to be quite reliable.

Pile Formulas

The bearing capacity Q_d of a point-bearing pile may, under some circumstances (Article 56), be approximately equal to the resistance Q_{dy} of the soil against rapid penetration of the pile under the impact of the falling ram of the pile driver. There is at least a theoretical possibility of estimating Q_{dy}, known as the *dynamic resistance*, from the average penetration S of the pile under the last few blows of the hammer, provided the weight W_H of the ram and the height of fall H are known. Therefore, many efforts have been made to compute the bearing capacity on the basis of this information. The results of these efforts are known as *pile formulas*. The following paragraphs deal with the fundamental concepts on which the pile formulas are based.

The work performed by the falling hammer is $W_H H$, and the work required to increase the penetration of the pile by S against a resistance Q_{dy} is $Q_{dy}S$. If the entire work of the falling hammer served to increase the penetration of the pile, we could write

$$W_H H = Q_{dy} S$$

whence

$$Q_{dy} = \frac{W_H H}{S}$$

This is Sanders' pile formula, published about 1850. The values obtained by means of this formula are too great, because part of the energy of the falling hammer is converted into heat and into elastic deformations.

If we assume that all the deformations and energy losses occur simultaneously upon application of the hammer blow; that is, if the existence of stress waves in the pile and soil is ignored, we can write

$$W_H H = Q_{dy} S + \Delta \tag{34.3}$$

where Δ represents the energy lost and therefore unavailable to cause penetration of the pile. If there were no penetration and all the driving energy were consumed in elastic compression of the pile, the energy expended would be

$$W_H H = \tfrac{1}{2} Q_{dy} S_e$$

where S_e is the elastic compression of the pile. Moreover,

$$S_e = \frac{Q_{dy} L}{AE}$$

whence

$$S_e = \sqrt{\frac{2W_H HL}{AE}} \tag{34.4}$$

If we assume that the energy loss consists only of the elastic deformation of the pile and is, moreover, not influenced by the penetration of the point of the pile, Eq. 34.3 becomes

$$W_H H = Q_{dy} S + Q_{dy} \frac{S_e}{2} = Q_{dy}\left(S + \frac{S_e}{2}\right)$$

whence

$$Q_{dy} = \frac{W_H H}{S + \frac{1}{2}S_e} \tag{34.5}$$

This expression is known as the Danish formula. Statistical studies show that it should be used with a factor of safety of 3 (Sörensen and Hansen 1957).

Numerous attempts have been made to take into account the remaining energy losses. Some of these attempts have resulted in very complicated expressions and procedures. However, inasmuch as all methods based on Eq. 34.3 are fundamentally unsound because they ignore the dynamic aspects of the phenomena (Cummings 1940), the complicated formulas possess no inherent advantage over the simpler ones. The relative merits and reliability of any of the pile formulas can be judged only on the basis of comparisons with the results of load tests.

The Danish formula possesses the merit of simplicity and has been found to be reliable over a wide range of conditions (Agerschou 1962). A slightly more refined form of the Danish formula was proposed earlier by Janbu (1953); Janbu's formula contains a semi-empirical adjustment to allow for variations in the ratio W_P/W_H of the weights of pile and ram of the pile hammer. It may be expressed as

$$Q_{dy} = \frac{1}{K_u}\frac{W_H H}{S} \tag{34.6}$$

where

$$K_u = C_d\left[1 + \sqrt{1 + \frac{1}{2C_d}\frac{S_e^2}{S^2}}\right] \tag{34.7}$$

In Eq. 34.7 the empirical coefficient

$$C_d = 0.75 + 0.15\frac{W_P}{W_H} \tag{34.8}$$

Statistical studies (Article 56) indicate that the Janbu formula should be used with a calculated factor of safety of 3, and that the real factor of safety is not likely to be less than 1.75 or more than 4.4 (Flaate 1964).

The *Engineering News formula* (Wellington 1888), widely used in North America, is similar in form to Eq. 34.5 except that the term containing the elastic compression of the pile is replaced by a constant c. Thus

$$Q_{dy} = \frac{W_H H}{S + c}$$

Wellington regarded the quantity c as an additional penetration of the point of the pile that would have occurred if there were no energy losses. He evaluated it on the basis of whatever empirical data he had at his disposal and concluded that c is approximately equal to 1 in. for piles driven by a drop hammer and 0.1 in. for piles driven by a steam hammer. Since he realized that his estimate involved uncertainties, he proposed that the allowable load Q_a per pile should not exceed one-sixth of the computed ultimate load Q_{dy}. By expressing H in feet and S in inches, he obtained

$$Q_a = \frac{1}{6} Q_{dy} = \frac{12}{6} \frac{W_H H}{S + c} = \frac{2W_H H}{S + c} \tag{34.9}$$

This equation is known as the *Engineering News formula.*

Studies to evaluate the degree of accuracy of Eq. 34.9 (Agerschou 1962, Flaate 1964) have conclusively demonstrated that no satisfactory relation exists between the capacity of piles as determined by load tests and as calculated by Eq. 34.9. For two of every one hundred piles the actual bearing capacity may be less than 1.2 or greater than 30 times the calculated value in contrast to the supposed factor of safety of 6. There is no way to predict for a given pile what will be the actual capacity within this range. In view of these conditions the continued use of the Engineering News formula can no longer be justified.

A fundamentally more satisfactory approach to the development of pile formulas is the adaptation of the theory of longitudinal impact on rods (Glanville et al. 1938, Smith 1960, Sörensen and Hansen 1957). The calculations are complex and cannot yet be condensed into relations simple enough for practical use; furthermore, the limitations of the procedure have not yet been established by sufficient comparisons between predicted and measured bearing capacities. Hence, for the present, the designer of a point-bearing pile foundation must choose among several alternatives. He may use one of the less objectionable dynamic pile formulas, such as the Danish or Janbu formulas, at the risk of driving two or three times more piles than the foundation requires; he may estimate the point resistance on the basis of the static formula (Eq. 34.1) at the risk of overestimating the capacity, particularly if the piles are long and are embedded in dense sand; or else he may go to the expense of making load tests on full-size test piles in the field. The latter alternative may involve special procedures to permit evaluating the point capacity separately from the skin friction (Article 56). Whether the load tests are justifiable depends on the available time and the relation between the cost of the tests and that of the entire foundation.

Problems

1. A reinforced-concrete pile with a cross section 16 by 16 in. was driven through a deposit of fine loose sand and soft clay 65 ft thick and into a stratum of dense sand for a distance of 2.5 ft. The water table was located near the ground surface. The loose sand and soft clay had a submerged unit weight of 45 lb/ft^3, and the angle of internal friction of the dense sand in a submerged state was 35°. Compute the point resistance of the pile.

Ans. 90 tons. By means of a loading and a pulling test, the point resistance was found to be 115 tons.

2. The pile referred to in the preceding problem was driven by means of a steam hammer having a weight $W_H = 4$ tons and a stroke $H = 2$ ft. The penetration of the pile under the last blow was $S = 0.056$ in. According to the Engineering News formula, what is the ultimate bearing capacity of the pile?

Ans. 616 tons. According to the load test, the real ultimate bearing capacity, equal to the sum of the point resistance (115 tons) and the skin friction (110 tons), was 225 tons.

3. A test pile of the type described in problem 1 was driven at another point of the area to be occupied by the structure. Soil conditions were identical, except that the sand encountered at a depth of 65 ft was loose ($\phi = 30°$). Compute the point resistance of the pile.

Ans. 19 tons. (No load test was made, but the pile penetrated the sand so easily under the blows of the hammer that it was decided to change the type of foundation over the entire area underlain by the loose sand.)

4. The pile referred to in problem 1 was 70 ft long. Its modulus of elasticity was 3.5×10^6 lb/in.^{2} What is its ultimate capacity according to the Danish formula? the Janbu formula?

Ans. 260 tons; 190 tons.

Selected Reading

One of the classic papers of soil mechanics is Cummings, A. E. (1940): "Dynamic pile driving formulas," *J. Boston Soc. Civil Engrs.*, **27**, pp. 6–27. It is reprinted in *Contributions to soil mechanics 1925–1940,* Boston Soc. Civil Engrs. 1940, pp. 392–413.

ART. 35 STABILITY OF SLOPES

Introduction

The failure of a mass of soil located beneath a slope is called a *slide*. It involves a downward and outward movement of the entire mass of soil that participates in the failure.

Slides may occur in almost every conceivable manner, slowly or suddenly, and with or without any apparent provocation. Usually, slides are due to excavation or to undercutting the foot of an existing slope. However, in some instances, they are caused by a gradual disintegration of the structure of the soil, starting at hair cracks which subdivide the soil into angular fragments. In others, they are caused by an increase of the porewater pressure in a few exceptionally permeable layers, or by a shock that liquefies the soil beneath the slope (Article 49). Because of the extraordinary variety of factors and processes that may lead to slides, the conditions for the stability of slopes usually defy theoretical analysis. Stability computations based on test results can be relied on only when the conditions specified in the different sections of this article are strictly satisfied. Moreover, it should always be remembered that various undetected discontinuities in the soil, such as systems of hair cracks, remnants of old surfaces of sliding, or thin seams of water-bearing sand, may completely invalidate the results of the computations.

Slopes on Dry Cohesionless Sand

A slope underlain by clean dry sand is stable regardless of its height, provided the angle β between the slope and the horizontal is equal to or smaller than the angle of internal friction ϕ for the sand in a loose state. The factor of safety of the slope with respect to sliding may be expressed by the equation,

$$F = \frac{\tan\phi}{\tan\beta} \tag{35.1}$$

No slope on clean sand can exist with a slope angle greater than ϕ, irrespective of its height.

Since very few natural soils are perfectly cohesionless, the remainder of this article deals with slopes underlain by cohesive materials.

General Character of Slides in Homogeneous Cohesive Soil

A cohesive material having a shearing resistance

$$s = c + p\tan\phi$$

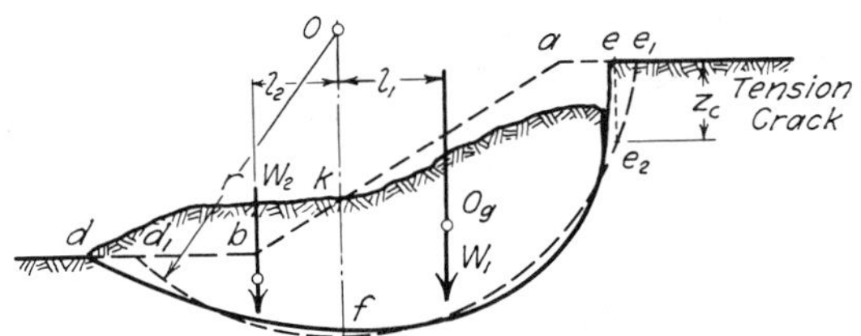

Fig. 35.1. Deformation associated with failure of slope.

can stand with a vertical slope at least for a short time, provided the height of the slope is somewhat less than H_c (Eq. 28.11). If the height of a slope is greater than H_c, the slope is not stable unless the slope angle β is less than 90°. The greater the height of the slope, the smaller must be the angle β. If the height is very great compared to H_c, the slope will fail unless the slope angle β is equal to or less than ϕ.

The failure of a slope in a cohesive material is commonly preceded by the formation of tension cracks behind the upper edge of the slope, as shown in Fig. 35.1. The force which produces the tension cracks behind the edge of a vertical slope is represented by the triangle *ace* in Fig. 28.3*b*. Sooner or later, the opening of the cracks is followed by sliding along a curved surface, indicated by the full line in Fig. 35.1. Usually the radius of curvature of the surface of sliding is least at the upper end, greatest in the middle, and intermediate at the lower end. The curve, therefore, resembles the arc of an ellipse. If the failure occurs along a surface of sliding that intersects the slope at or above its toe (Fig. 35.2*a*), the slide is known as a *slope failure*. On the other hand, if the soil beneath the level of the toe of the slope is unable to sustain the weight of the overlying material, the failure occurs along a surface that passes at some distance below the toe of the slope. A failure of this type, shown in Fig. 35.2*b*, is known as a *base failure*.

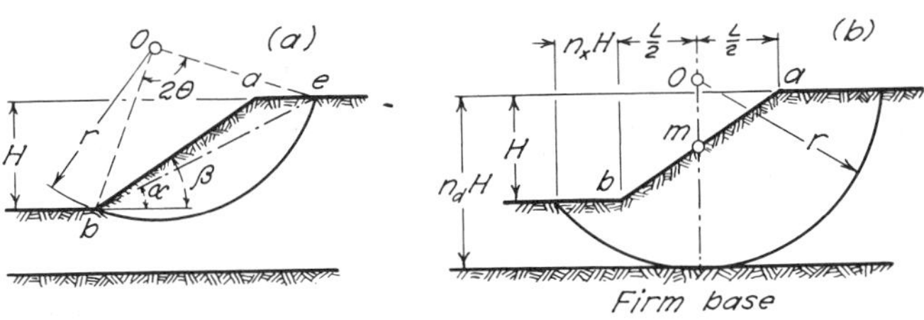

Fig. 35.2. Position of critical circle for (*a*) slope failure (after W. Fellenius 1927). (*b*) Base failure.

In stability computations the curve representing the real surface of sliding is usually replaced by an arc of a circle or of a logarithmic spiral. Either procedure is as legitimate as Coulomb's assumption of a plane surface of sliding in connection with retaining wall problems (Article 30). In the following discussions only the circle will be used as a substitute for the real surface of sliding.

Purpose of Stability Computations

In engineering practice, stability computations serve as a basis either for the redesign of slopes after a failure or for choosing slope angles in accordance with specified safety requirements in advance of construction.

Local failures on the slopes of cuts or fills are common during the construction period. They indicate that the average value of the minimum shearing resistance of the soil has been overestimated. Since such failures constitute large-scale shear tests, they offer excellent opportunities for evaluating the real minimum shearing resistance and for avoiding further accidents on the same job by changing the design in accordance with the findings. The general procedure is to determine the position of the surface of sliding by means of test borings, slope indicators, or shafts; to estimate the weights of the various parts of the sliding mass that tended to produce or to oppose the slide; and to compute the average shearing resistance s of the soil necessary to satisfy the conditions for equilibrium of the mass.

In order to design a slope in a region where no slides have occurred, the average shearing resistance s must be estimated or determined in advance of construction. Methods for evaluating the shearing resistance are discussed in Articles 17 and 18. After the value of s has been determined, the slope angle can be chosen on the basis of theory in such a manner that the slope satisfies the specified safety requirements. It is obvious that this method can be used only if the soil conditions permit a fairly reliable determination of s on the basis of the results of soil tests.

Computation of Shearing Resistance from Slide Data

The method for determining the average shearing resistance of soils on the basis of slide data is illustrated by Fig. 35.1. The depth z_c of the tension cracks and the shape of the surface of sliding are ascertained by field measurements. The line of sliding is then replaced by the arc of a circle having a radius r and its center at O. Equilibrium requires that

$$W_1 l_1 = W_2 l_2 + sr\,\widehat{d_1 e_2}$$

from which

$$s = \frac{W_1 l_1 - W_2 l_2}{r\, \widehat{d_1 e_2}}$$

where W_1 is the weight of the slice *akfe* which tends to produce failure, and W_2 is the weight of slice kbd_1f which tends to resist it.

If the shape of the surface of sliding is such that it cannot be represented even approximately by an arc of a circle, the procedure must be modified according to the methods described subsequently in connection with composite surfaces of sliding.

Procedure for Investigating Stability of Slopes

In order to investigate whether or not a slope on soil with known shear characteristics will be stable, it is necessary to determine the diameter and position of the circle that represents the surface along which sliding will occur. This circle, known as the *critical circle,* must satisfy the requirement that the ratio between the shearing strength of the soil along the surface of sliding and the shearing force tending to produce the sliding must be a minimum. Hence, the investigation belongs to the category of maximum and minimum problems exemplified by Coulomb's theory (Article 30) and the theory of passive earth pressure (Article 32).

After the diameter and position of the critical circle have been determined, the factor of safety F of the slope with respect to failure may be computed by means of the relation (Fig. 35.1)

$$F = \frac{sr\, \widehat{d_1 e_2}}{W_1 l_1 - W_2 l_2} \tag{35.2}$$

wherein r represents the radius of the critical circle and d_1e_2 the length of the surface of sliding.

Like the passive earth pressure of a mass of soil, the stability of a slope may be investigated by trial or, in simple cases, by analytical methods. To make the investigation by trial, different circles are selected, each representing a potential surface of sliding. For each circle, the value F (Eq. 35.2) is computed. The minimum value represents the factor of safety of the slope with respect to sliding, and the corresponding circle is the critical circle.

The analytical solutions can rarely be used to compute the factor of safety of a slope under actual conditions, because they are based on greatly simplified assumptions. They are valuable, however, as a guide for estimating the position of the center of the critical circle and for ascertaining the probable character of the failure. In addition, they may serve as a means for judging whether a given slope will be unquestionably safe, unquestionably unsafe, or of doubtful stability. If the stability appears doubtful, the factor of safety with respect to failure should be computed according to the procedure described in the preceding paragraph.

The analytical solutions are based on the following assumptions: Down to a given level below the toe of the slope, the soil is perfectly homogeneous. At this level, the soil rests on the horizontal surface of a stiffer stratum, known as the *firm base*, which is not penetrated by the surface of sliding. The slope is considered to be a plane, and it is located between two horizontal plane surfaces, as shown in Fig. 35.2. Finally, the weakening effect of tension cracks is disregarded, because it is more than compensated by the customary margin of safety. The following paragraphs contain a summary of the results of the investigations.

Slopes on Soft Clay

The average shearing resistance s per unit of area of a potential surface of sliding in homogeneous clay under undrained ($\phi = 0$) conditions (Article 18) is roughly equal to one-half the unconfined compressive strength q_u of the clay. This value of s is referred to briefly as the cohesion c. That is,

$$s = \tfrac{1}{2}q_u = c \tag{18.5}$$

If c is known, the critical height H_c of a slope having a given slope angle β can be expressed by the equation,

$$H_c = N_s \frac{c}{\gamma} \tag{35.3}$$

In this equation the *stability factor* N_s is a pure number. Its value depends only on the slope angle β and on the *depth factor* n_d (Fig. 35.2*b*) which expresses the depth at which the clay rests on a firm base. If a slope failure occurs, the critical circle is usually a *toe circle* that passes through the toe b of the slope (Fig. 35.2*a*). However, if the firm base is located at a short distance below the level of b, the critical circle may be a *slope circle* that is tangent to the firm base and that intersects the slope above the toe b. This type of failure is not shown in Fig. 35.2. If a base failure occurs, the critical circle is known as a *midpoint circle*, because its center is located on a vertical line through the midpoint m of the slope (Fig. 35.2*b*). The midpoint circle is tangent to the firm base.

The position of the critical circle with reference to a given slope depends on the slope angle β and the depth factor n_d. Figure 35.3 contains a summary of the results of pertinent theoretical investigations. According to this figure, the failure of all slopes rising at an angle of more than 53° occurs along a toe circle. If β is smaller than 53°, the type of failure depends on the value of the depth factor n_d and, at low values of n_d, also on the slope angle β. If n_d is equal to 1.0, failure occurs along a slope circle. If n_d is greater than about 4.0, the slope fails along a midpoint circle tangent to the firm base, regardless of the value of β. If n_d is intermediate in value between 1.0 and 4.0, failure occurs along a slope circle if the point representing the values of n_d and β lies above the shaded area in Fig. 35.3. If the point lies within the shaded area, failure occurs along a toe circle. If the point is below the shaded area, the slope fails along a midpoint circle tangent to the firm base.

If the slope angle β and the depth factor n_d are given, the value of the corresponding stability factor N_s (Eq. 35.3) can be obtained without computation from Fig. 35.3. The value of N_s determines the critical height H_c of the slope.

If failure occurs along a toe circle, the center of the critical circle can be located by laying off the angles α and 2θ, as shown in Fig. 35.2*a*.

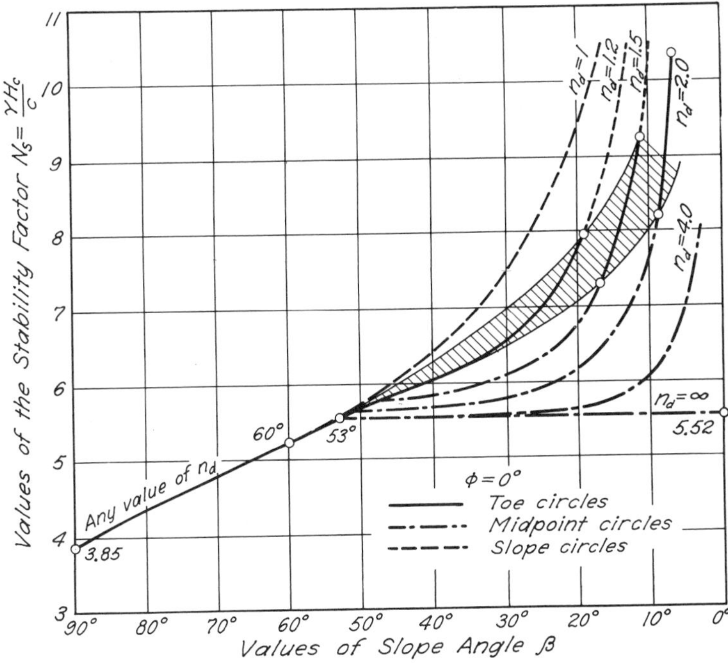

Fig. 35.3. Relation for frictionless material between slope angle β and stability factor N_s for different values of depth factor n_d (after Taylor 1937).

Values of α and θ for different slope angles β are given in Fig. 35.4*a*. If failure occurs along a midpoint circle tangent to the firm base, the position of the critical circle is determined by the horizontal distance $n_x H$ from the toe of the slope to the circle (Fig. 35.2*b*). Values of n_x can be estimated for different values of n_d and β by means of the chart (Fig. 35.4*b*).

If the clay beneath a slope consists of several layers with different average cohesion c_1, c_2, etc., or if the surface of the ground is irregular (Fig. 35.5), the center of the critical circle must be determined by trial and error. It is obvious that the longest part of the real surface of sliding will be located within the softest stratum. Therefore, the trial circle should also satisfy this condition. If one of the upper layers is relatively soft, the presence of a firm base at considerable depth may not

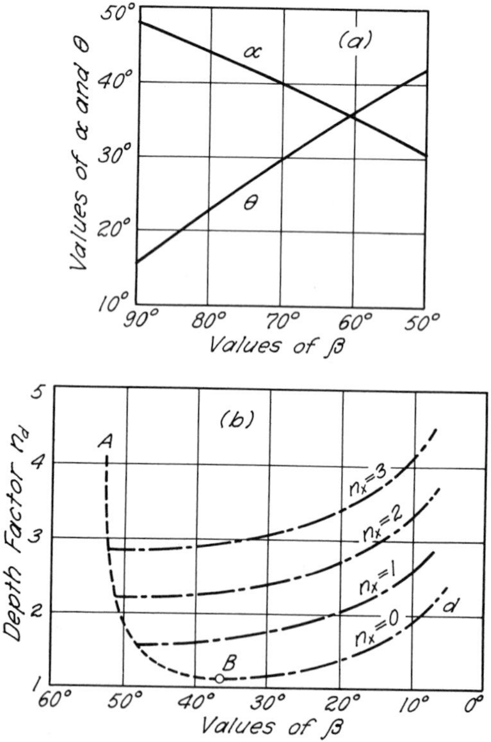

Fig. 35.4. (*a*) **Relation between slope angle β and parameters α and θ for location of critical toe circle when β is greater than 53°.** (*b*) **Relation between slope angle β and depth factor n_d for various values of parameter n_x (after W. Fellenius 1927).**

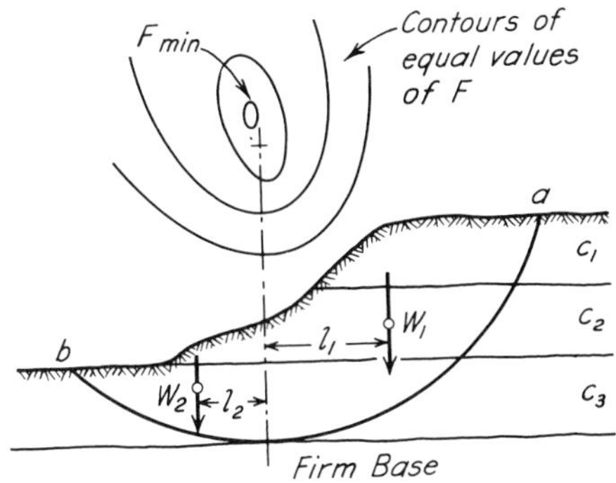

Fig. 35.5. Base failure in stratified cohesive soil.

enter into the problem, because the deepest part of the surface of sliding is likely to be located entirely within the softest stratum. For example, if the cohesion c_2 of the second stratum in Fig. 35.5 is much smaller than the cohesion c_3 of the underlying third layer, the critical circle will be tangent to the upper surface of the third stratum instead of the firm base.

For each trial circle we compute the average shearing stress t which must act along the surface of sliding to balance the difference between the moment W_1l_1 of the driving weight and the resisting moment W_2l_2. The value of t is

$$t = \frac{W_1l_1 - W_2l_2}{r\,\widehat{ab}}$$

Then, on the basis of the known values of c_1, c_2, c_3, etc., we compute the average value of the cohesion c of the soil along the sliding surface. The factor of safety of the slope against sliding along the circular trial surface is

$$F = \frac{c}{t} \tag{35.4}$$

The value of F is inscribed at the center of the circle. After values of F have been determined for several trial circles, curves of equal values of F are plotted (Fig. 35.5). These curves may be considered as contour lines of a depression. The center of the critical circle is located at the bottom of the depression. The corresponding value F_{min} is the factor of safety of the slope with respect to sliding.

If it is not obvious which of two layers may constitute the firm base for the critical circle, trial circles must be investigated separately for each possibility and the corresponding values of $F_{\min}$ determined. The smaller of the two values is associated with the firm base that governs the failure and is the factor of safety of the slope.

Slopes on Soils with Cohesion and Internal Friction

If the shearing resistance of the soil can be expressed approximately by the equation

$$s = c + p \tan \phi$$

the stability of slopes on the soil can be investigated by the procedure illustrated by Fig. 35.6*a*. The forces acting on the sliding mass are its weight W, the resultant cohesion C, and the resultant F of the normal and frictional forces acting along the surface of sliding. The resultant cohesion C acts in a direction parallel to the chord de and is equal to the unit cohesion c multiplied by the length L of the chord. The distance x from the center of rotation to C is determined by the condition that

$$Cx = cLx = c\,\overset{\frown}{de}\,r$$

whence $x = \overset{\frown}{de}\,r/L$. Therefore, the force C is known. The weight W is also known. Since the forces C, W, and F are in equilibrium, the force F must pass through the point of intersection of W and C. Hence, the magnitude and line of action of F can be determined by constructing the polygon of forces.

If the factor of safety against sliding is equal to unity, the slope is on the verge of failure. Under this condition each of the elementary reactions dF in Fig. 35.6*a* must be inclined at the angle ϕ to the normal to the circle of sliding. As a consequence, the line of action of each elementary reaction is tangent to a circle, known as the *friction circle*, having a radius

$$r_f = r \sin \phi$$

and having its center at the center of the circle of sliding. The line of action of the resultant reaction F is tangent to a circle having a radius slightly greater than r_f, but as a convenient approximation we assume that at a factor of safety equal to unity the line of action of F is also tangent to the friction circle. The corresponding error is small and is on the safe side.

For a given value of ϕ the critical height of a slope which fails along a toe circle is given by the equation,

$$H_c = N_s \frac{c}{\gamma}$$

which is identical with Eq. 35.3, except that N_s depends not only on β but also on ϕ. Figure 35.6*b* shows the relationship between β and N_s

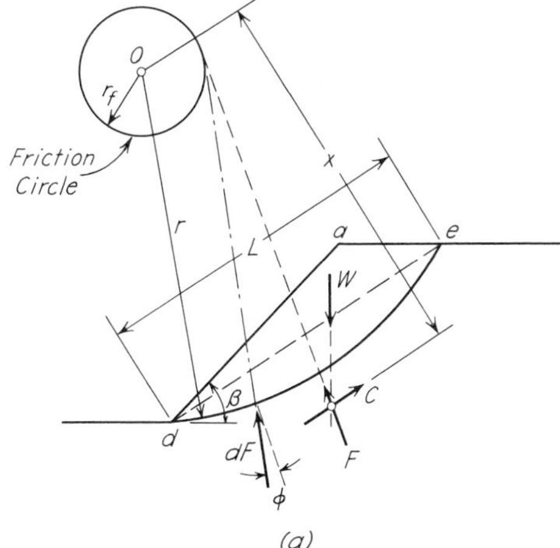

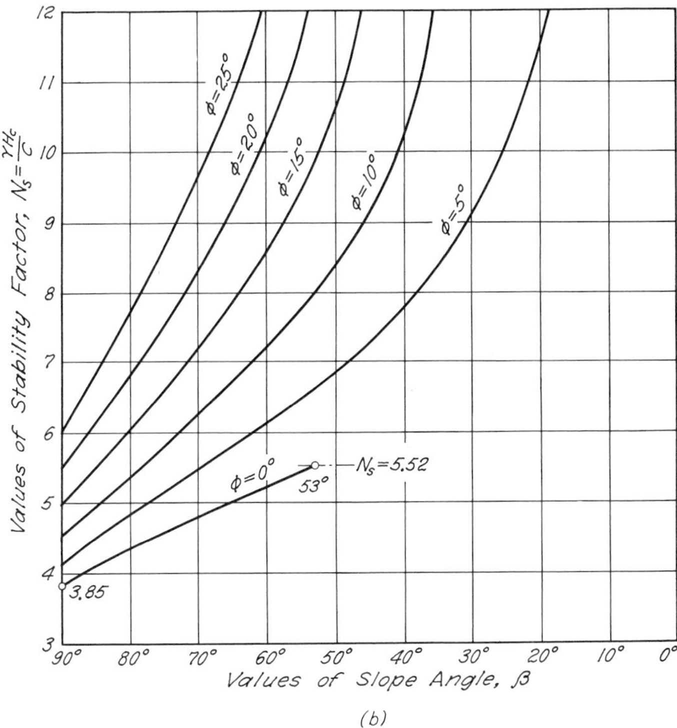

Fig. 35.6. Failure of slope in material having cohesion and friction. (*a*) Diagram illustrating friction-circle method. (*b*) Relation between slope angle β and stability factor N_s for various values of ϕ (after Taylor 1937).

for different values of ϕ. At a given value of the slope angle β, N_s increases at first slowly and then more rapidly with increasing values of ϕ. When $\phi = \beta$, N_s becomes infinite.

All the points on the curves shown in Fig. 35.6*b* correspond to failures along toe circles, because theory has shown that the possibility of a base failure does not exist unless ϕ is smaller than approximately 3°. Therefore, if a typical base failure has occurred in a fairly homogeneous soil in the field, it can be concluded that with respect to total stresses the value of ϕ for the soil at the time of the slide was close to zero.

Irregular Slopes on Nonuniform Soils

If a slope has an irregular surface that cannot be represented by a straight line, or if the surface of sliding is likely to pass through several materials with different values of c and ϕ, the stability can be investigated conveniently by the *method of slices*. According to this procedure a trial circle is selected (Fig. 35.7*a*) and the sliding mass subdivided into a number of vertical slices 1, 2, 3, etc. Each slice, such as slice 2 shown in Fig. 35.7*b*, is acted upon by its weight W, by shear forces T and normal forces E on its sides, and by a set of forces on its base. These include the shearing force S and the normal force P. The forces on each slice, as well as those acting on the sliding mass as a whole, must satisfy the conditions of equilibrium. However, the forces T and E depend on the deformation and the stress-strain characteristics of the slide material and cannot be evaluated rigorously. They can be approximated with sufficient accuracy for practical purposes.

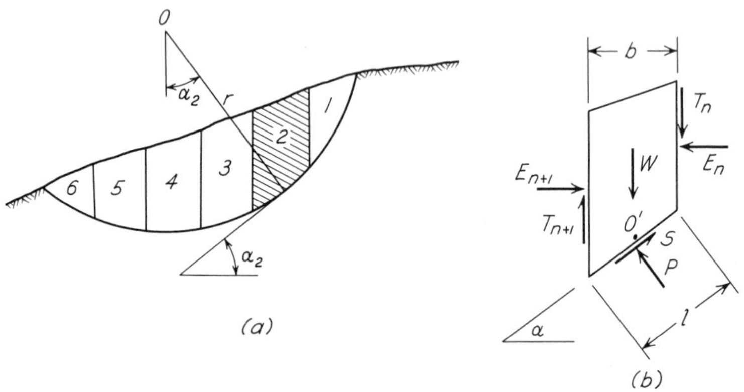

Fig. 35.7. Method of slices for investigating equilibrium of slope located above water table. (*a*) Geometry pertaining to one circular surface of sliding. (*b*) Forces on typical slice such as slice 2 in (*a*).

The simplest approximation consists of setting these forces equal to zero. Under these circumstances, if the entire trial circle is located above the water table and there are no excess pore pressures, equilibrium of the entire sliding mass requires that

$$r\Sigma W \sin \alpha = r\Sigma S \tag{35.5}$$

If s is the shearing strength of the soil along l, then

$$S = \frac{s}{F} l = \frac{s}{F} \frac{b}{\cos \alpha} \tag{35.6}$$

and

$$r \sum W \sin \alpha = \frac{r}{F} \sum \frac{sb}{\cos \alpha} \tag{35.7}$$

whence

$$F = \frac{\Sigma(sb/\cos \alpha)}{\Sigma W \sin \alpha} \tag{35.8}$$

The shearing strength s, however, is determined by

$$s = c + p \tan \phi$$

where p is the normal stress across the surface of sliding l. To evaluate p we consider the vertical equilibrium of the slice (Fig. 35.7*b*), whence

$$W = S \sin \alpha + P \cos \alpha$$

and

$$p = \frac{P}{l} = \frac{P \cos \alpha}{b} = \frac{W}{b} - \frac{S}{b} \sin \alpha \tag{35.9}$$

Therefore

$$s = c + \left(\frac{W}{b} - \frac{S}{b} \sin \alpha\right) \tan \phi = c + \left(\frac{W}{b} - \frac{s}{F} \tan \alpha\right) \tan \phi$$

and

$$s = \frac{c + (W/b) \tan \phi}{1 + (\tan \alpha \tan \phi)/F} \tag{35.10}$$

Let

$$m_\alpha = \left(1 + \frac{\tan \alpha \tan \phi}{F}\right) \cos \alpha \tag{35.11}$$

Then

$$F = \frac{\displaystyle\sum \frac{[c + (W/b) \tan \phi]b}{m_\alpha}}{\Sigma W \sin \alpha} \tag{35.12}$$

Equation 35.12, which gives the factor of safety F for the trial circle under investigation, contains on the right-hand side the quantity m_α (Eq. 35.11) which is itself a function of F. Therefore, Eq. 35.12 must be solved by successive approximations in which a value of $F = F_1$ is assumed and used for calculation of m_α, whereupon F is then computed. If the value of F differs significantly from F_1, the calculation is repeated. Convergence is very rapid. The calculations are facilitated by the chart (Fig. 35.8*a*) from which values of m_α can be taken (Janbu et al. 1956), and by a tabular arrangement of the computations (Fig. 35.8*b*).

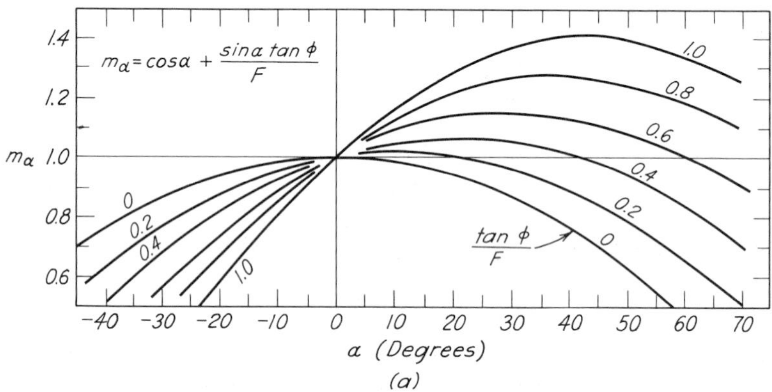

	Values from cross section							
	1	2	3	4	5	6	7	8
Slice No.	α°	$\sin\alpha$	W	$W\sin\alpha$	$c + \dfrac{W}{b}\tan\phi$	$(5)\cdot b$	m_α $F_a =$	$(6)/(7)$
								$\Sigma(8)$

For first trial, $F_a = \dfrac{\Sigma(6)}{\Sigma(4)}$ $\qquad F = \dfrac{\Sigma(8)}{\Sigma(4)}$

(*b*)

Fig. 35.8. Calculation of factor of safety for slope if surface of sliding is circular and forces between slices are neglected. (*a*) Chart for evaluating factor m_α. (*b*) Tabular form for computation.

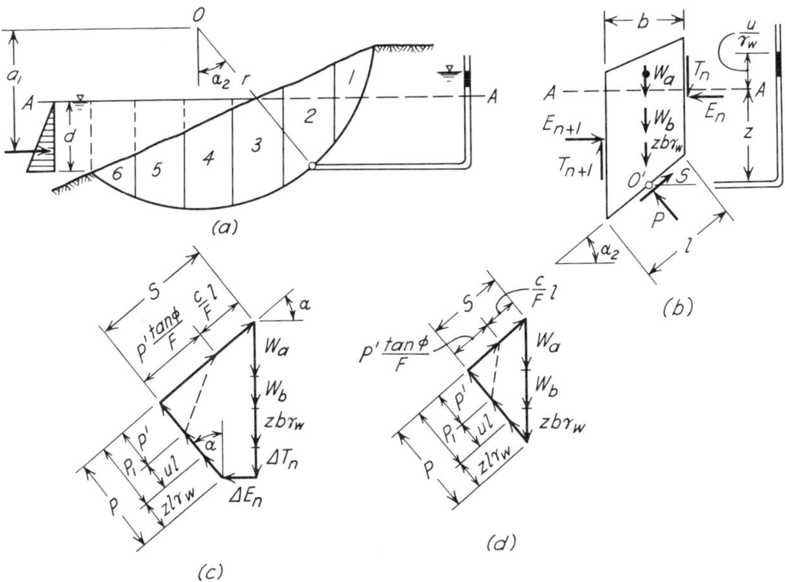

Fig. 35.9. Method of slices for circular surface of sliding if slope is partly submerged. (*a*) Geometry pertaining to one surface of sliding. (*b*) Forces acting on typical slice, such as slice 2 in (*a*). (*c*) Force polygon for slice 2 if all forces are considered. (*d*) Force polygon for slice 2 if forces *T* and *E* on sides of slice are considered to be zero.

Inasmuch as the calculations outlined in Fig. 35.8 refer to only one trial circle, they must be repeated for other circles until the minimum value of F is found.

In general, the slope may be partly submerged and there will be pore pressures acting along the trial circle (Fig. 35.9*a*). The magnitudes of the pore pressures depend upon the conditions of the problem. In some instances they may be estimated by means of a flow net (Article 23), by means of soil tests, or on the basis of field observations. If the level of the external water surface is denoted by $A - A$, the weight W of the slice (Fig. 35.9*b*) may be written as

$$W = W_a + W_b + zb\gamma_w \tag{35.13}$$

where W_a is the weight of that part of the slice above $A - A$, W_b is the submerged weight of the part below $A - A$, and $zb\gamma_w$ is the weight of a volume of water equal to the submerged portion of the slice. If the entire slice is located beneath water level, as slice 5 (Fig. 35.9*a*), the weight of the water above the slice must be included in $zb\gamma_w$. The pore pressure at the midpoint 0′ of the base of the slice is $z\gamma_w + u$, where u is the excess pore pressure with respect to the external water level. If the external water level A-A is located below O' on the base of the slice (Fig. 35.9*b*), the pore pressure at O' is $h\gamma_w$, where h is the height to which the water would rise in a piezometer at O'. If the pore pressure is due to capillarity, h is negative.

Since the forces acting on the slice are in equilibrium, they may be represented by the force polygon (Fig. 35.9*c*). The normal force P consists of an effective component P', the force ul caused by the excess pore pressure, and the force $zl\gamma_w$ caused by the hydrostatic pressure of the water with respect to $A - A$. The shearing stress t along the surface of sliding is

$$t = \frac{s}{F} = \frac{1}{F}(c + \bar{p}\tan\phi) = \frac{1}{F}\left[c + \left(\frac{P}{l} - z\gamma_w - u\right)\tan\phi\right] \quad (35.14)$$

whence

$$S = t \cdot l = \frac{1}{F}[cl + (P - zl\gamma_w - ul)\tan\phi] = \frac{1}{F}(cl + P'\tan\phi) \quad (35.15)$$

Equilibrium of the entire slide with respect to moments about the center of the trial circle requires that

$$\sum (W_a + W_b + zb\gamma_w)r\sin\alpha = \sum S \cdot r + \tfrac{1}{2}\gamma_w d^2 a_1$$

$$= \frac{1}{F}\sum (cl + P'\tan\phi)r + \tfrac{1}{2}\gamma_w d^2 a_1 \quad (35.16)$$

However, the water below level $A - A$ is in equilibrium, whence

$$\Sigma zb\gamma_w r\sin\alpha = \tfrac{1}{2}\gamma_w d^2 a_1 \quad (35.17)$$

Therefore,

$$\sum (W_a + W_b)r\sin\alpha = \frac{1}{F}\sum (cl + P'\tan\phi)r \quad (35.18)$$

and

$$F = \frac{\Sigma(cl + P'\tan\phi)}{\Sigma(W_a + W_b)\sin\alpha} \quad (35.19)$$

The value of F (Eq. 35.19) depends upon P' which may be determined for each slice from the force polygon (Fig. 35.9*c*). If the surface of sliding is circular, the influence of the forces T and E between the slices is relatively small and P' can usually be evaluated with sufficient accuracy on the assumption that the forces T and E are equal to zero. The force polygon then reduces to Fig. 35.9*d*, whence

$$W_a + W_b + zb\gamma_w = (zl\gamma_w + P' + ul)\cos\alpha + \left(P'\frac{\tan\phi}{F} + \frac{cl}{F}\right)\sin\alpha \tag{35.20}$$

and

$$P' = \frac{W_a + W_b - ub - \dfrac{cl}{F}\sin\alpha}{m_\alpha} \tag{35.21}$$

Substitution of Eq. 35.21 into 35.19 gives

$$F = \frac{\displaystyle\sum \frac{[cb + (W_a + W_b - ub)\tan\phi]}{m_\alpha}}{\Sigma(W_a + W_b)\sin\alpha} \tag{35.22}$$

Equation 35.22, like Eq. 35.12, must be solved by successive approximations because the factor of safety F is contained in m_α which appears on the right-hand side. It may be noted that the influence of the external water level is fully taken into account by the use of the submerged weight W_b, and that the excess pore pressure u is calculated for the base of each slice as explained in connection with Eq. 35.13.

The procedure described in the preceding paragraphs may be modified to take into account the forces T and E between the slices (Bishop 1955, Janbu 1954*a*). If the surface of sliding is circular, however, the improvement in accuracy is not likely to exceed 10 to 15% and the additional effort is not usually justified. On the other hand, if the surface of sliding is not circular the error may be significant. These circumstances will be considered in the next section. The procedures that will be developed may, if desired, be used to take into account the forces between slices for a circular surface of sliding as well.

Composite Surface of Sliding

In many instances the geometric or geologic conditions of the problem are such that the surface of sliding may not be even approximately circular. For these conditions, the method of slices can be extended (Janbu 1954*a*, Nonveiller 1965).

A sliding mass with a noncircular surface of sliding is shown in Fig. 35.10. The forces acting on any slice n are represented in the same manner as those shown in Fig. 35.9*b*, and the polygon of forces is identical to that in Fig. 35.9*c*.

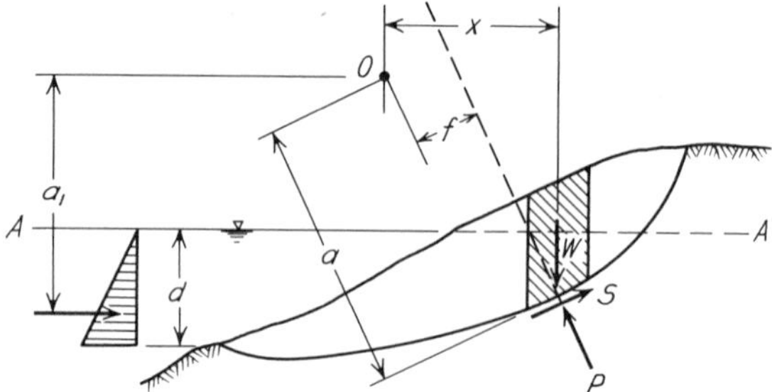

Fig. 35.10. Geometry of method of slices for investigating equilibrium of slope if surface of sliding is not circular.

The equilibrium of the entire sliding mass with respect to moments about the arbitrary pole O requires that

$$\Sigma Wx = \Sigma(Sa + Pf) + \tfrac{1}{2}\gamma_w d^2 a_1 \tag{35.23}$$

whence, from Eq. 35.15

$$\sum (W_a + W_b + zb\gamma_w)x = \frac{1}{F}\sum (cl + P' \tan \phi)a + \sum Pf + \tfrac{1}{2}\gamma_w d^2 a_1$$

and

$$F = \frac{\Sigma(cl + P' \tan \phi)a}{\Sigma(W_a + W_b + zb\gamma_w)x - \Sigma Pf - \tfrac{1}{2}\gamma_w d^2 a_1} \tag{35.24}$$

However, the water below level $A - A$ is in equilibrium, whence

$$\Sigma zb\gamma_w x - \tfrac{1}{2}\gamma_w d^2 a_1 = \Sigma zl\gamma_w f = \Sigma(P - P_1)f \tag{35.25}$$

where

$$P_1 = P - zl\gamma_w$$

Equation 35.24 then becomes

$$F = \frac{\Sigma(cl + P' \tan \phi)a}{\Sigma(W_a + W_b)x - \Sigma P_1 f} \tag{35.26}$$

This expression can be evaluated if P' and P_1 are known. These quantities may be determined from the force polygon (Fig. 35.9*c*). Summation of vertical components leads to

$$W_a + W_b + \Delta T_n + zb\gamma_w = zl\gamma_w \cos \alpha + (P' + ul) \cos \alpha$$
$$+ \frac{1}{F}(cl + P' \tan \phi) \sin \alpha$$

whence

$$P' = \frac{W_a + W_b + \Delta T_n - ub - (c/F)b \tan \alpha}{m_\alpha} \tag{35.27}$$

Moreover,

$$P_1 = P' + ul = \frac{W_a + W_b + \Delta T_n + (1/F)(ub \tan \phi - cb) \tan \alpha}{m_\alpha} \tag{35.28}$$

By substituting Eqs. 35.27 and 35.28 in 35.26 and combining terms, we find

$$F = \frac{\Sigma[cb + (W_a + W_b + \Delta T_n - ub) \tan \phi](a/m_\alpha)}{\sum (W_a + W_b)x - \sum \left[W_a + W_b + \Delta T_n + (ub \tan \phi - cb) \dfrac{\tan \alpha}{F}\right](f/m_\alpha)} \tag{35.29}$$

This equation must be solved by successive approximations because the factor of safety F occurs on the right-hand side explicitly as well as in the quantity m_α. Furthermore, the value of F depends on ΔT_n. As a first approximation, ΔT_n may be set equal to zero. The calculations are facilitated by the chart (Fig. 35.8*a*) and a tabular arrangement (Fig. 35.11). Inasmuch as the value of F determined in this manner refers to only one trial surface, the calculations must be repeated for other surfaces until the minimum value of F is found.

For most practical problems involving a noncircular surface of sliding, the assumption that ΔT_n is equal to zero leads to sufficiently accurate results. If the cross section of the surface of sliding departs significantly from a circular shape, the use of Eq. 35.29 with $\Delta T_n = 0$ is preferable to the assumption of a circular cross section and the use of Eq. 35.22. However, if greater refinement is justified, values of ΔT_n may be inserted in Eq. 35.29 and the factor of safety recalculated. The calculations are laborious.

If the values of T and E are not zero, they must satisfy the conditions for equilibrium of the entire sliding mass in vertical and horizontal directions. That is

$$\Sigma \Delta T_n = 0 \tag{35.30}$$

$$\Sigma \Delta E_n + \tfrac{1}{2}\gamma_w d^2 = 0 \tag{35.31}$$

Furthermore, for each slice, ΔT_n and ΔE_n are related in accordance with the requirements of the force polygon (Fig. 35.9*c*). By resolving the forces in the direction of S, we obtain

$$S = \Delta E_n \cos \alpha + (W_a + W_b + \Delta T_n + zb\gamma_w) \sin \alpha$$

Col. / Slice	1	2	3	4	5	6	7	8	9	10	11	12	13	14	15	16	17
	b	a	x	f	α	$\tan \alpha$	c	$\tan \phi$	cb	ub	W_a	W_b	ΔT_n	$W_a + W_b + \Delta T_n$	(14) $- ub$	(15) $\tan \phi$	(9) + (16)
1																	
2																	
:																	
n																	

Col. / Slice	18	19	20	21	22	23	24	25	26	27	28	29	30
	(17) $\cdot a$	$W_a + W_b$	(19) $\cdot x$	$ub \tan \phi$	(21) $- cb$	(22) $\tan \alpha$	F_1	$\frac{(23)}{F_1}$	(14) + (25)	(26) $\cdot f$	m_α	$\frac{(18)}{m_\alpha}$	$\frac{(27)}{m_\alpha}$
1													
2													
:													
n													

$\Sigma(20) =$ $\Sigma(29) =$

$\Sigma(30) =$

$$F_2 = \frac{\Sigma(29)}{\Sigma(20) - \Sigma(30)}$$

Repeat steps 24 to 30 incl.

Fig. 35.11. Tabular form for calculating factor of safety of slope by method of slices, if surface of sliding is not circular.

whence

$$\Delta E_n = S \sec \alpha - (W_a + W_b + \Delta T_n) \tan \alpha - zb\gamma_w \tan \alpha \quad (35.32)$$

However, it may also be seen from the force polygon that

$$S = \frac{1}{F} [cl + (P - zl\gamma_w - ul) \tan \phi] = \frac{1}{F} [cl + P' \tan \phi] \quad (35.33)$$

By substituting Eq. 35.27 into 35.33, we obtain

$$S = \frac{1}{F} \cdot \frac{cb + (W_a + W_b + \Delta T_n - ub) \tan \phi}{m_\alpha} = \frac{M}{F} \quad (35.34)$$

and, by using Eq. 35.32 and summing for all the slices,

$$\sum [\Delta E_n + zb\gamma_w \tan \alpha] = \sum \left[\frac{M}{F} \sec \alpha - (W_a + W_b + \Delta T_n) \tan \alpha\right] \quad (35.35)$$

But since

$$\Sigma zb\gamma_w \tan \alpha = \tfrac{1}{2}\gamma_w d^2$$

Eq. 35.31 requires the left-hand side of Eq. 35.35 to be zero. Hence the forces ΔT_n must satisfy not only Eq. 35.30, but also

$$\sum \left[\frac{M}{F} \sec \alpha - (W_a + W_b + \Delta T_n) \tan \alpha\right] = 0 \quad (35.36)$$

Because the problem is statically indeterminate, any set of values T_n satisfying Eqs. 35.30 and 35.36 will assure compliance with all conditions for equilibrium of the slide as a whole and for the horizontal and vertical equilibrium of each slice. However, not all such sets of values are reasonable or possible. For example, the values of T_n must not exceed the shearing strength of the soil along the vertical boundary of the corresponding slice under the influence of the normal force E_n. Moreover, tensile stresses should not occur across a significant portion of any vertical boundary between slices. In most instances it will prove satisfactory and expedient to assign arbitrary but reasonable values to the earth pressure E_n, and on the basis of these values and Eq. 16.5 to calculate approximate upper limiting values for T_n. By trial and error, smaller values of T_n are established that satisfy Eqs. 35.30 and 35.36. A systematic tabular arrangement (Fig. 35.12) is helpful. Values so obtained are substituted into Eq. 35.29. If F differs appreciably from the value determined previously, a revision by successive approximations is indicated. The revision may require alteration of the quantities T_n because of the dependence of M (Eq. 35.34) on F.

Col. / Slice		13	14	9	10	15	16	17	31			32	33	34	35	36
	T_n	ΔT_n	$W_a + W_b + \Delta T_n$	cb	ub	(14)–(10)	(15) tan ϕ	(9) + (16)	(14) tan α	F	m_α	$\frac{(17)}{m_\alpha} = M$	$\frac{M}{F}$	sec α	$\frac{M}{F}$ sec α	(35)–(31)
1																
2																
:																
n																

$\Sigma(13) \overline{} = 0$ $\qquad$ $\Sigma(36) \overline{} = 0$

Fig. 35.12. Tabular form for determining consistent set of shear forces T_n between slices, for substitution into Eq. 35.29, if values of ΔT_n are not considered to be zero.

There is, of course, no assurance that the value of F finally determined by this procedure is correct, because other consistent sets of T-values lead to other factors of safety. However, the values of F for different but reasonable sets of forces between the slices are not likely to differ to a great extent.

It may also be noted that the force polygon (Fig. 35.9*c*) presupposes that each slice is in equilibrium with respect to moments, whereas this condition will not generally be satisfied by the forces derived from the solution. This requirement can be added to those represented by Eqs. 35.30 and 35.36 but the difficulties of calculation are increased substantially. The use of electronic computation is virtually mandatory (Morgenstern and Price 1965).

If the subsoil contains one or more thin exceptionally weak strata, the surface of sliding is likely to consist of three or more sections that do not merge smoothly one into another. In stability computations such a surface cannot be replaced by a continuous curve without the introduction of an error on the unsafe side.

Figure 35.13 represents a slope underlain by a thin layer of very soft clay with cohesion c. If such a slope fails, the slip occurs along some composite surface *abcd*. In the right-hand part of the sliding mass, represented by the area *abf*, active failure must be expected because the earth stretches horizontally under the influence of its own weight. The central part *bcef* moves to the left under the influence of the active pressure on *bf*. The left-hand part of the sliding mass *cde* experiences passive failure due to the thrust of the advancing central part *bcef*.

The first step in investigating the conditions for the stability of the slope is to compute the passive earth pressure P_P of the soil located on the left side of a tentatively selected vertical section *ec* located near the toe of the slope. It is conservative to assume that P_P acts in the horizontal direction. The next step is to estimate the position of the right-hand boundary b of the horizontal part *cb* of the potential surface of sliding and to compute the active earth pressure P_A on a vertical section *fb* through b. The tendency for the mass *bcef* to move to the left

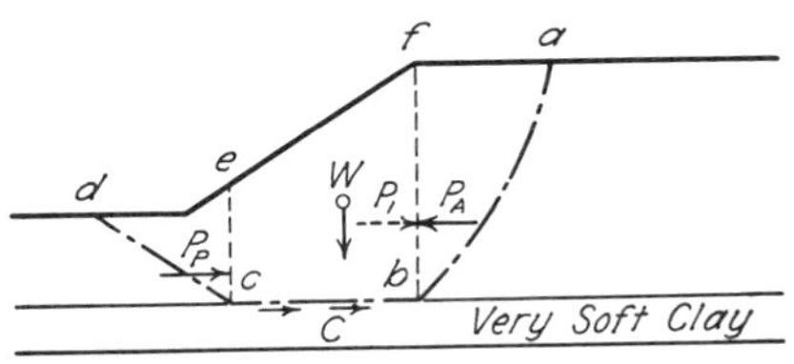

Fig. 35.13. Failure of slope underlain by thin layer of very soft clay.

is resisted by the passive earth pressure P_P and the total cohesion C along bc. If the slope is stable, the sum of these resisting forces must be greater than the active earth pressure P_A which is assumed to act in a horizontal direction. The factor of safety against sliding is equal to the ratio between the sum of the resisting forces and the force P_A. The investigation must be repeated for different positions of the points c and b until the surface of least resistance to sliding is found that corresponds to the least factor of safety.

Problems

1. A wide cut was made in a stratum of soft clay that had a level surface. The sides of the cut rose at 30° to the horizontal. Bedrock was located at a depth of 40 ft below the original ground surface. When the cut reached a depth of 25 ft, failure occurred. If the unit weight of the clay was 120 lb/ft^3, what was its average cohesive strength? What was the character of the surface of sliding? At what distance from the foot of the slope did the surface of sliding intersect the bottom of the excavation?

Ans. 500 lb/ft^2; midpoint circle; 18 ft.

2. The rock surface referred to in problem 1 was located at a depth of 30 ft below the original ground surface. What were the average cohesive strength of the clay and the character of the surface of sliding?

Ans. 450 lb/ft^2; toe circle.

3. A cut is to be excavated in soft clay to a depth of 30 ft. The material has a unit weight of 114 lb/ft^3 and a cohesion of 700 lb/ft^2. A hard layer underlies the soft layer at a depth of 40 ft below the original ground surface. What is the slope angle at which failure is likely to occur?

Ans. $\beta = 69°$.

4. A trench with sides rising at 80° to the horizontal is excavated in a soft clay which weighs 120 lb/ft^3 and has a cohesion of 250 lb/ft^2. To what depth can the excavation be carried before the sides cave in? At what distance from the upper edge of the slope will the surface of sliding intersect the ground surface?

Ans. 9 ft; 8 ft.

5. A bed of clay consists of three horizontal strata, each 15 ft thick. The values for c for the upper, middle, and lower strata are, respectively, 600, 400, and 3000 lb/ft^2. The unit weight is 115 lb/ft^3. A cut is excavated with side slopes of 1 (vertical) to 3 (horizontal) to a depth of 20 ft. What is the factor of safety of the slope against failure?

Ans. 1.2.

6. To what depth can the trench in problem 4 be excavated without bracing if the soil has, in addition to its cohesion, an angle of internal friction of 20°?

Ans. 14.2 ft.

Selected Reading

A detailed discussion of the method of slices and the assumptions on which it is based may be found in Taylor, D. W. (1948): *Fundamentals of soil mechanics,* New York, John Wiley and Sons, pp. 432–441. A condensed summary of the method from the point of view of effective stress and the use of pore-pressure coefficients is given in Bishop, A. W. (1955): "The use of the slip circle in the stability analysis of slopes," *Géot.*, **5,** pp. 7–17.

Charts for the solution of many cases of practical importance are contained in Bishop, A. W. and N. R. Morgenstern (1960): "Stability coefficients for earth slopes," *Géot.*, **10,** pp. 129–150. Solutions for many other cases are given by Janbu, N. (1954*b*): "Stability analysis of slopes with dimensionless parameters," *Harvard Soil Mech. Series No. 46,* 81 pp.

The most general analysis available, not restricted to a circular surface of sliding and considering the forces between slices, is developed mathematically by Morgenstern, N. R. and V. E. Price (1965): "The analysis of the stability of general slip surfaces," *Géot.*, **15,** pp. 79–93. An electronic computer is needed for the solution.

ART. 36 STABILITY OF EARTH DAMS

Critical States for Design

The factor of safety of an earth dam with respect to a slope or foundation failure depends to a large extent on the porewater pressures. In a dam of given cross section on a given foundation, the intensity and distribution of the porewater pressures vary with time between wide limits. For purposes of design it is convenient to distinguish among the porewater pressure conditions corresponding to three stages: during construction, and especially immediately after construction has been completed; after the reservoir has been full long enough to develop a state of steady seepage in the dam and its foundation; and during or immediately after lowering of the reservoir level. These three stages are briefly designated as the *construction, full-reservoir,* and *drawdown states.* The stability of the upstream slope may also be critical upon first filling of the reservoir, especially if the dam has a sloping core. Furthermore, in some instances, the state of the upstream slope may be more critical at an intermediate level, known as *partial pool,* than with the reservoir full.

Evaluation of Porewater Pressures in Critical Design States

During the construction stage, significant pore pressures can develop only in the cohesive portions of the dam and subsoil. They are associated with progressive consolidation. Therefore, the intensity and distribution

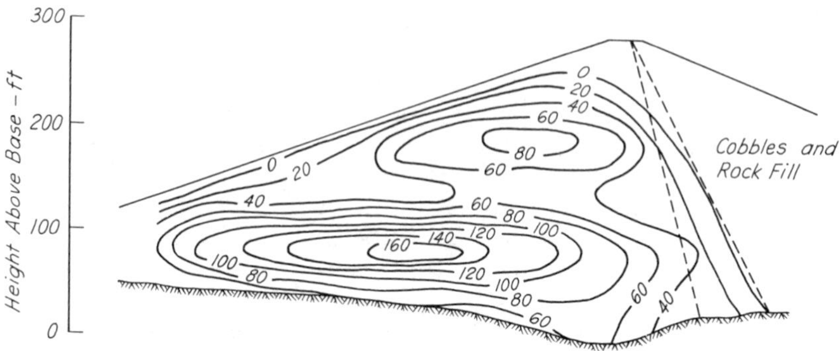

Fig. 36.1. Measured pore pressures in feet of water in impervious section of Green Mountain dam at end of construction (after Walker and Daehn 1948).

of the pressures depend not only on the characteristics of the materials and on the boundary conditions for drainage, but also to a large extent on the construction schedule. As an example, Fig. 36.1 shows the porewater pressures observed at the end of construction in the upstream portion of Green Mountain dam in Colorado (Walker and Daehn 1948). Placement of fill was discontinued for the winter when the dam had reached approximately half its final height; the influence of the interruption is evident. Attempts to predict the pore pressures under such conditions on the basis of the results of laboratory tests have been made (Hilf 1948) but the degree of reliability of the results has not yet been fully assessed.

After the dam has been completed and the reservoir has been full for some time, the dam is acted upon by the seepage pressures exerted by a steady flow of water percolating out of the reservoir through the dam toward the downstream slope. The corresponding porewater pressures can be estimated on the basis of the flow-net method (Fig. 23.4*a*) or an equivalent procedure, provided the subsoil conditions are simple enough to permit construction of a reasonably reliable permeability profile of the material upon which the dam is founded.

The porewater pressure conditions initiated by a drawdown depend to a large extent on the degree of compressibility of the different constituents of the body of the dam. In the semipervious and relatively incompressible portions, such as those constructed of well-compacted silty sand, most of the water which occupied the voids before drawdown is retained in the voids; the remainder drains out of the dam at practically unaltered void ratio. It emerges partly through the lower portions of the slopes and, if the subsoil is pervious, partly through the base of the dam. The drawdown condition is illustrated by Fig. 23.4*c* which shows the flow net for a homogeneous dam of fine, clean, well compacted sand, at the instant after a sudden, complete drawdown. The dam is assumed to rest on an impervious base. It can be seen that the seepage pressures on the downstream slope remain unchanged from those of the full-reservoir state (Fig. 23.4*a*), whereas those acting on the lower portion of the upstream slope tend to cause a failure of this portion by sliding. As time goes on, the uppermost flow line (the lower boundary of the capillary fringe) descends and all the seepage pressures decrease.

The impervious compressible portions of the dam, such as those consisting of clay, remain in a completely or almost completely saturated state even after the drawdown. The water is held by capillarity. However, the pore pressures in those portions of the clay located immediately beneath the uppermost flow line change from positive to negative whereas the total stresses remain almost unaltered. Consequently, the effective pressures in the clay increase and the clay starts to consolidate. Most of the excess water flows into the lower part of the clay portion of the dam, reducing its shearing resistance, and ultimately emerges from the lower part of the upstream boundary of the clay portion.

Whatever may be the cross section of a dam and the geological profile of its foundation materials, the importance of the undesirable consequences of a drawdown decreases with decreasing rate of drawdown. Hence, in order to determine the pore pressure conditions for the drawdown state, all the following factors need to be known: the location of the boundaries between materials with significantly different properties; the permeability and consolidation characteristics of each of these materials; and the anticipated maximum rate of drawdown. In addition, the pore pressures induced by the changes in the shearing stresses themselves (Article 15) need to be taken into consideration. In engineering practice, few of these factors can be reliably determined. The gaps in the available information must be filled by the most unfavorable assumptions compatible with the known facts.

Stability Computations

In every earth dam, except in a homogeneous dam resting on a rigid base, the potential surfaces of sliding pass through the weakest portions of the dam and subsoil (Fig. 36.2). For a given cross section and a given foundation, the position of these surfaces also depends on the intensity and distribution of the pore pressures. Therefore, the surfaces can rarely be defined as arcs of circles. They can in most instances be represented only by lines with a variable radius of curvature or by composite curves. Their position must be ascertained by trial and error,

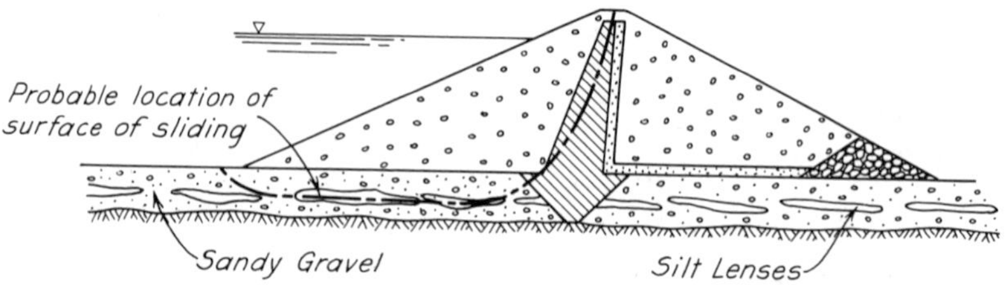

Fig. 36.2. Diagram illustrating likelihood of noncircular surface of sliding in zoned earth dam.

by starting with a curve which is believed to be located close to the surface of minimum resistance. The calculations may be carried out by the procedures outlined in Article 35.

The calculations for each of the three critical states for design require the determination of the corresponding pore pressures. According to the preceding section, this determination must be based on the adaptation of the theories of consolidation and of the flow of water through porous media to the internal and external boundary conditions, as well as on a knowledge of the stress-deformation pore-pressure characteristics of the materials. The latter characteristics are usually expressed in terms of estimated values of the pore-pressure coefficients $\bar{\mathbf{A}}$ and $\mathbf{B}$ (Article 15). If the pore pressures u_w have been estimated in accordance with these procedures, the values of c and ϕ to be inserted in the equations of Article 35 are the effective-stress values, and the analysis is said to be an *effective stress analysis.*

The values c and ϕ can be determined quite reliably on the basis of laboratory tests, whereas the choice of appropriate values of u_w requires much experience and judgment. Some investigators prefer to evaluate the shearing resistance more directly, on the basis of tests in which the influence of the pore pressures is taken into account in the test procedures. For example, in analyzing the stability of an upstream slope after a drawdown, the investigator may conclude that the materials will be fully consolidated under the conditions preceding the drawdown. He then may perform two sets of triaxial tests. One set is performed under consolidated-undrained conditions, in which the samples are consolidated under stresses corresponding to the conditions immediately preceding the drawdown. The other set is performed under fully drained conditions. The values of ϕ_{cu} and ϕ obtained from the two sets of tests represent limiting conditions, depending on whether the material beneath the upstream slope can experience practically no drainage during the drawdown, or can drain so rapidly as to dissipate practically all the excess porewater pressures as the drawdown progresses. On the basis of his knowledge of the rate of drawdown, and of the permeability and dimensions of the mass of soil affected by the drawdown, the investigator judges the most appropriate values of the shear-strength parameters between these limits. Analyses of this type are generally referred to as *total stress analyses.*

The experience and judgment required to choose the most appropriate values of the shearing strength between the limiting conditions imposed by the test procedures are fully as great as those needed to evaluate u_w for use in an effective stress analysis. From this point of view, neither procedure can be said to deserve the preference. Except for rather rare circumstances under which the simple $\phi = 0°$ conditions are applicable, there appears to be a growing tendency to estimate the pore pressures for use in effective stress analyses, partly because they can be compared more directly with the results of pore-pressure observations in the field.

Sources of Error in Stability Computations

The sources of error in stability computations can be divided into three categories: simplifying assumptions introduced into the computations; the assumption of simultaneous failure; and errors in the evaluation of the intensity and distribution of the pore pressures. By far the most important errors are those of the last category. For this reason, the design should be based on the most unfavorable assumptions regarding these pressures consistent with the known physical properties of the materials comprising the dam and its foundation.

Effect of Earthquakes on the Integrity of Earth Dams

If an earth dam is to be located in a region subject to earthquakes, the design must satisfy the condition that even the most severe earthquake anticipated in the region should not impair the integrity of the dam. In conventional methods of analysis, the severity of the earthquake is usually expressed by the ratio n_g between the maximum horizontal acceleration produced by the earthquake and the vertical acceleration produced by the force of gravity. During the earthquake all parts of the body of the dam are assumed to be acted on by a steady horizontal force γn_g per unit of volume, in addition to all the other forces to which the dam is subjected. The effect of the earthquake can then be considered in the stability computation by adding to the other forces acting on the mass above the potential surface of sliding a static horizontal force equal to the product of γn_g and the volume of the sliding mass. The surface of sliding corresponding to the minimum factor of safety must be found by trial and error.

Newmark (1965) has shown, in accordance with theory and model tests (Seed and Clough 1963), that the duration of the earthquake force is not long enough to justify its replacement by a static force γn_g, and that failures of the type indicated by the conventional stability analyses are not likely, provided the shearing resistance of the material in the dam is itself not reduced by the earthquake. Instead, the repeated pulses produce an S-shaped deformation of the slopes, or a moderate increase of the width of the dam, associated with a settlement of the crest. The settlement is essentially independent of n_g; it increases with increasing maximum ground velocity and with the total duration of the pulses. If the crest settles below the level of the tops of the highest waves which could occur during the earthquake at full reservoir, the dam could fail by overtopping. Therefore, in earthquake regions the design of the dam must satisfy the additional condition that the settlement of the crest of the dam should not be large enough to permit overtopping by the highest waves. including those caused by the earthquake or any induced landslides, at full reservoir.

Much more serious is the possibility that the shearing resistance of part of the dam, or especially of its foundation, may be radically reduced during the earthquake, particularly by liquefaction (Article 17). Earthquakes may also produce open cracks across the impervious portion of the dam; piping through the cracks involves the possibility of failure. However, such cracks may also be formed by unequal settlement without the assistance of earthquakes. Therefore, earth dams should always be designed in such a manner that piping through cracks cannot occur (Article 62).

Selected Reading

The effective-stress method of analysis, specifically in connection with rapid drawdown, is exemplified by Bishop, A. W. (1954): "The use of pore-pressure coefficients in practice," *Géot.*, **4,** pp. 148–152. The total-stress approach is described in detail in Manual EM 1110-2-1902: *Stability of earth and rockfill dams,* Corps of Engineers, U.S. Army, Dec. 27, 1960, 67 pp.

ART. 37 EARTH PRESSURE AGAINST BRACING IN CUTS

Deformation Conditions Imposed by Bracing

Figure 37.1 illustrates one of several methods for bracing an open cut. A row of H-piles is driven along each side of the proposed excavation to a depth of several feet below grade. The sides of the cut between H-piles are lined by horizontal boards placed directly against the soil as the cut is deepened. The two ends of each board are wedged against the inner flanges of the H-piles. The piles themselves are supported by horizontal steel or timber struts inserted as excavation proceeds. In order to design the struts, we must know the magnitude and the distribution of the earth pressure.

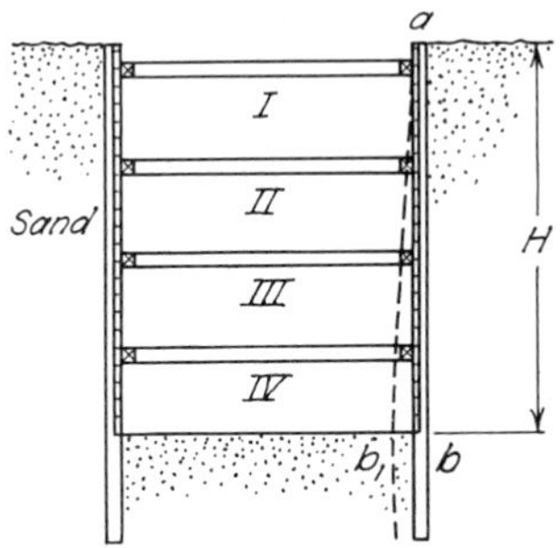

Fig. 37.1. Diagram illustrating deformation condition for lateral pressure against bracing in open cut.

In Article 27 it was shown that the earth pressure depends not only on the properties of the supported soil but also on the restrictions that the construction procedure imposes on the freedom of the support to yield. Hence, the first step in investigating the earth pressure against an open cut is to examine the nature of these restrictions. When the first row of struts *I* (Fig. 37.1) is placed, the amount of excavation is still so insignificant that the original state of stress in the soil is practically unaltered. Therefore, the first row of struts is in position before any appreciable yielding of the soil mass occurs. As excavation proceeds to the level of the next set of struts *II*, the rigidity of set *I* prevents further horizontal yielding of the soil located near the ground surface on each side of the cut. However, the H-piles are acted on by the lateral pressure of the soil outside the cut. Under the influence of this pressure, they yield inward by rotating about a line at the level of the uppermost set of struts. Hence, the placement of the second row of struts is preceded by a horizontal yielding of the soil located outside the cut at the level of this set. With increasing depth the yielding increases, because the height of the banks on either side of the cut increases. Therefore, while excavation proceeds, the vertical section ab (Fig. 37.1) advances into the position ab_1. Since the strut at the top of the cut prevents the stretching of the upper part of the sliding wedge, the soil can fail only as indicated in Fig. 27.3. Because of this, the active earth pressure against the bracing in the cut cannot be computed by means of Coulomb's or Rankine's theory. A method must be developed that takes into consideration the influence of the deformation conditions on the type of failure.

It has been shown that the deformation conditions represented by line ab_1 in Fig. 37.1 involve a failure of the type illustrated in Fig. 27.3. It has also been shown (Article 27) that the failure cannot occur unless the lower edge b of the lateral support (Fig. 37.1) yields more than a certain distance bb_1. This distance depends on the depth of the cut and on the physical properties of the soil. In the following discussion, we shall assume that this deformation condition is satisfied. The observations on which the assumption is based and the necessary qualifications are presented in Article 48.

Cuts in Dry or Drained Sand

Figure 37.2 shows a vertical section through one side of a cut with depth H in dry or drained sand. The initial position of the H-piles is indicated by the plain line ab and the final position by the dash line ab_1. The earth pressure on the bracing, per unit of length of the cut, is designated by P_a to distinguish it from the active earth pressure P_A exerted by a similar mass of sand against a retaining wall of height H. Since the upper part of the sliding wedge (Fig. 37.2*a*) cannot move laterally, the surface of sliding intersects the ground surface at a right angle (see also Fig. 27.3). The real curve of sliding can be closely approximated by a logarithmic spiral having the equation,

$$r = r_0 \epsilon^{\theta \tan \phi} \tag{37.1}$$

The center of the spiral is located on a straight line that passes through d and makes an angle ϕ with the horizontal. Since the yield of the lateral support causes the wedge to move downward along the back of the support, the resultant earth pressure acts at an angle δ to the horizontal. Theoretical investigations have shown that the point of application of the earth pressure is determined by the shape of the surface of sliding

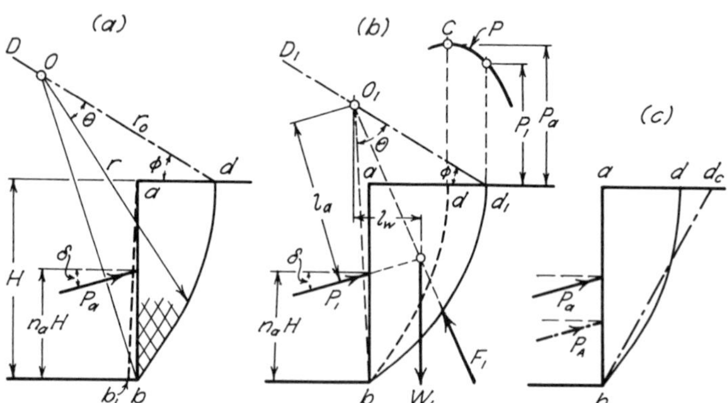

Fig. 37.2. Logarithmic spiral method for calculating earth pressure against bracing of open cuts. (*a*) Diagram illustrating assumptions on which computation is based. (*b*) Forces acting on sliding wedge. (*c*) Comparison of surface of sliding with that assumed in Coulomb's theory.

and vice versa. If the curve of sliding is similar to bd in Fig. 37.2, theory indicates that the distribution of the sand pressure against the bracing is roughly parabolic, as indicated in Fig. 27.3*b*, and that the elevation of the point of application n_aH should be between $0.45H$ and $0.55H$. This theoretical conclusion has been confirmed by pressure measurements in full-sized cuts. Therefore, in the following computation n_a is assumed to be known.

In order to determine the position of the surface of sliding, an arbitrary point d_1 (Fig. 37.2*b*) is selected on the horizontal surface adjoining the upper edge of the cut. Through this point and through the lower edge b of the bank, a logarithmic spiral bd_1 is traced with its center on d_1D_1. The reaction F_1 on the surface of sliding represented by bd_1 passes through the center O_1. Taking moments about O_1, we obtain

$$P_1 l_a = W_1 l_w$$

whence

$$P_1 = \frac{W_1 l_w}{l_a} \tag{37.2}$$

A similar computation is made for spirals through d_2, d_3 $\cdots$ (not shown). By plotting the values P_1, P_2 $\cdots$, etc., as ordinates above d_1, d_2, $\cdots$, the curve P is obtained. The active earth pressure P_a is equal to the maximum ordinate, corresponding to point C, and the surface of sliding passes through d. The width ad of the top of the wedge which exerts the maximum pressure P_a is always much smaller than the width of the top of the corresponding Coulomb wedge abd_c (Fig. 37.2*c*).

The value of P_a depends to a certain extent on n_a. It increases slightly with increasing values of n_a and is always greater than the corresponding Coulomb value P_A. For the values $\phi = 38°$ and $\delta = 0°$, an increase of n_a from 0.45 to 0.55 increases P_a from $1.03P_A$ to $1.11P_A$. If we assume $n_a = 0.55$, any error is on the safe side because this value is the greatest which has so far been obtained by field measurements. The angle δ has very little influence on the ratio P_a/P_A. Hence, for a preliminary estimate, it is sufficiently accurate to assume

$$P_a = 1.1P_A \tag{37.3}$$

The next step in the investigation is to determine the pressure in individual struts. The distribution of the lateral pressure against the bracing in cuts is roughly parabolic as shown in Fig. 27.3*b*, but from section to section in a given cut it deviates somewhat from the statistical average because of variations in soil conditions and in details of the construction procedure. As a consequence, for a given value of P_a the pressure in individual struts at a given elevation varies. The procedure

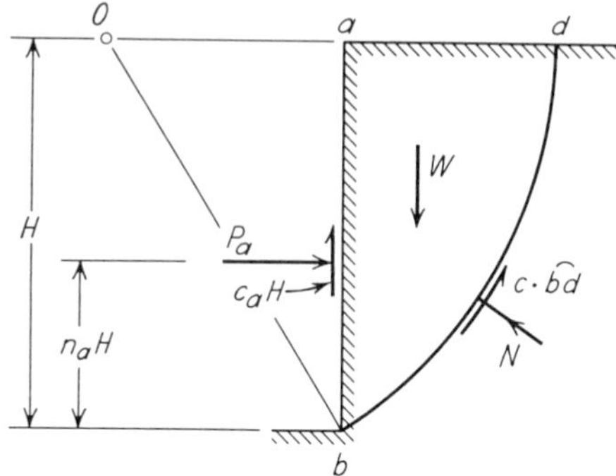

Fig. 37.3. Diagram illustrating assumptions upon which calculation of earth pressure is based for cuts in clays under $\phi = 0$ conditions.

for estimating the maximum pressure that may have to be resisted by the struts in any given row is described in Article 48.

Cuts in Saturated Clay

The time required to excavate and brace an open cut in an intact saturated clay is likely to be very short in comparison to that in which the water content of the clay may change significantly. Under these circumstances the $\phi = 0$ condition (Article 18) may be applicable, and Eq. 37.1 becomes identical with that of a circle having the radius $r = r_0$. Since the circle must intersect the ground surface at a right angle, its center must be located at the level of the surface (Fig. 37.3). The face *ab* of the cut is acted upon by the horizontal earth pressure P_a and the adhesion c_aH between the clay and the sheeting. The computation of P_a is made by taking moments about the center of the circle of the forces that act upon the sliding wedge. The driving moment is produced by the weight of the wedge. The resisting moment is equal to the sum of the moment of the resultant earth pressure P_a and the moments of the forces of adhesion c_aH and cohesion $c \cdot \widehat{bd}$ that act on the boundaries of the wedge. The value of P_a depends on the ratio c_a/c as well as on n_a. Measurements in full-sized cuts excavated in saturated soft to medium clays have shown that n_a varies between about 0.30 and 0.50 and is, on the average, about 0.39. The corresponding distribution of lateral pressure is, therefore, not usually triangular but, like that for sand, approaches a roughly parabolic shape. The method for estimating the maximum load which may have to be carried by an individual strut is described in Article 48.

Heave of the Bottom of Cuts in Soft Clay

In connection with open cuts in soft clay, we must consider the possibility that the bottom may fail by heaving, because the weight of the blocks of clay beside the cut tends to displace the underlying clay toward the excavation. Figure 37.4*a* represents a cross section through a cut in soft clay. The width of the cut is B, and the depth is H. The two strips *ab* and *cd* at the level of the bottom of the cut carry a surcharge due to the weight of the blocks of clay located above them. The strips *ab* and *cd*, therefore, act like footings. If the bearing capacity of the soil beneath the strips is exceeded, the bottom of the cut will fail by heaving. The bearing capacity of the soil, for $\phi = 0$ conditions, may be expressed as cN_c (Eq. 33.7). The factor of safety against heave is, therefore,

$$F = \frac{cN_c}{\gamma H} \tag{37.4}$$

The bearing-capacity factor N_c depends upon the shape of the cut in plan, and on the ratio of depth to width. If it is assumed that the excavated soil is analogous to a single large footing that exerts an upward load γH at the level *abcd*, the values of N_c may be taken equal to those for footings of the same ratios B/L of width to length, and H/B of depth to width (Bjerrum and Eide 1956). The calculations are facilitated by the chart, Fig. 37.4*b* (Janbu et al. 1956).

If sheet piles extend below the bottom of the cut, their stiffness reduces the tendency of the clay adjacent to the bottom to be displaced toward the excavation and, consequently, reduces the tendency toward

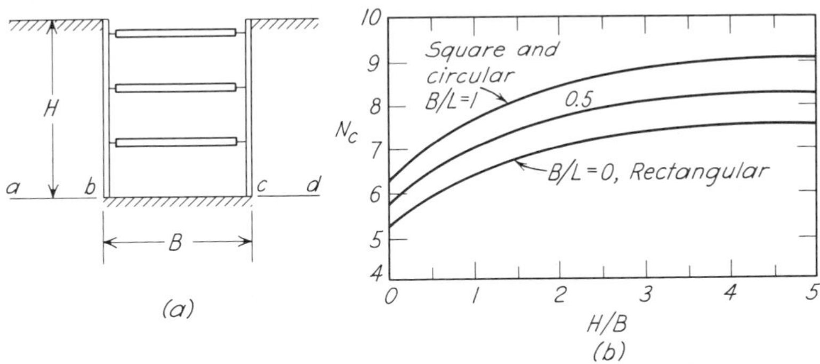

Fig. 37.4. (*a*) Section through open cut in deep deposit of clay. (*b*) Values of bearing capacity factor N_c for estimating stability of bottom of cut against heave (after Janbu et al. 1956).

heave. No satisfactory theoretical procedures have been developed for estimating the pressures that the sheeting must withstand. However, if the clay extends to a considerable depth below the cut, the beneficial effects of even relatively stiff sheeting have been found to be small. If the lower ends of the sheet piles are driven into a hard stratum, the effectiveness of the sheet piles is increased appreciably. Support of the lower edge of the piles reduces the maximum bending moment in the embedded portion of the piles; in addition, the vertical load on *ab* and *cd* (Fig. 37.4*a*) is reduced by the weight transferred by adhesion between the soil above base level and the sheeting. If the point resistance of the sheet piles is greater than the adhesion, the reduction is equal to the adhesion between the clay and the sheet piles. If it is smaller, the reduction is equal to the point resistance.

If the hard stratum is located a short distance below the bottom of the cut, the lower boundary of the zone of plastic equilibrium is tangent to the upper surface of the hard layer. The tendency for heave is greatly reduced even without sheet piles, and the effectiveness of sheet piles is substantially increased.

Problems

1. By means of the logarithmic-spiral method, determine the total pressure P_a against the bracing of a cut 30 ft deep in cohesionless sand for which $\gamma = 115$ lb/ft^3 and $\phi = 30°$. The value of δ is assumed to be zero. The center of pressure is 16 ft above the bottom of the cut. Determine also the Coulomb value P_A.

Ans. 18,400; 17,300 lb/lin ft.

2. An open cut is made to a depth of 40 ft in clay having a unit weight of 127 lb/ft^3 and a cohesion of 635 lb/ft^2. The values of ϕ and δ are assumed to be zero. The point of application of the resultant earth pressure against the bracing is 18 ft above the bottom. Find the value of the resultant earth pressure.

Ans. 59,700 lb/lin ft.

3. A braced open cut 30×150 ft in plan is to be made to a depth of 35 ft in a deep deposit of plastic clay having an undrained shear strength of 600 lb/ft^2 and a unit weight of 120 lb/ft^3. The sheeting extends only a few feet below the elevation of the bottom of the cut. What is the factor of safety against heave of the bottom?

Ans. 0.95. The bottom will heave when the excavation reaches a depth of 33 ft.

4. If the excavation in problem 3 consisted of a trench only 5 ft wide for a length of 150 ft, what would be the factor of safety against a heave of the bottom at a depth of 35 ft?

Ans. 1.07. The excavation could barely be made.

ART. 38 ARCHING IN SOILS

The earth pressure on the lateral support shown in Fig. 37.1 is greatest at about midheight of the sides of the cut. Yet, if a few of the horizontal boards supporting the soil at midheight are removed, the exposed part of the sides of the cut remains stable, provided the soil has at least a trace of cohesion. In order to explain this fact, we are compelled to assume that the pressure formerly exerted on the boards that were removed was transferred onto those that remained in place. This phenomenon of pressure transfer is known as *arching*.

The essential features of arching can be demonstrated by the test illustrated by Fig. 38.1. A layer of dry cohesionless sand with unit weight γ is placed on a platform that contains a trap door *ab*. The trap door is mounted on a scale (not shown) that permits measurement of the pressure on the door. The depth H of the layer of sand is several times greater than the width of the trap door.

As long as the trap door occupies its original position, the pressure on the trap door as well as that on the adjoining platform is equal to γH per unit of area. However, as soon as the trap door is allowed to yield in a downward direction, the pressure on the door decreases to a small fraction of its initial value, whereas the pressure on the adjoining parts of the platform increases. This is due to the fact that the descent of the prism of sand located above the yielding trap door is resisted by shearing stresses along its lateral boundaries, *ac* and *bd*.

Theory, as well as the results of tests and experience in tunneling, indicate that the ultimate pressure on the yielding trap door is practi-

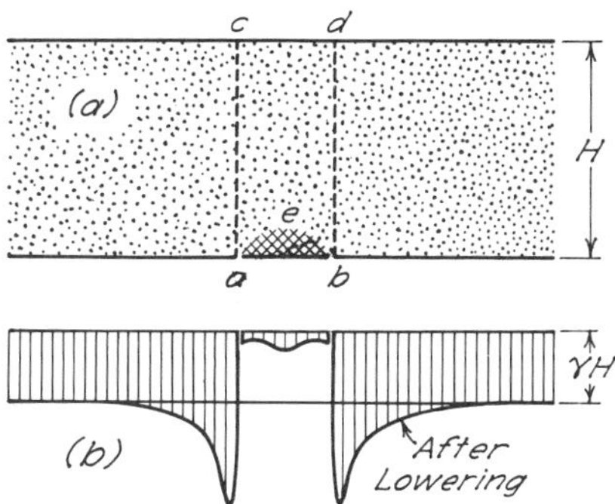

Fig. 38.1. (*a*) **Apparatus for investigating arching in layer of sand above yielding trap door in horizontal platform.** (*b*) **Pressure on platform and trap door before and after slight lowering of door.**

cally independent of the depth H of the layer of sand. It does not exceed the weight of a body of sand having approximately the dimensions indicated by the shaded area *abe* in Fig. 38.1. Hence, if the sand has a trace of cohesion, the trap door can be removed entirely, and the sand will not drop out of the gap.

Selected Reading

The "conventional" theory of arching over a conduit such as a culvert is detailed in Costes, N. C. (1956): "Factors affecting vertical loads on underground ducts due to arching," *Hwy. Res. Board Bull. 125,* pp. 12–57. The validity of the results depends, however, on the value of the lateral pressure considered to act on the assumed surfaces of rupture; no reliable means are available for predicting this pressure.

Chapter 6

SETTLEMENT AND CONTACT PRESSURE

ART. 39 INTRODUCTION

Purpose of Settlement Investigations

The term *settlement* indicates the sinking of a building due to the compression and deformation of the underlying soil.

The design of the framework of a building or other structure is, with rare exceptions, based on the assumption that the structure rests on an unyielding base. In reality, the weight of every structure compresses and deforms the underlying soil and, as a consequence, the design assumption is never strictly satisfied. If the base of the structure remains plane, the settlement is irrelevant because the stresses in the framework are not altered. On the other hand, if the weight of the structure causes the loaded area to warp, the base of the structure also becomes warped, and the entire structural framework is distorted. The supplementary stresses caused by this distortion are not considered in the design of the superstructure. Yet in many instances they are important enough to impair the appearance of a building or to cause permanent and irreparable damage.

Because of the complexity of the mechanical properties of soils and the disturbing influences of stratification, the settlement of buildings can be accurately predicted only under exceptional conditions. Nevertheless, a theoretical analysis of settlement phenomena is indispensable because the results permit the engineer at least to recognize the factors that determine the magnitude and the distribution of the settlement. Knowledge of these factors constitutes the prerequisite for converting construction experience into semiempirical rules for the design of foundations (Article 53).

Theoretical Approach to Settlement Problems

The theoretical methods for dealing with settlement problems must be chosen in accordance with the mechanical properties of the subsoil and the nature of the stratification. If a proposed structure is located above one or more layers of very compressible soil, buried beneath and separated by layers of relatively incompressible soil such as sand, the settlement depends only on the physical properties of the soft strata and on the intensity and distribution of the vertical pressure on these strata. Experience has shown that the vertical pressures can be computed with sufficient accuracy on the assumption that the subsoil of the building is perfectly elastic and homogeneous.

Similarly, if a structure rests on a fairly homogeneous subsoil, the distribution of the vertical stresses on horizontal sections can be estimated on the assumption that the subgrade is perfectly elastic. However, the intensity and distribution of all the other stresses are likely to be very different from those in an equally loaded perfectly elastic subgrade and, in addition, the determination of the stress-strain relations for the soil is commonly impracticable. Hence, in such instances it may be necessary to investigate the relation of intensity of loading, settlement, and size of loaded area by semiempirical methods.

Computation of Contact Pressure

After the designer has laid out the foundation in such a manner that the unequal settlement will not be great enough to injure the superstructure, he must design the foundation. The design requires a computation of the bending moments and shearing stresses in those parts of the foundation, such as footings or rafts, that transfer the weight of the building onto the subgrade. The pressure that acts on the base of footings or rafts is known as *contact pressure.*

The distribution of the contact pressure on the base of some foundations resembles that on the base of a similar foundation supported by an elastic isotropic material, but more often it is entirely different. Furthermore, if the supporting material is clay, the distribution of the contact pressure may change considerably with time. To simplify design, the computation of the bending moments in footings is commonly based on the arbitrary assumption that the footings rest on a uniformly spaced bed of springs. The procedure is described in Article 42. Experience has shown that it is usually accurate enough for practical purposes. Therefore, the designer needs to be familiar only with the general relationships between the type of soil and the character of the pressure distribution. If the difference between the computed and the real pressure distribution is likely to be large and on the unsafe side, the risk is eliminated by increasing the factor of safety.

ART. 40 VERTICAL PRESSURE IN SOIL BENEATH LOADED AREAS

Boussinesq's Equations

The application of a concentrated vertical load to the horizontal surface of any solid body produces a set of vertical stresses on every horizontal plane within the body. It is obvious, without computation, that the intensity of the vertical pressure on any horizontal section through the loaded soil decreases from a maximum at the point located directly beneath the load to zero at a very large distance from this point. A pressure distribution of this kind can be represented by a bell- or dome-shaped space, as indicated in Fig. 41.1*b*. Since the pressure exerted by the load spreads out in a downward direction, the maximum pressure on any horizontal section, represented by the maximum height of the corresponding bell-shaped pressure space, decreases with increasing depth below the loaded surface. Yet equilibrium requires that the total increase of pressure on any horizontal section must be equal to the applied load. Therefore, with increasing depth below the surface, the pressure bells become lower but wider.

Both theory and experience have shown that the shape of the pressure bells is more or less independent of the physical properties of the loaded subgrade. Therefore, in connection with practical problems, it is customary and justifiable to compute these stresses on the assumptions that the loaded material is elastic, homogeneous, and isotropic. On these assumptions, a concentrated vertical load Q (Fig. 40.1*a*) acting on the horizontal surface of a mass of very great extent produces at point N within the mass a vertical pressure having the intensity,

$$p_v = \frac{3Q}{2\pi z^2}\left[\frac{1}{1 + (r/z)^2}\right]^{5/2} \tag{40.1}$$

In this equation z represents the vertical distance between N and the surface of the mass, and r the horizontal distance from N to the line of action of the load.

Equation 40.1 is one of a set of stress equations, known as *Boussinesq's equations,* that determine the entire state of stress at point N (Fig. 40.1*a*). However, in contrast to the vertical pressure p_v, most of the other components of stress at point N depend to a large extent on the stress-deformation characteristics of the loaded material. Since soils are not even approximately elastic and homogeneous, the other

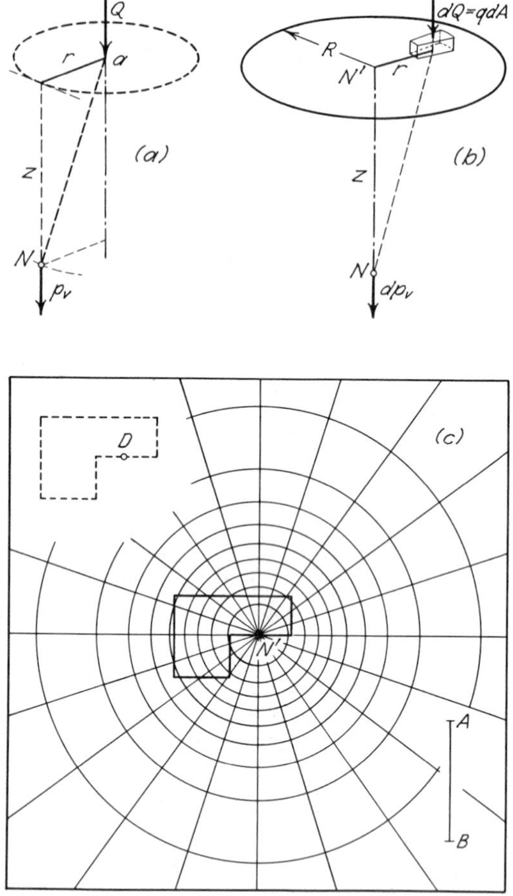

Fig. 40.1. **(*a*) Intensity of vertical pressure at point *N* in interior of semi-infinite solid acted on by point load *Q*. (*b*) Vertical pressure at point *N* beneath center of circular area acted on by load *q* per unit of area. (*c*) Diagram illustrating use of influence chart for computing vertical pressure (after Newmark 1942).**

stress equations of Boussinesq are not so generally suitable for the computation of stresses in soils.

Pressure Distribution on Horizontal Sections beneath Loaded Areas

In computing the vertical pressures in the soil beneath a building, it is commonly assumed that the building is perfectly flexible. If an area on the surface of a very large mass carries a uniformly distributed and perfectly flexible load of intensity q, the intensity of the vertical pressure at any point N (Fig. 40.1*b*) within the mass may be computed by dividing the loaded area into small parts dA, each of which sustains a load,

$$dQ = q\, dA$$

This load is considered to be concentrated at the centroid of the elementary area dA. According to Eq. 40.1, each concentrated load produces at point N a vertical pressure,

$$dp_v = \frac{3q}{2\pi z^2}\left[\frac{1}{1 + (r/z)^2}\right]^{5/2} dA \qquad (40.2)$$

The intensity of the vertical pressure at N due to the entire load is computed by integrating Eq. 40.2 over the loaded area. For example, if the point N is located at depth z beneath the center N' of a loaded area having the shape of a circle with radius R, the vertical pressure is found to be

$$p_v = q\left[1 - \left(\frac{1}{1 + (R/z)^2}\right)^{3/2}\right] \qquad (40.3)$$

If the load of intensity q is distributed over an area with a shape other than circular, the stress p_v at an arbitrary point N at depth z below this area can be computed readily with the aid of the chart (Fig. 40.2). The chart (Newmark 1942) represents a set of lines located on the ground surface. It is drawn to such a scale that the distance AB is equal to the depth z. The point N is located directly below the center of the concentric circles. The chart is so constructed that a load of intensity q distributed over any one of the smallest subdivisions bounded by two adjacent radial lines and two adjacent circles produces a pressure $p_v = 0.005q$ at point N. Each subdivision is, therefore, an *influence area,* with the value 0.005, for the stress p_v at point N.

To illustrate the use of the chart, we shall compute the value of p_v at a depth of 50 ft below point D of the building shown in plan in Fig. 40.1*c*. The weight of the building constitutes a uniformly distributed load of 3000 lb/ft^2 covering the area occupied by the building. The first step in the computation is to draw on tracing paper a plan of the building to such a scale that the depth, 50 ft, is equal to the distance AB on the chart. We then place the tracing over the chart so that point D is directly above point N' on the chart and count the number of influence areas enclosed by the outline of the loaded area. In this example the number of influence areas is 31.5, and the corresponding stress p_v at a depth of 50 ft below

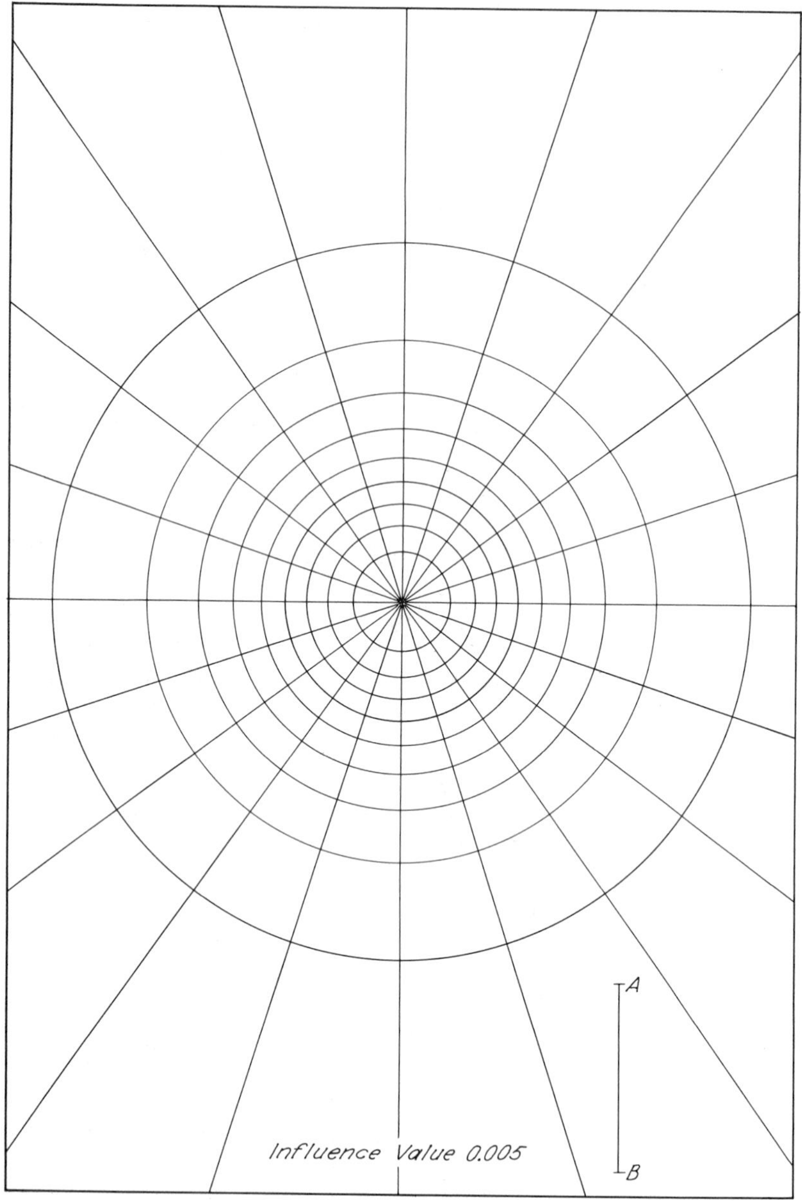

Fig. 40.2. Influence chart for vertical pressure (after Newmark 1942).

D is $31.5 \times 0.005 \times 3000 = 473$ lb/ft^2. The stress p_v at any other point at the same depth is obtained by the same procedure after shifting the tracing until the new point is directly above N'. In order to determine the stresses on a section at a different depth z_1, we draw a tracing to a different scale, such that the depth z_1 is equal to the distance AB on the chart.

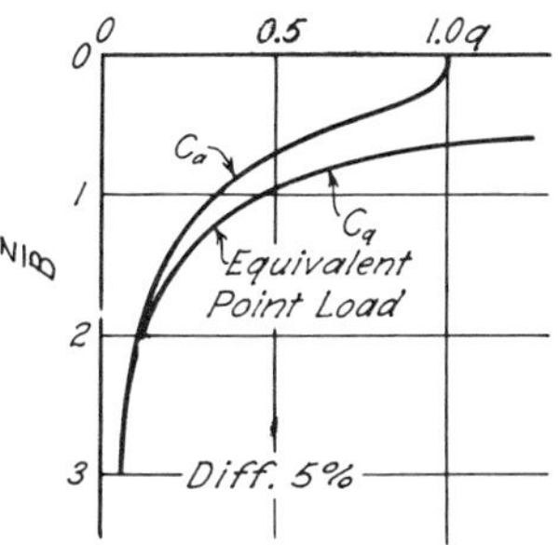

Fig. 40.3. Diagram illustrating effect on vertical pressure of replacing uniformly distributed load on square area by equivalent point load at center of square. Curves represent stress along vertical line beneath center of square.

Change of Pressure with Depth

The intensity of vertical pressure along any vertical line beneath a distributed load decreases with increasing values of the depth z below the surface. Therefore, if the compressible layer is very thick, the vertical pressure in the layer decreases appreciably from the top to the bottom. However, the compression of a thin layer depends merely on the average vertical pressure, which is roughly equal to the vertical pressure at midnight of the layer. Therefore, if the compressible layer is relatively thin, the change of pressure with depth can be disregarded and it may be sufficiently accurate to compute the intensity and distribution of the pressure on a horizontal plane at midheight of the layer.

In Fig. 40.3 the abscissas of the curve C_a represent the intensity of vertical pressure at different depths below the center of a square area $B \times B$ which carries a uniformly distributed load q per unit of area. If the total load B^2q is represented by a concentrated load $Q = B^2q$ acting at the center of the square area, the curve C_q is obtained instead of C_a. The figure shows that the curves become almost identical at a depth of about $3B$. At any depth greater than $3B$, the pressure on a horizontal section produced by loading a square area is practically the same as the pressure produced by an equivalent point load acting at the center of the loaded area. Hence, the stresses p_v on horizontal sections at a depth of more than $3B$ can be computed by means of Eq. 40.1.

The removal of soil from the space to be occupied by a basement reduces the vertical pressure at every point below the bottom of the excavation. In order to compute the resulting change in the stresses, it is assumed that the surface of the soil is located at the level of the bottom of the excavation, and that the weight of the excavated material acts in an upward direction at this level.

Problems

1. A point load of 5300 lb acts on the surface of an elastic mass of very great extent. What is the intensity of vertical pressure, due to the load, at a depth of 20 ft directly below the load? at a depth of 40 ft? at a depth of 200 ft? What is the intensity of vertical pressure at the same depths at a horizontal distance of 50 ft from the line of action of the point load?

Ans. 6.33, 1.58, 0.06; 0.045, 0.150, 0.054 lb/ft^2.

2. A circular area on the surface of an elastic mass of great extent carries a uniformly distributed load of 2500 lb/ft^2. The radius of the circle is 10 ft. What is the intensity of vertical pressure at a point 15 ft beneath the center of the circle? at a point at the same depth beneath the edge of the circle?

Ans. 1060; 640 lb/ft^2.

3. A building of very great length has a width of 120 ft. Its weight constitutes a practically uniform surcharge of 5000 lb/ft^2 on the ground surface. Between the depths of 70 and 90 ft there is a layer of soft clay. The rest of the subsoil is dense sand. Compute the intensity of vertical pressure due to the weight of the building, at the following points located in a horizontal plane at midheight of the compressible layer: directly below edge of building; 20 ft from edge toward center line; 40 ft from edge toward center line; directly below center line.

Ans. 2300; 2960; 3430; 3570 lb/ft^2.

4. If the building in problem 3 is 120 ft square, compute the stresses at the same points along a section midway between the ends of the building.

Ans. 1690; 2250; 2610; 2750 lb/ft^2.

5. The excavation for a rectangular building 200 by 120 ft in plan is 20 ft deep. The excavated material is a moist sand having a unit weight of 115 lb/ft^3. What is the reduction in vertical pressure, due to the removal of weight from the excavated area, at a point 70 ft below the original ground surface, at one corner of the building?

Ans. 560 lb/ft^2.

Selected Reading

The following references contain charts, tables, or influence values useful in the calculation of stresses in elastic materials:

Jurgensen, L. (1934). "The application of elasticity and plasticity to foundation problems," *J. Boston Soc. Civil Engrs.*, **21**, pp. 206–241. Reprinted in *Contributions to soil mechanics 1925–1940,* Boston Soc. Civil Engrs., 1940, pp. 148–183.

Newmark, N. M. (1942). "Influence charts for computation of stresses in elastic foundations," *Univ. of Ill. Eng. Exp. Sta. Bull. 338,* 28 pp.

Terzaghi, K. (1943*b*). *Theoretical soil mechanics,* New York, John Wiley and Sons, pp. 481–490.

Harr, M. E. (1966). *Foundations of theoretical soil mechanics,* New York, McGraw-Hill, pp. 55–116.

Burmister, D. M. (1956). "Stress and displacement characteristics of a two-layer rigid base soil system: influence diagrams and practical applications," *Proc. Hwy. Res. Board,* **35,** pp. 773–814.

Osterberg, J. O. (1957). "Influence values for vertical stresses in a semi-infinite mass due to an embankment loading," *Proc. 4th Int. Conf. Soil Mech., London,* **1,** pp. 393–394.

Mehta, M. R. (1959). *Stresses and displacements in layered systems,* Ph.D. thesis, Univ. of Illinois, 33 pp.

ART. 41 SETTLEMENT OF FOUNDATIONS

Foundations above Confined Strata of Soft Clay

The following paragraphs describe the procedure for estimating the settlement of a building located above a confined layer of soft clay. The weight of the building is transferred by a reinforced-concrete mat foundation (Fig. 41.1*a*) onto a stratum of sand that contains a layer of soft clay at a depth D below the mat. The weight of the building is assumed to be uniformly distributed over the area occupied by the mat.

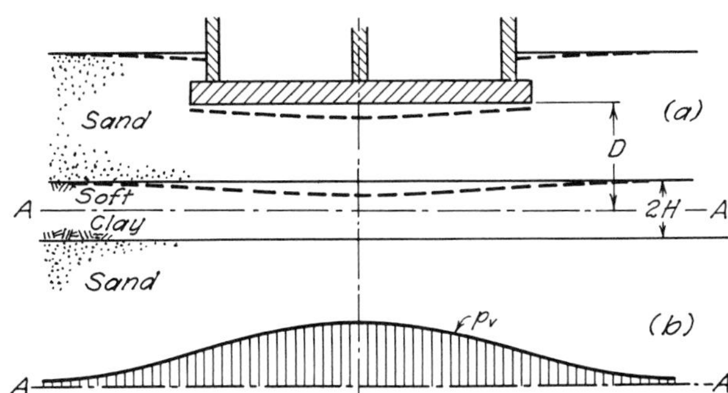

Fig. 41.1. **(*a*) Settlement of building having mat foundation supported by subsoil that contains compressible stratum at depth *D*. (*b*) Distribution of vertical pressure on horizontal plane at midheight of compressible layer.**

Inasmuch as sand is almost incompressible in comparison with soft clay, the settlement is caused almost entirely by the compression of the clay stratum. Since the clay is confined, the compression produced by a given pressure can be computed by the method described in Article 13. However, it is necessary to compute the settlement for several points on the base of the building, because the principal object of the settlement computation is to estimate the amount of warping that the base will experience. If the thickness of the clay stratum is small compared to the depth of the overburden, it can be assumed that the average intensity of vertical pressure p_v in the clay beneath a given point of the foundation is equal to the intensity of vertical pressure beneath this point at midheight of the stratum. This pressure may be evaluated by means of the chart (Fig. 40.2).

The next step is to compute the compression S of the clay layer below each of the selected points. According to Eq. 13.2, the change Δn in the porosity is given by the expression,

$$\Delta n = m_v \Delta p$$

The quantity m_v represents the average coefficient of volume compressibility (Eq. 13.3) for the range in pressure from the original value p_0 to the final value $p_0 + \Delta p$. The added pressure Δp is equal to the vertical pressure p_v computed as outlined in the preceding paragraph. Since the thickness of the compressible layer is $2H$, the change in thickness S due to the pressure p_v is

$$S = 2H\Delta n = 2Hm_v p_v \tag{41.1}$$

The value S represents not only the decrease in thickness of the stratum below the given point but also the settlement of the base of the foundation at that point. If the subsoil contains several compressible layers, the settlement of a given point on the foundation is equal to the sum of the compressions of each of the layers along the vertical line through the point.

If a clay stratum is relatively thick, or if p_v and m_v cannot be considered approximately constant through its entire thickness, we may resolve the stratum into several layers and determine p_v and m_v for each layer individually. On the other hand, we may replace Eq. 41.1 by the more general equation,

$$S = \int_0^{2H} m_v p_v \, dz \tag{41.2}$$

in which m_v and p_v are, respectively, the coefficient of compressibility and the added vertical pressure at any depth z below the point at

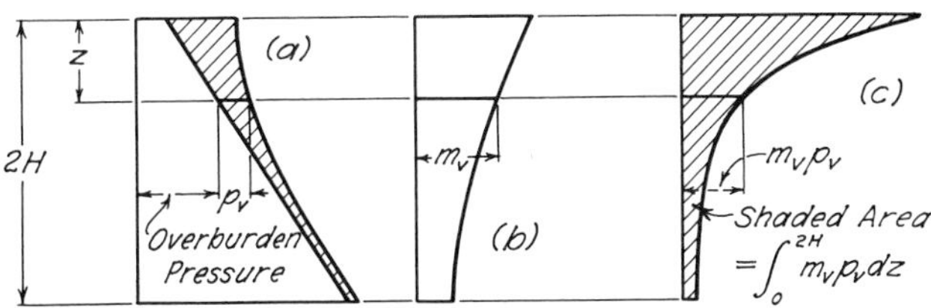

Fig. 41.2. Diagram illustrating graphical method for computing settlement of compressible layer if both vertical pressure p_v and coefficient of compressibility m_v vary with depth.

which the settlement is to be computed. The integration is performed graphically, as shown in Fig. 41.2. The added vertical pressure p_v at any depth z beneath the given point is represented by the width of the shaded area in Fig. 41.2*a*. In order to determine the right-hand boundary of the shaded area, the value of p_v must be computed for several values of z below each point. By plotting the values of m_v as abscissas and the depth as ordinates, the curve in Fig. 41.2*b* is obtained. The width of the shaded area in Fig. 41.2*c* at depth z is made equal to the product $m_v p_v$. Therefore, the entire shaded area in *c* represents the settlement S.

The compression of the clay stratum involves a decrease of the water content of the clay. Because of the low permeability of the clay, the excess water escapes very slowly and retards the compression (Article 14). The methods for computing the rate of settlement are presented in Article 25. At any given time, however, the settlement of a uniformly loaded area is trough- or dish-shaped, because the vertical pressure on the compressible layers is a maximum near the center and decreases toward the edges of the area (Fig. 41.1*b*).

Foundations on Unstratified Soil

If the subsoil of a foundation is fairly homogeneous, the weight of the building causes not only a compression of the underlying soil but also a lateral yield. Therefore, one part of the settlement may be regarded as a vertical shortening of the loaded stratum due to volume decrease, and the other as an additional shortening due to lateral bulging.

If the subgrade were perfectly elastic and homogeneous to a very great depth, the settlement due to bulging would be considerably greater than that due to volume decrease. At a given intensity of loading, the settlement of loaded areas having the same shape would

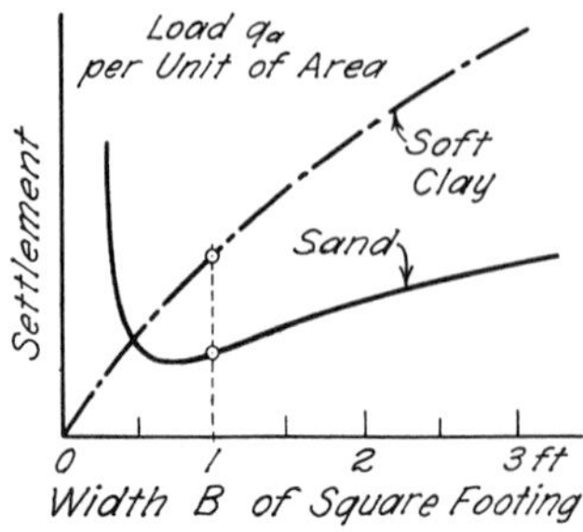

Fig. 41.3. Relation between width of square footing and settlement under same load per unit area (after Kögler 1933).

increase in simple proportion to the widths of the areas.

In connection with the settlement of loaded areas on soils, a distinction must be made between loads that rest on clay and those that rest on sand. If the subsoil consists of clay, the settlement due to lateral bulging is usually small compared to the total settlement. For this reason, even the settlement of foundations on thick strata of clay can be evaluated at least roughly by the method described under the preceding subheading. On the other hand, if the foundation rests on strata of inorganic silt or sand, the second part of the settlement is likely to be much greater than the first.

In order to determine the influence of the size of the loaded area and the position of the water table on the settlement of footings on cohesionless sand, we must consider the factors that determine the stress-strain properties of the sand (Article 15). Theoretical investigations of these relations as well as laboratory tests and field observations have led to the following conclusions (Kögler 1933).

The settlement of a footing with width B decreases with increasing average value of the initial tangent modulus E_i of the sand located between the base of the footing and a depth of about B below the base. According to Fig. 15.4, the initial tangent modulus of sand increases with increasing effective confining pressure. At any given depth below the surface of the sand, the effective confining pressure is roughly proportional to the effective overburden pressure. If the water table rises to the surface of the sand from a depth greater than B below the base of the footing, the effective confining pressure decreases by roughly 50% (Article 12). Therefore, the settlement increases approximately 100%.

At a given load per unit of area of the base of a footing, the depth of the body of sand subject to intense compression and deformation increases as the width of the footing increases. On the other hand, the ultimate bearing capacity of the footing and the average initial tangent modulus of the sand also increase. As a consequence of these several factors, the settlement varies with the width of the footing approximately as shown by the plain curve in Fig. 41.3.

In practice, the magnitude of the settlement of footings on sand cannot be predicted on the basis of the results of laboratory tests on soil specimens. However, it can be estimated roughly, by means of semiempirical rules based partly on the aforementioned general relations and partly on the observed relation between settlement and the results of simple field tests such as penetration tests (Articles 54 and 55).

Problems

1. The layer of soft clay referred to in problem 3, Article 40, has a natural water content of 45%. The unit weight of the solid matter of the clay is 2.70 gm/cm^3, and the unit weight of the dense sand is 130 lb/ft^3. The free-water level is at the ground surface. From the results of consolidation tests it has been ascertained that C_c is equal to 0.50. Compute the settlement of the edge and of the center of the building.

Ans. 8.5; 12.3 in.

2. A uniform load of 3000 lb/ft^2 is distributed over a very large area on the surface of the ground. The subsoil consists of a bed of dense sand containing two strata of clay each 10 ft thick. The top of the upper stratum is 20 ft deep, and of the lower 70 ft deep. The value of C_c is 0.35 for both layers; the natural water content and the unit weight of the solid constituents are 34% and 2.75 gm/cm^3, respectively. The sand weighs 125 lb/ft^3 and is completely submerged. What is the settlement of the ground surface due to the uniform load?

Ans. 15 in.

ART. 42 CONTACT PRESSURE AND THEORIES OF SUBGRADE REACTION

Contact Pressure on Base of Rigid Footings

Since the settlement of the base of a perfectly rigid footing is by necessity uniform, the distribution of the pressure on the base of such a footing is identical with the distribution of the load required to produce uniform settlement of the loaded area. If the subgrade consists of a perfectly elastic material, of clay, or of sand containing thick layers of soft clay, a uniformly loaded area assumes the shape of a shallow bowl or trough. In order to obtain uniform settlement it would be necessary to shift part of the load from the center of the loaded area toward the edges. Hence, the contact pressure on the base of a rigid footing resting on such a subgrade increases from the center of the base toward the rim. On the other hand, if a uniformly loaded area is underlain by sand, the settlement is greater at the edges than at the center. Uniform settlement can be obtained

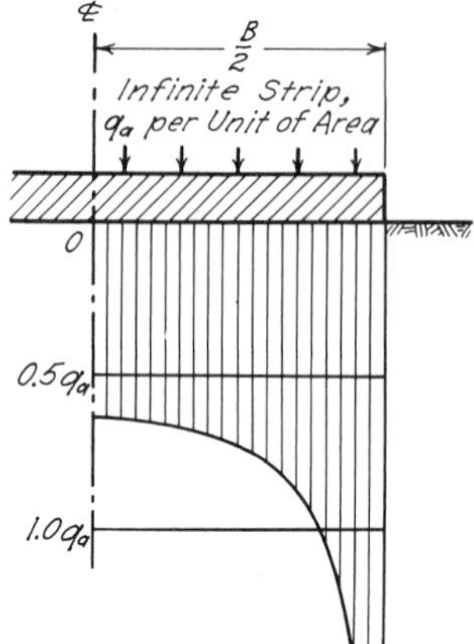

Fig. 42.1. Distribution of contact pressure on base of uniformly loaded rigid footing of very great length, resting on perfectly elastic, homogeneous, and isotropic subgrade.

only by distributing the load so that its intensity decreases from a maximum at the center to a minimum at the rim. Hence, the distribution of the contact pressure on the base of a rigid footing on sand has the same characteristics.

Figure 42.1 is a section through a rigid continuous footing with a width B resting on a perfectly elastic and homogeneous subgrade of very great depth. The load on the footing is q_aB per unit of length. Computations based on the theory of elasticity have shown that the contact pressure increases as shown in the figure from less than $0.7q_a$ at the center line to an infinite value at the edges. If the footing rests on a real elastic material, the pressure along the edges cannot exceed a certain finite value q_c at which the material passes from the elastic into a semiplastic or plastic state. The corresponding distribution of the contact pressure is shown in Fig. 42.2a by the curve C_1.

If the load on the footing in Fig. 42.2a is increased, the state of plastic equilibrium spreads from the edges, and the distribution of contact pressure changes. If the base of the footing is smooth, the

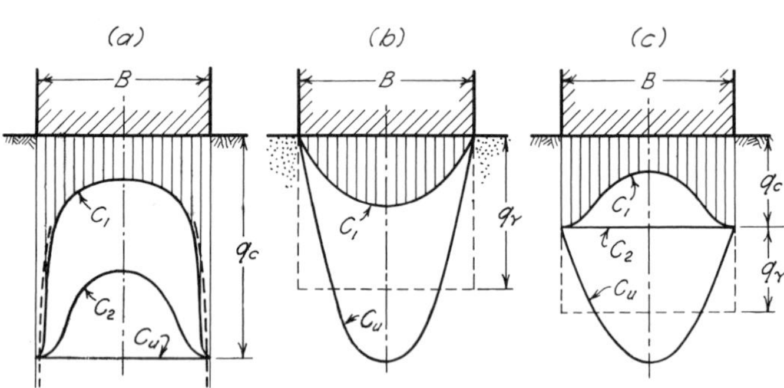

Fig. 42.2. Distribution of contact pressure on base of smooth rigid footing supported by (a) real, elastic material; (b) cohesionless sand; (c) soil having intermediate characteristics. Curves C_u refer to contact pressure when footing is loaded to ultimate value.

distribution becomes perfectly uniform at the instant when the subgrade fails by plastic flow. The curve C_u represents the distribution at this stage, and the curve C_2 at an intermediate stage.

If either a rigid or a flexible footing rests on the surface of a mass of dry cohesionless sand, theory indicates that the intensity of the contact pressure at any load decreases from a maximum at the center to zero at the edges, as shown in Fig. 42.2*b*. Experimental investigations have led to the same conclusion.

Figure 42.2*c* represents the distribution of contact pressure on the base of a footing supported by a subgrade intermediate in character between purely cohesive and purely cohesionless soils. At small loads the contact pressure increases from the center toward the edges of the footing (curve C_1). As the load increases, the pressure at the center increases whereas that at the edges remains unaltered. At the point of failure the pressure decreases from the center toward the edges, as indicated by curve C_u.

Definition of Subgrade Reaction

Figure 42.2 demonstrates that the relation between the stress-deformation characteristics of the subgrade and the contact pressure on the base of a perfectly rigid footing is by no means simple. If the footing is not rigid, the relation becomes even more complicated. Therefore, even a rough evaluation of the distribution of the real contact pressure is very cumbersome. Yet, without some knowledge of the contact pressure, footings or mats cannot be designed. Therefore, it is customary and necessary to estimate the contact pressure on the basis of simplifying assumptions and to compensate for the error due to these assumptions by an adequate factor of safety.

The simplified procedures are based on the arbitrary and incorrect assumption that the settlement S of any element of a loaded area is entirely independent of the load on the adjoining elements. It is further assumed, at variance with reality, that the ratio

$$K_s = \frac{p}{S} \tag{42.1}$$

between the intensity p of the pressure on the element and the corresponding settlement S is a constant K_s (grams per cubic centimeter). In contrast to the real contact pressure that acts on the base of the footing, the fictitious pressure p that satisfies Eq. 42.1 is called the *subgrade reaction.* In the following paragraphs of this article the symbol p is reserved strictly for the subgrade reaction.

It is not used with reference to the real contact pressure. The coefficient K_s is known as the *coefficient of subgrade reaction,* and the theories based on the aforementioned assumptions are the *theories of subgrade reaction.*

Subgrade Reaction on Rigid Foundations

In connection with a rigid foundation, Eq. 42.1 leads to the conclusion that the distribution of the subgrade reaction p over the base of the foundation must be planar, because a rigid foundation remains plane when it settles. Hence, in order to design a rigid foundation in accordance with Eq. 42.1, we merely assume that the subgrade reaction has a planar distribution. In addition, we must satisfy the requirements of statics that (1) the total subgrade reaction is equal to the sum of the vertical loads that act on the subgrade, and (2) the moment of the resultant vertical load about an arbitrary point is equal to the moment of the total subgrade reaction about that point.

As an example, the rigid gravity retaining wall shown in Fig. 42.3 is considered. The width of the base is B, and the resultant Q of the vertical loads on the base acts at the distance a from the toe. The subgrade reaction at the toe is p_a, and at the heel it is p_b. According to the previous paragraph, the distribution of the reaction is assumed to be linear between these two points. By statics we obtain the two equations,

$$Q = \tfrac{1}{2}B(p_a + p_b) \tag{42.2}$$

and

$$Qa = \tfrac{1}{6}B^2 p_a + \tfrac{1}{3}B^2 p_b \tag{42.3}$$

These equations can be solved for p_a and p_b.

It should be noted that Eqs. 42.2 and 42.3 do not contain the coefficient of subgrade reaction K_s. In other words, the distribution of subgrade reaction on the base of a rigid footing is independent of the degree of compressibility of the subgrade. This fact makes it easy to visualize the difference between the subgrade reaction and the real contact pressure. If the resultant Q of the load on a footing passes through the centroid of the loaded area A, the subgrade reaction is distributed uniformly over the base of the footing and is everywhere equal to Q/A. On the other hand, the distribution of the real contact pressure on the base of the same footing may be far from uniform, as shown by Fig. 42.2. It depends on the stress-deformation characteristics of the subgrade and on the intensity of the load.

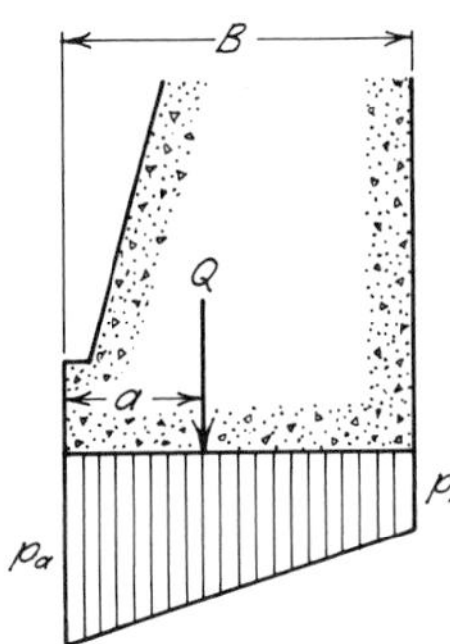

Fig. 42.3. Subgrade reaction on base of rigid gravity retaining wall.

In spite of these obvious discrepancies between theory and reality, the theories of subgrade reaction can be used safely in connection with the routine design of footings, because the errors are within the customary margin of safety and, as a rule, they are also on the safe side.

Subgrade Reaction on Flexible Foundations

If a footing or a mat is not rigid, the distribution of the subgrade reaction depends on both the numerical value of K_s and the flexural rigidity of the foundation. The influence of the latter is illustrated by Fig. 42.4, which represents a cross section through a long rectangular elastic slab. The longer axis of the slab carries a line load Q per unit of length. The slab rests on an elastic subgrade. Because of the flexibility of the slab, the settlement decreases from the center line toward the edges. Consequently, the subgrade reaction also decreases from a maximum at the center to a minimum at the edges. If the slab is very flexible, the edges may rise, and the subgrade reaction beneath the outer portions of the slab may become zero. In any event, for a given line load Q and a given width B of the slab the maximum bending moment in a flexible slab is very much smaller than that in a rigid one.

The subgrade reaction on the base of a relatively flexible member in a foundation can be computed by means of the *theory of elastic beams on a continuous elastic support*. The theory is based on the obvious fact that the vertical displacement of the loaded member due to settlement and bending must at every point be equal to the settlement of the ground surface at the same point. The computation of the settlement of the ground surface is based on Eq. 42.1. In contrast to Eqs. 42.2 and 42.3, which pertain to a rigid foundation, the equations for computing the subgrade reaction on an elastic foundation always contain the value K_s (Eq. 42.1).

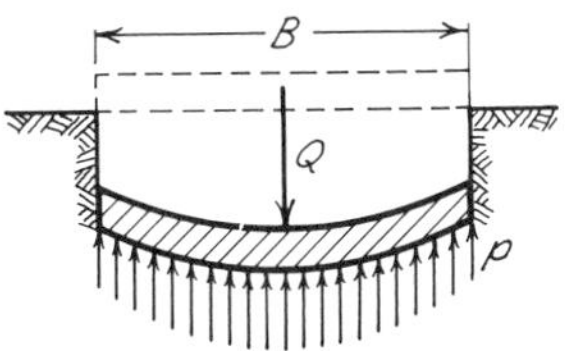

Fig. 42.4. Elastic footing of great length acted on by line load and supported by elastic subgrade. (*a*) Deformation of footing under load. (*b*) Distribution of subgrade reaction.

Since the theory of elastic beams on a continuous elastic support is based on

Eq. 42.1, it is no more accurate than the theory of subgrade reaction for rigid footings. It can be even less accurate, because it involves the error associated with evaluating K_s. Since the computations are always cumbersome, the investigation is not justified unless it leads to a considerable saving in the cost of the structure.

In all the theories of subgrade reaction, the coefficient K_s, which is the ratio between the intensity of load on the fictitious subgrade and the corresponding vertical displacement, is assumed to be a constant that depends only on the physical properties of the subsoil. However, the ratio between the average intensity of pressure on the surface of a given solid and the corresponding settlement is not a constant. For circular footings on an elastic isotropic base the ratio decreases as the radius of the footings increases. For a footing of given size resting on soil it also decreases with increasing values of the intensity of the load. Furthermore, it is different for different points at the base of the same footing. Therefore, the evaluation of K_s involves many uncertainties, and the customary procedure for determining K_s on the basis of small-scale load tests is subject to all the limitations of the load-test method described in Article 54.

Problems

1. A gravity retaining wall has a base width of 8 ft. The line of action of the resultant of the vertical and horizontal forces intersects the base at a point 3 ft from the toe of the wall. The vertical component of the resultant force is 12,000 lb/ft. What is the subgrade reaction at the toe? at the heel?

Ans. 2625 lb/ft^2; 375 lb/ft^2.

2. A footing with a trapezoidal base is 12 ft long, 3 ft wide at one end, and 6 ft wide at the other. It supports two columns along its center line, one at a distance of 2 ft from the narrow end and the other 3 ft from the wide end. The load on the first column is 18 tons, and on the second 36 tons. Assuming that the footing is rigid, what is the subgrade reaction at each end?

Ans. 2000 lb/ft^2.

Selected Reading

Hetényi, M. (1946). *Beams on elastic foundation,* Ann Arbor, Univ. of Michigan Press, 255 pp.

Terzaghi, K. (1955*b*). "Evaluation of coefficients of subgrade reaction," *Géot.*, **5,** No. 4, pp. 297–326.

PART III

Problems of Design and Construction

Part I contains a description of real soils as disclosed by laboratory tests. Part II consists of a condensed review of the theoretical procedures available for predicting the performance of ideal materials having properties approximating those of the real soils. Before the theories can be applied to the solution of problems of design and construction, two independent operations must be carried out. First, the significant properties of the subsurface materials must be determined by boring, sampling, and testing. Second, an idealized subsoil, consisting of a few homogeneous units with simple boundaries, must be substituted for the much more complex real one.

In the few instances in which the real soil profile is simple enough to be replaced without intolerable error by an appropriate idealization, theory combined with the results of soil testing makes possible a prediction of the performance of soil-supported structures on a mathematical basis. This procedure has been used successfully, for instance, to predict the magnitude and distribution of the settlement of structures founded above horizontal clay strata of fairly uniform thickness.

In all other instances the results of subsoil exploration inform the designer merely about the general characteristics of the subsurface materials and the location within them of potential sources of trouble. The detailed characteristics of these sources remain unknown, although even if they were known the time and labor involved in securing the data required for an accurate forecast of performance would be prohibitive. Under these circumstances, the designer can do no better than to construct idealized soil profiles showing approximately the outer boundaries of the potentially troublesome weak or compressible zones, and to assign to the materials located within these zones the most unfavorable properties compatible with the available data. The estimates of performance based on these profiles can furnish only upper limiting values for the undesirable consequences of the presence of the zones, but even the knowledge of these values enables the designer to avoid the undesirable consequences by appropriate design. Before the means for investigating the properties of subsurface materials were developed and before the theoretical principles of subsoil behavior were established, the significance or even the existence of troublesome zones remained undetected until they were disclosed by the unanticipated performance of the structures resting on the subsoil.

Part III contains a review of methods of subsoil exploration and of the inevitable uncertainties associated with the results. It then deals with the practice of subsurface engineering, with the potential sources of trouble that may be encountered in that practice, and with the means at our disposal to anticipate and avoid the detrimental consequences of the potential sources of trouble.

Chapter 7

SOIL EXPLORATION

ART. 43 PURPOSE AND SCOPE OF SOIL EXPLORATION

Definition of Soil Exploration

The design of a foundation, an earth dam, or a retaining wall cannot be made in an intelligent and satisfactory manner unless the designer has at least a reasonably accurate conception of the physical properties of the soils involved. The field and laboratory investigations required to obtain this essential information constitute the *soil exploration*.

Until a few decades ago soil exploration was consistently inadequate because rational methods for soil testing had not yet been developed. On the other hand, at the present time the amount of soil testing and the refinements in the techniques for performing the tests are often quite out of proportion to the practical value of the results. In order to avoid either of these extremes, it is necessary to adapt the exploratory program to the soil conditions and to the size of the job.

Influence of Soil Conditions on Exploratory Program

If the foundation of an important structure is to be established above a fairly homogeneous layer of clay, a considerable amount of soil testing by expert laboratory technicians may be justified because the test results permit a relatively accurate forecast of both the amount and the time rate of settlement. On the basis of such a forecast, it may be possible to eliminate the danger of harmful differential settlement at reasonable expense by appropriate distribution of the loads or by suitable adjustment of the depths of subbasements beneath different parts of the structure. On the other hand, if a similar structure is to be located above a deposit composed of pockets and lenses of sand, clay, and silt, the same amount of testing would add very little to the information that could be obtained merely by determining the index properties of several dozen representative samples extracted from exploratory drill holes. Additional data of far greater significance than those obtainable from extensive soil tests could be secured in a shorter time and at less expense by means of simple subsurface soundings along closely spaced vertical lines, because such soundings would disclose whatever weak spots might be located between drill holes. The discovery of such spots is more important than an accurate knowledge of the properties of random samples.

The preceding remarks demonstate that, if the soil profile is complex, an elaborate program of soil testing is likely to be out of place. Hence, the methods of soil exploration must be chosen in accordance with the type of soil profile at the site of the construction operations. The following paragraphs describe the significant characteristics of the principal types of soil profiles commonly encountered in the field.

The term *soil profile* indicates a vertical section through the subsoil that shows the thickness and sequence of the individual strata. The term *stratum* is applied to a relatively well-defined layer of soil in contact with other layers of conspicuously different character. If the boundaries between strata are more or less parallel, the soil profile is said to be *simple* or *regular*. If the boundaries constitute a more or less irregular pattern, the soil profile is called *erratic*.

From the ground surface to a depth of about 6 ft, and exceptionally to a greater depth, the physical properties of the soil are influenced by seasonal changes of moisture and temperature and by such biological agents as roots, worms, and bacteria. The upper part of this region is known as the A-horizon. It is subject primarily to the mechanical effects of weathering and to the loss of some constituents due to leaching. The lower part is referred to as the B-horizon, where part of the substances washed out of the A-horizon are precipitated and accumulate.

The properties of the soils in the A- and B-horizons are chiefly the concern of agronomists and road builders. Foundation and earthwork engineers are interested primarily in the underlying parent material. Beneath the B-horizon the character of the soil is determined only by the raw materials from which it is derived, by the method of deposition, and by subsequent geological events. The individual strata that constitute the soil profile beneath the B-horizon may be fairly homogeneous, or they may be composed of smaller elements having properties that depart more or less from the average. The shape, size, and arrangement of these smaller elements determine the *primary structure* of the deposit. Since most soils have been deposited under water, the most common primary structure is *stratification.* If the individual layers are not thicker than about 1 in. and are of roughly equal thickness, the soil is called *laminated.* For example, the varved clays described in Article 2 are laminated soils. The action of ice, landslides, torrential streams, and several other agents leads to the formation of deposits with an *erratic structure.* Such deposits have no well-defined pattern. The more the structure of a mass of soil approaches the erratic type, the more difficult it is to determine the average values of the soil properties and the more uncertain is the result.

In stiff clays and other soils with great cohesion the primary structure may be associated with a *secondary structure* that develops after the soil is deposited. Most important among the secondary structural characteristics are systems of hair cracks, joints, or slickensides. Hair cracks and joints occur commonly in flood-plain clays consisting of layers, each of which was temporarily exposed to the atmosphere after deposition. Shrinkage caused cracks to form during the period of exposure. Slickensides are smoothly polished surfaces that may be the result of volume changes produced by chemical processes or of deformations produced by gravity or tectonic forces involving slippage along the walls of existing or newly formed joints.

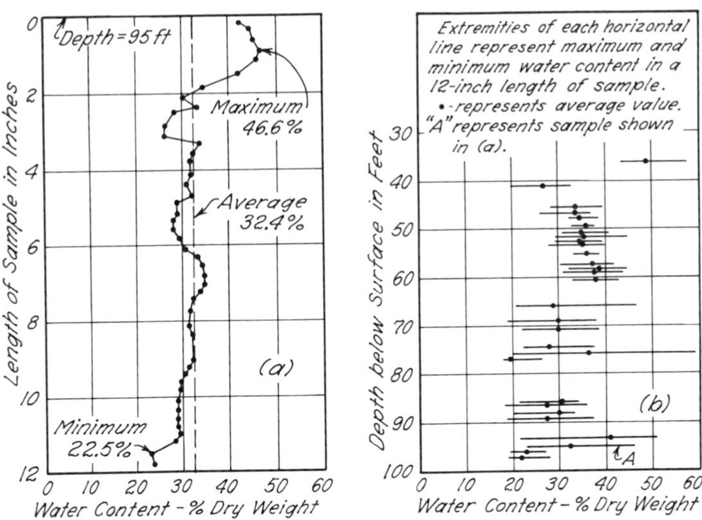

Fig. 43.1. Variation in natural water content of clay from one drill hole in Boston. (*a*) Variation within vertical distance of one foot. (*b*) Variation throughout entire boring (after Fadum 1948).

If a cohesive stratum has a well-developed secondary structure, the results of laboratory tests may give an erroneous conception of its mechanical properties. Therefore, in connection with such soils, the only guides on which the engineer can rely are his judgment based on experience with similar materials and, in some instances, on large-scale field tests.

Experience has shown that the physical properties of almost every natural soil stratum vary to a considerable extent in the vertical direction and to a smaller degree in horizontal directions. This fact is strikingly demonstrated by the variation in natural water content of clays that appear on visual inspection to be homogeneous. The results of an investigation of the variation in water content within a layer of clay in Boston are shown in Fig. 43.1. The variations within a 1-ft layer are shown in Fig. 43.1*a*, and those within a 60-ft layer in Fig. 43.1*b*. If a mass of clay appears to be nonhomogeneous, its water content is likely to vary with depth in a manner as erratic as that shown in Fig. 43.2.

If a soil stratum is of the erratic type, adequate information concerning the variations in the soil properties can be obtained only by securing continuous cores from top to bottom of the stratum and performing soil tests on every part of the core material, or else by performing suitable tests in the field. Field tests of one type, exemplified by subsurface soundings, furnish continuous records of the variations in penetration resistance of the stratum. Those of a second type, repre-

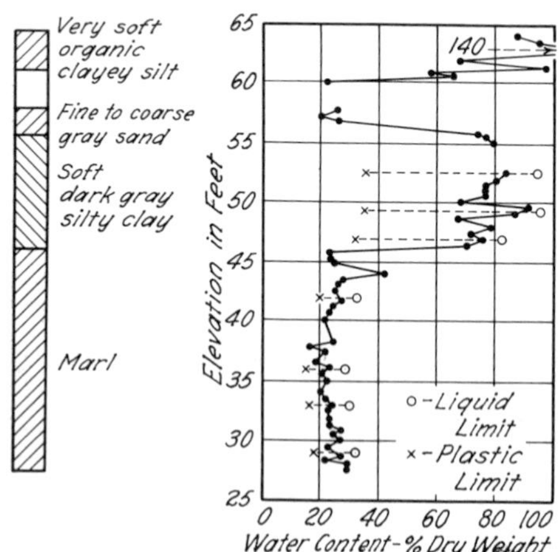

Fig. 43.2. Variation in natural water content of samples from boring in composite shore deposit.

sented by pumping tests for determining the coefficient of permeability, furnish average values of the soil property under investigation.

Influence of Size of Project on Exploratory Program

In the preparation of a program for soil exploration the magnitude of the job must also be considered. If the proposed construction operation involves only a small expenditure, the designer cannot afford to include more in the investigation than a small number of exploratory borings and a few classification tests on representative soil samples. The lack of accurate information concerning the subsoil conditions must be compensated by the use of a liberal factor of safety in design. On the other hand, if a large-scale construction operation of the same kind is to be carried out under similar soil conditions, the cost of even a thorough and elaborate subsoil investigation is usually small compared to the savings that can be realized by utilizing the results in design and construction, or compared to the expenditures that would arise from a failure due to erroneous design assumptions. Hence, on large projects extensive subsoil investigations are likely to be justified.

In order to adapt the exploratory program to the requirements of a given job and obtain the essential data at minimum expenditure of time and money, the engineer in charge must be familiar with the tools and processes available for exploring the soil, with the methods for analyzing and digesting the results of laboratory and field tests, and with the uncertainties involved in the results obtained by the different methods of soil exploration. These subjects are discussed in the following two articles.

Causes of Misjudgment of Subsoil Conditions

No matter what may be the subsoil conditions and the program for borings and soundings, the exploration furnishes information only concerning the sequence of materials along vertical lines, commonly spaced no closer than 50 ft, and concerning the significant physical properties of what are believed to be representative samples. On the basis of this rather fragmentary information, the designer is compelled to construct a soil profile by interpolation between drill holes and samples, to divide the subsoil into zones consisting of materials with approximately the same engineering properties, and to estimate for each zone the average values of the soil parameters that appear in his equations. Thereafter he forgets the real soils and operates with fictitious materials. Hence, the degree of reliability of the results of his computations depends entirely on the differences between the real and the ideal subsoil. If an unfavorable difference of an essential nature has escaped his attention, the design he has prepared on the basis of his data may turn out to be unsatisfactory in spite of conscientious subsoil exploration.

Experience has shown that the causes of fatal misjudgment of the subsoil conditions may be divided into three categories:

1. Influence on the test results of excessive sample disturbance or of significant differences between test and field conditions.
2. Failure to recognize or judge correctly the most unfavorable subsoil conditions compatible with the field data.
3. Inadequate contact between the design and construction organizations, resulting in failure to detect significant departures of conditions or of construction procedures from those the designer anticipated or specified (Terzaghi 1958*a*, 1963).

Observations during Construction

Design on the basis of the most unfavorable assumptions is inevitably uneconomical, but no other procedure provides the designer in advance of construction with the assurance that the soil-supported structure will not develop unanticipated defects. However, if the project permits modifications of the design during construction, important savings can be made by designing on the basis of the most probable rather than the most unfavorable possibilities. The gaps in the available information are filled by observations during construction, and the design is modified in accordance with the findings. This basis of design may be called the *observational procedure.*

The observational procedure has been practiced successfully throughout the ages in tunnel engineering, because the construction of permanent tunnel linings is usually preceded by the installation of temporary supports, and the observation of the performance of the temporary supports furnishes all the information required for adapting the design of the permanent lining to unanticipated unfavorable subsoil conditions. On the other hand, in earth-dam and foundation engineering the permanent structures are designed before the construction operations start, and the consequences of unanticipated sources of trouble do not appear until the structure is in an advanced state of construction.

In order to use the observational procedure in earthwork engineering, two requirements must be satisfied. First of all, the presence and general characteristics of the weak zones must be disclosed by the results of the subsoil exploration in advance of construction. Secondly, special provisions must be made to secure quantitative information concerning the undesirable characteristics of these zones during construction before it is too late to modify the design in accordance with the findings. These requirements could not have been satisfied until the mechanics of interaction between soil and water were clearly understood and adequate means for observation were developed. Depending on the nature of the project, the data required for practicing the observational procedure are obtained by measuring pore pressures and piezometric levels; loads and stresses; horizontal, vertical and angular displacements; and quantity of seepage. The means for making the measurements are described in Chapter 12. Some examples of the observational procedure are given in Chapters 8 and 11. More detailed accounts of modifications of design during construction are given in the list of suggested reading at the end of this article.

Selected Reading

Examples of the observational procedure, in which the design was modified as a consequence of observations during construction, are contained in the following references.

Graftio, H. (1936). Some features in connection with the foundation of Svir 3 hydro-electric power development. *Proc. 1st Int. Conf. Soil Mech.*, Cambridge, Mass., **1**, pp. 284–290. Note especially the means to adapt the design and construction to the elastic properties of the ground.

FitzHugh, M. M., J. S. Miller and K. Terzaghi (1947). "Shipways with cellular walls on a marl foundation," *Trans. ASCE,* **112,** pp. 298–324.

Zeevaert, L. (1957). "Foundation design and behavior of Tower Latino Americana in Mexico City," *Géot.*, **7,** No. 3, pp. 115–133.

Casagrande, A. (1960*d*). "An unsolved problem of embankment stability on soft ground," *Proc. 1st Panamerican Conf. Soil Mech. and Found. Eng.*, Mexico, **2,** pp. 721–746.

Terzaghi, K. (1960*d*). "Stabilization of landslides," Series of memoranda contained in *From theory to practice in soil mechanics,* New York, John Wiley and Sons, pp. 409–415.

Terzaghi, K. and T. M. Leps (1960). "Design and performance of Vermilion dam," *Trans. ASCE,* **125,** pp. 63–100.

Terzaghi, K. and Y. Lacroix (1964). "Mission dam, an earth and rockfill dam on a highly compressible foundation," *Géot.*, **14,** pp. 14–50.

Casagrande, A. (1965). "Role of the 'calculated risk' in earthwork and foundation engineering," *ASCE J. Soil Mech.*, **91,** No. SM4, July, pp. 1–40.

ART. 44 METHODS OF SOIL EXPLORATION

Principal Procedures

Every subsurface exploration should be preceded by a review of all available information concerning the geological and subsurface conditions at or near the site. In most instances this information must be supplemented by the results of more direct investigations. The first direct step is usually to drill a few holes into the ground by an expedient method, and to obtain fairly intact samples of soil from every stratum encountered by the drilling tools. In addition, more refined sampling operations, field tests, or both may be required. The samples provide material for an investigation of the soil properties by means of laboratory tests. Field tests such as subsurface soundings, in-place shear tests, or pumping tests supply direct information concerning the details of the soil profile and values for the physical properties of the soils *in situ*.

In recent years geophysical methods of exploration have been adapted to the purposes of civil engineering. By means of observations at the ground surface they provide information regarding the position of the boundary between soil and rock. If the rock is sound and its upper surface is not too uneven, the position and topography of the rock surface can be determined more cheaply and rapidly than by means of borings. Under favorable conditions geophysical methods have also been successfully used to determine the location of the boundaries between different soil strata and to obtain information about the physical properties of these strata. However, in a great many instances the results of such surveys have been utterly misleading; hence, geophysical methods should not be relied upon unless the findings are adequately checked by borings or other direct means of investigation.

The methods by which samples are obtained are adapted to the requirements of the project. On the other hand, the procedures for drilling the holes through which the samplers are inserted into and removed from the ground are determined largely by economy and site conditions. As a rule any one of several methods of drilling may be used in connection with a given sampling procedure. Therefore, in the following sections the methods of boring and sampling will be described separately.

Boring

METHODS OF DRILLING. The cheapest and most expedient procedures for making borings are wash boring, rotary drilling, and auger drilling. Shallow holes up to about 10 ft deep are commonly made with augers. To make deeper borings, any of the methods can be used.

WASH BORINGS. The most primitive equipment in common use for making a wash boring (Mohr 1943) usually includes a set of 5-ft lengths of pipe $2\frac{1}{2}$ in. in diameter, known as *casing*, which serves to support the walls of the hole; a weight for driving the casing into the ground; a derrick for handling the weight and casing; and wash pipe, 1 in. in diameter, in 5-ft or 10-ft lengths. A hose connection is made through a swivel head to the top of the wash pipe, and the lower end of the pipe is fitted with a chopping bit (Fig. 44.2) provided with water ports so that the wash water can be pumped down the wash pipe and forced out of the ports. The equipment also includes a tub to store the water and a hand- or power-operated pump.

In order to start a wash boring (Fig. 44.1), the derrick is erected and a 5-ft length of casing is driven about 4 ft into the ground. A tee is attached to the top of the casing with its stem in a horizontal position, and a short pipe is inserted horizontally in the stem. The tub is placed under the end of the short pipe and filled with water. The wash pipe is lifted to a vertical position by means of a hand

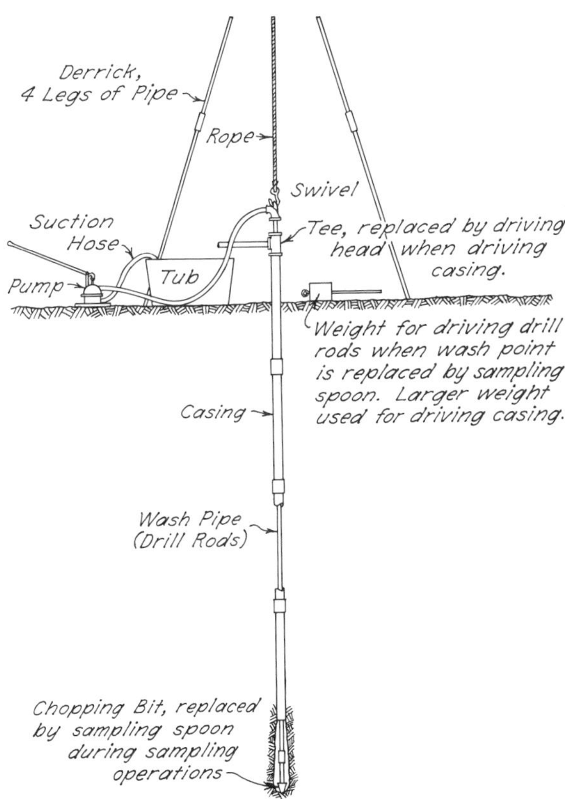

Fig. 44.1. Apparatus for making wash boring (after Mohr 1943).

rope that passes over a pulley at the top of the derrick and is lowered into the top of the casing. The pump is started and water is circulated from the tub through the swivel head into the wash pipe, whence it emerges at the chopping bit and rises in the annular space between the wash pipe and the casing. It returns to the tub, carrying cuttings of soil, through the tee and horizontal pipe at the top of the casing. As the water circulates, the wash pipe is churned up and down and is rotated at the bottom of each stroke to cut the soil loose. The hole is advanced by the churning and washing, and additional casing is driven as needed.

While drilling proceeds, the driller observes the color and general appearance of the mixture of soil and water that comes out of the hole. Whenever a conspicuous change is noticed, the wash water is turned off and a spoon sample (see page 303) is taken. Spoon samples are also secured, one for each 5 ft of depth, if the character of the subsoil appears to remain unaltered. Departures from this procedure should not be tolerated, because they may lead to serious misjudgment of the subsoil conditions. Even if the sampling is conscientiously done, the presence of clay strata with a thickness of several feet, located between sand strata, may remain unnoticed.

When the boring operations in pervious soils are discontinued to take a spoon sample, the water should be allowed to come to equilibrium in the casing. At this stage the elevation of the water table should be determined and recorded. It is not uncommon for water to rise from deeper strata to very much higher elevations than from the upper strata. Failure to recognize such a condition may be of serious consequence. In rare instances the reverse condition may be encountered.

The simple equipment described in the preceding paragraphs has the advantage that an experienced and conscientious driller, by the feel of the wash pipe as it is churned and rotated and by the color of the wash water, can usually detect changes in the character of the materials. Therefore, he can often establish the elevations of the boundaries between lenses or strata with reasonable accuracy and can stop drilling to take samples representative of all the materials penetrated. The other methods of drilling, or more elaborate wash-boring equipment, do not share this advantage. Nevertheless, they are widely used because of their economy and speed. Their shortcomings with respect to the detection of changes in the subsurface conditions must be compensated by more frequent or even continuous sampling.

Rotary Drilling. The essential features of rotary drilling are similar to those of wash boring except that the drill rods and cutting bit are rotated mechanically while the hole is being advanced. The cutting bit contains ports from which the circulating water emerges and lifts the cuttings as it rises in the annular space outside the drill rods. The rods while rotating are pressed downward mechanically or hydraulically. They can be withdrawn and the cutting bit replaced by a sampling spoon whenever a sample is required.

In rotary drilling the circulating fluid often consists not of water but of *drilling mud,* usually a suspension of bentonite of a creamy consistency and a specific gravity of 1.09 to 1.15. The higher unit weight of the fluid facilitates removal of the cuttings, and the slightly thixotropic character of the mud helps to prevent the accumulation of cuttings at the bottom of the hole in the interval of time between drilling and sampling. Moreover, the mud forms a thin layer of cohesive material on the walls of the hole which usually prevents the caving of those parts of the hole that are located in soil with little or no cohesion. Therefore, except for a short section at the top of the hole, casing may often be unnecessary.

The use of drilling mud eliminates the possibility of determining the piezometric levels corresponding to the various pervious strata through which the hole may pass.

AUGER BORINGS. Shallow borings are almost universally made by means of augers. The auger, usually of the type shown in Fig. 44.2*a*, is turned into the soil for a short distance and then withdrawn with the soil clinging to it. The soil is removed for examination, the auger again inserted into the hole and turned further. If the hole fails to stand open to permit the insertion of the auger because of squeezing from the sides or because of caving, it must be lined with a casing having an inside diameter somewhat larger than the diameter of the auger. The casing should be driven to a depth not greater than that of the top of the next sample and should be cleaned out by means of the auger. The auger is then inserted into the clean hole and turned below the bottom of the casing to obtain the sample. Auger borings cannot be made in sand below the water table because the material will not adhere to the auger.

Cohesive soil brought to the surface by the auger contains all its solid constituents, but the structure of the soil is completely destroyed, and the water content is likely to be greater than that of the soil in place. Hence, the use of augers as drilling tools does not eliminate the necessity for obtaining spoon samples whenever the drill hole reaches a new stratum. Only the spoon samples should be considered representative of the character of the undisturbed soil.

If a relatively firm stratum such as a layer of gravel is located

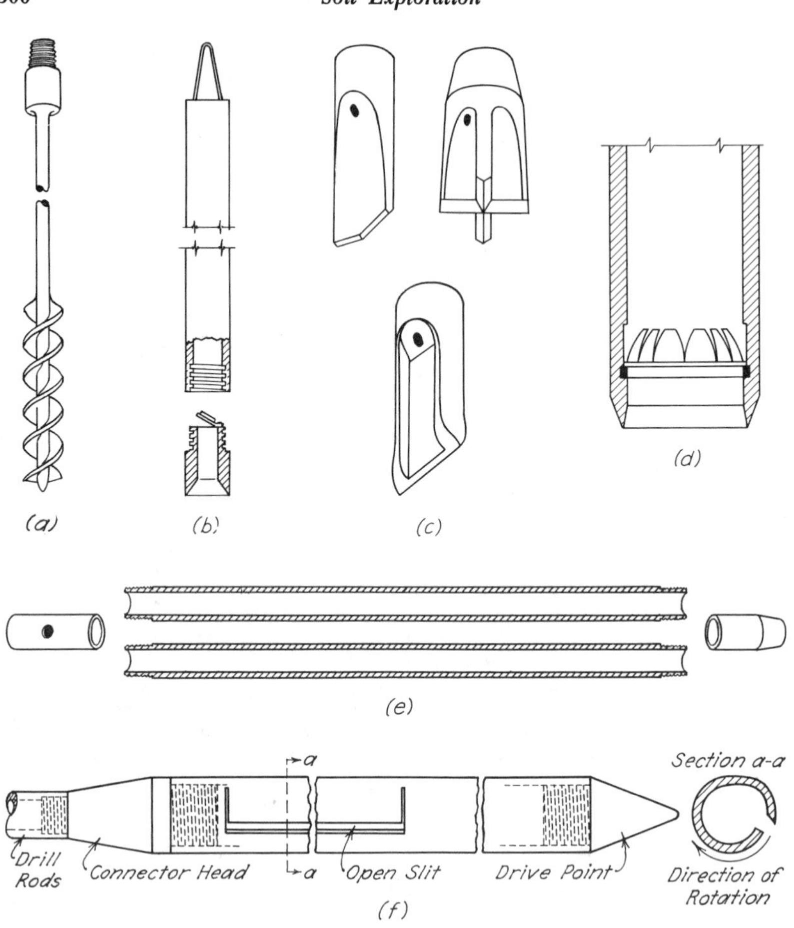

Fig. 44.2. Sampling tools for exploratory borings. (*a*) Earth auger. (*b*) Bailer. (*c*) Chopping bits. (*d*) Spring core catcher. (*e*) Split spoon sampler. (*f*) Scraper bucket.

above a very soft one, it is not uncommon for an auger boring to fail to disclose the real position of the boundary between the two strata. In one instance the existence of an 8-ft stratum of soft clay between two thick gravel layers was overlooked. In another the boundary between a bed of gravel and an underlying stratum of soft clay was reported 10 ft below its real position. Errors of this type are caused by driving the casing below the level at which the auger is operating. The casing pushes or drags stony material into the clay layer. The errors can be avoided by keeping the cutting tool as far in advance of the casing as the character of the soil permits.

By means of mechanized equipment, auger borings can also be made to depths greater than 100 ft and with diameters up to several feet. *Continuous flight augers* consist of segments that can be turned into the ground whereupon another segment is attached to the upper end, the assembly is again turned into the ground, another segment is attached, and the procedure repeated. The cuttings rise to the surface on the spirals, but the depth from which any given material comes cannot be ascertained. Therefore, the auger must be withdrawn frequently to permit examination of the material clinging to the bottom, or preferably to permit sampling. The *hollow-stem auger* (Fig. 44.3), a variation of the continuous flight auger, permits sampling below the bottom of the auger without removing the auger from the hole. It also eliminates the need for casing.

FIELD RECORDS OF EXPLORATORY BORINGS. Regardless of the procedure used for making an exploratory boring, the field notes kept by the foreman or the supervising engineer should contain the date when the boring was made, the location of the boring with reference to a permanent system of coordinates, and the elevation of the ground surface with respect to a permanent bench mark. They should include the elevations at which the water table and the upper boundary of each of the successive soil strata were encountered, the foreman's classification of the layers, and the values of the resistance obtained by means of the standard penetration test (page 304). The type of tools used in making the boring should be recorded. If the tools were changed, the depth at which the change was made and the reason for change should be noted. Incomplete or abandoned borings should be described with no less care than successfully completed drill holes. The notes should contain everything of significance observed on the job, such as the elevations at which wash water was lost from the hole.

If the base of a foundation is to be located below the water table, it is advisable to transform at least one drill hole into an observation well and to record the movement of the water table during construction. If concrete is to be placed beneath the water table, water samples of about 1 gal should be taken from several drill holes for a chemical analysis to determine whether the water contains detrimental constituents in sufficient quantity to attack the concrete. If there are any indications that the water contains gas, the analysis should be made at the site immediately after the samples are taken.

The information contained in the field notes should be assembled in the form of boring logs in which the boundaries between the strata are plotted at their correct elevation on a suitable vertical scale.

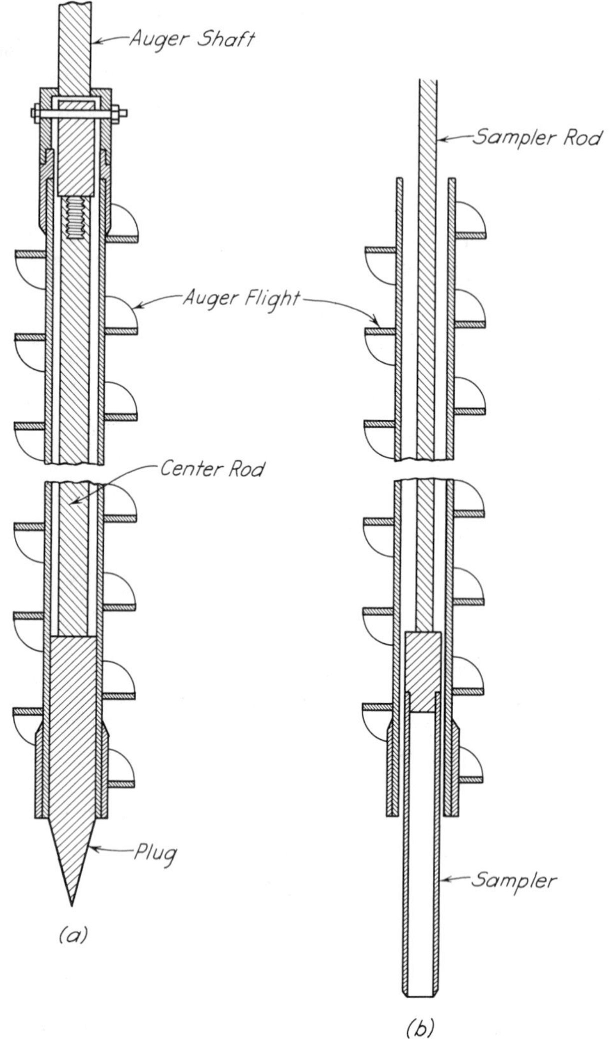

Fig. 44.3. Hollow stem auger. (*a*) Plugged while advancing auger. (*b*) Plug removed and sampler inserted to sample soil below auger.

Sampling

PURPOSE. The cuttings or washings from exploratory drill holes are inadequate to furnish a satisfactory conception of the engineering characteristics of the soils encountered, or even of the thicknesses and depths of the various strata. On the contrary, such evidence more often than not is grossly misleading and has been responsible for many foundation failures.

The proper identification of the subsurface materials requires that samples be recovered containing all the constituents of the materials in their proper proportions. Moreover, the evaluation of the appropriate engineering properties, such as the strength, compressibility, or permeability, may require the performance of laboratory tests on fairly intact or even virtually undisturbed samples. The expenditure of time and money increases rapidly as the requirements become more stringent with respect to the degree of disturbance that can be tolerated, and with increasing diameter of sample. Therefore, on small projects or in the initial exploratory stages of large or complex projects, it is usually preferable to obtain relatively inexpensive, fairly intact samples from the exploratory drill holes. On the basis of the information obtained from these samples, the necessity for more elaborate sampling procedures can be judged.

Split-spoon Sampling in Exploratory Drill Holes. In order to obtain soil samples from exploratory drill holes, a *sampling spoon* is attached to the bottom of the wash pipe or the drill rod in place of the bit and is lowered to the bottom of the hole. It is forced or driven into the soil to obtain a sample and is then removed from the hole.

Sampling spoons for exploratory borings commonly consist of a pipe with an inside diameter of about $1\frac{1}{2}$ in. and a length of 1 to 2 ft. The pipe is split lengthwise, as shown in Fig. 44.2*e*. Consequently the sampler is called a *split spoon*. While the sample is being taken, the two halves of the spoon are held together at the ends by short pieces of threaded pipe. One piece serves to couple the spoon to the wash pipe. The other, which has been sharpened, serves as the cutting edge while the spoon is driven into the soil.

According to the usual practice, the soil is extracted from the spoon by the foreman, who inspects and classifies the material and places a small portion of it in a glass jar that is covered tightly and shipped to the engineer for visual inspection. Preferably, fairly large samples should be removed from the spoon, sealed in airtight jars, carefully identified, and shipped to a laboratory for determination of the index properties. Only part of each sample should be used for the tests. The remainder should be retained in the jars, to be available for inspection by the bidders.

Clay samples obtained by means of a sampling spoon retain at least part of the characteristics of the undisturbed soil. On the other hand, samples of soils with a high permeability are almost always thoroughly compacted whether the soil *in situ* is loose or even fairly dense. Hence, the samples fail to inform the investigator on the relative density of the soil although, as a rule, this property is far more significant than the character of the soil grains themselves.

The simplest method of obtaining at least some information concerning the degree of compactness of the soil *in situ* consists of counting the number of blows of the drop weight required to drive the sampling spoon into the soil for a distance of 1 ft. A weight of 140 lb and a height of fall of 30 in. are considered standard. The spoon has the dimensions shown in Fig. 44.4. It is attached to the drill rods and lowered to the bottom of the drill hole after the hole has been cleaned by means of a water jet or an auger. After the spoon reaches the bottom, the drop weight is allowed to fall on the top of the drill rods until the sampler has penetrated about 6 in. into the soil, whereupon the penetration test is started, and the foreman records the number of blows required to produce the next foot of penetration. This procedure is referred to as the *standard penetration test.* Since this test furnishes vital information with very little extra effort, it should never be omitted.

In cohesionless or nearly cohesionless sand located below the water table, the sand is likely to drop out of the spoon while it is being lifted from the bottom of the drill hole. Bailers (Fig. 44.2*b*) are unsatisfactory, because the churning operation required to fill them washes the fine particles out of the sand. In order to secure sand samples which contain all their constituents, it is necessary to experiment with other devices such as a sampling spoon equipped with a core catcher made of spring steel (Fig. 44.2*d*). The core catcher is attached to the walls of the lower end of the sampling spoon. As the spoon is lifted, the springs bend toward the center of the sample and, if no coarse particle becomes caught between them, they join to form a dome-shaped bottom that supports the sample.

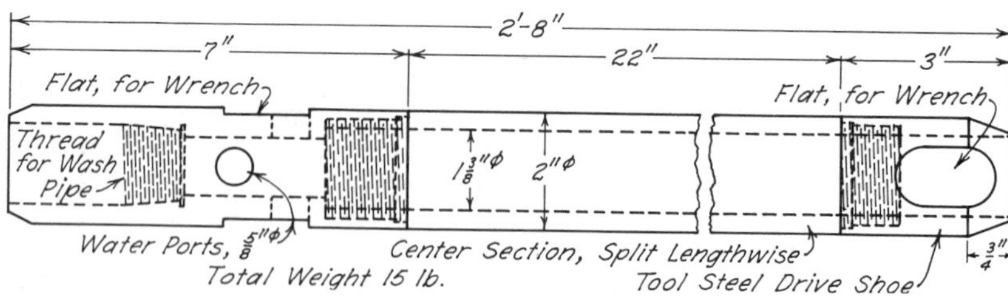

Fig. 44.4. Dimensions of sampling spoon for standard penetration test (courtesy Raymond Concrete Pile Co.).

If the sampling spoon equipped with core catcher fails to retain the sand, reasonably complete samples can be obtained from 4-in. holes by means of the scraper bucket shown in Fig. 44.2*f*. It has an internal diameter of 2½ in. and a length of 30 in. The lower end is plugged with a conical shoe. The upper half of the bucket is provided with a vertical slit. One side of the wall adjoining the slit is bent out and sharpened to form a cutting edge. The sampler is driven for its full length into the bottom of the hole and rotated in the direction shown in the figure, whereupon the cutting edge scrapes off the adjoining soil. The scraped-off material accumulates first in the lower half of the sampler and later in the upper part. The sample is thoroughly disturbed and partly segregated, but the loss of fines is very small.

If a stratum of gravel is encountered, no samples can be secured from exploratory drill holes with a diameter as small as 2½ in. It may even be impossible to drive the casing through the stratum, whereupon the hole must be abandoned. The next hole should be lined with a casing having a diameter of at least 4 in.

THIN-WALLED TUBE SAMPLES. If the project calls for reliable information concerning the shearing resistance or stress-deformation characteristics of a deposit, the degree of disturbance of the samples must be reduced to the minimum compatible with the benefits to be obtained from the information. Whatever type of sampler is used, a certain amount of disturbance of the soil is inevitable.

The degree of disturbance depends on the manner in which the sampler is forced into the soil and on the dimensions of the sampler. The greatest disturbance is caused by driving the sampler into the soil by successive blows of a hammer, and the best results can be obtained if the sampler is pushed into the ground at a high and constant speed. For samples of a given diameter forced into the soil by the same process, the degree of disturbance depends on the area ratio,

$$A_r(\%) = 100 \frac{D_e^2 - D_i^2}{D_i^2} \tag{44.1}$$

in which D_e is the external diameter, and D_i the internal diameter of the sampler (Hvorslev 1948). The area ratio of the split spoon for the standard penetration test is 112%, whereas the value should not exceed about 20% if disturbance is to be minimized.

If the exploratory borings are lined with casing having an internal diameter of 2½ in., the largest sampling device that can be used has an external diameter of 2 in. Reasonably satisfactory samples

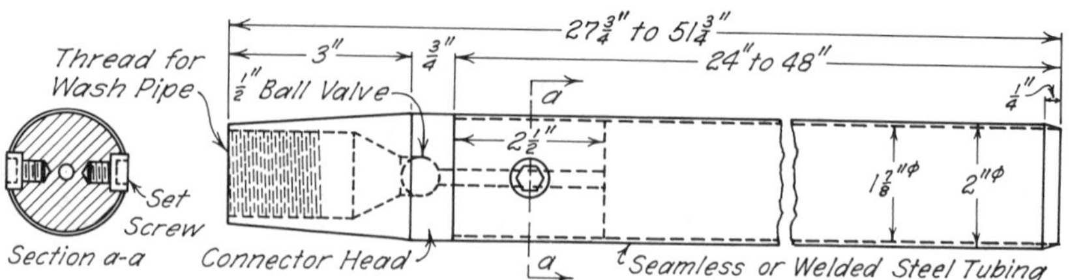

Fig. 44.5. Two-inch tube sampler.

can be obtained in 2-in. tube samplers made of no. 16 or no. 18 gage steel. Such samples have an area ratio of about 13%. The tubes commonly have a length of 30 or 36 in. The lower ends are beveled to a cutting edge, and the upper ends are fitted for attachment to the drill rods (Fig. 44.5).

In order to take a sample, the foreman attaches a tube to the bottom of the drill rods and lowers it into the hole that has previously been cleaned by a cleaning spoon or by washing. The sampler is then pushed downward from the bottom of the hole for a distance about 6 in. less than the length of the tube. Preferably, the sampler is forced down in one rapid continuous movement. Driving with a hammer should be avoided. When the sampler has been forced down, the drill rods are rotated to shear the end of the sample, and the sampler is removed. The material at each end of the tube is carefully cleaned out for a short distance and smoothed so that metal disks can be inserted to protect the faces of the soil sample. Paraffin is then poured against the metal disks to form a seal.

Usually, after two samples have been recovered, the casing is advanced to within a few inches of the bottom of the hole and is cleaned out by a spoon or a water jet. The next two samples are then taken. By repeating this procedure, an almost continuous core record of the clay strata can be obtained. During all these steps the hole should remain filled with water. The casing should not be driven into the clay below a given level until the sampling operations have been carried out for at least the length of one sampling tube below this level. Otherwise, the sample consists not of relatively undisturbed soil but of material that has been forced into the casing. If the clay is very soft, it may squeeze into the hole left by the sampler so rapidly that casing must be driven before the next sample can be obtained. If the soil is fairly stiff, several samples can be taken in succession before additional casing is needed.

If tube samples have been taken on a given job, it is always desirable to investigate the extent to which the consistency of the clay has been affected by the sampling operations. However, such information can be obtained only at sites where the clay is exposed, either in open excavations or on the bottom of shafts. Several sampling tubes are pushed into the clay on the bottom of the excavation and are allowed to remain in the soil while a bench containing the tubes is cut in the clay. A large sample is then carefully carved from the bench, and finally the filled sampling tubes are recovered.

Investigations of this kind were carried out in clays of various consistencies in tunnels of the Chicago subway (Peck 1940). The results of tests at one location are shown in Fig. 44.6, in which the plain curves *a* represent the stress–strain relations obtained by means of unconfined compression tests on the hand-carved samples, and the dash curves *b* those on the tube samples. The dash–dotted curve *c* represents the relation for one of the samples completely remolded at unaltered water content. On the basis of the results of a great number of tests of this kind, it was concluded that the unconfined

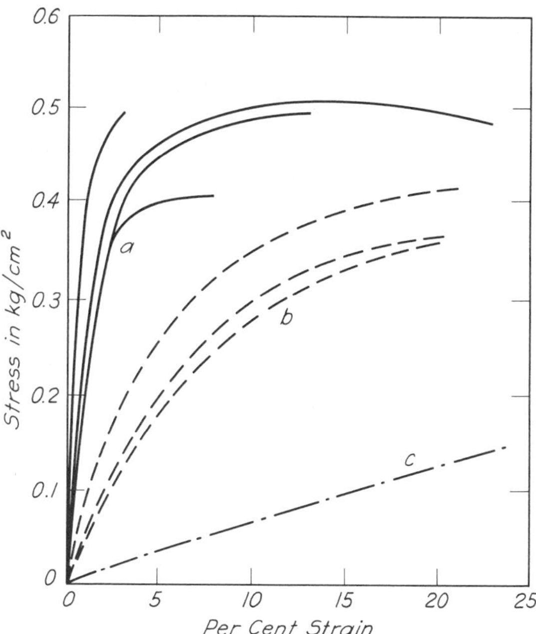

Fig. 44.6. Stress-strain curves obtained by means of unconfined compression tests on Chicago clay. (*a*) Undisturbed samples cut from bench in tunnel. (*b*) 2-in. tube samples of same clay. (*c*) Completely remolded samples (after Peck 1940).

compressive strength of 2-in. tube samples of the clay was roughly equal to 75% of that of the hand-cut samples, whereas complete remolding reduced the strength of the hand-cut samples to 30% of its original value.

Similar samplers with a diameter of 3 in. are also in common use. With tube samplers of larger diameter the difficulty of retaining the samples increases and samplers of other types are likely to be more satisfactory.

Piston Samplers. Part of the disturbance associated with tube sampling, especially in soft nonuniform cohesive soils, arises because the various portions of the soil *in situ* are not represented in the sample at their true thicknesses. When the empty sampler begins its downward thrust, the adhesion and friction on the outside of the tube combined with the tendency for the bottom of the drill hole toward instability may cause the soil to rise into the tube faster than the rate of descent of the tube. On the other hand, after the tube is partly filled, the adhesion and friction between the tube and the sample oppose the rise of the sample. Under extreme conditions the initial portion of the sample may act as a plug capable of displacing soft seams or layers so that they do not enter the sampler at all (Hvorslev 1948).

These conditions can be greatly improved by providing the sampling tube with a piston (Fig. 44.7) that closes the lower end of the tube until the sampler has arrived at the level of the top of the sample to be taken. The piston is then held at this elevation, in contact with the soil, while the tube is advanced around the piston and into the soil. In the early part of the stroke the presence of the piston prevents the entrance of a greater length of sample than the amount of penetration of the tube. In the later part of the stroke, the top of the sample cannot pull away from the piston without creating a vacuum; hence, at this stage the presence of the piston assists the rise of the sample into the tube. After the sampling tube has been advanced, the piston is fixed in its new position with respect to the tube, both elements are rotated to separate the sample from the underlying soil, and piston and tube are removed from the hole.

Piston samplers with small area ratios are capable of furnishing excellent samples of cohesive soils even if very soft and sensitive. The necessity for a separate piston rod rising through the drill rods to the ground surface can be eliminated by use of a hydraulically operated mechanism (Osterberg 1952).

Foil Sampler. Even in piston samplers, the length of sample that can be removed is limited to a few feet and the degree of disturbance

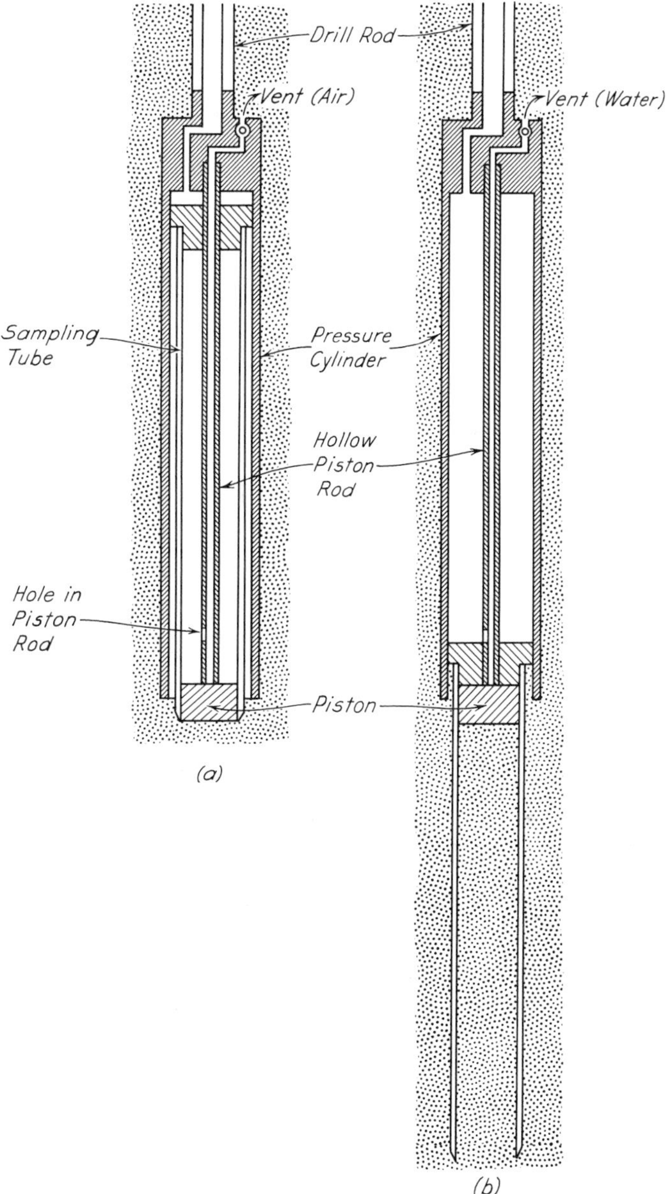

Fig. 44.7. Piston sampler of hydraulically operated type. (*a*) Lowered to bottom of drill hole, drill rod clamped in fixed position at ground surface. (*b*) Sampling tube after being forced into soil by water supplied through drill rod.

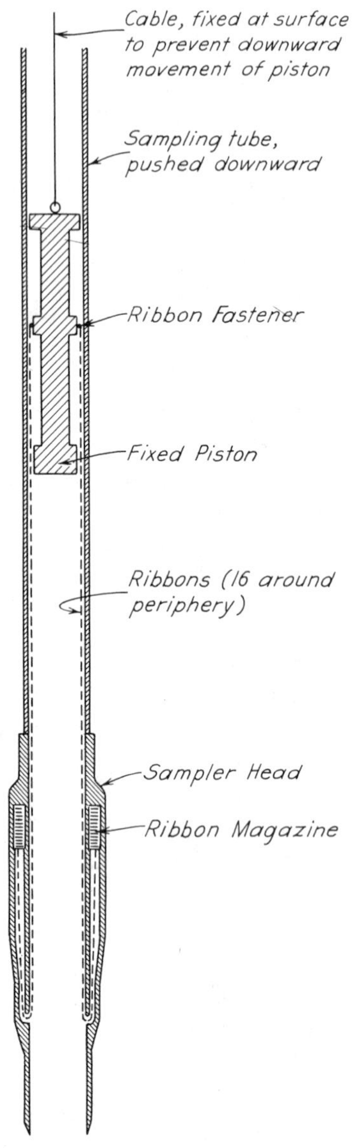

Fig. 44.8. Diagram showing principle of Swedish foil sampler (after Kjellman et al. 1950).

increases with increasing length. By elimination of the friction and adhesion between the sample and the tube, the sample can rise freely into the tube without disturbance of the soil below the cutting edge, and much longer continuous samples can be obtained. These objectives are accomplished in the *Swedish foil sampler* (Fig. 44.8) by lining the inside of the sampling tube with a series of thin vertical steel ribbons (Kjellman et al. 1950). The ribbons, fed from magazines near the bottom of the sampler, remain in contact with the soil after it enters the tube. They do not move vertically with respect to the soil, but remain stationary while the sampling tube slides downward around them. The magazines for the ribbons are located in an enlarged part of the cutting shoe far enough above the cutting edge to maintain a small area ratio. Although the apparatus is complex, virtually undisturbed continuous samples of extremely soft sensitive clays and silts have been obtained with lengths of as much as 60 ft.

Sampling Combined with Coring. Sampling by forcing thin-walled tubes into the soil cannot be done if the soil is too stiff or compact to permit penetration without damaging the cutting edge or buckling the tube. Moreover, even if a tube could be advanced by driving, the resulting disturbance, especially in brittle materials, would be excessive. In deposits containing successive layers of soft and hard consistency the likelihood of successful sampling by tube or piston samplers is very remote.

Under these circumstances the *Pitcher sampler*, in which rock-coring techniques are adapted to tube sampling, may prove satisfactory. The essential elements of the sampler are shown in Fig. 44.9. While the sampler is being lowered into the hole the thin-walled tube is suspended from the cutter barrel and drilling fluid circulates downward through the tube and flushes the cuttings from the bottom of

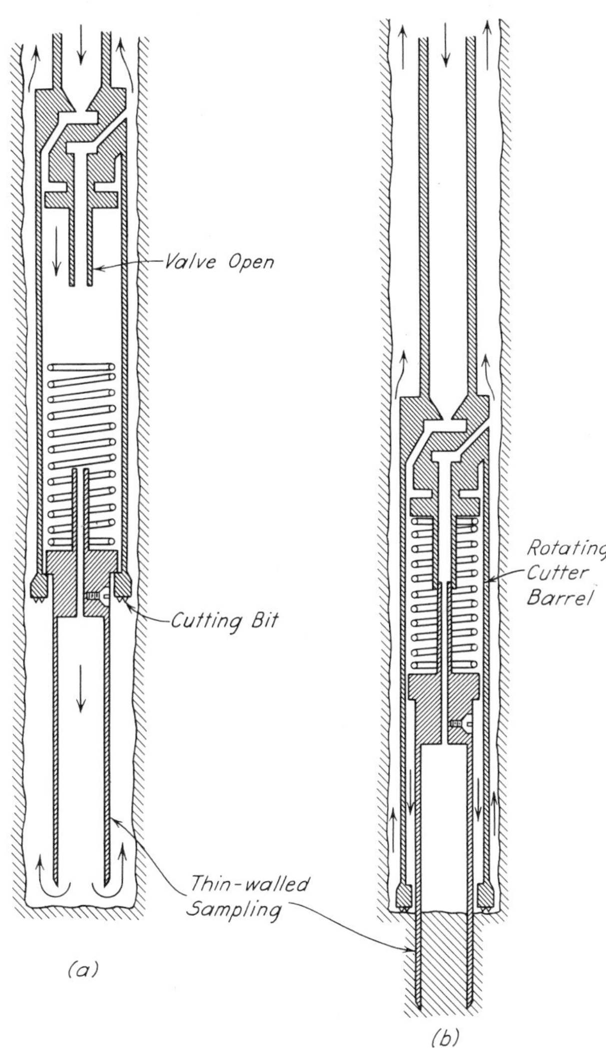

Fig. 44.9. Diagrammatic sketch of Pitcher sampler. (*a*) Sampling tube suspended from cutter barrel while being lowered into hole. (*b*) Tube forced into soft soil ahead of cutter barrel by spring.

the hole. When the tube encounters the bottom it is pushed upward with respect to the cutter barrel, whereupon circulation is diverted downward in the annular space between tube and barrel, beneath the rotating cutter barrel, and upward alongside the barrel. If the soil to be sampled is soft, the spring at the head of the tube keeps the cutting edge of the tube well below the cutter barrel and the tube is pushed into the soil in a manner similar to that in ordinary tube sampling. If the soil is hard, the spring is compressed until the cutting edge of the tube is forced above the level of the bottom of the cutter barrel. As the barrel rotates it cuts an annular ring leaving a cylinder of soil over which the tube sampler slides and protects the sample against further erosion by the circulating fluid. In this manner the sampler adapts itself to the consistency of the soil.

Another modification of rock-coring techniques, in which the sample enters a core barrel, has been used extensively by the United States Army Engineers and the United States Bureau of Reclamation in combination with a sampling tool known as the *Denison sampler* (Johnson 1940). Drilling mud is introduced into the hole in suspension through a set of hollow drill rods. The core barrel is located within a larger barrel (Fig. 44.10) provided with cutting teeth on the lower end. While drilling proceeds, the outer barrel rotates. The drilling mud flows downward through the annular space between the two barrels. It escapes through the openings between the cutting teeth and rises between the outer barrel and the walls of the hole into the upper part of the drill hole. The sampler has a length of 24 in. and an inner diameter of 6 in. It contains a thin cylindrical lining within which the sample may be removed from the tool, and it is equipped with a spring core catcher similar to that shown in Fig. 44.2*d*. While drilling proceeds, the sampler is pushed into the ground by means of jacks exerting a pressure between one and two tons. The jacks react against the drilling rig.

Stiff to hard cohesive soils including soft shales have been obtained successfully by means of the Denison sampler. In many instances even slightly cohesive or silty sands have been recovered with little disturbance. However, when used in clean sand below the water table, the sampler may come out of the hole empty. Strata of gravel may interfere with the drilling operations to such an extent that the drill hole must be abandoned.

HAND-CARVED SAMPLES IN CLAY. On construction projects involving shafts, cuts or tunnels in clays, the opportunity may exist to recover undisturbed samples without the necessity for drill holes. In some instances an exploratory shaft or pit may be preferable to borings.

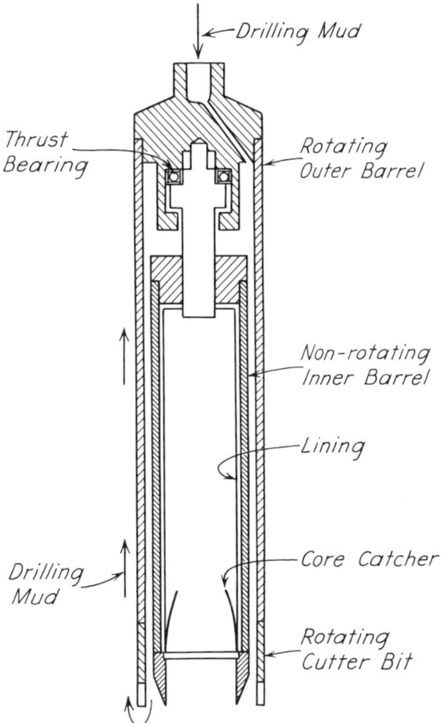

Fig. 44.10. Diagrammatic sketch showing principal features of Denison sampler.

Samples obtained from such excavations are, at least potentially, less disturbed than those recovered by any other procedure.

In order to obtain a large-diameter undisturbed sample in an open excavation or tunnel, the clay around the location of the proposed sample is carefully carved away, leaving a block somewhat larger than the sample standing in the form of a pedestal. Soft clay is usually carved with the aid of a tightly drawn piano wire or a loop of thin strap steel. In stiffer materials a knife or spatula may be more suitable.

The container for the sample consists of a thin-walled metal can with no projecting lugs or rim. When the pedestal has been carved into a size several inches larger than the final dimensions of the sample, the container, with top removed, is inverted and placed on top of the pedestal. The pedestal is carefully trimmed to the diameter of the container, a few inches at a time. As carving proceeds, the container is forced down. When it is full, the pedestal is cut off below the can by means of the piano wire. The soil is trimmed flush with the end of the container, and any voids left between the sample and the container are filled by pouring paraffin around the periphery of the sample. Finally a metal top is placed on the container and sealed.

SAMPLING IN SAND. A distinction must be made between sampling in sand above and below the water table. Above the water table the soil moisture imparts to the sand a trace of cohesion (Article 20). Samples of the slightly cohesive sand can be secured from drill holes for identification by means of sampling spoons equipped with core catchers (Fig. 44.2*d*). Less disturbed samples can be recovered by means of thin-walled piston samplers. Such samplers are adequate for study of the grain-size distribution and stratification. However, in most unsaturated sands the sampling operation is likely to cause a change in volume of the material that enters the sampler. Hence, if the nature of the job requires information concerning the natural void ratio of the sand, special procedures must be used. In many instances the most satisfactory procedure is to carve the samples from a shaft excavated for the purpose.

Before a sample is taken from a shaft, a bench is cut in the sand at the bottom. The surface of the sand is carefully leveled, and on it is placed a cylindrical metal shell with its axis in a vertical direction. The shell commonly has a diameter of 5 or 6 in. and a length of about 6 in. It consists of thin sheet steel. The shell is pushed gently into the sand for its full length so that it encases a cylinder of sand. The surrounding sand is trimmed away, and the top of the sample is sealed with a metal cap that fits over the shell. If the surface of the sample is not level with the top of the shell, the space is filled with paraffin before the cap is attached. A shovel is used to cut off the sample several inches below the shell, and the sample is inverted. The surplus sand is removed so that the surface now on top can also be sealed with a metal cap.

Sands from below the water table cannot ordinarily be retained in the types of samplers in common use. Moreover, their void ratio is likely to be markedly altered unless the area ratio of the sampler is small. Satisfactory samples can sometimes be recovered by means of piston or Denison samplers in holes stabilized by drilling mud. The mud may permeate very coarse sands but otherwise is not detrimental. Often, more elaborate procedures must be used.

Because of side friction a sample of saturated sand can be held in a tube provided at least a small capillary tension can be maintained in the porewater at the bottom of the tube. The tension cannot, of course, exist if the bottom of the tube is submerged. In the *Bishop sampler* (Bishop 1948) a thin-walled sampling tube is housed in a chamber similar to a diving bell at the bottom of the boring. After the tube has been pressed into the sand, the water is expelled from the bell by means of compressed air (Fig. 44.11) to form an air-filled chamber just above the sampler. The sampler, sealed at the top by a leak-proof valve, is raised into the chamber so rapidly that the sand does not escape; as soon as the bottom of the sample enters the air-filled chamber, capillary forces are created and the sample can be retained while sampler and chamber are together removed from the hole.

As an alternative the water table may be lowered to a level below the base of the sand stratum and a shaft excavated in the drained sand. If the shaft is unwatered by pumping from a sump, the water that flows toward the sump is likely to loosen the structure of the sand or, if the sand is already loose, the shaft may be invaded by a mixture of sand and water. For these reasons satisfactory results can be obtained only if the water table is lowered by pumping from well points (Article 21). The water level should be maintained several feet below the bottom of the shaft.

Finally, water-bearing cohesionless sand below the bottom of a drill hole may be transformed into a cohesive material whereupon it can be sampled by methods appropriate for clays. The transformation has been accomplished by injection of asphalt emulsion removed, after the sample has been recovered, by a solvent (Bruggen 1936), and by freezing a plug in the lower end of the sampling tube (Fahlquist 1941). These procedures are rather expensive and require elaborate equipment. Fortunately, in most problems in practice, sufficiently reliable information concerning the properties of sands below the water table can be obtained by indirect means such as penetration or pumping tests.

Subsurface Soundings

PURPOSE OF SUBSURFACE SOUNDINGS. Subsurface soundings are made for exploring layers of soil with an erratic structure. They are also used to make sure that the subsoil does not contain exceptionally soft spots located between drill holes and to get information on the relative density of soils with little or no cohesion.

Experience shows that erratic soil profiles are far more common than regular ones. The results of test borings in subsoils with an

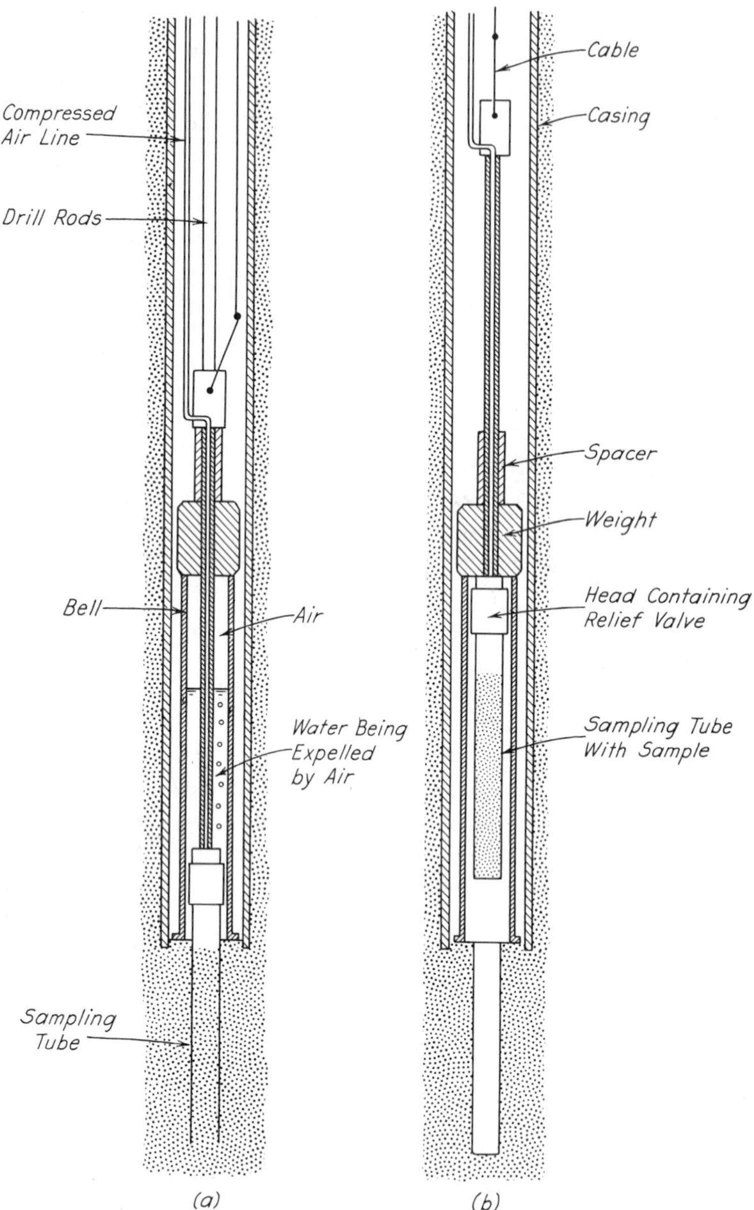

Fig. 44.11. Principle of Bishop sampler for sand below water table. (*a*) Sampler forced into sand by drill rods and water in bell being displaced by compressed air. (*b*) Sampler lifted by cable into air-filled bell (after Bishop 1948).

erratic structure leave a dangerously wide margin for interpretation unless the spacing between drill holes is very small, and the cost of a set of closely spaced drill holes is likely to be prohibitive unless the area under investigation is also very small. However, significant changes in the character of the subsoil are commonly associated with a change in the resistance of the soil against the penetration of a pile or of a pipe equipped with a drive point.

The effect of the relative density of sand on the penetration resistance is well known to every engineer experienced in pile driving. If the sand is very dense, a pile cannot be driven deeper than 10 or 15 ft. Driving is very hard, and the number of blows per foot increases rapidly with depth. If the sand is very loose, cylindrical piles can be driven to any depth, and the increase of resistance with depth is small.

The variation of the penetration resistance of a soil along vertical lines can be determined rapidly at moderate expense by tests known as *subsurface soundings*. The tool used to make the tests is the *penetrometer*. One of the most widely used procedures for measuring resistance to penetration is the *standard penetration test*, described on page 304, in which the penetrometer is the standard split spoon itself. Applications of the results of the standard penetration test will be discussed in Article 45 and in subsequent articles.

Whereas the standard penetration test furnishes only one value of resistance for each 5 ft of depth, or under special conditions one value for each $2\frac{1}{2}$ ft, many other types of subsurface soundings yield continuous or almost continuous penetration records.

Improvised Sounding Methods. For several generations engineers have made crude attempts to ascertain the consistency of the subsoil by driving rods, pipes, or railroad rails into the ground and recording the penetration produced by each blow of the drive weight. If this method is intelligently used in combination with at least a few exploratory borings, it can be very successful in spite of its simplicity. The following incident is an example.

The preliminary borings for a pile foundation disclosed an erratic deposit consisting principally of loose and medium sand with a few pockets of soft silt or clay. During construction of the foundation, it was noticed that the depth at which the piles met refusal varied between surprisingly wide limits. It was feared that the shorter piles might have met refusal in resistant deposits located above large pockets of soft silt or clay. To find out without undue loss of time whether or not this apprehension was justified, the sounding method was used. The only equipment available on short notice was a supply of 86-lb steel rails and a drop hammer weighing 2500 lb. The procedure consisted of driving the rails by dropping the hammer 30 in. per blow, and of recording the number of blows per foot of penetration. The soundings disclosed extremely erratic variations in the resistance of the soil against penetration of the rail. These variations are shown in Fig. 44.12 which represents the records of two soundings 42 ft apart. By means of the soundings it was possible to determine in a short time the boundaries of all the exceptionally soft pockets in the subsoil. After this information was secured, a few exploratory drill holes were made where the softest pockets were located. They

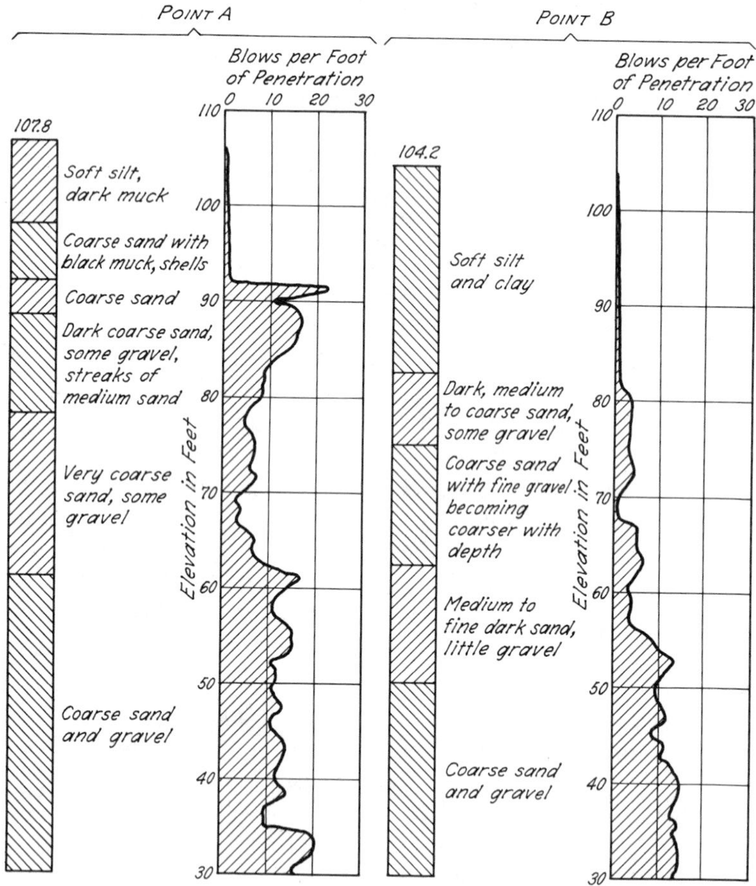

Fig. 44.12. Penetration record for steel rails driven at two points 42 ft apart through soft silt and clay into coarse-grained glacial outwash, Port Alberni, Vancouver, British Columbia.

showed that most of the pockets contained clean, well-graded, but very loose sand instead of compressible silt or clay. The variation in the lengths of the piles was due only to erratic and very large variations in the density of the sand.

If the sounding method is to be used to full advantage, the technique must be adapted to the subsoil conditions. For this reason a great many different procedures have been developed. They can be divided into two large groups, static and dynamic. In the static methods the sounding rod is pushed into the ground by static pressure. The dynamic methods consist of driving the rod by the impact of a drop hammer.

STATIC SOUNDING METHODS. Static sounding tools still in use were developed about 1917 by the Swedish State Railways (Fellenius et al. 1922), about 1927 by the Danish Railways (Godskesen 1936), and about 1935 by the Department of Public Works in the Netherlands (Barentsen 1936). Of these the latter, known as the *Dutch Cone* apparatus, has found wide application. In its primitive form it consists of a 60° cone with a diameter of 1.4 in. (Fig. 44.13*a*) attached to the lower end of a $\frac{5}{8}$-in. rod surrounded by a $\frac{3}{4}$-in. gas pipe (Fig. 44.14*a*). The cone is pushed 20 in. into the ground at a rate of 0.4 in./sec by one or two men who apply part of their weight to a crossbar attached to the upper end of the rod. The pressure exerted on the rod is registered by a Bourdon gage connected to a hydraulic cylinder located below the crossbar. After each downward stroke, the pipe is pushed down 20 in., and the stroke is repeated. The pressure exerted on the rod during each stroke is plotted against depth. The individual penetration records furnish the data for constructing consistency profiles (Fig. 44.14*b*).

The original Dutch Cone apparatus is still useful for rapid detailed surveys of erratic deposits of soft clays, silts, and peats. One sounding to a depth of 40 ft can be made in about 15 min. The equipment has been refined and mechanized, however, to permit rapid exploration of soft deposits as deep as about 100 ft, and to investigate the relative density of sands. It is extensively used, especially in Holland and Belgium, for estimating the lengths and capacities of long piles driven through compressible soils into sand. In the Dutch Cone penetrometers in current use (Sanglerat 1965), not only the resistance to penetration of the point is determined, but also the friction developed on the circumference of the casing.

The penetration resistance determined by the Dutch Cone in sands appears to be almost exclusively a function of the relative density or angle of internal friction. The depth of penetration below the sur-

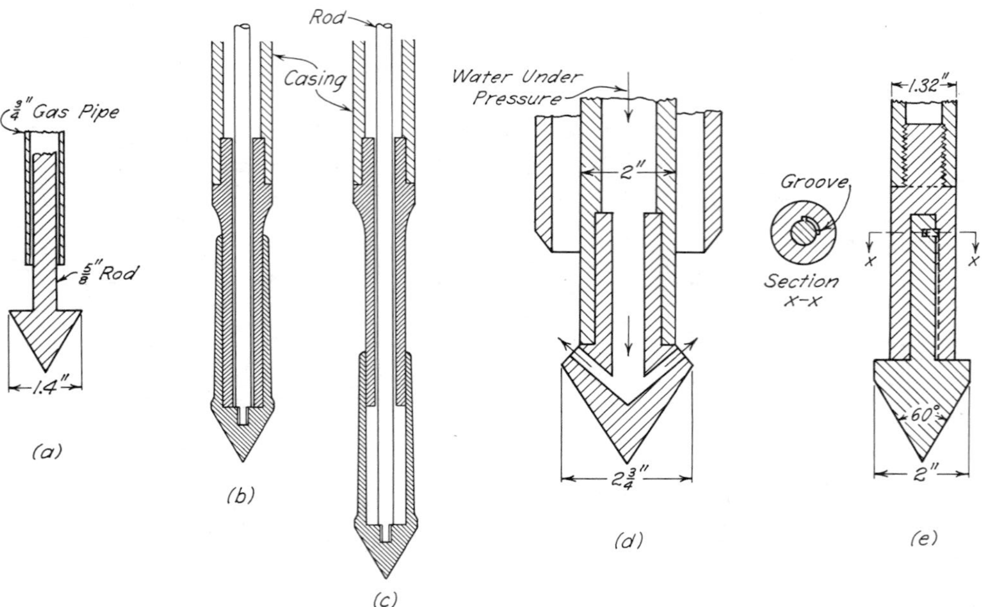

Fig. 44.13. Penetrometers. (*a*) Original Dutch Cone. (*b*) and (*c*) Refined Dutch Cone with point retracted while advancing casing and with point extended after measurement of resistance. (*d*) Wash-point penetrometer. (*e*) Conical drive point.

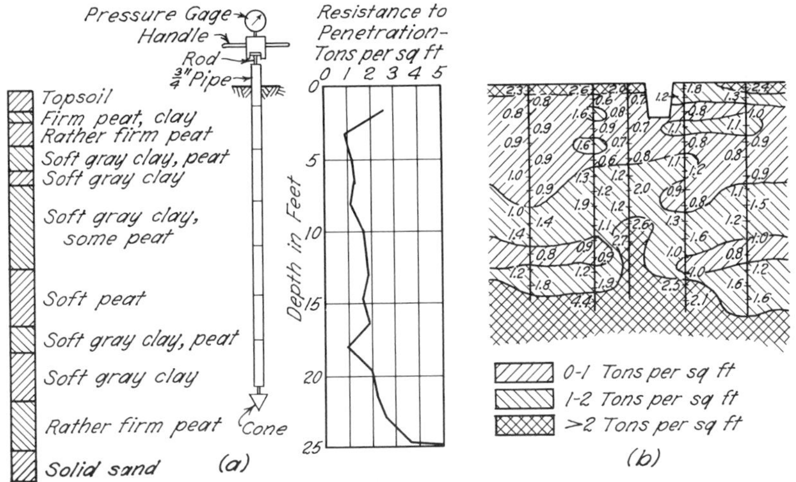

Fig. 44.14. (*a*) **Dutch penetrometer. Chart shows record obtained from single test hole.** (*b*) **Soil profile along route of highway, showing variation in penetration resistance (after Barentsen 1936).**

face has a small and usually negligible influence. To eliminate completely the influence of depth, as well as to permit investigation of deposits too dense to be penetrated by the Dutch Cone, a wash-point method was developed in 1928 for use on a subway job in New York. The subsoil consisted of clean medium to coarse sand. In the application of this method, a conical point (Fig. 44.13*d*) with a diameter of $2\frac{3}{4}$ in. is attached to the lower end of a heavy wash pipe with an external diameter of 2 in. The pipe, with the cone attached, is introduced into a casing with an inner diameter of 3 in. (Fig. 44.15*a*). The cone is forced into the soil to a depth of 10 in. by means of a hydraulic jack acting on the upper end of the pipe. The water is then turned on. It leaves the cone through holes pointing upward and transforms a cone-shaped body of soil (Fig. 44.15*b*) located above the top of the conical point into a semiliquid. Part of the soil is washed out through the space between the wash pipe and the casing. While the water circulates, a slight push is sufficient to press the casing down through a distance equal to the preceding stroke of the cone. Then the water is turned off, and the conical point is again forced down through a distance of 10 in. The pressure exerted by the jack during each downward stroke of the cone is read on a Bourdon gage attached to the oil conduit of the jack and is plotted on a diagram as a function of the depth. By using this procedure on the job in

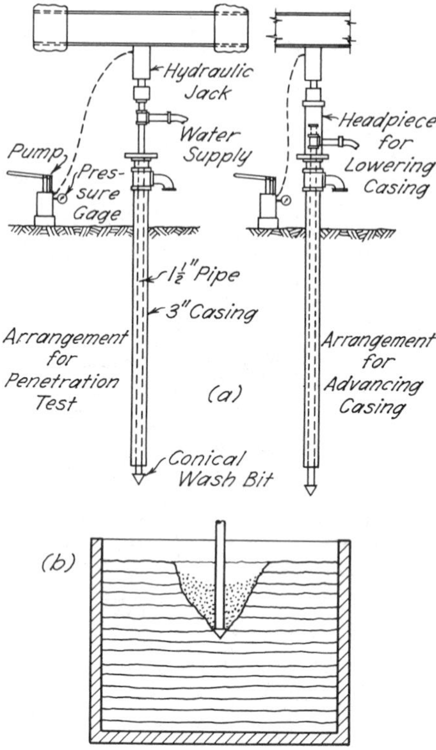

Fig. 44.15. **(*a*) Penetration device for investigating relative density of sand. (*b*) Sketch from photograph showing wash point at beginning of a downward stroke. In cone-shaped space above point, structure of sand was destroyed by jet action.**

New York, a great many soundings were made in a short time. The results of the observations were calibrated against the results of loading tests on bearing plates 1 ft square, resting on the bottom of an open shaft. The tests were made at different depths below the surface as the shaft was excavated. The results of the calibration tests are shown in Fig. 44.16. During both the penetration and the loading tests the reaction for the jack was furnished by the base of the foundations of existing buildings (Terzaghi 1930).

Dynamic Methods. The dynamic sounding methods consist of driving a rod with a drive point into the ground by means of a drop hammer and measuring the number of blows per foot of penetration. In addition to the standard penetration test, page 304, several other more or less standardized tests have been developed. In most of these the drive point is a retractable or expendable steel cone. The great variety of procedures that have come into use indicates that no one sounding method is equally suitable under all the soil conditions that may be encountered in the field. For a given site the method must also be chosen in accordance with the type of information called

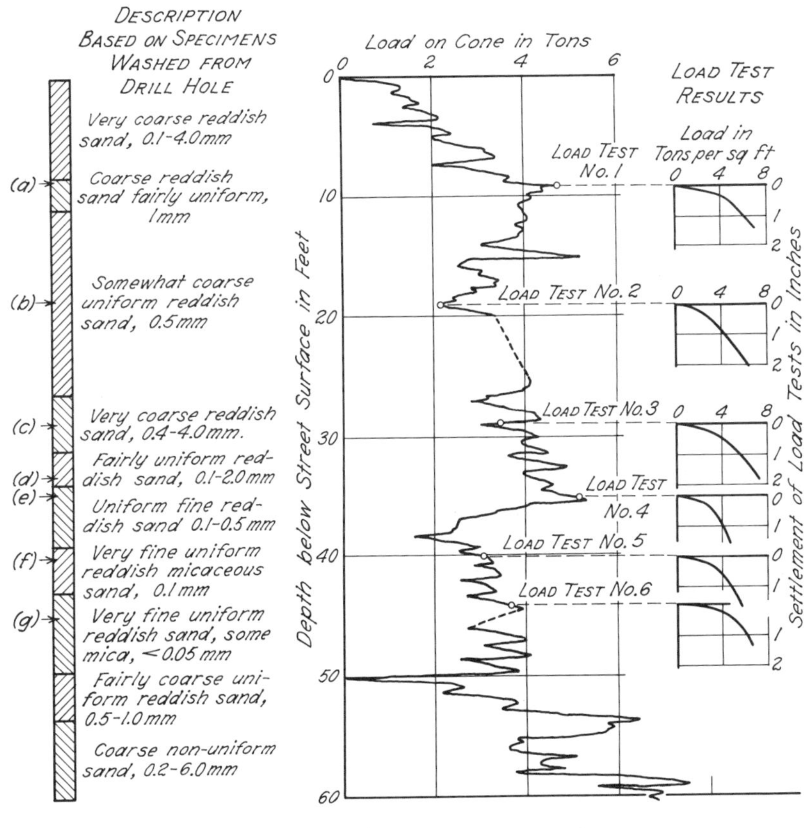

Fig. 44.16. Results of investigation of sand deposit at Houston Street subway, New York, by means of wash-point penetrometer and by means of load tests conducted on areas 1 ft square in test shaft after penetration record was obtained (after Terzaghi 1930).

for by the project. Whenever a new method is used, a certain amount of experimentation is required to adapt the procedure to local soil conditions.

The most common deposits with erratic structure are river or shore deposits consisting of lenses of silt or clay embedded in sand, or sand and gravel with variable relative density. General information on the structure of such deposits can be obtained by driving a 1-in. extra-strong steel pipe with a 2-in. conical drive point (Fig. 44.13*e*). The pipe is composed of 5-ft sections with flush joints. Each section of pipe weighs 11 lb. Attached to the cone is a short stem fitting a $\frac{1}{2}$-in. hole in a plug that is screwed into the lower end of the string of pipes. The pipe is driven into the ground by means of a 160-lb drive weight that falls 30 in., and a record is kept of the number of blows per foot. After the pipe is driven to refusal it is recovered, whereas the drive point remains in the ground.

By means of such a simple penetrometer, several soundings per day can be made to a depth of 60 or 80 ft. Since the diameter of the cone is larger than that of the pipe, the side friction is likely to be small compared to the point resistance. As the depth of the point increases, the weight of the pipe also increases. Hence, the relation between relative density and penetration resistance is to some extent dependent on the depth.

In-Place Shear Tests

In many practical problems it is necessary to determine the undrained shearing resistance and the sensitivity of soft deposits of clay. Inasmuch as both the strength and sensitivity of such materials may be radically altered by the processes of boring, sampling, and handling in the laboratory, various devices have been developed for measuring the undisturbed and remolded strengths *in situ*. Of these the most versatile and widely used is the *vane shear apparatus* (Carlson 1948; Cadling and Odenstad 1950). In its simplest form it consists of a four-bladed vane (Fig. 44.17*a*) fastened to the bottom of a vertical rod. The vane and the rod can be pushed into the soil without appreciable disturbance. The assembly is then rotated and the relation between torque and angular rotation is determined. Typical results for a sensitive soft clay are shown in Fig. 44.17*b*. Investigations have shown that the soil fails along a cylindrical surface passing through the outer edges of the vane, as well as along plane horizontal circular surfaces at the top and bottom of the blades. Hence, the shearing resistance can be computed if the dimensions of the vane and the torque are known. If the vane is rotated rapidly through

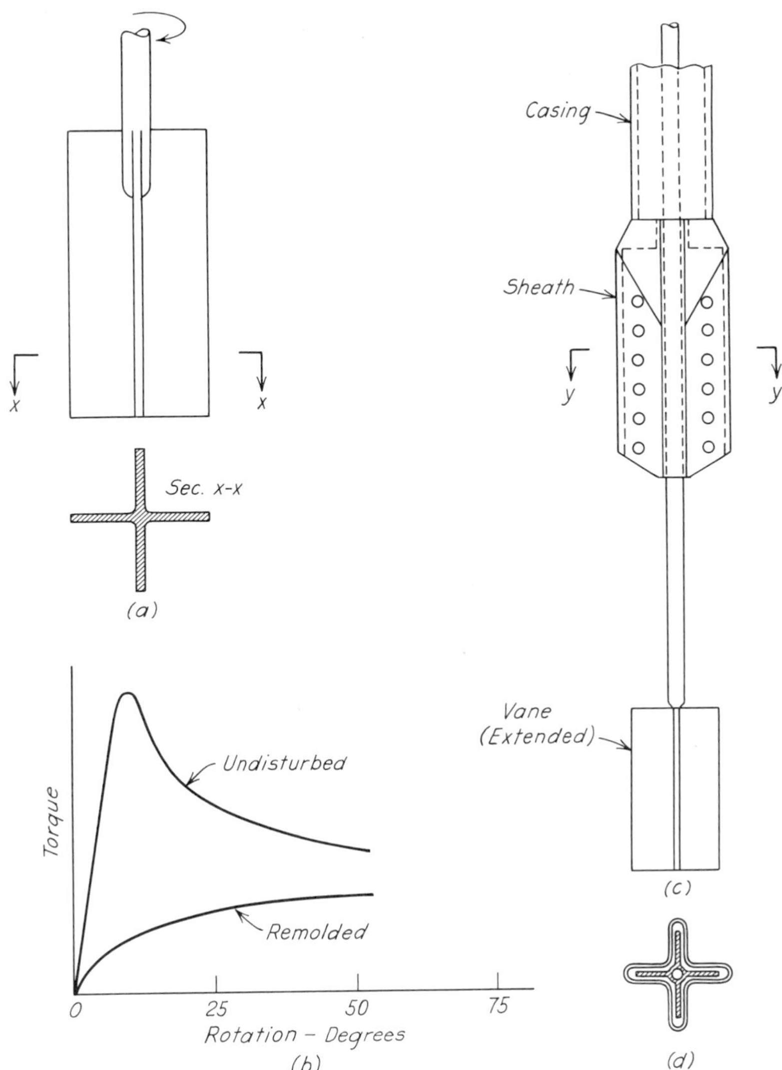

Fig. 44.17. Vane shear apparatus. (*a*) Simple four-bladed vane. (*b*) Typical torque-rotation curves for soft sensitive clay. (*c*) Sheath for advancing vane without drill-hole. (*d*) Section y-y through sheath before advancing vane (after Cadling and Odenstad 1950).

several revolutions, the soil becomes remolded and the shearing strength can again be determined. In this manner the sensitivity can be calculated. However, the degree of disturbance caused by rotating the vane differs from that obtained by kneading a sample in the laboratory; therefore, the numerical values of sensitivity determined by the two procedures are not strictly comparable.

The vane may be used to measure the shearing resistance of a clay beneath the bottom of a drill hole, and successive values may be determined as the drill hole is deepened. It may also, in soft soils, be pushed into the ground without first making a boring. Under these circumstances the rod is housed in a casing and the vane protected by a sheath until it has reached the depth where a test is to be made (Fig. 44.17*c*). The vane is then advanced out of its sheath and rotated.

If the soil contains even thin layers or laminations of sand or dense silt, the torque may be much greater than that required if the layers were not present. If these conditions prevail, the results of vane tests may be misleading.

In-Place Permeability Tests

Preliminary information about the order of magnitude and variability of the coefficient of permeability of a natural pervious stratum is obtained by permeability tests in the exploratory borings while drilling proceeds. The observations during drilling should also furnish information concerning the presence or absence of free communication between pervious strata encountered in the holes.

Most of the common procedures used in connection with drill holes are based on the principle of the falling head permeability test (Article 11). The hole is cased from the ground surface to the top of the zone to be tested and extends without support for a suitable depth below the casing. Usually the uncased part of the hole has a roughly cylindrical shape. If the pervious stratum is not too thick, the hole is preferably extended through the full thickness; otherwise the hole penetrates only part of the pervious material.

If the pervious zone is below the water table the test may be carried out by adding water to raise the water level in the casing and then allowing the water level to descend toward its equilibrium position. The elevation is measured as a function of time, and the coefficient of permeability is calculated by means of the expression

$$k = \frac{1}{C} \frac{A(\Delta h/\Delta t)}{r_0' h_m'} \tag{44.2}$$

where Δh is the drop in water level in the casing during an interval of time Δt, A is the inside cross-sectional area of the casing, h_m' is the mean distance during the interval Δt from the water level in the casing to the equilibrium water level in the pervious zone, and r_0' is the mean radius of the roughly cylindrical hole below the casing. The coefficient C is a dimensionless quantity depending on the shape of the cylindrical hole and the depth of penetration into the pervious layer. Values of C for various conditions are given in Fig. 44.18 (Zangar 1953).

In a falling head test in a drill hole it is likely that fines suspended in the water may form a filter skin over the walls and bottom of the hole in the pervious material; as a consequence, the observed permeability may be too small. The error may be avoided by bailing the water from the casing until the water level is below that of the pervious stratum and by measuring the elevation of the water level at various times as it rises toward its equilibrium position. The value of k can be calculated by Eq. 44.2 as before. However, if the permeable stratum is cohesionless the water level cannot be lowered too far or the hole will collapse and the cohesionless material may rise into the casing.

The results of such tests are little more than an indication of the order of magnitude of the coefficient of permeability. More reliable information is obtained from pumping tests on test wells.

The usual diameter of a test well is about 12 in. In a closed, fairly homogeneous aquifer, the water should be free to enter the well over the full thickness of the aquifer. Observation wells should be established in two lines, one in the direction of normal groundwater flow and the other perpendicular to it. At least two, and preferably four, observation wells should be established in each line. The observation wells should also permit entrance of water over most of the thickness of the aquifer. The initial groundwater level should be observed in all the wells for a period long enough to establish the amount and nature of any fluctuations that normally occur at the site. Pumping should then be started at a constant rate of discharge and the water levels measured in the observation wells until equilibrium is reached. The value of k can then be calculated by means of Eq. 23.9. The permeability can also be evaluated on the basis of the rate at which the water levels descend in the various observation wells. The procedures are known as nonequilibrium methods (Todd 1959).

If the pervious deposit is open (Fig. 23.6*b*), the observation wells may be used primarily for the purpose of estimating the radius of influence of the pumping well and k can be calculated on the basis

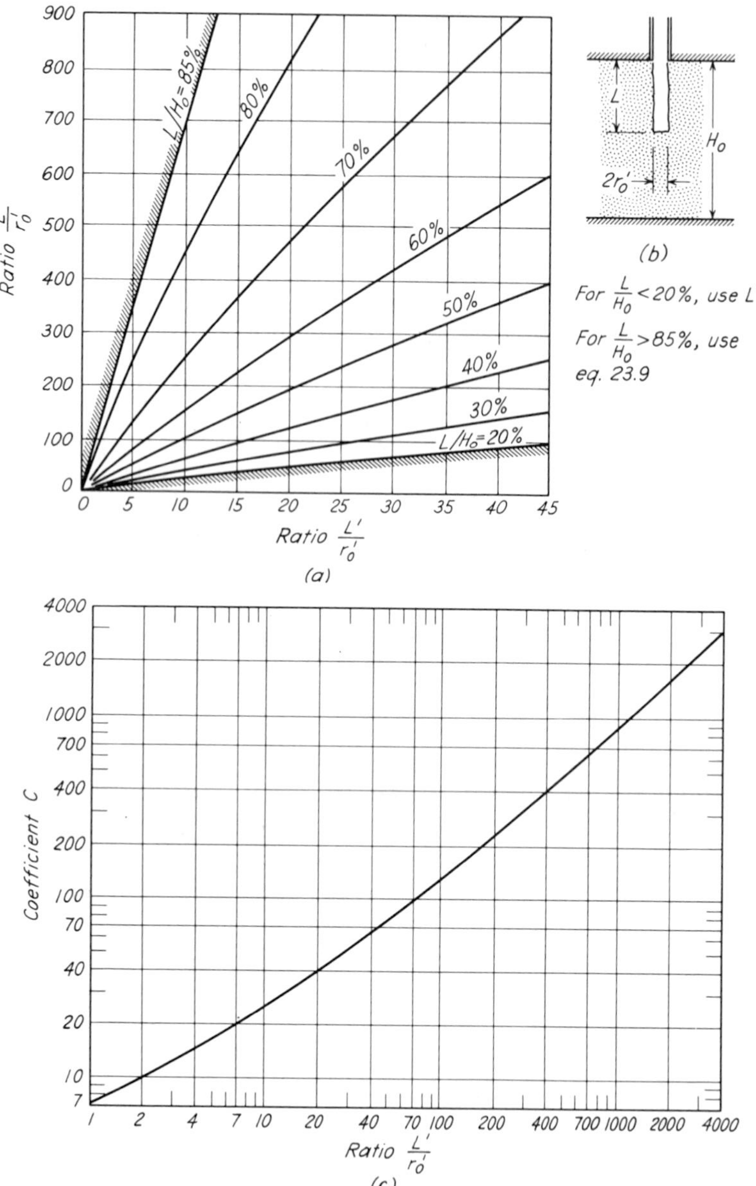

Fig. 44.18. Permeability test in open drill hole into pervious stratum. (*a*) Chart for determining ratio L'/r_o' for various penetrations L/H_o, (*b*). (*c*) Chart for determining coefficient *C* for use in Eq. 44.2 (after Zangar 1953).

of Eq. 23.12. Under these conditions, however, an allowance must be made for the loss of head experienced by the water as it enters the well screen (Petersen et al. 1955).

If more reliable results are to be obtained by measuring in observation wells the equilibrium levels associated with pumping at a constant rate from the test well, special precautions must be taken. If the calculations are to be based on Eq. 23.11, the nearest observation wells must not be closer to the pumping well than the thickness of the pervious layer below the original water table (Article 23). At these and greater distances the drawdown in the observation wells may be small. If the normal fluctuations in water level are significant fractions of those due to pumping from the test well, the resulting error in k may be intolerable. If the observation wells are located closer to the pumping well, the value of k can no longer be calculated by means of Eq. 23.11 because the real drawdown curve lies considerably above the Dupuit curve on which the equation is based (Article 23). Equations derived from more advanced theories are available (Boreli 1955). For the use of these theories, however, the observation wells must not extend too deeply into the pervious deposit because, within the radius to which the Dupuit and the real drawdown surfaces are significantly different, the piezometric levels are not the same along a given vertical line. Hence, to define the drawdown surface, the observation wells should not extend appreciably below the lowered position of the water table.

If a pumping well penetrates several aquifers separated from each other by impervious layers, the elevation of the original water table must be determined for each of the aquifers while the well is being drilled. If they are equal, k can be determined by a single pumping test, but H_0 in Eq. 23.9 must be modified in accordance with the boring records. In river valleys it is by no means uncommon that the upper, open aquifer is separated by a stratum of clay from a lower, closed aquifer with a much higher piezometric level. Two independent pumping tests are then required.

Geophysical Methods

At the beginning of this article it was mentioned that various kinds of information regarding subsoil conditions can be obtained by means of geophysical methods, without the aid of borings or soundings.

Some of the geophysical methods are based on the fact that the geometry of every field of force depends on the location of the boundaries between the substances that occupy the field. The field of force may already exist, as for example the gravitational or the magnetic field of the earth, or it may be created artificially, for instance by sending an electric current through the ground located between two electrodes embedded in the ground.

The geometry of any field of force in a perfectly homogenous medium is independent of the physical properties of the medium. It is simple and can be determined accurately by theory. The distortion of the field produced by the existence of an internal boundary depends on those physical properties of the substances located on either side of the boundary that create the field or that have a decisive influence on its intensity. Hence, the most suitable method for locating the boundary between two types of rock is determined by the type of field of force that will be distorted most conspicuously by the difference between the properties of these rocks. If their unit weights are very different, a gravitational method may be indicated. If their unit weights are almost equal, but their electric conductivities are very different, the electric potential or resistivity method may be used to advantage.

To locate the position of an internal boundary the pattern of the real field of force at the surface of the ground is determined by suitable surface observations. This pattern is compared with the one determined by computation based on the assumption that the seat of the field is perfectly homogeneous. The position of the internal boundary is ascertained on the basis of the difference between the real and the ideal patterns.

A second group of geophysical methods, known as seismic methods, is based on the fact that the rate of propagation of elastic waves is a function of the elastic properties of the media through which the waves travel. If a wave arrives at the boundary between two media with different elastic properties, part of it is reflected and another part is refracted. In order to determine the position of an internal boundary, for instance between hard and soft rock or between soil and rock, a small shock is produced at the ground surface by a hammer blow or by firing an explosive in a shallow hole, and the time is measured at which the reflected or the refracted waves arrive at different points on the surface. On the basis of the results of the observations the position of the internal boundary can be computed, provided the boundary is well defined and not too uneven.

In civil engineering, only the seismic and electrical resistivity methods are used to any significant extent, sometimes in conjunction with each other. The principal application of the seismic method is to locate the surface of the bedrock. If the thickness of the weathered top layer of the rock is small and the rock surface is not too uneven, the results are usually reliable. In fact, if a sedimentary overburden contains many boulders, a survey by boring may be almost impracticable, whereas the seismic survey may be as simple and reliable as if the boulders did not exist. In some instances the depth to the surface of a stiff or hard deposit of soil beneath soft overlying sediments can be determined. Since the velocity of seismic waves is much greater in saturated than in unsaturated soils, the method can also be used to locate the water table in pervious soils. On the other hand, the presence of a soft layer below a stiffer one cannot ordinarily be detected.

The resistivity method has been found useful in defining the boundaries between soils of low resistivity such as soft clays and soft organic deposits, and materials of higher resistivity such as sands, gravels, or bedrock. Materials having low resistivities can be detected even if they underlie those of high resistivities. The method can be used from the surface of a body of water. On the other hand, the boundaries cannot usually be detected between an organic soil and soft clay, or between stiff clay and soft clay shale, or between loose sand and coarse-grained sandstone. In all applications, the interpretation requires calibration of the equipment over known materials in the immediate area (Moore 1961).

Readily portable seismic and resistivity equipment has been developed for civil engineering purposes. With such equipment, exploration can often be carried out economically and rapidly over large areas. Under some circumstances the use of both types of equipment may facilitate interpretation. For instance, it may not be possible on the basis of a seismic survey to determine whether an inferred boundary is the water table or bedrock; a resistivity survey might permit the distinction because rock ordinarily has a high resistivity compared to that of waterbearing strata. It is, however, always advisable to check the results of a geophysical survey by at least a few borings.

Selected Reading

Hvorslev, M. J. (1948). *Subsurface exploration and sampling of soils for civil engineering purposes,* Waterways Exp. Sta., Vicksburg, Miss., 465 pp.

Cambefort, H. (1955). *Forages et sondages* (Borings and soundings), Paris, Eyrolles, 396 pp.

Lowe, J. (1960). "Current practice in soil sampling in the United States," *Hwy. Res. Board Special Rept. 60,* pp. 142–154.

Sanglerat, G. (1965). *Le pénétromètre et la reconnaissance des sols* (The penetrometer and soil exploration), Paris, Dunod, 230 pp.

ART. 45 PROGRAM FOR SUBSOIL EXPLORATION

Type and Sequence of Operations

Whatever the project may be, the engineer should never forget that most subsoils were formed by geological processes that changed at random in space and time. Because of the decisive influence of geological factors on the sequence, shape, and continuity of the soil strata, the first step in any subsoil exploration should always be an investigation of the general geological character of the site. The more clearly the geology of the site is understood, the more efficiently can the program for soil exploration be laid out. The second step is to make exploratory drill holes that furnish more specific information regarding the general character and the thickness of the individual strata. These two steps are obligatory. All others depend on the size of the job and the character of the soil profile.

On routine jobs, such as the design and construction of foundations for apartment houses of moderate size in districts with known foundation conditions, no further investigations are called for. The soil testing can be limited to the determination of the index properties (see Table 9.1, page 44) of spoon samples obtained from the exploratory borings. The test results serve to correlate the soils with others previously encountered on similar jobs. Hence, they make it possible to utilize past experience. The gaps in the information obtained from the exploratory drill holes are compensated by a liberal factor of safety. Wherever information can be obtained by inspection of existing structures in the vicinity, this opportunity should not be overlooked.

The soil exploration on large projects may call for the determination of one or several of the following: relative density of sand strata, permeability of sand strata, shearing resistance and bearing capacity of clay strata, or compressibility of clay strata. In every instance the program of the exploration should be prepared in accordance with the amount of useful information that can be derived from the results of laboratory tests. With increasing complexity of the soil profile the usefulness of elaborate laboratory investigations rapidly decreases. If the soil profile is erratic, the efforts should be concentrated not on obtaining accurate data regarding the physical properties of individual soil samples, but on obtaining reliable information regarding the structural pattern of the subsoil. Attempts to obtain this information by means of boring and testing are commonly wasteful. Since erratic soil profiles are far more common than simple and regular ones, the instances are relatively rare in which elaborate and large-scale soil testing is justified from a practical point of view. In the following discussion of the means for obtaining reliable information concerning the subsoil conditions, the influence of the degree of complexity of the soil profile on the practical value of soil testing is consistently emphasized.

Geological Considerations

Most natural soil deposits represent one of the following principal types: river-channel deposits, flood-plain deposits, delta deposits, shore deposits, glacial deposits, wind-laid deposits (dune sand or loess), deposits formed by sedimentation in standing water, and residual soils formed in place by weathering. The only ones likely to have a fairly regular structure are the flood-plain and wind-laid deposits and those formed in large bodies of standing water at a considerable distance from the shore. All the others are likely to be distinguished by important and erratic variations, at least in consistency or relative density, and usually in grain size as well.

In the upper reaches of river systems the *river-channel deposits* commonly occupy the bottoms of valleys carved out of rock. In the lower reaches they may be laid down in winding and interlaced channels eroded out of the broad sheet of fine-grained sediments that have previously been deposited by the river under different conditions of sedimentation. The average grain size decreases with increasing distance from the source, and at any one point it is likely to increase in a general way with increasing depth below the surface. However, the details of stratification are always erratic, and both grain size and relative density vary in an unpredictable manner. Still more abrupt and conspicuous are the variations in the so-called *glacial outwash* deposited by the melt waters along the rim of the continental ice sheets. The variations in the relative density of a fluvioglacial sand stratum, as indicated by the resistance to penetration, are illustrated in Fig. 44.16, and those of a fluvioglacial sand and gravel stratum capped by a blanket of soft silt in Fig. 44.12.

Flood-plain deposits are laid down during the high-water season on both sides of the lower courses of rivers. They commonly consist of continuous layers of silt or clay of fairly uniform thickness, separated from each other by equally persistent layers of coarser sediments. However, at any point or line the continuity of these strata can be broken by bodies of other sediments occupying troughs or abandoned river channels (Kolb and Shockley 1959). If such a body is located between two drill holes, its presence may escape attention. Several well-known foundation accidents have been due to this cause.

Delta deposits are formed at the points where water courses enter bodies of standing water. The main features of deltas are simple but the details of their structure can be very complex, as shown in Fig. 45.1, because the currents which transport the sediments shift continually.

Shore deposits are composed of sediments that were eroded by waves or carried into a body of standing water by rivers and transported

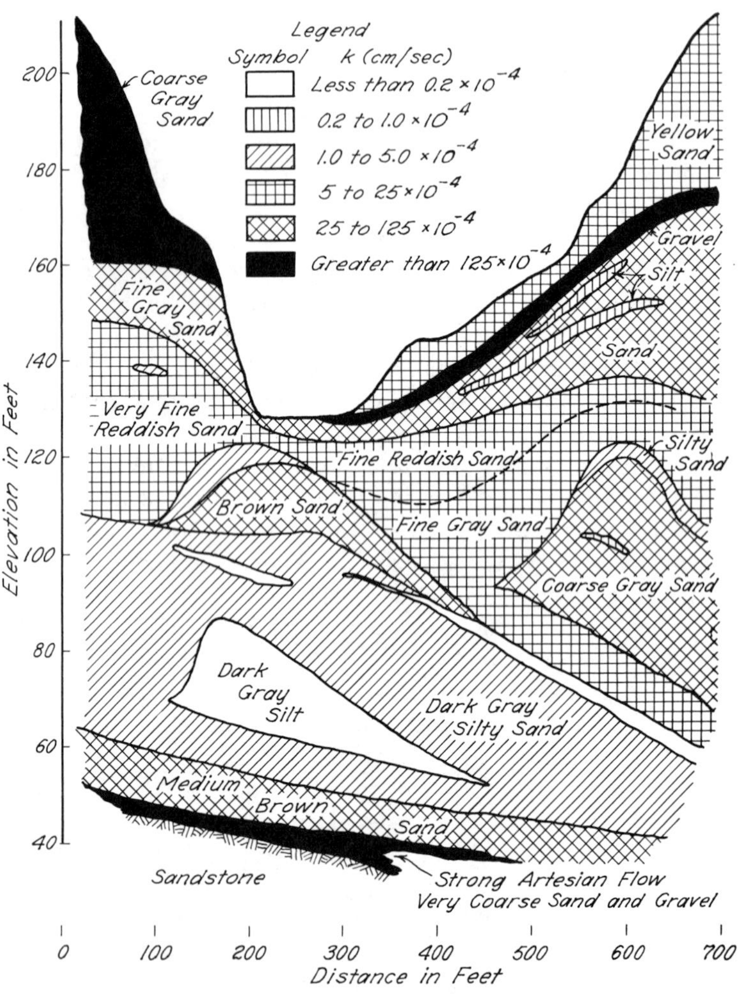

Fig. 45.1. Permeability profile of relatively homogeneous glacial delta deposit near Chicopee, Mass.

and deposited by shore currents. They commonly consist of sand and gravel. However, as a result of important fluctuations in the lake or sea level combined with the shifting of the water courses which cross the coastal belt, the sand and gravel deposits may alternate in an intricate manner with layers or pockets of silt, clay, or peat. Shore deposits of this kind are referred to as *composite shore deposits.* Figure 45.7 and the upper half of Fig. 43.2 illustrate the structure of deposits of this type.

The constituents of *glacial deposits* were picked up and transported by ice and laid down when the ice melted. The wasting away of ice sheets always alternates with periods of temporary growth and advancement. The advancing ice plows up or deforms previously deposited layers of glacial material. Furthermore, at the ice rim random sorting and shifting are carried on by the streams of water that emerge from beneath the ice. Hence, glacial deposits are among the most erratic with which the engineer has to deal. Irregular pockets and lenses of fine- and coarse-grained materials intermingled with boulders may follow each other in a chaotic manner.

In contrast to glacial deposits, *wind-laid sediments* invariably are remarkably uniform. However, the shape of their boundaries may be very irregular, because the wind may drop its burden in irregular heaps on very uneven surfaces. Furthermore, the fine-grained varieties known as loess (Article 2) may completely lose their original homogeneity on account of local leaching or weathering. Many faulty foundations on loess have been caused by the failure of the designers to recognize the existence of these partial alterations.

The various transporting agencies, running water, ice, and wind, deposit only part of their solid burden on their way or at the end of their path. The remainder is carried into large bodies of standing water such as lakes, bays, or the open ocean. Once they get beyond the narrow zone in which the shore currents travel, they are acted on by no force other than gravity. Therefore, in contrast to all other sedimentary deposits, those which are formed in large bodies of standing water commonly have a relatively simple structure. This structure reflects merely the periodic or progressive changes in the character of the material that enters the region of sedimentation. It is also influenced to some extent by the chemical composition of the water.

The effect of the seasonal changes in the character of the suspended material is disclosed by the water-content diagram (Fig. 43.1*b*). On account of this effect, the scattering of the water content from the average is as important for vertical distances as small as a few inches as it is for the entire depth. Still more conspicuous is the effect of

Fig. 45.2. Section through undisturbed sample of varved clay taken in well-designed sampler of 4-in. diameter (courtesy M. J. Hvorslev).

seasonal changes on the structure of sediments that were laid down in fresh-water lakes under arctic conditions such as those which prevailed in the northern United States and in Canada during the ice age. In the summertime, the suspended material in the offshore parts of the lakes consisted of silt and clay, because the coarser materials such as sand and gravel had already been laid down at the mouths of the rivers, building up delta deposits. The silt particles settled out during the summer. During the winter, however, no new material was carried into the lakes because the rivers were completely frozen. Hence, beneath the ice crust, only clay particles which did not settle during the summer were deposited. Therefore, the sediment is composed of light-colored summer layers consisting of silt with some clay, and dark-colored winter layers consisting chiefly of clay. Each double layer represents the deposit of one year. These deposits are the varved clays (Fig. 45.2) mentioned in Article 2. The thickness of the double layers is commonly less than 1 in., but exceptionally as much as several feet. It depends on the amount of material washed into the lake during the summer season. Deposits of such clay are very common in both North America and Europe north of the 40th parallel. They are a prolific source of serious construction difficulties.

If similar arctic rivers enter a bay of the ocean instead of a fresh-water lake, the segregation according to particle size is much less perfect because the salts contained in the sea water cause flocculation of the clay particles. As a consequence, most of the clay is deposited simultaneously with the silt.

The preceding review has shown that nature created an infinite variety of structural patterns ranging from the simple stratification of offshore deposits formed in large lakes to the utterly complex structure of masses of gravel, sand, and silt that were laid down, plowed up, distorted, locally eroded, and redeposited along the rim of a continental ice sheet. The character of residual soils, on account of variations in the parent material and in the degree of weathering, is no less complex (Article 49). If borings are made in a mass of soil at two points 100 or 200 ft apart, the engineer knows the character and the sequence of the strata along two vertical lines. Between these two lines the strata may be continuous. However, they may also wedge out at a short distance from each line, and the sequence of strata halfway between the two borings may not have the remotest resemblance to that at either one. An intelligent program for supplementary soil investigations can be prepared only by an engineer who is thoroughly familiar with the elements of physical geology and with the geology of the region in which the site is located.

A description of the geological history of the subsoil of large cities can usually be found in the publications of the local museum of natural history or of some similar institution. If the job is located in the open country, it is advisable to find out whether or not a geological study of the region has been made. General information concerning the geology of a particular locality can often be found in the following:

Fenneman, N. M. (1931), *Physiography of Western United States,* New York, 534 pp.

Fenneman, N. M. (1938), *Physiography of Eastern United States,* New York, 714 pp.

Atwood, W. W. (1940), *The Physiographic Provinces of North America,* New York, 536 pp.

More specific references should be looked for in the following bibliographies:

Geologic Literature on North America, bibliographic bulletins of the U.S. Geological Survey, published every two years. Cumulative bibliographies are available for the years 1785–1918 and 1919–1928.

Bibliography and Index of Geology Exclusive of North America, published annually since 1930 by the Geological Society of America.

Index to Geologic Mapping in the United States. Set of maps, one for each state, prepared by U.S. Geological Survey showing areas described in various publications and listing the publications.

R. F. Legget, "Geological Surveys of the World," Appendix B in *Geology and Engineering,* 2nd ed., New York, 1962. Contains brief discussions of the geological surveys of various countries, their mailing addresses, and their publications.

R. F. Legget, "Geological Societies and Periodicals," Appendix C in *Geology and Engineering.*

Catalogue of Published Bibliographies in Geology 1896–1920, Bulletin National Research Council, **6,** Part 5, No. 36, 1923.

Geologic maps and brief descriptions of a few regions have been published in the Folios of the U.S. Geological Survey. Descriptions and maps of many other regions are scattered throughout the periodical literature. These are listed in the bibliographic bulletins of the U.S. Geological Survey.

A large amount of useful information regarding regional geology is given in the *Water Supply Papers* published from time to time since 1896 by the U.S. Geological Survey.

Soil survey maps, prepared for agricultural use by the U.S. Department of Agriculture and the various state agricultural experiment stations, are commonly accompanied by at least brief descriptions of the parent materials and their geological origin.

If no specific information regarding the geology of the site of the job is available, the engineer must rely on his own capacities for geological observation and interpretation. On large projects a detailed geological survey of the site and its vicinity is imperative. It calls for the services of a professional geologist.

Spacing and Depth of Exploratory Borings

At the present time the spacing between exploratory drill holes is still governed primarily by convention and not by rational considerations. On building sites the borings are commonly spaced at about 50 ft in both principal directions. On subway or earth-dam projects a spacing of 100 ft is generally considered the minimum. However, if the line is very long or the site very large, it may be necessary to increase the spacing to 200 ft. Even at that spacing the required amount of drilling and testing may be very large, and it may cause undesirable delays in starting construction on the project.

The standardization of the spacing of exploratory drill holes has obvious disadvantages. If the soil profile is very simple, the customary spacing is too small, whereas, if the profile is erratic, the spacing is excessive. In order to avoid the loss of time and money due to drilling superfluous bore holes, the method of subsurface soundings can often be used to advantage. A sounding, which is cheaper and more expedient than a drill hole, may be made at each point where the conventional regulations call for a drill hole. If all the penetration diagrams are similar to each other, the soil profile is likely to be simple. Exploratory drill holes are required only at one or two locations where average conditions prevail, and near those few points where the penetration diagrams indicate maximum deviations from the statistical average. If the geology of the site involves the possibility that the continuity of the strata may be disrupted locally by channel fillings or other bodies of foreign material, supplementary soundings should be made wherever there is any surface indication of the presence of a compressible inclusion, such as a shallow depression on the surface of the ground. If a sounding strikes such an inclusion, an exploratory drill hole should be made near by to determine the type of soil of which the inclusion consists.

If the penetration diagrams obtained from the exploratory soundings are consistently very different, the soil profile is likely to be erratic and intermediate soundings should be made until the penetration data are complete enough to leave no doubt concerning the general shape and trend of the boundaries between the fine-grained and coarse-grained, and the loose and dense parts of the deposit. Yet, no more drill holes are required than those few needed to determine the types of soil located between the different surfaces of discontinuity or to find out whether a body of exceptionally resistant or nonresistant soil consists of sand or clay. Such a question arose when the sounding shown on the right-hand side of Fig. 44.12 was made. It was doubtful whether the soil between El. 80 and 60 consisted of very loose sand or of clay. To answer this question a boring was made next to the line of sounding. The boring record left no doubt that there was no clay below El. 80. The low resistance to penetration within this range of depth was due exclusively to the exceptionally loose structure of the sand.

The depth to which exploratory drill holes should be made is likewise more or less standardized. This practice is not only wasteful but also dangerous. Many buildings have been seriously damaged by settlement due to the consolidation of soft clay strata located below the depth to which the subsoil was explored. Yet, no general rules can be established for selecting this depth, because for a given weight and given dimensions of a structure the depth at which the seat of settlement is located depends to a large extent on the soil profile. The following paragraphs illustrate the factors that should be considered before the depth of the drill holes is specified.

If it is certain for geological reasons or from the results of previous borings in the vicinity that the subsoil of a group of buildings does not contain any strata of clay or soft silt, it is sufficient to explore the subsoil at the site of each building to a depth between 20 to 30 ft below subgrade, depending on the size and weight of the building. The size of the area occupied by the group does not require any consideration because the compressibility of sand strata decreases rapidly with increasing depth (Article 15) and, consequently, each building settles almost as if the others did not exist.

On the other hand, if the subsoil of a group of buildings contains soft strata, the seat of settlement may be located at a depth greater than the width of the entire area occupied by the buildings because, even at a depth of 150 or 200 ft, a moderate increase of the pressure on a thick stratum of soft clay may produce a settlement of more than 1 ft (Article 55). Hence, the depth to which the subsoil should be explored depends primarily on the absence or presence of compressible strata such as clay or plastic silt.

If the geology of the site indicates that clay or silt strata may be located at great depth below the surface, or if nothing whatsoever is known concerning the subsoil conditions, a rough estimate should be made of the intensity and distribution of the pressures that will be produced in the subsoil by the proposed group of buildings. The procedure has been described in Article 40. On the basis of this estimate, the greatest depth D_{max} can be evaluated at which the presence of a thick layer of soft clay with a high liquid limit may still have a significant influence on settlement. The first drill hole should be made to this depth. All the other borings and subsoil soundings can be discontinued at a depth of about 10 ft below the base of the lowest clay stratum that was encountered within the depth D_{max}. This procedure should be followed, regardless of whether the character of the upper soil strata may call for a footing, raft, or pile foundation.

The following example illustrates the possible consequences of disregarding the recommended procedure. A group of factory buildings was constructed on a tidal flat. None of the buildings was more than 40 ft wide. The subsoil was explored by borings to a depth of 90 ft. Within this depth there was a gradual transition from soft silt near the surface to sand with variable density at a depth of more than 65 ft. Because of the high compressibility of the top strata, it was decided to support the buildings on piles 70 to 90 ft long. To the surprise of the engineers in charge of the job, the buildings started to settle during construction, and in the course of three years the settlement increased to more than 2 ft. Subsequent soil investigations showed that the settlement was due to the consolidation of a stratum of soft clay 30 ft thick, located at a depth of about 115 ft below yard level.

If bedrock is encountered within the depth D_{max}, the topography of the rock surface must be determined at least approximately by sounding or boring, because the depressions in the rock surface may be filled with very compressible sediments that are encountered in only the deepest drill holes. The omission of this precaution also has repeatedly been the cause of important settlement.

The results of the exploratory borings and subsurface soundings should be assembled in a report containing all the information that was secured concerning the geology of the site, a list of the index properties of all the spoon samples that were taken, and a record of the results of the standard penetration tests. On the basis of this report it can be decided whether or not supplementary investigations are required concerning the relative density and permeability of the sand strata and the shearing resistance and compressibility of the clay strata.

Relative Density of Sand Strata

The relative density of sand strata has a decisive influence on the angle of internal friction of the sand (Article 17), on the ultimate bearing capacity (Article 33), and on the settlement of footings resting on the sand. If a submerged sand is very loose, a sudden shock may

Table 45.1
Relative Density of Sands according to Results of Standard Penetration Test

No. of Blows N	Relative Density
0–4	Very loose
4–10	Loose
10–30	Medium
30–50	Dense
Over 50	Very dense

transform it temporarily into a sand suspension with the properties of a thick viscous liquid (Article 17). In a dense state the same sand is insensitive to shock and is perfectly suitable as a base for even very heavy structures. For this reason the relative density of a sand is far more important than any of its other properties, except possibly its permeability.

While the exploratory borings are being made, some information regarding the relative density of the sand strata encountered in the drill holes can be obtained by performing the standard penetration test (page 304) whenever a spoon sample is taken. Considering the outstanding importance of the relative density, the standard penetration test should be considered an essential part of the boring operation. Table 45.1 gives an approximate relation between the number of blows N and the relative density.

The relation, Table 45.1, should be used with caution, and only if the penetration tests are carried out conscientiously. If the sand is located below the water table an inexpert driller may allow the water level in the casing to fall below the piezometric level in the sand at the location of the test, whereupon the sand may become quick and be transformed into a loose state; the N-value will then be too low. The mere removal of the drill rods at a rate too rapid to permit water to replace the volume of the rods may cause such a lowering of the water level. On the other hand, boulders or cobbles several inches larger than the diameter of the spoon may give excessively high N-values.

In saturated fine or silty sands of moderate to high relative density, and having an effective grain size between 0.1 and 0.05 mm, the number of blows may be abnormally great because of the tendency of such materials to dilate during shear under undrained conditions (Article 15). Hence, in such soils the standard penetration test should be supplemented by more reliable procedures, or the results should be interpreted conservatively.

On important jobs the information obtained from the standard penetration tests concerning the relative density of the sand should be supplemented by subsurface soundings. These soundings furnish continuous records, such as those shown in Figs. 44.12 and 44.16, of the variation of the penetration resistance with depth. However, the resistance against the penetration of a penetrometer into sand, or the energy required to produce a given penetration, depends not only on the relative density of the sand but also on the dimensions of the drive point and the stem and to some extent on the shape of the grains and the grain-size distribution. Hence, every new method of subsurface sounding and every use of the method in an unexplored locality call for a set of calibration tests that furnish the data for interpreting the penetration records.

A rough calibration can be achieved by making a subsurface sounding next to a drill hole in which standard penetration tests were performed. More cumbersome but also more reliable are surface load tests at different depths below a point close to the location of a subsurface sounding. The tests are made on a bearing plate 1 ft square, resting on a horizontal surface of the sand. No backfill or surcharge is placed within 3 ft of the plate. The relation between load and settlement for such tests on several different sands is shown in Fig. 45.3*a*. Curves 1 and 2 were obtained from tests on very dense sand, curve 4 on sand of medium density, and curve 5 on loose sand. With increasing relative density the bearing capacity increases rapidly, and the settlement under a given load decreases. Figure 45.3*a* shows, in accordance with field experience and in contrast to a widespread opinion, that the grain size has no influence on the relative density and bearing capacity of a sand.

On the right-hand side of Fig. 44.16 are shown the results of load tests made for the purpose of calibrating the wash-point penetrometer (Fig. 44.13*d*). The procedure was described on page 321.

On the basis of the results of standard load tests, such as those shown in Fig. 45.3*a*, the relative density of the sand can be determined by means of the diagram (Fig. 45.3*b*). For this purpose the curves obtained from the calibration tests are introduced into the diagram. Each curve corresponds to a definite penetrometer reading. The position of the curve with reference to the boundaries shown in the figure indicates the relative density of the sand penetrated by the drive point.

The standard load test may, however, furnish misleading results if the grain size of the sand is fine or very fine and if the sand contains appreciable soil moisture. On account of the apparent cohesion due to the capillary forces (Article 20), the sand may appear stronger and less compressible than it would if no soil moisture were present. The influence of the apparent cohesion decreases rapidly with

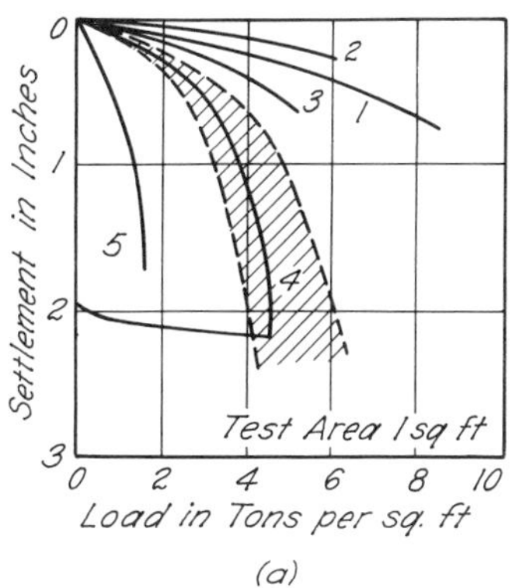

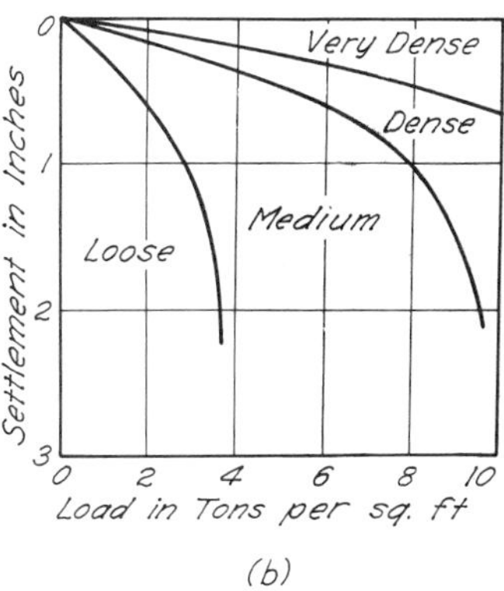

Fig. 45.3. (*a*) Relation between load and settlement of bearing plate 1 ft square resting on the surface of a sand. Curve 1 represents dense clean fine sand in caisson 26 ft below river bottom; 2 represents very dense very fine sand in open excavation 26 ft below ground surface in Lynn, Mass; 3 represents damp sand of medium density hand-compacted by tamping in layers; 4 represents medium-dense sand at bottom of shaft 30 ft deep in Houston Street, New York. Shaded area indicates range for curves obtained between depths of 20 and 60 ft; 5 represents loose, coarse, clean, and very sharp sand at bottom of open excavation near Muskegon, Mich. (*b*) Chart for estimating relative density of sand on basis of results of standard load test on bearing plate 1 ft square.

increasing width of the loaded area; it may be too great to ignore if the test area is as small as one foot square.

Still more accurate information regarding the relative density of sand strata can be obtained by laboratory tests on undisturbed samples hand-carved from test shafts or taken from drill holes by one of the procedures described in Article 44. The shafts or borings are made near points at which subsurface soundings were previously performed. By correlating the test results with the corresponding resistances to penetration, the data are obtained for accurate interpretation of the results of all the other subsurface soundings. However, the instances are very rare in which such refinements are warranted.

Permeability of Sand Strata

Reliable information on the permeability of sand strata may be required for either of two purposes. It may be necessary to estimate the quantity of water that will flow toward an excavation with specified dimensions, at a given position of the water table. Or it may be required to determine the depth to which the cutoff beneath a dam on a permeable foundation must be made in order to reduce the seepage losses from the reservoir to less than a specified amount.

The data for estimating the flow of water toward an excavation are obtained most conveniently by means of pumping tests (Article 44). The results of the tests make it possible to compute the average coefficient of permeability of the subsoil in horizontal directions. Once this coefficient is known, problems concerning the flow of water toward the proposed excavation can be solved on the basis of the laws of hydraulics. If the project calls for lowering the water table by means of filter wells (Article 47), the well system can be laid out and an estimate can be made of the capacity of the pumps required for maintaining the water table below the bottom of the excavation during construction.

In order to solve cutoff and seepage problems, it is necessary to determine not only the average permeability of the subsoil but also the major variations of the permeability within the sand strata located beneath and beside the water-retaining structure. This can be done by means of permeability tests on fairly continuous series of samples obtained from a considerable number of drill holes. However, natural deposits are never homogeneous. The water percolates through them along more or less tortuous lines, following lenses and layers composed of the coarsest constituents, and their permeability in a vertical direction is usually considerably smaller than that in a horizontal one. Therefore, laboratory investigations of any kind cannot be expected to disclose more than the order of magnitude of the permeability of the deposit even if the tests are performed by causing the water to flow through undisturbed samples separately in horizontal and vertical directions. Furthermore, the cores are never continuous. A single seam of silt located between two adjoining samples of sand may have a radical influence on the ratio of horizontal to vertical permeability. The occurrence of such seams is not uncommon (Fig. 45.4).

For these reasons the use of undisturbed samples for permeability tests is hardly justified. Results no less reliable can be obtained by testing samples recovered from drill holes by means of a sampling spoon equipped with a core catcher (Fig. 44.2*d*) or by a scraper bucket (Fig. 44.2*f*). If the samples were recovered in a spoon, their constituents should be thoroughly mixed before testing. After the technician has made 15 or 20 permeability tests on samples from a given stratum, he can estimate the coefficient of permeability of the others on the basis of their texture and general appearance. The estimates and test results can be adjusted to take into account the difference between the relative density of the remolded and the in-place material. The

Fig. 45.4. Silt seams in medium uniform sand. The presence of the seams could not be detected by ordinary test borings. Yet they reduce the permeability of the sand stratum in a vertical direction to a small fraction of that in horizontal directions.

ratio of horizontal to vertical permeability can be judged on the basis of Eqs. 11.10 and 11.11.

Elaborate investigations of this nature are rarely justified economically. The determination of the permeability of natural deposits below the water table by *in situ* permeability tests is always more reliable than that by laboratory tests.

Procedures have been developed to evaluate the permeability of sand strata located above the water table on the basis of the quantity of water that percolates into the soil from that part of a drill hole extending below the casing. The results are at best no more than crude estimates and may be unreliable, because the flow pattern into the soil remains unknown and the formation of a filter skin at the entrance surface can hardly be avoided. The procedure (Zangar 1953) is similar to that described in connection with *in-situ* permeability tests in drill holes below the water table.

Shearing Resistance of Saturated Clays

If a project involving clay soils calls for an investigation of the stability of slopes, the computation of the lateral pressure against the bracing of open cuts, or an estimate of the ultimate bearing capacity of footings or rafts, the shearing resistance of the clay must be determined. If the water content of the clay will not change significantly during the period when the slopes are unsupported or during the lifetime of the temporary bracing of the open cuts, or if the factor of safety of footings is a minimum before the water content can decrease on account of the loading, the $\phi = 0$ conditions are applicable (Article 18). The undrained shearing strength on the basis of total stresses is then equal to one-half the unconfined compressive strength q_u of undisturbed samples of the clay. The shearing strength can also be determined directly by means of the vane (Fig. 44.17) or torvane (Fig. 18.3). Inasmuch as many practical problems of outstanding importance fall into the $\phi = 0$ category, means for evaluating the undrained shearing strength of saturated clay soils deserve special consideration.

During the drilling of the exploratory holes the shearing resistance of the clay can be crudely estimated on the basis of the record of the standard penetration test. Table 45.2 shows the approximate relation between the unconfined compressive strength and the number of blows per foot of penetration of the sampling spoon. However, at a given number N of blows per foot, the scattering of the corresponding values of q_u from the average is very large. Therefore, compression tests should always be made on the spoon samples. The other

Table 45.2
Relation of Consistency of Clay, Number of Blows N on Sampling Spoon, and Unconfined Compressive Strength

	q_u in tons/ ft^2					
Consistency	Very Soft	Soft	Medium	Stiff	Very Stiff	Hard
N	<2	2–4	4–8	8–15	15–30	>30
q_u	<0.25	0.25–0.50	0.50–1.00	1.00–2.00	2.00–4.00	>4.00

routine tests on the spoon samples, listed in Table 9.1, are also obligatory because their results are required for comparing the clay with others previously encountered on similar jobs. The values of q_u obtained by means of compression tests are likely to be somewhat too low because spoon samples are appreciably disturbed. The supplementary investigations required on important jobs depend on the character of the soil profile.

If the soil profile is simple and regular, it is commonly possible to evaluate the average shearing resistance of the clay strata on the basis of the results of laboratory tests. The samples are secured by means of tube sample borings (Article 44) which furnish continuous cores. To obtain fairly reliable average values, the spacing between the sample borings should not exceed 100 ft. If it is known in advance that the soil profile is fairly regular and that tube sample borings will be required, continuous samples are taken in all those sections of the exploratory holes that are located within clay strata. In the sections located between clay strata spoon samples are extracted, and standard penetration tests are made.

The samples are delivered to the laboratory in sealed tubes commonly 30 or 36-in. long. Preferably, all the clay samples from one hole should be tested in the sequence in which they followed each other in the drill hole in a downward direction. Each sample is ejected from its tube by means of a close-fitting plunger in such a manner that the sample continues to move with respect to the tube in the same direction as it entered; if excessive side friction causes too much disturbance during ejection, the tube is cut into 6-in. sections, the soil itself is cut by means of a wire saw, and each section is ejected.

For routine testing each sample is cut into sections with lengths equal to about three times the diameter; that is, 2-in. tube samples should be cut into about 6-in. sections. If the uppermost clay-core section from a given tube appears to be relatively undisturbed, it is submitted to an unconfined compression test first in its natural state and then in a completely remolded state at the same water content. The ratio between the two values of the compressive strength is a measure of the sensitivity of the clay (Article 7). After the test, the sample is divided lengthwise into two parts. One half is used for a water-content determination, and the other is stored in a jar with an airtight cover. The same set of tests is made during subsequent operations whenever a sample is encountered that differs noticeably from its predecessor in consistency, color, or general appearance. A change in consistency is revealed by a noticeable change of the resistance of the clay to deformation between the fingers. The uppermost samples in each tube may be appreciably more disturbed than the others. If this is the case, the compression tests should be performed on one of the other less disturbed samples.

The samples following the first one are split lengthwise. One entire half is used for a water-content determination. The other half should be set aside in a fairly humid atmosphere with its plane surface facing upward, whereupon it starts to dry slowly. At an intermediate state of desiccation the details of stratification become clearly visible. At that state a record should be made of the details of stratification, indicating the color and approximate thickness of the individual layers, the degree of perfection of the stratification, and other visible features. The records are later used for preparing a general description of the characteristics of the stratification of the clay. A few representative specimens are photographed.

The following 6-in. sections are also used only for water-content determination and visual inspection. If the experimenter tests five or six samples in this manner without noticing a conspicuous change, the next section is submitted to an unconfined compression test in its natural state, as well as to the water-content determination. This procedure is continued until a sample is encountered that differs materially from its predecessor. This sample is submitted to the same tests as the very first one, whereupon the routine procedure is resumed.

If a more detailed record of the consistency is desired, each of the 6-in. sections mentioned in the preceding paragraph is sliced lengthwise into two halves and one or two small-diameter torvane tests are made on the cut face of one of the halves. The other half is used for determinations of water content, studies of stratification, and other appropriate tests. Figure 45.5 shows the results of such a detailed survey on samples of stratified highly sensitive silty clays

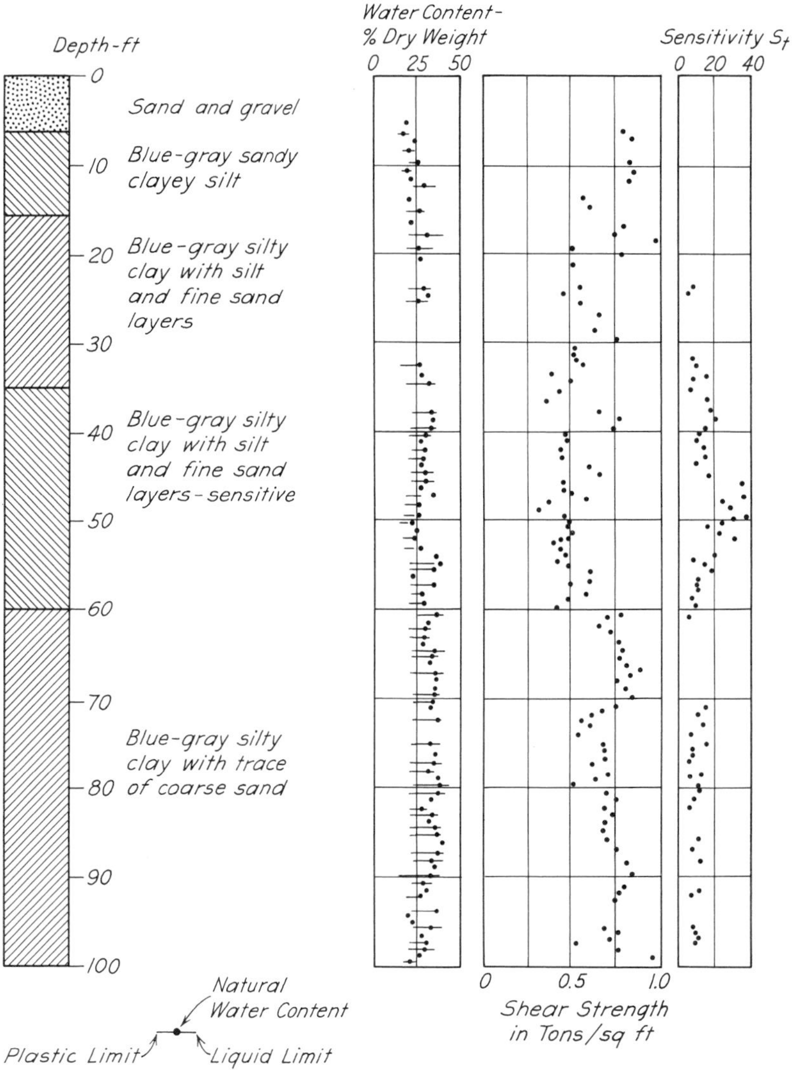

Depth-ft
0
10
20
30
40
50
60
70
80
90
100

Sand and gravel
Blue-gray sandy clayey silt
Blue-gray silty clay with silt and fine sand layers
Blue-gray silty clay with silt and fine sand layers - sensitive
Blue-gray silty clay with trace of coarse sand

Water Content-
% Dry Weight
0 25 50

Shear Strength in Tons/sq ft
0 0.5 1.0

Sensitivity S_t
0 20 40

Natural Water Content
Plastic Limit
Liquid Limit

Fig. 45.5. Results of detailed survey of strength, Atterberg limits and natural water content of boring in clays associated with landslides caused by Good Friday, 1964, earthquake in Anchorage, Alaska (after Shannon and Wilson 1964).

involved in landslides caused by a major earthquake (Shannon and Wilson 1964). The values of sensitivity were calculated on the basis of miniature vane tests on completely remolded portions of the samples at the locations where the torvane tests were made.

After all the tests on samples from one drill hole have been made, the Atterberg limits are determined on representative specimens of those samples that were submitted to compression tests in both the natural and remolded states. The results of the tests are represented in diagrams such as that shown in Fig. 45.5. The diagrams should be accompanied by a brief description of the characteristics of the stratification of the clay (not shown in the figure).

If the investigation is made for the purpose of estimating the factor of safety of slopes with respect to sliding or of fills with respect to spreading, the knowledge of the details of stratification is at least as important as that of the strength of the clay, because the major part of the potential surface of sliding may be located in one or more seams of fine sand or coarse silt and not in the clay. In such instances a detailed and well-illustrated description of the characteristics of the stratification should be prepared. A few typical samples of the stratified layers should be set aside for further investigation. This investigation consists of determining the natural water content and the Atterberg limits of each of the layers of which the sample is composed. Figure 45.6 shows the results of such an investigation.

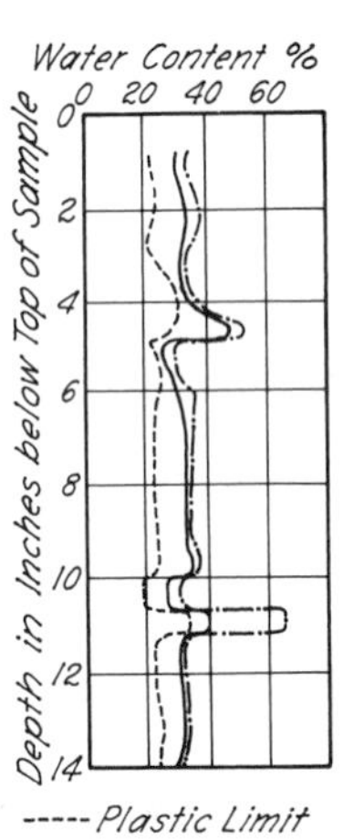

Fig. 45.6. Diagram showing variations of index properties within a 1-ft layer of a soft glacial clay.

In any event the opportunity should be sought to investigate the degree of disturbance of the tube samples as described in Article 44.

All the preceding discussions referred to the investigation of fairly homogeneous clay strata. If the clay strata contained in the subsoil have a variable thickness and consistency, the method of investigation must be modified. Instead of concentrating on soil tests, the engineer should make efforts to investigate the topography of the upper and lower boundaries of the clay layers and to locate the softest and hardest parts of the layers. The most expedient method to get this information is to make numerous subsurface soundings supplemented by exploratory drill holes. After the results of these investiga-

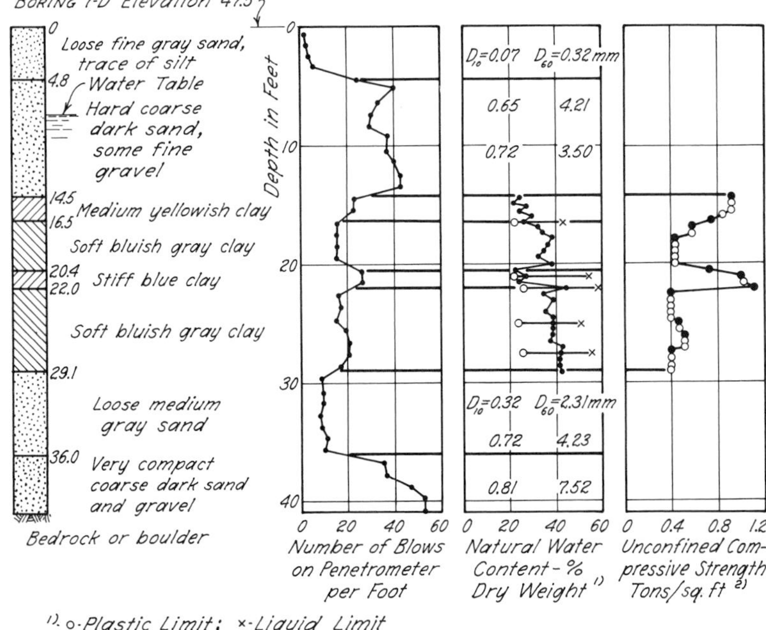

Fig. 45.7. Diagram representing boring record, penetration record, and results of soil tests on samples from drill hole through composite shore deposit.

tions are assembled, two or three tube-sample borings are made. These borings should be located at the best and the worst spots of the site. Within the bodies of soil located between the clay strata, spoon samples are taken and standard penetration tests are made, whereas within the clay strata continuous tube samples are secured. Figure 45.7 represents a boring of this type. The boring was made in a composite shore deposit located on one of the slopes of a drowned valley. On the left side is shown an abstract of the foreman's record. The first diagram is the penetration record of a subsurface sounding made a few feet from the drill hole. The last two diagrams contain the results of the soil tests.

Figure 45.8 represents the results of a survey of the unconfined compressive strength of a glacial clay deposit intermediate between regular and erratic. The individual clay strata were not homogeneous enough to justify assigning definite average values to their physical properties. Yet the project called for general information concerning the compressive strength of the clay and its variations in both hori-

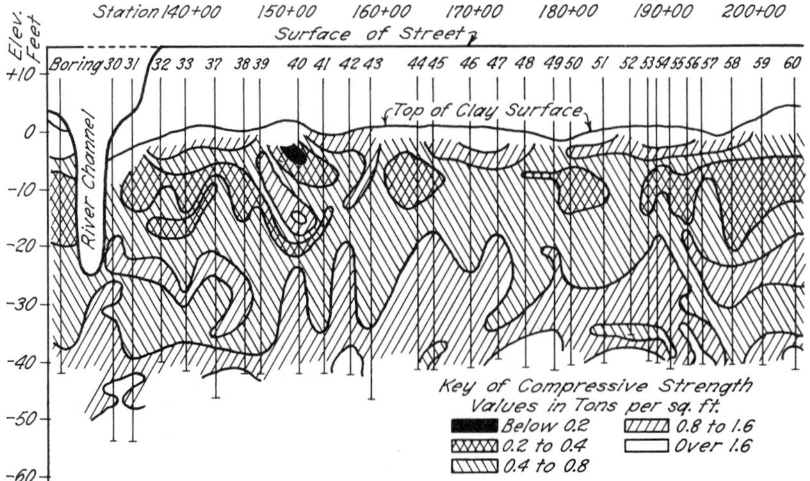

Fig. 45.8. Diagram showing variations in unconfined compressive strength of somewhat erratic glacial clay deposit in Chicago (after Terzaghi 1943*a*).

zontal and vertical directions. To satisfy these demands tube sample borings were made at points 200 ft apart, and the cores obtained from these borings were submitted to the same tests that are made on continuous cores from homogeneous strata. Subsequent tunneling operations showed that the profiles actually disclosed the general character of the clay strata encountered in the different tunnel sections. As would be expected, between drill holes the scattering of the properties of the clay from the average was important and called for continual vigilance during construction, but a more detailed subsoil investigation would have been impracticable and uneconomical (Terzaghi 1943*a*).

Compressibility of Clay Strata

The compressibility of clay strata is of interest as a source either of progressive settlement or else of delay in the increase of shearing resistance produced by superimposed load. Whatever the practical implications of the compressibility may be, a reliable forecast of its effects can be made only if the clay strata are continuous and fairly homogeneous.

If the subsoil contains a continuous and fairly homogeneous clay stratum, the settlement of the surface due to superimposed loads is at every point roughly proportional to the average pressure that the loads produce in the clay beneath this point. The intensity and distribution of the pressure in the clay can be computed by means of the methods described in Article 40. On the basis of the results of the computations and those of soil tests, the settlements due to the loads can be computed, and the curves of equal settlement can be constructed.

On minor jobs involving foundations above homogeneous clay strata, no soil investigations are required other than routine tests on the spoon samples. For clay these tests include the determination of the liquid and plastic limits. The statistical relation between the liquid limit and the compression index C_c is given by Eq. 13.11. For a normally loaded clay of ordinary sensitivity the value obtained by means of this equation is accurate enough for most practical purposes. However, if the clay is extrasensitive, the correct value of C_c is likely to be higher than the computed one and, if it is precompressed, the correct value is considerably lower. The degree of sensitivity is indicated by the effect of remolding on the compressive strength of the spoon samples. The existence of precompression can commonly be inferred from the geological character of the site.

On important jobs that call for accurate settlement forecasts, supplementary investigations are required. These consist, first of all, of tube-sample borings spaced not more than 100 ft. The continuous samples obtained from these borings are submitted to the same tests as those prescribed for the investigation of the shearing resistance of homogeneous clay strata. However, the unconfined compression or torsion shear tests need be performed on representative samples from only one hole in order to get reliable information concerning the sensitivity of the clay.

After the water-content profiles for all the tube-sample borings have been plotted in diagrams similar to Fig. 45.5, one representative boring is selected. Near this boring an undisturbed-sample boring is made that furnishes samples with a diameter of at least 3 in. to be submitted to consolidation tests.

Because of the great amount of time and labor involved in the performance of consolidation tests, these tests cannot be made on more than 10 or 15 samples without undue delay. Yet, even in relatively homogeneous clay strata the physical properties of the clay are likely to change to a considerable extent from point to point. As a consequence, the compressibility characteristics of a clay can be determined at a reasonable expense only on the basis of statistical relationships between the compressibility and the index properties of the clay.

Of all the tests listed in Table 9.1 as routine tests on clay, the cheapest and most convenient is the water-content determination. Furthermore, the natural water content is more closely related to the compressibility of the different parts of a clay stratum than any of the other index properties. Hence, the evaluation of the average compressibility of a clay stratum is most conveniently based on the statistical relation between the natural water content and the compressibility of the components of the stratum.

The settlement due to the consolidation of a normally loaded layer of clay with an average void ratio e_0 depends on the compression index C_c of the clay, provided all other conditions are equal. Experience has shown (Rutledge 1939) that the relation between the natural water content and the *compression ratio* $C_c/(1 + e_0)$ for such clays can be represented approximately by a linear equation. In order to take advantage of this relation, consolidation tests are performed on samples of the clay, and the values of $C_c/(1 + e_0)$ are plotted against the natural water content. Figure 45.9 shows such a diagram. All the points representing individual test results are located close to a straight line. The vertical distance between the dash lines represents the scattering of the values of $C_c/(1 + e_0)$ from the average corresponding to a given natural water content.

After the relation between the compression ratio and the natural water content has been determined, the next step is to make use of the relation to estimate the values of the compression ratio corresponding to the natural water content of all the tube samples that have been tested. Finally, the average value of $C_c/(1 + e_0)$ is determined by a suitable arithmetic or graphical procedure. This value can be used directly in Eq. 13.8 to compute the settlement.

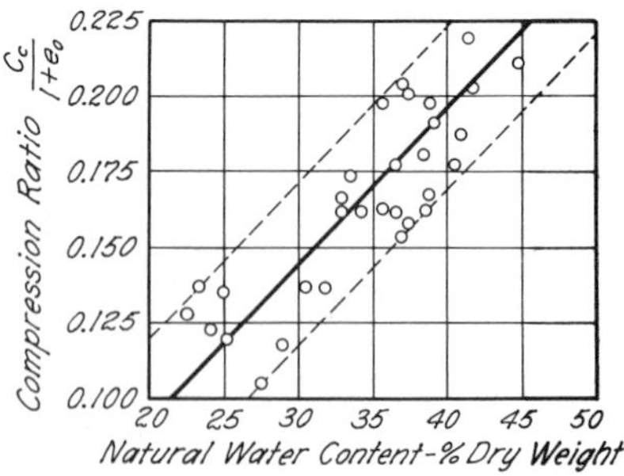

Fig. 45.9. Statistical relation between natural water content and compression ratio for samples of clay from boring in Boston, Mass. (after Fadum 1941).

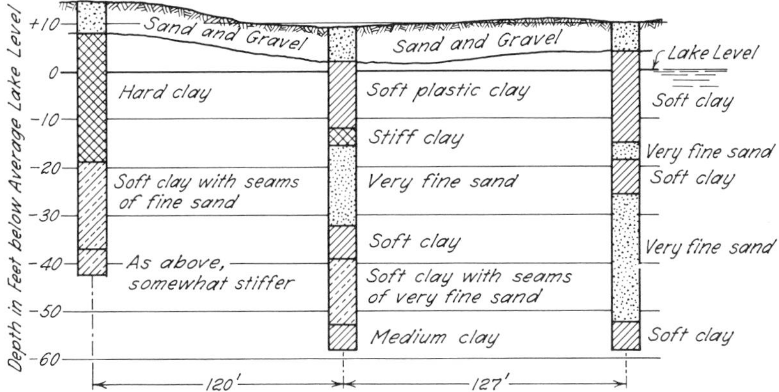

Fig. 45.10. Erratic shore deposit, Lake Erie near Cleveland.

If a clay is precompressed, Eq. 13.8 cannot be used, and the method of settlement computation must be adapted to the consolidation characteristics of the clay. If the project requires the highest degree of accuracy that can reasonably be achieved, the best possible undisturbed samples must be recovered and the field e–log p curve constructed by the procedure illustrated in Fig. 13.10. Nevertheless, for the reasons explained in Article 13, such a settlement forecast is much less reliable than that for normally loaded clays.

Figure 56.8 illustrates the degree of accuracy that can be achieved in computing the distribution of the settlement over the base area of a building located above fairly homogeneous clay strata. The real distribution is shown on the left-hand side, and the computed on the right. The structure itself is complex but symmetrical. No such results can be obtained if the profile of the subsoil of a building is erratic, because the settlement of buildings on such subsoils depends not only on the intensity and the distribution of the loads but also on the variations in the compressibility of the subsoil in horizontal directions. In addition, the rate of settlement depends on the degree of continuity of layers and pockets of cohesionless material present in the subsoil. As a consequence, it can change from place to place. Figure 45.10 illustrates such a profile. It represents the results of test borings in a composite shore deposit located at the south shore of Lake Erie. More than 100 borings were made, and the spacing between drill holes did not exceed 100 ft. Yet the boring records did not indicate whether or not the layers of clay encountered in the drill holes were continuous.

If the profile of the subsoil of a proposed structure is erratic, undisturbed sample borings and elaborate soil investigations are entirely out of place. Information of far more practical value can be obtained by means of numerous subsurface soundings supplemented by exploratory drill holes. The results of such investigations at least inform the designer of the location of the softest and the most resistant spots beneath the base of his building. At two or three points tube-sample borings may be made to get information on the details of stratification and on the sensitivity of the clay that was encountered in the exploratory holes. The maximum settlement is roughly estimated on the basis of the statistical relation between the liquid limit and the compression index C_c. In judging whether or not the proposed structure can stand the estimated settlement, the spacing between the hardest and the softest spots of the subsoil is taken into consideration. The results of even very elaborate soil investigations would hardly add anything to the information obtained by the recommended procedure.

Summary of Procedures in Subsoil Reconnaissance

According to the preceding discussions, a subsoil reconnaissance involves several successive stages. The first step is to decide on the depth and spacing of the exploratory borings.

If the proposed structure is a building, current practice calls for about one drill hole for every 2500 ft^2 of the area covered by the building. If a retaining wall is to be constructed or an open cut to be made, it is customary to make at least one drill hole for every 100 ft of length of the wall or the cut. However, these rules are based on convention rather than rational considerations. If the subsoil is erratic, more useful information can commonly be obtained in less time and at a smaller expense by combining the exploratory borings with subsurface soundings.

The depth to which the exploratory drill holes should be made depends on whether the subsoil may contain layers of soft clay. If the local geological conditions or the conditions revealed by earlier test borings in the vicinity exclude this possibility, the drill holes do not need to be made to a depth of more than 20 or 30 ft below subgrade. On the other hand, if the subsoil may contain layers of soft clay at an unknown depth, a sound decision regarding the minimum depth of the test borings can be made only on the basis of the results of a rough estimate of the maximum depth at which the presence of clay strata may still have a significant influence on the performance of the proposed structure.

The subsequent investigations depend on the size of the job, the nature of the design problems, and the subsoil conditions.

On routine jobs such as the foundations of ordinary buildings or bridges, no investigations are needed beyond the routine tests on the spoon samples (see Table 9.1). Large or unusual jobs may call for one or more of the supplementary investigations described under the preceding subheadings. After the results of these investigations are digested, the engineer must then judge whether the conclusions based on the data can be considered final or whether the remaining uncertainties require an observational procedure during construction. Because of the important practical implications of these uncertainties, they are discussed in detail at the end of this article.

The preceding summary of operations demonstrates that subsoil reconnaissance is seldom a simple procedure requiring only conscientious adherence to a set of hard and fast rules. Unless the investigator is guided by mature judgment and has had a varied practical experience in this field, much time and money may be wasted.

In connection with each step, a thorough knowledge of the geology of sedimentary and other unconsolidated masses is an asset of inestimable value, because factual knowledge is always limited to soil conditions along vertical lines spaced far apart. It has already been mentioned (Article 43) that the results of interpolation and the estimate of possible scattering can be very misleading, unless the investigator has a fairly clear conception of the anatomy of the body of soil under investigation. A knowledge of the geology of the region is also needed to determine whether clay beds beneath the building site have ever been subjected to greater loads than at present and, if so, to provide a basis for estimating the magnitude of the additional pressure.

The larger the job, the more necessary it is to supplement the results of the soil investigations by information obtained from purely geological sources, because on large jobs a detailed soil survey is likely to be a physical impossibility.

Discrepancies between Reality and Assumptions Based on Subsoil Reconnaissance

The results of the subsoil reconnaissance for every job, large or small, are ultimately condensed into a set of assumptions that constitute the basis for design. The steps that lead to this final result involve various processes of interpolation and correlation based on statistical relationships. Therefore, the assumptions are always to some extent at variance with reality. However, the importance of these inevitable discrepancies is very different for different types of assumptions. This fact is explained in detail in the following paragraphs.

Assumptions regarding the angle of internal friction of sandy soils, the relative density of sand strata, or the average compressibility of clay strata belong in one category. The errors associated with these assumptions depend chiefly on the number and quality of the field tests which furnish the basic data. Hence, faulty assumptions in this category can safely be blamed on inadequate soil reconnaissance, provided the soil profile is relatively simple. The dangerous character of some submerged or partly submerged very loose sands cannot be demonstrated reliably by tests of any kind (Article 17). Therefore, it should always be assumed that loose submerged sands may liquefy on slight provocation unless they are compacted by artificial means.

The accurate determination of the average coefficients of permeability k_I and k_{II} for soil strata of any kind on the basis of the results of soil tests is impracticable, because the values k_I and k_{II} depend on structural details of the strata that cannot be detected by any method of soil exploration. However, if the method for investigating the permeability is judiciously selected and intelligently used, fairly reliable limiting values can be obtained under almost any circumstances. The difference between the limiting values and the real average value cannot be determined, but for many practical purposes only a knowledge of the limiting values is needed.

By far the most unreliable information is obtained when attempts are made to predict the porewater pressures in stratified sand layers or in beds of clay containing thin seams or layers of more permeable material. This is due to the fact that the intensity and distribution of porewater pressures, under given hydraulic boundary conditions, depend on unexplorable structural details, even more than do the average coefficients of permeability of such strata. Hence, if the factor of safety of a foundation with respect to failure, or that of a mass of soil with respect to sliding, depends on porewater pressures, the fundamental assumptions regarding the porewater pressures should not be trusted under any circumstances, regardless of the care with which the subsoil has been explored.

In such instances the assumptions on which the design is based should be considered as nothing more than the expression of a working hypothesis, subject to revision on the basis of the results of observations made in the field during construction. Practically all the failures of the foundations of dams and other hydraulic structures can be attributed to unjustified confidence in assumptions of some kind, and many of the failures could have been avoided by appropriate field observations during the construction period. Considering the loss of life and capital involved in the failure of an important hydraulic structure, reliance on the assumptions on which the original design was based and omission of the field observations required for investigating the real conditions must, at the present state of our knowledge, be classified as unpardonable neglect.

In spite of the fact that computed values of the porewater pressure cannot be relied on, the computations should always be made because the results serve a vital purpose. They constitute the basis for evaluating the possible dangers, for preparing the program of field observations needed to detect impending dangers during construction, and for interpreting the results of these observations.

Selected Reading

The relationships among geological conditions, engineering properties and exploratory procedures are discussed or exemplified in the following publications.

Belcher, D. J. (1945). "The engineering significance of soil patterns," *Photogrammetric Engineering,* **11,** No. 2, pp. 115–148.

Lee, C. H. (1953). "Building foundations in San Francisco," *Proc. ASCE,* **79** (Separate 325), 32 pp.

Sowers, G. F. (1953). "Soil and foundation problems in the southern Piedmont region," *Proc. ASCE,* **80** (Separate 416), 18 pp.

Bjerrum, L. (1954). "Geotechnical properties of Norwegian marine clays," *Géot.,* **4,** No. 2, pp. 49–69.

Peck, R. B. and W. C. Reed (1954). "Engineering properties of Chicago subsoils," *U. of Ill. Eng. Exp. Sta. Bull. 423,* 62 pp.

Terzaghi, K. (1955*a*). "Influence of geological factors on the engineering properties of sediments," *Economic Geology, Fiftieth Anniversary Volume,* pp. 557–618.

Marsal, R. J. and M. Mazari (1962). *El subsuelo de la Ciudad de Mexico* (The subsoil of Mexico City). U. of Mexico, Faculty of Engineering, 2nd ed., 614 pp.

Woods, K. B., R. D. Miles, and C. W. Lovell, Jr. (1962). "Origin, formation, and distribution of soils in North America," Chapter 1 in *Foundation engineering,* G. A. Leonards, ed., New York, McGraw-Hill, pp. 1–65.

Lumb, P. (1965). "The residual soils of Hong Kong," *Géot.,* **15,** No. 2, pp. 180–194.

The following references contain examples of exploratory programs adapted to the conditions at the site and the requirements of the project.

Terzaghi, K. (1929*c*). "Soil studies for the Granville dam at Westfield, Mass.," *J. New Engl. Water Works Assn.,* **43,** pp. 191–223. Permeability survey of glacial outwash adjoining a reservoir site. The capillary rise method used in this study has been superseded by other procedures.

Peck, R. B. (1940). "Sampling methods and laboratory tests for Chicago subway soils," *Proc. Purdue Conf. on Soil Mech.*, pp. 140–150. Investigation of physical properties of somewhat erratic glacial clays in connection with tunneling operations.

Brown, F. S. (1941). "Foundation investigation for the Franklin Falls dam," J. Boston Soc. Civil. Engrs., **28**, pp. 126–143. Reprinted in *Contributions to soil mechanics 1941–1953*, Boston Soc. Civil Engrs., pp. 2–19.

Peck, R. B. (1953). Foundation exploration—Denver Coliseum," *Proc. ASCE*, **79** (Separate 326), 14 pp. Investigation of erratic sand and gravel deposits and location of boundary of filled area.

Peck, R. B., (1954). "Foundation conditions in the Cuyahoga River valley," *Proc. ASCE*, **80** (Separate 513), 20 pp.

Teixeira, A. H. (1960). "Typical subsoil conditions and settlement problems in Santos, Brasil," *Proc. 1st Panamerican Conf. on Soil Mech.*, Mexico, **1**, pp. 149–177.

Monahan, C. J. (1962). "John Day lock and dam: foundation investigations," *Proc. ASCE*, **88**, No. PO4, pp. 29–45. Exploration of five-mile stretch of river for selection of damsite, requiring geological, geophysical and engineering studies.

Chapter 8

EARTH PRESSURE AND STABILITY OF SLOPES

ART. 46. RETAINING WALLS

Design of Retaining Walls

The procedure for the design of retaining walls, like that of many other types of engineering structures, consists essentially of the successive repetition of two steps: (1) the tentative selection of the dimensions of the structure, and (2) the analysis of the ability of the selected structure to resist the forces that will act on it. If the analysis indicates that the structure is unsatisfactory, the dimensions are altered and a new analysis is made.

In making the first tentative selection of the dimensions of a retaining wall, the designer is guided by his experience and by various tables giving the ratio of base width to height of ordinary walls. In order to make the analysis, he first estimates the magnitude of all the forces that act above the base of the wall, including the pressure exerted by the backfill and the weight of the wall itself. Next, he investigates the stability of the wall with respect to overturning. He then estimates the adequacy of the underlying soil to prevent failure of the wall by sliding along a plane at or below the base, to withstand the pressure beneath the toe of the foundation without failing and allowing the wall to overturn, and to support all the vertical forces, including the weight of the backfill, without excessive settlement, tilting, or outward movement.

Soil mechanics may enter into the design of retaining walls in two of these operations: the estimate of the pressure exerted against the wall by the backfill; and the estimate of the adequacy of the foundation soil to support the structure. These two subjects are discussed independently.

Estimate of Pressure Exerted by Backfill

INTRODUCTION. Theoretical methods for computing the earth pressure against retaining walls are presented in Articles 28 and 30. These methods are based on three assumptions:

1. The wall can yield by tilting or sliding through a distance sufficient to develop the full shearing resistance of the backfill.
2. The pressure in the porewater of the backfill is negligible.
3. The soil constants appearing in the earth-pressure equations have definite values that can be determined reliably.

The use of earth-pressure theory to estimate the pressure of the backfill against a retaining wall is justified only if these three assumptions are satisfied. Every retaining wall not rigidly supported at its crest, unless it consists of a massive gravity section supported on rock, is able to yield far enough to comply with the first assumption. However, in order to satisfy the second, the drainage system in the backfill must be designed and constructed with the same care as the wall itself and, in order to satisfy the third, the backfill material must be selected and investigated before the wall is designed. Furthermore, it must be carefully placed, because the shearing resistance of a backfill which is merely dumped into position cannot be determined reliably by any practicable means.

If the latter two requirements are not satisfied, the wall will be acted on by various agents and forces beyond the scope of any earth-pressure theory. If a backfill is loosely deposited or inadequately drained, its properties change from season to season, and during the course of each year it passes through states of partial or total saturation alternating with states of drainage or even partial desiccation. All these processes cause seasonal changes in the earth pressure that receive no consideration in the classical earth-pressure theories. For example, pressure cell measurements on the back of a reinforced-concrete retaining wall 34 ft high indicated that within one year the pressure varied from the average value by $\pm 30\%$ (McNary 1925).

The maximum value of the earth pressure of backfills subject to seasonal changes is greater than the Coulomb or Rankine value. Yet, on routine jobs such as the construction of retaining walls along railroads or highways, it would be both uneconomical and impracticable to eliminate the seasonal pressure variations by design and construction in strict accordance with theoretical requirements. For the sake of economy and expediency, such walls are designed on the basis of simple semiempirical rules for estimating the backfill pressure. In their original form these rules were based chiefly on analyses of the stability of actual retaining walls, only a few of which had failed. Since the underlying causes of failure were not taken into consideration in formulating the rules, the design of walls by this procedure led on rare occasions to failure, but in the great majority of cases the walls were safer than necessary. The advent of soil mechanics has permitted improvements in the rules without loss of simplicity.

On the other hand, if a retaining wall constitutes the major part of a large job, or if the height of a wall exceeds about 20 ft, it is likely to be more economical to determine the properties of the backfill, to carry out such construction procedures as are necessary to satisfy the theoretical requirements for applying an earth-pressure theory, and to design the wall to withstand only the theoretical value of the earth pressure.

SEMIEMPIRICAL METHODS FOR ESTIMATING PRESSURE OF BACKFILLS. For many years most retaining walls have been designed by empirical or semiempirical methods. Perhaps the oldest of these methods is the use of charts or tables giving suitable values of the ratio of base width to height for various types of walls and backfills. The principal defect of this approach is that the foundation cannot be investigated adequately because the forces that act on it are unknown. A second procedure in common use is the *equivalent-fluid method,* in which the wall is designed to withstand the pressure of a liquid that is assumed to exert the same pressure against the wall as the real backfill. In spite of its widespread use, the equivalent-fluid concept has not led to generally accepted values for the unit weight of the equivalent fluid. Many designers prefer to use the theoretical equations for calculating the pressure of cohesionless earth and to substitute in them such values for the angle of internal friction as have usually led to satisfactory design in the past. However, a great diversity of opinion exists as to the proper values to use for ϕ under different circumstances, and attempts to use the procedure for cohesive backfills cannot be justified even on a theoretical basis.

Nevertheless, each of the various empirical or semiempirical procedures represents a body of valuable experience and summarizes much useful information. Today our knowledge of the physical properties of soils permits us to eliminate the most unreasonable values assigned to the soil parameters or to the unit weight of the equivalent fluid. Furthermore, a knowledge of earth-pressure theory can be used to take proper account of the cohesion and to estimate the influence of a surcharge carried by the backfill or of a backfill that has an irregularly shaped surface. A summary of all this information, in the form of an approximate design procedure for practical use, is given in the following paragraphs.

In the use of this procedure, it should be emphasized that every approximate method for the design of retaining walls involves two features: It must be based on more or less arbitrary assumptions,

Table 46.1
Types of Backfill for Retaining Walls

1. Coarse-grained soil without admixture of fine soil particles, very permeable (clean sand or gravel).
2. Coarse-grained soil of low permeability due to admixture of particles of silt size.
3. Residual soil with stones, fine silty sand, and granular materials with conspicuous clay content.
4. Very soft or soft clay, organic silts, or silty clays.
5. Medium or stiff clay, deposited in chunks and protected in such a way that a negligible amount of water enters the spaces between the chunks during floods or heavy rains. If this condition cannot be satisfied, the clay should not be used as backfill material. With increasing stiffness of the clay, danger to the wall due to infiltration of water increases rapidly.

and it cannot apply to all cases encountered in practice. Therefore, the following suggestions for the design of small retaining walls should serve merely as a basis for extrapolating from the stipulated simple conditions to the conditions encountered in specific instances in the field.

The first step in the design of a wall on a semiempirical basis is to assign the available backfill material to one of the five categories listed in Table 46.1.

If the wall must be designed before the nature of the backfill material can be learned, the estimate of the backfill pressure should be based on the most unsuitable material that may be used by the construction forces, or else alternative designs should be prepared. Each design should be accompanied by clear and simple statements indicating to which of the five soil types in Table 46.1 it pertains. The engineer in the field should then select the design appropriate for the existing field conditions.

The practical conditions likely to be encountered in the design of retaining walls can be divided into four categories, depending on the shape of the surface of the backfill and the surcharge which it carries. These are:

1. The surface of the backfill is plane and carries no surcharge.

2. The surface of the backfill rises on a slope from the crest of the wall to a level at some elevation above the crest.

3. The surface of the backfill is horizontal and carries a uniformly distributed surcharge.

4. The surface of the backfill is horizontal and carries a uniformly distributed line load parallel to the crest of the wall.

If the surface of the backfill is plane as in the first category, the backfill pressure may be estimated by means of the chart (Fig. 46.1). The first step in using the chart is to determine the height H of a vertical section passing through the heel of the wall, extending from the bottom of the base to the surface of the backfill. The total pressure against this section is $\frac{1}{2}k_hH^2$, and the total vertical force on it is $\frac{1}{2}k_vH^2$. Values of k_h and k_v are given on the right-hand side of Fig. 46.1, in terms of the slope angle β, for each of the given types of backfill material. The backfill pressure is assumed to increase in simple proportion to the depth below point a. Hence the point of application of the resultant backfill pressure is at the lower third-point of H. If the material consists of clay chunks, type 5, the value of H may be reduced by 4 ft for calculating the total pressure, but the re-

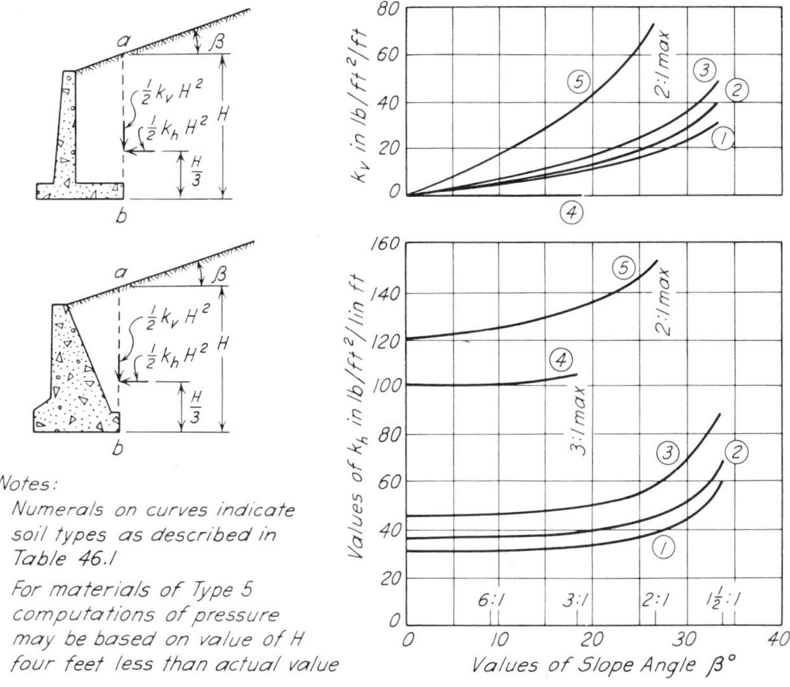

Fig. 46.1. Chart for estimating pressure of backfill against retaining walls supporting backfills with plane surface.

sultant pressure should still be considered to act at the lower third-point of the unreduced height H.

If the surface of the backfill rises at an angle β to the horizontal for a limited distance and then becomes horizontal (second category), values of k_h and k_v may be estimated from the curves in Fig. 46.2.

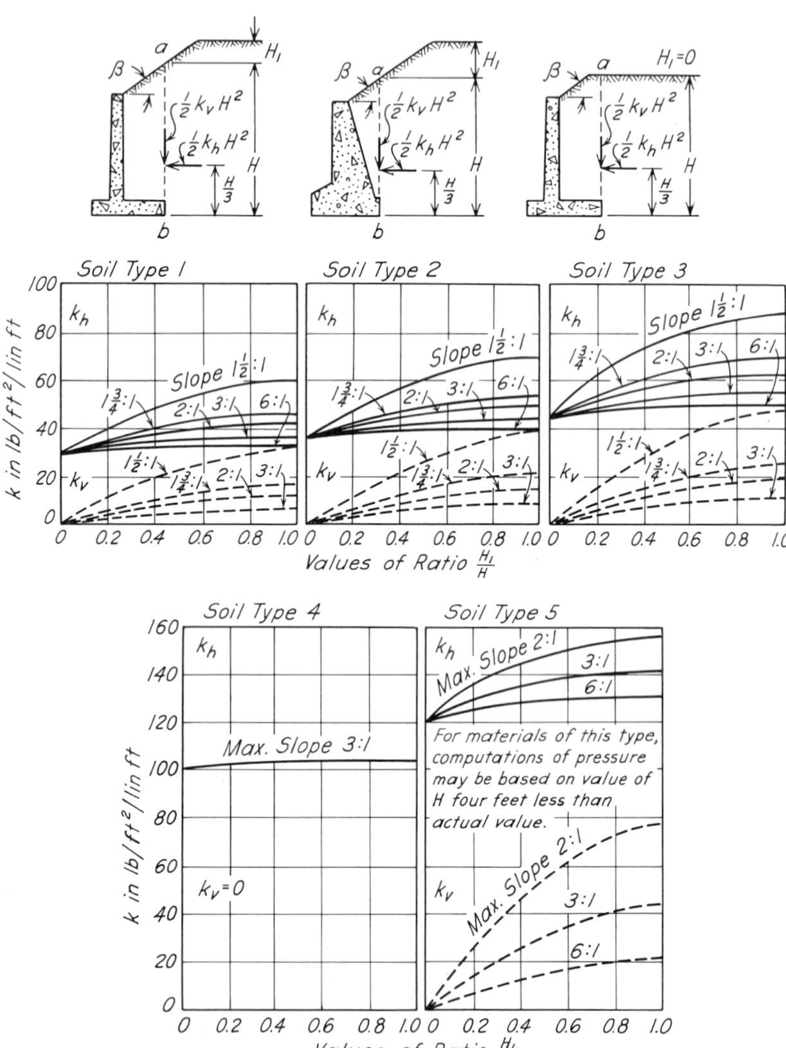

Fig. 46.2. Chart for estimating pressure of backfill against retaining walls supporting backfills with surface which slopes upward from crest of wall for limited distance and then becomes horizontal.

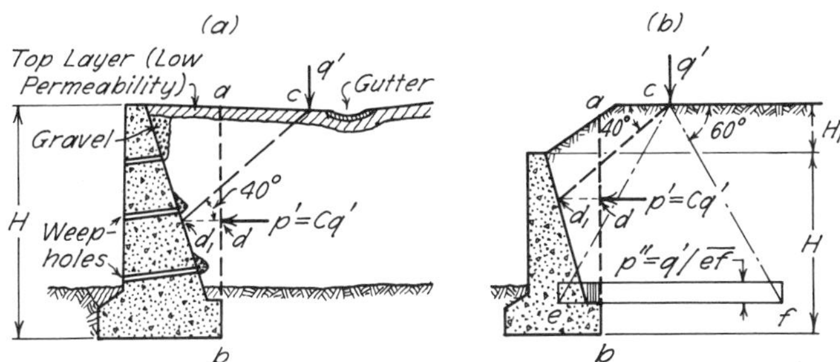

Fig. 46.3. Diagrams illustrating method for estimating magnitude and line of action of force against vertical section through heel of retaining wall, due to surcharge q' per unit of length parallel to crest of wall.

As before, the chart gives values of the pressure against a vertical section *ab* through the heel of the wall. The point of application of the resultant pressure is taken at the lower third-point of H. For materials of type 5, the value of H is reduced 4 ft in calculating the total pressure, but the resultant pressure should be considered to act at the lower third-point of the unreduced height H.

If the surface of the backfill is horizontal and carries a uniformly distributed surcharge q per unit of area (third category), the pressure per unit of area against the vertical section *ab* at any depth is increased because of the surcharge by an amount

$$p_q = Cq \tag{46.1}$$

where C is a coefficient depending on the type of soil. Values of C are given in Table 46.2.

If the surface of the backfill carries a line load, q' per unit of length, parallel to its crest (fourth category), the line is considered to exert against the vertical section *ab* a horizontal force,

$$p_q' = Cq' \tag{46.2}$$

per unit of length of the wall. The point of application d of the force p_q' (Fig. 46.3*a*) can be obtained by drawing a straight line from point c, which represents the point of application of the force q', at an angle of 40° to the horizontal until it intersects the back of the wall at point d_1. If point d_1 is located below the base of the wall, the influence of the line load on the surcharge can be disregarded. If point c is located to the left of the vertical plane *ab*, the rule remains unaltered.

The line load q' also produces an additional vertical pressure on the upper surface of the heel of the wall. It may be assumed that the pressure p'' at this level is uniformly distributed over the base ef of an equilateral triangle with apex at c. Hence, the intensity of the pressure is

$$p'' = \frac{q'}{\overline{ef}} \tag{46.3}$$

Only that part of p'' acting directly on the heel of the wall should be considered in the stability computation.

Table 46.2
Values of C in Equations 46.1 and 46.2

Type of Soil	C
1	0.27
2	0.30
3	0.39
4	1.00
5	1.00

The procedures described in the preceding paragraphs refer to walls on relatively unyielding foundations. Hence, the wall friction and adhesion tend to pull the wall down and to reduce the earth pressure. However, if a wall rests on a very compressible foundation, the settlement of the wall with respect to that of the backfill is likely to reverse the direction of these forces. This increases the earth pressure very considerably (Article 29 and Fig. 29.1). Therefore, if a wall rests on such a compressible foundation as soft clay, the values of backfill pressure computed for materials of types 1, 2, 3, and 5 should be increased by 50%.

The backfill pressures computed by means of the semiempirical procedures just outlined include the effect of seepage pressures and various time-conditioned changes in the backfill. In spite of this fact, however, provisions must be made to prevent the accumulation of water behind the wall, and to reduce the effect of frost action.

For removing water that seeps into the backfill during rainstorms, outlets known as weep holes or back drains are provided. *Weep holes* are commonly made by embedding 4-in. pipes in the wall, as shown in Fig. 46.3*a*. The vertical spacing between horizontal rows of weep holes should not exceed 5 ft. The horizontal spacing in a given row depends on the provisions made to direct the seepage water toward the weep holes. The cheapest but least effective method is to dump about one cubic foot of crushed stone or gravel at the intake end of each weep hole. If this method is used, the horizontal spacing of the weep holes should not exceed about 5 ft. The water that emerges from the weep holes seeps into the ground at the toe of the retaining wall where the soil should be kept as dry as possible. This undesirable condition can be avoided by substituting for every horizontal row of weep holes a longitudinal *back drain* that extends for the full length of the back of the wall. The outlets of the back drains are located beyond the ends of the wall. The most elaborate system of drainage in common use is the *continuous back drain,* which consists of a vertical layer of gravel covering the entire back of the wall. Outlets are provided at each end of the wall.

These drainage provisions prevent the accumulation of water behind the wall. However, regardless of which method is used, the water flows out of the backfill toward the drains. Theoretical studies based on the flow net have shown that the seepage pressure associated with this process of percolation may considerably increase the lateral pressure exerted by backfills with low permeability (Terzaghi 1936*a*). The values given in Figs. 46.1 and 46.2 take account of this temporary increase in earth pressure, because they are based on experience with retaining walls with the customary imperfect drainage provisions.

To prevent the saturation of backfills of types 2 and 3 (Table 46.1) during the wet season, the surface should be covered by a layer of soil with a permeability considerably smaller than that of the backfill. The surface should be given a slope toward a conveniently located gutter, as indicated in Fig. 46.3*a*.

If a water or sewer pipe is to be buried in the backfill, the pipe should be surrounded by a gravel drain with an outlet located in such a position that a break in the pipe cannot escape attention.

Since the semiempirical method given here takes into account the forces exerted by the earth, by water percolating toward the back drains, and by various other time-conditioned agencies, the only factor that requires independent consideration is frost action. If a backfill of types 2 and 3 in Table 46.1 is saturated, the freezing of the pore water adjoining the back of the wall draws more water out of the fill toward the zone of freezing, and ice layers may be formed parallel to the back of the wall (Article 21). If the backfill is permanently separated from the water table by a very permeable or very impermeable soil stratum, it constitutes a closed system. In such a backfill the formation of ice layers involves merely a migration of water from the central part of the backfill toward the zone of freezing, but the volume and shape of the backfill remain practically unaltered and the corresponding movement of the retaining wall is likely to be imperceptible. However, if the ground water rises into the backfill, the system is an open one, and the formation of ice layers produces an energetic outward movement of the wall, because no retaining wall is heavy enough to withstand the pressure of crystallization of ice. It has been suggested that the condition may be remedied by the installation of a continuous gravel drain at the intersection *b* of the back of the wall and the original ground surface (Fig. 46.4*a*). Such a drain would lower the water table into the position *bd*, but it would not prevent the water from being drawn by capillarity toward the zone of freezing, as indicated by the arrows in Fig. 46.4*a*. However, the backfill can be transformed into a closed system by covering the entire area of contact between the backfill and its base, up to a level a foot or two above the highest position of the water table, with a blanket of gravel or some other highly permeable material (Fig. 46.4*b*). The collector drain should be installed beyond the inner boundary of the zone of freezing, and its outlets should be protected against obstruction by ice. No serious frost action need be feared if the backfill constitutes a closed system, or if it consists of soils of the types 1, 4, or 5.

Estimate of Earth Pressure by Means of Theory. The magnitude of the earth pressure computed on the basis of theory is smaller than the backfill pressure estimated by means of the semiempirical procedure described in the preceding paragraphs. However, as stated previously, the design of a retaining wall on the basis of theory is justifiable only if the physical constants of the backfill material are reliably known, and if provisions are made to take care that the porewater pressure in the backfill will be permanently negligible. The cost of satisfying these requirements more than offsets the benefits from the use of the theoretical earth pressure in design, unless the

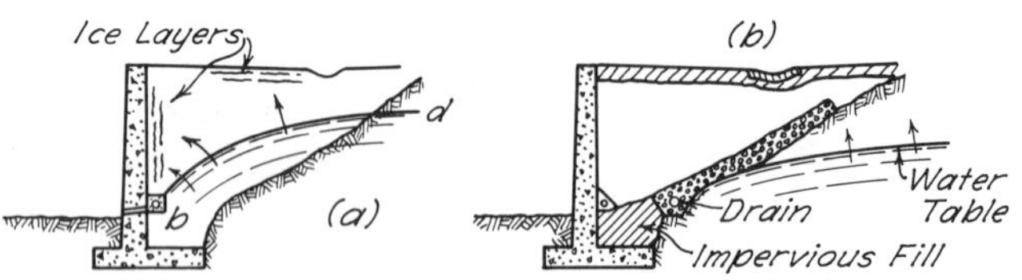

Fig. 46.4. (*a*) Frost action in backfill of retaining wall provided only with backdrain. (*b*) Method of draining backfill to prevent formation of ice layers.

retaining wall is of greater than ordinary height or length. In this event, however, it may prove more economical to investigate the properties of the backfill, to take suitable measures to assure that the properties will remain constant, to eliminate the possibility of excess porewater pressure, and to design the wall to withstand only the theoretical value of the earth pressure.

The physical soil properties that enter into the theoretical earth-pressure computations are the unit weight, the angle of internal friction, and the cohesion. No elaborate theoretical calculations are justified unless the values of these properties are determined by means of laboratory tests on representative samples of the backfill material compacted to a density equal to that after deposition and compaction in the field. The following three paragraphs summarize the procedure for getting the required information.

The unit weight of soils of types 1 to 3 in Table 46.1 should be determined by weighing samples that are first saturated and then allowed to drain by gravity for about 30 min through perforations in the bottom of their containers. The samples should be about 4 in. high. Clay is weighed at the water content at which it is placed in the fill.

The angle of internal friction of fairly permeable soils, such as types 1 to 3 in Table 46.1, can be determined by means of drained shear tests, because the void ratio of these materials in the field can adapt itself during construction to the change in stresses. The cohesion should be disregarded. The coefficient of wall friction tan δ can be assumed equal to two thirds of tan ϕ. If the fill will be subjected to traffic vibrations, or if it will sustain heavy surcharges of variable intensity, such as the loads on the floor of storage-warehouse docks, the values of tan ϕ and tan δ should be reduced by 20%. If there is a possibility that the wall may settle more than the backfill, the wall friction should be assumed to act against the wall in an upward direction.

The values of c and ϕ for clay soils, such as types 4 and 5 in Table 46.1, should be determined by means of undrained triaxial tests performed on samples at the density and water content anticipated in the field at the time the backfill is completed. The adhesion between the clay backfill and the back of the wall should be disregarded, and the value of δ taken equal to zero. The effect of traffic vibrations does not require consideration. Stiff clay should not be used as a backfill material unless conditions permit the complete and permanent exclusion of water from the fill; however, this can seldom be accomplished.

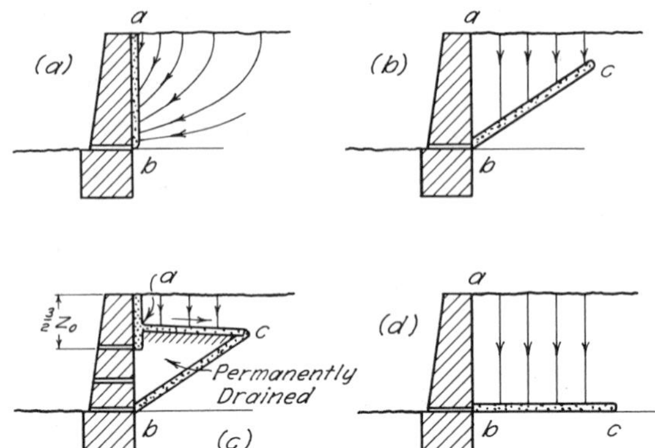

Fig. 46.5. Diagram showing provisions for drainage of backfill behind retaining walls. (*a*) Vertical drainage layers. (*b*) Inclined drainage layer for cohesionless backfill. (*c*) Horizontal drain and seal combined with inclined drainage layer for cohesive backfill. (*d*) Bottom drain to accelerate consolidation of cohesive backfill.

During rainstorms water percolates through the backfill toward the back of the wall as shown in Fig. 46.5*a*. The seepage pressure (Article 23) exerted by the percolating water increases the earth pressure exerted by backfills with medium permeability, types 2 and 3 (Table 46.1), as long as percolation continues. This should be prevented by means of inclined drainage layers, such as that shown in Fig. 46.5*b*. The drainage layers serve the double function of drains and protection against frost action. In addition, the surface of backfills of medium permeability should be covered with a well-compacted layer of less permeable soil, as shown in Fig. 46.3*a*.

A clay backfill is likely to pull away from the back of the wall to a depth of about z_0 (Eq. 28.7). In order to prevent the accumulation of water in the open fissure during rainstorms, a drainage layer should be inserted between the wall and the backfill to a depth of $1.5z_0$ below the crest. Since the uppermost layer of a clay backfill is likely to break up and become fairly permeable as the result of alternate wetting and drying, the vertical drainage layer should be connected to an inclined drainage layer by a gently sloping top filter (Fig. 46.5*c*). This filter collects the water that percolates through the top layer of clay. The physical properties of the wedge-shaped body of clay located between the top filter and the inclined drainage layer can be expected to remain fairly constant throughout the year.

The quantity of water that percolates through a well-constructed backfill is so small that there is no danger of the drains becoming obstructed by washed-out soil particles. Therefore, it is not necessary that the grain size of the materials in the drainage layers should satisfy the requirements for filter layers (Article 11).

EARTH PRESSURE AGAINST NONYIELDING RETAINING WALLS. Rigid walls in a fixed position, such as the front part of U-shaped bridge abutments or the side walls of deep basements, are acted on not by the active earth pressure but by the earth pressure at rest. The magnitude of the earth pressure at rest is greater than the active value. It depends not only on the physical properties of the backfill, but also, to a large extent, on the method of placing the fill. Hence, the intensity of the earth pressure against a fixed wall can be estimated only on the basis of experience. As yet, very little empirical information exists on this subject. The pressure exerted by a loose fill against a low fixed wall appears to be smaller than that of the same backfill in a compacted state (Terzaghi 1934*a*). The results of pressure-cell measurements on two U-shaped bridge abutments in northwestern Germany indicated that the pressure of the well-compacted medium-sand backfill was roughly equal to the Coulomb value at any depth plus a constant value of about 0.13 ton/ft^2 (Müller 1939).

Foundations for Retaining Walls

INTRODUCTION. Experience has shown that most retaining wall failures are caused by inadequacy of the foundations. Since an adequate foundation cannot be designed without at least some knowledge of the type of soil located beneath the base of the proposed wall, the subsoil must be investigated at least by primitive means. The minimum requirement for exploring the soil beneath the base of any retaining wall is to drill with a post-hole digger or some other convenient tool to a depth below the base equal to the height of the wall. If a firm stratum is encountered at a smaller depth, the boring can be discontinued after drilling about 2 ft into the stratum, provided local experience or readily available geological evidence leaves no doubt that soft strata do not exist at greater depth. On the other hand, if a soft stratum extends to a depth greater than the height of the wall, the boring should be continued until the bottom of the soft stratum is encountered, or until the stiffness of the soil increases perceptibly. The designer should also know the depth of frost penetration and the depth to which the soil is broken up by seasonal volume changes, so that he can establish the base of the foundation below both of these depths (Article 53). If no information can be obtained in advance, the dimensions for the foundation should not appear on the plans, and simple instructions should be given to the engineer on the job for selecting the dimensions after the required information is available.

Foundations for retaining walls must satisfy at least two conditions: The factor of safety against sliding must be adequate, and the soil pressure beneath the toe of the foundation should be equal to or smaller than the allowable soil pressure (Article 54). To prevent excessive tilting, it is quite properly considered good practice to require that the resultant of all the forces acting on the wall above its base should intersect the base within the middle third. In addition, if the subsoil is compressible, the further requirement must be satisfied that the differential settlement of the foundation should not be excessive.

Safety against Sliding. The sliding of a retaining wall on its base is resisted by the friction between the soil and the base and by the passive earth pressure of the soil in contact with the outer face of the foundation. It is commonly required that the factor of safety against sliding be at least 1.5.

The friction between the base and a fairly permeable soil such as a clean or silty sand is equal to the total normal pressure on the base times the coefficient of friction f between soil and base. For a coarse-grained soil containing no silt or clay the value of f may be taken as 0.55; for a coarse-grained soil containing silt $f = 0.45$.

If the wall rests on silt or clay, special precautions are required. Immediately before the footing is poured, about 4 in. of soil should be removed over the area to be covered by concrete and replaced by a 4-in. layer of well-compacted sharp-grained sand or sand and gravel. The coefficient of friction between the sand and the underlying soil can be assumed as $f = 0.35$. However, if the undrained shear strength s of the underlying soil is less than the frictional resistance beneath any part of the base, the slip will take place by shear in the soil at some distance below the base. If the normal pressure increases from zero at the heel to p at the toe, as shown in Fig. 46.6*a*, failure between a_2 and d would occur by sliding along the contact between the sand layer and the underlying soil, and between d and a_1 by shear in the soil itself. If the pressure on the base has a uniform value p per unit of area, the sliding resistance per unit of area of the base is equal to the smaller of the two values fp and s.

The second force that resists the sliding of the base is the passive

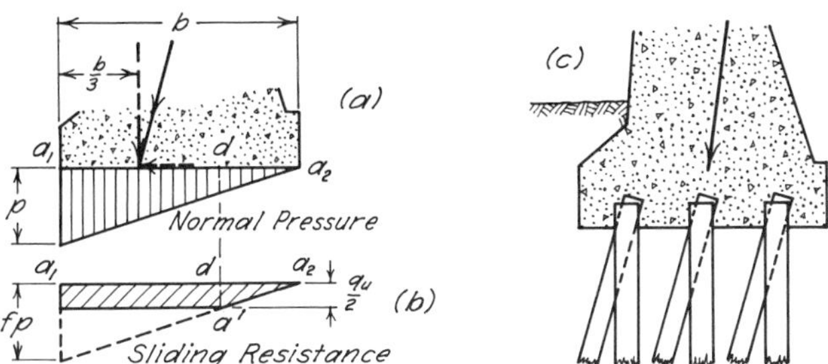

Fig. 46.6. (*a*) Approximate distribution of contact pressure on base of retaining wall if resultant force intersects base at outer third-point. (*b*) Diagram showing resistance to sliding if undrained shear strength of soil beneath base is less than the frictional resistance between base and underlying soil. (*c*) Base of retaining wall supported by vertical and batter piles.

earth pressure of the soil in front of the buried part of the wall. Within the zone of seasonal changes of moisture and temperature, the passive earth pressure is a rather unreliable resistance. The presence of root holes may make the soil so compressible that the passive resistance does not become effective until the wall has advanced through a considerable distance. If the subsoil contains silt and the water table is close to the surface, ice layers may be formed during the winter in the upper part of the soil (Article 21). During the subsequent thaw the soil may become so soft that its passive resistance is negligible. Because of these possibilities, the passive earth pressure should be disregarded unless local conditions permit reliable evaluation of its lower limiting value.

If the factor of safety with respect to sliding cannot be raised to 1.5 without the construction of an excessively heavy foundation, it is likely to be more economical to establish the wall on a pile foundation, as shown in Fig. 46.6*c*. The vertical forces are carried by vertical piles, and the horizontal ones by *batter piles,* driven at an angle to the vertical. The practice of driving some of the piles beneath the foundations of retaining walls on a batter is by no means universal, because vertical piles can be driven more cheaply. However, since the resistance against horizontal displacement offered by the upper part of vertical piles in soft soil is very small, the absence of batter piles is likely to be associated with a gradual outward movement of the wall. Some bridge abutments supported by pile foundations without batter piles have advanced in the course of time until the tension members have started to buckle owing to axial compression (Terzaghi 1929*b*). If the weight of the backfill exceeds about one-half the bearing capacity of the subsoil, the progressive movement of the retaining wall or abutment is likely to be excessive, even if the foundation is provided with enough batter piles to resist the backfill pressure (Peck, Ireland and Teng 1948). Under these circumstances, it may be necessary to substitute lightweight material for the ordinary types of backfill, or even to alter the layout of the entire project to eliminate the fill. Solid bridge abutments, for example, may be less desirable than open abutments through which the backfill extends on a slope. The roadway immediately behind the abutment is then carried on a structure rather than on a fill.

ALLOWABLE SOIL PRESSURE AND SETTLEMENT. If the resultant of all the forces acting on the wall above its base intersects the base of the foundation at the outer third-point, the contact pressure on the base increases roughly from zero at the heel to twice the average pressure at the toe. Therefore, the process of backfilling commonly produces an outward tilt of the wall. If the wall rests on a firm soil, such as a dense sand or a stiff sand-clay mixture, the tilt of the wall will be imperceptible provided the pressure beneath the toe does not exceed the allowable pressure for the given soil (Article 54). On the other hand, if the wall rests on a very compressible soil, such as soft clay, the tilt can become very large. Progressive consolidation of the clay beneath the toe may cause the tilt to increase for many years. The increased tilt causes the center of gravity of the wall to advance and the soil pressure under the toe to increase further, until finally the wall may fail by overturning. Hence, if a wall rests on a very compressible soil, the foundation must be designed in such a way that the point of application of the resultant pressure is located close to the midpoint of the base.

If a retaining wall serves as a bridge abutment, a tilt of the wall changes the clearance between the abutments. At some bridges the clearance decreases until the superstructure acts as a spacer, whereas at others the clearance increases and threatens to exceed the span of the superstructure. A movement of the second type can occur only if the subsoil of the backfill contains a fairly thick layer of compressible soil such as peat or soft clay. Under the weight of the fill the layer compresses, and the area beneath the backfill settles. Since the abutment is located near the edge of the loaded area, its base becomes inclined, and the wall tilts toward the backfill. The backward tilt due to this cause may be much greater than the forward tilt caused by the pressure of the backfill.

These considerations serve to indicate that the foundation of a retaining wall requires even more careful attention than that of an ordinary building. The general principles that govern the design of foundations are discussed in Articles 53, 54, and 56.

Field Observations

Further improvement in the design and construction of retaining walls cannot be anticipated until information is available concerning the actual performance of ordinary retaining walls backfilled in the customary manners, and concerning the effectiveness of drains intended to eliminate the porewater pressure in compacted backfills. Hence, field observations are needed to provide data of both kinds.

No empirical rules can be more reliable than the observations on which they are based. Yet the observational data which form the basis of the existing empirical rules for estimating backfill pressures are meager and inadequate. Records concerning the behavior of actual retaining walls seldom contain more than a vague description of the backfill material, and the data concerning displacements are generally limited to evidence every casual observer can see. Hence, the important field of estimating backfill pressures on a semiempirical basis still contains much room for improvement. Progress can be made only by observing retaining walls in the field for a number of years and by publishing and interpreting the results.

The records of observations made for the purpose of improving the semiempirical basis of design should contain an adequate description of the soil used as a backfill, the method of constructing the backfill, the provisions for drainage, the time of year when the backfill was placed, the average annual rainfall, and the depth of frost penetration. This information should be accompanied by a sketch that shows a cross section of the wall and a profile of the subsoil that leaves no doubt regarding the foundation conditions. The backfill samples can be obtained by means of a post-hole digger, and the description of the backfill material should contain the results of all of the pertinent identification tests listed in Table 9.1, page 44. The observations on the wall should include measurements of the tilt and of the horizontal displacement of the crest. The measurements should be made at least four times each year, at the end of every season.

The displacement of retaining walls due to frost action is practically an unexplored phenomenon. Yet, periodic measurements of tilt or displacement for a period of a few years could determine whether the observed movements are caused by frost action. If frost action is found to be the responsible agent, the structure of the ice in the frozen zone should be investigated by excavating along the back of the wall before the spring thaw.

The records of observations on large retaining walls designed on the basis of earth-pressure theory should also include the results of all soil tests made prior to construction and of the periodic measurement of the porewater pressure at several suitably located points in the backfill. Measurements of earth pressure against the back of the wall are desirable but not essential. The measured pressures should not be expected to agree with the values of active earth pressure calculated according to theory; the margin of safety provided by a well designed retaining wall would allow smaller movements than those necessary for reduction of the lateral pressures to the active value. Before such a wall could fail, however, the deformation conditions for active pressure would be satisfied. Hence, design on the basis of the active value is a rational procedure.

No satisfactory basis will be found for the design of rigid walls that cannot yield at the crest until numerous records are available of the earth pressure exerted against such walls in the field. The few available data have been obtained by means of pressure cells with a small area compared to that of the back of the wall and, as a consequence, the results are rather erratic. More reliable information could be secured by the use of devices that measure the average pressure on fairly large areas (Article 67).

Summary

In connection with the design of retaining walls, the planning of adequate drainage provisions and a careful consideration of the foundation conditions are more important than a correct evaluation of the earth pressure. The pressure exerted by the backfill can be estimated either on the basis of semiempirical rules or else by means of earth-pressure theory. The first method has the drawback that some walls are excessively safe, others are barely stable, and occasionally a wall fails. Nevertheless, for routine jobs the first method is cheaper and preferable. The second method requires that the backfill and the drainage system be constructed in strict compliance with the conditions imposed by the theory. The time and labor involved in this process are not justified unless the retaining wall constitutes a prominent part of an individual job or has a height exceeding about 20 ft.

Further progress in the design and construction of retaining walls cannot be expected without observations made on full-sized retaining walls in the field, to determine the seasonal variations in the condition of the backfill and their effect on the wall.

Selected Reading

A comprehensive treatise dealing with the classical earth-pressure theories and their application to the design of retaining walls is *Earth Pressures and Retaining Walls* by W. C. Huntington, New York, John Wiley and Sons, 1957. Convenient methods of analysis and design are presented for a wide variety of conditions.

Examples of the various semiempirical procedures mentioned in this article may be found in the following references:

Turneaure and Maurer (1913). *Principles of reinforced concrete construction,* 2nd ed., New York, pp. 370–373. Design by equivalent fluid method.

Trautwine (1937). *Civil Engineer's Reference-Book,* 21st ed., Ithaca, pp. 603–606. Design on basis of ratio of base width to height.

Other references containing useful information are the following:

Baker, Benjamin (1881). "The actual lateral pressure of earthwork," *Min. Proc. Inst. Civil Eng,* London, **65,** pp. 140–186, discussions pp. 187–241. This paper contains a graphic description of the causes and types of failure of retaining walls. The theoretical discussions and proposed design methods are out of date.

AREA (1933). "Use of portable cribbing in place of rigid retaining walls and the utility of the different kinds of cribbing," *Committee Report, Proc. Am. Rwy. Eng. Assn.,* **34,** pp. 139–148. Digest of maintenance experience.

Terzaghi, K. (1934*a*). "Large retaining-wall tests," *Eng. News Record,* **112,** pp. 136–140, 259–262, 316–318, 403–406, 503–508. Large-scale tests demonstrating the effect of the movement of a retaining wall on the intensity and distribution of earth pressure.

Terzaghi, K. (1934*b*). "Retaining-wall design for Fifteen-Mile Falls Dam," *Eng. News-Record,* **112,** pp. 632–636. Design of gravity retaining wall 170 ft high.

Kaufman, R. I. and W. C. Sherman, Jr. (1964). "Engineering measurements on Port Allen Lock," *ASCE J. Soil Mech.,* **90,** No. SM5, pp. 221–247. Measurements of lateral pressure of sand backfill against lock wall under several conditions of hydrostatic pressure.

ART. 47 DRAINAGE PRIOR TO EXCAVATION

Introduction

On many jobs, such as the installation of underground utilities, the construction of deep basements for buildings, and the preparation of foundations for dams, the soil must be excavated to a level beneath the water table, and the flow of water into the excavation must be eliminated or reduced to an inconsequential amount. To control the inflow of water, a system of drains must be established either during or, preferably, before removal of the soil. The sides of the excavation are given a slope adequate to maintain stability, or else they are made vertical and are braced with some type of lateral support (Article 48).

In an excavation with given dimensions, extending to a given depth below the water table, the quantity of water that must be disposed of and the time required for draining the surrounding soil depend on the permeability and the compressibility of the soil. On average jobs the planning of the drainage provisions does not require accurate information concerning the permeability of the subsoil. Hence, on such jobs no soil investigations need be made other than the routine tests (Table 9.1, page 44) on spoon samples obtained from the exploratory drill holes. On large jobs pumping tests are commonly made. On every job, regardless of size, the method of drainage and the location of the points at which water will be pumped require careful consideration.

Methods of Drainage

To obtain satisfactory results at least expense, the method of drainage should be adapted to the average permeability of the soil surrounding the site, to the depth of the cut with reference to the water table and, on small jobs, to the type of equipment most readily available at the site. The permeability of the soils that constitute most natural deposits, with the probable exception of those that are windlaid, varies considerably from point to point. The extreme limits within which the coefficient of permeability k has been found to vary in individual representative deposits of the most common types are given in Table 47.1.

According to their coefficients of permeability soils may be divided into five groups as indicated in Table 47.2. Soils of high permeability are rarely encountered and, when they are, they commonly alternate in the ground with less permeable layers. Practically impervious soils, such as clays, are very common.

Until the end of the last century the drainage of open excavations was generally accomplished by conducting the water that seeped into the excavation to shallow pits or timbered shafts called *sumps* and by pumping it out of these pits. On small jobs this method of sumping is still practiced. The principle of the method is illustrated by the left-hand side of Fig. 47.1, which represents a vertical section through a wide excavation with sloping sides. Most of the water emerges from the toes of the slopes. It is diverted through drainage ditches into

Table 47.1
Coefficient of Permeability of Common Natural Soil Formations

Formation	Value of k (cm/sec)
River Deposits	
Rhone at Genissiat	Up to 0.40
Small streams, eastern Alps	0.02 to 0.16
Missouri	0.02 to 0.20
Mississippi	0.02 to 0.12
Glacial Deposits	
Outwash plains	0.05 to 2.00
Esker, Westfield, Mass.	0.01 to 0.13
Delta, Chicopee, Mass.	0.0001 to 0.015
Till	Less than 0.0001
Wind Deposits	
Dune sand	0.1 to 0.3
Loess	0.001 ±
Loess loam	0.0001 ±
Lacustrine and Marine Offshore Deposits	
Very fine uniform sand, $U = 5$ to 2	0.0001 to 0.0064
Bull's liver, Sixth Ave., N. Y., $U = 5$ to 2	0.0001 to 0.0050
Bull's liver, Brooklyn, $U = 5$	0.00001 to 0.0001
Clay	Less than 0.0000001

one or several sumps S. At each sump a pump is installed that lifts the water into a discharge pipe.

The method of pumping from sumps has several disadvantages. First of all, it invites softening and sloughing of the lower part of the slopes, because in this region the seepage velocity and, as a consequence, the seepage pressures are greatest (Articles **23** and **24**). Second,

Table 47.2
Classification of Soils According to Their Coefficients of Permeability

Degree of Permeability	Value of k (cm/sec)
High	Over 10^{-1}
Medium	10^{-1} to 10^{-3}
Low	10^{-3} to 10^{-5}
Very low	10^{-5} to 10^{-7}
Practically impermeable	Less than 10^{-7}

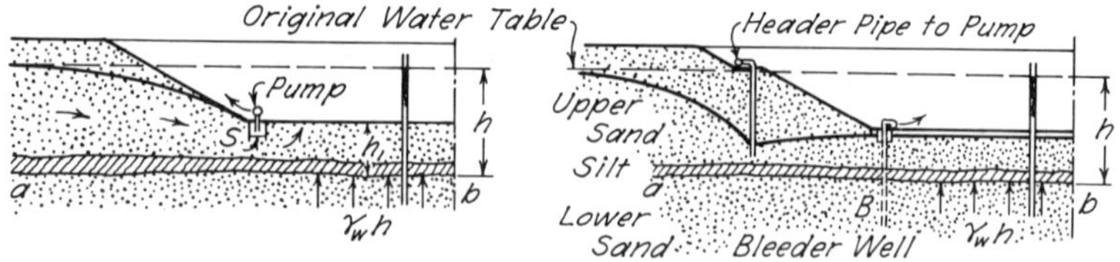

Fig. 47.1. Position of water table while pumping from sumps (left) and from well points (right). Soil conditions lead to failure by heave in spite of pumping unless bleeder wells *B* are installed.

since every natural soil stratum is more or less nonuniform, the water emerges in the form of springs. If the soil contains layers or pockets of fine sand or coarse silt, the springs are likely to discharge a mixture of soil and water instead of clear water. Springs of this type located on the bottom of the excavation are known as *boils*. Starting at the boils, underground erosion may work backward and form tunnels. The collapse of the roofs of these tunnels leads to subsidence of the ground surface surrounding the excavation, to slumping of the slopes, or to failure of the lateral supports (Article 63).

The probability of the formation of boils can be reduced by surrounding the excavation with a row of sheet piles driven to some distance below grade. The sheet piles intercept the seepage through all the strata located above the lower edge of the sheet-pile wall and reduce the hydraulic gradient at which the water rises toward the bottom of the excavation. Yet, if the soil conditions are unfavorable, even sheet piles may not prevent the formation of boils with all their undesirable consequences. On small jobs, such as the excavation of shallow timbered cuts in fine-grained water-bearing soils, attempts are often made to prevent the formation of boils by dumping gravel into the cut wherever the soil has a tendency to rise with the water, but this procedure is slow and hazardous. On a large job, such as the excavation for the foundation of a dam, it may be entirely impracticable.

Accidents and serious delays may also be caused by the hydrostatic pressure acting on the base of a continuous relatively impervious layer, such as *ab* in Fig. 47.1, located beneath the bottom of the excavation. The seepage toward the excavation lowers the piezometric level of only the body of water located above *ab*, whereas that below *ab* remains unchanged. If a piezometric tube is installed at a point located below *ab*, the water rises in this tube to the level of the original water table. If

h = vertical distance between ab and the original water table
h_1 = vertical distance between ab and the bottom of the excavation
γ_w = unit weight of water
γ = unit weight of soil, solid and water combined

the pressure on ab due to the weight of the overlying soil is γh_1, and the hydrostatic upward pressure is $\gamma_w h$. If $\gamma_w h$ is greater than γh_1, and if ab is fairly horizontal, the bottom of the excavation rises bodily. This phenomenon is known as a *heave*. On the other hand, if ab is very uneven, the ground heaves only in those places where h_1 is least. Such a local heave is sometimes referred to as a *blow*.

Historical Review of Drainage Techniques

The first attempts to replace the method of pumping from shallow pits by less hazardous procedures were made between 1870 and 1890 in England and Germany. The shallow sumps were replaced by filter wells with a diameter of 3 or 4 ft. Toward the end of the century it was realized that the efficiency of the new procedure could be improved by reducing the spacing between wells. This fact led to methods of drainage by pumping from rows of wells. The development of these methods took place along different lines in Europe and in the United States.

In Europe it became customary to install wells at a spacing of 20 to 40 ft. Each well was provided with a casing of 8 in.-diameter and a 6-in. suction tube connected through a horizontal header pipe to a centrifugal pump. The casing was perforated and screened in the pervious zone, and the screen surrounded by a filter. The procedure was referred to as the *Siemens method,* because it was developed by the Siemens Bau Union in Berlin (Kyrieleis and Sichardt 1930).

In the United States the well-point method was introduced about 1920. In contrast to the Siemens method, the well-point system consists of drawing the water from wells with a diameter of about 2 in., spaced at only about 3 to 6 ft. The well points are also connected to a header that leads to the pump.

In both methods the header pipe is commonly installed on a berm located close to the original water table. Because of the limited height to which water can be lifted in a suction tube, the water table cannot be lowered more than about 20 ft below its original position. Hence, if a project calls for draining the soil to a depth of more than 16 or 18 ft, either the water table must be lowered by stages, or else the pumping must be done by pumps that can lift water from any depth below the mouth of the well. Since about 1930 considerable use has been made of vertical turbine and submersible pumps installed inside deep-well casings with diameters ranging from 6 to 18 in. The spacing of such wells ranges from 20 to 200 ft. About 1960 the jet-eductor pump was adapted for use in wells of smaller diameter. The eductor wells usually have a diameter of 4 to 6 in. and are spaced at 5 to 25 ft.

Soon after the methods of drainage by pumping from batteries of wells came into general use, it was found that they were ineffective unless the soil had at least medium permeability. As the effective grain size D_{10} decreased below about 0.1 mm, the time required for draining the site of an excavation increased rapidly and, if D_{10} was less than 0.05 mm, pumping from wells did not accomplish its purpose. To prevent the rise of the bottom of excavations in cohesionless soils with effective size smaller than 0.05 mm, several different methods were devised.

In Germany, about 1930, there began the development of procedures to solidify the soil located below the bottoms of proposed excavations by the injection of chemicals that react in the voids of the soil to form an insoluble gel. The procedures are expensive and, if the soil contains layers with low permeability, are usually ineffective. As a consequence, their practical usefulness in connection with excavations is very limited. In the United States it was observed that fine-grained soils such as coarse silt could be consolidated by maintaining a vacuum in the riser pipes of well points. This observation led to the development between 1925 and 1930 of the *vacuum method*. Finally, in about 1934, fine-grained soils were consolidated successfully by the electro-osmotic process. This procedure is called the *electro-osmotic method*.

The following paragraphs contain brief discussions of the principal methods of drainage and of the conditions for their success. The effects of drainage on adjoining property are discussed in Article 59.

Well-Point Method

The term *well point* refers to the lower perforated end of a 2-in. or $2\frac{1}{2}$-in. pipe, commonly 40 in. long, that serves the double purpose of well casing and suction tube. The perforations are covered by a wire mesh. The well points are jetted into the ground at a spacing of 3 to 6 ft.

If a series of well points is located beneath a continuous stratum with relatively low permeability, the soil above this stratum is likely to remain undrained. To avoid such an incident and to improve the efficiency of well points in soils of low permeability the following procedure is commonly used. After a well point is jetted into the ground, the pressure in the jetting water is increased, whereupon the soil surrounding the riser pipe is scoured out, and a cylindrical hole is formed. During this process all the fine particles of the soil that formerly occupied the scoured space are washed out of the ground, but the coarser particles that remain and accumulate in the lower part of the hole form a cylindrical filter. If the wash water fails to produce a scouring effect, the hole is made by mechanical means, and the filter is constructed by shoveling sand into the hole.

Drainage of a narrow cut can usually be accomplished by pumping from a single row of well points located on one side of the cut, provided the depth of the cut is considerably less than the depth to which the water table can be lowered by the well points. Otherwise, two rows of well points are required, one on each side of the cut. The cost of pumping is usually small compared to that of transporting and installing the well points, unless the soil contains very permeable layers. If the exploratory borings indicate the presence of exceptionally permeable layers, a pumping test should be made for the purpose of estimating the capacity of the pumps that will be required. In all other instances it is justifiable to select the pumping equipment on the basis of empirical rules. Commonly, one 6-in. self-priming pump is installed for every 500 or 600 ft of the length of the row of well points. If the height to which the water has to be lifted above the level of the header pipe is not excessive, a 20-hp motor is sufficient. The drainage requires between about 2 and 6 days.

If the water table must be lowered more than 15 or 20 ft, a single-stage system of well points cannot be used. Figure 47.2, for example, shows a section through an open cut with a depth of 50 ft below the original water table. By means of the uppermost set of well points *a*, the water table can be lowered only to the level of point *b*, at a depth of less than 20 ft below *a*. In order to carry the excavation to a lower level, a second row of well points interconnected by a header pipe must be installed several feet above the level of point *b*, and so on. Such an arrangement is known as a *multiple-stage setup*. One row of well points is required for about every 15 ft of the depth, and an additional row may be needed along the toe of the slope.

Regardless of the number of stages, the average thickness of the inclined layer of soil that is drained cannot be increased to more than about 15 ft (Fig. 47.2*a*). Beneath this layer the soil is acted on by the seepage pressure of the percolating water. If the depth of the cut is many times greater than 15 ft, the drained layer is

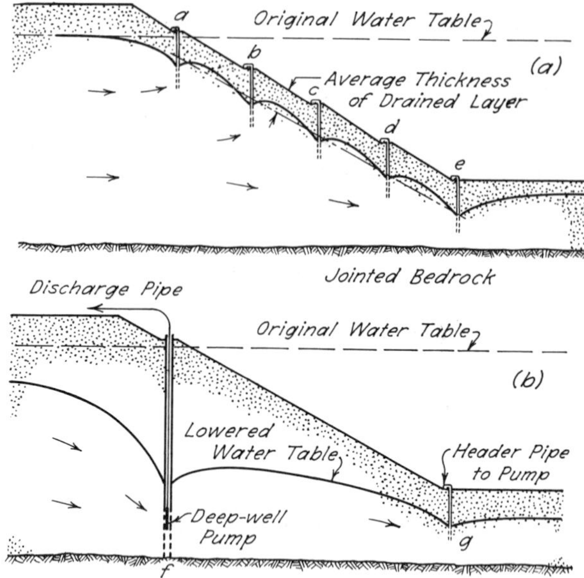

Fig. 47.2. Drainage of deep cut. (*a*) By multiple-stage setup of well points. (*b*) By means of deep-well pumps.

only skin deep compared to the thickness of the mass of soil adjoining the slope. The seepage pressures acting within this mass may compromise the stability of the slopes.

Deep-Well Drainage Method

The risk of reducing the stability of a slope because of the seepage pressure of the water flowing toward the thin skin drained by a multiple-stage well-point system can be eliminated by intercepting the flow of seepage by means of deep wells before it enters the space beneath the slope (Fig. 47.2*b*). Deep wells are also better suited than well points to the predrainage of sites in which a deep excavation is to be made with braced vertical sides. If the permeability of the soil increases with depth, and if the pervious soil extends below the intended position of the lowered water table far enough to assure submergence of the screen and pump, deep wells of large diameter may prove to be the most economical method of dewatering.

If the soil profile is fairly uniform, the spacing between the deep wells should be determined on the basis of a theoretical study of the flow of water toward the cut. Such an investigation is economically justified, because each unit of a deep-well installation is costly. The spacing usually ranges between 20 and 200 ft. The casing has a diameter from about 6 to 18 in. with a screen section 20 to 75 ft long usually surrounded by a filter of sand and gravel. The grain-size requirements for the filter material are indicated in Table 11.2. If the screen is slotted, the width of the slot should not exceed the diameter D_{70} of the material surrounding the screen. If the openings are circular, their diameter should not exceed D_{80} of the surrounding material. An electrically operated submersible or deep-well turbine pump, capable of lifting water to heights not restricted by suction lift, is installed in each well. The drilling of the holes and installation of the graded filters require special techniques (Mansur and Kaufman 1962).

A small quantity of water may flow into an excavation through the gaps between deep wells. To prevent sloughing of the toes of the slopes, it is advisable to remove the water through a row of well points g (Fig. 47.2*b*).

Eductor Well-Point Systems

If the water table must be lowered more than 15 or 20 ft but the permeability is relatively low, so that the quantity of water per well is too small for economical use of large-diameter deep-well pumps, a jet-eductor well-point system may be advantageous. The jet-eductor pump, located immediately above the well point, is operated by water furnished to the eductor under high pressure. The well point is established at the bottom of a casing at least 4 in. in diameter, in which are installed the pressure and discharge pipes for the eductor. The casing may be surrounded by a filter.

Eductor well points are usually spaced at 5 to 25 ft and discharge a maximum of about 10 to 15 gpm each. They are capable of lowering the groundwater from 50 to 100 ft. The efficiency of the eductors is considerably lower than that of centrifugal or turbine pumps.

Bleeder Wells

Since pumping from well points or filter wells lowers the water table to elevations everywhere below the slope or subgrade, the risk of slumps is eliminated. This is an important advantage over the method of pumping from open sumps. However, as previously explained, if the lower ends of the well points or filter wells are located above a relatively impervious stratum, such as ab in Fig. 47.1, a heave or a blow may occur in spite of the drainage achieved by pumping from the wells. To prevent such accidents outlets must be provided for the water located below the obstructing layer. These outlets are known as *bleeder wells*. The simplest method for constructing bleeder wells is to jet well points into the ground, to wash an annular space around the riser pipe, and to fill this space with coarse sand.

The saturated unit weight γ of most soils is roughly equal to twice the unit weight γ_w of water. Hence, as a rule, the condition for a heave or blow,

$$\gamma h_1 \lesseqgtr \gamma_w h$$

is not satisfied unless h (Fig. 47.1) is greater than $2h_1$. However, in some soil formations the water rises in piezometric tubes to greater elevations from deeper water-bearing strata than from strata nearer the ground surface. This is known as an *artesian condition*. If such a condition prevails, a heave or blow may occur even if h is considerably less than $2h_1$.

To detect the existence of artesian conditions the exploratory holes should be drilled to a depth of at least h and preferably $1.5h$ below subgrade level. Whenever a spoon sample is taken, the water should be allowed to rise in the casing until its level becomes stationary, and the elevation of the water level should be determined.

Vacuum Method

If the average effective grain size D_{10} of the soil is smaller than about 0.05 mm, the methods of gravity drainage described in the preceding paragraphs fail to produce the desired results, because the water is retained in the voids of the soil by capillary forces. However, the stabilization of very fine-grained soils can be accomplished at least gradually by maintaining a vacuum in the filters that surround the well points (Fig. 47.3). Before the vacuum is produced, both the upper surface of the fine-grained layer and the soil surrounding the filter are acted on by the pressure p_a of the atmosphere, approximately 1 ton per sq ft. After the vacuum has been produced, the pressure on the soil around the filters is almost equal to zero, whereas that on the surface of the layer remains equal to p_a. As a consequence water is gradually squeezed out of the soil into the evacuated filters, until the effective pressure in the soil adjoining the row of well points has increased by an amount equal to the atmospheric pressure. At the same time the shearing resistance of the soil increases by an amount equal to $p_a \tan \phi$, where ϕ is the angle of internal friction of the soil. This process is very similar to the stiffening of clay due to desiccation (Article 21).

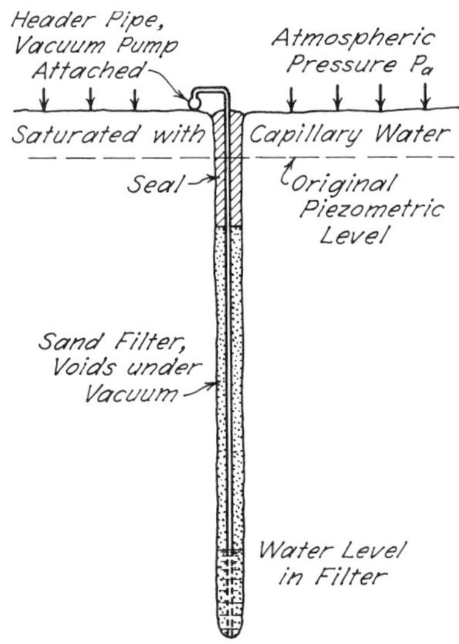

Fig. 47.3. Diagram illustrating principle of the vacuum method of drainage.

The following method is used to construct a filter that can be evacuated. After the well point is jetted into the ground, the pressure of the jetting water is increased until a hole with a diameter of 10 to 12 in. has been scoured out. While the water is still flowing, sand is shoveled into the hole until the top of the sand reaches an elevation a few feet below the surface of the fine-grained stratum. The water is then turned off, and the rest of the hole is filled with clay or silt which acts as a seal (Fig. 47.3).

The results that can be obtained by this method are illustrated by Fig. 47.4, which shows an open excavation in an organic silt with an average effective grain size less than 0.01 mm. Ninety-five per cent of the soil passed the 200-mesh screen (0.07 mm). The bottom of the excavation was about 16 ft below the original water table. Before pumping, the silt was so soft that the crane, visible in the background, had to be moved on a runway of heavy timbers. After two weeks of pumping the soil was so stiff that the sides of the excavation did not require lateral support. The distinct marks left by the excavating tools indicate the high degree of cohesion that the soil acquired during pumping.

When the vacuum method is used, the well points are commonly spaced at 3 ft. The pumping equipment is the same as that for draining soils of medium permeability. One 6-in. pump is used for every

Fig. 47.4. Open excavation in Camdem, N.J., in soft organic silt after consolidation by the vacuum method (courtesy Moretrench Corp.).

500 ft of the length of a row of well points. In addition, one or two vacuum pumps are attached to the header-pipe lines. One 20-hp motor is sufficient to operate the entire pump aggregate. Because of the low permeability of the soil, the water pump discharges for short periods only. The vacuum pumps operate continually. The success of the method depends to a large extent on the quality of the vacuum pumps and on the skill and experience of the foreman.

After a soil has been drained by the vacuum method, the soil particles are held together by an effective pressure equal to the unbalanced atmospheric pressure, but the voids are completely filled with water. Hence, if the structure of the soil is very loose, it is conceivable that a sudden shock produced by an occurrence such as pile driving or blasting may cause a collapse of the structure associated with spontaneous liquefaction (Article 17). However, no accident of this kind has yet to come to public attention.

Drainage by Electro-Osmosis

The principle of this method has been explained in Article 21. It has most often been applied in practice to the stabilization of slopes being excavated into cohesionless or slightly cohesive silts below the normal groundwater level. The time required to drain such materials by the vacuum method may be excessive, especially under emergency conditions. Yet, the materials readily become quick under the influence of the seepage pressures directed inward toward the face and upward toward the bottom of the excavation. By an arrangement of electrodes similar to that shown in Fig. 47.5 and the application of a suitable potential, seepage pressures due to electro-osmotic flow can be created in directions away from the faces of the excavation and toward the cathodes. The stabilizing influence of these pressures is in many instances spectacular and occurs as soon as the current is turned on. In addition there is a progressive decrease in the water content of the silt and a corresponding increase in strength (L. Casagrande 1949, 1962).

The anodes commonly consist of iron pipes although reinforcing bars or steel rails have also been used. Corrosion is likely to be concentrated at a few points of the anodes; consequently the anodes may become discontinuous whereupon the lower portions are no longer effective. If the anodes consist of pipes, smaller pipes or rods can be inserted to restore their continuity. The cathodes may also consist merely of iron rods along which the water flows as it escapes to the surface, but should preferably be perforated and screened for their full length to permit easier and more rapid escape of the water. The applied potential is usually on the order of 100 volts; the current required for stabilization of even a fairly small excavation is likely to be at least 150 amps. The actual power requirements depend on the resistivity of the soil and vary considerably. Potential gradients in excess of about 0.5 volt per cm may lead to excessive energy loss in the form of heat.

The process of electro-osmosis causes consolidation of compressible soils such as clays. The consolidation is accompanied by an increase

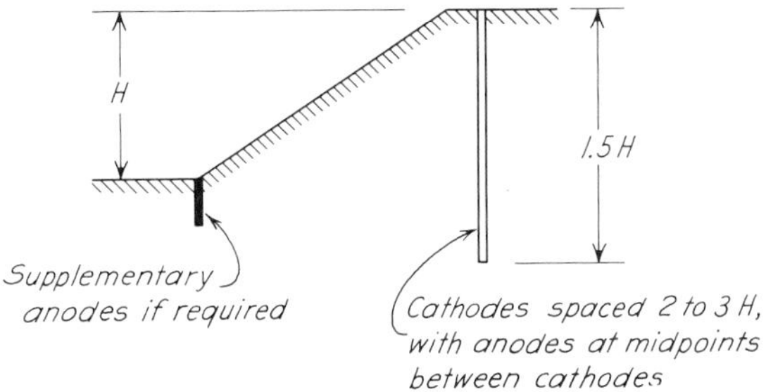

Fig. 47.5. One arrangement of electrodes suitable for stabilizing slope by electro-osmosis.

in strength and generally by a decrease in sensitivity. In addition, the clay becomes fissured. The use of electro-osmosis for altering the properties of clays in this manner has not been as frequent as that for the stabilization of slopes in silty materials.

Summary of Methods of Drainage

The quantity of water that flows toward an excavation with given dimensions and the methods that can be used to best advantage for draining the excavation depend primarily on the average permeability of the adjoining soil. On small jobs the permeability can be estimated with sufficient accuracy on the basis of the results of the routine tests performed on spoon samples from the exploratory borings. On large jobs pumping tests may be appropriate.

To determine whether or not bleeder wells are required, the exploratory borings should be drilled to a depth below subgrade equal at least to the vertical distance between the original water table and the subgrade level. As often as a spoon sample is taken, the water should be allowed to rise in the casing, and the elevation to which it rises should appear in the boring record.

Excavations in soils with high permeability (k greater than 0.1 cm/sec) or in very dense mixed-grained soils of medium permeability (k between 10^{-1} and 10^{-3} cm/sec) can as a rule be drained without undue risk by pumping from open sumps.

Under favorable conditions uniform soils of medium permeability can also be drained without mishap by pumping from sumps. However, this procedure involves the possibility of the formation of boils on the bottom of the excavation, associated with underground erosion and subsidence of the area surrounding the excavation. To avoid this risk it is preferable to drain soils of medium permeability by pumping from well points or filter wells. The drainage of the soil prior to excavation requires 2 to 6 days.

The greatest depth to which the water table can be lowered by drawing the water from one set of wells or well points is about 18 ft. If the bottom of the proposed excavation is located at a greater depth, a multiple-stage setup may be used. Two or more header pipes must be installed at a vertical spacing not exceeding about 15 ft. If limitations of space do not permit a multiple-stage installation, eductor well points may be suitable. If the depth of the excavation exceeds 50 or 60 ft, it is usually preferable to drain the soil adjoining the site by means of deep-well pumps operating within the casings of large-diameter wells.

Uniform soils of low permeability (k between 10^{-3} and 10^{-5} cm/sec) cannot be drained by pumping either from sumps or from wells. Such soils can be stabilized most successfully by the vacuum method. The quantity of water that can be drawn out is very small but, if the pumping is kept up for a period of several weeks, the soil may become so stiff that the sides of excavations up to 15 ft deep can be established at a slope angle of 60° to 70° without risk of failure.

Fine silts and uniform silty soils with a coefficient of permeability between about 10^{-5} and 10^{-7} cm/sec may be so soft that they will rise in the bottom of an excavation having even a moderate depth. They cannot be drained by gravity or by the vacuum method, but may be stabilized by electro-osmosis. Alternatively, excavations in soft soils of this category may be made by dredging or by the use of compressed air.

Soils with a coefficient of permeability less than about 10^{-7} cm/sec are with few exceptions very cohesive. Drainage is not practicable by any means except, in rare instances, by electro-osmosis. However, drainage is seldom necessary, because the shearing strength is normally great enough to maintain the stability of the bottom of an open excavation of moderate depth. The depth to which an excavation can be made in such a soil without the risk of a rise of the bottom can be increased only by reducing the side slopes or, if the sides of the excavation are vertical, by increasing the depth of penetration of the sheet piles that constitute part of the lateral support (Article 37).

Selected Reading

An excellent discussion of drainage for construction purposes, including details of the design and construction of dewatering systems, may be found in Mansur, C. I. and R. I. Kaufman (1962): "Dewatering," Chapter 3 in *Foundation engineering,* G. A. Leonards, ed., McGraw-Hill, New York, pp. 241–350.

Useful general and practical information concerning well points is contained in Griffin Wellpoint Corp., New York (1950): *The wellpoint system in principle and practice;* and in Moretrench Corp., Rockaway, N.J. (1954): *General instructions for the installation and operation of Moretrench pumps and wellpoint systems.*

References on specific aspects of drainage are included in the following:

Casagrande, L. (1949). "Electro-osmosis in soils," *Géot.*, **1,** No. 3, pp. 159–177.

Casagrande, L. (1962). "Electro-osmosis and related phenomena," *Revista Ingenieria,* Mexico, Suppl. 2, **32,** April, pp. 51–62. (Figures and Spanish text, pp. 1–50; English text, pp. 51–62), also published as *Harvard Soil Mech. Series No. 66.*

ART. 48 LATERAL SUPPORTS IN OPEN CUTS

Introduction

Open cuts may be intended to remain open permanently, like those for highways or railways, or they may be only temporary, to be backfilled after they have served their purpose. The sides of permanent earth cuts are commonly inclined at slopes not steeper than $1\frac{1}{2}$ to 1 (Article 49) or else they are supported by retaining walls (Article 46). On the other hand, the sides of temporary cuts are made as steep as the soil conditions permit without risk of slope failure (Fig. 47.4) or they are made vertical and are braced usually against each other. The choice depends on the relative costs and the restrictions imposed by the local conditions on the width of the cut.

This article deals with the design of the bracing in temporary cuts with vertical sides. If the bottom of a cut is to be located below the water table, the soil adjoining the cut is drained before or during excavation. Therefore, the design of the bracing can usually be made without considering the position of the water table.

The data needed as a basis for adequate design of the system of bracing depend primarily on the depth of the cut. Therefore, it is convenient to distinguish between *shallow cuts* with a depth less than about 20 ft and *deep cuts* with a greater depth. The bracing of shallow cuts such as trenches for the installation of sewers or water mains is more or less standardized. The customary systems can be used safely under very different soil conditions. Since refinements in the design of such systems would be uneconomical, only a general soil reconnaissance is needed in advance of construction, and no computations of earth pressure are required. On the other hand, in the design of the bracing of deep cuts such as those for subways, the dimensions of the cut and the character of the adjoining soil should be considered, because the savings that can be realized from such a procedure are likely to be very much greater than the cost of obtaining the data for design. In order to obtain adequate information concerning the character of the soil, tube-sample borings or penetration tests may be needed in addition to the standard exploratory borings.

In the past the design of the bracing of deep cuts was usually based on the assumption that the earth pressure increased like a hydrostatic pressure in simple proportion to the depth below the surface. However, both theory (Article 37) and experience have shown that this assumption is rarely justified. Hence, the discussion of deep cuts in the second part of this article includes the methods for designing bracing on the basis of the real pressure distribution.

Bracing of Shallow Cuts

In cohesive soils, cuts with vertical sides can theoretically be made to a depth H_c (Eq. 28.9) without bracing. The values of H_c for clays of various consistencies are approximately as follows:

	Very Soft	Soft	Medium
H_c (ft)	<8	8–16	16–32

Stiff and very stiff clays are likely to be jointed or fissured and, as a consequence, the value of H_c may be as low as 10 ft. The value of H_c for cohesive sand depends on the degree of cohesion; it commonly lies between 10 and 15 ft, but it may be considerably greater.

In reality, if a cut with entirely unsupported vertical sides is made in cohesive soil, tension cracks are likely to appear on the surface of the ground adjoining the cut a few hours or days after excavation. The presence of such cracks considerably reduces the critical height (Article 35), and sooner or later the sides cave in. To prevent such accidents the upper edges of narrow cuts are braced against each other, as shown in Fig. 48.1*a*. The horizontal cross-members are usually referred to as *struts* or *braces*. They may consist of timbers, or of extensible metal supports known as *trench braces*. They are tightened by wedges or screws and support horizontal timbers that usually consist of 3-in. planks. The braces are commonly spaced at about 8 ft, and the load they carry remains very small, unless the cut is located in stiff swelling clay.

If the depth of a narrow cut exceeds about $\frac{1}{2}H_c$, struts are usually inserted as excavation proceeds. They are wedged against short vertical timbers known as *soldier beams* that bear against horizontal

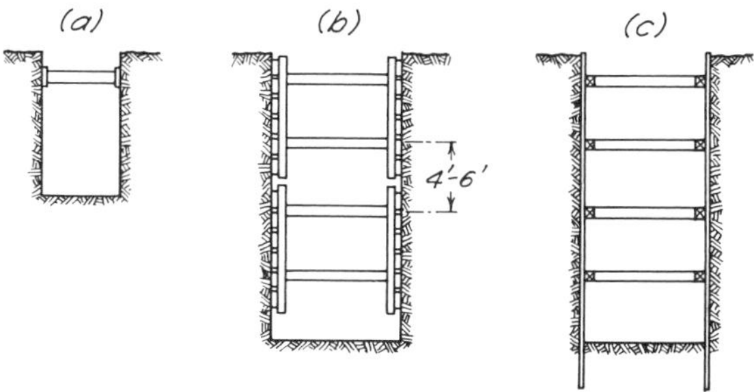

Fig. 48.1. Diagrams illustrating different methods for constructing shallow open cuts. (*a*) Single row of struts. (*b*) Lagging. (*c*) Sheeting.

boards known as *lagging* (Fig. 48.1*b*). It is usually unnecessary to fit the lagging boards tightly together; if space is left between them, they constitute *open lagging*. An alternative procedure is to wedge the struts against horizontal timbers known as *wales* that support vertical boards known as *sheeting*. The lowest part of the sides, with a height of about $\frac{1}{2}H_c$, can be left unsupported in order to furnish adequate working space, provided the soil does not have a tendency to slake or ravel. If it does, the sheets are carried down to the bottom of the excavation, but no struts are required to hold them in place.

In perfectly cohesionless sand or gravel only vertical sheeting can be used. One row of sheeting is usually driven on each side of the cut, and wales and struts are inserted as excavation proceeds. The sheets are commonly driven a few feet at a time, but their lower ends are always kept several feet below excavation level (Fig. 48.1*c*).

The dimensions of the bracing are fairly well standardized, regardless of the type of soil. Struts are spaced at about 8 ft horizontally and 4 to 6 ft vertically. Metal trench braces are available for cuts up to 5 ft in width. For narrow cuts wooden struts are usually 4 by 6 in. The dimensions increase to about 8 by 8 in. for cuts 12 ft wide. Sheeting or lagging usually consists of planks 6 to 10 in. wide. Bracing of these dimensions can be used safely in cohesionless sand to a depth of about 30 ft and in soft clay to a depth of about 5 ft in excess of $\frac{1}{2}H_c$.

Bracing of Deep Cuts

General Considerations in Design of Bracing. The most common methods for supporting the sides of deep cuts are illustrated in Fig. 48.2. When an open cut is excavated, the struts are inserted as the depth of the excavation increases. In Article 37 it is shown that this procedure is accompanied by an inward movement of the soil on each side of the cut. At the ground surface the movement is restricted to a very small amount, because the uppermost row of struts is inserted before the state of stress in the soil is appreciably altered by excavation. However, the movement that precedes the insertion of struts at lower levels increases as the depth of excavation increases. According to Article 27, this type of yielding is associated with a roughly parabolic distribution of pressure and the maximum intensity of pressure occurs near midheight of the cut, whereas the lateral pressure exerted against a retaining wall by a backfill with a horizontal surface increases like a hydrostatic pressure in simple proportion to the depth below the surface.

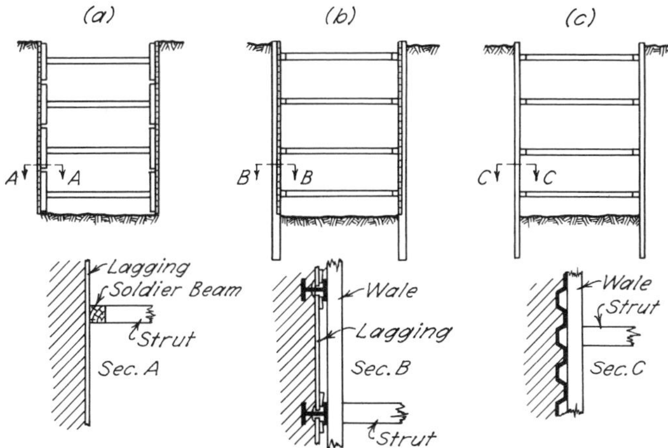

Fig. 48.2. Diagrams illustrating different methods for constructing deep open cuts. (*a*) Use of lagging and soldier beams. (*b*) Use of H-piles, lagging and wales. (*c*) Use of sheet piles and wales.

Another fundamental difference between a retaining wall and the bracing in a cut is the manner in which they fail. A retaining wall constitutes a structural unit, and it fails as a unit. Local irregularities in the magnitude of the backfill pressure are of little consequence. However, any strut in an open cut can fail as an individual. Since the failure of one strut involves an increase of the load on its neighbors, it may initiate a progressive failure of the entire system of bracing.

Finally, we should remember that the shearing resistance of the soil adjacent to a vertical face does not become fully active until the face has yielded through a certain distance (Article 27). It is impracticable to find out by laboratory tests or by any other indirect means whether or not the process of excavating and bracing a cut is actually associated with enough movement to reduce the total lateral earth pressure to the active value. Furthermore, at a given total pressure on the bracing system, the loads carried by the individual struts can be very different, because they depend on accidental factors such as the local variations in the adjoining soil, the rate and orderliness with which excavation proceeds, the time that elapses between excavation and insertion of the strut at a given point, and the extent and uniformity of prestressing. In view of these facts, no procedure for designing the bracing should be trusted until its reliability has been demonstrated by the results of measurements in full-sized cuts.

So far, comprehensive measurements of this kind in deep cuts have been made only in sands in Berlin, Munich, and New York; in soft to medium insensitive glacial clays in Chicago; and in soft to medium insensitive marine clays in Oslo. A few additional sets of observations are available from cuts in a wide variety of soils (Flaate 1966).

Most of the observations consist of measurements of the loads carried by the struts in a given vertical cross section or in several cross sections of a cut. In some instances the strut-load determinations are supplemented by measurements of deflections and settlement. Since reliable direct measurements of the earth pressure against the sheeting have rarely been made, the magnitude and distribution of the earth pressure against the sheeting must be inferred from the strut loads. Leading to a rough but reasonable approximation, the simplest procedure is to assume the load in each strut to be equal to the total earth pressure acting on the sheeting over a rectangular area extending horizontally half the distance to the next vertical row of struts on either side, and vertically half the distance to the horizontal sets of struts immediately above and below. The earth pressure is assumed to be uniformly distributed over the rectangular area. The rectangular area tributary to the uppermost strut in a cross section extends to the ground surface. For the purpose of the calculation, the bottom of the cut is assumed to be a strut. If the shear in the sheeting at the bottom of the cut has not been measured, the earth pressure per unit of area is assumed to have the same value as that tributary to the lowermost actual strut. The procedure is illustrated in Fig. 48.3.

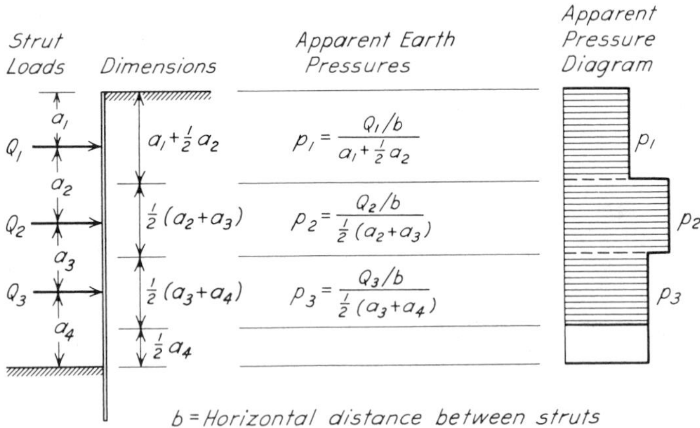

Fig. 48.3. Method of calculating apparent pressure diagram from measured strut loads Q in open cut.

The real distribution of the earth pressure against the sheeting may differ appreciably from that calculated in accordance with the foregoing procedure because of the continuity of the sheeting and because of the assumptions concerning the pressures near the bottom of the cut. Moreover, in cohesionless materials the earth pressure at the ground surface must be zero. For these reasons the pressure calculated in this manner is designated the *apparent earth pressure.* However, if the apparent earth pressure is known, the corresponding strut loads can be computed by following the inverse procedure.

Deep Cuts Excavated in Sand. Strut-load measurements were made during the construction of a subway in Berlin, for which an open cut was excavated to a depth of about 38 ft in fine, dense, fairly uniform sand. Before and during excavation, the ground-water level was lowered to a considerable depth below final grade by pumping from deep wells (Article 47). Hence, during construction the cut was located above the water table. The cut was braced as shown in Fig. 48.2*b*. The struts were arranged in vertical planes spaced uniformly along the length of the cut, and the loads in the struts were measured in six of these planes (Spilker 1937). The apparent earth pressure on four sets of struts is shown in Fig. 48.4*a*. The apparent pressure diagrams for the other sets lie within the range of those shown.

Although the sand at the site of this open cut was fairly uniform, the various diagrams representing the apparent earth pressure vary considerably from the statistical average. The variations were probably caused to some extent by local differences in the soil properties and to a greater extent by differences in the details of construction procedure at different locations. Yet the distance n_aH of the center of pressure from the bottom of the cut ranged between the rather narrow limits of $0.46H$ and $0.50H$. Similar results were obtained from measurements of seven sets of struts in an excavation for the Munich subway (Klenner 1941), for which n_a varied from 0.41 to 0.55, and for six sets of struts in an excavation for the New York subway (White and Prentis 1940) in which n_a varied from 0.46 to 0.54. Hence, the value of n_a was found, in all the cuts in sand, to be on the order of 0.5, corresponding to a roughly parabolic distribution of pressure, rather than 0.33, corresponding to a linear increase of pressure with depth.

According to Article 37, the total earth pressure for the deformation conditions associated with the excavation and bracing of an open cut in sand should correspond to that calculated on the assumption that the surface of sliding is a logarithmic spiral. For a cut of given depth H in a material of unit weight γ, the horizontal component of the total

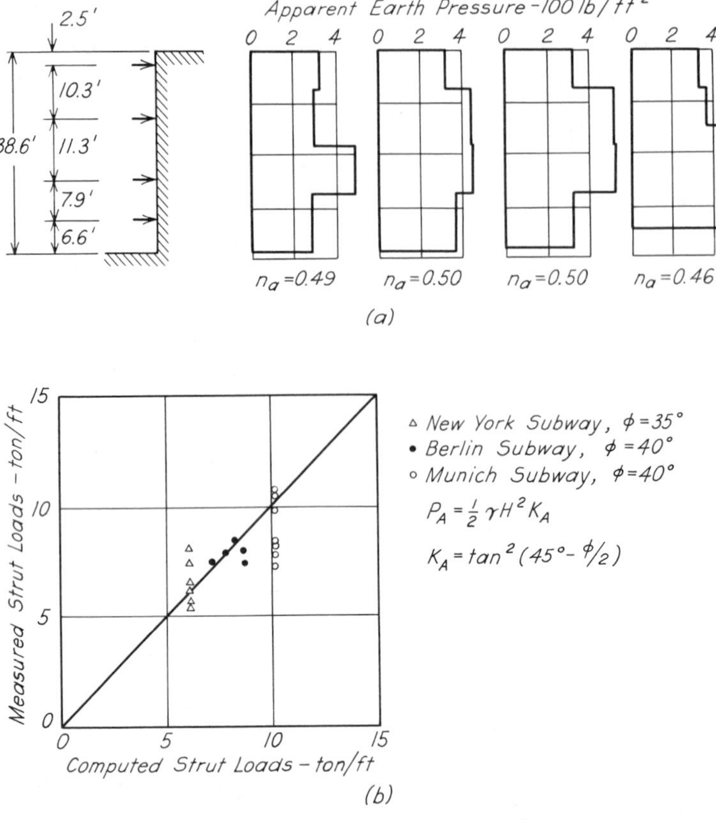

Fig. 48.4. (a) Apparent earth-pressure diagrams for four sets of struts in Berlin subway open cut. (b) Comparison of measured and calculated total loads in vertical sets of struts in various open cuts in sand.

active pressure depends on the values of n_a, ϕ, and the angle of wall friction δ between the horizontal and the direction of the earth pressure on the back of the sheeting. Depending on the material of which the sheeting consists and on the extent to which the sheeting can settle, δ may range between 0 and ϕ. Values of the coefficient of active earth pressure,

$$K_A = \frac{P_a}{\frac{1}{2}\gamma H^2} \tag{48.1}$$

calculated by the logarithmic spiral method for $\delta = \phi/2$, and for values of n_a ranging from 0.4 to 0.6, are given in Table 48.1. Also tabulated are the values of $K_A = \tan^2 (45° - \phi/2)$ corresponding to Rankine's theory, Eq. 28.1.

Table 48.1 indicates that, for a given value of ϕ and for a range of n_a from 0.4 to 0.6, the Rankine value does not differ from those obtained by the logarithmic spiral method by more than 15%. Moreover, for $n_a = 0.5$, the difference is not over 4%. On the other hand, a variation in the angle of internal friction ϕ of only 5° produces a change in K_A of nearly 50%. Inasmuch as the values of ϕ for the sands at the sites of the open cuts in Berlin, Munich, and New York were not determined by tests and can only be estimated from the descriptions of the materials, it is apparent that the strut-load measurements cannot be used as a basis for establishing the superiority of

Table 48.1
Values of Coefficient of Active Earth Pressure K_A for Open Cuts in Sand ($\delta = \phi/2$)

n_a	$\phi = 30°$	$\phi = 35°$	$\phi = 40°$
0.4	0.311	0.238	0.202
0.5	0.340	0.257	0.217
0.6	0.391	0.282	0.235
Rankine	0.332	0.270	0.220

the logarithmic spiral method over the simpler Rankine solution. Nevertheless, a general evaluation of the applicability of either procedure to the calculation of the total earth pressure against the sides of a cut can be made by comparing the sum of the loads in each vertical set of struts where measurements were made with the total earth pressure calculated on the basis of Eq. 28.1 and reasonable assumed values of ϕ. Such a comparison is shown in Fig. 48.4*b*. The measured strut loads include an allowance for the pressure transferred to the soil beneath the bottom of the cut, as shown in Fig. 48.3. The computed strut loads are based on a value of $\phi = 40°$ for the fairly dense sands at the cuts in Berlin and Munich, and 35° for the somewhat looser sands in New York. The excellent agreement indicates that use of Eq. 28.1 for calculating the total earth pressure against similar cuts in sand is justified.

On the other hand, the distribution of the apparent earth pressure at a given vertical section may resemble any of the diagrams in Fig. 48.4*a*. It changes from place to place. Since each strut should be designed for the maximum load to which it may be subjected, the design of the struts should be based on the envelope of all the apparent pressure diagrams determined from the measured strut loads. In Fig. 48.5*a*, the maximum apparent earth pressure for each of the three projects has been plotted. The pressure was computed by taking the highest individual strut load at each level and converting it into apparent earth pressure. The apparent earth pressure was then expressed in terms of the quantity $K_A\gamma H$, where K_A is the Rankine coefficient $\tan^2(45° - \phi/2)$. The simplest envelope of best fit corresponds to a uniform pressure equal to 0.65 $K_A\gamma H$ for the entire depth of the cut.

Hence, for a similar cut in dense sands, the struts should be designed for loads determined from the apparent pressure diagram (Fig. 48.5*b*). This procedure should provide for the highest strut loads that can occur. The most probable value of any individual strut load is about 25% lower than the maximum.

The apparent pressure diagram for design (Fig. 48.5*b*) has been established on the basis of a rather limited number of cuts ranging in depth from about 28 to 40 ft. Hence, it should be used with caution for cuts of substantially greater depth. Moreover, it is emphasized

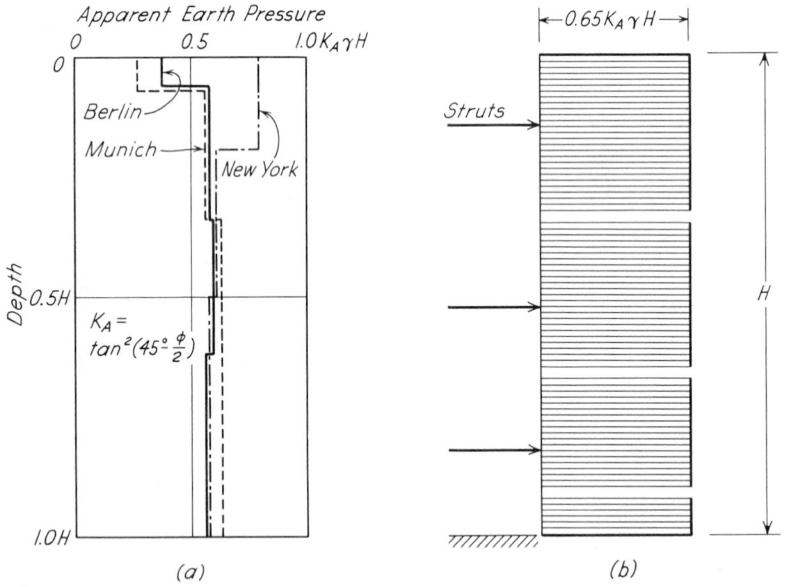

Fig. 48.5. (*a*) Envelope of apparent earth-pressure diagrams for measured strut loads in cuts in sand. (*b*) Suggested apparent earth-pressure diagram for design of struts in open cuts in sand.

that the apparent pressure diagram for design does not bear any resemblance to the real distribution of earth pressure against the sheeting of the sides of the cut; it is merely an artifice for calculating values of the strut loads that will not be exceeded in any real strut in a similar open cut. In general, the bending moments in the sheeting or soldier piles, and in wales and lagging, will be substantially smaller than those calculated from the apparent earth pressure diagram suggested for determining strut loads.

If the water table is lowered by pumping from open sumps in the cut, ample allowance should be made for seepage pressures against the lower part of the bracing. Drainage through spaces between the lagging boards is not sufficient to eliminate seepage pressures. The effect of this type of drainage is similar to that of the vertical drainage layer behind the retaining wall shown in Fig. 46.5*a*.

DEEP CUTS IN SATURATED SOFT TO MEDIUM CLAYS. In contrast to the relatively few measurements of strut loads in cuts in sand, numerous observations have been carried out in cuts in soft to medium clays. Although most of the information was obtained in Chicago and Oslo, several sets of observations have also been made in England and Japan. At all sites the undrained shearing strength of the clays has been investigated.

The apparent earth-pressure diagrams exhibit a variety of shapes, of which those in Fig. 48.6*a* are representative. The average value of n_a for 42 sets of struts, representing all the localities where measurements were made, is 0.39. The values range from 0.30 to 0.50; at one cut the exceptionally high value of 0.59 was found. The measurements leave no doubt that minor and inevitable variations in construction procedure, such as differences in the interval of time between excavation of the clay and placement of the strut, are of paramount importance in determining the load that will be carried by the strut. This fact is illustrated by Fig. 48.6*b*, in which each horizontal bar represents the average load in all eight struts at the same level and at the same excavation stage in one open cut in Chicago, as well as the maximum and minimum values of the strut loads at that level and stage. The cut contained 5 levels of struts. Excavation was carried out systematically from one level to the next, and after each stage of excavation the eight new struts were carefully installed and prestressed to 10 tons each. Yet, in spite of the unusually uniform construction procedure, individual strut loads in each level varied as much as $\pm 60\%$ from the average. Similar variations are characteristic for all the cuts in which enough strut loads were measured to provide statistically significant data.

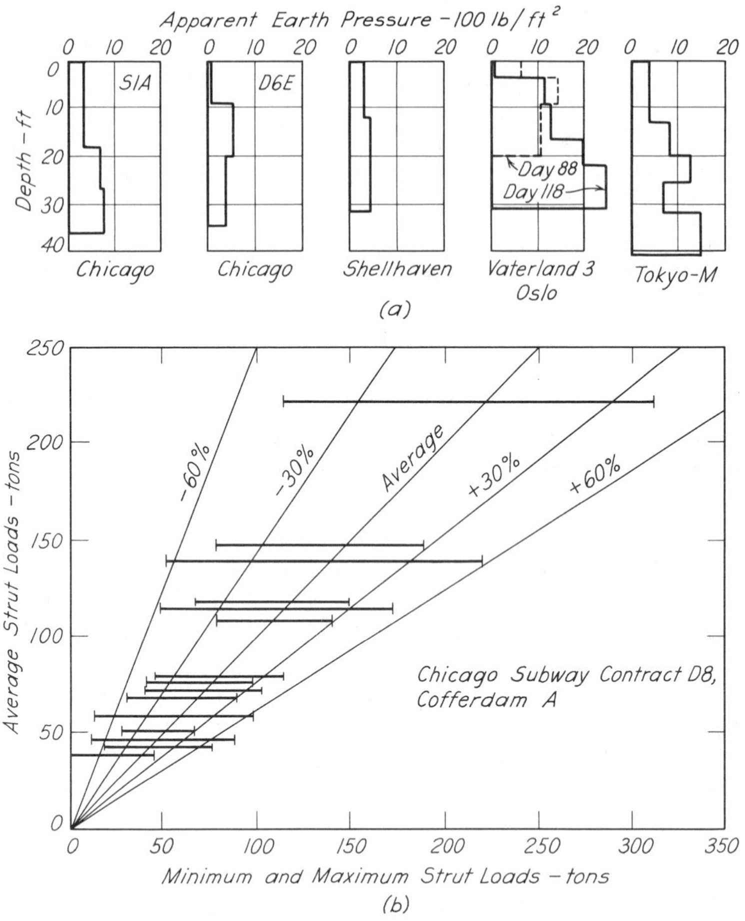

Fig. 48.6. (*a*) Representative apparent earth pressure diagrams for loads in sets of struts at various localities with soft to medium clay deposits. (*b*) Variation of loads in struts in one open cut in Chicago; each horizontal bar represents average, minimum and maximum loads in 8 struts at same level and at same excavation stage.

The total loads carried by vertical sets of struts in a given open cut, provided the sets are equally spaced in the horizontal direction, vary much less than the loads in individual struts. Nevertheless, even the variation in total loads is considerable. This fact is illustrated in Fig. 48.7, in which each horizontal bar indicates the range in total load as well as the average load in identical vertical sets of struts in one open cut in Chicago. The data for 5 different cuts, containing from 5 to 17 sets of struts, are included. For some of the cuts the variation from the average is as much as ±30%. There is no indication that the variation is smaller in the other localities where measurements have been made.

These findings are of outstanding practical importance. They demonstrate that erroneous conclusions concerning the validity of theories for the earth pressures against the bracing of open cuts are likely to be drawn if strut-load measurements have been made on only one or two sets of struts in a given cut. Moreover, the results of any theoretical calculations for estimating the loads that must be carried in the individual struts in a cut must take into account the inevitable scatter in both the total load on sets of struts and in the load on struts at the same level in the sets.

According to Article 37, the resultant earth pressure P_a against the bracing of an open cut in clay under $\phi = 0°$ conditions can be calculated on the assumption that the surface of sliding is an arc of

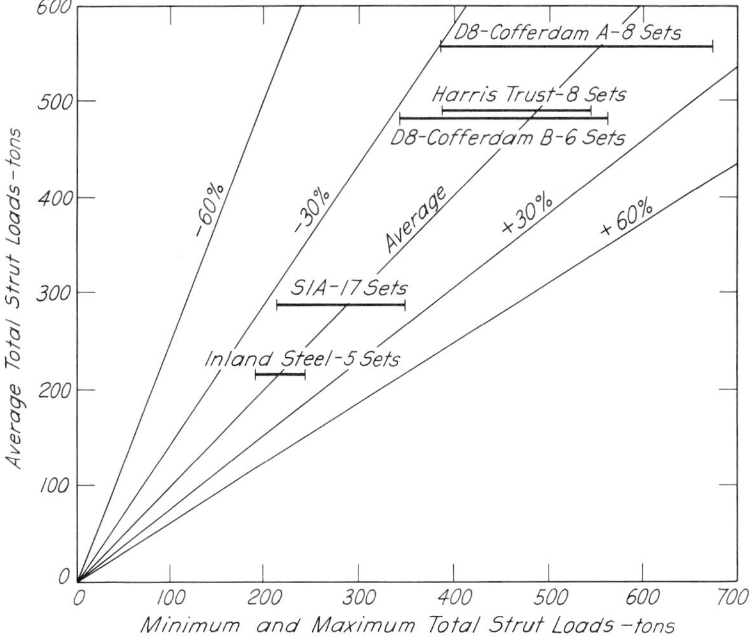

Fig. 48.7. Variation of loads in identical vertical sets of struts in various open cuts in soft to medium clay in Chicago; each horizontal bar represents average, minimum and maximum sum of loads in the sets of struts in one open cut when that cut had reached its maximum depth.

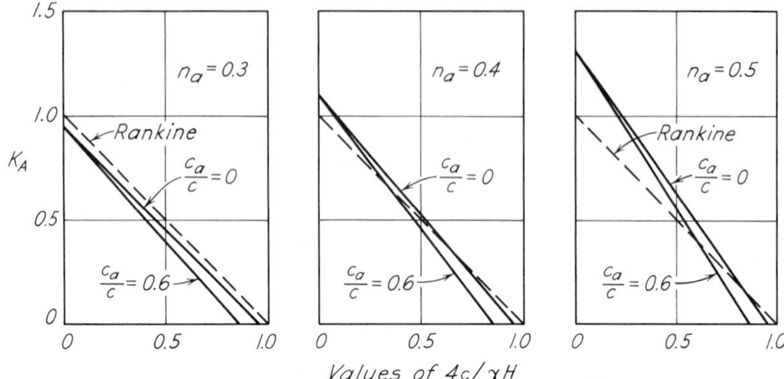

Fig. 48.8. Values of earth-pressure coefficients $K_A = P_a/\frac{1}{2}\gamma H^2$ for clay soils calculated on assumption that surface of sliding is circular (solid lines), and by means of Rankine's theory (dash lines).

a circle. Moreover, the value of P_a depends not only on the ratio n_a, which defines the center of pressure, but also on the ratio c_a/c, in which c_a is the adhesion developed between the sheeting and the clay. The theory indicates, however, that the influence c_a/c is small compared to that of n_a, and negligible in comparison with that of the shearing strength c itself. This can be seen in Fig. 48.8, in which the earth pressure, expressed in terms of $K_A = P_a/\frac{1}{2}\gamma H^2$, has been calculated, on the assumption that the surface of sliding is circular, for various values of c_a/c and n_a. The figure also demonstrates that, for values of n_a not greater than about 0.5, the value of K_A can be approximated with reasonable accuracy by the Rankine value

$$K_A = 1 - \frac{4c}{\gamma H} \tag{48.2}$$

represented in the figure by the dash lines. For the value $n_a = 0.4$, representing closely the average for all the observed cuts, the approximation is excellent. Hence, for practical purposes, it is appropriate to compare the total lateral pressure against the various cuts, as determined from measurement of the loads in vertical sets of struts, with that computed by means of Eq. 48.2.

Such a comparison is shown in Fig. 48.9. It is evident that all the observed pressures are within about 30% of the computed ones with the exception of those measured in certain cuts in Oslo. Inasmuch

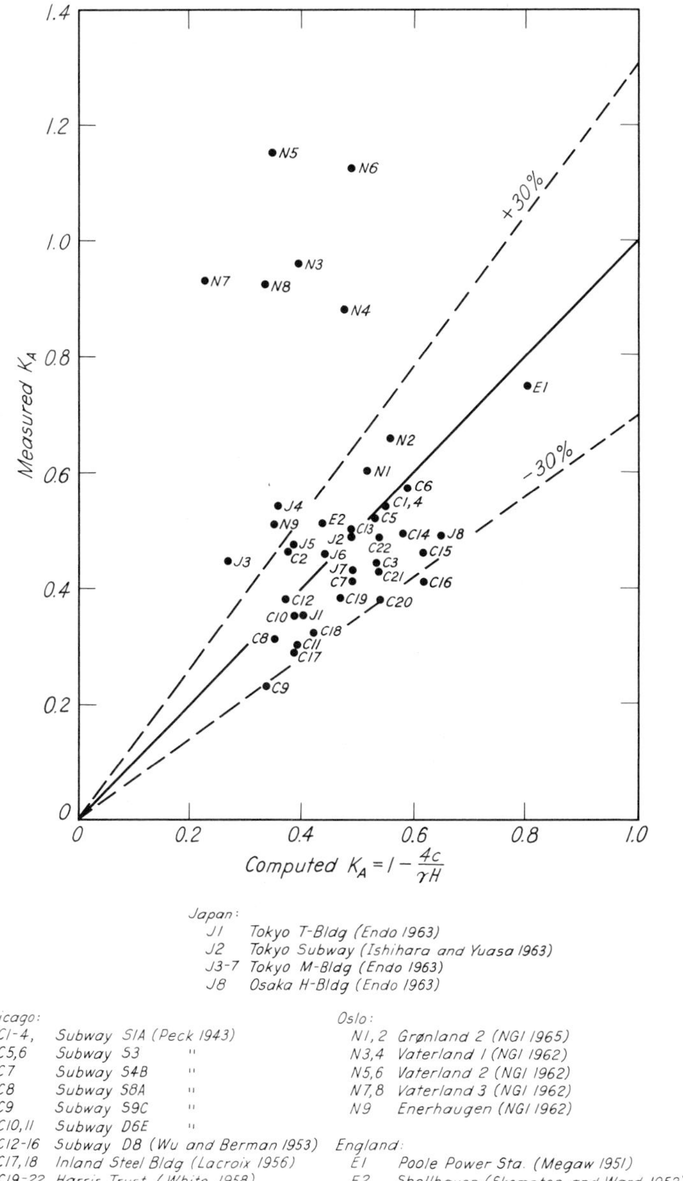

Fig. 48.9. Comparison of measured earth pressures against bracing of open cuts in soft to medium clays with pressures computed by Rankine theory (after Flaate 1966).

as the calculated pressures for the latter cuts are far on the unconservative side, the conditions leading to the discrepancy require definition. They are believed to be related to the stability conditions at the bottom of the cut.

As the depth of an open cut increases, the weight of the blocks of soil on either side of the cut acts like a surcharge on the clay at the level of the bottom of the cut and tends to displace the underlying clay laterally toward the cut and to cause failure of the bottom by heave (Article 37). If the average shearing strength of the clay beneath the bottom of the excavation is c, base failure is likely to occur when the depth of the cut reaches a critical value determined by the relation

$$\gamma H = N_c c \tag{48.3}$$

where N_c is the stability number. According to Fig. 37.4, N_c is on the order of 6 to 7. As an index to the extent that the excavation is approaching a complete base failure, the dimensionless number N may be used, where

$$N = \frac{\gamma H}{c} \tag{48.4}$$

It has been found (Article 58) that movement of the sheeting and settlements of the ground surface adjacent to an open cut in clay become significant for values of N on the order of 3 to 4. At about this value, a plastic zone begins to form in the clay near the lower corners of the excavation, and as N increases the plastic zone enlarges. Under these conditions, the simple assumption that the surface of sliding extends as the arc of a circle from the ground surface to the lower corner of the cut becomes increasingly in error, as the wedge behind the cut merges with the plastic zone at the bottom to form a larger plastic zone bounded by a surface of sliding that extends much farther from the edge of the cut and much deeper into the subsoil. Correspondingly, the earth pressure increases.

In most of the cuts on which observations have been made, the depth to which the plastic zone could extend was limited by the presence of bedrock, or by materials of increasing stiffness with depth at or near the bottom of the excavation. In such instances, agreement between measured and computed earth pressures (Fig. 48.9) was satisfactory. On the other hand, at three of the cuts in Oslo, represented in Fig. 48.9 by points $N3$ to $N8$ inclusive, an extensive body of soft clay extended below the cut, and the values of N ranged, at final depth, from 6.3 to 8.5. Accordingly, the plastic zones were able to develop without restriction below the bottom of the cut, and the surface of sliding no longer resembled that corresponding to the theory upon which Fig. 48.8 is based. The corresponding earth pressures were greatly in excess of those predicted on the basis of Eq. 48.2.

No consistent or satisfactory theory has yet been developed for calculating the earth pressure against the bracing of a cut under the conditions described in the preceding paragraph. As a rough approximation Eq. 48.2 can be modified empirically to incorporate a reduction factor m to be applied to the shear strength c. Thus

$$K_A = 1 - m \frac{4c}{\gamma H} \tag{48.5}$$

For those Oslo cuts beneath which the plastic zone could form freely and for which N exceeded about 4, the value of m was found to be 0.4. Values of m for other clays can be determined only on the basis of measurements of strut loads or lateral pressures in those deposits.

It seems likely that the value of m depends on the stress-strain characteristics of the clay. In several of the cuts in Chicago at intermediate depths the value of N exceeded 4 and the depth to stiff clay below the bottom of the excavation was great enough to permit the plastic zones to develop fully. Thus the conditions for increased earth pressure against the bracing appeared to be satisfied. Nevertheless, the measured strut loads corresponded to those indicated by Eq. 48.2, or to a value $m = 1$ in Eq. 48.5. The most conspicuous difference between the Oslo and Chicago clays is the extent of preloading. Except for an upper crust the Oslo clays appear to be truly normally loaded, whereas even the soft Chicago clays have been slightly preloaded by glacial ice. Although the preloading of the Chicago clays did not significantly alter their strength, it was sufficient to increase the initial tangent modulus (Article 15) to a value appreciably greater than that of a truly normally loaded clay. Hence, the deformations accompanying the reduction in pressure due to excavation, and the corresponding spread of the plastic zones, may have been smaller than those for the Oslo clays. Inasmuch as truly normally loaded clays are rather rare, the value of m is likely usually to be approximately 1.0.

It appears, therefore, that Eq. 48.2 provides a reliable estimate of the total pressure against the bracing of open cuts in soft to medium clay unless the value of N exceeds about 4 and unless, in addition, the clay beside and for a considerable depth beneath the cut has an unusually low initial tangent modulus, such as that of a truly normally loaded clay.

As was found for cuts in sand, the distribution of the apparent earth pressure varies from cut to cut, and from section to section in the same cut. Since each strut should be designed for the maximum load to which it may be subjected, the design of the struts should be based on the envelope of all the apparent pressure diagrams determined from the measured strut loads. In Fig. 48.10 are plotted representative maximum apparent earth-pressure diagrams for cuts in several different localities. The apparent earth pressure in each instance is expressed in terms of the quantity $K_A \gamma H$, where K_A is defined by Eq. 48.5. In diagrams (*a*) through (*h*) the reduction factor m has been taken as 1.0, and the maximum values rarely exceed $1.0 K_A \gamma H$. However, as discussed in connection with Fig. 48.9, much larger maximum apparent pressures, represented by the diagrams in (*i*), are indicated for those cuts in Oslo having values of N exceeding 4 and being underlain by a considerable depth of clay with a very low initial tangent modulus. If m for these cuts is assigned the value 0.4, the diagrams in (*j*) are obtained; they are similar in all respect to those for the other cuts.

The apparent pressure diagram shown in Fig. 48.11*a* is considered to be a reasonably conservative basis for estimating the loads for design of struts in cuts in soft to medium saturated clay. In a few instances the maximum apparent pressure diagrams (Fig. 48.10) exhibit somewhat greater pressures over limited vertical distances, but the influence of the excess on the strut loads is small and well within the factor of safety for which the struts should be designed. The method of calculation of strut loads is the same as that previously described for sands. The reduction factor m should be taken as 1.0 unless the stability factor N exceeds 4 and the cut is underlain by a deep deposit of clay with a very low initial tangent modulus.

Some of the cuts in Oslo and Chicago were subjected to extended periods of below-freezing weather. The apparent earth pressures corresponding to these conditions have not been included in Fig. 48.10. In some instances they reached magnitudes several times those that had developed at the same stages of excavation before freezing.

Cuts in Other Materials. Measurements have been made in two cuts in stiff fissured clays. One, a test trench in Oslo, was only 14 ft deep (DiBiagio and Bjerrum 1957). The other, at Park Village East in London (Golder 1948) had a depth of as much as 52 ft. For both cuts the quantity $1 - 4c/\gamma H$ is negative if c is determined

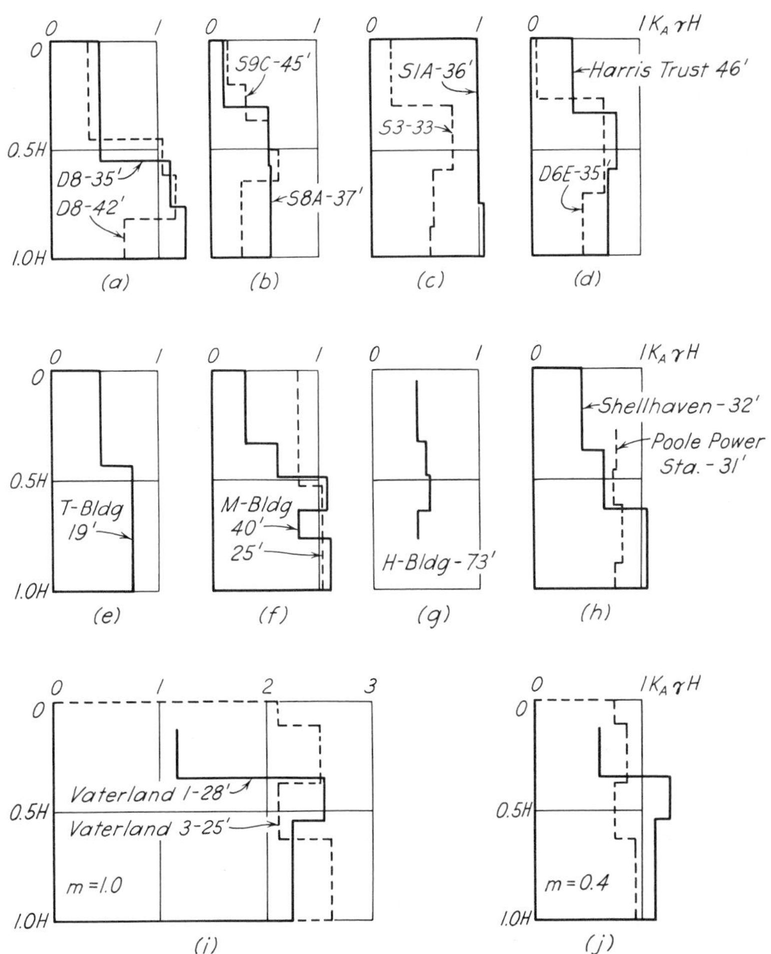

Fig. 48.10. Maximum apparent pressure diagrams for representative open cuts in soft to medium clays. (*a* to *d*) Cuts in Chicago. (*e* and *f*) Cuts in Tokyo. (*g*) Cut in Osaka. (*h*) Cuts in England. (*i*) Cuts in Oslo above deep deposit of normally loaded clay on assumption reduction factor $m = 1.0$. (*j*) Data from same cuts in Oslo if $m = 0.4$.

by means of undrained tests on intact specimens. Nevertheless appreciable pressures developed. On the basis of the meager information available, the very tentative maximum apparent earth-pressure diagram for design of struts shown in Fig. 48.11*b* is suggested. The maximum design pressure is $0.2\gamma H$ to $0.4\gamma H$. The lower values may be used if the movements of the sheeting can be kept to a minimum

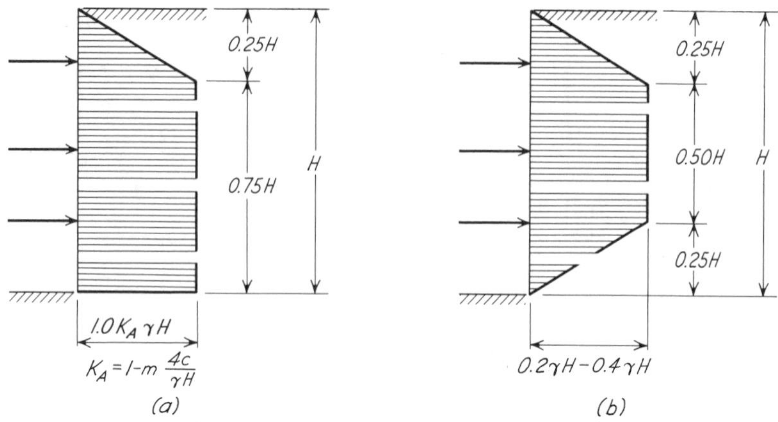

Fig. 48.11. Apparent pressure diagrams for design of struts in cuts excavated in clay soils. (a) Diagram for soft to medium clays; value of m to be taken as 1.0 except for truly normally loaded clays when $N = \gamma H/c$ exceeds about 4, under which conditions $m < 1.0$. (b) Tentative diagram for stiff-fissured clays; lower pressure to be used only when movement can be kept to a minimum and construction time is short.

and the construction time will be short. Otherwise, the higher values are applicable.

Measurements have been made in two narrow trenches for the cutoff of a dam through residual soil and weathered rock to a depth of more than 80 ft (Humphreys 1962). Because the width of the trenches, and consequently the weight of the excavated soil, were small, the original state of stress in the ground may not have been greatly altered. The pressure increased almost linearly with depth, in accordance with the relation $p_a = 0.35\gamma H$, where p_a is the intensity of the apparent earth pressure.

No data are as yet available for cuts in stiff intact clays, or for soils that are likely to exhibit values of both c and ϕ. Included in the latter category are sandy clays, clayey sands, cohesive silts, and a variety of other soils of common occurrence. Design rules for such materials cannot be worked out until adequate observations have been made.

Summary of Procedures and Problems

The excavation and bracing of cuts with a depth of less than about 20 ft merely require conscientious adherence to existing empirical rules. The earth pressure against the bracing of such cuts is a factor of secondary importance, because it is more economical to use one of the standard systems of bracing at the price of some excess material than to adapt the bracing to local soil conditions.

On the other hand, the bracing of deep and wide cuts accounts for a considerable part of the total cost. Furthermore, substantial savings can often be realized by various departures from the standard methods of bracing, such as providing for a large unobstructed working space between the bottom of the cut and the lowest row of struts. In order to comply with the requirements of both safety and economy, it is necessary to make a thorough soil survey and to prepare the plans on the basis of the results of earth-pressure computations.

Experience has shown that the magnitude of the total earth pressure against the bracing of cuts in sands and in soft to medium saturated clays can be computed reliably on the basis of earth-pressure theories, providing the cut is not underlain by a deep deposit of truly normally loaded clay with a low initial tangent modulus in which large strains are induced by the excavating operations. On the other hand, the distribution of load among struts arranged in vertical sets cannot be predicted by earth-pressure theory. The loads for which the struts should be designed may be computed by the methods described under the preceding subheadings. The application of theory or of similar procedures to the design of bracing systems for cuts in other types of soil should be practiced with caution until the reliability of the results is demonstrated by field measurements.

The need for additional field measurements in soils of other types, together with adequate descriptions of the soils, is apparent.

Selected Reading

Among the more general papers dealing with the earth pressures against the bracing of open cuts are:

Terzaghi, K. (1941*b*). "General wedge theory of earth pressure," *Trans. ASCE*, **106,** pp. 68–97.

Peck, R. B. (1943). "Earth-pressure measurements in open cuts, Chicago subway," *Trans. ASCE,* **108,** pp. 1008–1036.

An early paper by J. C. Meem (1908): "The bracing of trenches and tunnels, with practical formulas for earth pressures," *Trans. ASCE,* **60,** pp. 1–23, with discussions pp. 24–100, contains interesting records of observations in trenches in very different soils. The theoretical parts of the paper and the discussion can claim only historical interest, and the soil data are inadequate.

Full details of the Oslo measurements, without interpretations, are reported in *Technical Reports* Nos. 1–9 of the Norwegian Geotechnical Institute.

ART. 49 STABILITY OF HILLSIDES AND SLOPES IN OPEN CUTS

Causes and General Characteristics of Slope Failures

Every mass of soil located beneath a sloping ground surface or beneath the sloping sides of an open cut has a tendency to move downward and outward under the influence of gravity. If this tendency is counteracted by the shearing resistance of the soil, the slope is stable. Otherwise a slide occurs. The material involved in a slide may consist of naturally deposited soil, of man-made fill, or of a combination of the two. In this article only slides in natural soil are considered. The other types are discussed subsequently.

Slides in natural soil may be caused by such external disturbances as undercutting the foot of an existing slope or digging an excavation with unsupported sides. On the other hand, they may also occur without external provocation on slopes that have been stable for many years. Failures of this nature are caused either by a temporary increase in porewater pressure or by a progressive deterioration of the strength of the soil.

In spite of the variety of conditions that may cause a slide, almost every slide exhibits the general characteristics illustrated by Fig. 49.1. The failure is preceded by the formation of tension cracks on the upper part of the slope or beyond its crest. During the slide the upper part of the slide area, known as the *root,* subsides, whereas the lower part, known as the *tongue,* bulges. Hence, if the original surface of the slope is a plane, the profile of the ground surface along the axis of the slide becomes distorted into an S-shaped curve (Fig. 35.1). The shape of the tongue depends to a certain extent on the type of sliding material. Homogeneous clay with a low degree of sensitivity to disturbance is likely to bulge, as shown in Fig. 49.1. On the other hand, clay with a very sensitive structure or clay with sand pockets is likely to flow like a liquid.

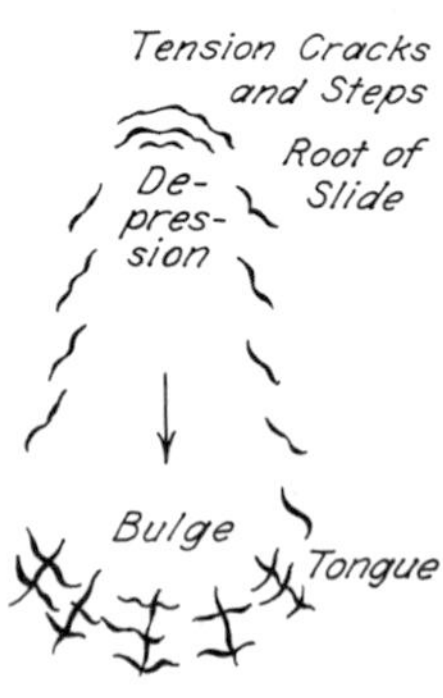

Fig. 49.1. Plan of typical slide in cohesive material.

Even on uniform slopes of great length and approximately uniform height, slides seldom occur at more than a few places, separated from each other by considerable distances. For example, the well-known slides in the Panama Canal appear in plan as isolated scars separated by long stretches of intact slope. Slides in long railroad cuts of fairly uniform cross section exhibit similar characteristics.

One important class of slides, however, provides an exception to the general rule that slides do not occur over a broad front. If the geological conditions are such that the major part of the surface of sliding is located within a horizontal layer of coarse silt or sand that separates two layers of clay, the width of the slide measured parallel to the crest of the slope is likely to be very much greater than the length of the slide. Slides of this type are commonly caused by an excess porewater pressure in the sand or silt layer. In contrast to slides of the other types, they are not preceded by readily detectible symptoms of impending danger, and the failure occurs almost suddenly.

Engineering Problems Involving the Stability of Slopes

Most of the problems involving the stability of slopes are associated with the design and construction of unbraced cuts for highways, railways, and canals. The necessity for excavating deep cuts did not arise until early in the 19th century, when the first railways were built. Since that time, however, countless cuts with increasingly greater depth and length have been excavated.

Experience has shown that slopes of $1\frac{1}{2}$ (horizontal) to 1 (vertical) are commonly stable. As a matter of fact, the sides of most railway and highway cuts less than 20 ft deep rise at that slope, as do the sides of many deeper perfectly stable cuts. Therefore, a slope of $1\frac{1}{2}$ to 1 can be considered the standard for highway and railway construction. The standard slopes for flooded cuts such as those for canals range between 2:1 and 3:1. Steeper than standard slopes should be established only on rock, on some residual soils, on dense sandy soil interspersed with boulders, and on true loess.

Slopes on rock are beyond the scope of this book. On dense residual soils and mixtures of sand and gravel with boulders, slopes of 1:1 have been permanently stable. Slopes in loess are discussed subsequently.

Preliminary estimates of the quantity of excavation required for establishing a new line of transportation are commonly based on the assumption that all the cuts in earth will be provided with standard slopes. However, experience has shown that the standard slopes are stable only if the cuts are made in favorable ground. The term *favorable ground* indicates cohesionless or cohesive sandy or gravelly soil in a moist or dry state. In soft clay or in stiff fissured clay the excavation of even a very shallow cut with standard slopes may cause the soil to move toward the cut, and the movement may spread to a distance from the cut equal to many times the depth. Clay soils containing layers or pockets of water-bearing sand may react to a disturbance of their equilibrium in a similar manner. Deposits with properties of this type constitute *troublesome ground.*

Experienced engineers always locate new lines of transportation with a view of avoiding cuts in troublesome ground as far as conditions permit. If a project requires long cuts in potentially troublesome soil, estimates are likely to show that the project is uneconomical unless the margin of safety is reduced to considerably less than the margin of error in stability computations. As a consequence, in cuts through troublesome soil local slides are commonly and justly considered inevitable. At the same time sound engineering requires that the slides should not involve loss of life or serious damage to property. This requirement can be satisfied only by means of extensive and conscientiously executed field observations during and after construction. Such observations and no other means make it possible to detect the symptoms of impending slides and to take appropriate measures for avoiding fatal consequences.

The methods for dealing with unstable slopes depend primarily on the nature of the soils involved. Hence, for practical purposes it is most suitable to classify slides in accordance with the types of soil in which they occur. The most common types of troublesome soils and soil formations are layers of weathered schists or shales *in situ,* talus deposits, very loose water-bearing sand, homogeneous soft clay, stiff fissured clay, clay with sand or silt partings, and bodies of cohesive soil containing layers or pockets of water-bearing sand or silt. In the following text the causes of slides are described, and present practice in dealing with the engineering aspects of the problems is summarized. Because of the complexity of the subject, the information serves merely as an introduction to a study of the stability of slopes in natural soil strata.

Stability of Slopes and Cuts in Sand

Sand of any kind, permanently located above the water table, can be considered stable ground in which cuts with standard slopes can be made safely. Dense and medium sands located below the water table are equally stable. Slides can occur only in loose saturated sand. They are caused by spontaneous liquefaction (Article 17). The disturbance required to release a sand slide can be produced either by a shock or by a rapid change in the position of the water table. Once

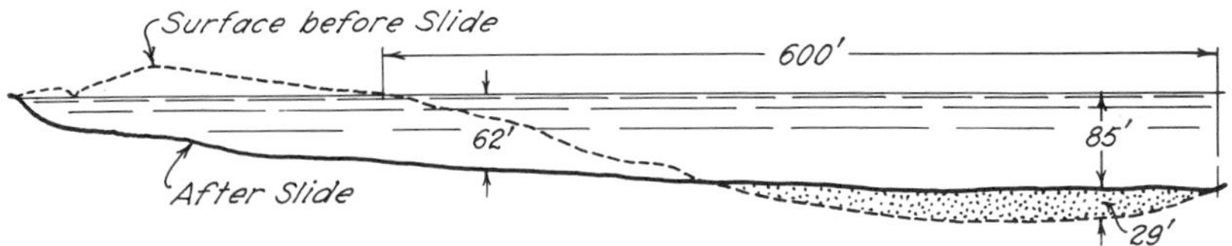

Fig. 49.2. Section through flow slide in sand at coast of Zeeland (after Müller 1898).

the movement has started, the sand flows as if it were a liquid and does not stop until the slope angle becomes smaller than 10°.

The sand slides along the coast of the island of Zeeland in Holland belong in this category (Müller 1898). The coast is located on a thick stratum of fine quartz sand that consists of rounded grains. The slope of the beach is only about 15°. Yet, once every few decades after exceptionally high spring tides, the structure of the sand breaks down beneath a short section of the coastal belt. The sand flows out and spreads with great speed in a fan-shaped sheet over the bottom of the adjacent body of water. The tongue of the slide is always very much broader than the root. Figure 49.2 shows a section through one such slide. The final slope of the ground surface was less than 5°. A slide that occurred at Borssele in 1874 involved nearly 2,000,000 yd^3.

Since flow slides in sand occur only if the sand is very loose, the tendency toward sliding can be reduced by increasing the density of the sand. This can be accomplished by several different means, such as pile driving or exploding small charges of blasting powder at many points in the interior of the mass (Article 50). On slopes of marginal stability, these means may, however, induce a slide.

Stability of Cuts in Loess

Real loess is a cohesive wind-laid soil with an effective grain size between about 0.02 and 0.006 mm and a low uniformity coefficient. It consists chiefly of angular and subangular quartz grains that are slightly cemented together. Furthermore, it always contains an intricate network of more or less vertical root holes. The cohesion of the loess is due to thin films of slightly soluble cementing material that cover the walls of the root holes. Since the root holes are predominantly vertical, the loess has a tendency to break by splitting along vertical surfaces, and its permeability in a vertical direction is very great compared to that in a horizontal direction. Its porosity may be as great as 52%.

When loess is located permanently above the water table, it is a very stable soil except for the fact that it is readily attacked by erosion. To reduce the erosion as much as possible, cuts in loess must be given a nearly vertical slope (Turnbull 1948). The foot of the vertical faces requires careful protection against temporary saturation during rainstorms. In spite of this precaution, slices inevitably break down from time to time, again leaving vertical faces that remain stable for years. To prevent blocking of traffic by the debris it is customary to make cuts in loess with a width greater than that called for by the traffic requirements.

On the other hand, permanently submerged loess is likely to be very unstable because of its high porosity and because of the leaching effects of submergence. Leaching removes the cementing substance and transforms the loess into an almost cohesionless material that is not stable unless its porosity is less than about 47% (Scheidig 1934).

The effect of submergence is illustrated by the results of a large-scale experiment performed on a plateau of loess in Soviet Turkestan. The loess has an average porosity of 50%. In dry cuts it stands with unsupported vertical faces for a height of more than 50 ft. The experiment was made to find out whether the material would remain stable if an unlined canal were excavated across the plateau and filled with water for irrigation. An open pit, 160 by 60 ft in plan, was dug 10 ft deep with sides sloping at 1.5 (horizontal) to 1 (vertical). The pit was then filled with water, and the water level kept constant by replacing the seepage losses. After a few days the slopes started to slough, and the bottom began to subside. This process continued at a decreasing rate for a period of about 6 weeks. At the end of this period the surface surrounding the excavation had cracked and subsided within a distance of about 20 ft from the original edge of the pit, and the bottom had subsided about $2\frac{1}{2}$ ft. Within the area of subsidence and sloughing the loess was so soft that it was not possible to walk on it.

Slides in Fairly Homogeneous Soft Clay

If the sides of a cut in a thick layer of soft clay rise at the standard slope of 1.5:1, a slide is likely to occur before the cut reaches a depth of 10 ft. The movement has the character of a base failure (see Article 35 and Fig. 35.2*b*) associated with a rise of the bottom of the cut. If the clay stratum is buried beneath stable sediments, or if it has a stiff crust, the heave occurs when the bottom of the cut approaches the surface of the soft material.

On the other hand, if the soft clay is underlain by bedrock or

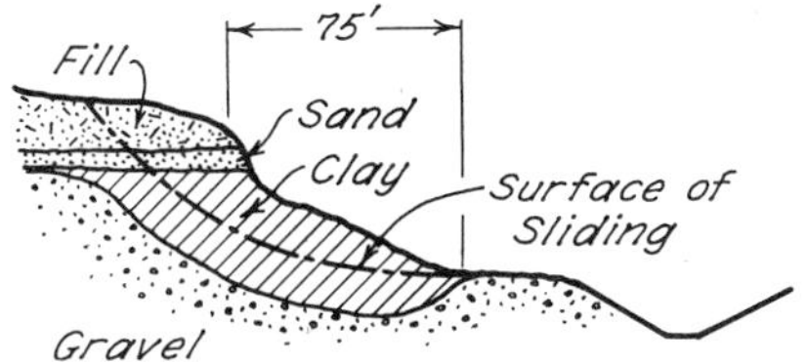

Fig. 49.3. Section through toe-circle slide in soft clay on Södertalje canal in Sweden (after Fellenius et al. 1922).

a layer of stiff clay at a short distance below the bottom of the cut, failure occurs along a toe or slope circle tangent to the surface of the stiff stratum, because the bottom cannot heave (Article 35).

If a mass of soft clay has an irregular shape, the location of the surface of sliding is likely to be determined by that shape. Figure 49.3 illustrates this statement. It represents a section through a slide that occurred during the construction of the Södertalje canal in Sweden. If the soft clay had extended to a considerable depth, base failure would have occurred approximately along a midpoint circle. However, the presence of the gravel below the soft clay excluded the possibility of a base failure, and the slide occurred along a toe circle. The movement was so rapid that several workmen were killed (Fellenius et al. 1922).

Experience has shown that sliding failures during construction in masses of homogeneous saturated soft clay take place under undrained conditions. Therefore, $\phi = 0°$ conditions prevail (Article 18) and the shearing resistance c may be taken as half the unconfined compressive strength. The factor of safety of the slopes of proposed cuts in such clay with respect to sliding can be estimated in advance of construction by the method described in Article 35. However, it should be emphasized that discontinuities in the clay, consisting of sand or silt partings, may invalidate the results of the computation. The reason is explained in the paragraphs dealing with nonhomogeneous clay.

Clay Flows

After a slope on soft clay fails, the movement commonly stops as soon as the tongue of the slide (Fig. 49.1) has advanced to a moderate distance from its original position. There is, however, one notable exception to this rule. In quick clays (Article 7), such as those that occur in the St. Lawrence River valley in Quebec and in Norway and Sweden, extensive progressive slides occur from time to time, often without apparent provocation. The movement begins as a small slide, usually at the bank of a stream, but the deformation of the sliding material transforms the clay into a thick slurry which flows out and deprives the new escarpment of its support, whereupon another slip occurs. The disturbance propagates rapidly away from the initial point as the clay is transformed into a fluid matrix of remolded material flowing toward the break in the river bank and carrying with it floating chunks of still intact clay. The principal features of such a flow on one of the northern tributaries of the St. Lawrence River are shown in the block diagram (Fig. 49.4). During the flow a roughly rectangular depression was created with a depth of 15 to 30 ft, a length of about 1700 ft parallel to the river, and a width of 3000 ft (Sharpe 1938). Within about half an hour 3,500,000 yd^3 of clay moved into the river channel through a gap 200 ft wide. The channel was blocked for over 2 miles, and the upstream water level was raised 25 ft. In the great clay flow of Vaerdalen, Norway, in 1893, over 60,000,000 yd^3 flowed out of a narrow gap in less than an hour (Holmsen 1953).

Considerable evidence indicates that the clays may have acquired their extreme sensitivity and may have gradually experienced a decrease in strength as a consequence of a reduction of salt content in the porewater since their deposition in a marine environment. The material consists largely of finely divided quartz and inactive clay minerals. Although as much as 40% of the material may be of clay size (<0.002 mm), the liquid limit and plasticity index are low, on the order of 26 and 7% respectively. The natural water content, on the other hand, is well above the liquid limit and the sensitivity S_t often exceeds 40. The salt content of the porewater is likely to be only from 1 to 3 grams per liter in contrast to a normal content of about 36 grams per liter for sea water (Article 4).

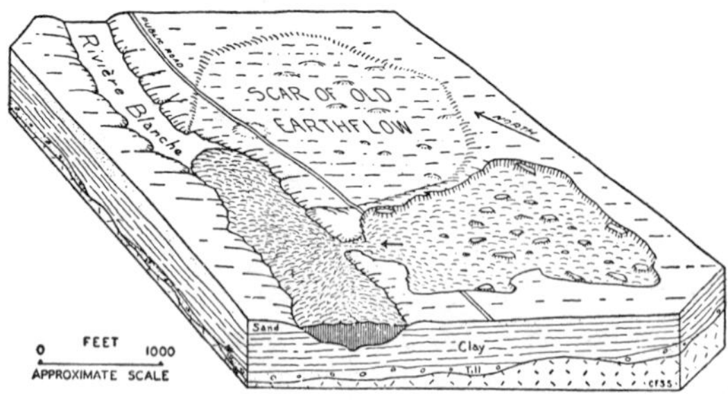

Fig. 49.4. Block diagram showing principal features of flow in quick clay near St. Thuribe, Quebec (from Sharpe 1938).

The low salt content of the porewater may be a consequence of leaching by fresh water that has percolated through the clay since the land surface has been uplifted with respect to the sea level prevailing at the time of deposition. Hence, the process may still be going on. If this hypothesis is correct, the frequency of clay flows may increase in those areas where geological conditions are favorable.

Stability of Slopes on Clay Containing Layers or Pockets of Water-Bearing Sand

In the preceding text we have considered only the stability of more or less homogeneous soils. The most important nonhomogeneous soil formations are stratified deposits consisting of layers of sand and clay and masses of cohesive soil containing irregular lenses or pockets of sand or silt.

In a sequence of layers of clay and sand or coarse silt, at least some of the latter are commonly water-bearing during part or all of the year. If a cut is excavated in such a soil, water seeps out of the slopes at various points or along various lines. Therefore, such cuts are commonly referred to as *wet cuts*. They require special attention, particularly if the strata dip toward the slope. The springs that issue along the base of the sand outcrops are likely to cause sloughing, and frost action may also lead to deterioration. Therefore, it is common practice to intercept the veins of water by means of drains that follow the base of the water-bearing layers at a depth of at least 5 ft measured at right angles to the slope. If the clay strata are soft or fissured, they may constitute an additional source of structural weakness. Hence, if the cut is deep, a stability investigation should be made to learn whether or not it is advisable to adhere to the standard slope.

Masses of cohesive soil containing irregular lenses or pockets of cohesionless soil are common in regions of former glaciation where the sediments were deposited by melting ice and then deformed by the push of temporarily advancing ice sheets. They have also been encountered at the site of old landslides that took place in stratified masses of sand and clay.

The sand pockets within the clay serve as water reservoirs. During wet weather they become the seat of considerable hydrostatic pressure that tends to cause the outward movement of the masses in which they are located. As the soil masses move outward, they disintegrate into a mixture of saturated silt, sand, and chunks of clay that flows like a glacier or like a thick viscous liquid.

Since the source of instability is the pressure of the water trapped in the sand pockets, stabilization can be accomplished by means of drainage. However, the geological profile is likely to be very irregular, and the spacing of drains may be very difficult to determine in advance even after the soil and hydraulic conditions have been thoroughly investigated by boring, testing, and periodic surveys of the water table. Under these circumstances an expedient and effective procedure may be the insertion of *horizontal auger drains* (Smith and Stafford 1957). Such drains commonly consist of perforated or slotted metal or plastic pipe of about 2-in. diameter, inserted into holes drilled nearly horizontally into the soil beneath the slope. The lengths of the drains range from a few feet to more than 200 ft. Their horizontal spacing depends on the local conditions; it often ranges from about 15 to 50 ft. Several rows at different elevations may also prove effective. The drains are usually given a small downward slope toward the face of the cut to facilitate the removal of the water by gravity.

The holes for the drains are commonly made by continuous-flight hollow-stem augers (Fig. 44.3) which permit insertion of the drain pipe without collapse of the hole. In some materials a filter may be required to prevent underground erosion and clogging; the filter material under favorable circumstances may be transported into the hole around the drain by means of the auger upon reversing its direction of rotation and gradually withdrawing it from the hole. Holes have also been formed successfully by a modification of rotary drilling wherein a casing terminating in a hollow bit is advanced by rotation while water is supplied through the interior of the casing and returns around the outside of the casing. The bit is abandoned when the hole reaches its final length, the drain is inserted, and the casing withdrawn.

The technique of installation of horizontal auger drains requires adaptation to local conditions, but such drains can often be installed so rapidly and economically that the length and spacing are established on the basis of trial. Several drains may be nonproductive, but those that encounter the pervious pockets may be remarkably effective. Once drainage has been accomplished, the terrain may become so stable that the cut can be made with standard slopes.

Slides in Stiff Clay

Almost every stiff clay is weakened by a network of hair cracks or slickensides. If the surfaces of weakness subdivide the clay into small fragments 1 in. or less in size, a slope may become unstable during construction or shortly thereafter. On the other hand, if the spacing of the joints is greater, failure may not occur until many years after the cut is made.

Slides in clay with closely spaced joints occur as soon as the shearing stresses exceed the average shearing resistance of the fissured clay. Several slides of this type took place in a long railway cut at Rosengarten, near Frankfurt in Germany. The slope of the sides was 3:1. The greatest depth of the cut was 100 ft, and the average shearing stresses along the surfaces of sliding adjoining the deepest part of the cut were roughly 10 tons/ft^2. The clay was very stiff, but large specimens broke readily into small angular pieces with shiny surfaces. Slides started immediately after construction, and continued for 15 years (Pollack 1917).

If the spacing of the joints in a clay is greater than several inches, slopes may remain stable for many years or even decades after the cut is made. The lapse of time between the excavation of the cut and the failure of the slope indicates a gradual loss of the strength of the soil. Present conceptions regarding the mechanics of the process of softening are illustrated by Fig. 49.5. Before excavation, the clay is very rigid, and the fissures are completely closed. The reduction of stress during excavation causes an expansion of the clay, and some of the fissures open. Water then enters and softens the clay adjoining these fissures. Unequal swelling produces new fissures until the larger chunks disintegrate, and the mass is transformed into a soft matrix containing hard cores. A slide occurs as soon as the shearing resistance of the weakened clay becomes too small to counteract the forces of gravity. Most slides of this type occur along toe circles involving a relatively shallow body of soil, because the shearing resistance of the clay increases rapidly with increasing distance below the exposed surface. The water seems to cause only the deterioration of the clay structure; seepage pressures appear to be of no consequence.

Fig. 49.5. Section through fissured stiff clay mass. (*a*) Old fissures closed before relief of stress by excavation. (*b*) Relief of stress causes fissures to open, whereupon circulating water softens clay adjoining the walls.

Fig. 49.6. Photograph of slide in very stiff fissured clay.

Figure 49.6 shows a slide in very stiff fissured clay beside a railroad cut having side slopes of 2.5:1. The height of the slope was 60 ft. The characteristic S-shape of the slope after failure is apparent. Failure occurred about 80 years after the cut was excavated. No springs or other indications of percolating water were present.

A study of the records of several delayed slides in stiff clays with widely spaced joints has shown that the average shearing resistance of the clay decreases from a high initial value at the time of excavation to values between 0.20 and 0.35 ton/ft^2 at the time of the slide. Since the process of deterioration may require many decades, it would be uneconomical to select the slope angle for the sides of cuts in such clays on the basis of the ultimate value of the shearing resistance. However, it is desirable to delay the deterioration as much as possible by draining the strip of land adjoining the upper edge of the cut for a width equal to the depth of the cut and by treating the ground surface of the cut area to reduce its permeability. Should local slides occur at a later date, they can be remedied by local repairs. If delayed slides would endanger life or cause excessive property damage, the slope should be provided with reference points and periodic observations should be made, inasmuch as slides of this type are always preceded by deformations that increase at an accelerated rate as a state of failure is approached. When the movement becomes alarming, the slopes in the danger section should be flattened.

Hard core drains have also been successfully used to prevent movements at danger sections. These drains consist of ribs of dry masonry installed in trenches running up and down the slope at a spacing of about 15 or 20 ft. The trenches are excavated to a depth somewhat greater than that to which the clay has been softened. A concrete footwall supports the lower ends of all of the ribs. The beneficial effect of this type of construction is commonly ascribed to the action of the ribs as drains, but it is more likely that the principal function of the ribs is to transfer part of the weight of the unstable mass of clay through side friction to the footwall.

The behavior of poorly bonded clay shales is governed by many of the same considerations as that of stiff clays. Hence, further information concerning slides in heavily overconsolidated clays is contained in the next section.

Slopes on Shale

From an engineering point of view, shales are of outstanding importance because they constitute about 50% of the rocks that are either exposed at the earth's surface or are buried beneath a thin veneer of sediments. All the rocks of this category consist of deposits of clay or silt that have acquired their present characteristics under the influence of relatively moderate pressures and temperatures.

As the thickness of the overburden increases from a few tens of feet to several thousands, the porosity of a clay or silt deposit decreases; an increasing number of cohesive bonds develops between particles as a result of molecular interaction, but the mineralogical composition of the particles probably remains practically unaltered. Finally, at very great depth, all the particles are connected by virtually permanent, rigid bonds that impart to the material the properties of a real rock. Yet, all the materials located between the zones of incipient and complete bonding are called shale. Therefore, the engineering properties of any shale with a given mineralogical composition may range between those of a soil and those of a real rock.

The most conspicuous differences among the shales produced by the compaction of identical sedimentary deposits have their origin in the number of permanent interparticle bonds per unit of volume of shale. A relative measure of the degree of bonding is provided by the performance of intact specimens obtained from a depth of several hundred feet. Upon submersion, all of these gradually break up into fragments. However, depending on the degree of bonding, the sizes of the fragments may be as great as a large fraction of an inch or as small as the individual mineral particles themselves. Between these limits, shales may be said to range from those categorized as well-bonded, of which the extreme types are rock-like shales, and those described as poorly bonded, of which the extreme types are heavily overconsolidated silt- to clay-shales. In addition to an estimate of the degree of bonding, descriptions of shales should indicate whether the prevalent constituent is clay or silt, as, for example, "poorly bonded clay-shale." Nevertheless, within the limits of any given description the engineering properties of the shale may range as widely as those of clays or silts.

During the removal by geological processes of the load that was responsible for the transformation of a silt or a clay into shale, the shale expands at practically constant horizontal dimensions. Many of the interparticle bonds, which were formed during or after compression, are stressed by the subsequent expansion to or beyond the point of failure. Therefore, shales are commonly weakened by a network of joints. Below a depth on the order of 100 ft the joints are completely closed and are spaced several feet apart. However, as the depth of overburden is decreased further, the joints open because of unequal expansion of the blocks located between them. The water content of the blocks then increases and their strength decreases like that of any clay or silt during reduction in pressure from the preconsolidation load. During this process, new fissures are formed. The final result depends, like that of the immersion of a shale specimen, on the degree of bonding.

Within the depth of seasonal variations of moisture and temperature the shale may undergo additional alterations such as further mechanical disintegration or slight mineralogical changes revealed by discoloration. These alterations are manifestations of weathering. However, because the boundary between the hard or intact shale and the shale which is weakened by load removal may be located many tens of feet below the base of the weathered top layer, the presence or absence of such a layer has very little influence on the properties of a shale deposit. Nevertheless, it has become customary among engineers to designate as "weathered" the entire body of shale located above the hard shale, even where the top layer is absent. In other words, the term is applied to materials that owe their present engineering properties to two entirely different processes. To avoid misunderstanding, the term weathered will not be used in the following discussion.

On shales of any kind, the decrease of the slope angle to its final equilibrium value takes place primarily by intermittent sliding. The scars of the slides give the slopes the hummocky, warped appearance known as *landslide topography*. The details of the performance of shales underlying the slopes depend primarily on the mineral constituents and the degree of bonding. Extreme representatives of the well-bonded shales can be found in the Allegheny region and of the poorly

Fig. 49.7. Photograph of slide on gentle slope above well-bonded shale, near Barboursville, W. Va. (from Ladd 1935).

bonded ones among the shales underlying large areas located southwest and northwest of the Great Lakes.

In the Allegheny region of West Virginia, southern Pennsylvania and eastern Ohio, many slopes are underlain by well-bonded, more or less silty shales. The increase in water content of the blocks between joints, associated with load removal, is very small and slides seldom cut to a depth of more than about 15 ft. The resulting landslide topography is illustrated by Figs. 49.7 and 49.8. Steep slopes in side cuts can remain stable for many years. Slope failures, either on hillsides or cuts, occur only during the rainy season. If a failure occurs, the slide material flows for a short time like a viscous liquid and then comes to rest. On account of its relatively high permeability the slide material can be stabilized by simple means such as the installation of horizontal auger drains.

Whenever an attempt has been made to account for such slides

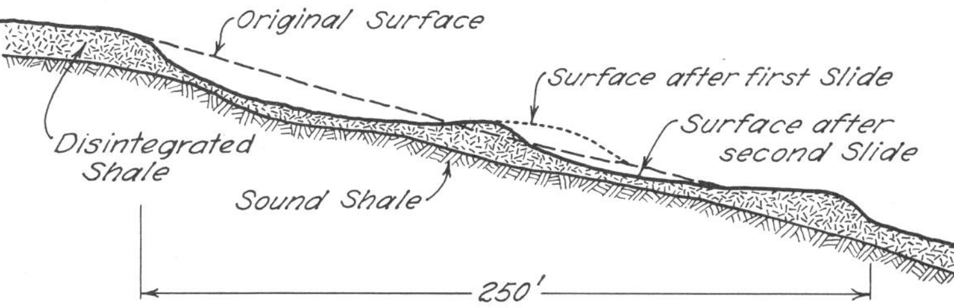

Fig. 49.8. Profile of double slide in well-bonded shale (after Ladd 1935).

in fairly well-bonded shales by stability computations based on the results of laboratory tests, it has been found that the factor of safety of the slope with respect to sliding should have been more than adequate even on the assumption that the water table at the time of failure was located at the ground surface. The striking and persistent discrepancy between forecast and field performance can most logically be explained by assuming that the slides are preceded by a sudden, but temporary and local, increase of the porewater pressure in the zone of sliding. The shale within this zone consists of macroscopic fragments which are in the process of progressive deterioration. Because of the slope the accumulation of fragments is acted on by shearing forces and the joints between the fragments open. During wet spells the open spaces are filled with water. As soon as the deterioration reaches a critical stage, which occurs in different places at different times, the fragments break down during a wet spell under the combined influence of the overburden and seepage pressures. The weight of the overburden is temporarily transferred to the water, whereupon the effective pressure and the corresponding shearing resistance along a potential surface of sliding are reduced and a slide occurs. It may stop rather abruptly, because the excess porewater pressure dissipates rapidly on account of the relatively high permeability of the accumulation of shale fragments.

The other extreme members of the shale family are the poorly bonded clay shales such as those that prevail in parts of the Dakotas, Montana, and the western prairie provinces of Canada. These shales, too, owe their present engineering properties to intense consolidation under overburdens with a thickness on the order of a thousand feet and to subsequent removal of the load. Beneath a depth measured in tens of feet these shales, too, are very hard. In tunnels extending below this depth it can be seen that they are transected by widely spaced, tight joints, like the well-bonded shales. On the other hand, the changes in water content that have occurred within the uppermost tens of feet as a consequence of unloading are radically different. Whereas in the well-bonded shales they may be almost imperceptible, in the poorly bonded clay shales they may amount to 10% or more. As a consequence, the final result of the removal of load is also very different. Well-bonded shales turn into fairly pervious aggregates of angular, macroscopic rock fragments, but the poorly bonded clay shales become stiff plastic clays. The mechanics of the transition from hard shale into a clay-like material is essentially the same as the process responsible for the slides in stiff, fissured clay, illustrated by Fig. 49.5. However, since the permeability of a clay shale is much lower than that of a less heavily preloaded clay, the softening proceeds much more slowly.

On account of the large changes in volume associated with the unloading of poorly bonded clay shales under conditions of no lateral strain, they become the seat of severe horizontal residual stresses. While the overburden is being gradually removed, the ratio between the horizontal and vertical normal stresses increases. Even in less heavily preconsolidated clay deposits the ratio may approach the coefficient K_p of passive pressure of the clay (Skempton 1961*b*, Terzaghi 1961*a*). These stresses can contribute to the mechanical disintegration of the shales adjacent to the slopes of river valleys or behind man-made cuts.

Natural slopes on poorly bonded clay shales recede primarily by intermittent sliding, and thereby become flatter and flatter. As the slope angle decreases, the average shearing stresses also decrease along potential surfaces of sliding. Nevertheless, slides continue to occur, at increasing intervals of time, until the slopes are reduced to 1 vertical on 10 horizontal or even less. These observations indicate that the loss of strength due to unloading is extremely slow and cannot be predicted reliably on the basis of laboratory tests. Ultimately (Skempton 1964), along surfaces where shearing strains become very large, the resistance may approach the residual strength of the soil (Article 18). Every slide is preceded by accelerated creep to a depth much greater than that of seasonal variations of moisture and temperature. As soon as the rate of creep attains a value of several inches per year a slide occurs. During the slide the shale located above the surface of sliding remains almost intact and retains the character of a fairly stiff and intensely fissured clay.

Because of the low permeability of poorly bonded clay shales at all depths below their present surface, failures of newly cut slopes or of the foundation material beneath new fills take place under $\phi = 0$ conditions. Yet, all attempts to determine the shearing resistance of the shale by means of undrained triaxial tests so far have failed (Peterson et al. 1960). The strengths determined in the laboratory are consistently several times greater than those at which the shales failed in the field. Opinions concerning the causes of these striking discrepancies are divided. Undoubtedly, progressive failure is a significant factor (Bjerrum 1966). However, the existence of the discrepancies still precludes the possibility of securing reasonably reliable information concerning the shear characteristics of a given shale from any source other than the analysis of slides that have occurred in the same shale under similar conditions.

Beneath the slopes of river valleys the shales have been subjected to severe shearing stresses for a long period. Consequently, they have reached a more advanced stage of expansion than those beneath the adjacent, less dissected uplands. Therefore, on such slopes the conditions for the stability of new cuts are much less favorable than at a greater distance from the deep valleys. As time goes on, however, the shearing resistance of the clay adjacent to new slopes, even away from the valleys, will decrease. At any location, excavation should be started at the top of a new slope to reduce to the inevitable minimum the opening up of existing joints in the adjacent shale.

A great deal of quantitative information regarding the movement of slopes on poorly bonded shales has been obtained in the Bearpaw formation at the site of the South Saskatchewan River Dam. The natural slopes rise to a height of about 200 ft above the river valley, at 1 vertical on 8 to 12 horizontal, and exhibit typical landslide topography. The gentleness of the slopes and the long periods of time between slides indicate that the slopes are already in an advanced stage of development. During the period of observation (1944–1964), no slides occurred outside the area affected by construction operations. Yet, cracks across trails along the edge of the uplands gradually became wider. Wherever the equilibrium of a slope was disturbed, for instance by the excavation of side cuts, a slide occurred and the shearing resistance was always found to be as small as that which had resisted similar slope failures due to natural causes. Exceptionally unfavorable stability conditions were encountered where the shales contained seams of bentonite, and in the proximity of minor faults or shear zones. Since the permeability of the shale is very low, drainage is almost ineffective.

Slopes on Weathered Rock

Rock slopes rising at an angle of less than about 40° are commonly covered with a layer of products of rock weathering that may vary in thickness over short distances between zero and tens, or even hundreds of feet. It has been pointed out in Article 2 that the physical properties of these soils *in situ* can be very different from those of transported soils with similar mineralogical composition and grain-size characteristics. Furthermore, these properties are likely to change over short distances in every direction on account of erratic variations in the degree of weathering. Therefore, the consequences of undercutting existing slopes, the effects of seepage toward such slopes from artificial sources, or the degree of stability of slopes to be produced by excavation can never be predicted with any degree of confidence on the basis of the results of boring and testing. Soil mechanics provides us only with the knowledge required for a correct interpretation of what can be observed in the field before and during construction, for anticipating the performance of the materials in a general way, and for taking full advantage of the existence of precedents.

As time goes on, slopes made by nature on rock of any kind become flatter; the process does not stop until the inclination becomes as flat as 15 horizontal to 1 vertical or even less. This fact indicates that the products of rock weathering are removed more or less continuously and descend toward the base of the slope where they accumulate or are carried away by erosion. The removal takes place most commonly by creep, an imperceptible, glacier-like movement of the material located within the depth of seasonal variations of moisture and temperature. However, on some types of rock much of the removal occurs by intermittent sliding. The continuity of slopes subject to sliding is disrupted by the remnants of numerous slide scars (*landslide topography*). Therefore, this type of waste removal is often disclosed by the details of the slope topography.

The character of the residual materials, as well as the mechanism of waste removal, reflect the type and the mechanical properties of the underlying weathered rock. For example, in soluble rocks such as limestone, there is commonly a very sharp but extremely uneven boundary between intact and completely weathered rock. Any zone of transition is conspicuously absent. The weathered rock consists of the insoluble, commonly very fine-grained, mineral constituents of the parent rock. Waste removal occurs exclusively by creep. Below the creep layer, the residual soil is likely to possess considerable cohesion, so that slopes of as much as 1 horizontal to 5 vertical in side cuts may be stable.

In contrast, gradual transitions from weathered to intact rock are characteristic of intrusive igneous rocks such as granite, and high grade metamorphic rocks such as gneiss, with a low percentage of micaceous constituents and a high percentage of chemically unstable materials such as feldspar. Removal of the insoluble waste takes place, as for the soluble rocks, only by creep. Side slopes rising 3 vertical on 2 horizontal are by no means uncommon even in deep cuts. Occasional slides take place at points where the slope cuts across weathered shear zones or where the orientation and location of relict joints are especially unfavorable. The spots where they may occur are seldom known in advance of construction, but the cost of avoiding them by reducing the inclination of the entire slope would be prohibitive. In most instances it is more economical to cut the slopes as steeply as seems appropriate for the intact material and to provide adequate width at the base of the cut to permit material from the occasional falls to accumulate and to be cleaned out from time to time without blocking the drainage ditches at the foot of the slope.

The sites of ancient landslides are commonly indicated by the topography of the slopes. Cuts in the landslide material are always troublesome, because the material came to rest as soon as its factor of safety with respect to further movement became equal to unity. If these locations cannot be avoided, construction should be preceded by radical and permanent drainage.

Within the zone of weathering of the insoluble types of rock, it is by no means uncommon for the coefficient of permeability of the weathered rock to increase from very small values near the ground surface to maximum values close to the boundary between weathered and sound rock. Thus, the zone of rock weathering forms a relatively impervious skin resting on a pervious layer. If water enters the pervious layer through a gap in the skin or through open fissures in the sound rock, artesian conditions may develop in the pervious zone and the impervious top layer may slide down the slope even if the inclination is very gentle. This possibility is illustrated by the following example.

Figure 49.9 is a section through a saddle in a series of hills separating a storage reservoir on the left from a deep valley on the right. A small earth dike was constructed in the saddle. The hills consisted of gneiss with the planes of foliation dipping about 60° in the direction of the axis of the dike. At the crest the depth of rock weathering was about 80 ft, at El. 1300. Borings in this vicinity encountered a top layer consisting chiefly of clay resting on soft to medium decom-

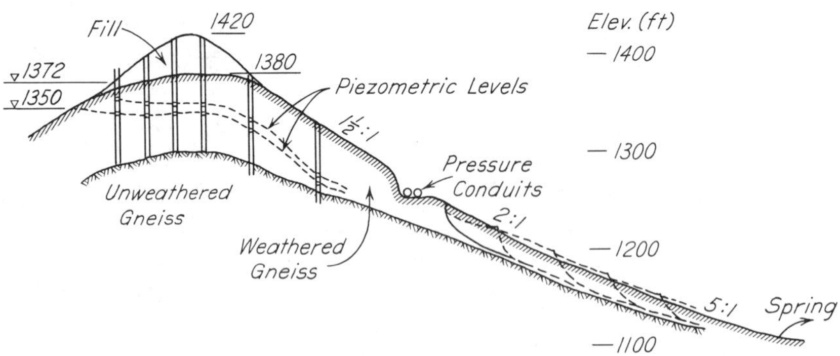

Fig. 49.9. Section through saddle in weathered gneiss separating reservoir at left from deep valley on right.

posed gneiss containing less clay than the top layer, and resting in turn on compact to hard decomposed gneiss in which there was a marked loss of wash water. As indicated in the figure, the inclination of the long right-hand slope decreased from 1.5 horizontal to 1 vertical near the top to about 5 to 1 at El. 1100. Before the reservoir was filled, slide scars were conspicuously absent; waste removal had obviously taken place by creep only.

At El. 1250 two pressure conduits were established on concrete piers in a side cut with a width of about 25 ft. The excavation for the foundation of each pier was discontinued at a depth where no further progress could be made without blasting. When the water level in the reservoir reached El. 1305, about 5 ft above the elevation of the top of the unweathered rock, the pressure conduits sagged at two points and had to be underpinned. When the reservoir level rose to El. 1320, the underpinned piers as well as additional ones settled, and a small spring emerged at El. 1100, at a distance of about 800 ft from the saddle. When the reservoir rose to El. 1380, more springs appeared above El. 1100, and the entire lower part of the slope between the pressure conduits at El. 1250 and the spring at El. 1100 moved downhill along a surface close to the upper boundary of the weathered rock. Yet, the nearly vertical slope of the side cut and the steep natural slope above it did not move perceptibly.

As the reservoir level rose, the piezometric levels also rose in the pervious contact layer of the zone of weathering (Fig. 49.9). The sequence of events recounted in the preceding paragraph indicates that the hydrostatic pressures in the pervious zone increased in a downhill direction, and that the slide started at the foot of the slope at about El. 1100 where the hydrostatic pressure first exceeded the weight of the overlying top layer.

Talus Slopes

The term *talus* refers to a loose aggregation of pieces of rock that accumulate at the foot of a rock cliff. The slope of the talus pile is commonly between $1\frac{1}{4}$ to $1\frac{3}{4}$ horizontal to 1 vertical. Ordinarily the slopes are stable. Slides occur most commonly when snow is melting and less often during heavy rainstorms. The nature of the fragments does not seem to be significant. After a slide starts, the saturated material rushes in a swift torrent down the valley, transporting rock fragments up to several cubic yards in size, removing bridges in its path, and spreading at the mouth of the valley like a fan. These slides, known as *mud spates* or *mud rock flows*, are common in high mountain chains in every part of the world. On the western slope of the Wasatch Mountains in Utah every canyon contains the remnants of at least one mud rock flow (Sharpe 1938). Since slides of this type occur irrespective of the relative density or petrographic character of the talus, and only on steep slopes, it is probable that they are caused exclusively by the seepage pressure of percolating water.

In the Alps it has been observed that mud rock flows are commonly preceded by the drying up of springs emerging from the root area of the flows. This phenomenon indicates a temporary increase in the pore space of the talus material prior to the slide, similar to the increase of the void ratio of a test specimen of dense sand before failure by shear (Article 15).

Since no slide in talus can take place without an abundance of water, the danger of such slides can be eliminated by preventing temporary saturation. This can be accomplished by installing a deep drain along the upper boundary of the area to be protected and by covering the surface of the area with a layer of relatively impermeable soil. In many instances the drain alone will have the desired effect.

Sudden Spreading of Clay Slopes

Experience has shown that failures of clay slopes by sudden spreading tend to occur in cycles with periods of maximum frequency at more or less regular intervals. It is characteristic of this type of failure that a gentle clay slope, which may have been stable for decades or centuries, moves out suddenly along a broad front. At the same time the terrain in front of the slide heaves for a considerable distance from the toe. On investigation, it has invariably been found that the spreading occurred at a considerable distance beneath the toe, along the boundary between the clay and an underlying water-bearing stratum or seam of sand or silt. The probable causes of these sudden and frequently catastrophic slope failures are illustrated by Fig. 49.10*a*.

Figure 49.10*a* represents a section through a valley located above a thick stratum of soft clay that gradually merges toward the left into sand. The clay, which has an average cohesion c, contains thin horizontal layers of fine sand or coarse silt, such as the layer *S–S*. The porewater in *S–S* communicates with the water in the large body of sand on the left side of the diagram. The plain lines *Ad* and *Be*, respectively, represent the water table in the sand during a dry and an exceptionally wet season. The dash lines *Ab* and *Bg* represent the corresponding piezometric levels for the porewater in *S–S*.

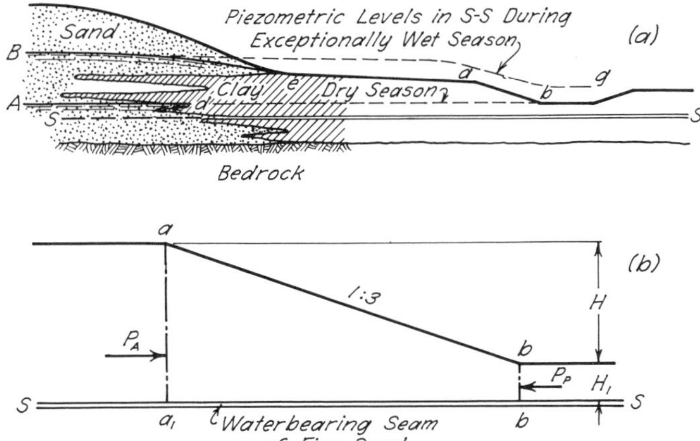

Fig. 49.10. **(*a*) Geological conditions involving danger of slope failure by spreading. (*b*) Diagram of forces which act on soil beneath slope *ab*.**

A cut *ab* has been excavated in the clay to a depth H. Every horizontal section beneath the cut, including that through S–S, is acted on by shearing stresses, because the overlying clay tends to settle vertically and to spread horizontally under the influence of its own weight. If the porewater pressure in the layer S–S is low, corresponding to the piezometric line Ab, the shearing resistance along S–S is likely to be considerably greater than the sum of the shearing stresses. When this is true, the stability of the slope depends only on the cohesion c of the clay. For any slope angle less than 53° the critical height H_c of the slope is

$$H_c = 5.52 \frac{c}{\gamma} \tag{49.1}$$

where γ is the unit weight of the clay (Article 35). If a firm base underlies the clay stratum at a shallow depth below the bottom of the excavation, corresponding to a low value of the depth factor n_D (Fig. 35.2), the critical height is even greater, and it increases with decreasing slope angles up to values of $9c/\gamma$ for slopes of 20°, as shown in Fig. 35.3.

However, because of a protracted wet spell or the melting of snow on the surface of the ground above the large body of sand, the piezometric levels for the stratum S–S may rise to the position indicated by the line Bg. During the rise the total load on S–S, p per unit of area, remains unchanged, but the porewater pressure u_w increases.

Since the layer S–S consists of almost cohesionless soil, its shearing resistance is determined by the equation,

$$s = (p - u_w) \tan \phi \qquad (17.1)$$

Hence, the increase of the piezometric levels for this layer corresponds to a decrease of the shearing resistance on any horizontal section through the layer. As soon as the average shearing resistance decreases to the value of the average shearing stress, the slope above S–S fails by spreading, in spite of the fact that it may still possess an adequate factor of safety against sliding along any curved surface located above or cutting across S–S.

The critical height of the slope above S–S can never be less than the value obtained on the assumption that the porewater pressure u_w is equal to p (Eq. 17.1), whereupon the shearing resistance along S–S becomes zero. The implications of this condition are illustrated by Fig. 49.10*b* which represents a vertical section through the slope *ab* to a larger scale. According to Eq. 28.10, the active earth pressure on the vertical section aa_1 is

$$P_A = \tfrac{1}{2}\gamma(H + H_1)^2 - 2c(H + H_1)$$

and, according to Eq. 28.17, the passive earth pressure on bb_1 is

$$P_P = \tfrac{1}{2}\gamma H_1^2 + 2cH_1$$

If the shearing resistance on a_1b_1 is zero, the slope will be on the verge of failure when $P_A = P_P$, whence

$$H = H_c = 4\frac{c}{\gamma} \qquad (49.2)$$

This value is approximately equal to $3.85c/\gamma$ which, according to Fig. 35.3, is the critical height of a vertical slope. Hence, if the porewater pressure is great enough to eliminate the friction in the seam S–S, it reduces the critical height of the slope located above the seam to slightly more than the critical height of a vertical slope, regardless of what the actual slope angle may be. For gentle slopes, this effect of the porewater pressure may involve a reduction of the critical height by almost 50%.

During exceptionally wet years or during the melting of an exceptionally thick snow cover, the water table rises everywhere. As a consequence, the shearing resistance of every water-bearing seam decreases, and slopes may fail that were previously always stable. In 1915 a slide occurred on a very gentle slope about 40 ft high, within the boundaries of the Knickerbocker Portland Cement Company on

Claverack Creek near Hudson, N.Y. The slope was located on varved clay consisting of alternate layers of clay and silt, each about $\frac{1}{2}$ in. thick. Suddenly, without any visible provocation, the slope moved out over a length of 1200 ft, and the surface of the flat in front of the toe heaved for a distance of about 300 ft. Over a length of about 600 ft, the creek bottom was lifted above the level of the surrounding ground, and the heave occurred so rapidly that fish remained stranded on the gentle ridge that occupied the former site of the creek. The powerhouse located on the premises was wrecked, and the occupants perished. This slide was only one of many that have occurred in the varved clays of the Hudson River valley since its settlement (Newland 1916). The history of the valley indicates quite clearly that the slides occurred most frequently at intervals of roughly 20 or 25 years, corresponding to years of maximum rainfall.

The difference between plain gravity slides due to inadequate cohesion of the clay and slides due to spreading of the clay strata is illustrated by Figs. 49.11*a* and *b*. In contrast to slides of type *a*, those of type *b* occur almost suddenly. It is probable that they are not even preceded by measurable deformations of the mass of soil that ultimately fails, because the seat of weakness is located not within the clay but only at the boundary between the clay and its base. Furthermore, the critical height for slopes on homogeneous clay depends only on the slope angle and the average cohesion *c*, whereas the critical height for slopes on clay located above water-

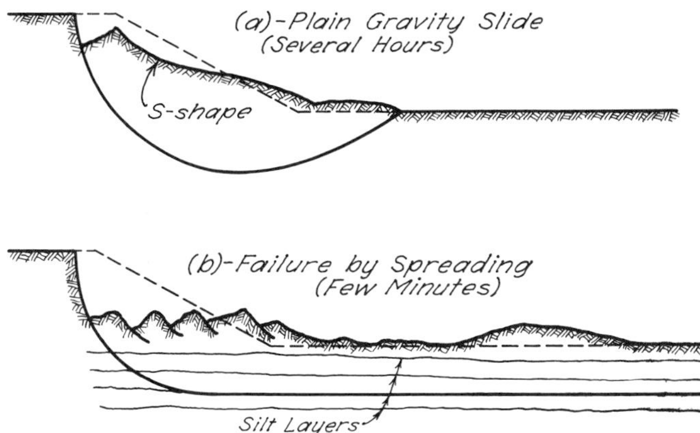

Fig. 49.11. Cross section through typical slide in varved clay. (*a*) If porewater pressure in silt layers is inconsequential. (*b*) If porewater pressure in silt layers is almost equal to overburden pressure.

bearing seams or strata of cohesionless soil depends to a large extent on the porewater pressure u_w in the seams. As the porewater pressure increases, the critical height decreases and approaches the value H_c (Eq. 49.2) regardless of what the slope angle may be. Hence, if the height of a slope on a clay with water-bearing seams of sand or silt is greater than H_c, no reliable opinion can be formed concerning the factor of safety of the slope with respect to sliding unless the porewater pressure u_w is known.

A rough estimate of the maximum possible value of the porewater pressure in the water-bearing seams can be made on the basis of the general geology and physiography of the region in which the slope is located. However, the real value of the porewater pressure can be computed neither by theory nor on the basis of laboratory tests. It can be determined only in the field, by means of piezometric observations. Hence, if the prerequisites for a slide of type *b* (Fig. 49.11) may conceivably exist, the engineer must judge what the practical consequences of such a slide would be. If it could do no more than block traffic, the engineer might be justified in proceeding with construction without special precautions, in full knowledge that a slide might occur a few years or decades after construction. On the other hand, if a slide could cause loss of life or serious damage to valuable property, the installation and periodic observation of porewater pressure gages are imperative. Whenever a stability computation based on the results of the gage readings indicates that the margin of safety for the slope is narrow, sound engineering requires that the danger be eliminated by the installation of drains to keep the porewater pressure in the water-bearing constituents of the subsoil within safe limits.

Summary of Problems and Procedures

In selecting the route for a highway or railway, or the site for a project that requires the excavation of open cuts, the degree of engineering skill required depends to a considerable extent on the nature of the terrain. The layout and the construction of open cuts in favorable ground is fairly well standardized, but, if the project involves troublesome ground, the engineer must possess the highest qualifications. This is due partly to the infinite variety of combined soil and hydraulic conditions that may lead to slides and partly to the fact that economic considerations commonly require a radical departure from the customary standards of safety. The engineer in charge of location must be capable of identifying favorable, troublesome, and very troublesome ground on the basis of surface indications and occasional exploratory borings. He must also be able to visualize the worst construction difficulties that may arise at the various possible sites, and to evaluate the corresponding expenditures and delays.

If troublesome ground cannot be avoided, the engineer must successively perform the following tasks:

1. Locate the most critical sites and explore them by sampling and testing.
2. Select the slope angles on the basis of a reasonable compromise between the requirements of economy and safety.
3. Design drainage systems, if they are needed.
4. Prepare the program of observations that must be made during construction to remove the uncertainties in the knowledge of the site and to eliminate the risk of accidents.
5. Stabilize those slopes that begin to move out, at a minimum of expense and delay.

The preceding sections of this article have made it plain that no hard and fast rules can be established for performing any one of these tasks. The theory of the stability of slopes (Article 35) can be used to advantage only in those rare instances when a cut is to be made in a fairly homogeneous mass of soft or medium clay. When dealing with other soils or with combinations of soils the engineer must depend entirely on his capacity to recognize the factors that determine the stability of the soil deposit under consideration, on his ability to visualize the implications of the uncertainties that will still exist while the project is in the design stage, and on his ingenuity in devising means for eliminating these uncertainties as construction proceeds.

The development of these vital attributes requires a working knowledge of geology and a thorough acquaintance with the laws that govern the interaction between water and the different types of soil. These laws are set forth in Part I of this book. They must be supplemented by a broad knowledge of construction experience dealing with excavations and slides. Personal experience can supply only part of this knowledge; equally important is the experience summarized in published case histories.

Selected Reading

Several publications deal with the description, mechanism or correction of landslides from a comprehensive point of view. Among the more useful are:

Ladd, G. E. (1935). "Landslides, subsidences and rock-falls," *Proc. Am. Rwy. Eng. Assn.*, **36**, pp. 1091–1162.

Sharpe, C. F. S. (1938). *Landslides and related phenomena.* New York, Columbia Univ. Press. The classification of landslides is not suitable for purposes of the civil engineer, but the descriptions of the phenomena are valuable.

Terzaghi, K. (1950). "Mechanism of landslides," *Geol. Soc. Am., Engineering geology, Berkey Vol.*, pp. 83–123. Reprinted in *From theory to practice in soil mechanics,* New York, John Wiley and Sons, 1960, pp. 202–245.

HRB (1958). "Landslides and engineering practice," Committee on landslide investigations, *Hwy. Res. Board Special Rept. 29,* 232 pp.

Among many excellent papers concerning individual landslides, the following have been selected to represent widely different conditions:

Newland, D. H. (1916). "Landslides in unconsolidated sediments," *N.Y. State Museum Bull. 187,* pp. 79–105. Slides in varved clays on Hudson River.

Close, U. and E. McCormick (1922). "Where the mountains walked," *Nat. Geog. Mag.*, **41,** pp. 445–464. Flow slides in loess in China.

Bjerrum, L. (1955). "Stability of natural slopes in quick clay," *Géot.* **5,** No. 1, pp. 101–119.

Skempton, A. W. and D. J. Henkel (1955). "A landslide at Jackfield, Shropshire, in a heavily overconsolidated clay," *Géot.*, **5,** No. 2, pp. 131–137. The mechanism has been reinterpreted in Skempton, A. W. (1964): "Long-term stability of clay slopes," *Géot.*, **14,** No. 2, pp. 77–101.

Deere, D. U. (1957). "Seepage and stability problems in deep cuts in residual soils, Charlotte, N.C.," *Proc. Am. Rwy. Eng. Assn.*, **58,** pp. 738–745. Failures caused by surface and subsurface erosion and presence of relict joints.

Gould, J. P. (1960). "A study of shear failure in certain Tertiary marine sediments," *Proc. ASCE Research Conf. on Shear Strength of Cohesive Soils,* pp. 615–641. Slides along coastal area near Los Angeles, California.

Terzaghi, K. (1960*b*). "Memorandum concerning landslide on slope adjacent to power plant, South America," *From theory to practice in soil mechanics,* New York, John Wiley and Sons, pp. 410–415. Details of investigations to control movement of slope on tropically weathered residual soil.

Kjaernsli, B. and N. Simons (1962). "Stability investigations of the north bank of the Drammen river," *Géot.*, **12,** No. 2, pp. 147–167. Circular slip in soft silty clay.

ART. 50 COMPACTION OF SOILS

Purposes and Methods of Soil Compaction

The preceding article dealt with the stability of soil masses in their natural state. By excavating such soil masses and redepositing them without special care, the average porosity, permeability, and compressibility of the soil are increased, and the capacity to resist internal scour by water veins is greatly reduced. Therefore, even in ancient times it was customary to compact fills to be used as dams or levees. On the other hand, no special efforts were made to compact highway embankments, because the road surfaces were flexible enough to remain unharmed by the settlement of the fill. Until very recently railroad fills were also built up by loose dumping and allowed to settle under their own weight for several years before placement of high-quality ballast.

The settlement of uncompacted fills did not result in any serious inconveniences until the beginning of the 20th century, when the rapid development of the automobile created an increasing demand for hard-surfaced roads. It soon became apparent that concrete roads on uncompacted fills were likely to break up and that the surfaces of other types of high-grade pavements had a tendency to become very uneven. The necessity for avoiding such undesirable conditions fostered the development of methods of soil compaction that would satisfy the requirements of both economy and efficiency. Simultaneous increase of activity in the field of earth-dam construction provided an additional incentive for the development of compaction methods.

The investigations that were made have led to the conclusion that no one method of compaction is equally suitable for all types of soil. Furthermore, the extent to which a given soil is compacted as a result of a given procedure depends to a large extent on the water content of the soil. The greatest degree of compaction is obtained when the water content has a certain value known as the *optimum moisture content*, and the procedure for maintaining the water content close to the optimum value during the compaction of a fill is known as *moisture-content control.*

In the following review of the subject, the current methods for compacting artificial fills are divided into three groups, those suitable for cohesionless soils, those for sandy or silty soils with moderate cohesion, and those for clays. Finally, methods are discussed for compacting masses of natural soil in their original position.

Compaction of Cohesionless Soils

The methods for compacting sand and gravel, arranged in order of decreasing effectiveness, are vibration, watering, and rolling. In practice, combinations of these methods have also been used.

Vibration can be produced in a primitive manner by tamping with hand or pneumatic tools or by dropping heavy weights on the soil from a height of several feet. However, the compacting effect of these procedures is extremely variable and usually quite small because a state of resonance is rarely approached (Article 19). The best results are achieved by machines vibrating at a frequency f_1 close to the resonant frequency f_0 for the soil and vibrator. If f_1 is approximately equal to f_0, the settlement is likely to be 20 to 40 times greater than that produced by a static load equivalent to the pulsating force.

Effective compaction has been obtained of coarse sand and gravel, and of rockfill consisting of particles of comparable sizes, by means of 5- to 15-ton rollers equipped with vibrators operating at a frequency between about 1100 and 1500 pulses per minute (Bertram 1963). The material is spread in layers from 12 to 14 in. thick; in a few instances even thicker layers have been compacted successfully, but segregation during placement is difficult to avoid. The maximum size of the particles is limited only by the thickness of the layers. From 2 to 4 passes of such rollers, drawn at a speed not exceeding about 1.5 mi per hr, are usually adequate to achieve a high degree of compaction. Moisture-content control is not necessary. Such materials have also been compacted by means of pneumatic-tired rollers drawn by heavy track-mounted diesel tractors. Water may be added to the fill during the process. Much of the compaction under these circumstances is produced by the tractor rather than the roller. Six to eight passages of the equipment over a given spot are usually required to attain a satisfactory degree of compaction, provided the material is deposited in layers not more than 1 ft thick.

In limited areas small self-propelled, hand-operated vibratory compactors may be useful. The weight of the compactors varies from a few hundred to a few thousand pounds, and the pulsating force is delivered to the ground, at approximately the resonant frequency for the compactor and soil, through a flat plate or through a roller. The thickness of layers that can be compacted effectively ranges from about 4 to 8 in.

Compaction by watering is based on the facts that the seepage pressure of percolating water breaks down unstable groups of grains, and temporary flooding at least briefly eliminates capillary forces. It is much less effective than compaction by vibration. Two different methods of watering have been used. In one the sand is deposited in ridges along both sides of the working surface and washed toward the center with jets of water under a pressure at the nozzle of 60 to 75 lb/in.^{2} The deposit that is built up has somewhat the character of a hydraulic fill. In the second method water is ponded in shallow pools on the working surface, so that it seeps into the previously placed sand and escapes through the toes of the fill. Both methods require about 1.5 yd^3 of water per cubic yard of sand. By a comparison of the porosity of fills before and after treatment it has been found that the degree of compaction obtained by either method is relatively low (Loos 1936). The practice should be discouraged.

The use of rollers to compact cohesionless soils is relatively ineffective. The best results are obtained if the sand is practically saturated. However, in clean sand the water flows away rapidly, and it may be impracticable to maintain a state of saturation.

Compaction of Sandy or Silty Soils with Moderate Cohesion

With increasing cohesion, the compacting effect of vibrations decreases greatly, because even a slight bond between the particles interferes with their tendency to move into more stable positions. Furthermore, the low permeability of these soils makes watering ineffective. Compaction in layers by rollers, on the other hand, has given very satisfactory results.

Two types of rollers are in general use: pneumatic-tired rollers, and tamping or sheepsfoot rollers. Pneumatic-tired rollers are most suitable for compacting slightly cohesive sandy soils, mixed-grained soils ranging from gravels to silt in size, and nonplastic silty soils. Sheepsfoot rollers are most effective on plastic soils.

Pneumatic-tired rollers usually consist of a load cart supported by a single row of 4 wheels equipped with tires inflated at pressures ranging from 50 to 125 psi. The wheels are mounted in such a manner that the weight from the cart is transmitted to all wheels even if the surface of the ground is uneven. Fills for buildings are commonly compacted in lifts having compacted thicknesses of 6 to 12 in. with 25-ton rollers and comparatively low tire pressures. For embankments and dams, higher tire pressures, 50-ton rollers and compacted layers of 6 to 12-in. are usual practice, whereas 100-ton rollers and compacted lift thicknesses of 12 to 18 in. are sometimes used. From 4 to 6 passes commonly achieve the required compaction. On large jobs, or on jobs at which unusual materials are encountered, the number of passes should be determined by means of field tests at the beginning of the job.

The surface of sheepsfoot rollers is covered with prismatic attachments or feet, one for approximately each 100 $in.^2$ of area of the roller. The rollers commonly used in connection with earth-dam construction have a diameter of about 5 ft and a length of about 6. When loaded they weigh about 17 tons. The feet extend a distance of at least 9 in. from the roller and have areas of 5 to 14 $in.^2$ Depending on the size of the feet, the contact pressure varies from about 300 to 600 lb/in.[2] Somewhat smaller and lighter rollers are used extensively for compaction of highway fills. With the ordinary equipment the thickness of the layers after compaction should not exceed about 6 in. The required number of passes should be determined in the field by means of tests on small experimental embankments. Satisfactory compaction is usually obtained after 6 passes of the roller (Turnbull and Shockley 1958).

Regardless of the type of compacting equipment or the degree of cohesion of the soil, the effectiveness of the compaction procedure depends to a large extent on the moisture content of the soil. This statement applies especially to almost nonplastic uniform fine-grained soils. If the water content is not almost exactly equal to the optimum, these soils cannot be compacted at all.

If a test embankment is constructed of a soil of uniform properties under carefully controlled conditions in the field, and if the layer thicknesses, type of compacting equipment and number of passes are all kept constant, it is found that the effectiveness of compaction depends only on the water content of the soil in the layer at the time compaction is started. The effectiveness of compaction is measured by the weight of solids per unit of volume, known as the *dry density*. The relation between dry density and placement water content has the characteristic shape shown by the solid curve *a* in Fig. 50.1. For the conditions of the test, the dry density at the peak of the curve is called the *maximum dry density* or the *dry density at 100% compaction*, and the corresponding water content is designated the *optimum moisture content*. Neither of these quantities is a property of the soil itself. If, for example, all conditions are kept unchanged except that a lighter roller is used, the value of the maximum dry density, as indicated by curve *b*, is lower and the optimum moisture content higher than for the heavier roller. An increase in the number of passes of the lighter roller may increase the maximum dry density, but even if it should reach a value comparable to that for curve *a*, the optimum moisture content corresponding to the new value of $\gamma_{d\,\max}$ would be likely to exceed that for the heavier roller. Similar changes in the moisture-density relations for a given soil accompany variations in thickness of layers and in type or weight of compaction equipment. Hence, the terms *100% compaction* and *optimum moisture content* for a given soil have specific meaning only in connection with a specific compaction procedure. Nevertheless, for any potential borrow material, it is essential to know in advance of construction whether, for the compaction procedure likely to be

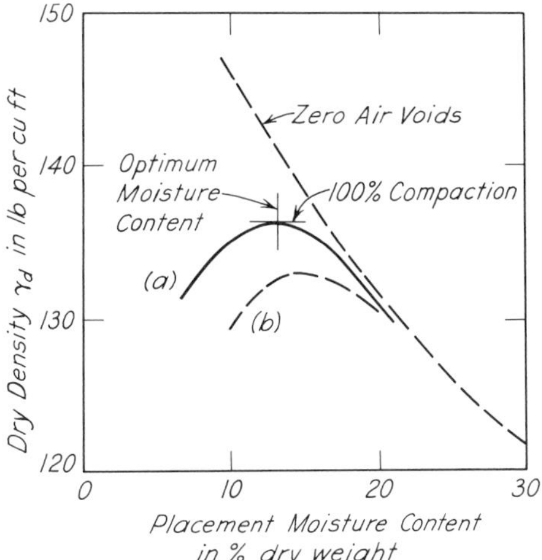

Fig. 50.1. Relation between dry density and placement moisture content for a particular soil (*a*) under a specific compaction procedure using a given roller. (*b*) Under the identical compaction procedure but with a lighter roller. Both curves approach zero-air-voids line representing relation for fully saturated soil.

specified, the moisture content in the field is excessive or deficient with respect to the optimum value for that procedure. Moreover, during placement of a fill the engineer must have the means for determining whether the specified compaction is being achieved even if the character of the borrow materials changes from time to time. These requirements have led to the development of laboratory compaction tests.

The purpose of every laboratory compaction test is to determine in the laboratory a moisture-density curve comparable to that for the same material when compacted in the field by means of the equipment and procedures likely to be used. Most of the current methods are derived from one developed by the California Highway Department in the early 1930's when compaction equipment was of relatively light weight. According to this procedure, now known as the *Standard Proctor test* (Proctor 1933, ASTM D-698-58T), a sample of soil is dried, pulverized, and separated into two fractions by a no. 4 sieve. About 6 lb of the finer fraction are moistened with a small quantity of water and thoroughly mixed to produce a damp aggregate that is then packed, in three equal layers, into a cylindrical container of specified dimensions. Each layer is compacted by 25 blows of a standard tamper that is allowed to fall from a height of 1 ft. When the cylinder has been filled and struck off level at the top, the weight and the water content of the damp soil in the container are determined. By means of these quantities, the dry density can be computed. In a similar manner the dry density is determined for successively damper mixtures of the aggregate, until the dry density after compaction decreases conspicuously with increasing water content. A curve is plotted showing the relation between dry density and water content. The optimum moisture content according to the Standard Proctor test is the value of water content at which the dry density is a maximum.

Because of the influence of the method of compaction on the moisture-density curve, no standard test of any kind, including the Proctor test, should be expected to produce results of general validity. Conclusive information on the optimum moisture content can be obtained only by making large-scale field tests with the compacting equipment to be used on the job.

For some time efforts have been made to develop laboratory methods that imitate the common types of field equipment more closely than does the Standard Proctor test. These efforts have led to various modifications of the original procedure. For the heavy equipment in modern use, especially in connection with the construction of earth dams or of aprons and taxiways for heavy aircraft, the Modified Proctor Test (ASTM D-1557-58T) is likely to be more appropriate. Various types of kneading compactors (Johnson and Sallberg 1962) have been developed that lead to more realistic moisture-density curves, but so far no such tests have been widely accepted.

Typical moisture–density curves for several soils are shown in Fig. 50.2. They were obtained by the Standard Proctor method. Curve *a* represents the moisture–density relation for a sand–clay mixture, *b* for a clay soil with low plasticity, *c* for a uniform coarse silt with low compressibility, and *d* for a clay with high plasticity.

If the moisture content of the soil in the field is greater than the optimum, the soil should be given an opportunity to dry out in storage or after spreading. If it is less, water should be added in the borrow pit or by sprinkling before compaction. With reasonable care it is usually possible to maintain the water content within 2 or 3% of the optimum. However, for uniform slightly cohesive nonplastic soils a closer approach to the optimum water content may be required.

The unit weight and the water content of the soil are checked in the field by routine sampling and testing. To determine the unit weight, a hole having a volume of at least 1/20 ft^3 is excavated in the compacted soil, and the excavated material is carefully recov-

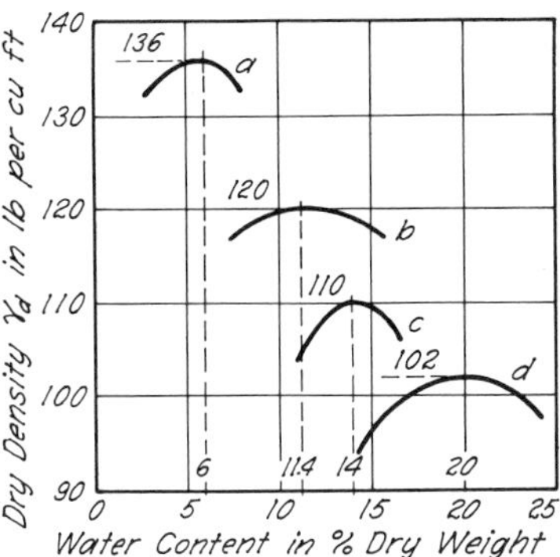

Fig. 50.2. Typical moisture-density curves for various soils. (*a*) Well graded sand with small percentage of clay. (*b*) Clay of low plasticity. (*c*) Inorganic nonplastic silt. (*d*) Clay of high plasticity.

ered and weighed before any moisture is lost by evaporation. The volume of the excavated material may be measured by one of several methods. In one of the older and more common procedures, the volume is measured by filling the hole with dry sand in a loose state, after the unit weight of the sand in this state has been established. The sand is poured from a container that is weighed before and after the hole is filled. According to a second procedure, a rubber balloon is placed beneath a horizontal cover and forced by the injection of water to accommodate its shape to that of the hole; the volume of the hole is determined by measuring the volume of water injected. An approximate value of the water content can be obtained quickly by determining the loss of weight due to drying the sample in a pan resting on a hot plate. However, after a moderate amount of experience on a given job, an inspector can estimate the water content quite accurately from the appearance and texture of the material. If the material to be used for a fill is quite variable in character, or if the job is located in a region subject to frequent rainy spells, compliance with the moisture-content requirements may considerably increase the cost of constructing the fill.

The water content at which a soil is compacted has an effect on all the physical properties of the compacted soil, including the permeability. Experience indicates that an increase in initial water content from a value somewhat below the optimum to a value somewhat above is likely to cause a large decrease in the coefficient of permeability. The decrease seems to increase with increasing clay content of the soil. In connection with the core material for Mud Mountain

Dam, which contained as much as 3% of clay with a high montmorillonite content, it was found that an increase in the water content from 2% below optimum to 2% above decreased the coefficient of permeability about 10,000 times (Cary et al. 1943). An influence of this magnitude is probably a rare exception, but even much less important effects deserve consideration.

Compaction of Clay

If the natural water content of a clay located in a borrow-pit area is not close to the optimum water content, it may be very difficult to change it to the optimum value. This is particularly true if the water content is too high. Therefore, the contractor may be compelled to use the clay in approximately the state in which it is encountered.

Excavating machinery removes the clay from a borrow pit in chunks. An individual chunk of clay cannot be compacted by any of the procedures previously mentioned, because neither vibration nor pressure of short duration produces more than an insignificant change in the water content. However, use of a sheepsfoot roller is effective in reducing the size of the open spaces between the chunks. The best results are obtained if the water content is slightly greater than the plastic limit. If it is considerably greater, the clay tends to stick to the roller, or the roller tends to sink into the ground. If it is considerably less, the chunks do not yield, and the spaces remain open.

Compaction of Natural Masses of Soil and of Existing Fills

Natural strata and existing fills cannot be compacted in layers. This fact excludes the application of most of the methods previously described, because in order to be effective the compacting agent must act in the interior of the soil mass. The method most suitable for a given job must be selected in accordance with the nature of the soil.

Cohesionless sand can be compacted most effectively by vibration. The simplest method for producing vibrations at considerable depth is to drive piles. When piles are driven into loose sand, the ground surface between the piles commonly subsides in spite of the displacement of the sand by the piles. In one instance, after cast-in-place concrete piles 45 ft long were driven at 3-ft centers into fine sand below the water table, the ground surface subsided as much as 3 ft, although the volume of the piles was equivalent to a 1-ft layer. The pile driving reduced the porosity of the sand from about 44 to about 38%.

Deep deposits of sand may also be compacted by *vibroflotation* (Steuermann 1939, D'Appolonia 1953). The tool that produces the compaction consists of a vibrator combined with a device for forcing water into the surrounding sand. The vibrator is first jetted into the sand to the depth within which the sand stratum is to be compacted and then is gradually lifted out again. The compaction is produced during the upward trip by the vibrations combined with the action of the water jets. The operation compacts the sand within a cylindrical space having a diameter of 8 to 10 ft at moderate expense. However, the method is most successful in clean sand. If the sand contains an admixture of silt or clay, the results are likely to be disappointing.

Satisfactory compaction of thick strata of very loose sand has also been accomplished by detonating small charges of dynamite at many points in the interior of the strata. The prerequisites for successful application of this method are the same as those for the vibroflotation process. In one such stratum, extending from the ground surface to a depth ranging between 15 and 30 ft, 8-lb charges of 60% dynamite were fired at a depth of 15 ft. The vibrations produced by the blasts reduced the porosity of the sand from its original value of 50 to 43% (Lyman 1942). At Karnafuli dam a large scour-hole having a volume of 60,000 yd^3 was filled by dumping into the water a uniform clean sand ($D_{10} = 0.18$ mm, $U = 2$) and compacting the sand by the explosion of a series of charges, usually of 8 lb each, at depths of 15, 33, and 50 ft below the surface of the sand. The holes were spaced 20 ft horizontally. The lowermost charges were fired first, followed at intervals of 4 hours by the middle and upper charges; a fourth series was then installed and fired at a depth of 25 ft. The porosity of the sand was reduced from about 47 to 41% (Hall 1962).

Sandy soils with some cohesion and existing cohesive fills can also be compacted by pile driving. However, the compaction of such soils is caused not by the vibrations associated with the driving, but by static pressure which decreases the size of the void spaces. If the soil is located above the water table and the voids are largely filled with air, the compacting effect of pile driving is commonly very satisfactory. However, if the soil is located below the water table, this effect decreases rapidly with decreasing permeability of the soil. In order to facilitate removal of the water, gravel drains may be installed. Thus, for instance, the following procedure was successfully used to compact a loose marl fill placed within the cells of a sheet-pile cofferdam (FitzHugh et al. 1947). Steel pipes with a diameter of 12 in. were driven into the fill. The lower end of each pipe was closed with a loosely attached steel disk that remained in the ground when the pipe was pulled. After a pipe had been driven to the base of the marl, it was filled with a mixture of gravel and sand and was provided with an air-tight cap. The pipe was then removed by pumping air into it under a pressure of 20 to 30 lb/in.^{2} The air pressure held the soft soil in place and prevented it from squeezing in while the gravel was falling out of the pipe into the hole. The consolidation of the surrounding soil was accelerated by bailing the water from the drains.

Compressible soils such as clays, loose silts, and most organic soils may also be compacted by *surcharging* or *preloading*. The area to be treated is covered by a fill having a weight per unit of area great enough to consolidate the soil sufficiently to increase its strength or reduce its compressibility to the required extent within the time available for the surcharging operation. Silty soils containing lenses or layers of sand are likely to consolidate almost as rapidly as the surcharge can be applied, but much longer times may be required for more impermeable soils. The rate of consolidation can be calculated by means of the theory in Article 25, but the estimates may be very unreliable because the spacing and degree of continuity of the more pervious drainage layers cannot usually be accurately assessed. If the estimated rate of consolidation is too slow, the natural drainage layers are often supplemented by the installation of *sand drains* similar to those described in the preceding paragraph. The drains commonly have diameters of at least 12 in. and are spaced in triangular or square patterns at 6 to as much as 15 ft. The required spacing may be calculated by theory, but the reliability of the conclusions is subject to the limitations of the knowledge of the actual permeability of the deposits in horizontal and vertical directions. The techniques for the installation of sand drains have been perfected to a high degree of efficiency (Carpenter and Barber 1953). Before the surcharge fill is placed above the area occupied by the drains, a pervious *drainage blanket* must be laid down to permit escape of the water emerging from the drains. Whether sand drains are present or not, the surcharge fill should not be built up so rapidly or with such steep slopes that a slide or base failure occurs; if drains are present, they are likely to become discontinuous and to be rendered ineffective. To avoid such slides, surcharge and sand-drain installations are provided with means for observing the settlement of the surface of the soil supporting the surcharge, the porewater pressures in the subsoil, and heave or lateral movement of the natural ground beyond the limits of the surcharge (Chapter 12).

Soft silt below the water table is transformed by pile driving into a semiliquid state. Hence, instead of inducing compaction, the process of pile driving weakens the soil at least temporarily. The compaction of such strata can be accomplished only by some process of drainage, by surcharging, or by a combination of the two.

Selected Reading

Compaction equipment, procedures, and control for earth dams are well described in Sherard, Woodward, Gizienski and Clevenger (1963): *Earth and earth-rock dams,* New York, John Wiley and Sons, 725 pp.

ART. 51 DESIGN OF FILLS AND DIKES

Principal Types of Earth Embankments

Earth embankments can be divided into three large groups: railway and highway fills, levees, and earth dams. The embankments in each group are similar not only in the purpose they serve, but also in the factors that should be considered when the side slopes are selected. In the following discussion of the choice of slopes it is assumed that the embankments rest on stable subsoil. The conditions for the stability of the base and the effect of unfavorable subsoil conditions on the stability of the embankments are discussed in Article 52.

Early Practice in Construction of Railway and Highway Fills

Until about the 1930's, railway fills were usually constructed by dumping the borrow material from timber trestles or over the end of the completed portion of the fill. Such fills were considered satisfactory if they were permanently stable. Since artificial compaction was not used, hard ballast was not placed beneath the track until the fills had "seasoned" for several years. During this period the fills settled under their own weight. The settlement amounted to about 3% of the height of rock fills, 4% of the height of fills of sandy materials, and about 8% of the height of fills with a considerable clay content. To prevent the development of sags in the track between the ends of a fill, the crest was customarily established at a distance above theoretical grade equal to the expected settlement.

The standard slope of railway fills constructed in this manner was 1.5 (horizontal) to 1 (vertical). However, it was noticed that fills with heights greater than 10 or 15 ft were likely to fail either during construction or after a few wet seasons if they contained a high percentage of clay. Therefore, it became the practice to reduce the slope angle of such fills from 1.5:1 at the crest to about 3:1 at the base. The decision whether the character of the clay required flattening of the slopes was commonly left to the engineer in charge of construction. However, even the most experienced engineers occasionally misjudged the character of the soil and, as a consequence, sections of the fills failed. The slopes were then repaired and their stability increased either by constructing low fills along the toes that moved out or by means of dry masonry footwalls, possibly supplemented by hard core drains.

Under the influence of increasing and heavier traffic, the upper portions of fills consisting of clay or silty clay often became soft, especially in the presence of water, whereupon the ballast tended to work downward into the fill and form ballast pockets in the shape of shallow troughs in which water collected and further softened the subgrade. The progressive deterioration of the roadbed led to costly maintenance. In the 1940's various methods of stabilization were attempted; among the most successful was the injection of sand and cement in the form of grout into the lower part of the ballast section (Smith and Peck 1955).

The earliest highway fills were similarly constructed by loose dumping over the ends of the completed portions. The standard slopes varied in different parts of the United States from 1.5:1 to 1.75:1. Differential settlements of the fills led to unsatisfactory riding characteristics and, frequently, to failure of the road surfaces. Unlike railroad tracks, which could be raised routinely by tamping additional ballast beneath the ties, highway pavements could be leveled only by resurfacing or, to some extent beneath concrete pavements, by mud-jacking. Moreover, no time was ordinarily available for seasoning of a highway fill before it was made available to high-speed traffic. Hence, within a decade after the introduction of the modern "hard road," spreading in layers and compaction by hauling equipment and rollers was becoming common practice.

It was recognized that the behavior of fills compacted in this manner depended primarily on the physical properties of the fill material. Consequently, efforts were made by the various state highway departments and the Bureau of Public Roads to correlate the behavior of compacted embankments with the index properties of the fill material. These efforts led to a generally accepted practice of judging the quality of the soil on the basis of its Atterberg-limit values and its maximum compacted dry density as determined by the Standard Proctor test or its local equivalent. Requirements such as those in Table 51.1

Table 51.1
Embankment Soil Compaction Requirements

Abstract of 1946 Construction and Material Specifications of the Department of Highways, State of Ohio

Condition I		Condition II	
Fills 10 ft or less in height, and not subject to extensive floods		Fills exceeding 10 ft in height, or subject to long periods of flooding	
Maximum Laboratory Dry Weight* (lb/ft³)	Minimum Field Compaction Requirements (Per cent of laboratory maximum dry weight)	Maximum Laboratory Dry Weight* (lb/ft³)	Minimum Field Compaction Requirements (Per cent of laboratory maximum dry weight)
89.9 and less	†	94.9 and less	‡
90.0–102.9	100	95.0–102.9	102
103.0–109.9	98	103.0–109.9	100
110.0–119.9	95	110.0–119.9	98
120.0 and more	90	120.0 and more	95

* Maximum laboratory dry weight is determined by standard Proctor test, as described in Article 50.

† Soils having maximum dry weights of less than 90.0 lb/ft³ will be considered unsatisfactory and shall not be used in embankment.

‡ Soils having maximum dry weights of less than 95.0 lb/ft³ will be considered unsatisfactory and shall not be used in embankment under condition II requirements or in the top 8-in. layer of embankment which will make up the subgrade for a pavement or subbase under condition I requirements.

Soil, in addition to the above requirements, shall have a liquid limit not to exceed 65, and the minimum plasticity index number of soil with liquid limits between 35 and 65 shall not be less than that determined by the formula 0.6 liquid limit minus 9.0.

were adopted in conformity with experience in some localities and were, unfortunately, often copied in other localities without the benefit of similar experience. Restriction of the water content to values within a few per cent of the optimum moisture content was rarely specified or required, provided the specified minimum percentage of the maximum Proctor dry density was achieved.

Modern Practice for Railway and Highway Fills

Inasmuch as most new railroad fills in industrially developed countries are made in connection with improvements in alignment or gradient to permit higher speeds or heavier loads, there is no longer any essential difference in construction procedures for railway or highway fills. Where possible, stable granular materials are used, but economy commonly dictates the placement of the closest available materials, regardless of their composition, unless they contain highly compressible organic constituents. Under most circumstances the fill is spread by bulldozers in layers about one foot thick and is compacted by rollers until a specified dry density is achieved. Moisture-content control is rarely specified. Standard slopes remain as steep as 1.5:1 for granular materials, and for cohesive soils vary from about 2:1 for fills 10 ft high to as flat as 3:1 for fills 100 ft high.

This procedure has led to generally satisfactory results if the water content of the material in the borrow pit does not exceed the appropriate Standard Proctor optimum value by more than a few per cent. If the borrow material is too wet, serious difficulties and delays may be encountered. Hence, the most important aspect of the soil survey for fine-grained fill materials is to ascertain the relation between the field moisture content and the optimum value. This information should be supplemented by determination of the liquid and plastic limits which serve as a basis for judging the likelihood that the borrow material can be dried to the optimum moisture content under the prevailing climatic conditions.

If the moisture content is excessive and the climate too humid to permit effective drying, no amount of compactive effort can satisfy a requirement in the specifications that 90 or 95% of a standard maximum dry density must be achieved. Under these circumstances, the engineer must investigate the strength that will be developed by the fill material after placement at its natural moisture content and sufficient manipulation by hauling and compaction equipment to eliminate any large voids. Ordinarily only light equipment can work satisfactorily on such a fill. The engineer must then select the slopes of the fill to provide an appropriate factor of safety against exceeding the strength of the material as placed. In many instances, satisfactory fills have been built at moisture contents such that no more than 50 to 70% of Standard Proctor density can possibly be achieved (Jimenez-Quiñones 1963).

In the humid tropics, the degree of saturation of residual soils is often close to 100% and reduction of the moisture content is imprac-

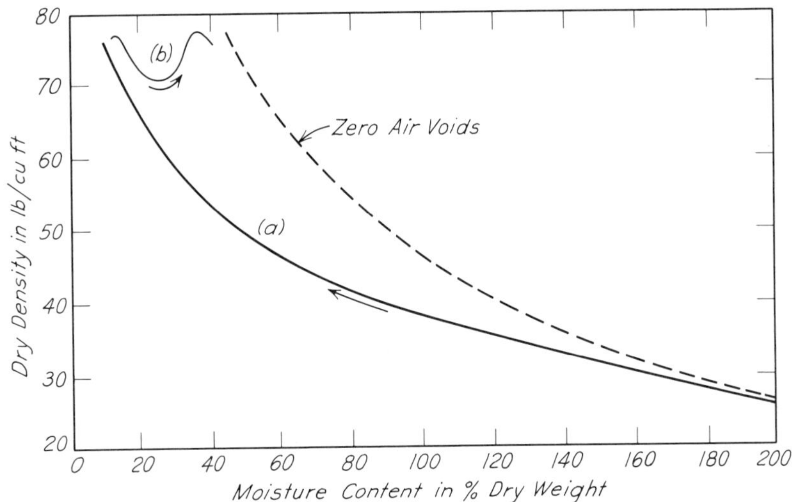

Fig. 51.1. Moisture-density curves obtained by Standard Proctor procedure for Peepeekeo ash from Hawaii. (*a*) Curve obtained by drying each of successive samples from natural water content to moisture content at which sample is compacted. (*b*) Curve obtained if soil is dried to $w = 10\%$, whereupon water is added and samples compacted in usual manner for performing Proctor test (after Willis 1946).

ticable. Furthermore, if the soils are dried for the purpose of making compaction tests, the characteristics of the soils may undergo drastic irreversible changes and the laboratory moisture-density curve may have no relation to the field conditions.

An extreme example of the error that may be introduced by drying the soil before performing the compaction test is shown in Fig. 51.1, which refers to the Peepeekeo ash, a weathered volcanic soil occurring in the most humid part of the island of Hawaii (Willis 1946). The liquid limit of the ash is about 240% and the plastic limit about 130%. Air-drying renders the soil nonplastic. The natural water content is close to 200%. If a sample is compacted according to the Standard Proctor procedure at this moisture content, the dry density is about 25 lb/ft^3. If each of a series of samples is allowed to dry to a different water content and then compacted, the relation shown by curve *a* is obtained. However, if the soil is dried to a moisture content of 10% and a moisture-density curve obtained in the usual manner by adding moisture to the dried sample, curve *b* is obtained. It exhibits a peak at a moisture content of 35%, corresponding to a maximum dry density of 77 lb/ft^3. Attempts to dry the material in the field to the optimum moisture content determined in this manner are futile. Moreover, use of a roller for compaction turns the soil into a fluid that flows from the embankment. Yet, it has been found that this unpromising material can be used to construct stable highway fills up to 90 ft high with slopes of 1.5:1 by placing it as gently as possible in 4-ft layers with a very light bulldozer (Hirashima 1948). Although the behavior of the Peepeekeo ash is unusual, similar experiences have been observed to a lesser degree in other tropically weathered soils containing hydrated oxides of iron and alumina, or the clay mineral halloysite (Terzaghi 1958*b*, Jiminez-Quiñones 1963). A marked drop in the liquid limit as a consequence of air-drying constitutes grounds for suspicion.

The use of fairly stiff clay for the construction of an embankment may involve the danger of subsequent expansion by swelling in contact with water. If the expansion is unequal, fissures are likely to form, whereupon the structure of the clay may disintegrate and the slopes may start to slough.

The extent of swelling in a fill depends upon the inherent swelling capacity of the material and upon such factors as the moisture content at which the fill is placed, the degree and method of compaction, and the pressure due to the weight of the overlying part of the embankment. The combined effect of these influences on a fill composed of rolled chunks of a stiff clay can be investigated by preparing samples representative of the clay in its initial state in the fill. Each sample is introduced into a consolidation ring and subjected to a pressure equal to that which will act on the clay at some point in the fill. Water is then admitted to the porous stones that cover the top and bottom of the sample, and the increase in volume is measured. The suitability of the material is judged by its tendency to swell. If the increase in volume under the anticipated vertical pressure in the fill exceeds about 5%, the fill is likely to perform unsatisfactorily.

The expense of carrying out swelling tests and the uncertainties involved in interpreting the results justify attempts to identify those borrow materials that exhibit excessive inherent swelling capacity. In a general way, it has been found that the inherent potential to swell depends on the plasticity index (Holtz and Gibbs 1956*a*, Seed et al. 1962), as indicated in Table 51.2.

If a clay with high to very high inherent swelling capacity must be used, the effects of swelling may be minimized by placing the clay at the highest practicable moisture content and by utilizing any available non-swelling materials in the outer portions of the fill. The weight of even a few feet of surcharge over a swelling material substantially reduces the amount of expansion and consequent loss of strength of the clay.

Levees or Dikes

Levees serve to protect lowlands against periodic inundation by high water, storm floods, or high tides. They differ from earth storage dams in three principal respects: Their inner slopes are submerged during only a few days or weeks per year; their location is determined by flood-protection requirements regardless of whether or not the foundation conditions are favorable; and the fill material must be derived from shallow borrow pits located near the site of the levees. These

Table 51.2
Approximate Relation Between Plasticity Index and Inherent Swelling Capacity

Plasticity Index	Inherent Swelling Capacity
0–15	Low
10–35	Medium
20–55	High
35–	Very high

(After Seed et al. 1962)

conditions introduce a considerable element of uncertainty into the design of such structures. Nevertheless, the necessity for levees existed in some regions during the earliest days of human civilization, and as a consequence the art of levee construction was developed in these regions to a high degree of perfection.

If the soil conditions in the borrow-pit area change from place to place, the cross section of a levee is customarily chosen in accordance with the requirements of the most unsuitable materials that will have to be used. Consideration is also given to the degree of freedom permitted the contractor in choosing the time and method of construction. In some levee districts the method of placing the soil is rigidly controlled, whereas in others the contractor is free to choose between widely different methods of construction. The influence of the method of construction on the cost of a levee depends chiefly on the ratio between the cost of hand and machine labor. Since this ratio is very different in different countries, efforts to build satisfactory levees at minimum expense have led to different rules in different parts of the world. For example, before World War II in countries such as Germany and Holland, where hand labor was cheap, levees were carefully compacted and built with steep slopes. On the other hand, in the Mississippi Valley no efforts were made to compact levees, because uncompacted levees with gentle slopes were commonly cheaper than carefully compacted ones with much smaller cross sections (Buchanan 1938). In Europe and Asia many levees of clay were constructed with side slopes of 2:1, whereas clay levees along the Mississippi River were commonly given an inner slope of 3:1 and an outer slope of 6:1. Both types of construction grew out of a slow process of trial and error, and both served their purpose equally well under the conditions that prevailed in the regions where they originated.

Even in the United States, however, in areas of high land and property values, steep slopes may be economically justified. While levees along the lower Mississippi River were being constructed with flat slopes, those in the industrialized Ohio River valley were much steeper. This trend has become more pronounced in recent years, and has justified increased use of theoretical methods in the design of levees even in regions where levee systems are already in existence. On the other hand, if in such regions economic factors have not changed significantly, soil mechanics can be used to advantage only for correlating construction and maintenance experience with the index properties of the soils that serve as the construction materials. The information obtained in this manner leads to the elimination of guesswork in classifying the soils encountered in new borrow-pit areas.

The use of theoretical methods in the design of levees on stable subsoil can also be justified if a levee is to be located in a region where few levees have previously been built. Under such circumstances the method of trial and error is too slow and expensive, and experience based on existing levee systems can hardly be used as a guide, because very few of the available construction records contain adequate data concerning the properties of the construction materials. Therefore, the designer is compelled to use the methods practiced in connection with the design of earth dams.

The influence of the subsoil conditions on the stability of levees and other earth embankments is discussed in Article 52.

Selected Reading

Casagrande, A. (1949). "Soil mechanics in the design and construction of the Logan airport," *J. Boston Soc. Civil Engrs,* **36,** No. 2, pp. 192–221. Reprinted in *Contributions to soil mechanics 1941–1953,* Boston Soc. Civil Engrs. (1953), pp. 176–205. Hydraulically placed clay fill.

AREA (1955). "Soil engineering in railroad construction," *Proc. Am. Rwy. Eng. Assn.*, **56**, pp. 694–702. Railroad embankment design to avoid common shortcomings.

ART. 52 STABILITY OF BASE OF EMBANKMENTS

Types of Base Failure

Whenever possible, embankments and earth dams are constructed on firm relatively incompressible subsoils. However, in many regions railway or highway embankments must be built on broad swampy flats or buried valleys filled with soft silt or clay. Levees must be constructed near the flood channels, irrespective of subsoil conditions. Even earth dams must occasionally be located at sites underlain by undesirable materials. In all these instances the design of the embankment must be adapted not only to the character of the available fill material, but also to the subsoil conditions.

Base failures may occur in several different ways. The fill may sink bodily into the supporting soil. Such an accident is referred to as *failure by sinking*. On the other hand, the fill together with the layer of soil on which it rests may spread on an underlying stratum of exceptionally soft clay or on partings of sand or silt containing water under pressure (Article 49 and Fig. 49.11*b*). This is known as *failure by spreading*. If the embankment retains a body of water, it may also fail by *piping*, as a consequence of backward erosion from springs that emerge from the ground near the toe of the fill. Finally, base failures may occur beneath fills located above strata of very loose sand because of liquefaction of the sand. This type of failure has rarely occurred except during major earthquakes (Ambraseys 1960). The likelihood of liquefaction beneath fills of moderate height can be considerably reduced by compacting the sand by one of the methods described in Article 50. Failure by piping is discussed independently in Article 63. Hence, the present article deals only with base failures by sinking or spreading.

Methods for Investigating Stability

The design of a fill to be constructed above clay strata should always be preceded by a thorough soil exploration involving boring, sampling, and testing. The results of the exploration inform the designer about the soil profile and the physical properties of the subsoil. The next step is to compute the factor of safety of the fill with respect to a failure of its base (Article 35). Under normal conditions, the foundation conditions are not considered satisfactory unless the factor of safety with respect to a base failure during or immediately after construction is at least equal to 1.5.

The conditions for the stability of the base of fills and the methods for preventing base failures are discussed in the following sequence: fills on very soft or marshy ground, fills on thick strata of fairly homogeneous soft clay, fills on stratified ground containing fairly homogeneous layers of soft clay, and fills on clay containing sand or silt partings. Subsoil conditions of the first two types are likely to be associated with failures by sinking, and those of the last two with failures by spreading.

Fills on Very Soft Organic Silt or Clay

Natural deposits of this type are common in regions formerly occupied by shallow lakes or lagoons. The fringes of such shallow deposits are likely to support growths of peat moss or other types of marsh vegetation. The silt or clay brought into the lakes in suspension intermingles with the decayed organic constituents washed in from the fringes. Hence, the fine-grained sediments in such bodies of water are likely to have high organic content. The natural void ratio of the sediments is very commonly greater than 2. The deposits may contain layers of peat or be buried beneath a bed of peat.

If the surface of such a deposit has never before carried an overburden, the subsoil is likely to be unable to sustain the weight of a fill more than a few feet in height. In many regions soft marshy ground is covered with a mat several feet thick that is stiffer than the deeper layers and is effectively reinforced by a dense network of roots. The mat acts like a raft and may be able to carry, at least temporarily, the weight of a fill several feet high. However, fills on such foundations continue to settle excessively for many years or decades, and maintenance records show that they may even suddenly break through the mat long after construction. Hence, if a fill is to be permanent, the continuity of the mat should be destroyed before the fill is placed, to facilitate the penetration of the fill material into the softer layers.

Inasmuch as the cost and the relative merits of the various methods for constructing fills across marshy flats depend on the depth of the soft stratum, construction should be preceded by the preparation of a contour map of the firm bottom. If the thickness of the soft stratum does not exceed 5 or 6 ft, it may be economical to excavate the soft material. If the thickness is greater, it is usually preferable to permit the sinking fill to displace the soft material. This procedure is known as the *displacement method*.

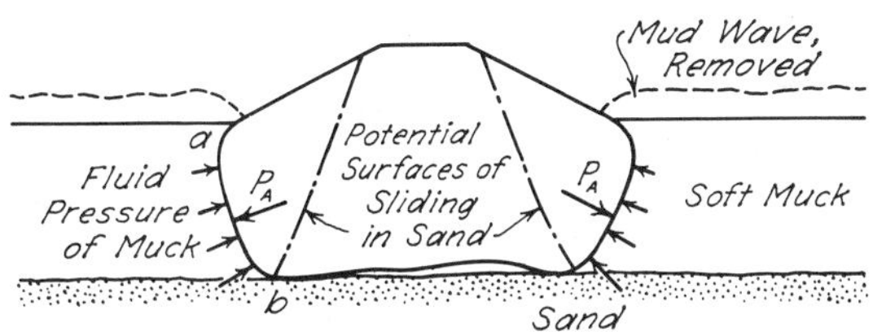

Fig. 52.1. Diagram showing forces that act on soil adjoining buried part of a fill constructed by the displacement method.

To accelerate the penetration of the fill material and to shorten the subsequent period of settlement, the fill may be built up to a height of 15 or 20 ft above final grade, and the excess material removed later. As an alternative the penetration of the fill material may be facilitated by blasting the soft subsoil. If the position of the bottom of the soft stratum is known, the quantity of material required for constructing the fill can be estimated fairly accurately in advance of construction.

The conditions for the equilibrium of a fill with its base established by displacement are illustrated by Fig. 52.1. The contact face *ab* is acted on by the active earth pressure exerted by the fill material. The displacement of *ab* toward the left is resisted by the sum of the liquid pressure of the soft material and the force required to overcome its cohesion. If the penetration of the fill is aided by a temporary surcharge or by blasting, the force that produces the corresponding displacement is very much greater than the force that acts on *ab* after construction. Furthermore, after the fill is completed, the soft material regains a part of the strength lost during the process of displacement (Article 4). Therefore, if the fill has a cross section similar to that shown in Fig. 52.1, the progressive settlement of its crest is likely to become inconsequential shortly after construction.

An outstanding example of the successful application of the displacement method is the Kiel canal, built during the years 1887–1895. For a distance of about 12 miles the canal had to be established on a layer of peat and very soft organic clay with a thickness up to 30 ft. In some sections the soil was so soft that a man could not walk on it. The method of building the canal in these sections is illustrated by Fig. 52.2. On the inner side of the center line of each of the future dikes sand fills were constructed as indicated by the dash line. These fills displaced the soft material over a broad belt almost down to firm ground. They served as a base for the dikes and formed the uppermost part of the slopes of the finished canal.

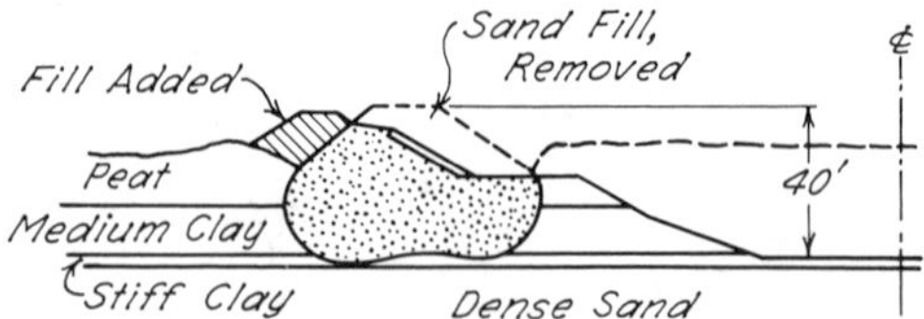

Fig. 52.2. Typical cross section of Kiel Canal (after Fülscher 1898).

To reduce the danger of slides during construction, excavation was not started until 6 months after the fill was placed. Nevertheless, slides did occur at a few points.

One of the slides is illustrated in Fig. 52.3. This figure represents four successive stages in the excavation of the canal. The second stage *b* was followed by slides, stage *c*, during which the sand fills moved toward the center line of the canal. In order to finish construction, it was necessary to dump still more sand (stage *d*), whereupon the excavation was completed without further mishap (Fülscher 1898).

In connection with the construction of railway and highway fills, the displacement method has been reduced to a routine procedure. It was even proposed as a means for the construction of a rock-fill

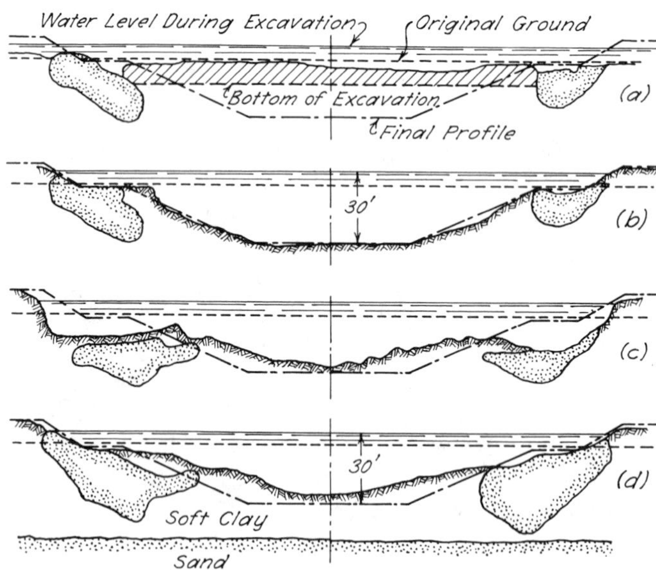

Fig. 52.3. Four successive stages in the excavation of the Kiel Canal in exceptionally soft ground (after Fülscher 1898).

dam with a height of about 100 ft across Cobsock Bay on the Passamaquoddy project in Maine (Hough 1938).

Soft strata with relatively high average permeability in horizontal directions can be made suitable for supporting the weight of superimposed fills by the use of sand drains, possibly supplemented with surcharge fills (Article 50).

Fills on Soft Homogeneous Clay

In the following discussion, it is assumed that the surface of the clay is very close to the base of the fill, that the thickness of the clay stratum is at least half as great as the base width of the fill, and that the stratum is fairly homogeneous.

The failure of a fill on such a base has the general character of a base failure along a midpoint circle (Article 35). However, the uppermost part of the surface of sliding is located within the artificial fill, and the shearing resistance per unit of area along this part is different from that along the lower part. The first step in making a stability computation is to ascertain the average undrained shearing resistance s along the lower part, on the basis of a strength survey of the clay stratum. The second step is to determine the average shearing resistance s_2 along the part of the surface of sliding located within the fill. This shearing resistance may consist of cohesion and friction or of friction alone. In the stability analysis the real fill is replaced by an ideal clay ($\phi = 0°$) that has a cohesion equal to s_2. As a first approximation it is assumed that failure occurs along a midpoint circle; however, the real critical circle must be determined by trial and error. Because of progressive failure, the average shearing strength along the surface of sliding may be less than the weighted average of the peak strengths s and s_2 (Article 16).

It is commonly required that the factor of safety with respect to a base failure should be at least 1.5. Considering the unavoidable errors in estimating the average shearing resistance of the clay, this factor is very low. Nevertheless, in order to satisfy the requirement, high fills on soft clay must be provided with very gentle slopes. Hence, if a high fill is also very long, it may be economical to reduce the factor of safety still further, to 1.2 or 1.1, and to rely on the results of observations during construction to detect impending slides and to prevent the slides by local modifications in the design.

The failure of the base of an embankment on clay is commonly preceded by the gradual lateral displacement of the material beneath the toes and by the gradual heave of broad belts located one on each side of the fill. The rate of displacement increases as failure

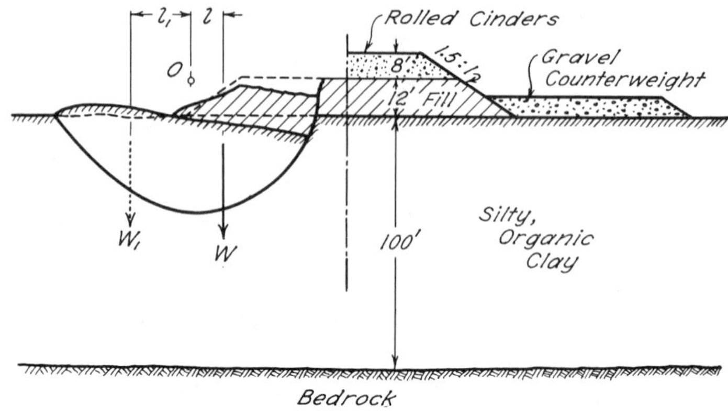

Fig. 52.4. Section through gravel fill on deposit of uniform soft clay. Left side shows principal features of failure during construction; right side shows reconstructed fill stabilized by means of gravel counterweight (after Gottstein 1936).

is approached. If the movement is detected in its initial state by means of suitable observations (Chapter 12), the failure can be prevented by covering these areas with a thick layer of fill.

Slides caused by the failure of a soft clay base generally occur during or immediately after construction, because thereafter the strength of the base gradually increases on account of consolidation. If a slide has already occurred, it is usually possible to ascertain the position of the surface of sliding by means of test shafts or slope indicators and to compute the average shearing resistance of the clay with considerable accuracy. The value thus obtained serves as a basis for redesign. Figure 52.4 illustrates the procedure. It shows a section through a highway fill of well-compacted gravel placed on a deposit of organic silty clay (Gottstein 1936). Failure occurred when the top of the fill was 8 ft below finished grade. A mass of soil, having an effective weight W (Article 12), failed by rotating about the point O. The driving moment was Wl. To complete the fill a gravel counterweight W_1 was added with its line of action at a horizontal distance l_1 from O. The counterweight was given dimensions such that its moment W_1l_1, plus that due to the shearing resistance along the surface of sliding, exceeded the driving moment of the complete fill by 50 per cent. The right side of Fig. 52.4 shows a section through the finished fill. The upper 8 ft were made of rolled cinders to keep the weight of the fill as small as possible. After construction of the counterweight there was no movement other than a slight subsidence due to consolidation of the base.

After a fill has been successfully constructed on the surface of a homogeneous mass of clay, its base gradually settles on account of the consolidation of the underlying clay. The magnitude of the settlement can become very great. It should be estimated by means of the procedure outlined in Article 41, and the crest of the fill established a corresponding distance above final grade. As consolidation progresses, the bearing capacity of the fill increases.

Observations on rock fills serving as bases for breakwaters suggest that the settlement of such fills depends not only on the properties of the underlying clay but also to a large extent on the method of construction. In the last century the fills were made by dumping large rocks into the water. This procedure completely destroyed the structure of the uppermost layer of clay and caused great local stress concentrations in the underlying material. The settlement of these fills was very large. The older part of the breakwater in the harbor of Spezia, Italy, is an example. Figure 52.5*a* is a section through the fill. The depth of water was 33 ft, and the water content of the soft clay was close to 100%. The results of load tests indicated that the deeper layers had an unconfined compressive strength of about 0.5 ton/ft^2. Construction was started in 1862. To maintain the crest of the fill at approximately constant elevation in spite of the rapid settlement, it was necessary to add new material to the fill. This in turn accelerated the settlement. During a period of 50 years the material that had to be added was equivalent to a layer 60 ft thick. As subsidence increased, the base of the fill assumed the shape shown in Fig. 52.5*a*.

In 1912 construction of a new section of the breakwater was started. To prevent excessive settlement of the new section, the mud was re-

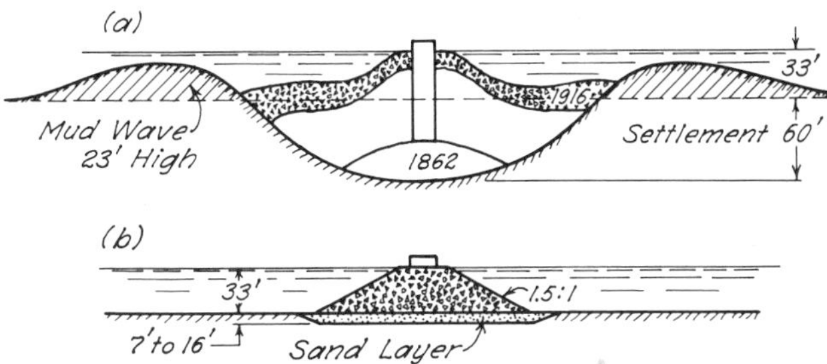

Fig. 52.5. Rock fill breakwater on soft clay in harbor of Spezia, Italy, constructed (*a*) By dumping rock directly onto clay. (*b*) By dumping rock onto sand layer which was previously deposited in shallow dredged cut (after Barberis 1935).

moved by dredging to a depth between 7 and 16 ft below its original surface and replaced by sand with a grain size between 0.2 and 0.4 mm (Fig. 52.5*b*). Hence, when the fill was constructed, the rocks came to rest on the sand instead of penetrating into the clay, and no local stress concentrations were established in the clay. Probably as a result of this condition, the settlement of the new fill was insignificant compared to that of the old one. At the end of the construction period the settlement was 1.7 ft; 9 years later it had reached only 2.7 ft. Similar procedures were successfully used in constructing breakwaters in the harbors of Valparaiso, Chile, and Kobe, Japan (Barberis 1935).

Varieties of Failure by Spreading

Failures by spreading have been observed only in connection with fills located above stratified deposits that contain layers of soft clay. Such fills are commonly safe with respect to sinking, but they may fail by spreading.

During the last 40 years half a dozen major and several minor dam failures have occurred in this manner. Hence, the stability of fills above clay strata deserves special attention. Outstanding failures due to spreading are those of Lafayette Dam in California in 1928 (ENR 1929), Marshall Creek Dam in Kansas (ENR 1937), and the Hartford flood-protection dike in Connecticut (ENR 1941).

A study of the records reveals two different types of failure by spreading. One type is distinguished by a relatively slow subsidence of the crest of the fill. The slope, if originally plane, assumes a gentle S-shape, as shown in Fig. 49.11*a*, and the heave of the ground surface extends only a short distance beyond the foot of the slope. The failure of Chingford Dam near London, England (Cooling and Golder 1942), and of Lafayette Dam are instructive examples of this type. Failures of the other type occur very rapidly, and the heave extends to a great distance from the foot of the slope. During the failure of Lafayette Dam, which was 120 ft high, the crest subsided 15 ft in about 3 days, over a length of about 500 ft. The toe moved out about 20 ft, and the heave was confined to a short distance from the foot of the slope. On the other hand, the Hartford dike, only 30 ft high, failed in less than 1 min. The crest subsided 15 ft over a length of more than 1000 ft. A row of sheet piles at the foot of the slope moved laterally for 60 ft, and the heave extended about 150 ft from the foot.

Analysis of case records and study of the causes of failure have demonstrated that the catastrophic rapid type of failure does not occur unless the clay stratum contains continuous layers or extensive lenses of coarse silt or sand. Therefore, the details of stratification of the clay layer are of decisive importance, and a distinction must be made between clay strata with and without highly permeable partings. In the following discussion we shall first examine the causes of failure in each of the two types of clay strata and subsequently consider methods for improving the stability of fills located above such strata.

Spreading of Fills above Fairly Homogeneous Layers of Soft Clay

The clay stratum below the fill shown in Fig. 52.6*a* is assumed to be perfectly homogeneous. Shortly after filling starts, the clay begins to consolidate and the stratum becomes stiffer near its upper and lower boundaries. Near midheight, however, the weight of the fill is still carried by excess hydrostatic pressure, indicated by the piezometric levels shown on the left side of the figure. In this part of the stratum the shearing resistance of the clay remains equal to its initial value. Hence, if failure occurs, the surface of sliding follows some layer of minimum shearing resistance located near midheight

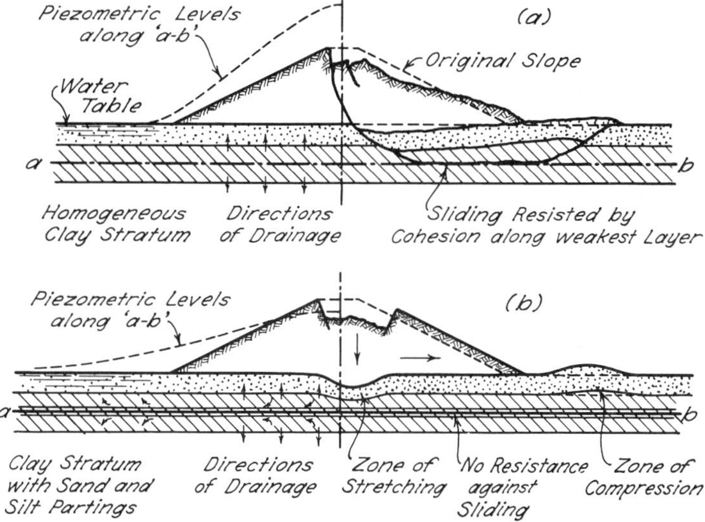

Fig. 52.6. Type of failure of base of fill containing thin clay stratum (*a*) If clay stratum contains no horizontal pervious partings. (*b*) If clay stratum contains pervious sand or silt parting.

of the clay stratum. In order to estimate the value of the minimum shearing resistance, a survey of the undrained shearing strength must be made (Article 45). Since the strength of the clay is likely to vary in both the horizontal and vertical directions, the selection of a representative value requires experience, mature judgment, and a thorough investigation of the character of the stratification of the clay bed. It is also essential to make sure that the clay actually does not contain any continuous seams of sand or silt.

After the appropriate value for the shearing resistance has been selected, the factor of safety with respect to sliding can be evaluated by the method described in Article 35 in connection with composite surfaces of sliding. Since there is an appreciable resistance along the horizontal portion of the surface of sliding, the slope assumes the characteristic S-shape shown in Fig. 52.6*a*.

Spreading of Fills above Clay Strata with Sand or Silt Partings

If the clay contains fairly continuous seams of sand or silt, the excess water from the central part of the stratum drains not only vertically through the top and bottom of the stratum, but also horizontally through the highly permeable seams as shown in Fig. 52.6*b*. Therefore, the seams become the seat of high excess hydrostatic pressure. The difference between the excess porewater pressure and the weight of the overlying soil and fill is greatest near the toes of the slopes. In these regions the shearing resistance of the cohesionless seams is likely to be reduced to zero, and the only resistance to the spreading of the fill is offered by the passive pressure of the earth located above and beyond the surface of sliding. If this pressure is exceeded, the outer parts of the fill move bodily away from the center, and the central part subsides leaving a troughlike depression, as indicated in Fig. 52.6*b*. Since soil conditions are never exactly symmetrical with respect to the center line of the fill, failure occurs on one side only, but it is hardly possible to predict on which side. The troughlike subsidence characteristic of this type of failure has been observed repeatedly.

The factor of safety against sliding depends on the distribution of the excess hydrostatic pressure within the pervious seams, which in turn depends on unknown local variations in the permeability and on other unknown geologic details. The practical implications of these uncertainties are illustrated by Fig. 52.7. Test borings were made along the center line of the proposed fill shown in the figure. Since no pervious seams were encountered in any of the holes, the designers assumed that during construction the hydrostatic conditions would

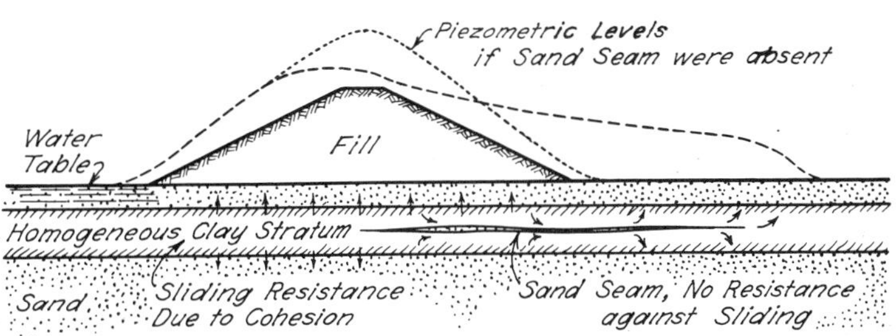

Fig. 52.7. Diagram showing effect on hydrostatic pressure conditions of pervious seam in clay stratum below fill.

be as indicated by the dotted piezometric line. These conditions are normal and do not compromise the stability of the base of a fill. In reality, the clay contained a seam of fine sand located beneath the right-hand section of the dam. Since hydrostatic pressures are freely transmitted through such seams, the real pressure conditions assumed the character indicated by the dash line, and the fill failed as shown in Fig. 52.6*b*.

Hence, if the geology of the stratum indicates that the clay may contain highly permeable seams, the risk of failure can be avoided only by providing the fill with very flat slopes, at the cost of excessive yardage, or else by using one of the construction expedients described in the following paragraphs.

Means for Increasing Stability of Fills above Thin Strata of Soft Clay

If the bottom of the soft clay stratum is located at a depth less than 5 or 6 ft below the ground surface, it is advisable to remove the clay over the full width of the base of the fill. Otherwise, the designer may choose between two alternatives. He may specify that the fill be constructed more slowly than the rate of consolidation of the clay at midheight of the stratum, or he may make provisions to accelerate the process of consolidation by means of sand drains. Each of these methods deserves consideration, regardless of whether or not the clay strata contain thin permeable seams.

To use the first procedure, the designer must know the rate of consolidation of the inner part of the stratum. Computation alone should not be relied on, because the result may be invalidated by some unnoticed geological detail such as the presence of highly colloidal seams. It should be used only to make a preliminary estimate of the maximum rate at which the fill can be constructed. To eliminate the risk of failure, the progress of consolidation must be observed in the field during construction by means of porewater pressure gages, and the rate of construction must be adapted to the findings. This is a decided disadvantage, because it does not eliminate the possibility that construction may be intolerably delayed.

If the results of computation show that the normal process of consolidation is too slow to be utilized as a means for strengthening the base of the fill, acceleration of the process by means of vertical sand drains should be considered. The procedure has already been described.

Summary

On very soft ground high fills can be established by one of two methods. The first consists of displacing the soft ground by the weight of the fill. To avoid excessive settlement after construction, the fill should be built up to a height of 15 or 20 ft above final grade, and the excess material removed after the fill has subsided. The second method consists of accelerating consolidation by means of sand drains extending to the bottom of the stratum. The drains discharge the water into drainage conduits located at the base of the fill. To determine the most economical procedure, it is necessary to prepare a contour map of the firm base of the soft layer. Wherever the depth of the layer is less than 5 or 6 ft, it may be advantageous to remove the soft soil by excavation.

The design of fills to be built above thick beds of soft clay should be preceded by a stability computation. Under normal conditions a factor of safety of 1.5 with respect to sinking should be specified. However, if the fill is very long, it may be more economical to base the design on a factor of safety of 1.2 or 1.1, to locate the weakest spots in the subsoil by means of heave observations during construction, and to cover the heave areas by a counterweight consisting of a thick layer of fill.

Special vigilance is required if a fill is to be built on stratified soil containing layers of soft clay. A number of catastrophic accidents have occurred because the stability of subsoils of this type has been overestimated. If the clay strata contain no sand or silt partings, the resistance against spreading depends on the average shearing resistance of the weakest layers in the subsoil. Since exceptionally weak layers are not necessarily continuous, their presence may escape the attention of even a conscientious investigator. If the clay contains sand or silt partings, the resistance against spreading depends chiefly on the porewater pressure in the partings. This pressure changes during construction, and an accurate forecast of its magnitude is impracticable. Only two reliable safeguards are known against failure due to spreading along such a parting. These are the periodic measurement of the porewater pressure during construction for the purpose of detecting impending danger and the elimination of the pressure by adequate drainage provisions.

Selected Reading

Porter, O. J. (1936). "Studies of fill construction over mud flats including a description of experimental construction using vertical sand drains to hasten stabilization," *Proc. 1st Int. Conf. Soil Mech.*, Cambridge, Mass., **1**, pp. 229–235.

Moran, Proctor, Mueser and Rutledge (1958). *Study of deep soil stabilization by vertical sand drains*, U.S. Dept. Commerce, Office Tech. Serv., Wash., 192 pp.

Casagrande, A. (1960). "An unsolved problem of embankment stability on soft ground," *Proc. 1st Panamerican Conf. of Soil Mech., Mexico,* **2**, pp. 721–746. Railroad fill across Great Salt Lake.

Stamatopoulos, A. C. and P. C. Kotzias (1965). "Construction and performance of an embankment in the sea on soft clay," *Proc. 6th Int. Conf. Soil Mech.*, Montreal, **2**, pp. 566–570.

Chapter 9

FOUNDATIONS

ART. 53 FOUNDATIONS FOR STRUCTURES

Types of Foundations

The *foundation* is that part of a structure which serves exclusively to transmit the weight of the structure onto the natural ground.

If a stratum of soil suitable for sustaining a structure is located at a relatively shallow depth, the structure may be supported directly on it by a *spread foundation*. However, if the upper strata are too weak, the loads are transferred to more suitable material at greater depth by means of *piles* or *piers*. Spread foundations are of two types. If a single slab covers the supporting stratum beneath the entire area of the superstructure, the foundation is known as a *mat* or *raft*. If various parts of the structure are supported individually, the individual supports are known as *spread footings*, and the foundation is called a *footing foundation*. A footing that supports a single column is called an *individual footing;* one that supports a group of columns is a *combined footing*, and one that supports a wall is a *continuous footing*.

The *depth of foundation* D_f is the vertical distance between the base of the footing or pier and the ground surface, unless the base is located beneath a basement or, if the structure is a bridge, beneath the surface of the river. In these instances the depth of foundation is referred to the level of the basement floor or to that of the river bed. The principal difference between footings and piers lies in the value of the ratio D_f/B, where B is the width of the base. For footings D_f/B commonly ranges between 0.25 and 1, whereas for piers it is usually greater than 5 and may be as great as 20. However, monolithic supports for bridges are commonly called piers, irrespective of the value of D_f/B. Depending on this value, bridge piers are designed according to the same principles as those governing the design of footings or piers for buildings.

Minimum Depth of Building Foundations

The conditions that determine the minimum depth of building foundations are illustrated by Fig. 53.1, which represents a cross section through part of a building. The outer portion of the structure does not have a basement but the inner part does.

The first requirement is that the base of every part of the foundation should be located below the depth to which the soil is subject to seasonal volume changes caused by alternate wetting and drying. This depth usually does not exceed 4 ft, but there are notable exceptions to this statement. One of them was mentioned in Article 21 in connection with the seasonal swelling and shrinking of certain clays in central Texas. Although these clays are stiff enough to sustain a load of 2 or 3 tons/ft^2 without perceptible settlement, the seasonal volume changes make it necessary to provide even light structures with pier foundations that extend to a depth of more than 20 ft (Simpson 1934). Similar seasonal volume changes extending to great depth have been observed in Canada, South Africa, and many other parts of the world (Bozozuk 1962, Jennings 1953). Withdrawal of water from the ground by the root systems of large trees located close to buildings has also been responsible for important and detrimental differential settlement.

The base of each part of the foundation should also be located below the depth to which the structure of the soil is significantly weakened by root holes or cavities produced by burrowing animals or worms. The lower boundary of the weakened zone can be discerned readily on the walls of test pits.

In regions with cold winters the foundations of the outside columns or walls should be located below the level to which frost may cause

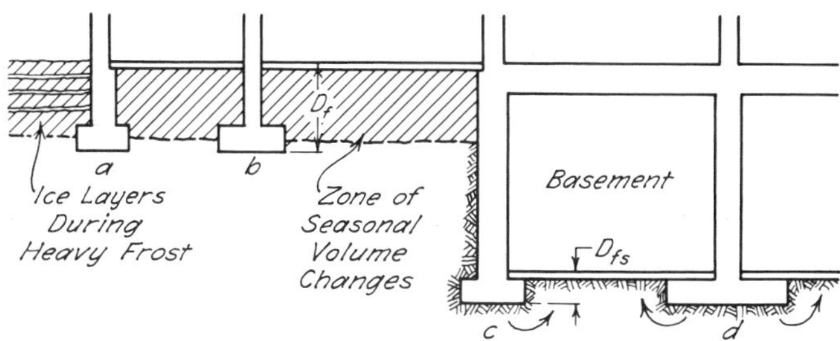

Fig. 53.1. Simplified section through footing foundation of building with basement beneath its central part.

a perceptible heave (Article 21). In the northeastern part of the United States this depth may be as great as 5 ft. Hence, outside walls or columns may require deeper foundations than interior columns.

Basement floors are commonly located below the minimum depth required for footings of buildings without basements. Hence, under normal conditions the minimum depth of foundations located within the boundaries of a basement (*c* and *d* in Fig. 53.1) is governed solely by structural requirements. Exceptions to this statement need be considered only if conditions may arise that could subsequently affect the integrity of the soil beneath the footings. In one instance large unequal settlement of a building resting on medium clay was caused by the gradual desiccation of the clay surrounding a deep boiler room. On account of the low humidity and high temperature of the air in the boiler room, the water in the clay evaporated through the concrete walls of the room. In another instance the footings of a building on fine sand settled because of the leakage of water through open joints in a defective sewer located beneath the base level of the footings. The water washed sand into the sewer, and settlement occurred because of the loss of ground. Hence, before the minimum depth of foundation for a building with a basement is decided upon, the possibilities for subsequent artificial changes in the conditions of soil support should be considered.

Minimum Depth of Bridge Foundations

Whenever the water level in a river rises, the soil that constitutes the river bed starts to move throughout the greater part of the length and width of the river, and the bottom of the river goes down. This process is known as *scour*. The minimum depth for the foundation of a bridge pier is determined by the condition that the base of the foundation should be several feet below the level to which the river may scour during high water.

In those sections of a river where flood water is prevented by high banks or dikes from spreading over a wide area, scour can be very deep, even in a channel unobstructed by bridge piers. Figure 53.2 illustrates this possibility. Figure 53.2*a* is a section through the Colorado River near Yuma, Arizona. The river bed consists of fine silty sand and silt. As the river level rose 14 ft, the level of the bottom of the river channel went down as much as 36 ft (Murphy 1908). Figure 53.2*b* is a section through a mountain stream confined between the abutments of a bridge. The river bed consists of coarse sand and gravel with a high percentage of large cobblestones. A rise of

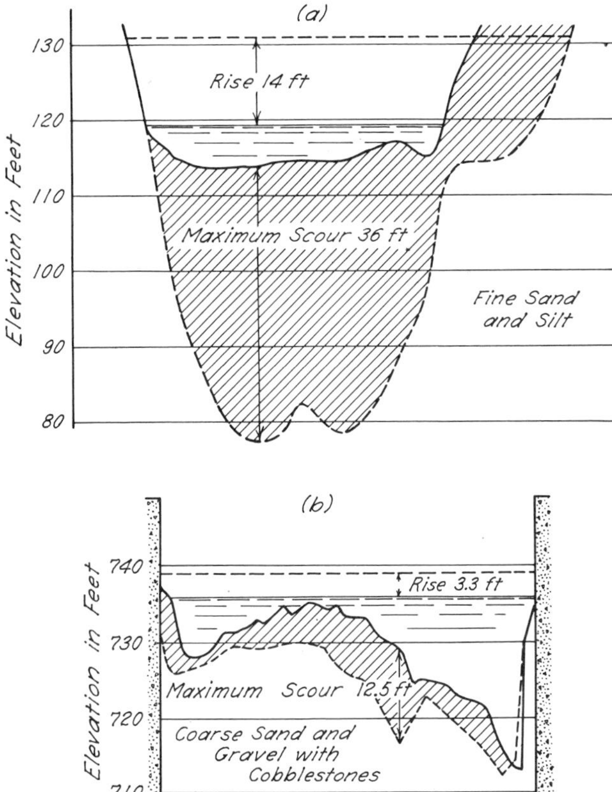

Fig. 53.2. Scour due to high water (*a*) in unobstructed river bed of Colorado River near Yuma, Ariz. (after Murphy 1908). (*b*) Between abutments of bridge over Drau River in eastern Alps. Horizontal scale 10 times vertical scale.

the river level of 3.3 ft was associated with a scour ranging between 2 and 12.5 ft.

Obstruction of the flow by bridge piers increases the amount of scour, particularly in the proximity of the piers. The influence of the shape of the piers on the topography of the depression formed by the scour is illustrated in Fig. 53.3. The information is based on the results of model tests (Rehbock 1931).

Scour does not always receive the attention it deserves and, as a consequence, failure of bridge piers due to this cause is not uncommon. Failure may occur even under conditions that seem to exclude the risk of scour. In a torrential river in Colorado the base of a bridge pier was established at a depth of 10 ft below the bottom of the river channel. At that depth the river bed contained boulders, of a size up to 8 ft^3, so tightly wedged that further excavation would have been impracticable without blasting. Therefore, the base of the pier was established at that depth. Yet, the first high water after construction caused the pier to fail.

Near the east coast of the United States, a bridge pier was founded 2 ft below the surface of a stratum of gravel 7 ft thick. The gravel was covered with 8 ft of soft mud. During exceptionally high water the pier settled appreciably. After the water level dropped the gravel was still buried beneath mud. From the records of the failure, it appeared likely that the settlement was due to scour in the gravel, preceded by complete removal of the overlying mud layer. While the high water was receding, a new layer of mud was deposited.

In those parts of a river where the high water has an opportunity to spread over a wide area, the scour may be imperceptible. Locally, the river bed may even be raised. However, bridges are commonly located at points where these conditions are not satisfied. Furthermore, at any given cross section of the river the point of deepest scour may shift from year to year in an unpredictable manner.

Since reliable scour forecasts require mature and varied experience in the hydraulics of rivers, they can be made only by specialists in this field. On account of the inevitable uncertainties involved in the forecasts, a large margin of safety is required. If no scour investigation has been made by a river specialist and if, in addition, the depth to bedrock or a scourproof stratum is very great, it is advisable to establish the base of the foundation at a depth below the bottom of the low-water channel equal to not less than four times the greatest known rise of the river level.

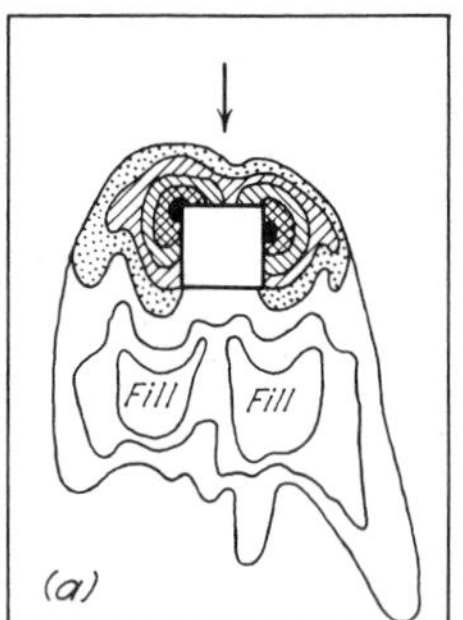

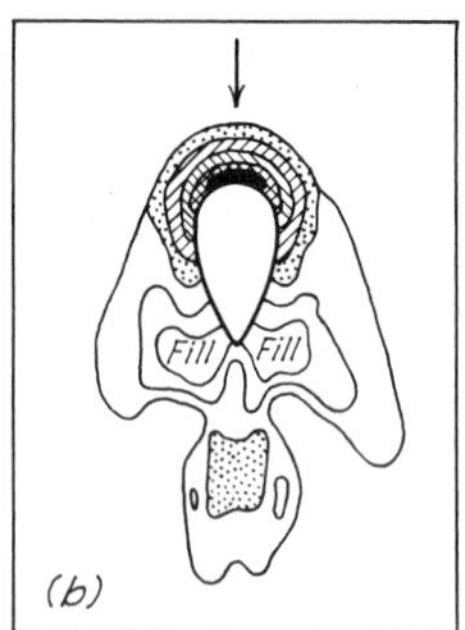

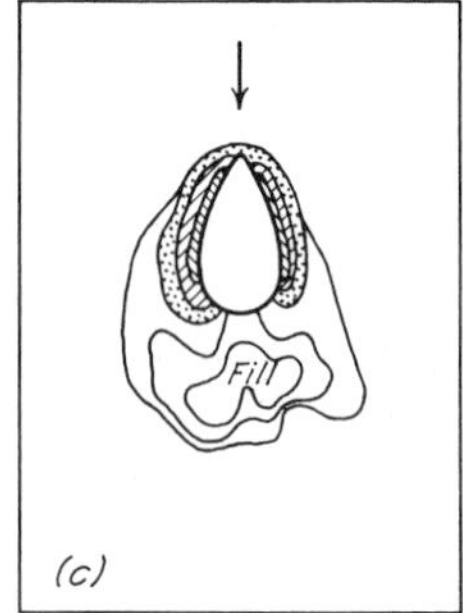

Fig. 53.3. Results of hydraulic model tests for investigating effect of shape of bridge piers on scour (after Rehbock 1931).

Allowable Pressure on the Subsoil

As the load on a given foundation increases, the foundation settles. At low loads the settlement may increase in direct proportion to the load, but with increasing load the rate of increase of settlement usually increases. If the load becomes sufficiently great, the increment of settlement may be excessively or uncontrollably large and the foundation is said to have broken into the ground or to have experienced a bearing-capacity failure. Obviously, the distinction between excessive settlement and failure by breaking into the ground is, in many instances, quite arbitrary. Nevertheless, it is convenient to consider the two conditions independently, especially because the settlement may increase on account of the consolidation of underlying compressible soils even if the loading remains constant.

It is appropriate, therefore, to require that every foundation satisfy two independent conditions. First, the factor of safety of the foundation with respect to breaking into the ground should not be less than about 3, which is the minimum factor of safety customarily specified against ultimate failure for the design of the superstructure. Second, the deformation of the base of the structure due to unequal settlement should not be great enough to damage the structure. Since the theoretical methods for evaluating the factor of safety of foundations with respect to breaking into the ground (Article 33) are simple and fairly reliable, they can be used without essential modification in connection with the design of foundations. On the other hand, the methods for evaluating the magnitude and distribution of the settlement are cumbersome and in many instances very unreliable. This fact determines the procedure for estimating the allowable soil pressure whenever the design must be based on settlement considerations.

Since all substances including soil and rock are compressible, every foundation settles. If the base of a structure remains plane while settlement proceeds, the magnitude of the settlement may be irrelevant. However, if the base becomes warped during the settlement, the structure may be damaged. For this reason, the distribution of the settlement over the base of the structure is far more important than the maximum value. At the same time, it is also far more difficult to evaluate.

According to Article 41, the magnitude and the distribution of the settlement of a loaded area depend on the physical properties of the soil located beneath the area, on the size of the area, on the depth of foundation, and on the position of the water table. If a building rests on footings, the estimate of the settlement is further complicated by the fact that the soil conditions prevailing under the different footings are likely to be different (Article 45). An accurate evaluation of the effect of all these factors on the settlement is impracticable. Therefore, under normal conditions the designer is compelled to estimate the settlement on the basis of simple semiempirical rules. The theory of settlement (Article 41) serves merely as a basis for a rational interpretation of the results of soil and load tests and for determining the limits of validity of the semiempirical rules. Refined settlement computations are justified only if the subsoil contains strata of soft clay located below the base of the foundation or the points of the piles (Articles 54 to 56).

Semiempirical rules for determining settlement are based on observed relations between the results of simple field tests such as penetration tests, the load per unit of area, and the behavior of existing structures. Every relation of this type is a statistical one involving more or less important scattering from the average. Experience shows that a relation developed within a geologically well-defined region always involves less scattering than the corresponding relation for all deposits of a given kind, irrespective of their geological origin and environment. In this book only relations of the latter type can be considered. On account of the great scattering, they represent a very conservative basis for design. Therefore, whenever extensive construction operations are carried out within a limited area, such as that occupied by a large city, the rules given in the following articles should be checked against local experience. If they are found to be too conservative for this particular region, they should be modified accordingly.

For example, according to one of the general relations discussed in the next article, a sand having an N-value of 25 as determined by the standard penetration test (page 304) is of medium density and, if the groundwater level is near footing level, it should be assigned an allowable soil pressure beneath a large footing of about 2400 lb/ft.^{2} Yet local investigations have demonstrated that the sand deposit at the south end of Lake Michigan near the Illinois–Indiana border, which has an N-value of 25, is actually a dense sand that can safely be subjected to a load of 3200 lb/ft^2 beneath large footings.

Until local rules are established, the requirements of safety call for design on the basis of the more conservative general rules. Because of the additional expense involved in this procedure, the accumulation of the observational data needed for establishing local rules is a very good investment and should be encouraged. Only in this manner can the engineer exploit the desirable characteristics of the principal local soil types to the fullest extent.

The following articles deal with the methods for adapting the four principal types of foundations to the subsoil conditions.

Selected Reading

Foundation movements due to seasonal volume changes and methods for their prevention are discussed in the following references:

Jennings, J. E. (1953). "The heaving of buildings on desiccated clay," *Proc. 3d Int. Conf. on Soil Mech.*, Zurich, **1,** pp 390–396.

Dawson, R. F. (1959). "Modern practices used in the design of foundations for structures on expansive soils," *Colo. School of Mines Quarterly,* **54,** No. 4, pp. 67–87.

Means, R. E. (1959). "Buildings on expansive clay," *Colo. School of Mines Quarterly,* **54,** No. 4, pp. 1–31.

Considerable information is contained in the papers of the Symposium on Expansive Soils, South Africa, 1957–1958.

The following references contain data on the observed depth of scour around bridge piers under field conditions:

Schneible, D. E. (1924). "Some field examples of scour at bridge piers and abutments," *Better Roads,* **24,** Aug. p. 21.

Hubbard, P. G. (1955). "Field measurement of bridge-pier scour," *Proc. Hwy. Res. Board,* **34,** pp. 184–188.

Laursen, E. M. (1955). "Model-prototype comparison of bridge-pier scour," *Proc. Hwy. Res. Board,* **34,** pp. 188–193.

Neill, C. R. (1964). "A review for bridge engineers," *Canadian Good Roads Assn., Ottawa, Tech. Publ. No. 23.*

ART. 54 FOOTING FOUNDATIONS

Origin and Shortcomings of Conventional Design Methods

The most important step in the design of a footing foundation is the evaluation of the greatest pressure that can be applied to the soil beneath the footings without causing either failure of the loaded soil or excessive settlement. Before the advent of soil mechanics, the methods for choosing this pressure were based on experience and inadequate knowledge of the properties and behavior of soils. Although the methods contained many shortcomings, they were expedient. As a consequence, their general form has been retained and suitable modifications have been introduced to take account of the findings of soil mechanics. Intelligent use of the modified procedures presupposes a familiarity with the methods prevalent during the first half of the century.

Before the 19th century the framework of most large buildings consisted of strong but somewhat flexible main walls interconnected by massive but equally flexible partition walls intersecting each other at right angles. Since such buildings could stand large settlements without damage, their builders gave little consideration to foundations other than to increase the wall thickness at the base. If the ground was obviously too soft to support the loads, the walls were established on piles. When exceptional structures were built with large domes, vaults, or heavy individual columns, the designers tended to underdimension the foundations because they had neither rules nor experience to guide them. As a consequence, many important buildings either collapsed or were disfigured by subsequent reinforcements.

The development of highly competitive industry during the 19th century led to a demand for large but inexpensive buildings. The types that developed were more sensitive to differential settlement than their predecessors. Furthermore, many of the most desirable sites for industrial buildings were located in regions that had previously been avoided because of notoriously bad soil conditions. Hence, designers found themselves in need of a reliable procedure, applicable under all soil conditions, for proportioning the footings of a given building in such a manner that they would all experience nearly the same settlement.

To satisfy this need the concept of an "allowable soil pressure" was developed during the 1870's in several different countries. The concept was based on the obvious fact that, under fairly similar soil conditions, footings transmitting pressures of high intensity to the subsoil generally settled more than those transmitting pressures of low intensity. With this fact in mind designers began to observe the condition of buildings supported by footings that exerted various pressures against the subsoil. The pressures beneath the footings of all those buildings that showed signs of damage due to settlement were considered too great for the given soil conditions. The maximum pressure not associated with structural damage was considered a satisfactory basis for design and was accepted as the *allowable soil pressure* or *allowable bearing value*. The values obtained for each type of soil in a given locality by this purely empirical procedure were assembled into a table of allowable soil pressures that was subsequently incorporated into the building code governing construction in that locality. Excerpts from pre-1930 building codes of several American cities are given in Table 54.1.

Table 54.1
Soil Pressures Allowed by Various Building Codes

Character of Foundation Bed, Loads in tons/ft²	Akron, 1920	Atlanta, 1911	Boston, 1926	Cleveland, 1927	Denver, 1927	Louisville, 1923	Minneapolis, 1911	New York, 1922	St. Paul, 1910	Jacksonville, 1922
1 Quicksand or alluvial soil	$\frac{1}{2}$	–	–	$\frac{1}{2}$	–	–	–	–	–	–
2 Soft or wet clay, at least 15′ thick	1	1	–	2	–	–	1	1	–	1
3 Soft clay and wet sand	$1\frac{1}{2}$	–	–	$1\frac{1}{2}$	–	–	–	–	1	–
4 Sand and clay mixed or in layers	–	2	–	–	–	–	2	2	2	2
5 Firm clay	–	–	–	–	–	–	–	2	–	–
6 Wet sand	–	–	–	–	–	–	–	2	–	–
7 Fine wet sand	2	–	–	2	–	–	–	–	–	–
8 Soft clay held against displacement	–	–	2	–	–	–	–	–	–	–
9 Clay in thick beds, mod. dry	–	–	–	–	2–4	–	–	–	–	–
10 Dry solid clay	–	–	–	–	–	–	–	–	–	3
11 Loam, clay or fine sand, firm and dry	–	–	–	–	–	$2\frac{1}{2}$	3	–	–	–
12 Firm dry loam	$2\frac{1}{2}$	2–3	–	–	1–2	–	–	–	–	–
13 Firm dry sand	3	2–3	–	–	2–4	–	–	3	–	3
14 Quicksand when drained	–	–	–	3	–	–	–	–	–	–
15 Hard clay	–	3–4	–	3	–	4	4	–	4	–
16 Fine-grained wet sand	–	–	3	–	–	–	–	–	–	–
17 Very firm coarse sand	–	3–4	–	–	4–6	4	4	–	4	4
18 Gravel	–	3–4	–	–	–	4	4	6	–	4
19 Dry hard clay	–	–	–	–	–	–	–	4	–	–
20 Clay in thick beds always dry	4	–	–	–	4–6	–	–	–	–	–
21 Fine dry clay	–	2–3	–	–	–	–	–	–	–	–
22 Fine-grained dry sand	–	–	4	4	–	–	–	–	–	–
23 Compact coarse sand and gravel	–	–	–	–	–	–	–	–	–	4

Table 54.1 (Continued)

Character of Foundation Bed, Loads in tons/ft²	Akron, 1920	Atlanta, 1911	Boston, 1926	Cleveland, 1927	Denver, 1927	Louisville, 1923	Minneapolis, 1911	New York, 1922	St. Paul, 1910	Jacksonville, 1922
24 Gravel and coarse sand in thick beds	5	–	–	8	–	–	–	–	–	–
25 Wet or dry, med. or coarse sand	–	–	5	–	–	–	–	4	–	–
26 Firm coarse sand and gravel	–	–	–	–	–	–	–	–	6	–
27 Gravel, compact sand, and hard yellow clay	–	–	6	–	8–10	–	–	–	–	–
28 Hardpan	–	–	–	–	–	–	–	10	–	–
29 Hard shale, unexposed	6	–	–	6	–	–	–	–	–	–
30 Shale and hardpan	–	–	10	–	–	–	–	–	–	–
31 Soft rock	–	–	–	–	–	–	–	8	–	–
32 Rock	10	15	100	10	10–200	–	–	40	–	–

From Kidder-Parker: *Architects' and Builders' Handbook*, 1931.

Although most building codes contained tables of allowable soil pressures, they did not offer any hint regarding the origin of the values or any explanation of the meaning of the term "allowable soil pressure." These omissions fostered the belief that the settlement of a building would be uniform and of no consequence if the pressure on the soil beneath each footing were equal to the allowable soil pressure. The size of the loaded area and the type of building were believed to be immaterial. Some engineers were even under the delusion that a building with footings that exert the allowable soil pressure would not settle at all. To a considerable degree, these misconceptions still prevail today.

Many foundations designed on the basis of the allowable-soil-pressure tables were entirely satisfactory, but from time to time the results were disappointing and structures settled excessively. Since engineers believed that footings would not settle noticeably if the allowable pressure was not exceeded, they attributed the failures to faulty classification of the soil. They assumed that the wrong allowable pressures had been selected because the terms used to describe the soil in the field and in the building codes did not have the same meaning. In order to avoid this difficulty, it gradually became customary to select, or at least to verify, the allowable soil pressure on the basis of the results of load tests.

A load test is made by increasing the load on a bearing plate by small increments and measuring the corresponding settlements. The bearing plate rests on the bottom of a pit at the level of the base of the footings. Depending on the preference of the engineer who makes the test, the plate may be surrounded by a box and the pit backfilled to final grade (Fig. 54.1*a*), or the pit may be made so large that the plate rests in the middle of a level area. The test

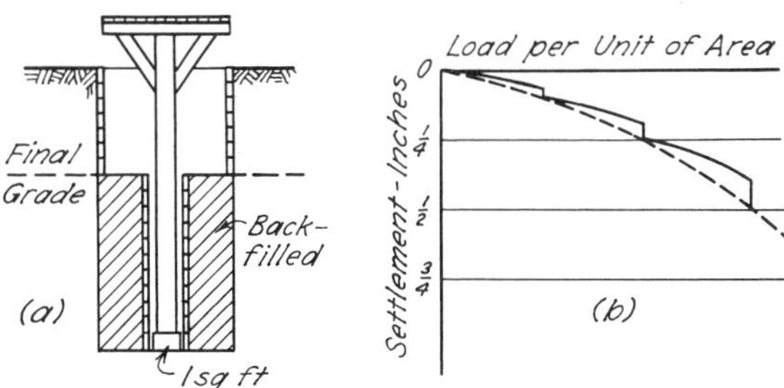

Fig. 54.1. (*a*) **Test arrangement for determining relation between load per unit of area and settlement of test plate, as basis for selecting allowable soil pressure.** (*b*) **One of several customary methods for plotting results of load test.**

results are represented by load-settlement curves similar to the one shown in Fig. 54.1*b*. In the following paragraphs two of the most common methods for performing the tests and interpreting the results are described.

The first method consists of loading a square or circular bearing block of any dimensions chosen by the investigator. The allowable load q_a per unit of area is taken as some fraction, such as one half, of the average pressure on the block at the time of failure. This procedure is objectionable for several reasons. In the first place, if the load-settlement curve resembles C_2 (Fig. 33.1), there is no definite failure load. Second, the size of the loaded area, which is optional, may have a large influence on the ultimate bearing capacity per unit of area (Article 33). Hence, by using this first procedure two different investigators can obtain very different values of q_a for the same soil.

The second method consists of loading a bearing block covering an area of 1 ft^2. The allowable load q_a is arbitrarily defined as one-half the load at which the settlement of the bearing block is 0.5 in. (In countries using the metric system the area of contact is customarily taken as 0.1 m^2 or 1.08 ft^2, and the settlement as 1 cm or 0.4 in.) This procedure, although arbitrary, is preferable because two different investigators will at least obtain the same value of q_a for the same soil.

There are many other methods for performing load tests and many other rules for interpreting the results. Yet, whatever the method may be, the test results reflect the character only of the soil located within a depth of less than twice the width of the bearing plate, whereas the settlement of the footings depends on the properties of a much thicker soil stratum. Consequently, if the character of the soil changes below a depth of about twice the width of the bearing plate, as it commonly does, the test results are certain to be misleading. In the past, it was almost universal practice to select the allowable soil pressure without regard for the size of the footings, the type of superstructure, and other vital characteristics of the proposed foundation; therefore, it is not surprising that increasing recourse to load tests did not significantly reduce the frequency of faulty footing design. In fact, several complete foundation failures occurred in spite of the conscientious performance of load tests. To reduce the risk of faulty design, the allowable soil pressure must be chosen in accordance not only with the results of load tests or their equivalent, but also with the character of the soil profile and of the foundation itself. Part of the necessary information can be obtained from the theories given in Articles 33, 40, and 41. The rest is derived from construction experience.

Because of the great variety of soils and combinations of soils encountered in practice, no single method for determining the allowable soil pressure can be developed that would be suitable under all circumstances. The procedure must always be adapted to the soil conditions revealed by the exploratory borings. In particular, the procedure depends on the *significant depth.* This term refers to the depth within which the load on the footing alters the state of stress in the soil enough to produce a perceptible contribution to the settlement.

The significant depth depends not only on the size of the footing and on the load it supports, but also to a high degree on the soil profile and the physical properties of the soils that constitute the individual strata. If the initial tangent modulus of the soil (Article 15) increases as the depth below a footing increases, the significant depth may not exceed the width B of the footing. On the other hand, if the soil beneath the footing becomes softer with depth, the significant depth may be equal to several times the width B.

In the following discussion, four principal types of soil conditions are considered:

1. The footings are supported by sand or sand and gravel that do not contain any layers of soft clay or other highly compressible soil within the significant depth.
2. The footings are supported by clay that is fairly homogeneous within the significant depth.
3. The footings rest on soil with properties intermediate between those of sand and clay, such as silt, some types of fill, or loess. The soil is fairly homogeneous within the significant depth.
4. The footings are supported by soil that contains one or more soft layers within the significant depth.

Footings on Homogeneous Sand

Conceptions prevalent in building codes before 1930, and even still retained in some, are exemplified by Table 54.2. As the first step in discussing a rational basis for selecting the allowable pressure, we shall examine the shortcomings of this table. The numerical values are likely to be inappropriate, because the soil classification is based on properties that are largely irrelevant, whereas significant properties are ignored. For example, the term quicksand (1) is not descriptive of a type of sand. It does not even indicate a sand that was necessarily in a loose state before construction started. This fact is illustrated by the ill-deserved reputation of a very fine uniform sand located beneath the water table near Lynn, Mass. Curve 2 in Fig. 45.3*a* represents the results of a load test made on this sand after the water table was lowered by well points; it indicates that the sand is firm and dense. Yet, among the construction men in its locality it once had the reputation of a dangerous quicksand because on former jobs, when more primitive methods of drainage were used, it became soft on the bottom of excavations and started to boil on slight provocation. Description 6 does not state whether the sand is above or below water table, although this factor is decisive. The grain size, mentioned in descriptions 11, 17, and 24, has no direct influence on the bearing

Table 54.2
Allowable Bearing Values for Sand From Pre-1930 Building Codes*

Soil	q_a in tons/ft^2
1 Quicksand	0.5
6 Wet sand	2
11 Fine sand, firm and dry	2.5–3
14 Quicksand when drained	3
17 Very firm coarse sand	3–6
24 Gravel and coarse sand in thick beds	5–8

* Abstract from Table 54.1.

capacity. The poorest of the sands represented in Fig. 45.3*a*, indicated by curve 5, was clean, coarse, mixed-grained, and dry. The best one, represented by curve 1, was uniform, fine, and wet. In order to establish more reliable criteria for the design of footings on sand, the allowable soil pressure must be correlated not with irrelevant properties of the sand, but with properties and conditions that have a significant influence on the behavior of the sand under load. These conditions are the relative density of the sand and the position of the water table with reference to the base of the footings.

The relative density has a decisive influence on the angle of internal friction ϕ and the shape of the load-settlement curve. Depending on the relative density, the value of ϕ for a sand may vary over as wide a range as 27.5 to 45° (Article 17) and the load-settlement curve may have any shape intermediate between C_1 and C_2 in Fig. 33.1. If standard penetration tests are made, the relative density can be judged by means of Table 45.1, page 341. More comprehensive data can be obtained expediently by making subsurface soundings.

The position of the water table with reference to the base of the footings has an influence on both the ultimate bearing capacity of the sand and the settlement. If the water table rises from beneath the seat of settlement toward the base of a footing, it reduces the effective unit weight of the soil located between the two positions of the water table by roughly 50% (Article 12). As a consequence the effective pressures are reduced at all depths below that of the new water table, the factor of safety of the footing with respect to breaking into the ground is correspondingly reduced (Article 33) and the settlement is substantially increased (Article 41).

Computations based on the theory presented in Article 33 lead to the following conclusions regarding the factor of safety F of footings designed on the basis of the customary allowable pressures on sand: If the base of a footing rests on loose sand at or below the water table; if, in addition, the width B of the footing is less than about 6 ft; and if the depth of foundation is less than B below the ground surface or basement floor, the value of F may be smaller than the required minimum of 3. In the rare instances when these conditions are simultaneously satisfied, a stability computation should be made to find out whether or not the safety requirement is met. In all other instances the factor of safety is greater and commonly much greater than 3. Hence, under normal conditions, the allowable soil pressure on sand is determined exclusively by settlement considerations.

The distribution of the settlement over the base of a building supported by footings with width B is determined chiefly by the variations in the compressibility of the layer of sand with thickness B located immediately below the footings (Article 45). The practical importance of these variations is illustrated by Fig. 54.2, which shows the settlement of several uniformly loaded continuous footings of con-

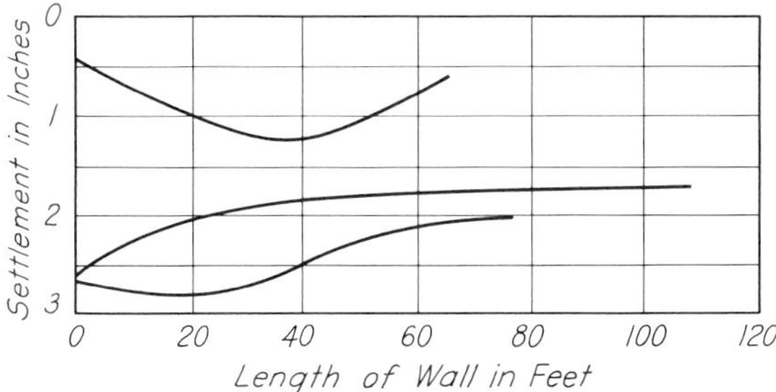

Fig. 54.2. Diagram showing settlements of long narrow continuous footings supporting brick walls (after Terzaghi 1938*b*).

stant width resting on sand or gravel. If the subsoils had been uniform, each footing would have settled almost uniformly. The unequal settlement was due to local variations in the compressibility of the soil (Terzaghi 1938*b*).

A study of available settlement records leads to the conclusion that the differential settlement of uniformly loaded continuous footings and of equally loaded spread footings of approximately equal size is unlikely to exceed 50% of the maximum settlement. However, in practice, the size of footings supporting the different columns of a building may be very different because the loads on the columns are likely to be different. This is a source of additional differential settlement.

According to the discussion in Article 41, based on theoretical considerations and on the stress-deformation characteristics of sands, the settlement of square footings exerting equal soil pressure on a homogeneous sand should increase with increasing width in the general manner shown by the plain curve in Fig. 41.3. In accordance with this conclusion, the results of observations on structures and of various field and laboratory experiments indicate that the settlement increases with the width B of the footing approximately as shown by curve a in Fig. 54.3. In this figure, S_1 is the settlement of a loaded area 1 ft square under a given load q per unit of area, and S is the settlement at the same load per unit of area of a footing with a width B.

In the construction of curve a (Fig. 54.3), the greatest weight has been given to the empirical data derived from the observed differential

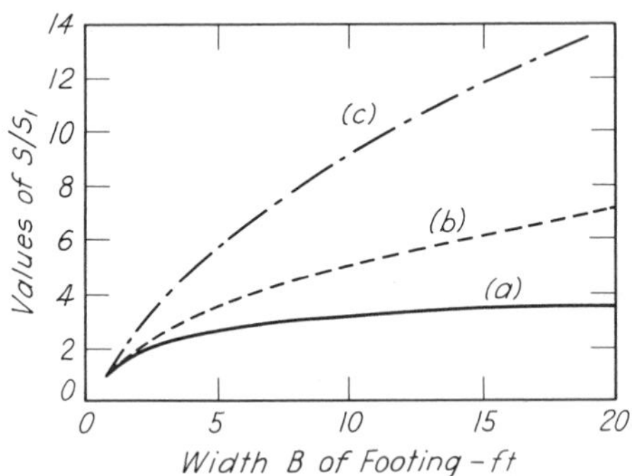

Fig. 54.3. Approximate relation between width B of footing on sand and the ratio S/S_1, wherein S represents the settlement of a footing with width B and S_1 the settlement of a footing 1 ft wide subject to the same load per unit of area. Curve (a) refers to usual conditions. Curve (b) represents possible relation for loose sands. Curve (c) refers to sand with a small organic content.

settlement of structures established on footings of different sizes on the same sand deposit. Small-scale field experiments, including load tests on plates as small as 1 ft square, are likely to lead to the erroneous conclusion that relatively greater settlements should occur with increasing width of footing because the apparent cohesion due to capillarity in the sand has a disproportionately great influence on the settlement of small loaded areas as compared to large areas. The presence of even a small organic content may radically increase the influence of the size of the loaded area, as indicated by curve c in Fig. 54.3. There is some indication that the ratio S/S_1 may increase more rapidly with increasing width, as in curve b, if the sand is loose than if it is medium or dense (Bjerrum and Eggestad 1963).

For curve a, the relation between S, S_1 and B is given approximately by the equation

$$S = S_1\left(\frac{2B}{B+1}\right)^2 \tag{54.1}$$

in which S and S_1 are expressed in inches and B in feet. There is no significant difference between the settlements of square and continuous footings having the same width B, because the effect of stressing the sand to a greater depth below a continuous footing is compensated by the restraint that keeps the sand from being displaced in directions parallel to the footing. According to curve a (Fig. 54.3), the settlement of a large footing, greater than about 20 ft square, exceeds that of a small footing 4 or 5 ft square by roughly 30%, provided the soil pressures are equal. At a given width B of the footing, the settlement decreases to some extent with increasing values of the *depth ratio* D_f/B, wherein D_f is the depth of foundation (Article 53). Yet, even under extreme conditions involving a foundation on footings with very different sizes and depth ratios (Fig. 53.1), the differential settlement is unlikely to exceed 75% of the maximum settlement. Normally it is very much smaller.

Most ordinary structures, such as office buildings, apartment houses, or factories, can withstand a differential settlement between adjacent columns of three quarters of an inch. As indicated in the preceding paragraph, this settlement will not be exceeded if the soil pressure is selected such that the largest footing would settle 1 in. even if it rested on the most compressible part of the sand deposit. Therefore, the allowable soil pressure for the design of the footings of such structures can be assumed equal to the pressure that will cause the largest footing to settle 1 in. if located above the loosest part of the deposit. The following paragraphs contain a description of an approximate method for selecting the allowable soil pressure on sand in accordance with this assumption. If a differential settlement ΔS of more than $\frac{3}{4}$ in. can be tolerated, the allowable soil pressure can be multiplied by $4\Delta S/3$. However, in such instances it may be advisable to investigate whether the stability condition is satisfied (Terzaghi 1935).

Allowable Pressure on Dry and on Moist Sand

The settlement of a footing on dry or moist sand depends primarily on the relative density of the sand and the width of the footing. According to Article 45 the direct determination of the relative density of sands is difficult and time-consuming. As a consequence, the relative density is judged in practice on the basis of indirect means such as penetration tests or plate-bearing tests. The results of such tests depend not only on the relative density of the sand, but also on numerous other factors such as the shape of the grains and the grain-size distribution. Calibration tests to establish the relation between the results of penetration or load tests and the relative density are seldom practicable on routine jobs and, even if conscientiously performed, may be inconclusive on account of the variability of most natural sand deposits. Therefore, semiempirical procedures have been developed for estimating the settlements of footings on sand on the basis of the results of the penetration or load tests themselves.

In the United States the most commonly used procedure for investigating the characteristics of sand deposits is the standard penetration test (page 304). Although the procedure is very crude and involves many uncertainties, the results are a far more reliable basis for judging the allowable soil pressure than the soil-pressure tables or the results of a few conventional load tests.

In order to select the allowable soil pressure on the basis of the results of standard penetration tests, it is necessary to estimate very roughly the width B of the largest footings. Between the level of the base of the footings and a depth B below this level one standard penetration test should be performed for every $2\frac{1}{2}$ ft of depth. The average value of N for this depth is at least a crude indication of the relative density of the sand within the seat of settlement of the footing. If the tests in different drill holes furnish different average values of N, the lowest value should be used for estimating the allowable soil pressure.

The value of the allowable soil pressure is then obtained by means of the chart (Fig. 54.4) in which the curves represent the relation between the width B of a footing, in feet, and the soil pressure required to produce a settlement of the footing of 1 in., provided the footing rests on a sand for which the number of blows N has the value inscribed on the curve. If N has a value other than those for which the curves are drawn, the allowable soil pressure is obtained by linear interpolation between curves.

The chart (Fig. 54.4) was prepared on the basis of present knowledge concerning the relation between the standard penetration resistance N, the behavior of actual footing foundations on sand, surface load tests, and Eq. 54.1. If B is the width of the largest footing supporting a structure, and if all of the footings are proportioned in accordance with the allowable soil pressure corresponding to B, the maximum settlement of the foundation should not exceed 1 in., and the differential settlement $\frac{3}{4}$ in.

If the subsoil consists of gravel or of sand containing large pieces of gravel, the number of blows on the sampler cannot be considered indicative of the degree of compactness of the soil. Yet, the bearing properties of such soils are as variable as those of sand. A well-packed mixture of sand and gravel is less compressible than a very dense sand, whereas the compressibility of a loose gravel may be as great as that of a sand of only medium density. In order to avoid overesti-

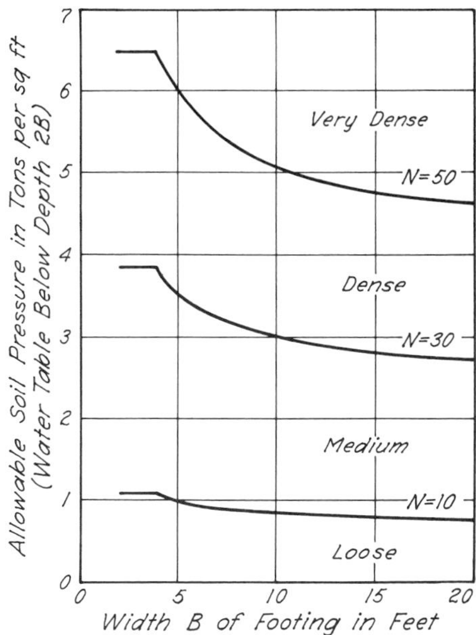

Fig. 54.4. Chart for estimating allowable soil pressure for footings on sand on the basis of results of standard penetration test.

mating the allowable soil pressure on gravel, several test pits should be excavated into the layers that will constitute the seat of settlement of the footings, and the degree of compactness of the exposed soil should be evaluated at least on the basis of its appearance, stability, and resistance to excavation. If the allowable soil pressure for the gravel is assumed equal to that of a sand with the same relative density, conservative values are obtained by means of the chart (Fig. 54.4). On some projects it may be expedient to use a penetrometer having a diameter large enough to be only slightly influenced by the size of the gravel and to calibrate the penetrometer against the results of standard penetration tests performed in materials of smaller maximum size (Peck 1953).

In Europe, static penetration tests such as the Dutch-Cone test (page 319) are used more commonly than the standard penetration test for investigating sand deposits. The procedures described in the preceding paragraphs and the chart (Fig. 54.4) may be used in connection with the results of the Dutch-Cone test by means of the following crude but simple correlation between the cone penetration resistance q_p (kg/cm^2) and the standard penetration resistance N (Meigh and Nixon 1961, Schultze and Melzer 1965, Meyerhof 1956)

$$q_p = KN \tag{54.2}$$

where K is on the order of 5 to 10.

Attempts have been made to evaluate the compressibility of the sand directly or on the basis of such indirect procedures as the Dutch Cone, to calculate the increase in vertical pressure at various depths below the base of the footing, and to compute the settlement by the use of the methods described in Article 41 (Buisman 1943). Such a procedure implies that the settlement of a footing on sand is associated primarily with a decrease in volume of the material whereas, in reality, a substantial part is a consequence of lateral displacements in the subsoil, particularly if the sand is dense (Eggestad 1963). Moreover, the procedure fails to emphasize the implications of the inevitable variations in compressibility of the sand within the few feet immediately beneath the footings. Therefore, there seem to be no substantial grounds for preferring this procedure over the empirical approach represented by Fig. 54.4. For very large footings or rafts, however, the method may lead to useful results (Article 55).

Even if very low soil pressures are used in design, footings on sand are likely to settle excessively if the sand is subject to high-frequency vibrations. The statement applies to saturated as well as to moist or dry sands. Foundations intended to support vibrating machinery must be designed on the basis of the theory of vibrations (Barkan 1962, Richart 1960). They are beyond the scope of this book.

Allowable Soil Pressure on Saturated Sand

If a saturated sand located beneath a footing is very loose, a shock of any kind may cause spontaneous liquefaction (Article 17) followed by a sinking of the footing. It has even been observed that a rapid change of the ground-water level in loose sand occasionally causes a large subsidence. Hence, if the sand is very loose (N equal to 5 or less), the footings should be supported by piles, or else the sand should be compacted (Article 50). During a prolonged series of earthquake vibrations, such as those that occurred in Niigata, Japan, in 1964, catastrophic subsidence of footings may occur above uniform sands with N-values up to approximately 15 (IISEE 1965). Hydraulically deposited uniformly graded sand fill appears to be particularly vulnerable.

If the value of N for the sand in its natural state is greater than 5, or if the sand has been compacted, the allowable soil pressure q_a on the sand should be chosen for ordinary structures such that the maximum settlement will not exceed 1 in. In order to determine q_a by means of the chart (Fig. 54.4), the effect of submergence on settlement must be considered.

According to theory, the submergence of the sand located beneath the base of a footing should approximately double the settlement, provided the base is located at or near the surface of the sand (Article 41). If this fact is taken into consideration, the load per unit of area required to cause a footing on saturated sand to settle 1 in. can be estimated by means of the chart (Fig. 54.4) in the following manner: if the depth ratio D_f/B for the footings is small, like that for the basement footings in Fig. 53.1, the values obtained from the chart should be reduced 50%. On the other hand, if the depth ratio is close to unity, the values from the chart need be reduced by only one-third, because the effect on the settlement of the weight of the soil surrounding the footing partly compensates for that of the increase due to saturation.

The procedure outlined in the preceding paragraph leads to conservative, and quite probably to overconservative, results (Meyerhof 1965). The influence of submergence on the results of the standard penetration or Dutch cone test has not yet been adequately investigated. Submergence may, at least under some circumstances, tend to reduce the penetration resistance; if this occurs, use of the field values inherently involves a correction for submergence. In view of the inadequate state of present knowledge in this connection, estimates of settlement should be corrected for submergence unless experience in the locality has conclusively demonstrated the procedure to be excessively conservative.

The conditions favoring a bearing-capacity failure of a footing on submerged sand have been discussed on page 487. Where these conditions prevail, a stability computation is imperative. The calculations can be made by means of the equations given in Article 33 and the chart (Fig. 33.4). For loose sand with an N-value of 5 the dash curves in Fig. 33.4 should be used, and for dense sand, with an N-value equal to 30 or more, the full curves. The bearing-capacity factors for values of N between 5 and 30 can be estimated by linear interpolation between the two curves. If the investigation shows that the factor of safety of these footings is smaller than 3, either the size of the footings or the depth of foundation should be increased sufficiently to satisfy the safety requirement.

Prerequisites for the Success of Load Tests on Sand

The procedure for determining the allowable soil pressure on sand by means of the chart (Fig. 54.4) eliminates much of the guesswork involved in the use of soil-pressure tables such as Table 54.1, because it furnishes values that are related to significant and not to irrelevant soil properties and conditions. Moreover, it allows the designer to adapt the soil pressures at least in a crude way to the differential settlement that he feels can be tolerated, and the method lends itself to progressive improvement as knowledge and experience increase. At present, more reliable data regarding the allowable soil pressure on sand can be obtained only by means of load tests, at a considerable expenditure of time and money.

Year after year in almost every country a great number of load tests are performed. Yet, most of them are worthless, if not misleading, because the results are unfit for rational interpretation. Therefore, the engineer should be familiar with the prerequisites for obtaining trustworthy results.

Each load test should be made on a bearing plate 1 ft square located at the bottom of a test pit at least 5 ft square. The bearing surface should be at the level of the base of the footing. The load should be applied in increments of about 10% of the estimated allowable soil pressure and increased to at least 1.5 times the estimated allowable soil pressure. The apparatus for measuring the settlement should permit direct readings to 0.01 in. Load tests that satisfy these conditions are referred to as *standard load tests.*

The results of each test should be represented graphically by a load-settlement curve. The load per unit of area at which the settlement of the largest footing would be equal to the predetermined allowable value can be approximated by means of the relations shown in Fig. 54.3. For example, if the sand contains no organic matter and is not extremely loose, and if the design is to be based on a maximum settlement $S = 1$ in., the allowable soil pressure is equal to the load per unit of area at which the settlement of the bearing plate in inches is

$$S_1 = \left(\frac{B + 1}{2B}\right)^2 \tag{54.3}$$

where B is the width of the footing in feet.

If load tests are made at different points of the same site, the results are usually more or less different. This is due to local variations of the relative density of the sand in horizontal directions. Similar variations in vertical directions are observed whenever penetration tests are made in drill holes (Figs. 44.16 and 45.7). These omnipresent variations are a potential source of serious errors. If, for instance, a load test is made on a layer of dense sand 2 ft thick that rests on a stratum of loose sand, the test result will be the same as if the dense sand extended to a very great depth. However, the full-sized footing will settle much more than would be anticipated on the basis of the load test. The reason is illustrated by Fig. 54.5.

This figure represents a vertical section through a stratified subgrade. A is a bearing block covering an area 1 ft square, and B is a full-sized footing. The load on both A and B has the same intensity q. Beneath A and B are shown curves of equal vertical pressure in the subsoil. The pressures were computed by means of the chart

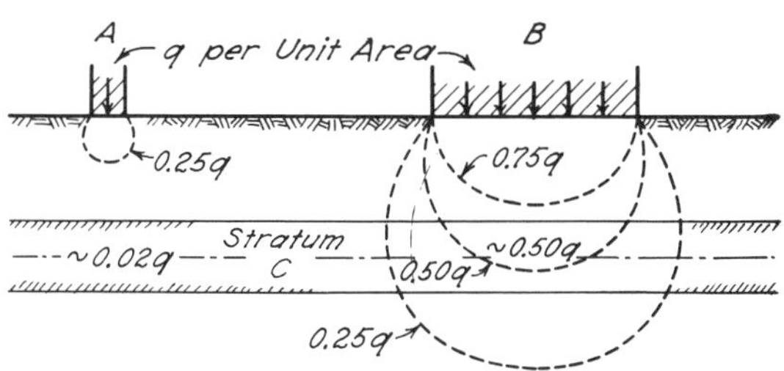

Fig. 54.5. Section through stratified subgrade showing stress produced in stratum *C* by load *q* per unit of area transmitted to surface of ground by (*A*) bearing block one ft square; (*B*) full-sized footing.

(Fig. 40.2). The load on A increases the average vertical pressure in stratum C beneath the loaded area by about $0.02q$, whereas the footing B increases it by $0.50q$. If stratum C is very compressible, the settlement of B may be very large. If C is hard, the settlement of B may be very small. Yet, the result of the load test is practically independent of the compressibility of C, because the increase of the pressure in stratum C due to the load on the bearing plate is negligible.

Because of the fact illustrated by Fig. 54.5, it is necessary to make sure, by means of penetration tests, whether the variations in the density of the subsoil are purely erratic or whether the density of the subsoil within the significant depth for the proposed footings increases or decreases conspicuously and consistently in a downward direction. If the variation is entirely erratic, it is sufficient to make at least six load tests in different locations at the level of the base of the proposed footings. If the density varies consistently with depth, additional load tests must be made at one or two different levels within the significant depth. The allowable load should be selected on the basis of the most unfavorable test results.

The preceding description of the technique for making load tests is based on the assumption that the water table is located at a considerable depth below the base of the footings. If it is located at or slightly above this level, the bearing plate should be established at the water table, on the bottom of a pit 5 ft square. On the other hand, if the water table is located at a considerable height above the level of the base of the footings, it must be lowered by pumping from well points or open sumps before the load tests are made. If well points are used, the pit need not be wider than 5 ft. The bearing plate should be located at the lowered water table. The allowable soil pressure can be computed by means of Eq. 54.3.

Even if the water table is located 3 or 4 ft below the base level of the footings, the load tests should be made at the water table. Otherwise the apparent cohesion imparted to the sand by the soil moisture may introduce an error on the unsafe side. As a matter of fact, even if the water table is at considerable depth, the influence of the apparent cohesion on the load-settlement curve for a bearing plate one foot square underlain by very fine or fine sand may be intolerably large, whereas the influence on a full-size footing may be much less or negligible. Under these circumstances, use of the standard load-test procedure may be inadvisable.

If the water level is lowered by pumping from open sumps, the test pit must be at least 10 ft wide. As soon as the excavation level arrives at the water table, a drainage ditch must be dug all the way around the floor of the pit. During further excavation the ditch must be maintained deep enough to prevent water from flowing up through the central part of the bottom. These requirements call for great care and close supervision. If they are not strictly satisfied, the results of the load tests can be very misleading, because the seepage pressure of water rising toward the bottom of the excavation may greatly increase the settlement.

In any event, the load-test method is very expensive and cumbersome because of the elaborate preparations and the great number of tests required. If the program is not expertly planned and executed, the results may be misleading. Therefore, use of the method should be considered only on very important jobs where the cost of the tests is a small fraction of the total expenditure.

Allowable Pressure on Clay

Customary values for the allowable soil pressure on clay are given in Table 54.3. This table, like Table 54.2 that applies to sands, is

Table 54.3
Customary Allowable Bearing Values for Clay*

	Soil	q_a in tons sq/ft
2	Soft or wet clay, at least 15 ft thick	1–2
3	Soft clay and wet sand	$1\text{–}1\frac{1}{2}$
5	Firm clay	2
8	Soft clay held against displacement	2
9	Clay in thick beds, moderately dry	2–4
10	Dry solid clay	3
15	Hard clay	3–4
19	Dry hard clay	4
20	Clay in thick beds always dry	4–6

*Abstract from Table 54.1.

open to the criticism that the terminology is vague, and the soil properties on which it is based are irrelevant. A satisfactory procedure for design can be developed only by correlating the allowable soil pressures with well-defined mechanical properties of the clay.

The allowable soil pressure on clay, as well as that on sand, should satisfy the two requirements that the factor of safety against breaking into the ground should be adequate, and the settlement produced by the load should be within tolerable limits.

The factor of safety against the breaking of a footing into clay depends on the shearing resistance of the clay. As long as its water content is not appreciably altered by consolidation, the clay behaves in the field with respect to total stresses as if ϕ were equal to zero and as if the cohesion c were approximately equal to one-half the unconfined compressive strength q_u of fairly undisturbed samples (Article 18). Hence, the net ultimate bearing capacity $q_{d\text{ net}}$ (Article 33) per unit of area of a footing resting near the surface of the clay may be calculated from

$$q_{d\text{ net}} = 5c\left(1 + 0.2\frac{D_f}{B}\right)\left(1 + 0.2\frac{B}{L}\right) \qquad (33.17)$$

where B and L are the width and length of the footing, and the depth of foundation D_f does not exceed $2.5B$. For a circular footing the diameter D may be taken as $D = B = L$.

For soft clays the values computed by means of these equations are only slightly greater than the allowable soil pressures given in Table 54.3. Therefore, it is not surprising that complete failures of footings on soft clay were by no means infrequent when the design of footings was customarily based on tables of empirically selected allowable soil pressures.

Under normal conditions the factor of safety of footings on clay, like that of footings on sand, should not be smaller than 3. If the loads for which a footing is designed are very unlikely to develop, the value of $F = 2$ can be tolerated. For example, this value might be appropriate if the design load for the footing of an office building included the maximum live load combined with maximum snow and wind load.

To compute the bearing capacity of a clay it is necessary to determine the undrained shear strength of the clay below the proposed footings. One expedient procedure for obtaining the information consists of making test borings at the site of several footings and securing continuous 2-in. or, preferably, 3-in. tube samples between the level of the base of the footings and a depth below the base equal to the width of the footing. The samples may be ejected from the tubes, sliced longitudinally, and tested at intervals of 4 to 6 in. by a portable torsion shear device as illustrated in Fig. 18.3, or the unconfined compressive strength may be determined at intervals of about 6 in. by means of laboratory tests. The average value of shear strength c from the torsion shear tests, or of $c = q_u/2$ from the unconfined compression tests, is computed for each of the borings, and the smallest of these average values is introduced into Eq. 33.17. The value of the ultimate bearing capacity is then calculated and divided by a factor of safety of 3.

This procedure is valid providing the subsoil does not contain within the significant depth a layer of clay softer than the clay defined by the value c on which the estimate of the allowable soil pressure is based (Skempton 1951). Hence, it cannot be used for computing the allowable soil pressure on a stiff clay crust that rests on softer clay.

The determination of the values of q_u by means of unconfined compression tests or of c by means of a portable torsion vane is so expedient that the ultimate bearing capacity should not be estimated on the basis of the results of the standard penetration test described in Article 44. If tube samples cannot be obtained, the tests should be performed on the spoon samples, although the disturbed condition of the samples may lead to an appreciable underestimate of the shearing strength.

If the clay does not contain numerous sand or silt partings that would invalidate the results, the value of c can also be determined by the performance of vane tests in the field (Article 45). The tests should be made at intervals not greater than one foot along several vertical lines. The average value of shear strength along each vertical line should be computed, and the smallest of these averages used for calculation of the ultimate bearing capacity.

Some stiff clays consist of small angular fragments separated from each other by hair cracks. The presence of the cracks makes it impracticable to determine the unconfined compressive strength of the clay, because the test specimens are likely to disintegrate while they are being prepared. Furthermore, the hair cracks invalidate Eq. 33.17, because they change the stress conditions for failure. The ultimate bearing capacity of such clays should be determined by the load-test method. The tests should be made at the level of the base of the footings, on bearing plates 2 ft square at the bottom of test pits 6 ft square. If the consistency of the clay varies considerably between this level and a depth B (square footings) or $2B$ (continuous footings), load tests must be made at two or three different levels within this depth. The number of load tests or sets of tests that are required depends primarily on the degree of homogeneity of the clay stratum and the number of footings. The load should be applied in increments until the load-settlement curve (Fig. 33.1) indicates that the bearing capacity of the soil $q_{d\ \mathrm{net}}$ has been reached, or until the pressure is at least three times the value of the maximum pressure that would be transmitted to the ground by the foundation.

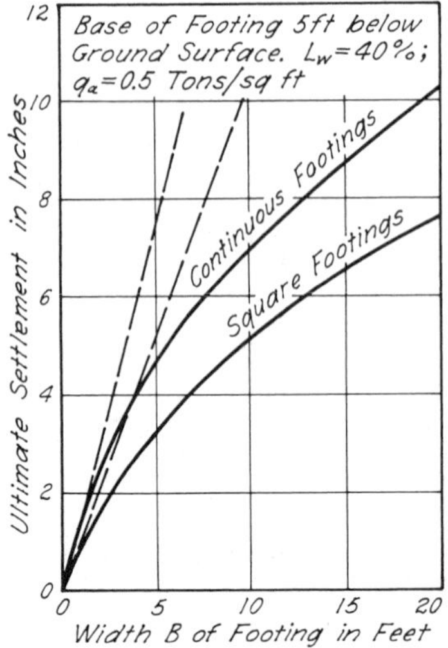

Fig. 54.6. Approximate relation between width *B* and ultimate settlement of footing on normally loaded clay.

The allowable soil pressure q_a on a clay can be assumed equal to one-third the value of $q_{d\,\mathrm{net}}$ determined by Eq. 33.17 or by means of the load tests described in the preceding paragraph, provided the soil conditions justify the assumption that the settlement of the foundation will be tolerable. Whether this requirement is satisfied depends primarily on whether the clay is normally loaded or precompressed.

If the footings rest on normally loaded clay, the magnitude of both the total and the differential settlement can be very large. This can be demonstrated by computing the ultimate settlement of continuous footings of different widths resting on soft normally loaded clay. The results of such a computation are shown in Fig. 54.6. The soil pressure on the base of the footings was taken as 1000 lb/ft^2. In addition, it was assumed that the depth of foundation was 5 ft, that within this depth the effective unit weight of the soil was 100 lb/ft^3, that the liquid limit of the clay was 40%, and that the settlement of the footings was caused solely by consolidation. The compression index of the clay was estimated by means of Eq. 13.11, and the settlement by means of Eq. 13.8. The curve that represents the relation between the settlement and the width of the footing resembles the dash–dotted line in Fig. 41.3. The trend of the curve indicates that the settlement of footings on clay, in contrast to that of footings on sand, increases in almost direct proportion to the width of the footings.

The diagram (Fig. 54.6) shows that the settlement of continuous uniformly loaded footings of constant width on a uniform deposit of normally loaded clay can be very large and that the settlement of footings with different widths can be very different. Furthermore, the settlement of footings with the same width can also be very nonuniform, because the compressibility of natural clay strata may vary considerably in horizontal directions. As a matter of fact, in those parts of such cities as Istanbul or Mexico City that are underlain by normally loaded clays, the unequal settlement of the house fronts can be discerned with the naked eye. Fortunately, footing foundations on normally loaded clays are rare exceptions. In most localities even soft clays are precompressed to some extent, either by desiccation or temporary lowering of the water table.

In the few regions where structures must be built above normally or almost normally loaded clays, differential settlements of several inches or even half a foot are commonly considered unavoidable. Attempts to reduce the settlement by reducing the allowable soil pressures to values smaller than those determined on the basis of

$$q_a = \tfrac{1}{3} q_{d\,\text{net}}$$

(Eq. 33.17) are ineffective and wasteful. Hence, the designer must choose between two alternatives. Either he designs the footings on the basis of Eq. 33.17 at the risk of large unequal settlements, or else he provides the structure with another type of foundation (raft, pile, or pier foundation). The characteristics of the alternative types of foundations are discussed in subsequent articles.

Medium and stiff clays beneath a shallow overburden are always precompressed. The allowable soil pressures q_a corresponding to a factor of safety of 3 against a bearing-capacity failure are almost always less than the preconsolidation pressure. As a consequence, the differential settlements of footing foundations on such clays seldom exceed those of adequately designed footing foundations on sand. The maximum settlements, although likely to be greater than those of comparable foundations on sand, are also generally moderate. They can be estimated on the basis of the results of carefully conducted consolidation tests on undisturbed samples. The field e-log p relation should be determined in accordance with the procedures outlined for undisturbed precompressed clays (Article 13). This relation is valid, however, only on the condition that lateral strains in the clay are prevented as they are in the consolidation test. If in the field lateral strains are free to develop, the initial pore pressure produced by the applied load is a function of the pore-pressure coefficient $\bar{\mathbf{A}}$ (Eq. 15.3) which for overconsolidated clays is likely to have a value considerably less than 1.0. As a consequence, the actual settlements due to consolidation are smaller than those calculated from the results of the e–log p curves. The calculated settlement may, with reasonable approximation, be multiplied by a correction factor ranging from about 0.4 to 0.7 for moderately overconsolidated clays to about 0.2 to 0.6 for heavily overconsolidated clays (Skempton and Bjerrum 1957).

Although the settlement of structures supported by footing foundations on soft clay is likely to be excessive if the clay is normally loaded, even a slight precompression may radically reduce the settlements. Unfortunately, even the best techniques of sampling and testing may be inadequate to detect or to permit reliable evaluation of very small preconsolidation pressures (Simons 1963) and the calculated settlements may be much greater than the real ones. On the other hand, the consequences of over-estimating the degree of precompression may be very detrimental. A careful study of the behavior of existing structures on the same deposit may permit a sound appraisal of the conditions. Otherwise, a conservative approach must be adopted.

Allowable Pressure on Soil Intermediate between Sand and Clay

The most important soils intermediate in character between sand and clay are silt and loess. Preliminary information on the nature of a silt can be obtained by means of the standard penetration test. If the number of blows required to drive the sampler (Article 44) is smaller than 10/ft, the silt is loose. If it is greater than 10, the silt is medium or dense.

Loose silt is even less suitable for supporting footings than normally loaded soft clay. This fact is demonstrated by the results of settlement observations on nine structures resting on silt deposits in Germany. Although the soil pressure ranged between the relatively low values of 1.1 and 2.0 tons/ft^2, the settlement ranged between 8 and 40 in. A reduction of the soil pressure by 50% would have greatly increased the cost of the foundations without reducing the settlements to tolerable amounts (L. Casagrande 1936).

Medium or dense silts can be divided into two categories: those with the characteristics of a rock flour, and those that are plastic (Article 2). The allowable pressure on silts of the rock-flour type can be selected crudely by means of the procedures for sand, and that on plastic silts by the methods used for clay.

On important projects for which more refined analyses are justified, the ultimate bearing capacity may be calculated by means of the appropriate equations in Article 33. The evaluation of these equations requires a knowledge of c and ϕ obtained from triaxial tests on undisturbed samples. The samples should be consolidated under the cell pressure before the axial load is increased to failure. The subsequent procedure depends on the rate at which the pore pressures in the soil beneath the footing can be expected to dissipate in relation to the rate at which the load is expected to be applied to the footings. If the silt is relatively impermeable and the rate of loading rapid, consolidated-undrained tests are appropriate. On the other hand, if the silt is permeable and the rate of construction and loading very slow, drained conditions may be approached (Article 15). With silty soils and usual rates of loading, intermediate conditions are likely to prevail. Considerable judgment, aided by estimates of the rate of pore-pressure dissipation based on the theory of consolidation (Article 25), is required for selection of appropriate values of c and ϕ under these circumstances.

Reliable procedures for estimating the settlement of footing foundations above nonplastic silts are not yet available; hence, recourse to the semiempirical methods developed for sands is unavoidable. Settlement forecasts for the plastic varieties can be based on the results of laboratory consolidation tests on undisturbed samples.

The second important soil intermediate in character between sand and clay is loess (Article 2). It covers large areas in the central part of each of the five continents.

Because of the cohesive binder and the root holes typical of every true loess, the properties of loess are very different from those of other soils with similar grain-size characteristics. The bearing capacity of a normally loaded silt is commonly very low, whereas that of loess may be very high. If a stratum of true loess is located permanently above the water table, it may be capable of supporting footings at a soil pressure of 2 or 3 tons/ft^2 without perceptible settlement.

Nevertheless, loess cannot always be trusted, because in some localities its bearing capacity changes considerably with the seasons. The changes are caused by variations in the strength of the cohesive bond due to changes in moisture content. Thus, for example, the foundation for a coal bin in central Russia was designed on the basis of the results of load tests made during the summer. The bin was also constructed in summertime. Before construction was finished the autumn rains began, whereupon the bin started to settle unequally and the walls cracked. In central Germany a boilerhouse was built on a stratum of loess located partly below water table. Here also the designers were deceived by the apparent strength of the soil. The footings were designed on the basis of a soil pressure of 1.2 tons/ft^2, but under very much smaller loads the settlement had already become excessive. Some of the footings had to be underpinned, whereas others were redesigned during construction for a soil pressure as low as 0.35 tons/ft^2 (Scheidig 1934).

Because of the extraordinary variety of the physical properties of loess soils, no simple empirical rules comparable to those for sand or clay can be established for evaluating the allowable soil pressure.

Hence, if a footing foundation is to be constructed on loess in a region where there are no precedents, the designer must resort to the load-test method combined with an investigation of the effect of moisture on the bearing capacity of the soil. In some instances he will learn that a footing foundation is impracticable in spite of the apparent solidity of the loess. (Clevenger 1958, Peck and Ireland 1958).

Footing Foundations Located on Firm Soil above Soft Layers

The values given for the settlements of footings designed in accordance with the rules discussed under the preceding subheadings are based on the assumption that the soil does not become softer with depth. If this condition is not satisfied, the values cease to be reliable. The reason for this fact is illustrated by Fig. 54.7.

Figure 54.7 shows the stress conditions beneath a footing that rests on a firm stratum A located above a soft stratum B. If the upper boundary of the soft stratum is located close to the base of the footing, the footing may break through the firm layer into the soft deposit. Failures of this type are not uncommon (Skempton 1942). They can be avoided by giving the footing such dimensions that the pressure on the upper boundary of stratum B does not exceed the allowable bearing value for the soil in that stratum. The pressure at the boundary can be computed by the method described in Article 40. Less accurately, the total footing load can be assumed uniformly dis-

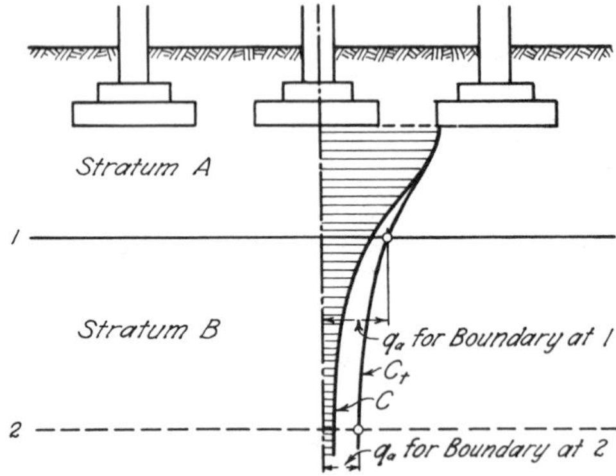

Fig. 54.7. Diagram illustrating method of calculation to ascertain whether allowable soil pressure is exceeded for members of stratified clay subsoil. Curve C represents variation with depth of vertical pressure below single footing neglecting influence of adjacent footings. Curve C_t represents vertical pressure below same footing, considering influence of adjacent footings.

tributed over the base of a truncated pyramid whose sides slope from the edges of the footing to the upper surface of B at an angle of 60° with the horizontal.

If the upper boundary of the soft stratum B is located at a considerable depth below the base of the footings, failure by breaking into the ground cannot occur because stratum A acts like a thick raft that distributes the entire weight of the building almost uniformly over the surface of B. The flexural rigidity of this natural raft prevents the surface of B from heaving beyond the loaded area. Nevertheless, the settlement may be very large. For example, the weight of the building represented in Fig. 54.8 is transmitted by continuous footings onto a stratum of dense sand and gravel that rests, at a depth of 23 ft below the footings, on a layer of soft clay 50 ft thick. The footings were designed for a soil pressure of 2.5 tons/ft^2, a conservative value for dense sand and gravel. The greatest pressure on the surface of the clay due to the weight of the building was 1.1 tons/ft^2. During the construction period, which lasted 1 year, the footings settled between 1 and 4 in. During the following 40 years the maximum settlement increased to about 3 ft. Since the basement floor, which rested on the sand between the footings, neither cracked nor moved with respect to the footings, it is evident that the layer of sand and the footings settled together.

Ten years after construction the deterioration of the building was such that the owners decided to strengthen the foundation. In spite of the symptoms previously mentioned, it was not suspected that the seat of settlement was located below the sand. Hence, the "strengthening" was accomplished by increasing the width of the footings so that the intensity of the pressure exerted by the footings was reduced about 30%. However, since the pressure on the clay remained unchanged, the expensive alterations did not have the slightest effect on the trend of the time-settlement curves shown in Fig. 54.8*c*.

At a later date undisturbed samples were taken from the clay at some distance from the building. On the basis of the results of consolidation tests the average rate of settlement for the building as a whole was computed. The theoretical trend of the settlement, represented by the dash curve in Fig. 54.8*c*, is very similar to the real one except for the secondary time effect which cannot yet be estimated reliably (Article 14). On account of the secondary time effect, the real settlement approaches a constant rate ranging for different parts of the structure from 0.12 to 0.32 in. per year, whereas the curve of computed settlement approaches a horizontal asymptote (Terzaghi 1935).

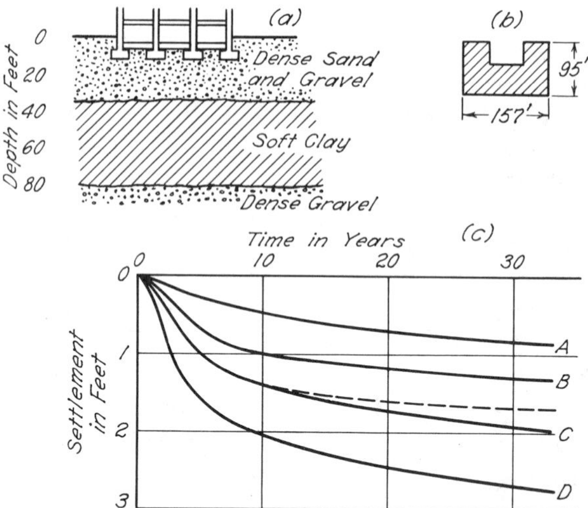

Fig. 54.8. (*a*) **Cross section through foundation of structure supported by dense sand underlain by soft clay.** (*b*) **Plan of structure.** (*c*) **Observed time-settlement curves. Dash curve represents time-settlement relation computed from results of consolidation tests (after Terzaghi 1935).**

The observations illustrated by Fig. 54.8 show very clearly that the settlement due to consolidation of soft layers located at a considerable depth below the footings is in reality practically independent of the pressure on the base of the footings, because the firm stratum supporting the footings acts like a natural raft that distributes the load from the footings over the softer strata. Procedures for computing the settlements caused by the consolidation of the lower layers and methods for reducing them are discussed in connection with raft foundations (Article 55). After the foundations are laid out in such a manner that the settlement due to consolidation of the soft layers will be tolerable, the footings can be designed as if the soft strata did not exist. Hence, the presence of the soft strata may compel the designer to change the entire layout of the foundation, but it has no bearing on the allowable soil pressure for the footings.

Summary of Rules for Selecting Allowable Soil Pressure

1. Except for narrow footings on loose saturated sand, the allowable bearing values for sand are governed only by settlement considerations, because it can be taken for granted that the factor of safety with respect to a base failure is adequate. The rules suggested for choosing

these values satisfy the condition that the maximum settlement is unlikely to exceed 1 in. and the differential settlement $\frac{3}{4}$ in. On routine jobs the allowable soil pressure on dry and moist sand can be determined by means of the chart (Fig. 54.4) on the basis of the results of standard penetration or static cone tests. If the water table is located close to or above the base of the footings, the depth ratio D_f/B must also be considered. If the depth ratio is very small, the values obtained from the chart should be reduced by 50%; if it is close to unity, the values need be reduced by only one third. The most important sources of error in this procedure and the means for avoiding them have been discussed. On large jobs the load-test procedure may be used. However, it is expensive and cumbersome and, if it is not expertly planned and executed, the results may be very misleading. If the sand is very loose and saturated, it should be compacted.

2. The allowable bearing value for clay is usually determined by the condition that the factor of safety with respect to breaking into the ground should be at least 3. The ultimate bearing capacity can be calculated on the basis of the equations given in Article 33 and the results of unconfined compression tests or undrained shear tests of the clay beneath the proposed footings. After the allowable soil pressure has been selected, it is necessary to find out whether the settlement will be tolerable. If the clay is normally loaded, the settlement is likely to be excessive, and a type of foundation other than a footing foundation may be indicated. On the other hand, if the clay is precompressed, the differential settlement is likely to be tolerable. In doubtful cases the load-test method should be used. The allowable soil pressure on stiff fissured clays can be determined only by this method.

3. Loose saturated silt of any kind is unsuitable for supporting the weight of a foundation on footings. The allowable bearing value of medium or dense silt of the rock-flour type can be estimated crudely by means of the procedures proposed for sand. That of medium or stiff plastic silt can be approximated by the procedures for clays. More refined estimates require the performance of triaxial tests and mature judgment concerning the rate of dissipation of pore pressure in the silt as the loads on the footings increase. No general rules can be established for ascertaining the allowable soil pressure for loess.

4. If the area occupied by the footings exceeds half the total area covered by the building, it is commonly more economical to provide the building with a raft foundation.

Design of Footings

Steps in Design. The first step in designing a footing is to compute the total effective load that will be transferred to the subsoil at the base of the footing. The second step is to determine the allowable bearing value for the soil. The area of the footing is then obtained by dividing the total effective load by the allowable bearing value. Finally, the bending moments and shears in the footing are computed, and the structural design of the footing is carried out.

Design Loads. The total effective or *excess load* Q_t transferred to the subgrade may be expressed by the equation,

$$Q_t = [Q - W_s] + Q_l = Q_{dn} + Q_l \tag{54.4}$$

in which

Q = permanent or *dead load* on the base of the footing, including the weight of the footing and the soil located above the footing. If the water table is higher than the base of the footing, the hydrostatic uplift (Article 12) on the submerged part of the body of soil and concrete should be deducted.

W_s = effective weight of the soil (total weight of soil reduced by hydrostatic uplift) that was located above the base of the footing prior to excavation. However, in connection with basement footings such as *c* and *d* in Fig. 53.1, the weight of the soil previously located above the basement floor should not be deducted, because the soil was removed not only above the base but also above the area adjoining at least one side of the base.

Q_{dn} = $Q - W_s$ = net dead load

Q_l = live load on footing, including that due to wind and snow

In any discussion of live load, a distinction must be made between the *normal live load* and the *maximum live load*. The normal live load Q_{ln} is that part of the live load which acts on the foundation at least as often as once a year; the maximum live load $Q_{\max}$ acts only during the simultaneous occurrence of several exceptional events. For instance, the normal live load in a tall office building includes only the weight of the equipment and furniture, of the persons who normally occupy the building on weekdays, and of a normal snow load. The maximum live load is the sum of the weights of the furniture and equipment, of the maximum number of persons who may crowd into the building on exceptional occasions, combined with the maximum snow and wind load. The total excess load on a footing at normal live load will be indicated by the symbol

$$Q_{tn} = Q_{dn} + Q_{ln} \tag{54.5}$$

and at maximum live load by

$$Q_{t\max} = Q_{dn} + Q_{l\max} \tag{54.6}$$

Because of the exceptional character of the maximum live load and the low probability that the foundation will ever be called on to sustain it, it is customary to design footings in such a manner that the soil pressure produced by the normal total load Q_{tn} is the same for all the footings. However, sound engineering also requires that even the maximum load $Q_{t\max}$ should not cause irreparable damage to the structure. The procedure for complying with this requirement without excessive expenditure depends on the type of subsoil.

If the footings rest on sand, an increase of load produces an almost simultaneous increase of settlement, but it can be assumed that the factor of safety with respect to a foundation failure remains adequate. In order to eliminate the possibility of serious damage due to the maximum live load, the designer should estimate the greatest differential settlement ΔS in excess of the normal value of $\frac{3}{4}$ in. that, in his judgment, the structure can stand without serious injury. An additional differential settlement of ΔS would correspond to a maximum settlement of 1.33 ΔS plus the normal maximum value of 1 in.

If all of the footings were designed on the basis of a maximum settlement of 1 in. at normal live load, the maximum live load would increase the maximum settlement to

$$S_{\max} = 1'' \times \frac{Q_{t\max}}{Q_{tn}} \tag{54.7}$$

If $S_{\max}$ is smaller than the tolerable maximum of $(1.33\Delta S + 1'')$, the maximum live load can be disregarded. On the other hand, if $S_{\max}$ is larger than $(1.33\Delta S + 1'')$, the footings should be designed so that the soil pressure at normal live load is

$$q_a' = q_a \frac{1.33\Delta S + 1''}{S_{\max}} \tag{54.8}$$

The value of q_a' is commonly different for different footings. The smallest value should be used for proportioning all the footings; it corresponds to the footing for which the ratio $Q_{t\max}/Q_{tn}$ is greatest.

If the footings of a building rest on clay, the allowable soil pressure is determined by the conditions that under the normal total load the factor of safety against failure should be equal to 3, but under no circumstances should it be less than 2. If the factor of safety F at normal total load is equal to 3, the factor of safety F' at maximum total load is

$$F' = 3 \frac{Q_{tn}}{Q_{t\max}} \tag{54.9}$$

If F' is equal to 2 or more, the maximum live load can be disregarded, and all the footings can be proportioned for normal live load on the basis of $F = 3$. On the other hand, if F' is less than 2, the allowable soil pressure must be so chosen that the factor of safety at normal live load is equal to $6/F'$.

Reduction of Settlement by Adjusting Footing Size. In the discussion of allowable soil pressure, it was mentioned that the settlement of loaded areas with similar shape but different size increases at a given intensity of load with increasing width of the area. If the footings of a structure differ greatly in size, the differential settlement due to this cause can be important. In such instances it may be justifiable to adapt the pressure on the base of the footings to some extent to the size of the footings. If the subsoil consists of sand, the differential settlement can be reduced by decreasing the size of the smallest footings, because even after the reduction the factor of safety F of these footings with respect to breaking into the ground is likely to be adequate. The application of this procedure to footing foundations on clay would reduce the value of F for the smallest footings to less than 3, which is not admissible. Hence, the differential settlement of footing foundations on clay can be reduced only by increasing the size of the largest footings beyond that required by the allowable soil pressure. However, sound judgment is required to make such adjustments with prospects for success, because periodic and exceptional changes in the loading conditions must be considered.

Layout of Footings and Computation of Moments. It is customary to lay out each footing so that the resultant load Q_{tn} (Eq. 54.5) passes through the centroid of the area covered by the footing. The bending moments are then computed on the assumption that the soil pressure is distributed uniformly over the base. In reality, the contact pressure against footings on sand decreases from the center toward the rim (Fig. 42.2*b*) and the real bending moments are usually less than the computed ones. On the other hand, if the footings are very rigid, and they rest on soft or medium clay, the contact pressure may increase toward the rim (Fig. 42.2*a*) and the real moments may exceed the computed ones. However, the difference is amply covered by the margin of safety customarily provided in structural design.

The columns that support crane runways in industrial buildings are subject to large eccentric loads whenever the crane operates near by, but during the rest of the time they carry ordinary dead and live loads. It is customary to design the connections between the columns and the footings for the eccentric loads. Consequently, the moments are transmitted to the base of the footings. If the footings rest on clay, the allowable soil pressure q_a should not be exceeded under the toe of any footing when all the loads, including that due to the crane, are acting. The centroid of the base of every footing should be made to coincide with the resultant of the net dead load, the normal live load, and a small fraction, such as 25%, of the crane load; and all of the footings should be proportioned for the same soil pressure under this resultant load. On the other hand, if the footings rest on sand, they should be laid out so that the soil pressure is uniform and equal to q_a under the net dead load, the normal live load, and the maximum crane load that can be expected under ordinary operating conditions. Under no conceivable combination of loads should the pressure $1.5q_a$ be exceeded.

PRECAUTIONS DURING CONSTRUCTION. All footing foundations are inevitably designed on the assumption that the soil beneath the footings is in approximately the same state as that disclosed by whatever borings or load tests were made. If the soil contains soft pockets not encountered by the borings, or if the soil structure is disturbed during excavation, the settlement will be larger and more unequal than the designer anticipated. To avoid such a risk a simple penetration test should be made at the site of each footing after the excavation is completed. One of several practicable methods is merely to count the number of blows per foot required for driving a sounding rod into the ground by means of a drop weight. If exceptionally soft spots are encountered within the seat of settlement of any one footing, this footing should be redesigned. Such a procedure is more economical than subsequent repair.

Disturbance of the structure of the subsoil during construction is especially likely to occur under two conditions commonly encountered in the field. If the subsoil consists chiefly of silt or fine sand, it can be radically disturbed by pumping from open sumps. The disturbance is likely to be associated with serious damage to adjoining property due to loss of ground. Hence, if footings on such soils require excavation below the water table, the site should be drained by pumping from well points and not from open sumps (Article 47). Pumping from well points occasionally causes a noticeable settlement of the adjoining ground surface. However, if this does occur, it is certain that the detrimental effects of pumping from open sumps would have been far greater.

If the subsoil consists of clay, the top layer of the exposed clay is likely to become soft because of the absorption of moisture from puddles and the kneading effect of walking on it. Therefore, footings on clay should be concreted and backfilled immediately after the excavation is completed. If this cannot be done, the last 4 to 6 in. of clay should not be removed until preparations for placing the concrete are complete.

Selected Reading

Peck, R. B. (1948). "History of building foundations in Chicago," *U. of Ill. Eng. Exp. Sta. Bull. 373,* 64 pp.

Sowers, G. F. (1962). "Shallow foundations," Chapter 6 in *Foundation engineering,* G. A. Leonards, ed., New York, McGraw Hill, pp. 525–632.

Aldrich, H. P. (1965). "Precompression for support of shallow foundations," *ASCE J. Soil Mech.,* **91,** No. SM2, pp. 5–20.

Meyerhof, G. G. (1965). "Shallow foundations," *ASCE J. Soil Mech.,* **91,** No. SM2, pp. 21–31.

ART. 55 RAFT FOUNDATIONS

Comparison between Raft and Footing Foundations

If the sum of the base areas of the footings required to support a structure exceeds about half the total building area, it is usually preferable to combine the footings into a single mat or raft. Such a raft is only a large footing and, like a footing, it must satisfy the requirements that the factor of safety with respect to a base failure should be not less than 3 and that the settlement should not exceed an amount acceptable to the designer of the superstructure.

The factor of safety of raft foundations depends on the nature of the subsoil. If the soil consists of very loose sand in a saturated state, it should be compacted by artificial means before the raft is constructed (Article 50). If the sand is medium or dense, the factor of safety of a raft is considerably greater than that of footings, and its adequacy can be taken for granted without any computation.

The factor of safety of raft foundations on clay is practically independent of the size of the loaded area. It is commonly very low, and several failures have occurred. One of these is illustrated by Fig. 55.1. The structure, a grain elevator near Winnipeg, Canada, was 77 by 195 ft in plan and 102 ft high. It rested on a stratum of "firm" clay overlying rock. On the basis of the results of load tests it was estimated that the ultimate bearing capacity of the clay was between 4 and 5 ton/ft², and the design was based on an allowable soil pressure of 2.5 ton/ft². When the excess or net load on the raft approach this value, one side of the structure settled 29 ft, whereas the opposite side rose 5 ft. The movements took place within less than 24 hr (Peck and Bryant 1953, White 1953). To avoid the risk of such a failure, a raft foundation on clay soil should be designed so that the excess load divided by the loaded area does not exceed one-third the value of $q_{d\ \mathrm{net}}$ determined by Eq. 33.17.

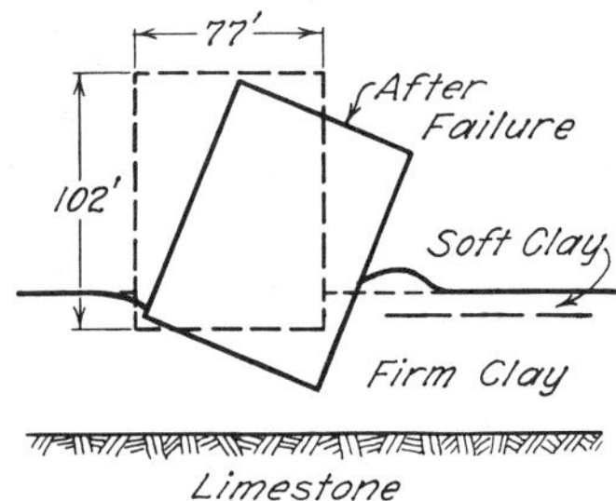

Fig. 55.1. Diagram illustrating failure of grain elevator near Winnipeg, Canada, by breaking into clay stratum.

The excess load on the base of a raft is computed in the same manner as that on the base of a spread footing (Article 54). If the raft is located beneath a basement (Fig. 55.3), it constitutes with the basement walls a large hollow footing. Since the loaded soil can rise, as indicated by an arrow, only outside the area occupied by the raft, the depth of the overburden is equal to D_f, measured from the ground surface, and not to D_{fs} as for spread footings below basements (Fig. 53.1*c* and *d*). Hence, the excess load Q_t on the base of the raft is equal to the difference between the total effective load $Q + Q_l$ at the base of the raft, reduced by the total effective weight W_s of the soil replaced by the basement, or

$$Q_t = (Q + Q_l) - W_s \tag{55.1}$$

If q_a is the allowable pressure on the subsoil, and A the area covered by the raft, the foundation must satisfy the condition,

$$\frac{Q_t}{A} \lesseqgtr q_a \tag{55.2}$$

The relation expressed by Eq. 55.1 indicates that the excess load on the base of a raft can be reduced by increasing the depth of the basement. This reduction increases the factor of safety of the foundation with respect to breaking into the ground and reduces the settlement. The existence of such a relation was recognized by a few engineers over a century ago, and they used it to advantage in establishing heavy structures on soft ground without the use of piles.

Although the rules governing the factor of safety of rafts and footings are quite similar, the general character of the settlement of these two types of foundations is very different. The causes of the difference are illustrated in Fig. 55.2. This figure represents a vertical section through each of two structures, one of which rests on footings and the other on a raft. The footings and the raft both exert on the subsoil the same load per unit of area, as indicated by the rectangular diagrams at the base of the foundations. In addition, the figure shows the intensity and the distribution of the vertical pressure at different depths below the base level of each foundation.

The footings shown in Fig. 55.2*a* are so far apart that each one settles much as if the others did not exist. If the soil were homogeneous, the footings would settle almost equally; in reality, they settle by different amounts because no natural soil stratum is homogeneous. Since the seat of settlement is located within the uppermost soil stratum, the distribution of the settlement reflects the variations in the compressibility of the soil located within this stratum (Fig. 54.2). It is always erratic and cannot be predicted by any practicable means. This fact determined the rules that were established for evaluating the allowable soil pressures for footing foundations (Article 54).

The seat of settlement of the raft foundation (Fig. 55.2*b*) extends to a very much greater depth than that of the footing foundation. Within this depth weak spots are scattered at random, as shown in Fig. 55.3, and their effects on the settlement of a loaded area partly cancel each other. Therefore, the area settles as if the loaded soil

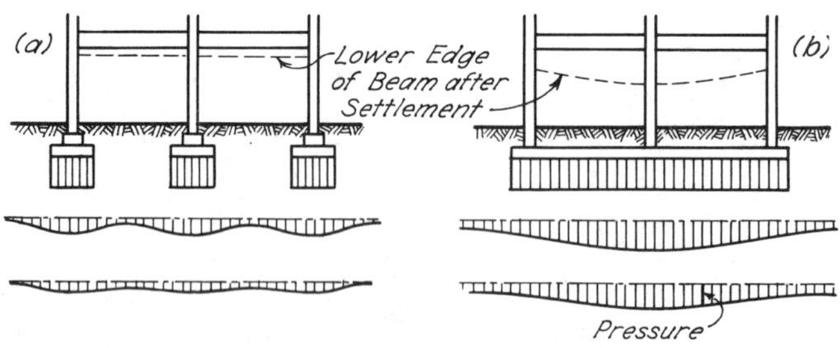

Fig. 55.2. Distribution of pressure in soil beneath buildings supported by (*a*) widely spaced spread footings and (*b*) concrete mat. The load per unit of area is the same beneath the footings and the mat.

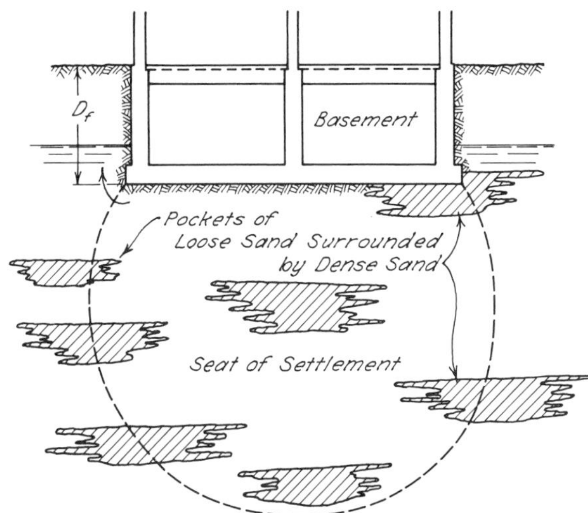

Fig. 55.3. Diagram representing erratic distribution of pockets of loose sand throughout a stratum of dense sand located beneath base of a building.

were more or less homogeneous. The settlement is not necessarily uniform, but it follows a fairly definite instead of an erratic pattern. The pattern differs, however, depending on whether the soil located within the seat of settlement consists of sand or clay.

Settlement of Raft Foundations

Both theory and experience indicate that the settlement of a uniformly loaded area on sand is fairly uniform, provided the area is located at a depth of more than about 8 ft below the adjacent ground surface. If the depth is smaller, the outer parts of the loaded area are likely to settle somewhat more than the central part unless lateral yield of the sand is prevented within a depth of 8 or 10 ft from the ground surface.

The differential settlement of the area covered by the raft reflects in a general way the variations in the compressibility of the subsoil. However, because of the random distribution of compressible zones in the subsoil (Fig. 55.3) combined with the stiffening effect of the raft and building frame, it can safely be assumed that the differential settlement of a raft foundation per inch of maximum settlement is not more than one half the corresponding value for buildings on footings. Hence, if a differential settlement of $\frac{3}{4}$ in. can be tolerated, the allowable soil pressure can be so chosen that the maximum settlement is 2 in. instead of 1 in. as specified for spread footings. The width B of rafts commonly lies between 40 and 120 ft. Within this range the value of B has very little influence on the maximum settlement (Fig. 54.3). Therefore, the width can be disregarded in selecting the allowable soil pressure. Finally, at least the major part of the sand located within the seat of settlement is likely to be saturated, because the vertical distance between the base of the raft and the water table is commonly small compared to the width of the raft.

The preceding conditions determine the allowable soil pressure, provided the average compressibility of the sand is also taken into account. This property is closely related to the relative density. At present, the most expedient methods for investigating the relative density are the standard penetration test or the Dutch-Cone test (Article 44).

If the standard penetration test is used, one test should be made for every $2\frac{1}{2}$ ft of the depth of the drill hole from the level of the base of the raft to a depth B below this level. The N-value for the hole is equal to the average of all the N-values within this depth. At least 6 drill holes are required, and the allowable soil pressure should be chosen on the basis of the smallest N-value furnished by the tests.

Allowable soil pressures corresponding to different N-values are given in Table 55.1. The values are based on the assumption that

Table 55.1
Proposed Allowable Bearing Values for Rafts on Sand

Relative density of sand	Loose	Medium	Dense	Very dense
N	Less than 10	10–30	30–50	Over 50
q_a	Requires compaction	0.7–2.5	2.5–4.5	Over 4.5

Values are based on maximum settlement of 2 in.

Depth of sand stratum is presumed to be greater than the width B of the raft, and water table to be close to or above base of raft. If depth of bedrock is much less than $B/2$, or if water table is at depth greater than $B/2$, the allowable bearing values can be increased.

The loads are presumed to be distributed fairly uniformly over the base of the building. If different parts of a large raft on sand carry very different loads per unit of area, it is advisable to establish construction joints at the boundaries between these parts.

N = number of blows per foot in standard penetration test
q_a = proposed allowable bearing value in tons/ft^2

the allowable soil pressure on the base of a raft is twice the bearing value for saturated sand obtained by extrapolation from the chart (Fig. 54.4). This assumption is based on the conclusion that the tolerable maximum settlement of rafts is 2 in., in contrast to 1 in. for buildings on spread footings.

If the subsoil contains gravel, appropriate check tests or corrections are required (Article 54) which may lead to lower values than those given in Table 55.1. On the other hand, if the sand rests at a depth less than $B/2$ on sound rock, or if the water table is located permanently below this depth, somewhat higher pressures can be tolerated.

If the investigation is carried out by means of the Dutch Cone, at least 6 soundings are required with essentially continuous records of penetration resistance q_p as a function of depth. For each sounding the average value of q_p should be determined for a depth B below the level of the base of the raft. On the basis of the smallest average q_p value, the corresponding average N-value may be estimated crudely by means of Eq. 54.2, whereupon the allowable soil pressure may be taken from Table 55.1.

The results of the Dutch-Cone test may also be used for evaluating an index of compressibility C for the sand and for calculating the settlement of the raft in a manner analogous to that used for clay. The index of compressibility is estimated by means of the statistical relation (Buisman 1943)

$$C = 1.5 \frac{q_p}{\bar{p}_0} \tag{55.3}$$

where $\bar{p}_0$ is the effective overburden pressure at the level where the penetration resistance was measured. The increase Δp in vertical pressure at depths z below the base of the footing, due to the load on the footing, is calculated on the assumption that the subsoil is elastic (Article 40) and the settlement is estimated from the equation

$$S = \int_0^\infty \frac{1}{C} \log_\epsilon \frac{\bar{p}_0 + \Delta p}{\bar{p}_0} \, dz \tag{55.4}$$

The results of such calculations appear to be reasonable and generally on the conservative side (DeBeer and Martens 1957, Meyerhof 1965, Bogdanovič et al. 1963).

The allowable soil pressures determined on the basis of the preceding discussion are based on the tacit assumption that the distribution of the loads over the raft is fairly uniform. If the structure supported by the raft consists of several parts with very different heights, it may be advisable to provide construction joints at the boundaries between these parts.

The maximum permissible value for the soil pressure beneath rafts on clay, like that beneath footings on clay, is obtained by dividing the ultimate net bearing capacity $q_{d\ \mathrm{net}}$ (Eq. 33.17) by a factor of safety F equal to 3 for dead load and not more than 2 for dead load plus extreme combinations of live load. However, because of the large dimensions of the area covered by a raft and the rapid increase of settlement of clay with increasing size of the loaded area (Fig. 54.6), it is always necessary to find out, at least by a crude estimate, whether the settlement will be tolerable. The computation can be based on the assumption that the loaded clay is laterally confined. The results of the computations show, in accordance with experience, that the base of a uniformly loaded area on clay assumes the shape of a shallow bowl, because the consolidation pressure decreases from the center toward the edges (Fig. 55.2*b*). However, the slopes of the bowl are so gentle that the difference between the settlement of two adjacent columns never exceeds a small fraction of the difference between the maximum and the minimum settlement. For rafts on sand the differences can be almost equal. Therefore, the tolerable differential settlement for rafts on clay is very much greater than that for rafts on sand.

Design of Raft Foundations

The average gross load per unit of area on the base of a raft is equal to the total effective weight of the building, $Q + Q_l$, divided by the total area A of the base. Since the area occupied by the raft can only be equal to or slightly greater than the area occupied by the building, the designer has no opportunity to change the soil pressure by adjusting the size of the raft. Hence, to satisfy Eq. 55.2 he is compelled to increase W_s (Eq. 55.1). This can be done only by providing the structure with one or more basements with adequate depth. The required depth can be computed by trial.

After the depth of basement has been determined, the next step in the design is to compute the forces that act on the raft. In this operation, the designer must depend to a large extent on the soundness of his judgment. The factors and conditions that need to be considered are illustrated by Fig. 55.4.

Figure 55.4*a* shows a vertical section through a structure consisting of a heavy tower and two wings. The water table is located below the base of the raft. On this assumption the total soil reaction is equal to the full weight $Q + Q_l$ of the building including the weight

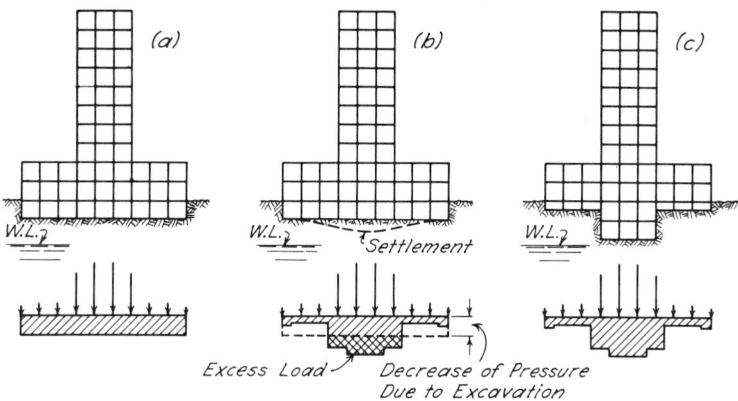

Fig. 55.4. Diagram illustrating three different methods for the design of raft foundations on very compressible subsoil. (*a*) Rigid superstructure, capable of enforcing uniform settlement. (*b*) Flexible superstructure, capable of sustaining large deflections without damage. (*c*) Flexible superstructure, uniform settlement produced by adapting depths of basements to weight of structure located above them.

of the raft, whereas the excess load Q_t (Eq. 55.1) which determines the settlement is equal to the difference between the weight of the structure and the weight W_s of the soil that has been excavated. If the excess load Q_t is zero and if, in addition, the structure is rigid, there will be practically no settlement even if the soil reaction is very large. As a rough approximation the soil reaction on the base of a rigid structure may be considered uniform, as indicated by the shaded rectangle in Fig. 55.4*a*. Yet, the loads are concentrated on the central part of the base of the building. Therefore, the frame of the structure is acted on by very severe bending moments. The cost of the reinforcement required to carry these moments may be prohibitive.

If the building is flexible, the soil reaction on every part of the raft is roughly equal to the load that acts on it (Fig. 55.4*b*). The corresponding bending moments are relatively small. However, because of the heavy concentration of loads on the middle part of the raft, this part carries an excess load whereas the excess load on the outer parts is negative. Consequently, the tower will settle more than the wings, as shown in the figure. A difference in settlement is inovitable even if the total excess load on the subsoil is zero. If the building rests on sand, the difference between the settlements of the tower and the wings is likely to be too small to have an injurious effect on the superstructure, and the raft can be designed as if it were acted on by the forces shown in Fig. 55.4*b*. On the other hand, if the raft rests on clay, the differential settlement due to the nonuniform pressure distribution may be very large. Construction joints between the tower and the wings may slightly improve the stress conditions in the members of the superstructure, but they cannot prevent the settlement of both wings from increasing toward the tower. Hence, it is necessary to make a settlement computation to determine whether the differential settlement is likely to exceed what the structure can stand without injury. If it is, the designer must choose between two alternatives. Either he specifies a pile or a pier foundation for the structure, or else he provides the tower and the wings with basements of different depth (Fig. 55.4*c*). The depth of each basement must be determined in such a manner that the settlement of the tower and the wings would theoretically be equal. If this condition is satisfied, the designer can be fairly certain that the differential settlement will be tolerable.

In computing the thickness of the raft and the amount of reinforcement, it is commonly assumed that the raft is a continuous slab freely supported at every point and along every line at which load is transferred onto the slab from above. A distributed load acts on the slab from below. It is equal to the total soil reaction which, in turn, is equal to the full weight of the building without any deduction for hydrostatic uplift or basement excavation. Since the difference between the theoretical and the real distribution of the bending moments in the raft can be very large, it is commonly advisable to provide the raft with considerably more than the theoretical amount of reinforcement and to allow for the possibility that even the signs of the bending moments may differ from those calculated.

In the preceding discussion, it has been tacitly assumed that a rigid raft does not settle until the load on the raft becomes equal to the weight of the excavated soil. In many instances, the error due to this assumption can safely be ignored. However, if the subsoil is soft and the excavation is deep, the settlement that occurs before the effective load on the raft becomes equal to the effective weight of the excavated soil may be large enough to require consideration. The cause of this settlement is discussed under the following subheading.

Heave during Basement Excavation

The excavation for a basement or a subbasement involves the complete removal of the pressure originally exerted against the soil at the base level of the raft. Consequently, the bottom of the excavation rises. During the subsequent period of construction, the weight of the building becomes equal to and generally exceeds the original overburden pressure; hence, the heave disappears, and the building settles. If the building has a greater weight than the excavated soil, the settlement passes through two stages. The first lasts until the load per unit of area at the base of the raft becomes equal to the original overburden pressure, and the second begins when this pressure is exceeded. The characteristics of the settlement during the second stage have already been described. Those of the first stage may be very different.

At the end of the first stage, when the building load becomes equal to the weight of excavated material, the settlement is equal to or slightly greater than the preceding heave, which is commonly very small. If the building load is not further increased, the settlement stops shortly after construction is finished. It has been mentioned that this fact was utilized long ago, in the design of buildings on soft soils, but it is not generally realized that the progressive settlement of buildings on stiffer soils can also be eliminated by excavating enough soil to compensate for the weight of the building. As a matter of fact, some buildings with basements deep enough to satisfy this requirement have actually been provided with expensive pile support; it is obvious that the money spent for the piles was wasted.

The amount of the heave and subsequent settlement depends on the nature of the subsoil and the dimensions of the excavation. If the excavation is made in sand above the water table, the heave is so small that it can usually be disregarded. A soft clay subsoil deforms at practically constant water content like an incompressible elastically isotropic material. Hence, the heave can be computed on the basis of the theory of elasticity provided the modulus of elasticity is determined by means of soil tests. Unfortunately, the value of the initial tangent modulus E_i (Article 15) is extremely sensitive to the degree of disturbance of the samples. Consequently, the heave may be grossly overestimated.

The samples for evaluating E_i should be obtained with the least possible disturbance. They should be tested under undrained conditions in a triaxial device under a cell pressure approximately equal to the effective overburden pressure. The axial load should alternately be increased moderately and then reduced to zero several times, and the value of E_i taken as the initial slope of the stress-strain curve under the last increase of axial load. If these precautions are followed, the heave is not likely to be greatly overestimated unless the soil in its natural state possesses a rigidity that cannot survive the operation of sampling.

If the clay beneath an excavation contains a great number of continuous layers or seams of coarse silt or sand, the water content of the clay may increase to such an extent that the major part of the heave is caused by swelling. Predictions of the rate of swelling, based on the results of laboratory consolidation tests, are likely to be very inaccurate, because the degree of continuity of the pervious strata cannot be learned by sampling in advance of construction.

If the depth of the basement is increased by open excavation beyond a certain value, the bottom of the excavation becomes unstable and fails by heaving, regardless of the strength and nature of the lateral support for the sides (Article 37). However, the critical depth can be almost doubled by performing the excavation under compressed air. In excessively soft ground, raft foundations have been successfully established by constructing the side walls and floor of the basement as a unit near the surface of the ground and lowering the entire unit to grade by washing or pumping through holes in the floor.

Footing Foundations on Natural Rafts

If the footings of a building rest on a thick firm stratum underlain by considerably more compressible ones, the firm stratum acts like a natural raft that distributes the weight of the building over the soft layers. The footings are designed as if the soft strata did not exist, because the settlement due to consolidation of the soft strata is practically independent of the pressure on the base of the footing.

The load responsible for the settlement due to consolidation is equal to the total effective weight of the building reduced by the effective weight of the excavated soil. In the computation of the magnitude and distribution of the consolidation pressure within the soft layers, the weight of the excavated soil is assumed to represent a negative load uniformly distributed over the bottom of the basement. The weight of the building is a positive load that acts on the bases of the footings. At any point in the soft layers the consolidation pressure is equal to the difference between the pressures produced by these two loads. The settlement due to consolidation is estimated on the assumption that the soft soil is laterally confined. The importance of the settlement that may ensue is illustrated by Fig. 54.8.

If the computation shows that the settlement conditions are unacceptable, the design of the foundation must be changed. This can be done, for instance, by providing the different parts of the building with basements of different depths (Fig. 55.4*c*) or by supporting the structure on piles or piers.

Footings on Sand in Basements below the Water Table

A basement located below the water table must be provided with a watertight floor slab interconnecting the footings. If the load on the footings is applied after the slab is concreted, the footings together with the slab constitute a raft whose base is acted on not only by water pressure but also by a more or less uniformly distributed soil reaction.

In order to avoid the necessity for making the floor slab strong enough to withstand both pressures, the slab between the footings should not be concreted until the footings carry the full dead load. The load on the base of the footings will then be equal to the full weight of the building reduced by the full hydrostatic uplift on the cellar floor, and the interconnecting slab will be acted on by water pressure only. However, the footings must be designed on the assumption that the hydrostatic uplift is inactive, because the water table is not allowed to rise above the cellar floor until the footings carry the full dead weight of the structure. The postponement of the construction of the floor slab requires the continuation of pumping until the superstructure is completed. The sequence of operations is shown in Fig. 55.5. To prevent the floor slab from floating, it must be anchored either to the columns or, preferably, to the footings.

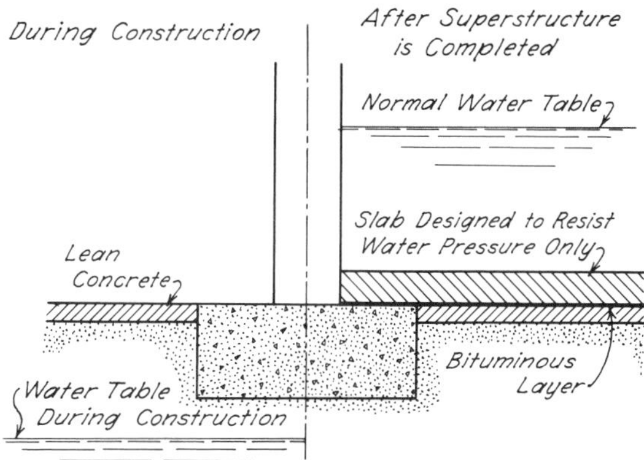

Fig. 55.5. Details of footing on sand in basement located below water table.

Summary of Rules for the Design of Raft Foundations

1. If a structure resting on a sand stratum can stand a differential settlement of $\frac{3}{4}$ in. between adjacent columns without injury, a maximum settlement of 2 in. can be tolerated. The corresponding allowable soil pressures are given in Table 55.1.

2. The allowable soil pressure for rafts with a width B cannot be determined reliably by means of load tests unless several sets of tests are made, at several depths within a distance B below the base of the raft. Such tests are economically justified only under exceptional conditions.

3. If different parts of a large raft on sand carry very different loads per unit of area, it is advisable to establish construction joints at the boundaries between these parts.

4. A raft foundation on clay should satisfy the conditions that the factor of safety with respect to a failure of the loaded clay should not be less than 3 and that the differential settlement should not be large enough to damage the superstructure. Both the factor of safety and the settlement depend not on the total weight of the structure but on the difference between the weight of the structure and that of the excavated soil. Therefore, the design requirements can usually be met by appropriate selection of the depth of the basement.

5. The differential settlement of a uniformly loaded flexible raft on clay is chiefly due to dishing. It is roughly equal to one half the maximum settlement. If the building itself is flexible, the differential settlement can be eliminated by providing the building with a very stiff substructure. If different parts of a large raft on clay carry very different loads, the bending moments in a stiff substructure are so great that the cost of the substructure is likely to be prohibitive. Another alternative is to vary the depths of the basements in accordance with the loads in such a manner that the difference between the building load and the weight of the excavated soil per unit of area has approximately the same value for every part of the raft. Whichever alternative is adopted, the design requires at least a rough settlement computation.

6. Layers of stiff clay or of dense sand located above layers of soft clay act like natural rafts. The footings of buildings resting on such layers are designed as if the soft layers did not exist. Since the settlement due to the consolidation of the soft layers may be very large, a settlement computation is required. The means for reducing the settlement due to consolidation are the same as those described in connection with rafts on homogeneous beds of clay.

Selected Reading

Golder, H. Q. (1965). "State-of-the-art of floating foundations," *ASCE J. Soil Mech.*, **91**, No. SM2, pp. 81–88.

ART. 56 PILE FOUNDATIONS

Function of Piles

A structure is founded on piles if the soil immediately below its base does not have adequate bearing capacity or if an estimate of costs indicates that a pile foundation may be cheaper than any other.

Piles are made in many forms and of a variety of materials. A description of the principal types and of the methods for installing them can be found in Chellis (1961). In this discussion we shall consider only piles of the more common types that are driven into the ground by a mechanical device known as a pile driver. However, the general principles are applicable, with minor modifications, to the design of foundations on other types of piles, installed in a different manner.

With respect to the manner in which they function, piles may be divided into three categories:

1. *Friction piles in coarse-grained very permeable soil.* These piles transfer most of their load to the soil through skin friction. The process of driving such piles close to each other in groups greatly reduces the porosity and compressibility of the soil within and around the groups. Therefore, piles of this category are sometimes called *compaction piles.*
2. *Friction piles in very fine-grained soils of low permeability.* These piles also transfer their load to the soil through skin friction, but they do not compact the soil appreciably. Foundations supported by piles of this type are commonly known as *floating pile foundations.*
3. *Point-bearing piles.* These piles transfer their load onto a firm stratum located at a considerable depth below the base of the structure.

In nature, homogeneous soil strata are very rare. Therefore, no sharp boundaries can be established between the three principal categories of piles. The same pile may displace part of the mass of soil through which it is driven without changing the relative density, whereas the remainder of the soil may undergo compaction. The point of a pile may be embedded in a firm sand stratum capable of supporting the pile by point bearing but, nevertheless, a considerable part of the load is likely to be carried by skin friction. Because of the wide variety of soil conditions encountered in practice, any attempt to establish rules for the design of pile foundations necessarily involves radical simplifications, and the rules themselves are useful only as guides to judgment. For the same reason, theoretical refinements in dealing with pile problems, such as attempts to compute the distribution of load among the piles in a group by means of the theory of elasticity, are completely out of place and can safely be ignored. Even conclusions based on the results of small-scale model tests may be far from reliable.

Design of Pile Foundations

Historical Development. Before the 19th century almost all buildings were established on continuous footings. Piles were used as a means of support wherever the ground appeared incapable of sustaining the pressure exerted by the footings. Since timber was abundant and labor cheap, as many piles were driven as the ground would take. Settlement caused no concern, because the prevalent type of structures could stand a considerable amount of unequal settlement without injury.

In the 19th century, when industrial development created a demand for heavy but inexpensive structures in locations underlain by soft ground, the cost of pile foundations became an item of consequence, and engineers were expected to specify no more piles than were necessary to provide adequate support for the buildings. This could not be done without at least some knowledge of the ultimate load that an individual pile could carry. Efforts to obtain the necessary information at a minimum expenditure of money and labor led to theoretical speculations that resulted in an impressive assortment of pile formulas. However, the realization slowly grew that the pile formulas had inherent shortcomings, and it became more and more customary to determine the allowable load per pile on all but the smallest jobs by making load tests on test piles.

The number of piles required to support a given structure was determined by the simple procedure of dividing the total load by the allowable load per pile. Many foundations designed in this manner were satisfactory, but now and then excessive and unexpected settlements occurred. These incidents indicated that the settlement of an entire pile foundation was not necessarily related to the settlement of a single test pile, even at the same load per pile. They led to the obvious conclusion that a knowledge of the bearing capacity of a single pile constitutes only part of the information necessary for the design of a satisfactory pile foundation. To find out whether the settlement of a pile foundation will remain within tolerable limits, the designer must consider the stresses produced in the soil by the entire load assigned to the foundation, and he must estimate the settlement produced by these stresses. This estimate requires a knowledge of the fundamental principles of soil mechanics. If the results of the investigation show that the settlement may exceed an acceptable value, the design must be changed.

STEPS IN DESIGN OF A PILE FOUNDATION. The first requirement for the preliminary design of a pile foundation is a soil profile representing the results of exploratory borings. The factors that determine the depth to which the subsoil should be explored are discussed in Article 45. Usually, the soil profile provides all the information required to decide whether the foundation can be supported by friction piles entirely embedded in sand, by point-bearing piles driven through soft strata into a firm one, or by a floating pile foundation.

The next step in the preliminary design is to select the length and type of pile. If point-bearing piles are appropriate, it may be possible to judge the required length with reasonable accuracy on the basis of the soil profile. Methods for estimating the length of friction piles in sand are in a rudimentary state of development, and reliable determinations can be made only by driving test piles. The length of friction piles in soft clay can be determined by making an estimate of the factor of safety of the pile groups against complete failure (see page 538). The selection of the type of pile is governed at least partly by practical considerations (Chellis 1961).

After the length and type of pile have been tentatively chosen, the ultimate bearing capacity of a single pile is either estimated or else determined by means of load tests. This value is divided by an appropriate factor of safety to obtain the "safe design load" per pile. The total number of piles required to support the structure is determined by dividing the total weight of the structure by the "safe design load" per pile.

After the number of piles has been determined, the next step is to choose their spacing. It is generally agreed that the distance D between the centers of piles with a top diameter d should not be less than $2.5d$. This rule is based on practical considerations. If the spacing is less than $2.5d$, the heave of the soil is likely to be excessive, and the driving of each new pile may displace or lift the adjacent ones. On the other hand, a spacing of more than $4d$ is uneconomical, because it increases the cost of the footings without materially benefiting the foundation. The most suitable value of D between these limits must be selected in accordance with the soil conditions, as explained subsequently.

When the spacing has been decided on, the piles are laid out in either a square or a triangular pattern. By multiplying the number of piles by D^2 (square pattern) or by $\frac{1}{2}D^2\sqrt{3}$ (triangular pattern), the total area required for the pile-supported parts of the foundation is obtained. If this area is considerably smaller than half the total area covered by the structure, the structure is established on pile-supported footings; if it is considerably greater, the structure is founded on a pile-supported raft, and the spacing of the piles is increased so that the pile layout forms a continuous pattern. If the intensity of loading on different parts of the raft is very different, the spacing between piles is adapted to the intensity on each of the parts. Finally, if it is doubtful whether the structure should be established on footings or on a raft, the decision is made after a comparison of the costs of the two alternatives.

If the foundation is supported by friction piles in soft clay or plastic silt, an estimate must be made of the ultimate bearing capacity of the pile groups, and the load on the groups must not be allowed to exceed one half, or preferably one third of the ultimate value. The consequences of ignoring this condition can be catastrophic. In several instances, structures together with the supporting piles and the soil located between the piles have sunk suddenly into the ground, although the load per pile did not exceed the "safe design load." The procedure for estimating the bearing capacity of pile groups is described subsequently.

If the load per pile is such that the bearing capacity of the pile groups is not exceeded, the foundation will not fail suddenly by sinking into the ground. However, adequate bearing capacity does not exclude the possibility of excessive settlement, because the settlement of an entire pile foundation has no relation whatsoever to the settlement of a single pile under the load per pile assigned to the foundation. The settlement of the foundation may range between a fraction of an inch and several feet, depending on the soil conditions, the number of piles, and the area covered by the structure. Settlements of less than about 2 in. are commonly not harmful, but settlements of 6 in. or more may have very undesirable effects on the superstructure. Hence, if a foundation rests on friction piles driven into soft clay, or if the points of point-bearing piles are located above soft strata, a settlement computation is imperative. Failure to make such a computation has been responsible for many unsatisfactory pile foundations.

The final step in the design of the foundation is the structural design of the pile-supported footings or raft. The computations of bending moments and shears are commonly based on the assumption that each pile carries the same load. Theoretical considerations and the results of field tests (Swiger 1941) both lead to the conclusion that this assumption is usually far from correct. If the substrata are fairly horizontal and the points of the piles do not rest on bedrock, the load per pile in a group supporting a rigid footing increases from the center piles toward the edges. The error involved in the commonly accepted assumption, however, is well within the customary margin of safety for reinforced-concrete design.

The details of the successive steps in the design of pile foundations are presented under the following subheadings.

Ultimate Loads and "Safe Design Loads" for Single Piles

Skin Friction and Point Resistance. In a general way, the term *ultimate load* or *bearing capacity* of a single pile indicates the load at which the settlement of the pile increases continuously with no further increase in load, or at which the settlement begins to increase at a rate far out of proportion to the rate of increase of the load. Whatever the load may be, it is carried partly by skin friction and partly by the resistance of the soil directly beneath the point, as indicated in Fig. 56.1*a*. Therefore, the ultimate bearing capacity Q_d can be considered rather arbitrarily as resolved into two parts: Q_f which is due to skin friction, and Q_p which is due to point resistance. Hence,

$$Q_d = Q_f + Q_p \tag{56.1}$$

In Fig. 56.1*b*, *ab* represents a horizontal section through the point of the pile, and the shaded areas indicate the pressure on this section. The total pressure is obviously equal to Q_d. Various refined theoretical methods have been used to compute the distribution of this pressure, but the results of the computations cannot be relied on, because all the methods are based on the assumptions that the soil is perfectly homogeneous and elastic. Authoritative informa-

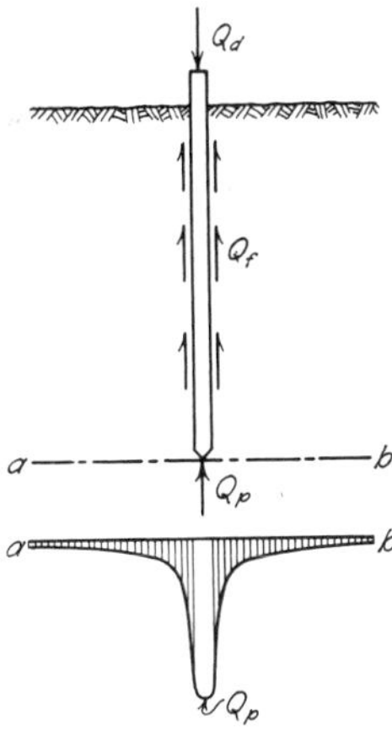

Fig. 56.1. (*a*) Loaded friction pile in soft clay. (*b*) Distribution of pressure on horizontal section through point of pile.

tion on the pressure distribution can be obtained only by direct measurements, and so far no such measurements have been made. However, there is no doubt that the distribution depends not only on the dimensions of the pile, but also on the load, the nature of the soil, and the conditions of stratification. It is also likely to change appreciably with time.

SKIN FRICTION ON SINGLE PILE IN SAND. When a pile is driven into very dense sand, refusal against further penetration is met at a depth of a very few feet whereas, in very loose sand, piles can be driven to great depth without encountering appreciable resistance. In every sand the average skin friction per unit of area of contact and the point resistance both increase with increasing depth. The total skin friction that resists further penetration of a cylindrical or prismatic pile into a homogeneous stratum of sand is considerably greater than one half the total ultimate bearing capacity Q_d of the pile, but the resistance against pulling is considerably less than half of Q_d. The difference between the two values of skin friction is due to the fact that a downward movement of the pile increases the pressure against its sides, whereas an upward movement decreases it. After the pile is driven to refusal, the average skin friction resisting further penetration under static load is of the order of 500 lb/ft^2 for loose sand (long piles), and 2000 lb/ft^2 for very dense sand (short piles).

The ultimate capacity of a single friction pile in sand depends not only on the relative density of the sand, but also on the taper of the pile, the roughness and configuration of the surface of the pile, and the volume of sand displaced. The relative density itself depends on the number and spacing of the neighboring piles. Attempts to predict the ultimate bearing capacity of such piles on a semi-empirical basis appear promising but as yet the procedures are in a formative stage (Nordlund 1963).

Occasionally, it has been observed that the bearing capacity of piles in sand decreases conspicuously during the first 2 or 3 days after driving. Although this phenomenon is rather exceptional, the possibility of its occurrence should never be ignored. It is probable that the high initial bearing capacity is due to a temporary state of stress that develops in the sand surrounding the point of the pile during driving. This state of stress is associated with a temporary excess point resistance.

SKIN FRICTION ON SINGLE PILE IN SATURATED CLAY. The point resistance of a friction pile embedded in soft clay is negligible compared to the skin friction. The skin friction per unit of contact area is more or less independent of the depth of penetration and the method of installing the pile. It depends almost entirely on the properties of the clay. The resistance against pulling is commonly, but not always, nearly equal to the resistance against further penetration under load. All these relations are much simpler than the corresponding ones for a friction pile in sand. However, by contrast, the relations between skin friction and time are much more complex and as yet are unpredictable. The skin friction usually increases during the first month after the pile is driven, but the amount of increase varies considerably with the nature of the soil.

The curve in Fig. 56.2 represents the increase in bearing capacity of a friction pile with time. The pile was driven into soft brown clay with streaks of silt. The liquid limit of the clay was between 37 and 45%, the plastic limit between 20 and 22%, and the natural water content slightly below the liquid limit. During the pile-driving operations the soil turned almost liquid, and the skin friction had a very small value. Although the piles penetrated 12 in. under a single blow, they rose 10 in. when the hammer was lifted, and a special device had to be used to prevent the piles from rising. Yet, within a month the skin friction increased to more than three times its initial value.

When a pile is driven into soft clay, the clay in the path of the penetrating pile is displaced and becomes severely distorted. After the pile is driven, the disturbed clay surrounds it like a shell (Cummings et al. 1950). Beyond the outer boundaries of this shell the disturbance of the soil structure is quite small. If a large number of piles is driven in a small area, the effects of disturbance may

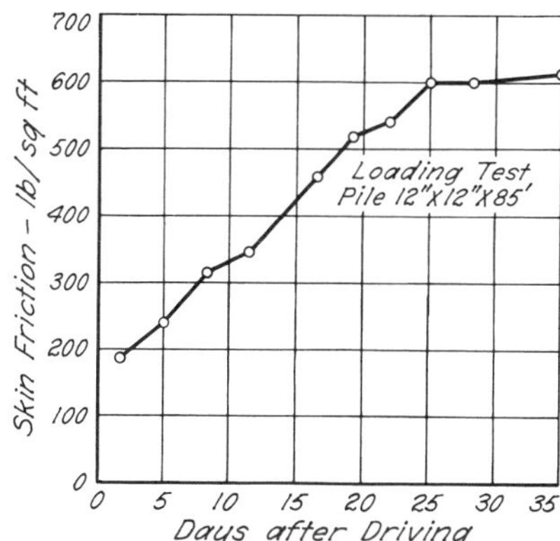

Fig. 56.2. Diagram illustrating increase of ultimate bearing capacity of friction pile with time.

be accumulative and may extend well beyond the limits of the pile-driving operations. Consequently, settlements of adjacent structures have in some instances been induced (A. Casagrande 1947, Lambe and Horn 1965). On the other hand, the shell of badly disturbed clay around each pile usually consolidates rapidly and becomes stiffer than the undisturbed clay; it is likely to adhere to the pile if the pile is pulled. If a pile is driven into an extremely sensitive or quick clay, the shell of disturbed material is likely to be only an inch or two thick and to behave like a liquid during pile driving. The displaced material flows upward along the pile to the surface (Legget 1950) where it accumulates, but the soil outside the liquid shell remains virtually undisturbed. Within the shell the liquefied material regains strength by consolidation and thixotropy, but the final strength is not necessarily equal to that of the undisturbed soil.

Driving piles into saturated clay or silt produces a temporary increase in porewater pressure (Lambe and Horn 1965). In a silt, the excess pressures may temporarily liquefy the silt for a considerable distance. Yet, within a few days or weeks the silt appears to become as solid and stable as it was originally.

The effects of pile driving in soft clays or silts are obviously complex and not well understood. They should always be given consideration, but in many instances they are not detrimental. Hence, in the following discussions it is assumed that the physical properties of the soils do not experience significant permanent alteration.

In spite of the influence of disturbance and the diverse time effects that take place immediately after a pile is driven into soft clay or soft plastic silt, the ultimate value of the skin friction is commonly equal to about the undrained shear strength or one half the unconfined compressive strength of the clay (Peck 1958). In a few unusual instances, however, significantly smaller skin-friction values have been experienced (Peck 1961). Thus far, the abnormally low values have been associated only with varved clays, but the conditions for the reduced strength are not yet known. Therefore, final decisions concerning the skin friction of piles in soft saturated clays and silts should be based on pile load tests.

In stiffer clays the ultimate value of the skin friction is likely to be substantially smaller than half the unconfined compressive strength of the undisturbed clay, and the discrepancy increases with increasing strength of the clay (Tomlinson 1957, Peck 1958, Woodward et al. 1961). The skin friction also becomes more dependent on the material comprising the outer surface of the pile. Present knowledge is summarized in Table 56.1. It may be emphasized that

Table 56.1
Ultimate Values of Skin Friction on Piles Embedded in Cohesive Soils*

Material of Pile	Unconfined Compressive Strength of Clay tons/ft²	Ultimate Skin Friction between Pile and Clay lb/ft²
Concrete and timber	0–0.75	0–700
	0.75–1.5	700–1000
	1.5–3.0	1000–1300
	Over 3.0	1300
Steel	0–0.75	0–700
	0.75–1.5	700–1000
	1.5–3.0	1000–1200
	Over 3.0	1200

* (After Tomlinson 1963.)

this table as well as more elaborate ones serve only as a guide for making preliminary estimates. Reliable information cannot be obtained without performing loading and pulling tests on full-sized piles in the field.

ACTION OF POINT-BEARING PILES. In contrast to friction piles, point-bearing piles are assumed to transfer the load through their points onto a firm stratum. Nevertheless, a considerable part of the load is carried at least temporarily by skin friction. This has been demonstrated by load tests in both the laboratory and the field (Vey 1957, D'Appolonia and Romualdi 1963, D'Appolonia and Hribar 1963). However, if the pile passes through a very compressible soil such as soft silt or clay, the pressure transferred to this soil by skin friction gradually consolidates it and, as a consequence, the pile tends to settle. The settlement is resisted only by the soil in which the point is embedded and, as time goes on, the pressure on the point increases. This process continues until the major part of the load on the pile is carried by the point. If the load assigned to the pile in the foundation exceeds the point resistance, the resulting settlement can be very large. Yet, the danger is not revealed by the results of a load test on a single pile, even if the load test is applied for several weeks. Hence, it is more important to know the point resistance than the total bearing capacity of a point-bearing pile.

RELATIONS BETWEEN DRIVING RESISTANCE AND DEPTH. If the depth to which a pile has penetrated is plotted against the number of ham-

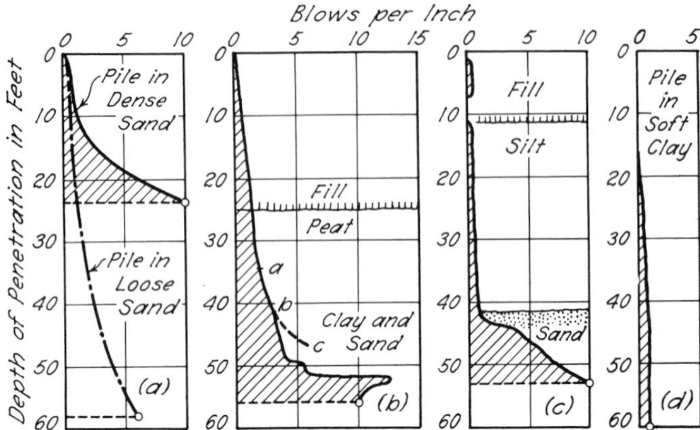

Fig. 56.3. Relation between blows per inch of penetration and total depth of penetration for wood piles driven into subsoils of various characteristics.

mer blows per inch of penetration, a resistance diagram is obtained. Typical diagrams are shown in Fig. 56.3. The shape of the penetration curve indicates almost unmistakably into which of the three main categories the pile belongs. Figure 56.3*a* shows curves typical for piles driven into loose and into dense sand. In both types of sand the penetration resistance increases with depth. On the other hand, the pile represented by Fig. 56.3*d* was driven through soft clay, and the penetration resistance became practically constant. The sharp break in the curve in Fig. 56.3*c* indicates that the pile point passed from soft silt into a fairly dense sand. Such a break is typical for point-bearing piles. By correlating the resistance diagrams with the soil profile on a given job, the engineer can usually obtain a reliable conception of the material in which each pile is embedded. In particular, he can determine whether the point of the pile has reached a suitable bearing stratum.

Use of Pile Formulas for Estimating Ultimate Bearing Capacity. When a point-bearing pile encounters a firm stratum, the penetration resistance increases sharply (Fig. 56.3*c*). In a general way, the greater this increase, the greater is likely to be the point resistance of the pile. This observation has led to various attempts to express the relationship between the bearing capacity of a pile and the resistance to penetration just before driving is discontinued. The results are known as pile formulas (Article 34).

In every such formula in common use, such as that developed by Janbu (Eq. 34.6) or the Engineering News formula (Eq. 34.9), the calculated ultimate bearing capacity depends upon the penetration S under the last blow of the hammer. According to Fig. 56.3*d*, the value of S that appears in the formulas is, for friction piles in clay, practically independent of depth. As a result, application of any of the formulas leads to the conclusion that the ultimate bearing capacity of such piles is also independent of depth. However, experience has shown that the ultimate bearing capacity of friction piles in clay increases approximately in direct proportion to the length of the piles. This fact strictly excludes the application of any pile formula to friction piles in soft silt or clay. As a matter of fact, in several cities, including Shanghai and New Orleans, where the prevalence of thick deposits of soft soil calls for extensive use of friction piles, no experienced engineer would even consider using a pile formula. On small jobs the bearing capacity is estimated on the basis of empirical values for the average skin friction per unit of area, and the point resistance is disregarded. On large jobs load tests are made.

However, even for point-bearing piles and other piles for which the penetration resistance increases with depth, the agreement between the real ultimate bearing capacity and that computed on the basis of the best of the pile formulas is only moderately good, whereas that computed on the basis of the widely used Engineering News formula is so poor and so erratic as to offer no justification for its continued use. The reasons are explained in Article 34. Hence, on small jobs, the ultimate capacity of the piles may be estimated by Janbu's formula with a factor of safety of at least 3, at the risk that the real factor of safety may be as little as about 1.75 or, conversely, that it may be as great as about 4.4 and unnecessary piles may be driven. On large jobs good engineering practice calls for determining the ultimate bearing capacity by means of load tests on full-sized piles.

Load Tests on Piles. It has been pointed out that the bearing capacity of all piles except those driven to bedrock does not reach its ultimate value until a certain time has elapsed. Hence, the results of load tests are not conclusive unless the tests are made after the period of adjustment. For piles in permeable ground this period amounts to 2 or 3 days, and for piles partly or wholly surrounded by silt or clay to about a month.

Pile load tests are sometimes made by constructing a platform on top of the pile, loading it with sand or pig iron, and observing the settlement of the pile by means of a level. This procedure is cumbersome because of the large weights to be handled and the time required. A more expedient procedure is known as a *bootstrap test*. In preparation for such a test three piles are driven in line at a spacing of about

5 ft. The upper ends of the outer piles are connected by a strong and stiff yoke. The middle pile is the test pile, and it is loaded by jacking against the yoke. The pull on the anchor piles slightly reduces the settlement of the test pile, but this disadvantage is more than offset by the ease with which the process of loading can be repeated once every few days. The curve shown in Fig. 56.2 was obtained by means of such a test. In another widely used procedure, the test pile is loaded by jacking against the center of a platform loaded by dead weights to provide a reaction and supported by a pile at each corner. The supporting piles should be at least 5 ft from the test pile.

In order to design foundations on point-bearing piles driven through clay strata into sand, information is needed concerning the bearing capacity of that part of the pile located in the sand. For the sake of brevity this is referred to as the point resistance, although it includes the skin friction at the surface of contact between the pile and the sand. Unless it is absolutely certain that the "safe design load" is considerably less than the point resistance, the latter should be determined by load tests in the field. For this purpose two test piles may be driven at a distance of about 5 ft from each other. One of them is driven to refusal within the bearing stratum. The other is driven until its point is about 3 ft above the surface of the bearing stratum. Since the point resistance of a pile embedded in sand approaches its ultimate value quite rapidly, the load tests can be made as soon as 3 days after driving the test piles. The effect of time on the skin friction can be eliminated by loading both piles simultaneously and at equal rates. The point resistance is approximately equal to the difference between the ultimate bearing capacity of the two piles.

EVALUATION OF "SAFE DESIGN LOAD." The term *"safe design load"* Q_a indicates the load at which the factor of safety with respect to a downward plunging or sinking of a single pile has a value consistent with the customary safety requirements.

If the "safe design load" for the pile is estimated by means of the Janbu formula (Eq. 34.6), a factor of safety not less than 3 should be used. If the ultimate bearing capacity is determined by means of a load test, the customary factors of safety range between 1.5 and 2 (Chellis 1961). Even the smaller value is fully adequate unless the subsurface conditions are very nonuniform. The principal uncertainties in the load-test method lie in the assignment of a value for the ultimate bearing capacity based on an interpretation of the load-settlement curve.

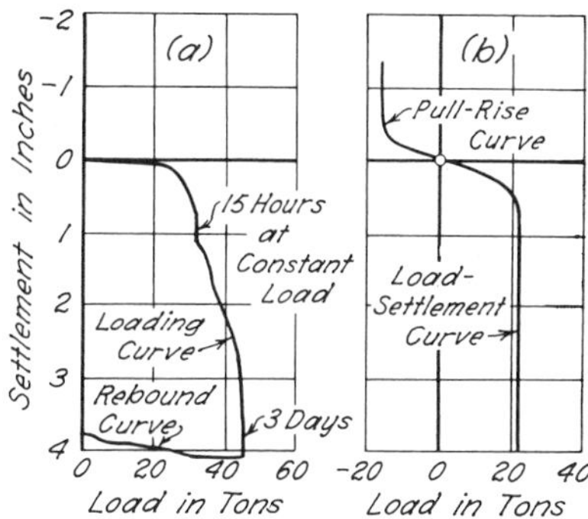

Fig. 56.4. Typical load-settlement curve for (*a*) point-bearing pile. (*b*) Friction pile.

The general character of load-settlement curves ranges between the two extremes shown in Fig. 56.4. The curve in Fig. 56.4*a* is typical for friction piles embedded in coarse-grained soils and for point-bearing piles that transfer their load onto sand strata. Since the settlement curve gradually approaches an inclined tangent as the load on the pile increases, no definite value can be assigned to the ultimate bearing capacity. The "safe design load" on such piles should not exceed the load required to produce a penetration of 1 in. divided by a factor of safety of not less than 1.5.

The load-settlement curves for friction piles differ considerably from each other in character. An extreme possibility is represented in Fig. 56.4*b*. The test pile was driven 37 ft through soft silt and clay with layers of peat. The point did not reach a firm stratum. Under loads of less than 22 tons the settlement of the pile was quite small but, when the load became equal to this value, the pile suddenly sank several feet and did not stop until the loading platform hit the ground. The pull-rise curve, also shown in the figure, is similar to the load-settlement curve. The "safe design load" for such piles can be considered equal to the ultimate bearing capacity Q_d divided by a factor of safety of not less than 2.0.

Ultimate Bearing Capacity of Pile Groups

Both theory and experience have shown that pile groups may fail as units by breaking into the ground before the load per pile becomes equal to the "safe design load." Such a failure is illustrated in Fig. 56.5. Hence, the computation of the "safe design load" must be supple-

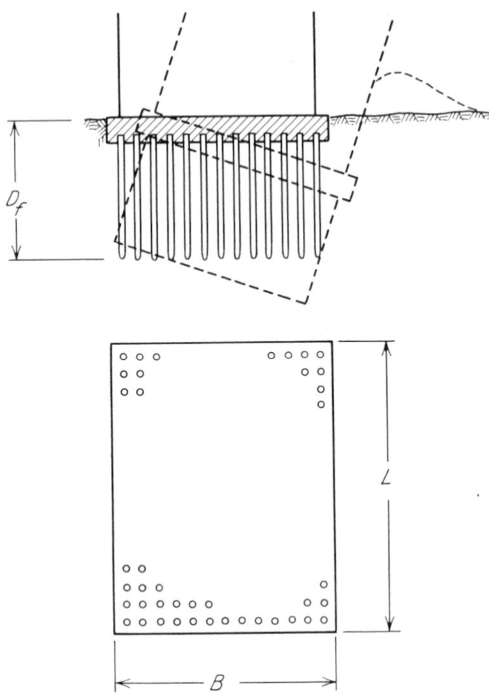

Fig. 56.5. Diagram illustrating failure due to breaking into ground of entire pile cluster, including the soil located between the piles.

mented by a computation of the ultimate bearing capacity of the entire group. Let

B = width of pile group
L = length of pile group
s = average shearing resistance of soil, per unit of area, between ground surface and depth D_f
q_d = ultimate bearing capacity, per unit of area, of a rectangular loaded area with dimensions $B \times L$ and depth D_f. This quantity may be estimated by means of the appropriate equations (Article 33).

If the piles and the confined mass of soil sink as a unit like a pier, the ultimate bearing capacity Q_g of the group is given with sufficient accuracy by

$$Q_g = q_d BL + D_f(2B + 2L)s \tag{56.2}$$

Computations based on this equation have shown that a base failure can hardly occur unless the pile group consists of a large number of friction piles embedded in silt or soft clay, as shown in Fig. 56.5, or else of point-bearing piles that transfer their load onto a firm but thin stratum underlain by a thick deposit of silt or soft clay. A pile group can be considered safe against such a failure if the total design load (number of piles times the "safe design load" per pile) does not exceed $Q_g/3$. If this condition is not satisfied, the design of the foundation must be changed.

Settlement of Friction Pile Foundations in Sand

Dense sand is an excellent subsoil that does not need any reinforcement by piles. If piles have to be installed in dense sand for some purpose such as to transfer the weight of a bridge pier to a level below that of the deepest scour, it is usually necessary to aid the penetration of the pile by jetting. Hence, in the following paragraphs we consider only piles driven into loose sand. Furthermore, we assume that the sand within which the piles are driven is not underlain by any material more compressible than the sand itself.

If all other conditions are the same, the skin friction against the piles increases with increasing relative density of the sand. While a pile is being driven, the density of the surrounding sand increases (Plantema and Nolet 1957). Large-scale experiments have shown that the compaction caused by driving one pile influences the bearing capacity of any other pile located within a distance equal at least to five times the diameter of the pile (Press 1933). Nevertheless, if a single pile is driven and loaded, its settlement S_0 under a given load is likely to be less than that of a group of similar piles subjected to the same load per pile, because the beneficial influence of the compaction does not extend significantly below the points of the piles, whereas the depth of highly stressed sand increases with the size of the group. On the basis of the few available field observations, Skempton (1953) suggested the empirical relation shown in Fig. 56.6. It is evident that the settlement increases significantly with increasing width of the pile group or foundation, on the assumption that the spacing of the piles is approximately the same for all the groups.

The ultimate bearing capacity of piles in sand increases roughly with the square of the depth of penetration, whereas the cost of the piles increases with depth at a much smaller rate. Therefore, it is economical to drive piles in sand to such a depth that further penetration becomes difficult and slow. The number of blows per inch at which driving must be discontinued, regardless of other considerations, is determined by the condition that the pile should not be injured by the process of driving (Bruns 1941).

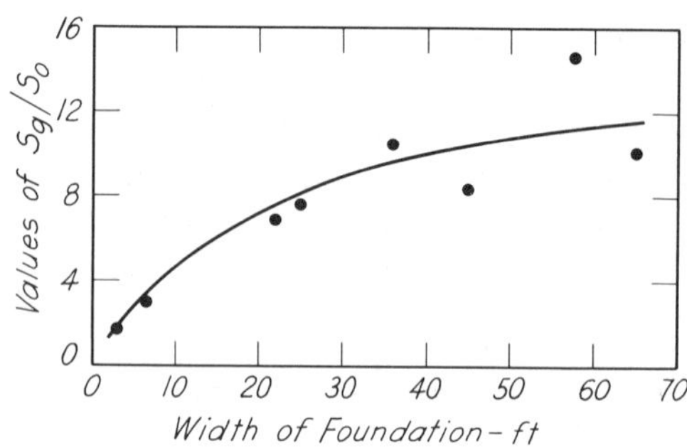

Fig. 56.6. Approximate empirical relation between width B of group of piles in sand and the ratio S_g/S_0, wherein S_g represents the settlement of a group of piles with width B and S_0 the settlement of a single isolated test pile subject to the same load as each of the piles in the group (after Skempton 1953).

The most suitable spacing D between the centers of piles with a top diameter d appears to be $3d$. In clusters each pile should be driven until the number of blows per inch becomes equal to that at which the driving of the test pile was discontinued. Driving should proceed outward from the center of the cluster; otherwise, the inner piles cannot be driven to as great a depth as the others.

After the piles have been installed, each cluster constitutes the core of a column of compacted sand embedded in loose sand. If the load per pile of such a foundation does not exceed the "safe design load," the settlement will not be greater than that of a similar structure supported by footings on dense sand. However, if the sand stratum containing the piles is interspersed with pockets or layers of silt or clay, the settlement may be almost as large as that of a floating pile foundation, because the pressure transmitted through the skin of the piles onto these layers causes them to consolidate.

Settlement of Point-Bearing Pile Foundations

INTRODUCTION. Point-bearing pile foundations may be divided into five principal categories, depending on the nature of the bearing stratum. The following possibilities are considered under separate subheadings:

1. The points rest on sound bedrock.
2. The points have been driven into decomposed bedrock.
3. The points are embedded in dense sand underlain by equally incompressible strata.

4. The points are embedded in stiff clay underlain by even more incompressible strata.
5. The points are embedded in a stratum of dense sand or stiff clay located above a stratum of soft clay.

POINTS RESTING ON SOUND BEDROCK. Under ideal conditions piles driven to sound bedrock act like piers, and the settlement does not exceed the elastic shortening of the piles. However, unless the points of wood piles are adequately protected, they are likely to be injured by brooming, whereupon the beneficial effect of the rigid point support is lost. If the points encounter a smooth but inclined rock surface, they may travel down the slope without giving any visible indication of their progressive deflection. When the weight of the building is added, the deflection may increase still further and the foundation may fail. Under such conditions wood piles should not be used. Even reinforced-concrete piles may break.

POINTS DRIVEN INTO DECOMPOSED BEDROCK. Decomposed rocks, particularly of metamorphic origin, may be as compressible as medium clay. Yet, they usually contain fragments of fairly intact rock that interfere with driving the piles through the compressible zone. Under these circumstances reliable information on the probable settlement can be obtained only by securing undisturbed cores of the decomposed material and making a settlement forecast on the basis of the results of consolidation tests. If the settlement may conceivably exceed the tolerable value, some method must be used for punching through the zone of decomposed rock.

POINTS DRIVEN THROUGH COMPRESSIBLE STRATA INTO SAND. In the preceding discussion of the ultimate bearing capacity of a single pile of this type it was shown that the settlement of the pile depends primarily on the ratio between the point resistance and the pile load. The same general statement can be made with regard to the settlement of the entire foundation. If the load per pile is equal to or less than the point resistance, the settlement is likely to be of no consequence. On the other hand, if it is greater, the settlement may be large and detrimental. However, in any event, the average settlement of the foundation will be many times greater than the settlement of a single pile acted on by the "safe design load." These statements are illustrated by the following example (Terzaghi 1938*b*).

An apartment house in Vienna, Austria, was constructed on continuous footings 40 in. wide, supported by cast-in-place concrete piles driven through about 20 ft of loose fill into fairly dense gravel. Each pile carried a load of 24 tons. In Fig. 56.7*b*, curve C_0 shows the result of a load test on a single pile, and C the load-settlement curve for the same pile during construction. When the load due to the weight of the building reached its ultimate value of 24 tons, the settlement of the pile was very much greater than that of the same pile during the load test. Curves of equal settlement for the entire foundation 11 weeks after the building was completed are shown in Fig. 56.7*a*. Their complete lack of symmetry suggests that the seat of settlement was in the upper part of the firm stratum and that the settlement reflected primarily the local variations in compressibility of this stratum. If the walls had been perfectly flexible, the curves of equal settlement would extend without any break across the expansion joint at midlength of the building. The discontinuities indicate that the walls acted as semirigid beams that bridged the weakest spots in the supporting stratum. However, from a practical point of view the settlement was irrelevant, because the maximum differential settlement did not exceed ½ in. The foundation was successful because the load per pile was smaller than the point resistance.

In the example referred to in the preceding paragraph the point resistance was large, because only a small amount of energy was consumed by the soft upper strata during driving. On the other hand,

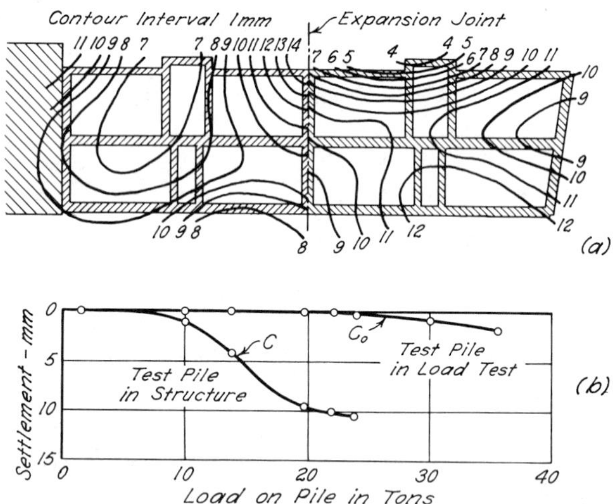

Fig. 56.7. (*a*) Settlement contours one year after construction of brick buildings supported by continuous footings on conical piles bearing on dense gravel stratum. (*b*) Load-settlement relations for single pile under test load and under same load beneath structure.

if some of the top strata are very firm, most of the driving energy is used in overcoming the side friction in the upper strata, and the resistance against further penetration becomes excessive while the point resistance is still very low. Reliable information on the point resistance of such piles can be obtained only by making load tests on two piles of different length, as described on page 536 or, on very large jobs where the installations may be sensitive to settlement, by providing the test piles with strain gages at various depths in order to permit a determination of the load actually reaching the points during the load tests. Such tests, although costly, have been carried out on several important projects (Stevens et al. 1965).

If the load per pile does not exceed two thirds of the point resistance, the settlement of the foundation will be unimportant regardless of the spacing of the piles. A spacing of $3d$ satisfies all practical requirements. The center piles in a cluster should be driven first to insure that their points will have adequate penetration into the bearing stratum.

In some localities the bedrock is covered by a composite stratum consisting of irregular pockets of sand or sand and gravel alternating with pockets of more compressible material such as clay, or rock fragments embedded in clay. This composite base is buried beneath soft sediments. The conventional exploratory borings do not necessarily reveal its composite character. However, while a test pile is being driven, the variations in the character of the subsoil come into prominence. For example, Fig. 56.3*b* represents the depth-resistance curve for a pile driven through fill and peat into a sand stratum containing layers and pockets of clay. The curve consists of inclined sections such as *ab* followed by vertical ones. These abrupt transitions may indicate the passage of the point of the pile from a firm soil into soft silt or clay; if the soil below the level of point *b* had been similar to that above it, the curve would have continued as indicated by the dash line *bc*. Since the depth-resistance curve has several steps, the point of the pile evidently passed through several firm strata alternating with soft ones. However, quite similar abrupt transitions may be caused by the passage of the point of the pile from a dense layer of sand into a loose one. This possibility is illustrated by Fig. 44.12. In any event, the individual piles of a cluster driven into a subsoil with an erratic soil profile are likely to meet refusal at very different depths. For instance, one of two adjacent piles driven at a distance of 2.5 ft from each other into the subsoil shown in Fig. 44.12 met refusal at a depth of 60 ft, whereas the other penetrated to a depth of 85 ft. If the soil located between the points of piles with very different length consists only of loose sand, the performance of the cluster under load may be perfectly satisfactory. On the other hand, if it contains pockets of soft clay or silt, the settlement of the pile-supported footing may be excessive. Hence, if adjacent foundation piles meet refusal at very different depths, a boring should be made near by to determine the cause of the difference. If the boring shows that the soil contains pockets of highly compressible material below the level of the shortest piles, it is necessary to enforce the penetration of all the piles to a level below the bottom of the zone that contains the pockets. If this cannot be done by jetting, spudding may be required. All those piles that met refusal above the level of the lowest soft pockets should be pulled, or disregarded and replaced by piles with adequate length.

On flood plains and along the seacoast the construction of pile foundations is often preceded by placing a fill over the site of the proposed structure. If the subsoil consists of loose sand or other highly permeable and relatively incompressible soils, the effect of the fill on the piles can be disregarded. On the other hand, if the subsoil contains layers of soft silt or clay, the presence of the fill considerably increases the load on the piles and, as a consequence, also causes an increase in settlement. This fact was first recognized in Holland, where many buildings located in the coastal plains rest on point-bearing piles driven through about 60 ft of very soft strata to refusal in a bed of sand. Wherever the site was covered by a thick layer of fill shortly before the piles were driven, it was found that the buildings supported by the piles settled excessively. Once this fact was noticed, the cause of the settlement became obvious.

Before such piles are driven, the compressible strata gradually consolidate under the weight of the newly applied fill, and the fill settles. As soon as the piles are installed, the fill material located in the upper part of the pile clusters can no longer settle freely because its downward movement is resisted by skin friction between the fill and the piles. An imperceptible downward movement of the fill with respect to the piles is sufficient to transfer onto the piles the weight of all the fill located within the clusters. If A represents the area of a horizontal section included within the boundaries of a cluster, n the number of piles, H the thickness of the fill, and γ its unit weight, the load Q' which acts on each pile due to the weight of the fill within the cluster is

$$Q' = \frac{A}{n} \gamma H \tag{56.3}$$

In the spaces between clusters, the weight of the fill causes progressive settlement. If the clusters consist of point-bearing piles, the piles do not participate in the downward movement and, as a consequence, the soil that surrounds the clusters moves down with reference to the clusters. It tends to drag each cluster down.

The drag increases as the consolidation of the clay surrounding the cluster proceeds. The minimum value depends on the distance through which the surface of the clay subsides. It is almost zero for very small subsidences, and it increases with increasing subsidence. It cannot become greater than the product of the thickness H of the clay stratum, the circumference L of the cluster, and the average shearing resistance s of the clay. If n is the number of piles in the cluster, the maximum value of the drag is

$$Q_{max}'' = \frac{LHs}{n} \tag{56.4}$$

The real value Q'' ranges between zero and Q_{max}''. At the present state of our knowledge it can be evaluated only by judgment.

The forces that produce the loads Q' and Q'' are known as *negative skin friction*. With increasing spacing between piles both Q' and Q'' increase. Hence, to reduce the effects of negative skin friction the spacing between piles should be reduced to $2.5d$, the minimum compatible with practical requirements.

If Q is the load per pile exerted by a building on piles driven through new fill and soft clay into a sand stratum, the lower ends of the piles will ultimately receive a load

$$Q_t = Q + Q' + Q'' \tag{56.5}$$

If this load is greater than the point resistance of the pile, the settlement of the foundation will be excessive, regardless of what ultimate bearing capacity a load test may indicate. Hence, if a foundation on point-bearing piles is to be established at the site of a recent fill, both the point resistance and the value of Q_t (Eq. 56.5) should be determined.

Various expedients are used to reduce the effects of negative skin friction. For example, in Holland it has become customary to use precast concrete piles with shafts of small cross section as compared to that of the points (Plantema and Nolet 1957). Under extreme conditions, piles have been driven inside casings, the spaces between piles and casings filled with a viscous material, and the casing pulled (Golder and Willeumier 1964).

Points Driven through Compressible Strata into Stiff Clay. Under these conditions most of the load on the piles is ultimately carried by the points. This produces a large concentration of stress in the clay near the point of each pile. The results of a load test on a single pile may be perfectly reassuring, because, first of all, the major part of the load during the test is carried by skin friction and, second, the consolidation of the clay near the pile points develops very slowly. However, as time goes on, the settlement due to this consolidation may become very large. To obtain reliable information concerning this possibility, it is advisable to drive a steel pipe with a conical loosely attached point through the soft strata into the stiff clay and to make a load test by inserting a column in the pipe to transfer the load directly to the point. The diameter of the pipe should be roughly equal to that of the tip of the proposed piles. The point should preferably be made of a permeable material such as artificial porous stone. The load should be left on the point for at least a month. The settlement should be measured once a day during the first week and twice a week thereafter. The shape of the time-settlement curve obtained by plotting the settlement data permits at least a crude estimate of the ultimate settlement of the pile.

The spacing of the piles should not be less than $3d$, to reduce the disturbance of the clay constituting the bearing stratum as much as possible. A spacing of $3.5d$ is preferable. The difference between the ultimate settlement of the single test pile and of the entire foundation is likely to be unimportant.

If the area to be occupied by the foundation has been covered by a recent fill, the foundation should be designed for a load Q_t per pile (Eq. 56.5) on account of the negative skin friction.

Points Embedded in Firm Stratum Underlain by Soft Clay. If the bearing stratum, such as a thick layer of dense sand, is located above a layer of soft clay, the settlement of the pile foundation is the sum of two independent items. The first part is equal to the settlement that would take place if the sand stratum were not underlain by compressible material. The factors that determine this part of the settlement are discussed under the preceding subheadings. The second part is due to the gradual consolidation of the compressible layer located beneath the stratum in which the points of the piles are embedded. Whereas the first part is negligible if the foundation is properly designed, the second may be very large and detrimental. This possibility has often been overlooked even in recent years.

In one instance about 5000 wood piles 80 ft long were driven to firm bearing in dense sand through fill and through 50 to 65 ft of loose fine sand containing thin layers of silt and soft clay. The piles were arranged in clusters and capped by footings. The load per pile was about 16 tons, less than one quarter of the ultimate bearing capacity as determined by load tests, and no measurable settlement was anticipated. Yet, the foundation actually settled more than 2 ft. The seat of settlement was a layer of clay 30 ft thick, located 25 ft below the points of the longest piles. The water content of the clay was close to the liquid limit.

The settlement of a pile foundation due to the consolidation of a soft layer below the bearing stratum can be computed by the procedures given in Articles 13 and 41, on the assumption that the structure is perfectly flexible and that the loads act directly on the surface of the bearing stratum. The total load that produces the consolidation is equal to the difference between the total effective weight of the building and the effective weight of the excavated soil (Article 55). The reliability of this procedure is illustrated by Fig. 56.8. Figures 56.8*a* and *b* represent, respectively, a simplified plan and vertical section through a steel frame building with cut-stone facing. The structure rests on about 10,000 wood piles 26 ft long, driven so that their points came to bearing in the upper part of a layer of dense sand. The load per pile is 15 tons. Since the average settlement of the test pile was only $\frac{1}{4}$ in. under 30 tons, the designers did not expect that the maximum settlement of the entire foundation would

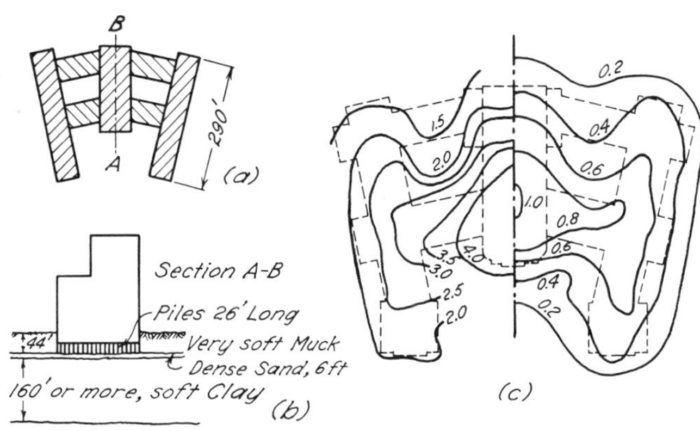

Fig. 56.8. (*a* and *b*) Plan and cross section of structure supported by piles driven into dense sand layer above deep clay deposit. (*c*) Settlement contours for structure. Contours on left side represent observed settlement in inches at completion of structure; contours on right side represent lines of equal relative settlement based on computation and results of consolidation tests.

exceed this value. The real maximum settlement, however, had already exceeded 1 ft within 2 years after construction. The contours of observed settlement, in inches, at the end of the construction period are shown on the left side of Fig. 56.8*c*. The right side shows curves of equal computed settlement, plotted as a fraction of the maximum settlement. In spite of the simplifying assumptions made in the calculations, the computed differential settlement is in good agreement with the actual settlement. According to the results of the settlement analysis, the ultimate maximum settlement will be about 18 in., but the real settlement will be considerably greater because of the secondary time effect (Article 14).

In order to get information about the magnitude of the settlement due to consolidation of compressible strata located beneath the pile points, the exploratory borings must be supplemented at least by several tube-sample borings from which continuous samples of all the highly compressible strata are secured. If an accurate settlement forecast is called for, an undisturbed sample boring must also be made. The program for testing the samples and the method of computation are identical with those outlined for the settlement of raft foundations located above soft clay strata (Article 55). If the computation indicates that the settlement may exceed a tolerable value, other methods for constructing the foundation must be considered.

If the computation indicates that the settlement will be tolerable, the spacing between piles may be determined by means of the same rules that are used for foundations on point-bearing piles embedded in sand.

Redriving of Point-Bearing Piles. If a pile is driven through silt or clay, the neighboring piles may rise as much as several inches, and their points lose contact with the point-supporting soil. Subsequent application of a load on these piles causes a settlement equal to the preceding rise. Hence, if the soil conditions are conducive to a rise, reference points should be established on the heads of the piles and observed by means of a level from time to time. If a rise is observed, the piles must be redriven before the footings are constructed (Klohn 1961). If the piles consist of non-rigid steel shells, tell-tales should be installed to permit detection of heave of the points. If heave occurs, the points should be redriven before the piles are concreted.

Settlement of Floating Pile Foundations

In some types of soft ground, piles of any kind can be driven to great depth without appreciable resistance against further penetration. The depth-penetration diagrams for such piles resemble that shown in Fig. 56.3*d*. These conditions call for a floating pile foundation, wherein the minimum length of the piles is determined not by a specified resistance against further penetration under the blows of a hammer, but by the requirement that the factor of safety of the pile groups with respect to a base failure should be equal to at least 2 or 3. The ultimate bearing capacity Q_g of each group can be estimated by means of Eq. 56.2. The value of s in this equation can best be determined by loading to failure several test piles of different length. However, before the computation can be made, the spacing between piles must be decided on.

According to Eq. 56.2, the ultimate bearing capacity of a friction pile group increases with increasing spacing. Furthermore, at a given load per pile the settlement of a cluster consisting of a given number of piles decreases as the spacing increases. Hence, it would appear that a fairly large spacing is advantageous. However, as yet, empirical data concerning the effect of spacing on the settlement are very scarce. In 1915 two groups of friction piles embedded in soft silty clay were loaded with 240 tons per group (Staniford 1915). Each group contained 16 piles 77 ft long. In one the piles were spaced at $2\frac{1}{2}$ ft, and in the other at $3\frac{1}{2}$ ft. After 40 days both groups had settled $4\frac{1}{2}$ in. but after 270 days the settlement of the group with the closer spacing was 11 in., and that of the other group only 8 in. Inasmuch as such an advantage must be paid for by constructing very much larger footings, it is doubtful that a spacing in excess of 3.5*d* is economical.

If the number of piles in a group is increased at a given spacing and at a given load per pile, both the intensity of the greatest stress in the soil and the depth of the highly stressed zone increase. This can be seen by comparing the right-hand sides of the diagrams in Figs. 56.9*a* and *b*. Hence, the settlement of a pile-supported foundation covering a large area is greater than that of a smaller foundation supported by equally loaded piles of the same length and driven at the same spacing. Similarly, the settlement of a foundation covering a given area and supporting a given total load decreases with increasing length of the piles, in spite of the fact that fewer piles are needed to carry the load. These conclusions are confirmed by experience in every city where the soil conditions call for floating pile foundations (Clarke and Watson 1936).

On the left sides of Figs. 56.9*a* and *b* are shown the intensity and the distribution of the stresses in the soil computed on the assumption that no piles are present. The ultimate settlement of the pile foundations shown on the right side of the diagrams can be roughly estimated

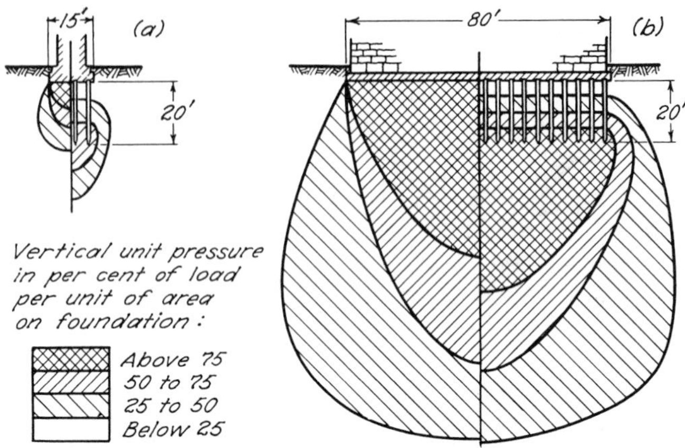

Fig. 56.9. Diagram illustrating increase of vertical pressure in soil beneath friction pile foundations having piles of equal lengths carrying equal loads. In (*a*) width of foundation is small compared to pile length. In (*b*) width of foundation is large compared to pile length.

on the following simplifying assumption. Above the level of the lower third-point of the length of the piles the water content of the clay remains unchanged, and below this level consolidation proceeds as if the building were supported on a flexible raft located at that level. The presence of the piles is disregarded. According to this assumption, the benefit derived from the piles is equivalent to replacing the subsoil by a practically incompressible material that extends from the base of the foundation to a depth equal to two thirds of the length of the piles. If this depth is several times greater than the width of the footings, and the footings are widely spaced, the settlement of the pile foundation will be small, no matter how bad the subsoil may be. On the other hand, if this depth is considerably smaller than the width of the loaded area, and the loaded area is large, the ultimate settlement may be excessive even under a moderate load. These conclusions have been consistently confirmed by experience. Both experience and theory have also shown that raft foundations supported by uniformly loaded and equally spaced friction piles, like simple raft foundations, always tend to assume the shape of a shallow bowl.

If the structure contains a basement, the load that produces consolidation is equal to the difference between the effective weight of the building and the effective weight of the soil that was excavated to form the basement (Article 55).

Efficiency Equations

The preceding discussions have demonstrated that there is no general relation between the settlement of a pile foundation and that of a single pile under a load equal to the load per pile in the foundation. The growing realization of this fact has led to various attempts to express the influence of the number and spacing of the piles on the settlement of the foundation by so-called *efficiency equations* (Seiler and Keeney 1944, Masters 1943, Feld 1943). However, the extraordinary variety of soils encountered in piling practice excludes the possibility of establishing a limited number of sufficiently accurate efficiency equations of general validity. The effect of the number and spacing of the piles on the ratio between the settlement of a single pile under a given load and that of a group under the same load per pile depends to a large extent on the sequence and properties of the soil strata. Furthermore, at a given length and spacing of the piles the ratio changes to a considerable extent with the load per pile. Nevertheless, in none of the existing efficiency theories are these vital facts given adequate consideration. Because of the great number and diversity of the factors involved, it seems very doubtful, to say the least, whether the efficiency equations represent a step in the right direction.

At the present state of knowledge and for many years to come it seems preferable to consider every case individually and to evaluate the probable settlement of a proposed pile foundation on the basis of the physical properties of the soils onto which the load is transmitted by the piles. Examples of the use of this procedure have been given under the preceding subheadings. If the probable settlement exceeds the tolerable maximum, the design must be modified. The maximum tolerable settlement of pile foundations is determined by the same factors as those that govern the permissible settlement of footing and raft foundations (Articles 54 and 55).

If the distribution of the loads over the area to be occupied by a structure is very uneven, the secondary stresses in the structure due to unequal settlement can be appreciably relieved by dividing the structure into blocks separated from each other by continuous vertical joints.

Selection of Type of Pile

The designer of a pile foundation can choose among several different types of piles, any one of which may provide adequate support for the proposed foundation. The final choice is governed by economic considerations and by conditions imposed by the character of the job.

Until the end of the last century untreated timber piles were used almost exclusively. This type of pile is relatively cheap, but it has two major disadvantages. First, a wood pile must be cut off below the lowest water table; if the water table is subsequently lowered on account of a permanent change in groundwater conditions, the uppermost parts of the pile disintegrate within a relatively short time. Second, a wood pile may break if it is driven too hard, although the foreman may not detect anything unusual. The risk of deterioration may be reduced by impregnation with wood preservatives, but the risk of breakage can be reduced only by stopping the driving of the pile while its bearing capacity is still relatively low. Since concrete or steel piles can be driven harder than wood piles without risk of damage, the "safe design load" for such piles is considerably greater than that for wood piles. Recognition of this fact in practice is exemplified by the values that represent the loads commonly assigned to piles of various types. Such values are given in Table 56.2. However, under many circumstances the design loads differ widely from those in the table.

Although the "safe design loads" for piles of different types are different, the spacing between piles of all types is practically the same. Therefore, the footings required to transfer a given load onto wood piles are considerably larger and more expensive than footings with equal bearing capacities supported by concrete or steel piles. Furthermore, the bases of footings resting on concrete or steel piles can be established at any convenient elevation, whereas those of footings on wood piles must be located below the lowest position of the water table. These advantages in many instances compensate for the fact that the cost of a concrete or steel pile is several times that of a wood pile.

Before the beginning of the 20th century all concrete piles were of the precast reinforced type. During the following decade cast-in-place piles became widely used, and the manufacture of concrete piles developed into a highly specialized industry. More recently, prestressed concrete piles have also entered the field. Structural steel sections and steel pipe have similarly become commonplace. The piles from which the designer may choose differ in their method of installation, their shape, the texture of their surface, and several other aspects. Almost every type of pile has features that make it exceptionally suitable under certain soil conditions and less suitable or inapplicable under others. For example, if piles are expected to carry their load by skin friction, conical types are preferable to prismatic ones, and bulb piles are not applicable at all. On the other hand, if the piles are expected to derive their support by point bearing, conical piles offer no advantage, whereas bulb piles may be very suitable. In any particular set of circumstances, economy may dictate the final choice among those piles that are technically acceptable. In order to meet the wide variety of soil conditions encountered in practice, every large pile company offers its clients several very different types of piles.

Table 56.2
Customary Design Loads for Piles

Type of Pile	Allowable Load (tons)
Wood	15–30
Composite	20–30
Cast-in-place concrete	30–50
Precast reinforced concrete	30–50
Steel pipe, concrete-filled	40–60
Steel H-section	30–60

The choice of the type of pile may also be influenced by special requirements imposed on the designer by the character of the job. Precast reinforced-concrete piles require heavy pile drivers with leads high enough to handle the longest piles on the job. They also require a large vacant space to serve as a casting yard. If these requirements are not met, precast piles cannot be used. If vibrations due to pile driving cannot be tolerated for some reason, a pile must be adopted which can be jacked down or else installed in a drill hole.

These and similar factors must be considered by the designer in connection with every pile job. Proper choice of a pile type requires judgment, experience in pile driving, and thorough grounding in the principles discussed in this article.

Summary of Principles in Design and Construction of Pile Foundations

Design of a pile foundation requires first the selection of the type, length, and spacing of the piles, and the "safe design load" per pile.

Selection of the type of pile is governed chiefly by economic and practical considerations. The choice among point-bearing piles of different types should be based on point resistance, not on total ultimate bearing capacity.

The length of point-bearing piles is determined by the location of the bearing stratum. Friction piles in any type of soil should be made as long as economically possible. By increasing the length of friction piles, the number of piles required to carry a given load is reduced, the ultimate bearing capacity of the entire pile foundation is increased, and the settlement is decreased.

The spacing D between the centers of wood piles with a butt diameter d should conform roughly to the following rules: For point-bearing piles driven to rock, or driven through soft clay strata to sand a short time after the ground surface has been covered by a fill, $D = 2.5d$. For point-bearing piles driven through less compressible strata into dense sand and for friction piles in loose sand, $D = 3d$. For point-bearing piles driven to bearing in stiff clay, and for friction piles in soft clay, $D = 3d$ to $3.5d$.

The "safe design load" can be determined either by means of a pile formula or a load test. The load-test method is more accurate. However, a pile foundation is not necessarily satisfactory even if the load per pile is less than the "safe design load." It may settle excessively, or if it is a floating pile foundation it may fail completely. To avoid these risks the group action of the piles must be considered.

Attempts to evaluate the group action by efficiency equations are likely to be misleading. All the existing equations claim a wide range of validity, whereas the variety of actual soil conditions precludes the possibility that any such equation could have more than a very limited range of applicability. Therefore, efficiency equations should not be used. General instructions for judging group action on the basis of soil profiles have been given in the text.

Point-bearing piles driven through highly compressible strata into sand may settle excessively unless the load per pile is considerably smaller than the point resistance. If a large part of the energy of the blows of the hammer is consumed by the skin friction in the uppermost strata, the point resistance may be smaller than the "safe design load." In case of doubt the point resistance should be determined. If the individual piles in clusters of point-bearing piles meet refusal at very different depths, a boring should be made near the cluster to determine the cause of the variation. If the boring shows that the soil located between the levels of the shortest and the longest piles contains pockets or layers of soft clay or silt, only those piles should be considered satisfactory that extend below the level of the lowest pockets. The others should be disregarded and replaced by substitute piles that should be driven to the necessary depth with the aid of jetting or spudding.

Clusters of point-bearing piles driven through clay strata supporting newly deposited fill will be acted on not only by the weight of the structure but also by the weight of the new fill located between the piles in each cluster, and by the negative skin friction along the vertical boundaries of the clusters.

If the points of the piles in large clusters are driven into a sand stratum located above soft clay layers, or if such clusters are entirely embedded in soft clay, appreciable progressive settlement is inevitable and should be estimated in advance of construction.

Large clusters of friction piles embedded in soft clay may not have adequate safety with respect to a base failure of the group as a whole. Hence, the factor of safety with respect to such a failure should always be estimated.

If piles are driven into sand without the aid of a water jet, driving should proceed from the centers of the clusters toward the edges. Friction piles in soft silt or clay should be driven to the same depth, regardless of the number of blows for the last foot. Piles of any other category should be driven until the number of blows for the last inch becomes equal to that of the test piles that furnished the information for evaluating the design load. If point-bearing piles are to be driven through firm strata underlain by or alternating with soft compressible ones, jetting or spudding may be required.

Selected Reading

A comprehensive treatment of the subject is found in *Pile foundations* by R. D. Chellis (1961), New York, McGraw-Hill. The chapters on "Piled foundations" in *Foundation design and construction* by M. J. Tomlinson (1963), New York, John Wiley and Sons, contain much useful information for the designer, with special emphasis on British practice. Economic and practical aspects, including the choice of type of pile, are discussed extensively in *Foundation design and practice* by J. H. Thornley (1951), New York, Columbia Univ. Press.

A summary of the current state of research and practice may be found in the General Report on deep foundations presented by A. Kezdi at the Sixth International Conference on Soil Mechanics and Foundation Engineering, Montreal, 1965. It is published in Vol. 3 of the *Proceedings*, pp. 256 to 264.

The article by J. D. Parsons, "Piling difficulties in the New York Area," *ASCE J. Soil Mech.*, **92**, No. SM1 (1966), pp. 43–64, is an exemplary addition to the meager documentation of the behavior of piles and pile foundations under adequately described conditions.

ART. 57 PIER FOUNDATIONS

Function of Piers

Piers are prismatic or cylindrical columns that have essentially the same function as piles or pile clusters. If piers are constructed to support a bridge, their sole purpose may be to transfer the loads to a level below that of the deepest scour. In some semiarid regions piers are used to transfer the loads to a level below the zone of periodic desiccation of highly plastic clays (Article 21). However, in all other instances, piers serve, like point-bearing piles, to transfer the loads onto or into a firm stratum located beneath softer ones.

The principal difference between piers and piles lies in the method of installation. The relative merits of piers in comparison with piles depend not only on economic but also on several technical factors. These include the influence of the method of construction on the load that can be assigned to the foundation, and the influence of the soil conditions on the ease or difficulty of construction and on the integrity of the completed foundation. These factors, moreover, are interdependent. The following examples illustrate the possibilities.

If a pile is driven through soft ground into a stratum of dense sand, the point of the pile displaces and compacts the sand. The point resistance of such a pile is likely to be many times greater than that of a cylindrical pier with equal diameter, because the process of installing the pier does not compact the sand but, instead, gives it an opportunity to expand. On the other hand, if the layer of dense sand is located beneath a sequence of thin layers of soft clay and thick layers of sand, most of the energy of pile driving is likely to be used up by skin friction, and driving may have to be discontinued while the point resistance is still very small. Under such conditions piers would probably be safer and more economical than point-bearing piles.

It may be intended to transfer the weight of a structure onto sound bedrock overlain by a thick layer of decomposed rock which, in turn, is buried beneath soft sediments. The decomposed rock may be as compressible as medium or even soft clay, but it may contain large fragments of less decomposed material. These fragments would prevent the points of piles from being driven into sound rock, whereas they could easily be removed from the excavation for a pier.

If a structure is underlain by a medium clay which rests in turn at comparatively shallow depth on a deep deposit of stiff clay, it may be possible to support the entire load from each column on a single machine-drilled pier with an enlarged base resting just below the upper surface of the stiff clay. Such a foundation may be far more economical than friction piles in the stiff clay. On the other hand, if the clay deposits contain seams of waterbearing silt and sand, it may not be possible to enlarge the bases of the shafts without collapse or without inflow of loose wet material that would deprive the piers of firm support and might make the placement of sound concrete impracticable.

Types of Piers

Since piles and piers serve the same purposes, no sharp distinction can be made between the two. Cast-in-place piles installed in drill holes might preferably be considered piers of small diameter because they are constructed by procedures identical to some of the methods used for constructing larger piers. Drilled-in caissons are made by driving a heavy steel pipe with a cutting shoe down to bedrock and as far into the rock as it will go. In this respect the caissons are piles. After refusal is met, the soil encased in the pipe is removed, a hole is drilled through the decomposed top layer into sound rock by means of a churn drill, and the hole and the pipe are filled with concrete. These procedures are characteristic for piers.

The great variety of types intermediate between typical piles and typical piers involves a similar diversity in the methods of installation. If the diameter of a pier is small enough to justify the use of well-drilling methods, the pier can be installed in almost any type of soil. On the other hand, the most suitable method for constructing piers of large diameter is chiefly determined by the soil conditions. If an attempt is made to build such a pier by a method that is not practicable under the given soil conditions, the contractor will be compelled to change the method during construction. An emergency change of procedure always involves a considerable loss of time and money. Therefore, the engineer who chooses the method for constructing piers should be familiar with the prerequisites for success. The most common methods of construction are discussed under the following subheading.

Methods for Constructing Piers

The methods for constructing piers can be divided into two general classes, the sinking of caissons, and the excavation of open shafts. Caissons are rarely used for piers with diameters less than about 10 ft, whereas the dimensions of open shafts range from as little as one foot to the size of the largest foundation units. Strictly speaking, a caisson is a shell within which the excavation is made. The shell descends into the ground to the level of the base of the foundation and eventually becomes an integral part of the pier. The oldest type of caisson is the drop shaft or open caisson (Figs. 57.1*a* to *c*). The shell sinks under its own weight as the soil at the bottom of the caisson is excavated. If the bottom is located above the water table, or if the water is removed by pumping from an open sump, the excavation can be made by hand (Fig. 57.1*a*); otherwise, the soil must be removed by dredging (Fig. 57.1*b* and *c*) and the bottom of the caisson sealed with underwater concrete when grade is reached. Obstacles in the path of the cutting edge, such as buried logs or boulders, may delay the sinking of the caisson by several days or weeks. If they cannot be removed within a reasonable length of time, the work must be continued by the compressed-air method (Fig. 57.1*d*). As the caisson descends, air pressure is maintained in the working chamber at the value of the hydrostatic pressure in the porewater at the level of the cutting edge. For physiological reasons the use of air pressure is limited to a depth of about 120 ft below the water

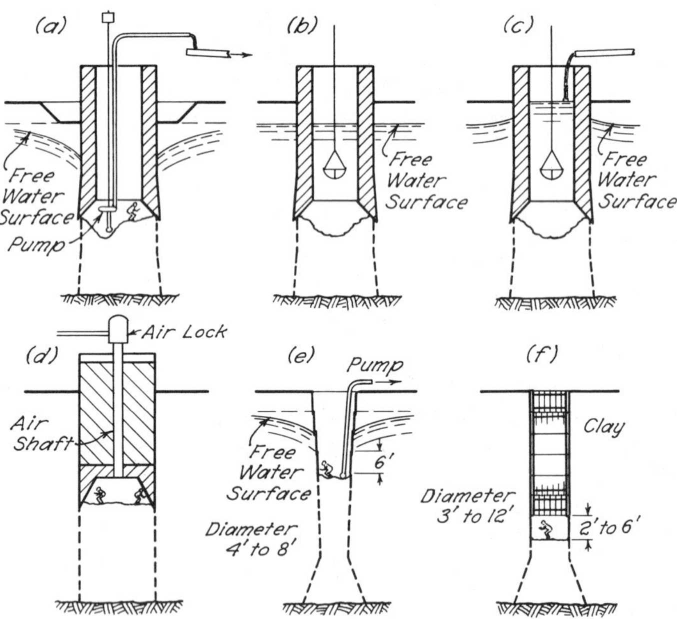

Fig. 57.1. Diagrams illustrating methods for constructing piers. (*a* to *c*) Open-caisson method. (*d*) Compressed-air caisson. (*e*) Open shaft with telescoping steel-shell lining (Gow method). (*f*) Open shaft lined with wood lagging and steel rings (Chicago method).

table. Beyond a depth of 40 ft the cost increases rapidly. The compressed-air method must also be used as a substitute for pumping if the specifications call for cleaning the bottom of the pier excavation before placing the concrete.

Until about the 1950's, open-shaft methods for establishing piers were almost without exception associated with hand excavation. Subsequently, methods for mechanically drilling open shafts have been developed to a high degree of efficiency and have largely replaced hand methods. Nevertheless, hand excavation is still required in locations not accessible to the drilling equipment, and is often used in conjunction with mechanical drilling, especially for enlarging or cleaning the bottoms of the shafts or for coping with unexpected conditions.

Two commonly used open-shaft methods for establishing piers by hand excavation are the *Gow method* (Fig. 57.1*e*) and the *Chicago method* (Fig. 57.1*f*). Neither procedure can be used unless the water can be removed by pumping or bailing. In the Gow method the sides of the excavation are supported by a series of steel cylinders, each of which is 2 in. smaller in diameter than the one above. The cylinders are driven with a light hammer while the soil is being excavated by hand. The lowest part of the shaft is usually belled out. After the excavation has been completed, the shaft is filled with concrete, and the cylinders are recovered one by one (Mohr 1964).

The Chicago method is used exclusively in clay. A cylindrical hole is dug by hand for a depth which varies from as little as 2 ft in soft clay to 6 ft in stiff clay. The sides of the excavation are accurately trimmed and lined with vertical boards held against the clay by two or more steel rings. The hole is then deepened for several more feet and lined in the same manner. When the bottom reaches final elevation, the hole is filled with concrete. In homogeneous clay, water causes no difficulty but, if water-bearing strata of sand or silt are encountered, special construction expedients may be required (Peck 1948).

A wide variety of equipment and techniques has been developed for the mechanical excavation of pier shafts. In their simplest form the machines consist of rotating vertical shafts equipped with augers or with buckets provided with cutting vanes. Reamers attached to a special bucket may be used to form a bell at the bottom of the shaft. The augers or buckets can be raised from the hole quickly for removal of the spoil and quickly reinserted. Hence, under favorable circumstances, progress may be very rapid in comparison to that which can be achieved by hand. The diameters of the shafts may vary from as little as one foot to 10 ft or more, and the depths may reach more than 100 ft. Heavy equipment is required for the larger diameters and greater depths.

In firm dry ground without obstacles, machine drilling may be extremely economical. Such conditions are encountered in many localities in southwestern United States, for example, where piers are needed to establish even light structures below the depth of seasonal volume change (Article 21). If the ground contains lenses or layers of slightly cohesive soils, even below the water table, it may remain stable in the walls of the shaft at least briefly after excavation; because of the speed of the mechanical excavation it may be possible under these circumstances to complete the hole before an objectionable amount of material falls in, and then to insert a temporary casing to protect the walls while the bottom is being prepared for concreting. If the materials will not remain stable long enough to permit the use of this procedure, casing may have to be inserted before the final depth is reached. The cutting tool must then have retractable blades in order that it may be raised and lowered inside the casing, or else the diameter of the hole must be decreased beneath the casing. In any event, the necessity for casing appreciably reduces the rate of progress. Alternatively, the shaft may be kept filled with a slightly thixotropic drilling mud (Article 44) that provides support for the walls of the shaft until the hole reaches its final depth. A casing is then inserted and the drilling mud pumped out to permit inspection of the bottom. In cohesionless soils, particularly below the water table, the use of drilling mud is a prerequisite to success. Obstacles such as boulders encountered in the mud-filled hole can usually be removed or broken up by equipment especially devised for the purpose (Gaunt 1962). If the shafts are to be excavated through soft cohesive soils, the use of mud may also be necessary to reduce loss of ground (Article 58).

For economy, the temporary casing is often withdrawn during concreting. The procedure requires the utmost care and most expert supervision in order to prevent the incursion of the surrounding soil into the space that should be occupied by the concrete. The top of the fresh concrete must be maintained high enough above the bottom of the casing to counterbalance the pressure of the surrounding soil which would otherwise invade the concrete; at the same time the column of concrete in the bottom of the casing must not be long enough to form a plug that would rise with the casing and create a space below the casing into which the surrounding soil would be drawn. In numerous instances, investigations of severe settlements of pier foundations have disclosed that the shafts were completely severed by soil (Peck 1965).

In all methods of pier excavation, the stability of the bottom is of outstanding practical importance during excavation and, particularly, during the final preparations for placing the concrete on the material that will support the pier. These matters will be considered under the following subheading.

Stability of Bottom and Preparation for Concreting

The stability of the bottom of a pier excavation is determined by the same factors that govern that of the bottom of open cuts (Articles 47 and 48). In very dense sand it may be possible to pump water from the caisson or open shaft without destroying the stability of the material below the bottom, because the deformation produced by the seepage pressure does not cause an increase in the neutral stress. In loose sand, however, the deformation causes an excess hydrostatic pressure that tends to liquefy the sand. In one instance the Gow method (Fig. 57.1*e*) was used successfully to establish the first members of a group of piers. The excavations were carried through a stratum of fine dense sand to rock located about 10 ft below the water table. However, beneath one part of the building site a mixture of sand and water rose into the shafts as soon as the excavation was carried to a depth of a few feet below the water table. All efforts to stop the flow failed, and the remaining piers had to be constructed by a method that did not require pumping. The most probable reason for this unexpected development is that the latter shafts happened to penetrate a large pocket of loose sand. The occurrence of such pockets surrounded by dense sand with essentially the same grain-size characteristics is by no means uncommon.

If pumping from sumps seems impracticable, alternative methods must be considered. They include predrainage of the soil by pumping from well points or deep wells, excavation under compressed air, dredging, and excavating in a slurry-filled hole. The soil investigations required to learn whether a given method of unwatering is practicable on a given job are described in the discussions of the methods for unwatering open cuts (Article 47). If the soil consists of fine silt, even the vacuum method may fail to stabilize the soil. Because of the expense and other limitations of the compressed-air method (page 558), it is not frequently used. The most common alternatives for pumping are dredging and the use of slurries.

Dredging in sand usually removes a volume of sand greater than the volume of the caisson. If the sand is loose, the quantity excavated may be twice that displaced by the caisson. The excess excavation is associated with loss of ground and settlement of the near-by ground surface. However, the loss can be almost entirely prevented by maintaining the water level within the caisson several feet above the water table, as shown in Fig. 57.1*c*. The excess head produces a flow of water from the caisson into the sand located below the bottom of the excavation, and the corresponding seepage pressure counteracts the tendency of the sand to rise. To maintain the flow, the excavating tools must be hoisted out of the caisson slowly.

In choosing between the dredging and compressed-air methods for sinking a caisson, several factors should be considered. If obstacles in the soil are encountered when the dredging method is being used, they may cause unpredictable delays. The compressed-air method avoids this risk, because the workmen have access to the obstacle. It has the added advantage that the base for the pier can be prepared carefully, and all loose material removed. On the other hand, the use of compressed air is inherently much more expensive. In constructing a bridge pier it is not uncommon to lower a caisson by dredging until the bearing stratum is approached or reached, whereupon the caisson is converted to the compressed-air type. In this manner the economy of open dredging is combined with the greater certainty of the pneumatic method during preparation of the bottom for placement of concrete.

The use of a slurry to stabilize the walls of a drilled shaft simultaneously prevents instability of the soil below the bottom as excavation proceeds. However, when the shaft has reached its final depth and casing has been inserted for support of the walls, the bottom may blow up when the slurry is pumped out unless the shaft has been carried to a material firm enough to remain stable under the influence of the upward seepage pressures. If the material below the bottom consists of rock overlain by potentially unstable cohesionless materials, it may not be possible to pump out the slurry without causing an inflow of water and soil through the gaps between the bottom of the casing and the irregular surface of the rock. It is sometimes necessary to resort to expensive and elaborate procedures, such as providing the bottom of the casing with hardened steel teeth and grinding the casing into the rock to obtain a seal, in order to be able to expose the rock and place the concrete in the dry (Peck and Berman 1961).

Evaluation of Skin Friction during Sinking of Caissons

While excavation is being carried on within a caisson of the drop-shaft type, the soil adjoining the caisson is supported laterally by the walls of the shaft. Sections are added to the shaft above the ground surface and, as excavation proceeds, the caisson slides down. The downward movement is resisted by skin friction. In order to overcome the skin friction, lightweight drop shafts, such as those of steel, must be loaded with dead weights. On the other hand, heavy caissons, such as those of reinforced concrete, may descend under their own weight.

Adding weights to the top of a caisson is a cumbersome procedure that considerably increases the cost of construction. Therefore, concrete caissons are commonly designed in such a manner that their weight exceeds the skin friction at every stage of construction. Hence, the design requires evaluation of the skin friction. Experience has demonstrated that methods for evaluating the skin friction on the basis of soil tests are unreliable. The principal source of existing information regarding skin friction is the record of loads required to start caissons that had become stuck. These records suggest that for a

Table 57.1
Values of Skin Friction During Sinking of Caissons

Type of Soil	Skin Friction f_s (lb/ft²)
Silt and soft clay	150–600
Very stiff clay	1000–4000
Loose sand	250–700
Dense sand	700–1400
Dense gravel	1000–2000

given soil the skin friction per unit of contact area reaches a fairly constant value below a depth of about 25 ft. Table 57.1 gives values that have been obtained for caissons ranging in depth from 25 to 125 ft. For each material the range of values is fairly close to that for skin friction on piles in the same material. However, no perfect agreement should be expected, because in a given material f_s depends on the shape of the lowest part of the caisson, on the method of excavation, and on the diameter of the caisson. Values from other jobs in the vicinity should not be relied on unless all of the circumstances attending the caisson sinking are known. In clay the skin friction is likely to increase with time.

The friction between the walls of concrete caissons and fine-grained soils such as silt or clay can be considerably reduced by providing the outside of the caissons with a coating that has a smooth oily surface and that, in addition, is tough enough not to be rubbed off while the caisson descends. Such a coating was used on the caissons for the piers of the San Francisco-Oakland Bay bridge. The results of friction tests made prior to construction indicated that it reduced the friction between the concrete and a fairly stiff clay by roughly 40%.

Piers on Sand

Piers commonly serve to transfer the weight of a structure onto a firm stratum covered by soft and compressible soil. If piles are driven into such a stratum, almost the entire load on the pile is ultimately carried by the point resistance (Article 56). For similar reasons practically the entire load on a pier is ultimately carried only by its base. Hence, the allowable load on piers surrounded by relatively compressible soil should not include any allowance for skin friction.

The buried part of a bridge pier may be completely surrounded by sand that has a low compressibility and is capable of carrying a considerable part of the load on the pier by skin friction. However, the base of such a pier is commonly located at a moderate depth below the maximum depth of scour (Article 53). During exceptionally high water, most of the sand surrounding the pier is temporarily removed. Hence, even in connection with bridge piers entirely surrounded by sand, it should be assumed that the entire load on the pier is carried by the base.

The ultimate bearing capacity of a pier on sand beneath compressible deposits can be calculated with sufficient accuracy by means of Eqs. 33.7, 33.13, or 33.14, in which the term $\gamma D_f N_q$ is considered to be the effective weight of the soil between the ground surface and the level of the base of the pier. Because of the influence of this term, the ultimate bearing capacity increases rapidly with increasing depth of foundation. Hence, unless the pier is comparatively shallow and has a small width, it can usually be taken for granted that it will not experience a base failure. Therefore, the allowable bearing value is determined exclusively by considerations of settlement.

In Article 34 it was pointed out that the ultimate bearing capacity of piers entirely embedded in sand does not increase with depth as rapidly as indicated by Eqs. 33.13 or 33.14. Nevertheless, in most practical problems involving subsurface conditions of this nature, settlement is likely to govern the allowable soil pressure and the ultimate bearing capacity is likely to be irrelevant. The ultimate capacity may have to be evaluated, however, in connection with such projects as large bridges which exert very small live loads onto the foundation in comparison to the dead loads, and for which the settlement during construction is largely irrelevant because adjustments can readily be made during erection of the spans.

The settlement of a loaded area on sand depends to a large extent on the stress conditions that exist in the sand before the load is applied. The construction of a pier is always preceded by the excavation of a shaft. This process is associated with a relaxation of all the stresses in the sand adjoining the walls and the bottom of the shaft. If the depth of the shaft exceeds four or five times the diameter, the state of stress in the sand near the bottom of the shaft is practically independent of the depth of the shaft. Therefore, it is to be expected that the influence of the depth of foundation on the settlement of the piers will be relatively small compared to its influence on the ultimate bearing capacity. This conclusion is corroborated by the following observations.

Load tests were made at the same depth on two circular bearing plates each covering an area of 1 ft^2. One plate was located on the bottom of a large open shaft, and the other on the bottom of a drill hole with a diameter of 1.15 ft. At a load of 2 tons/ft^2, the settlement of the plate in the shaft was 0.90 in., and that in the drill hole was 0.52 in.

Similar experiments were made in the shaft to which Fig. 44.16 refers. After the shaft was excavated to a depth of about 50 ft, a load test was made on a bearing plate 1 ft square. The settlement under a load of 2 tons/ft^2 was 0.25 in. A second bearing plate 3.3 by 3.3 ft was installed on the bottom of the shaft, and the narrow space between the edges of the plate and the sides of the shaft was filled with concrete to prevent even a local heave of the loaded sand. The settlement of the plate at a load of 2 tons/ft^2 was 0.47 in. According to Eq. 54.1, the settlement of a plate of the same size on the surface of a similar sand deposit with no confinement or surcharge would be 0.59 in. (Terzaghi 1930).

These and various other observations indicate that the settlement of the base of a pier on sand at any depth is likely to be about one half the settlement of an equally loaded footing covering the same area on sand of the same characteristics. Therefore, the allowable bearing values for piers on sand can be assumed to equal twice the value that would be admissible for footings resting on the same sand in the same state (Article 54). If the excess unit pressure on the base of the piers does not exceed this value, the maximum settlement will not exceed 1 in. Furthermore, if the bases of all the piers have approximately the same width, the differential settlement between piers will not exceed $\frac{1}{2}$ in. If the designer feels that he can tolerate larger settlements, he may increase the bearing values accordingly.

A modification of this procedure may be required if the base of a bridge pier is located fairly close to the level to which scour may remove the sand. The scour temporarily reduces the depth of foundation of the pier to much less than 4 or 5 times the width of the base. Hence, the pressure on the base of such piers should not exceed that appropriate for footings of the same area resting on the same sand in a saturated condition.

Piers on Clay

The ultimate bearing capacity of a pier founded on a stratum of stiff clay located beneath soft compressible deposits is determined by Eq. 33.17, in which D_f is taken as the vertical distance between the top of the stiff clay and the level of the base of the pier. The value of N_c is not increased above the value corresponding to that for a shallow footing, because the low strength and compressible character of the overlying materials prevent the development of the zones of plastic equilibrium characteristic of a homogeneous cohesive material (Article 34). Consequently, the procedure is conservative because the strength of the overlying material increases the ultimate bearing capacity of the pier to some extent.

In some localities, including London and parts of Southern California, the subsoil beneath the surficial deposits consists of stiff, often fissured, clay to great depths. Drilled piers, either straight-shafted or with enlarged bases, have been found expedient and economical for transferring the loads from columns into the clay. A substantial part of the load, even on the piers with enlarged bases, is carried by skin friction. Large-scale tests (Skempton 1959, Whitaker and Colman 1965, Woodward et al. 1961) indicate that the ultimate bearing capacity is given approximately (Eq. 34.1) by

$$Q_d = Q_p + Q_s = Q_p + 2\pi r f_s D_f \qquad (57.1)$$

The ultimate base resistance Q_p may be considered equal to cN_cA_p, where c is the undrained shear strength of the undisturbed clay, N_c has the value 9 corresponding to a deep foundation in a homogeneous cohesive soil, and A_p is the area of the base. The value of c of intact samples is likely to be greater than that of the undisturbed fissured clay, but the influence of the fissures is usually small at the depths of the bottoms of the piers.

The ultimate capacity of the shaft Q_s may be expressed as $\alpha c A_s$, in which α is a reduction factor to be applied to the average undrained shear strength of the clay adjacent to the shaft with area A_s. The factor α must be evaluated on the basis of full-scale tests. So far, extensive experiments have been made in only a few localities. For the London clays, α appears to be on the order of 0.45 (Skempton 1959), whereas values in stiff clays in southern California ranged from 0.49 to 0.52 (Woodward et al. 1961). The results are more or less independent of the presence or absence of an enlarged base, provided any skin friction on the enlargement is ignored.

On the other hand, at a given factor of safety, the immediate settlements of piers with bells are always larger than those of straight-shafted piers in the same material. This condition arises because the skin friction reaches its ultimate value at very small settlements, on the order of 1% of the diameter of the shaft for the London clays, and continues to exert a resistance almost equal to the ultimate value during further settlement. In contrast, the point resistance develops slowly with increasing load and does not reach a maximum until settlements have been reached on the order of 10% of the diameter of the base (Whitaker and Colman 1965).

The total load that can safely be applied to the clay beneath the pier is equal to the sum of the allowable load on the base of the pier and the effective weight of the soil excavated during construction. Hence, the design load on large piers, at a given allowable load on the base, can be considerably increased by making the piers hollow. This fact has been utilized many times in the design of bridge piers.

On clay the settlement of piers, like that of footings, depends to a large extent on the load history of the clay. Pier foundations on normally loaded clay are uneconomical, and their settlement is prohibitive. Therefore, piers are established only on precompressed clay. Yet, if the area covered by a pier is large, the precompressed state of the underlying clay does not necessarily preclude important progressive settlement. This statement is illustrated by the following observation. Near the end of the last century bridge piers were established by the compressed-air method on a thick stratum of very stiff precompressed clay beneath the Danube River. The base of each pier was 75 ft long and 20 ft wide. The effective load on the bases of the piers ranged between 3.3 and 4.8 tons/ft^2. For a very stiff clay this load is well below the critical value for a base failure. Yet within half a century the difference between the settlement of the piers became equal to 3 in. The value of the maximum settlement could not be ascertained, but there is no doubt that it was much greater than the differential settlement. Hence, if the base of a pier on stiff clays covers a large area, a settlement computation should be made. The uncertainties involved in the determination of the settlement of loaded areas on precompressed clay are discussed in Article 13.

Pier Foundations on Natural Rafts

Pier foundations on natural rafts differ in no essential way from footing foundations on such rafts. The piers are designed as if the raft rested on a rigid base, but the design must be supplemented by a settlement computation (Article 55).

Summary of Rules for Estimating Allowable Bearing Value for Bases of Piers

1. The allowable bearing value for the base of a pier on sand is equal to twice the allowable bearing value for footings covering the same area on sand of the same characteristics (Article 54). The settlement of piers designed on the basis of these values is unlikely to exceed 1 in. If larger settlements are acceptable, the bearing values can be increased accordingly. If the piers are surrounded by sand that may be removed by scour during high water, the allowable bearing value should be taken equal to that for footings of the same size resting on sand of the same characteristics.

2. If the excavation for a pier is made by dredging, the water level within the dredging shaft should be maintained several feet above the outside water level. This procedure reduces the tendency of the sand to flow toward the bottom of the excavation. Yet, even if the flow is prevented, the bottom of the excavation will be very uneven and partly covered with a layer of loose sand. Hence, if the sand is excavated by dredging, an allowance must be made for the unavoidable disturbance of the sand. No such allowance is required if the excavation is made under compressed air.

If excavation is carried through sand by mechanical drilling methods, the stability of the sides and bottom of the hole during the drilling can be maintained by means of a slurry. Use of a casing will protect the sides when the slurry is pumped out, but the bottom is likely to become unstable unless a firm cohesive stratum has been encountered.

3. Normally loaded clay is unsuitable for sustaining loaded piers. The allowable bearing values for piers on precompressed clay can be selected on the basis of Eq. 33.17 if the bearing stratum is located below weak, compressible deposits, or on the basis of Eq. 34.1 if the stiff clay is thick and fairly homogeneous. If the width of the

area covered by the base of each pier exceeds 10 ft, a settlement computation should be made.

4. Pier foundations on natural rafts call for the same investigations as footing foundations on such rafts.

Selected Reading

General descriptions of types of foundations and methods of construction, with emphasis on bridge piers, are included in Jacoby, H. S. and R. P. Davis (1941): *Foundations of Bridges and Buildings*. New York, McGraw-Hill, 3rd ed., 523 pp.

Drilled piers are extensively treated in Chapter 9 of Teng. W. C. (1962). *Foundation design*. New Jersey, Prentice-Hall, pp. 254–286.

Chapter 10

SETTLEMENT DUE TO EXCEPTIONAL CAUSES

ART. 58 SETTLEMENT DUE TO CONSTRUCTION OPERATIONS

Extraneous Causes of Settlement

In Chapter 9 we discussed the settlement of buildings and other structures under the influence of their own weight. Although this is the most common type of settlement, other types are important enough to deserve consideration. They include settlement due to increasing the load on the surrounding soil, to excavation in the vicinity, to lowering the groundwater table, and to vibrations. In this article we shall consider only the first two categories.

Settlement Due to Increasing Load on Surrounding Soil

The application of a load to one portion of the ground surface above any type of soil causes the surface of the adjacent soil to tilt (Fig. 58.1*a*). The distance within which the tilt is of any practical importance depends, however, upon the soil profile as well as the dimensions of the loaded area. If the subsoil contains soft clay, the magnitude and distribution of the settlement can be roughly estimated on the basis of the results of soil tests. If the subsoil is sand, the settlement cannot be computed and estimates can be based only on the records of precedents.

If rafts on sand are designed in accordance with the rules contained in the municipal building codes, they are likely to settle as much as 2 in. Exceptionally, they may settle even more (Article 55). Since the greatest part of this settlement occurs during construction, the structure itself will not be damaged unless it is very sensitive. However, the tilt of the adjoining ground surface toward the loaded area may be great enough to damage neighboring structures. In New York, for example, a 20-story building was constructed on a lot between two 7-story buildings supported by spread footings on a deposit of fine sand. The new building rested on a raft at a depth of 20 ft below the ground surface. The soil pressure was 2 tons/ft^2 in ex-

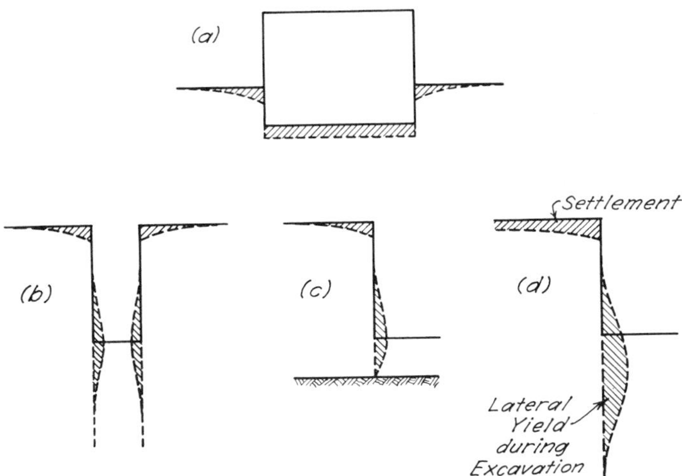

Fig. 58.1. Diagrams indicating settlement of ground surface adjacent to areas in which construction operations are carried out. (*a*) Settlement due to weight of structure. (*b*) Settlement due to lateral yield of clay beside narrow deep cut. (*c*) Settlement due to lateral yield of clay beside wide deep cut above stiffer soil. (*d*) Settlement due to lateral yield of clay beside and below wide deep cut in soft clay of great depth.

cess of the weight of the soil removed. Since the building itself settled only 1.8 in., and the settlement was fairly uniform, the building remained intact. Yet, the neighboring buildings were damaged by shear cracks and by distortion of door and window frames.

If the subsoil consists of soft clay, the effect of the weight of a new building on its neighbors can be much greater, although not necessarily more detrimental. In Istanbul a tall building was erected on a site separated from that of its equally tall neighbor by a narrow alley. The new structure caused such a large tilt of the old one that the cornices of the two structures came into contact with each other. Yet, neither structure was damaged.

Settlement Due to Excavation

FORECAST OF SETTLEMENT DUE TO EXCAVATION. If all other conditions are the same, settlement due to excavation depends to a large extent on the type of bracing used to support the adjoining soil and on the care with which the bracing is installed. Therefore, the magnitude of the settlement cannot be computed. A forecast can be based only on reliable well-documented case records.

The most common types of large excavations are open cuts, in which entire structures or basements can be built, and individual shafts, in which single piers are constructed. Tunnels constitute a third type which, however, is beyond the scope of this book.

Open Cuts in Sand. Even if the ground surface adjacent to an open cut in sand carries shallow heavily loaded footings, the settlement due to excavating the cut does not extend beyond a distance equal to the depth of the cut. If the adjacent ground surface carries no load, the settlement does not extend beyond half this distance. If the cut is properly braced, the maximum settlement is not likely to exceed about 0.5% of the depth of the cut. However, even this amount may be enough to cause damage, as shown in Fig. 58.2. The excavation responsible for the settlement was made in water-bearing gravel. The bracing was carefully placed, according to the method illustrated by Fig. 48.2*c*, and the sheet piles were driven through the gravel into a stratum of stiff clay. This made pumping unnecessary. Nevertheless, the footings of the adjacent building settled 1 to 2 in., and cracks developed in the walls as indicated in the figure. Part of the settlement occurred while the sheet piles were being driven.

Open Cuts in Soft Clay. When an open cut is excavated in soft clay, the clay located at the sides of the cut acts like a surcharge. Under this surcharge the clay near the bottom of the cut yields laterally toward the excavation, and the bottom of the cut rises. As a consequence of these movements, the ground surface located above the yielding clay settles. An additional lateral yield occurs above

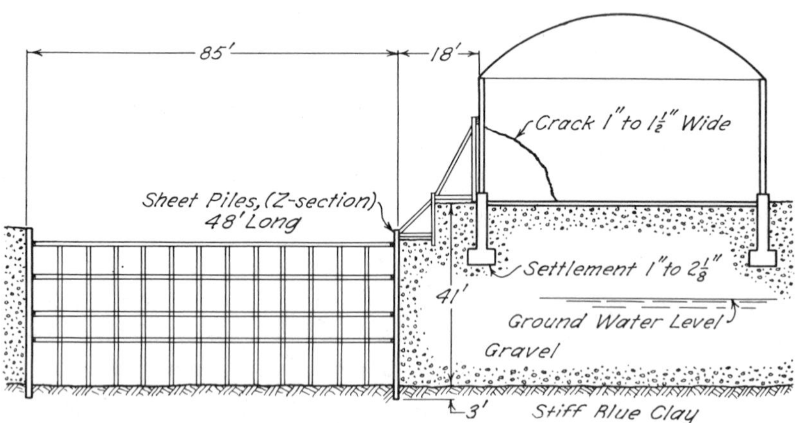

Fig. 58.2. Cross section through open cut in gravel, showing method of bracing and damage to adjacent structure due to settlement.

the bottom of the cut during the intermission between excavation and installation of the struts. The magnitude of these lateral movements and of the corresponding settlement depends primarily on the width-depth ratio of the cut, on the construction procedure, and on the depth and stress-strain characteristics of the soft clay beneath the bottom of the cut.

If the cut is very narrow (Fig. 58.1*b*) or if the bottom of the cut is located close to the surface of a firm stratum (Fig. 58.1*c*), the lateral yield spreads only a short distance from the sides of the cut. Therefore, the settlement of the ground surface is restricted to relatively narrow belts located on each side of the cut. The width of these belts does not exceed the depth of the cut. Beyond this distance the settlement is inconsequential. By careful bracing the inward yield of the clay can be kept within 0.5% of the depth of the cut, and the greatest surface settlement is of the same order of magnitude. Appreciably greater settlements are usually due to poor workmanship.

The soil deformations that lead to the settlement of the surface adjacent to wide cuts in clay have been observed and measured in several instances in Chicago and Oslo. At the site of the cut illustrated in Fig. 58.3 (Peck 1943) the uppermost 12 ft consisted of sand underlain by a stratum of soft clay which, in turn, rested on a moderately stiff clay at the depth of about 14 ft below final grade. The sides of the cut were supported by sheet piles that were driven through the soft clay into the stiff stratum before excavation was started. The curves on the left side of Fig. 58.3 represent successive positions of the sheet-pile wall on the dates shown. On the right side of the figure are inscribed the dates at which the struts were placed. The dash lines indicate the corresponding positions of the bottom of the cut. The diagram shows that the lateral yield spread to the base of the soft clay layer at an early stage of the excavation. Since the sheet piles penetrated into the stiff clay, the inward movement of the buried section decreased toward the lower ends of the piles. As a consequence, the heave of the bottom was unimportant, and the small tunnel shown in the figure rose only 1 in. The exceptionally great movement of the sheeting at a depth of 10 ft below the ground surface was caused by a delay in installing the top strut; otherwise, it would not have occurred. At a distance from the edge of the cut equal to its final depth, the settlement amounted to 0.7 in., but settlement was noticeable up to distances of 86 ft.

If the cut is wide, and the clay is soft to a great depth below the bottom, the lateral yield involves a wide and deep body of clay (Fig. 58.1*d*). The corresponding settlement may extend to a distance

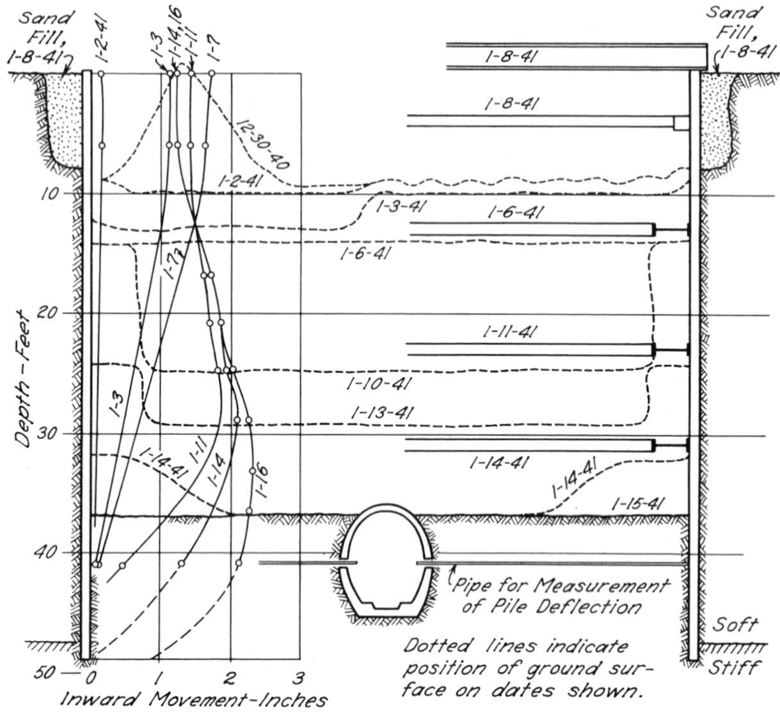

Fig. 58.3. Diagram showing results of measurements of lateral yield of sheet piling at side of open cut in soft clay underlain at shallow depth by stiff clay. Positions of deflected sheeting shown at left side of diagram. Dash lines indicate stage of excavation on dates shown. Dates shown on struts indicate time of placement (after Peck 1943).

considerably greater than the depth of the cut. With increasing depth of excavation, the stability number $N = \gamma H/c$ (Eq. 48.4) also increases. When it reaches about 4, the settlement begins to increase rapidly and spreads to a great distance from the edge of the cut, regardless of the care with which the sides are braced. At values of N approaching 7 or 8, base failure becomes inevitable and the bottom of the cut rises (Article 37).

The cut illustrated in Fig. 58.4*a* was excavated in a deep deposit of medium clay (S_t = 3 to 7) for a subway in Oslo (NGI 1962*e*). The average undrained shearing resistance was about 0.4 ton/ft^2. Heavy steel sheet piles were driven on both sides of the cut to a depth of 6 to 8 ft below final grade; bedrock was at a depth of more than 35 ft below the bottom of the excavation. While the excavation progressed the sheet piles moved inward, in spite of the successive insertion of struts, as shown in the figure. Correspondingly, the soil beneath excavation level heaved, as indicated by the dash lines representing the upward movements of reference points $H1$ and $H2$, and the surface of the ground settled adjacent to the cut. Up to 80 days after the start of excavation, the stability number N did not exceed about 3 (Fig. 58.4*b*), and all the movements were small. Between days 80 and 109, N increased to more than 6 and the heave of the bottom and the inward movements of the sheet piles increased markedly. During the same interval, the settlement spread from the edge of the cut to a distance of more than twice the depth of excavation, and cracks developed in the 3-story brick building. The rapid increase of the settlement with increasing N is shown in Fig. 58.4*b*, in which the settlement is represented by the volume of subsidence adjacent to one side of the cut, per lineal foot measured along the cut.

Although the settlements adjacent to the cut represented in Fig. 58.4 might have been reduced by increasing the depth of the sheet piles, the reduction would have been very small. This conclusion is based on the large changes in curvature experienced by the comparatively stiff sheet piles even at intermediate stages of excavation.

The inevitable movements associated with the excavation of wide and deep cuts in deep deposits of soft clay are in some instances too great to be tolerated. They can be reduced by adopting a procedure that does not involve removing the weight of the excavated soil from the entire area at one time. In the *trench method,* the first step is to excavate around the periphery of the site a trench in which the permanent outer walls of the substructure can be built. The sides of the trench are usually supported by lagging and struts (Fig. 48.1*b*). Because the trench is narrow, the movements adjacent to the excavation are much smaller than those that would occur adjacent to a wide cut of the same depth. In some instances the trench is excavated without bracing by means of the slurry method and the permanent wall constructed by displacing the slurry by underwater concrete. After the exterior walls have been completed, cross-trenches are excavated in a similar manner. Bracing is installed in these trenches to support the exterior walls. The bracing at the level of the bottom usually consists of concrete struts that will be incorporated into the base slab of the finished substructure. In this fashion the site is subdivided by trenches and struts into small rectangular units. The soil in these units is excavated, one unit at a time, and the base slab concreted before adjacent units are excavated. By this procedure the reduction in stress in the soil beneath the substructure, and the result-

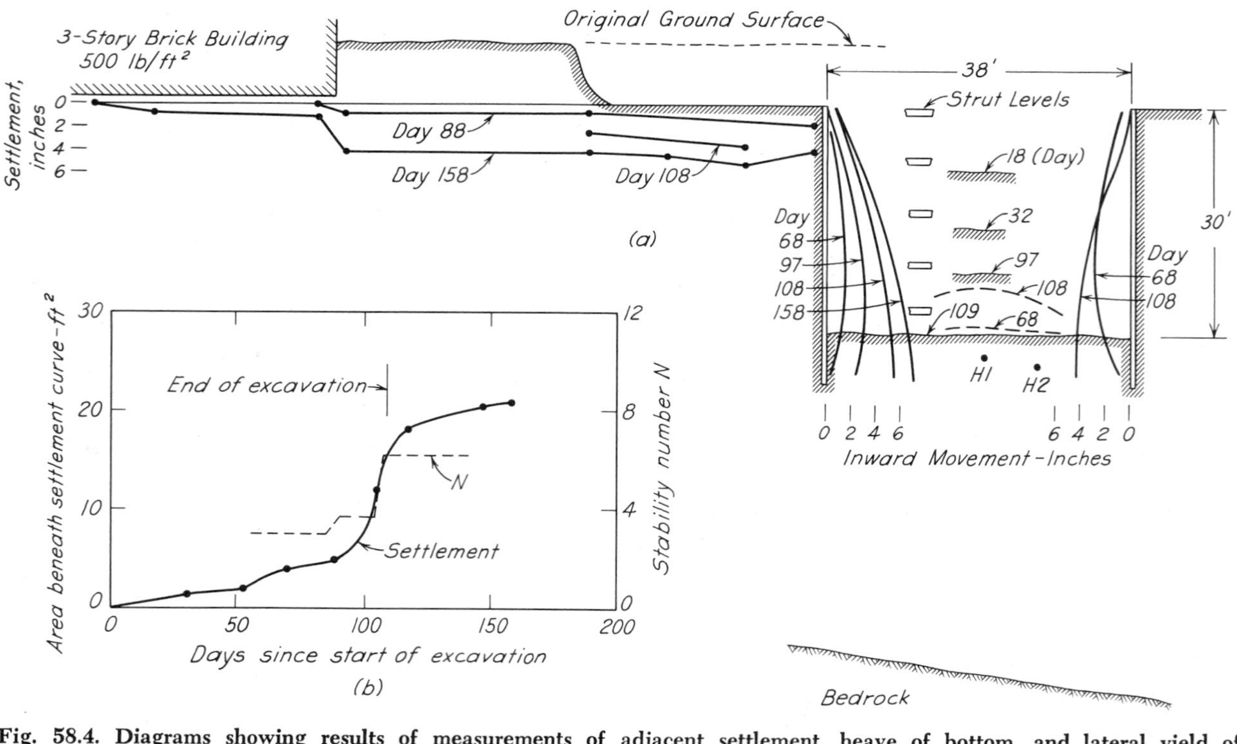

Fig. 58.4. Diagrams showing results of measurements of adjacent settlement, heave of bottom, and lateral yield of sheet piling at sides of open cut in deep deposit of medium clay in Oslo. (*a*) Successive positions of ground surface and sheeting on designated days. (*b*) Progress of settlement and increase of stability factor *N* as function of time since start of excavation (after NGI 1962*e*).

ing movements, are substantially reduced. In some instances the completed base slabs have been temporarily backfilled until the weight of the superstructure has become great enough to compensate for the unloading. In Japan, several large substructures for buildings have been installed to depths of more than 100 ft by the caisson method (Mason 1952).

OPEN SHAFTS OR CAISSONS IN SOFT CLAY. During the excavation of an open shaft or the sinking of an open caisson in soft clay, the clay beneath the bottom also heaves. Furthermore, if the lower parts of the walls of the shaft are unsupported, as in the shafts excavated by the Chicago method (Fig. 57.1*f*) the lateral squeeze can be appreciable. Because of these movements the volume of excavated soil is greater than that of the shaft or caisson. The difference is known as *lost ground;* loss of ground always involves settlement of the ground surface.

The physical processes associated with the lateral squeeze are represented in Fig. 58.5*b*. The figure shows a vertical section through a shaft being excavated according to the Chicago method. Installation of each new set of lagging with height h is preceded by excavation to depth h below the lower edge of the lined section. Figure 58.5*a* represents a horizontal section through the unlined part of the shaft. Before excavation, the cylindrical surface with diameter d_0 is acted on by a radial pressure p_0. Excavation reduces this pressure to zero, whereupon the cylindrical shell of clay surrounding the shaft is acted on by an unbalanced external radial pressure. This pressure reduces

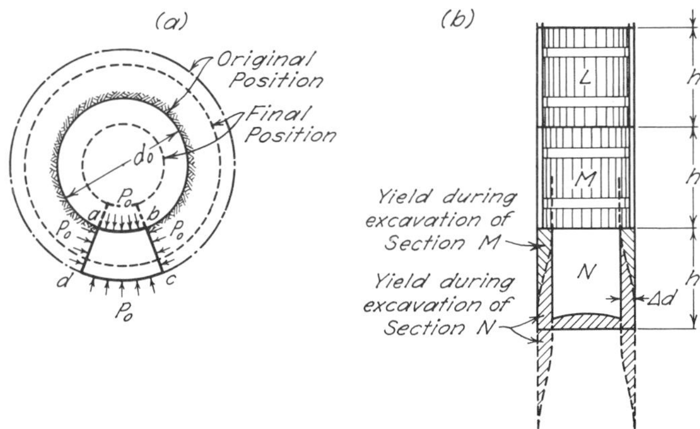

Fig. 58.5 **(*a*) Diagram illustrating cause of loss of ground during excavation by Chicago method. (*b*) Lost ground associated with excavation.**

the inner diameter of the shell, and the clay advances toward the shaft, as indicated by shaded areas along the walls of the excavation in Fig. 58.5*b*. Because of the squeeze, every wedge-shaped element *abcd* (Fig. 58.5*a*) is compressed circumferentially and stretched radially. For similar reasons the bottom of the shaft heaves, as indicated in Fig. 58.5*b*. Each of the shaded areas represents the ground lost during the excavation of one section. The total volume of lost ground is roughly equal to the entire area of the walls of the shaft times the width Δd of the squeeze area adjoining the walls in Fig. 58.5*b*. This loss causes a settlement of the ground surface surrounding the top of the shaft.

When a single shaft is excavated, the effect of the loss of ground may not be noticeable at the ground surface. However, if many shafts are sunk close to each other, the subsidences accumulate and affect the entire vicinity. Such a subsidence occurred on the job illustrated by Fig. 58.6. On a building lot 190 by 150 ft, 120 shafts with diameters ranging from 5 to 8 ft were excavated through soft glacial clay to hardpan. Sinking of the shafts required 3 months and involved the removal of 17,000 yd^3 of clay. Immediately after the excavation was started, the area surrounding the lot started to subside and finally settled to the position indicated in Fig. 58.6*b*. The adjoining buildings had to be provided with temporary shoring and underpinning to maintain them at their original elevation. Figure 58.6*c* shows the progress of the excavation of the shafts with time, and also the corresponding settlements of two reference points P_1 and P_2 located, respectively, at midlength of one side and at the corner of the lot. The similarity between the curves shows very clearly that the settlement was due chiefly to the loss of ground associated with the excavation of the shafts.

Several means are available for reducing the settlement caused by excavating shafts for piers in soft clay.

1. Use of sheet piles or cylindrical shells that eliminate the vertical part of the working face. One such method is illustrated in Fig. 57.1*e*.
2. Use of mechanical drilling tools and the slurry or heavy liquid method (Article 57).
3. Use of compressed air. Since the air pressure may compensate for only part of the pressure at the level of the working face, some loss of ground is inevitable. However, the settlement is reduced to a small fraction of that associated with the Chicago method (Fig. 57.1*f*).

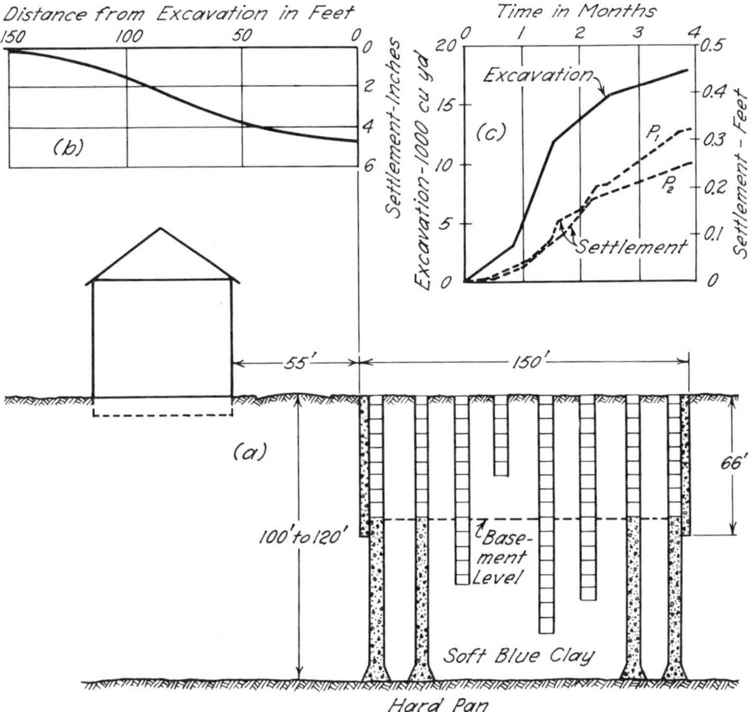

Fig. 58.6. **(*a*) Cross section through foundation of structure during excavation for piers by Chicago method. (*b*) Relation between settlement of ground surface and distance from edge of foundation area. (*c*) Relation between quantity of soil excavated from shafts, settlement of ground surface, and time.**

4. Use of heavy steel shells driven to final grade and left in the ground. After a shell is driven, the soil is removed by dredging or by means of an air or water jet, and the shell is cleaned with suitable tools, such as a mechanically operated brush, and is filled with concrete. This method has frequently been used with success for the construction of cylindrical piers in very soft soils.

Practical Value of Settlement Observations during Excavation Period. The preceding review of the various causes of the settlement of the area surrounding an excavation leaves no doubt that a certain amount of settlement is inevitable. For example, unless the entire construction procedure is changed to such an alternative as the trench method, nothing can be done to prevent the settlement due to the lateral yield of the soil toward the zone of heave below the bottom of an open cut. Neither can the bulging of the sides of the cut be prevented during the process of excavating from the level of one set of struts to that of the next set. However, in contrast to the yield that occurs below the bottom, the amount of inward yield of the sides depends to a large extent on the vertical spacing between struts, on the speed of excavation, and on various details of the construction procedure. Hence, the corresponding settlement can be considerably reduced by appropriate modifications of the construction procedures.

On a given job reliable information on the relative importance of the lateral yield below and above the bottom of the excavation can be obtained only by measuring the settlement and keeping a record of all the circumstances that might have influenced it. On the basis of the results, the engineer is able to decide whether or not the settlement can be reduced substantially by practicable changes in the procedure. In addition to serving their purpose on the job, the settlement records are of great assistance in planning the construction procedure for other excavations to be made in similar soils, and in predicting the effects of the excavation on structures and public utilities located near the site.

Selected Reading

The results of investigations made to determine the settlement during construction and of the factors influencing it are contained in the following:

Terzaghi, K. (1938*b*). "Settlement of structures in Europe and methods of observations," *Trans. ASCE,* **103,** pp. 1432–1448. Effect of filling of oil tanks on settlement of neighboring tanks.

Terzaghi, K. (1942). "Shield tunnels of the Chicago subway," *J. Boston Soc. Civil Engrs.* **29,** pp. 163–210. Record of heave and settlement due to shoving a shield through soft clay.

Peck, R. B. (1943). "Earth-pressure measurements in open cuts, Chicago subway," *Trans. ASCE,* **108,** pp. 1008–1036. Record of settlements due to excavation of open cuts in soft clay.

Terzaghi, K. (1943*a*). "Linerplate tunnels of the Chicago subway," *Trans. ASCE,* **108,** pp. 970–1007. Settlements caused by the construction of linerplate tunnels in soft clay.

Ireland, H. O. (1955). *Settlements due to foundation construction in Chicago, 1900–1950,* Ph.D. thesis, Univ. of Illinois, 128 pp.

Norwegian Geotechnical Institute, *Technical Reports* Nos. 1–8, Oslo, 1962–1966. Series of reports on measurements made in connection with excavations for braced open cuts in Oslo, including observations of settlement, movement of sheeting, and heave of bottom.

ART. 59 SETTLEMENT DUE TO LOWERING THE WATER TABLE

Causes of Settlement

Whenever a large open excavation is to be made below the water table by any process other than dredging or the sinking of a caisson by the compressed-air method, the water table must be temporarily lowered by pumping (Article 47). By lowering the water table the effective load on the subsoil is increased by an amount equal to the difference between the drained weight (solid and soil moisture combined) and the submerged weight of the entire mass of soil located between the original and the lowered water table. The increase of the effective overburden pressure causes additional compression. This, in turn, produces a settlement that at every point is roughly proportional to the descent of the piezometric level at that point. For a given descent the settlement depends on the compressibility of the subsoil.

Effect of Lowering the Water Table in Sand Strata

Pumping from sand that does not contain any clay strata increases the effective pressure, but the corresponding settlement is usually small unless the sand is very loose. However, if the water table is raised and lowered periodically, the settlement may become important, because every temporary increase of effective pressure increases the settlement by a certain amount. This fact can be demonstrated by means of laboratory tests on laterally confined sand. The magnitude of the increment of settlement decreases with increasing number of cycles and approaches zero, but the final total settlement is many times greater than the settlement produced by the first cycle. The looser the sand, the greater is the settlement.

During construction in an open excavation fluctuations of the lowered water table are usually insignificant. Therefore, if pumping causes large settlements in any but very loose sand, the settlements are probably due to causes other than the increase of the effective weight of the drained portion of the sand. The most common cause is careless pumping from an open sump (Article 47). Several instances of settlement due to this cause are described in Article 63. In all these instances one or more subsurface conduits were formed by backward erosion from springs that discharged into a pit. The settlement produced by the erosion led to the formation of shallow and narrow troughs located above the conduits. The width and depth of the troughs increased with increasing distance from the springs, and the troughs terminated in sink holes. Settlement of this type can be avoided by pumping from well points or by providing the sump with a filter lining.

Loss of ground can also occur on one or both sides of an open cut lined by watertight sheet piles. The loss is caused by the erosive action of water as it rises toward the bottom of the cut along the inner face of the sheet piles. It can be avoided by providing the sides of the cut with a permeable instead of an impermeable lining (Prentis and White 1950). The following observation demonstrates the efficacy of this procedure. A subway cut was made in New York through fine sand and coarse silt close to buildings founded on pile-supported footings. The points of the piles did not rest on a hard stratum. In one section of the cut the bracing was of the type shown in Fig. 48.2*b*. The sheeting consisted of boards placed horizontally with spaces between them, as shown in Fig. 59.1*a*. The spaces were packed with marsh hay to permit free flow of water into the cut without allowing sand to enter. In a second section the sheeting consisted of steel sheet piling driven along the sides of the cut. The sheeting compelled the water to enter the cut by percolating under the sheet piles, as indicated in Fig. 59.1*b*. Therefore, conditions were favorable for the development of erosion by springs, and the footings of the adjacent buildings settled about 6 in. Excavation of the section with permeable sheeting, on the other hand, produced no noticeable settlement.

Effect of Pumping on Clay Strata

If the subsoil contains layers of soft clay, silt, or peat, lowering of the groundwater table may cause large settlements. In Mexico

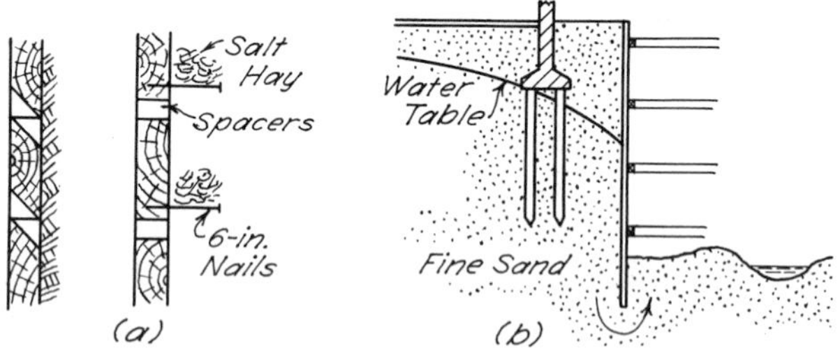

Fig. 59.1. (*a*) Louvre-type sheeting successfully used in part of open cut in waterbearing sand to prevent loss of ground. (*b*) Continuous steel sheeting used in other sections of the same cut. Adjacent foundations settled on account of loss of ground by erosive action of rising water veins (after Prentis and White 1950).

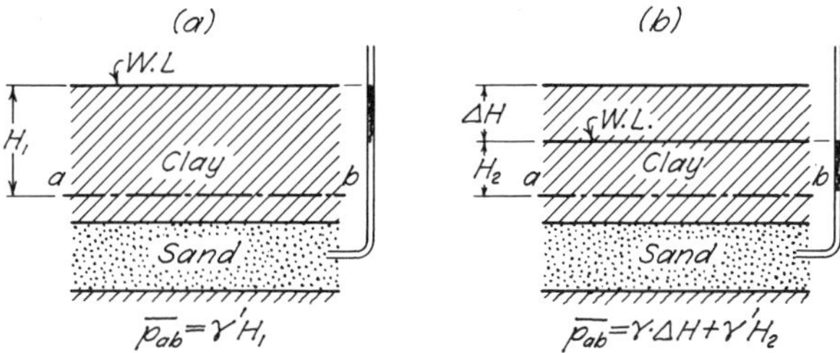

Fig. 59.2. Diagrams illustrating cause of settlement of clay surface due to pumping from underlying waterbearing sand.

City, for example, where the subsoil consists of highly compressible clays with horizontal layers of water-bearing sand, the withdrawal of water by drainage and by pumping from sand layers has been accompanied by a general irregular subsidence of the whole area. Between 1900 and 1956 the surface in some places settled more than 20 ft (Marsal and Mazari 1962). Similarly, in the Santa Clara Valley in California, the operation of 2000 wells to provide water for irrigation initiated a process of progressive settlements. The floor of this valley is underlain by a thick bed of marine clay that contains layers of water-bearing sand and gravel at a depth of 100 to 200 ft. In 1918 the withdrawal of water began to exceed the natural supply, and the piezometric levels began to descend. By 1956 the corresponding settlement had reached locally as much as 8 ft (Poland 1958). Even localities underlain by stiff clays have experienced significant subsidences. Lowering of the water levels beneath Houston by some 250 ft between 1905 and 1951 was accompanied by a subsidence of as much as 2 ft associated with local faulting and sharp differential settlements of ordinary structures (Lockwood 1954). Similarly, subsidences in London on the order of 7 in. occurred between 1865 and 1931; during the same period the piezometric levels declined roughly 200 ft (Wilson and Grace 1942). Similar phenomena have occurred above oil fields, notably in Long Beach, California (Berbower 1959) and at Lake Maracaibo, Venezuela (Collins 1935).

The physical causes of this phenomenon are illustrated in Fig. 59.2, which represents a section through a bed of saturated clay overlying a pervious sand layer. In Fig. 59.2*a* the piezometric level is assumed to be at the ground surface; in *b* it has been lowered through the distance ΔH by pumping from the layer of sand. Before pumping, the effective pressure on a section *ab* is

$$\bar{p}_{ab} = \gamma' H_1$$

where γ' is the submerged unit weight of the clay (Article 12). During and after pumping the effective pressure gradually increases and approaches a final value

$$\bar{p}_{ab} = \gamma\Delta H + \gamma' H_2$$

where γ is the unit weight (soil plus water) of the saturated clay. The change in effective pressure due to the lowering of the piezometric level is

$$\gamma\Delta H + \gamma' H_2 - \gamma' H_1 = \Delta H(\gamma - \gamma') = \gamma_w \Delta H$$

Therefore, lowering the water table by a distance ΔH ultimately increases the effective pressure on a horizontal section through the clay by an amount equal to the weight of a column of water ΔH in height. This increase involves a progressive settlement of the surface of the clay due to consolidation. The rate and magnitude of the settlement can be computed on the basis of the theory of consolidation and the results of soil tests (Article 25). However, in areas of large regional subsidence due to the consolidation of great thicknesses of stiff soils, the results are not likely to be reliable because the compressibility of the soils is altered substantially by the sampling procedures, and the location and degree of continuity of drainage layers often cannot be assessed.

If the clay strata are soft and thick, and if the water table is lowered through a considerable distance, the settlement due to pumping is likely to be very great and to spread over a large area. A record of settlement of this type was obtained during the construction of the Vreeswijk locks in Holland. At the site of the locks the subsoil consisted of 20 to 23 ft of clay and peat underlain by a thick layer of water-bearing sand. The bottom of the pit was 21 ft below the ground surface and covered an area 170 ft wide by 900 ft long.

Before construction the water level was 8 in. above the ground. During excavation it was lowered, by pumping from filter wells that extended into the sand, to the position indicated in Fig. 59.3. As a result of the pumping, the total effective vertical pressure on any horizontal section such as *ab* was gradually increased by an amount equal to the height of the shaded area above *ab* times the unit weight of water. Since the total height of the shaded area was a maximum adjacent to the excavation, the settlement was greatest at the edge of the cut. Even at a distance of 130 ft, the settlement amounted to 2 ft, and it was noticeable as far as 2500 ft (Brinkhorst 1936).

On several construction projects the subsidence adjacent to the dewatered zone has been prevented or greatly reduced by surrounding the site by sheet piles and injecting water outside the sheeting by

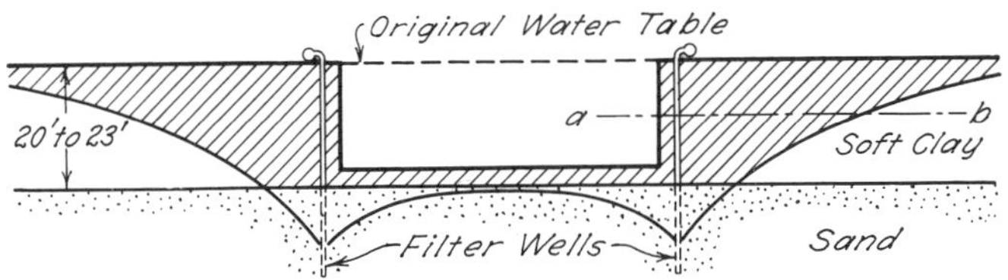

Fig. 59.3. Simplified cross section through excavation for Vreeswijk Locks, Holland, showing position of water table lowered by filter wells during excavation (after Brinkhorst 1936). Vertical scale greatly exaggerated.

means of well points or filter trenches (Zeevaert 1957, Parsons 1959). The injected water is commonly obtained from the dewatering system. Provision should be made for cleaning the injection wells because of their tendency to become clogged, particularly by bacterial action. The subsidence in the vicinity of Long Beach was virtually stopped after a program of repressurization of the oil-bearing formations by the injection of water was put into effect in 1959.

Selected Reading

Wilson, G. and H. Grace (1942). "The settlement of London due to underdrainage of the London clay," *J. Inst. Civil Engrs., London,* **19,** pp. 100–127.

Zeevaert, L. (1953). "Pore pressure measurements to investigate the main source of surface subsidence in Mexico City," *Proc. 3d Int. Conf. Soil Mech.,* Zurich, **2,** pp. 299–304.

Lockwood, M. G. (1954). "Ground subsides in Houston area," *Civ. Eng.,* **24,** No. 6, pp. 48–50.

Poland, J. F. (1958). "Land subsidence due to ground-water development," *ASCE J. Irr. and Drainage Div.,* **84,** Paper 1774, 11 pp.

ART. 60 SETTLEMENT CAUSED BY VIBRATIONS

Factors Determining Magnitude of Settlement

Any structure founded on cohesionless soil is likely to settle excessively if the soil is subject to vibrations from such sources as moving machinery, traffic, pile driving, blasting, or earthquakes. On the other hand, the settlement caused by vibration of a foundation on clay is usually so small that it is unlikely to cause serious damage under any circumstances. This conspicuous difference between the effect of vibrations on sand and on clay has already been emphasized in the discussion of the methods for the compaction of fills (Article 50). On account of its sensitivity to vibrations, sand can most effectively be compacted by vibratory equipment, whereas clay can be compacted only by static forces. Therefore, only the effects of vibrations on sand will be considered.

In Article 19 it is shown that the settlement of the surface of sand due to a pulsating load is many times greater than that produced by static action of the peak value of the load. At a given peak value the settlement depends on the frequency of the pulsations. The greatest settlements occur within a range of about 500 to 2500 impulses per min. Inasmuch as the frequency of the unbalanced forces in many types of machinery, such as steam turbines, diesel power units and air or gas compressors, lies within this range, the effect on settlement of the operation of these machines is particularly conspicuous.

Examples of Settlement Due to Vibrations

The following examples demonstrate the magnitude of the settlements that may be caused by the vibration of machinery. In Germany a coal-handling plant, 170 by 66 ft in plan, contained coal crushers mounted on concrete blocks 10 ft square. The building rested on footings supported by a bed of fairly dense sand 60 to 130 ft deep. Although the allowable soil pressure of 1.4 tons/ft^2 was very conservative, the unequal settlement assumed such proportions that the building was severely damaged and had to be underpinned. In another locality turbogenerators were installed in a powerhouse founded on fairly dense sand and gravel. The number of revolutions was 1500 per min. The maximum settlement of the foundations exceeded 1 ft within a year after the power plant started to operate.

Traffic may generate vibrations of a periodic character. Experience has shown that continued exposure to such vibrations is likely to produce considerable settlement. In Holland it has been observed that new buildings adjoining old main highways commonly tilt away from the highways. The cause of the tilt is the fact that traffic vibrations had compacted the subsoil beneath and next to the highway, whereas the sand supporting the rear part of the buildings was still in its original condition. In Berlin some of the foundations of the elevated railway settled as much as 14 in. during 40 years of operation. They rested on fairly dense sand and were designed on the basis of an allowable soil pressure of 3.5 tons/ft^2. In Munich where most buildings rest on 20 ft of dense sand and gravel overlying rock, truck traffic of increasing intensity caused settlements of such magnitude that several streets had to be entirely closed to trucking. Within a 10-year period, the damage to adjacent structures rose to about $1,500,000.

Pile driving may also be responsible for the settlement of adjacent areas. The frequency of the hammer-blows of conventional drivers is far below that corresponding to resonance but each blow gives rise to a series of vibrations of the soil at the resonant frequency. In one instance about 100 piles were driven into a deposit of sand and gravel so loose that piles as long as 50 ft could be driven without jetting. Within the area occupied by the piles the ground surface settled 6 in. The settlement decreased with increasing distance from the edge of the area to a value of $\frac{1}{8}$ in. at a distance of 50 ft.

Prolonged intense earthquakes have caused spectacular settlements of the surface of cohesionless deposits even if liquefaction was not a factor. The floors of the deeply filled alluvial valleys of the Kenai peninsula subsided as much as 5 ft during the Good Friday earthquake of 1964 in Alaska; the differential settlements severely damaged highways, railways and buildings. Nevertheless, even during a very mild earthquake in Vienna, it was observed that a grain bin with a width of 50 ft and a height of 80 ft settled 1.7 in. more on one side than on the other. The absolute maximum settlement is unknown. The bin was supported by short conical piles embedded in very fine fairly dense water-bearing sand. The load was 4 tons/ft^2 over the total area. When the bin was filled for the first time, the settlement was practically uniform and amounted only to about 0.2 in.

The effect of blasting is somewhat similar to that produced by a mild earthquake. Most damage attributed to blasting arises, however, not from settlement but from the transient ground motions and the air blast associated with the shock.

Occasionally, pile driving and blasting give rise to complaints or suits for damage, whereupon the engineer may be called on to determine whether or not the complaints are justified. One method of investigation that eliminates the personal equation is illustrated by the following examples.

In the first instance the owner of a house complained that vibrations due to pile driving were causing damage to his structure. To check the validity of his complaint, a fully loaded truck of the heaviest type was driven past the house at maximum legal speed, while seismographic observations were made in the house at points where the owner claimed the vibrations were strongest. During pile-driving operations the seismic observations were repeated. The results indicated that the vibrations caused by pile driving were milder than those caused by the truck. Since the owner was not justified in objecting to vibrations smaller than those caused by trucks passing his house at the maximum legal speed, his claim to damages was disallowed.

In the second instance an owner also protested against blasting in the vicinity of his house, whereupon a similar truck experiment was made. After the experiment charges differing in size were fired, and the corresponding vibrations were observed in the house. The contractor was given permission to blast with charges not greater than those that caused vibrations equivalent to the ones produced by the truck.

The settlements of machine foundations can be reduced most effectively by avoiding frequencies at which resonance occurs. Present knowledge (Barkan 1962, Lysmer and Richart 1966) permits fairly reliable estimates of the resonant frequencies under the simplest conditions, but the complexities of actual installations are usually so great that unrealistic simplifying assumptions have to be made and, as a consequence, undesirable conditions of resonance may occur. Various procedures have been tried to alter the resonant frequency of such systems, including permanently lowering the groundwater level, adding weight to the foundation base, or injecting chemical grout into the granular subsoil. In some instances these procedures have been highly successful, but in many they have not served their purpose. Careful design of the machines to reduce unbalanced forces to a minimum is mandatory.

No generally applicable procedures are available for reducing the influence of vibrations reaching the subsoil of a structure from an external source. One method is to surround the structure with a deep ditch. The sides of the ditch should preferably be unsupported. If space is so limited that the ditch must be provided with vertical sides braced against each other, the bracing must be designed so that it does not transmit the vibrations from one side of the ditch to the other. Observations suggest that protective ditches are most effective if the frequency of vibration is high. The detrimental effects of blasting can be considerably reduced by the introduction of millisecond delays in the sequence of firing.

Selected Reading

Crandell, F. J. (1949). "Ground vibration due to blasting and its effect upon structures," *J. Boston Soc. Civil Engrs.*, **36**, pp. 245–268. Reprinted in *Contributions to soil mechanics 1941–1953*, Boston Soc. Civil Engrs., 1953, pp. 206–229.

Gnaedinger, J. P. (1961). "Grouting to prevent vibration of machinery foundations," *ASCE J. Soil Mech.*, **87**, No. SM2, pp. 43–54.

Barkan, D. D. (1962). *Dynamics of bases and foundations*. Translated from the Russian by L. Drashevska. New York, McGraw-Hill, 434 pp.

Chapter 11

DAMS AND DAM FOUNDATIONS

ART. 61 EARTH DAMS

Requirements for Satisfactory Design

The design of an earth dam must be adapted to the available construction materials. At most sites both pervious and impervious materials can be obtained. Under these circumstances the dam is made up of a relatively impervious central zone known as the core, and of outer zones that provide the structure with the required stability. Such dams are called *zoned earth dams*. The relative quantities in the different zones are determined chiefly by economic considerations. A dam consisting almost exclusively of clay may be provided with a thin filter core that maintains the downstream portion of the dam in a permanently drained state. On the other hand, a dam made entirely of pervious materials must contain a core wall. The cores of many of the older earth dams of this type were thick walls of unreinforced concrete. Experience has shown that few of them, if any, served their purpose because they cracked on account of unequal lateral yielding of the supporting soil. Only thin, adequately reinforced concrete walls can be expected to remain reasonably intact.

If the dam is to be constructed above pervious sediments, the design requires an estimate of the upper limiting value of the loss of water through the subsoil. The estimate may be made by means of the flow-net method, on the basis of the results of pumping tests and laboratory tests. If the estimated loss exceeds an amount consistent with the functions of the project, the dam is supplemented by some type of cutoff or by an impervious upstream blanket.

An earth dam may fail on account of overtopping, slope failure, spreading, or piping. Failure as a consequence of overtopping can be avoided by conservative spillway design, attention to the possibility of large rapid landslides into the reservoir, and generous freeboard. Slope failures and failures by spreading can be avoided by design in accordance with Articles 36 and 52 supplemented during construction by field observations, principally by measurement of porewater pressures. Furthermore, failures of these types are most likely to occur during construction; they cause unanticipated expenditures and delays but they are not catastrophic. Really catastrophic failures are those due to piping by subsurface erosion, because they occur without any warning, at full reservoir, sometimes many years after the reservoir is first put into operation. They devastate the valley downstream from the dam. Hence, the first and foremost condition to be satisfied by design is to exclude the possibility of failure by piping.

Subsurface erosion leading to failure always starts at springs fed by seepage. It proceeds in an upstream direction toward the reservoir, following a line or lines of least resistance against scour. The mechanics of the process are described in greater detail in Article 63. Lines of least resistance may be located along outlet conduits, through cracks across the impervious core of the dam, in inadequately compacted core material at its contact with uneven surfaces, or in zones susceptible to erosion within the subsoil. Scour along outlet conduits can be reliably prevented by collars, flexible joints, and thorough compaction of the fill in contact with the conduits. Piping through cracks across the core can be prevented by the use of thick transition layers along both the upstream and downstream faces of the core; and piping along uneven surfaces of contact with bedrock can be avoided by adequate compaction. The only lines along which resistance against scour cannot be increased by adequate design and construction of the dam are those located in the subsoil, and the resistance along these lines depends on unknown details of the pattern of stratification. The possibility of a failure by piping due to seepage through the subsoil can be eliminated only by adequate subsurface drainage and other provisions that prevent the removal of solid particles from the subsoil. If these conditions are not satisfied, the theory of piping (Article 24) can be very misleading.

Cutoffs and Impervious Blankets

If part or all of the base of a dam is located on pervious sediments, water will escape from the reservoir by underseepage. If the loss is estimated to be excessive, it must be reduced by such artificial means as constructed or grouted cutoffs or impervious blankets. If a dam forms part of a high-head power development, it may be economically justifiable to collect the seepage and to pump it back into the reservoir during periods of low demand.

Depending on the depth and permeability of the sediments and on the loss of water considered tolerable, constructed cutoffs are extended over their full length to impervious materials (complete cutoffs), or they are terminated at a depth between the base of the dam and the top of the impervious formation (partial cutoffs). They may consist of clay-filled trenches with sloping or vertical sides, of plain concrete walls, of sheet piles, or of skin-tight rows of cast-in-place concrete piles. As a result of constructing a cutoff, the piezometric elevation immediately upstream from the cutoff exceeds the corresponding elevation immediately downstream by an amount h'. The efficiency of a cutoff can be expressed conveniently by the ratio

$$E = \frac{h'}{h} \tag{61.1}$$

wherein h is the total head, equal to the difference between the elevation of the water level at full reservoir and the tailwater level. In the design stage E is estimated with the aid of the flow net (Article 23) on the assumption that the cutoff is perfectly watertight. For tight, complete cutoffs, $E = 100\%$.

The real efficiency cannot be determined until the reservoir is filled for the first time, and then only on the basis of the results of readings on observation wells located immediately upstream and downstream of the cutoff. In many instances the observations have shown that the real efficiency is very much less than the designer anticipated. The magnitude of the difference between the estimated and the real values of E depends on the type and depth of the cutoff, the subsoil conditions, and the workmanship as it reflects the qualifications of the contractor's personnel. Hence, in the selection of the type of cutoff all these factors should be considered.

The only type of cutoff of which the efficiency can be controlled positively by conscientious supervision is the clay cutoff established in an open cut with sloping sides. However, before construction the subsoil conditions at the bottom of the trench are known at only a few points. Therefore, the specifications for preparing the base for the cutoff and placing the lower part of the cutoff material may require radical modification after the trench has been opened. Such modifications were needed, for example, at Mammoth Pool Dam (Terzaghi 1962). The depth to which a clay cutoff deserves the preference over other types depends to a considerable extent on economic considerations.

All other types of cutoffs may turn out to be defective in spite of conscientious supervision. For any given type, the difference between the computed and real efficiency increases with increasing depth. Under unfavorable conditions the real efficiency may be so low as to make the cutoff practically ineffective.

For example, even if sheet-pile cutoffs are intact they are not watertight because of leakage across the interlocks. In addition, the locks may break because of defects in the steel or when a pile hits an obstacle. Once the lock is split, the width of the gap increases rapidly with increasing depth and may assume dimensions of many feet. Such gaps have often been encountered in open excavations carried out within sheet-pile enclosures. At the site for two shipways surrounded by a cellular cofferdam, sheet piles 75 ft long were driven to a depth of 30 ft into a stiff calcareous clay containing no obstacles. Yet, when the shipways were unwatered it was found that four piles had driven out of interlock within a depth of less than 5 ft below the surface of the clay. The number of lock failures at a greater depth is unknown (FitzHugh et al., 1947). The measured efficiency of complete sheet-pile cutoffs beneath several of the large dams on the Missouri River was found to be as low as about 10% in spite of the fact that the sheet piles were driven into shale (Lane and Wohlt 1961).

If steel sheet piles are driven to hard rock with a very uneven surface, a continuous row of triangular gaps may be present between their lower edges and the rock, or the piles may curl if they are driven too hard. Defects of this type can be avoided by constructing a cutoff consisting of a row of intersecting cast-in-place concrete piles installed in drill holes with diameters of 20 to 24 inches and drilled several feet into the rock. The deepest cutoff of this type installed up to 1963 is located beneath the upstream cofferdam of the Manicouagan V project, Quebec. It extends as deep as 250 ft through river sediments containing large boulders (Jacobus 1963). Piezometric observations demonstrated it to be practically impervious. On this project the maximum tolerance for deviation of any drill hole from verticality was 6 in. However, in such construction it becomes with increasing depth more and more difficult to prevent drill holes from getting out of plumb. If the specified distance between two adjacent piles is increased to more than about one foot, the piles may be separated by a gap.

The successful construction of a cutoff of any type other than a clay-filled trench at one site is no indication that the same type will not be almost ineffective at another site, for such reasons beyond the control of the designer as the qualifications of the foreman assigned to the job by the contractor. Engineering literature contains descriptions of a great many constructed cutoffs of all types, but very few of the descriptions contain reliable information concerning the E-values. Indeed, if the results of an E-determination were unfavorable, it is rather unlikely that they would have been published.

Therefore, the reader may get the erroneous impression that most of the cutoffs were successful and that the failures he may have heard about were caused only by inadequate supervision.

Grouted cutoffs are produced by filling, within the zone assigned to the cutoff, the voids of the sediments with cement, clay, chemicals, or a combination of these materials. Until about 1925 the injected substance consisted almost invariably of neat Portland cement. However, neat cement does not penetrate the voids of a granular material unless the effective size D_{10} exceeds 0.5 mm if the material is loose, or 1.4 mm if it is dense. These conditions are seldom satisfied. Therefore, although a few cement cutoffs were successful and were advertised, most were completely unsuccessful and were kept, if possible, confidential. In 1925 Joosten patented a procedure for solidification and impermeabilization by the successive injection of solutions of sodium silicate and calcium chloride. The procedure is still used, although the cost is commonly prohibitive in connection with large cutoffs. There followed a period of experimentation, chiefly in France, culminating in the practice of injecting mixtures of cement and clay in varying proportions, with occasional admixtures of chemicals usually acting as deflocculants. More recently, solutions such as AM-9 have been developed which polymerize in the voids and plug them; all of these are very expensive. An essential feature of all grouting procedures is successive injection, usually from the same grout holes, of progressively finer zones in the deposit. Inasmuch as grout cannot be made to penetrate the finer materials as long as more pervious zones are available, the coarser materials are treated first, usually with the less expensive and thicker grouts, whereupon the finer portions are penetrated with less viscous fluids.

Large scale tests performed in river sediments at Aswan, Serre Ponçon, and Mangla Dams have led to the conclusion that the coefficient of permeability of the grouted sediments is in the range 10^{-4} to 10^{-5} cm/sec, irrespective of the coefficient of permeability of the untreated sediments (Wafa 1961, Guelton et al. 1961, Skempton and Cattin 1963). On the basis of these values, the thickness of the grout curtain can be adapted to the upper limiting value of seepage loss that can be tolerated.

A recent example of the successful application of injection techniques is the grout curtain at Mission Dam*, British Columbia. Five

* On September 8, 1965, at the opening session of the Sixth International Conference on Soil Mechanics and Foundation Engineering, Mission Dam was formally renamed the Karl Terzaghi Dam by the owners, the British Columbia Hydro and Power Authority.

rows of grout holes were drilled, to a maximum depth of about 500 ft, at a spacing of 10 ft. The average deviation of the drill holes from the vertical was 1.7%, and the maximum 4.2%. In the outer rows the injected material was neat cement. The second and fourth rows were injected with cement and clay in various proportions, and the middle row with mixtures of cement, clay, and a small quantity of chemicals. The coefficient of permeability of the sediment after grouting was 2×10^{-4} cm/sec, and the measured value of E exceeded 90%.

Grout curtains of any kind have one or more shortcomings. The sizes and locations of those portions of the sediments not penetrated by the grout are unknown. Yet, if a layer of very fine untreated sand, for instance, crosses a grout curtain, water percolates through it under a high gradient and may scour out a gap. Most chemical grouts are extremely compressible; therefore, seepage pressures of long duration may puncture the curtain. The result of grouting operations depends to an uncomfortably large degree on the skill and experience of the grouting personnel. For these reasons, no opportunity should be missed for collecting information concerning the performance of existing grout curtains. On important projects large-scale grouting tests should be performed in spite of the fact that they do not furnish any information on the long-time performance of the curtain.

As an alternative to a grout curtain as a means for reducing loss of water by leakage through sediments of great depth, an upstream blanket should be considered. Its only function is to increase the length of the path through which water must travel from the reservoir to the nearest exit, and thereby to decrease the average hydraulic gradient. The effectiveness depends to a large extent on the ratio between the coefficients of permeability of the sediments in the horizontal and vertical directions. This ratio is almost always unknown and can be quite large (Articles 11, 45). However, if the loss of water is estimated on the assumptions that the ratio is unity and the value of k is equal to that in the horizontal direction, the estimated quantity will represent an upper limiting value.

The floor of many reservoirs is covered by an upper layer which is less pervious than the underlying sediments but which has been locally removed by stream erosion or must be excavated near the dam to permit construction. Frequently the effectiveness of the upstream blanket can be greatly enhanced by repairing the known defects in the upper layer and by connecting it to the natural blanket. Such a procedure proved to be highly beneficial at Vermilion Dam in California (Terzaghi and Leps 1960). In many reservoirs, loss of water decreases with time because of deposition of silt on the floor.

Drainage Provisions

The water that escapes from the reservoir through the subsoil and through gaps in the cutoff comes out of the ground in the form of springs downstream from the impervious portion of the dam. The location of the springs is unknown before the reservoir is filled for the first time. Subsurface erosion starting at the springs may lead to failure by piping. To eliminate this risk, the pervious downstream portion of the dam should be established on an inverted filter, and any springs that emerge downstream from the dam should be covered by such a filter. Nevertheless, if the subsoil should contain an impervious layer terminating beneath the reservoir, the pressure of the water beneath this layer might lift the overlying sediments near the toe, whereupon the dam would also fail by piping. This possibility can be investigated by systematic pore-pressure observations in the subsoil near the toe during an early stage of filling of the reservoir. If it exists, or if the pore-pressure observations cannot be made before the reservoir is filled, relief wells should be installed in the vicinity of the toe. They serve the double purpose of relieving the hydrostatic excess pressures beneath the valley floor and of drying up or at least reducing the discharge of the springs.

Relief wells are designed to discharge water without suspended solids. They are commonly spaced initially at 50 to 100 ft and are equipped with provisions for measuring discharge. An observation well is established between each adjacent pair of relief wells. As time goes by, the discharge from the relief wells may decrease for one of several reasons: the reservoir may be silting up; the wells may be plugging with silt; or the well screens may be becoming obstructed by chemical deposits or products of corrosion. If the decrease in the discharge is due to silting of the reservoir, the water levels in the observation wells at full reservoir go down; in all other circumstances they go up. Excessive discharge of silt should be prevented by sealing off any silt layers or lenses during installation of the wells. Minor accumulations of silt should be flushed out periodically. For this reason, and to permit replacement of deteriorated screens, the heads of the wells should be readily accessible. If the observations during the first years after filling the reservoir indicate that the dam would be safe without the relief wells, their supervision and maintenance can be discontinued.

Relief wells for the control of excess pressures beneath permanent dams have been developed to a high degree by the Corps of Engineers, U.S. Army, in projects on the Mississippi-Missouri River system. They have formed an integral part of the design of major structures such as Randall Dam (Thorfinnson 1960).

Consequences of Settlement

The construction of an earth dam is always associated with and followed by a differential settlement of its crest and slopes. If the foundation material is rock, the settlements have their seat almost exclusively within the body of the dam. Their importance and their effects on the performance of the dam depend on the design and on the construction procedure. Normally they are inconsequential. However, under unfavorable conditions they can be associated with the formation of open cracks across the impervious section of the dam. In narrow valleys, for example, such cracks have been produced by the tendency of the core to arch between the valley walls. In wide valleys, a section of the dam is often omitted to permit passage of the river while the dam is completed on either side; cracking has been caused by the tendency of the closure section to arch between the adjacent finished portions. Cracks have also developed near a sharp break in the slope of an abutment.

The Stockton Creek Dam in California appears to have failed immediately after construction in 1950 as a result of piping through a crack caused by differential settlement (Sherard 1953). The dam (Fig. 61.1) was a homogeneous fill of residual clay derived from the weathering of amphibolite schist, compacted in 6-in. layers to a height of 80 ft. The downstream rock toe was supplemented by a horizontal filter. In November 1950, six months after completion, when the reservoir was filled for the first time, the water broke out of the reservoir and created a gap at the left abutment. The gap was 40 ft deep, 40 ft wide at the top and 20 ft wide at the bottom.

Once a crack has started to form below reservoir level, water enters it under pressure and widens it by wedge action and by scour. Although the formation of cracks cannot always be avoided, particularly in earthquake regions, the possibility of destructive consequences can be eliminated by establishing generously dimensioned transition zones on both sides of the impervious portion of the dam.

After a dam has been completed the crest continues to settle at a decreasing rate. The settlement decreases from a maximum above the highest portion of the dam to zero at both ends. To preserve the freeboard and to enhance the appearance of the dam the crest

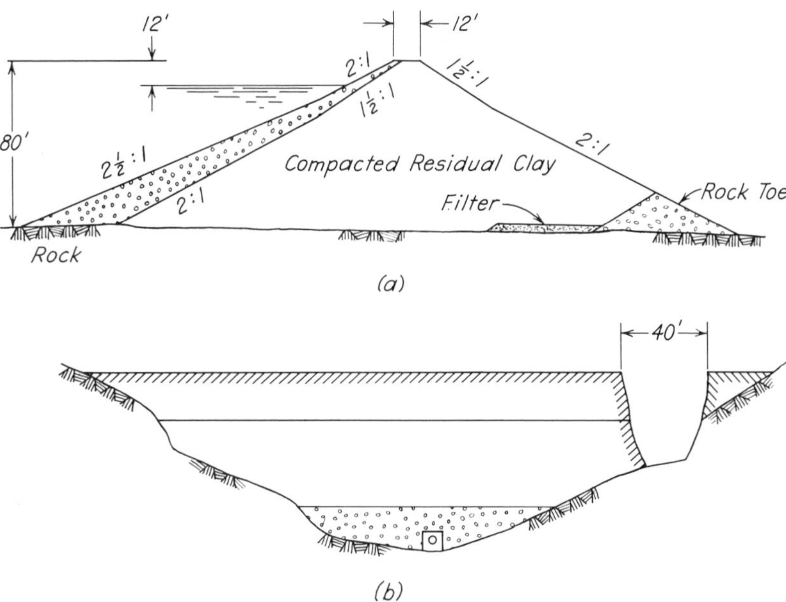

Fig. 61.1. (*a*) Cross section and (*b*) view looking upstream of Stockton Creek Dam, California. Failure by piping through crack caused by differential settlement over steep rock abutment (after Sherard 1953).

is established at a height above the design elevation equal to the anticipated settlement. For well compacted dams of moderate height, the settlement of the crest after completion of the embankment is rarely more than 0.25% of the height of the dam.

If the dam rests on sediments, the settlement of the crest and slopes is increased by the compression of the foundation materials produced by the weight of the dam and of the impounded water. The most severe conditions are encountered on sedimentary deposits containing thick lenses or layers of clay. At sites of this kind, settlements in excess of 10 ft are by no means uncommon. Under these circumstances the essential steps in the design, in addition to a conservative stability analysis, are the determination of the consolidation characteristics of the clay strata underlying the site and the construction of curves of equal anticipated ultimate settlement of the base of the dam. The computations should be based on the most unfavorable assumptions compatible with the results of the laboratory tests.

In reality the consolidation characteristics of natural clay deposits always vary between fairly wide limits. Therefore, the real curves of equal ultimate settlement cannot be expected to be identical with the calculated ones. However, the results of the computations inform the designer in a general way about the distortions he must expect at the site, and help him to visualize where the most serious effects on the proposed structure may develop. It may become evident that the impervious core will inevitably crack, that cutoffs other than wide clay-filled trenches may be deformed or even damaged by unbalanced lateral water pressures, or that the crest may settle by many feet.

It is obvious that the owner cannot expect a dam constructed at such a site to be perfectly watertight. However, it is always possible in advance of construction to estimate the maximum quantity of water that may escape through the dam. If this quantity is compatible with the purpose of the project, the dam can always be designed in such a manner that it cannot fail. The Karl Terzaghi Dam, B.C., with a height of 170 ft, is an instructive example. It is located above very pervious sediments containing a lens of exceptionally compressible normally consolidated clay with a maximum thickness of 80 ft. Two years after the reservoir was filled, the settlement of the base of the dam already ranged from zero to more than 15 ft. Yet the loss of water by leakage did not exceed 8 ft^3/sec, and the dam was designed in such a manner that no local failure could impair its integrity (Terzaghi and Lacroix 1964).

Exceptional settlement conditions are encountered at sites located above loess deposits (Clevenger 1958). In semiarid regions where the foundation material consists of relatively cohesionless sediments that have never been completely saturated, severe settlements associated with the formation of cracks across the impervious section of the dam have occasionally been experienced after filling the reservoir (Marsal 1960).

Selected Reading

The following general treatises contain much information concerning details of design and construction:

U.S. Bureau of Reclamation (1960). *Design of small dams,* Wash., D.C., 725 pp.

U.S. Bureau of Reclamation (1963). *Earth manual,* Wash., D.C., 783 pp.

Sherard, Woodward, Gizienski and Clevenger (1963). *Earth and earth-rock dams,* New York, John Wiley and Sons, 725 pp.

In addition, the proceedings of the Congresses on Large Dams contain a great variety of valuable papers. These were held in Stockholm, 1933; Washington, D.C., 1936; Stockholm, 1948; New Delhi, 1951; Paris, 1955; New York, 1958; Rome, 1961; and Edinburgh, 1964.

A useful discussion of injection methods from a historical point of view is presented by Glossop, R. (1960): "The invention and development of injection processes," Part 1:1802–1850, *Géot.*, **10,** No. 3, pp. 91–100; Part 2:1950–1960, *Géot.*, **11,** No. 4, pp. 255–279.

ART. 62 ROCKFILL DAMS

Varieties of Rockfill Dams

The term *rockfill dam* refers to a dam in which the major portion of the pressure exerted by the impounded water is transmitted onto the foundation through a rockfill. The fill material consists of fragments of sound rock obtained from quarries by blasting, from spillway excavations or from tunnels. The weight of the largest fragments may range between thirty or forty pounds and twenty tons. The material may be deposited in lifts with a height up to many tens of feet or it may be spread and compacted in layers with a thickness of a few feet.

The first rockfill dams were constructed in the mid-19th century under frontier conditions in the western United States, where few construction materials other than broken rock and timber were available. Such a structure usually consisted of a rockfill dike with side slopes equal to the angle of repose of the rockfill, and a timber skin that covered the upstream slope and was sealed to the bedrock. At the present time a great variety of construction materials is being used in combination with the rockfill and, as a consequence, the cross sections of rockfill dams range between that of the ancestral type and that of a modern zoned earth dam. Three categories can be recognized: membrane-faced rockfill dams, rockfill dams with inclined cores, and rockfill dams with central cores.

The membrane, which is a substitute for the timber facing of the ancestral type, is commonly a reinforced-concrete facing. Some dams are provided with composite reinforced-concrete decks containing layers of asphaltic concrete or a drainage layer, or with facings made entirely of asphaltic concrete. Exceptionally, steel membranes are used. The volume occupied by an inclined earth core may be as small as 1% of that of the entire dam, whereas that of a central core may be as large as 50%. Examples of the numerous varieties of rockfill dams are shown in Fig. 62.1.

From the designer's point of view, the principal difference between a modern rockfill dam and a zoned earth dam lies in the magnitude and type of the uncertainties associated with predictions of the performance of the dam, and in the amount of personal attention the dam

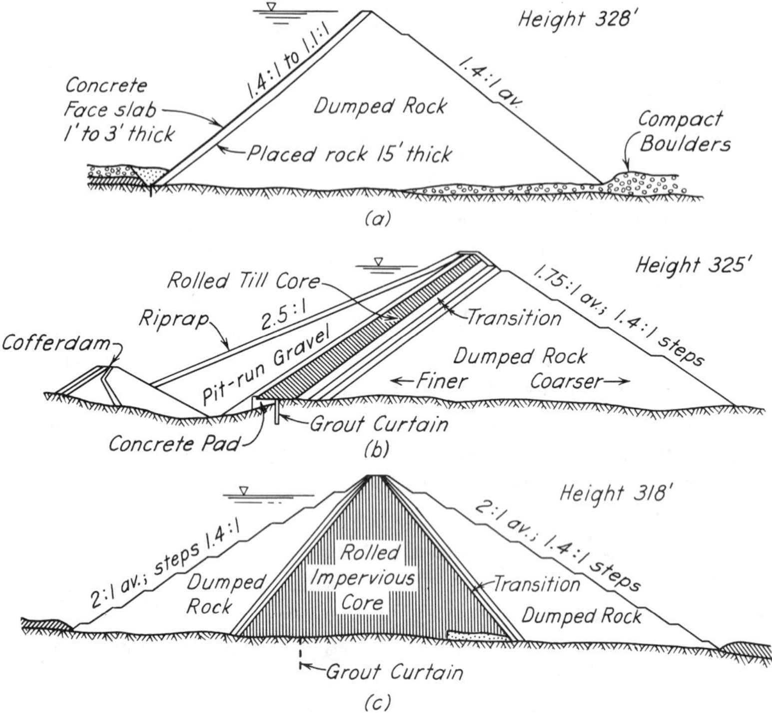

Fig. 62.1. Types of dumped rockfill dams. (*a*) Salt Springs Dam, California, with concrete face slab. (*b*) Kenney Dam, British Columbia, with thin sloping earth core. (*c*) Watauga Dam, Tennessee, with wide central earth core.

requires during construction. In the design of zoned earth dams the significant physical properties of all the materials can be determined by laboratory tests before construction and the required properties can be adequately defined in the specifications for control of the job in the field. By contrast, the significant properties of dumped rockfills, particularly their compressibility, cannot be determined by laboratory tests. They are known only in a general way and may vary significantly among materials obtained at short distances from each other in the borrow area. Yet, they may have a very important influence on the performance of the impervious portion of the dam. Furthermore, the hydraulic gradients at which water percolates through the contact zone between the natural ground and the impervious membrane or core can be very high, and the consequences of this condition depend on geological details that are never known in advance of construction.

Stress-Deformation Characteristics of Dumped Rockfill

Dumped rockfills commonly rest either on bedrock or on a layer of dense, more or less cohesionless waterlaid sediments. Experience has shown that the compressibility of dumped rockfill is much greater than that of such sediments. On the other hand, rockfill compacted in layers, although somewhat more compressible than compacted granular earthfill, behaves essentially like earthfill. Hence, compacted rockfill dams may be designed according to the principles in the preceding article, and in this article only the stress-deformation characteristics of dumped rockfills will be considered.

The deformation of the slopes of dumped rockfills consists of two parts, one of which is caused by the weight of the fill itself, and the other by the water pressure that acts on the impervious portion of the dam after the reservoir has been filled. The instantaneous deformation produced by either of these loads is followed by a slight additional deformation which proceeds at a decreasing rate under constant stress. The mechanics of this process is similar to that of the compression of a layer of cohesionless sand under constant load (Article 13). When the load is applied, some pieces of rock move into more stable positions or suffer corner breakage, whereupon the conditions change for the equilibrium of all the other pieces. Such a chain reaction requires a considerable amount of time. Since the number of pieces experiencing breakage decreases with time, the rate of delayed compression also decreases with time.

Most of the settlement of rockfill due to the weight of the dam occurs during construction and is usually inconsequential, but the filling of the reservoir is always associated with a deformation of the upstream face of the impervious portion of the dam. This deformation is illustrated by the contour lines of the upstream slope of Salt Springs Dam (Fig. 62.2) several years after the reservoir was filled for the first time (Steele and Cooke 1960). The dam, 328 ft high, was constructed in 1929 of blocks of hard sound granite supporting an upstream facing of reinforced concrete. Filling the reservoir for the first time required two years. The maximum deflection of the upstream slope, at the lower third-point, was 4 ft at the end of filling. During the following 25 years it increased at a decreasing rate to 5.5 ft. The maximum settlement of the crest during first filling was 1.5 ft; during the following 25 years the settlement, including the delayed settlement due to the weight of the fill, increased to 3.5 ft. Steele and Cooke concluded from their observations on this and numerous other granite dams that the maximum deflection D of the

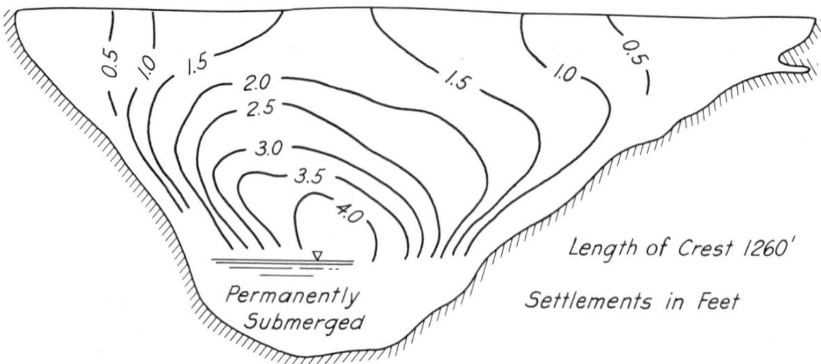

Fig. 62.2. Contours of settlement, measured perpendicular to slope, of concrete face slab of Salt Springs Dam (Fig. 62.1*a*) during first filling of reservoir and subsequent two-year period (after Steele and Cooke 1960).

upstream slope increases approximately in proportion to the square of the height of the dam.

The value D determines the average radius of curvature R of the deformed slope. However, the performance of impervious surface membranes or of sloping earth cores depends on the minimum value R_{min} of the radius of curvature rather than the average, and there is no consistent relation between R and R_{min}. Near the abutments the ratio R_{min}/R for a given fill depends primarily on the details of the topography of the lateral boundaries of the fill, and in the central portion of the slope on the degree of structural homogeneity of the rockfill. Neither of these influences can be evaluated reliably. The practical consequences depend on the design of the dam.

There does not appear to be any evidence that earth cores located between adequate transition zones have ever developed significant leakage because of unequal settlement of supporting rockfills. On the other hand, detrimental effects of even moderate differential settlements on concrete or other semiflexible slope coverings are unavoidable and are usually rather conspicuous. Since the percentage of dams with concrete decks resting on dumped rockfill is rather high, much thought and effort have been spent on reducing the differential settlements of dumped rockfill by enforcing adherence to more or less rigid specifications for selecting and placing the fill material.

Specifications for Dumped Rockfills

The deformation characteristics of dumped rockfills depend on the strength of the rock fragments, the range of particle size, the process of deposition, and the degree of structural homogeneity. The importance of each of these factors increases with increasing height of dam.

A large percentage of the total compression of rockfills under load is caused by corner breakage. Therefore, many designers specify that the unconfined compressive strength of the rock to be used in dumped fills should not be less than 10,000 lb/in.^{2}; if it is smaller, the rock should be spread and compacted in layers. This is an arbitrary but conservative rule.

The range in particle size is commonly set forth in the specifications, but the requirements can seldom be met because the grading of the quarry run is likely to change from place to place, and the sizes are too large to permit performance of mechanical analyses. Consequently, the specifications concerning particle size mean very little. The rockfill of the Lower Bear River Dam (Steele and Cooke 1960) serves as an example. According to the specifications the fill should consist of blocks weighing from one to 10 tons and should not contain more than 5% by weight of particles smaller than 4 in. As constructed, the fill contains rocks weighing up to 20 tons, and many loads went into the fill with more than 5% of fines. Yet, the performance of the dam has been satisfactory. Some designers specify that the largest blocks should be dumped on the downstream slope, or that materials with too many fines can be deposited near the center of the fill where they must be spread and compacted. In any event, the degree of structural homogeneity of dumped rockfills depends much less on the specifications than on the way the rock breaks in the quarry and on the qualifications of the inspectors.

The degree of structural homogeneity of dumped rockfills depends markedly on the height of the lifts. The dumping of the rock down the slope from a great height is inevitably associated with a certain amount of segregation, with the result that in each lift the average particle size increases with increasing vertical distance below the top of the lift. Moreover, the large particles that arrive on the lower part of the slopes have been subjected to a greater amount of corner breakage during deposition than those that come to rest near the top; this circumstance probably reduces the delayed compression of the lower layers. At the top of each lift the traffic of the hauling equipment crushes many of the larger particles and produces an upper layer with low permeability and compressibility. Therefore, it is always specified that this top layer be broken up before the next lift is started. It is evident that the deformation characteristics of each lift change in an unknown manner in the vertical direction, and that the horizontal boundaries between lifts are planes of discontinuity associated with abnormal stress conditions in the adjacent skin or membrane covering the upstream slope. These stresses cannot be evaluated. Therefore, opinions concerning the most desirable height of the lifts are divided (Terzaghi 1960, Steele and Cooke 1960). At the present time, dumped rockfills are commonly placed in lifts with a height ranging from about 50 to 150 ft.

All designers of rockfill dams agree that dumped rockfills should be generously watered ("sluiced") while they are being placed. However, there is wide divergence of opinion about the reasons for the beneficial results, or the quantity of water that should be added to the fill (Terzaghi 1960*a*, Steele and Cooke 1960). According to current practice, the prescribed volume of sluicing water ranges between 2 and 7 volumes per volume of rockfill. There is no tangible evidence that anything can be gained by increasing the ratio to more than 2. On the other hand, the consequences of dry dumping can, under exceptional conditions, be catastrophic. This was demonstrated by the performance of Cogswell Dam in southern California, a rockfill dam 280 ft high with a thin concrete facing. The fill was built by dry dumping in lifts with a height of 25 ft. It consisted of "sound granite" and contained blocks weighing up to 7 tons. However, the average unconfined compressive strength of 212 samples of the granite was only 6630 lb/in.^{2}, which for granite is very low (Baumann 1960). When the dam was 80% complete, a heavy rainstorm caused the crest of the fill to settle 8 ft. Subsequent watering through infiltration wells increased the settlement by another 8 to 12 ft.

Irrespective of the specifications for selecting and placing the fill materials, the compressibility characteristics of dumped rockfill depend to a large extent on factors beyond the knowledge and control of the designer. Nevertheless, adherence to specifications derived from experience reduces the importance of the difference between the real fill and the fill the designer hopes to achieve. Therefore, the specifications serve a useful purpose in spite of the facts that they cannot be rigidly enforced and that, particularly in the design stage, the benefits that will be derived from adhering to them cannot actually be assessed. Yet, experience has demonstrated that conspicuous deviations from them have had, on occasion, very detrimental effects.

Core Materials and Transition Zones

In earth dams with thick impervious sections, the core material is adequately defined by its shear and permeability characteristics. In rockfill dams with very narrow earth cores, the resistance of the material against scour and its degree of brittleness must also be taken into consideration. From the point of view of scour resistance the most undesirable materials are inorganic nonplastic or slightly plastic silts because they go into suspension readily and it is almost impracticable to construct a transition zone that prevents coarse silt particles from migrating through it into the rockfill.

The degree of brittleness determines the amount of warping that the core can stand before it cracks. Some clays remain relatively plastic indefinitely, whereas others become more brittle as they get older. However, the loss of water by leakage through cracks across a clay core is seldom a matter of serious concern provided the core is located between adequately designed and constructed transition zones.

Figure 62.3*a* is a section through an inclined clay core containing a tension crack. Transition zones *a* and *b* are located upstream, and *c, d,* and *e* downstream of the core. The deformation conditions responsible for the crack also prevail in the adjacent transition layers. The water flows from the upstream transition layer *b* into the crack and from the crack into layer *c*. Layer *b* is saturated. Ideally, it should consist of moderately compacted sand that will migrate into the crack in the core and fill it. Even if layer *b* contains some silt and therefore possesses slight cohesion, the cracks in *b* will not remain open because of the seepage pressures. However, if *b* is well graded and thoroughly compacted, a natural filter will develop and prevent the desirable migration of finer particles into the crack in the core.

From the crack the water percolates into layer *c* in which it spreads radially outward. Before the formation of the crack the *c*-material is only moist. It possesses slight cohesion which is greatly increased by intense compaction or a slight admixture of silt. A crack in such a material can stay open. If water invades the crack, it flows into the adjacent material and stabilizes the walls of the crack; it thus maintains free communication with the pervious layer *d*. Because of this highly undesirable possibility, it is essential that the *c*-layer should contain no silt and should be compacted only by the hauling equipment. It may be necessary to remove the silt from otherwise suitable *c*-material. This can be done economically by a rudimentary washing procedure such as that used at Kenney Dam (Bleifuss and Hawke 1960).

The rules for establishing the grain-size criteria for selecting the *c*-material are given in Article 11. They were derived from the results of laboratory tests, and their validity presupposes that the grain-size characteristics of each filter material are the same throughout the layer. In reality, the placement of filter layers is inevitably associated with a certain amount of segregation. Inasmuch as the pockets of excessively coarse material are scattered more or less at random, the adverse effects of segregation can be minimized by increasing the thickness of the *c*-layer. For ease of construction the smallest horizontal dimension that should be assigned to any layer is about 8 ft, equal to the width of the hauling equipment, but at least the *c*-layer should have a considerably greater width.

On account of the vital functions of the *b*- and *c*-layers, the specifications for the transition materials should preferably be based on the results of laboratory tests designed to secure information regarding the action of water while flowing from *b*-material through a narrow slit in a layer of core material into *c*- and *d*-materials. Such a procedure was used at Kenney Dam (Huber 1960).

If the rockfill has an inclined core (Fig. 62.3*a*), the placement of

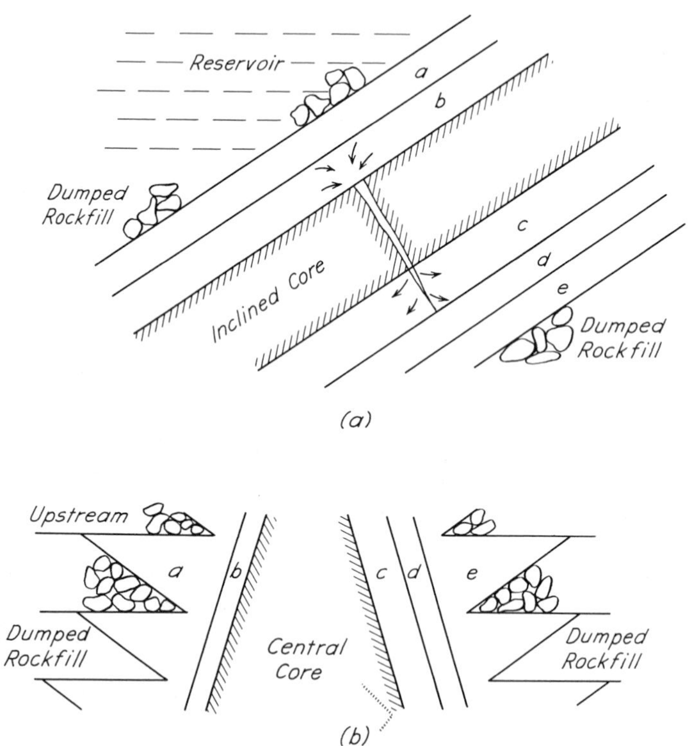

Fig. 62.3. Cores and transition zones in dumped rockfill dam. (*a*) Section through inclined clay core containing tension crack. (*b*) Section through central core showing slopes on successive lifts of rockfill.

the transition materials always proceeds simultaneously with the construction of the core. The working surface of the core should have a gentle slope toward one of the transition zones to prevent the accumulation of rainwater on the core material. If the rockfill has a central core, the inner slope of each new lift dips toward the core (Fig. 62.3*b*), whereas the outer surface of the core dips in the opposite direction. The boundaries of the transition zones must, therefore, be laid out as shown in the figure. The compressibility of the outermost transition layers *a* and *e* can be very different from that of the adjacent dumped rockfill and, as a consequence, the lateral support for the core may be very nonuniform. This condition precludes the use of very thin central earth cores in combination with dumped rockfill. The bases of the earth cores of all such dams have thus far been at least one third the height of the dam, unless the core and adjacent transition zones have been separated from the dumped rockfill by broad zones of well-compacted rockfill.

On account of the great difference between the deformation characteristics of earth cores and dumped rockfill, the contact zones between central cores and rockfill are the seat of intense shearing stresses. These stresses are responsible for large differential settlements of the crest at right angles to the axis of the dam and for the formation of cracks parallel to the axis. Such cracks, however, have no serious detrimental effects.

Seepage Control

Provisions for seepage control have two independent functions: reduction of the loss of water to an amount compatible with the purpose of the project, and elimination of the possibility of a failure of the structure by piping. Many dams have been in successful service for decades in spite of losses of water amounting to more than 100 ft^3/sec. Therefore, the rational design of measures for reducing seepage should be initiated by an estimate of the largest quantity of water that may escape from the reservoir if no attempt were made to intercept percolation through any but the most conspicuously pervious strata encountered in the borings. In many instances it will be found that the cost of reducing the loss of water further would be far in excess of the value of the additional water that can be retained. The means for reducing leakage were described in Article 61.

The safety of a dam with respect to a failure by piping has no relationship to the amount of water that escapes from the reservoir. Large losses of water may be associated with a high degree of safety against piping. Hence, the means for eliminating the danger of piping require independent consideration. The danger of failure of a dam by piping increases rapidly with increasing values of the hydraulic gradient at which the water percolates through the "impervious" portion of the dam and along the contact between this portion and the natural ground. In rockfill dams, hydraulic gradients up to 10 are by no means uncommon. Piping through the core can be eliminated reliably by adequately designed and constructed transition zones. However, the prevention of piping along the contact between the core and the natural ground requires more than the application of routine procedures. The designer's attention must, therefore, be concentrated on this contact. The necessary precautions are described in the following paragraphs.

At most rockfill dams, the site for at least the earth core and the transition layer *c* (Fig. 62.3) can be stripped to bedrock. The core is then keyed into the rock to a depth not exceeding that to which the rock can be excavated without the use of explosives. After the key trench has been excavated, the sound but fissured rock located below the bottom of the trench should be grouted under low pressures, not appreciably exceeding at any depth the weight of the overlying rock. The orientation and spacing of the grout holes may be laid out to take advantage of the known geological features of the bedrock, but ordinarily the grout holes may be shown in the contract drawings to have a depth equal to the width of the bottom of the trench, and a spacing of about 10 ft both ways. As the real rock conditions are gradually revealed by the observations of the grouting personnel, the depth and spacing are modified during construction (Terzaghi 1962).

Before the grouting operations are started, the fissures encountered in the grout holes should be cleaned, and the open fissures exposed on the bottom and sides of the trench should be sealed with gunite or slush grout to reduce the loss of grout by surface resurgences. The rock surface, downstream of the trench, which is to be covered by the *c*-layer should also be cleaned and open fissures sealed with gunite to prevent subsequent removal of the vital *c*-material by springs emerging from the rock.

After the key trench has been excavated and the underlying rock grouted, the trench should be backfilled with hand-tamped core material to a depth such that the remainder of the fill can be compacted reliably by conventional compacting equipment. As the construction of the core in the contact zone proceeds, many situations will be encountered that were not anticipated by the writer of the specifications. Therefore, the vital contact zone between core and natural ground cannot be expected to be satisfactory unless it is established under the continual personal supervision of well qualified inspectors who know when and where the intervention of the designer may be indicated.

When the site for the lowest, more or less horizontal, portion of the earth core is stripped, the rock surface may turn out to be so uneven that the lowest portion of the core cannot be compacted with conventional equipment. Adequate compaction of this portion may require a prohibitive amount of hand tamping under close supervision. Under such circumstances it is safer, and probably more economical, to fill the grooves and potholes with concrete and to start placing the core material on a concrete pad, the top surface of which rises slightly in a downstream direction. This was done, for instance, at Kenney Dam (Huber 1960).

Similarly, after the abutments in the core contact area are stripped, joints or bedding planes may be found through which water could flow and cause erosion of the core and adjacent transition layers. In horizontally stratified rock with vertical joints the abutments may exhibit a series of steps; at the bottom of each step there is likely to be a more or less continuous right-angled groove into which fill cannot be reliably compacted. There may also be overhangs. The overhangs should be trimmed back, preferably without blasting, or the gaps beneath them filled with concrete. The grooves should be provided with fillets of concrete or gunite, and the joints gunited to prevent the water flowing in them from attacking the fill. Great care is also required to prevent segregation of the transition materials in the crucial corners between the core and the abutment.

Design of Concrete Facings on Dumped Rockfills

The performance of concrete facings depends primarily on the deformation characteristics of the rockfill, especially the amount and rate of deformation after construction, and on the pattern of the deviations of these characteristics from the average. These factors cannot be evaluated reliably in advance of construction by tests of any kind. Therefore, the only suitable guide in the design of the facings consists of adequately documented performance records such as those assembled in the American Society of Civil Engineers Symposium in 1960 (see page 610).

Concrete facings are not only more sensitive to differential settlements than are earth cores, but they are also subject to severe stresses produced by daily and seasonal variations of temperature. Hence, they are always divided into square or rectangular panels with dimensions of about 40 ft, separated from each other by expansion joints. The problem of adequately sealing the joints at their points of intersection has not yet been satisfactorily solved. Near the abutments, many cracks may develop across the panels, and the water that enters the cracks escapes freely into the adjacent rockfill. Furthermore, if the normal operational cycle does not involve emptying the reservoir every year, the repair of defective concrete decks can be very inconvenient and expensive. Therefore, at equal cost, earth cores are preferable to concrete facings although the provisions for eliminating the failure of concrete-faced rockfill dams by piping are much less exacting than those described in the preceding section for earth cores.

Selected Reading

In June 1958 the American Society of Civil Engineers conducted a symposium on rockfill dams. The papers were planned to present data on the design, construction and performance of most of the world's higher rockfill dams. Together with their discussions, they were collected as the *Transactions ASCE*, **125,** Part 2, 1960. This volume is an invaluable source of information on all aspects of the subject.

Additional references of general interest include the following:

Baumann, P. (1942). "Design and construction of San Gabriel dam No. 1," *Trans. ASCE,* **107,** p. 1607.

Fucik, E. M. and R. F. Edbrooke (1960). "Ambuklao rockfill dam, design and construction," *Trans. ASCE,* **125,** Part 1, pp. 1207–1227.

Sherard, Woodward, Gizienski and Clevenger (1963). *Earth and earth-rock dams,* New York, John Wiley and Sons, 725 pp.

ART. 63 CONCRETE DAMS ON SEDIMENTS

Varieties of Concrete Dams on Sediments

Concrete storage dams resting on sediments are composed of rigid units separated from each other by sealed joints which, to compensate for unequal settlement, permit moderate displacement of the units with respect to each other. The conditions calling for such dams are usually encountered in large rivers flowing on alluvial deposits of great depth, either where the major part or all of the dam serves as a spillway, or where during high water it is desired to lower the crest of the dam to the level of the bottom of the river. The principal categories are massive concrete gravity dams with attached stilling basins (Fig. 63.5), buttressed reinforced-concrete dams of the Ambursen type (Fig. 63.1), gated dams consisting of sluice gates guided by piers resting on a heavy concrete sill, or movable dams such as the beartrap type.

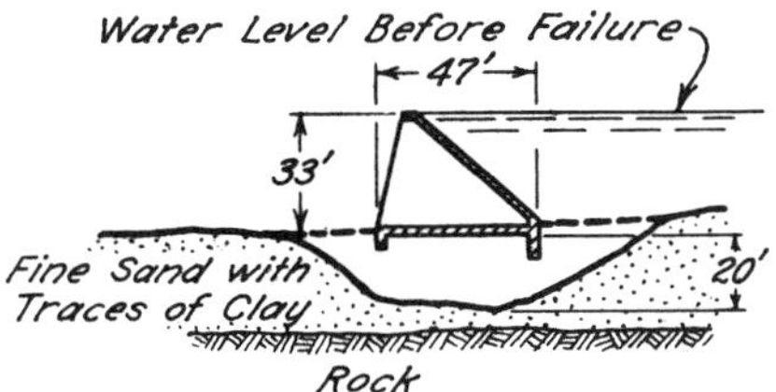

Fig. 63.1. Diagram illustrating failure of dam foundation due to piping.

Dams of any of these categories may fail by piping or sliding on their bases; they may be damaged by unequal settlement; or they may fail to serve their purpose on account of excessive loss of water by leakage out of the reservoir. Leakage out of the reservoir can have only embarrassing financial consequences, and damage due to unequal settlement can commonly be repaired. However, failures due to piping or sliding always cause catastrophic floods and may claim human lives. Furthermore, inasmuch as they may occur without warning after many years of satisfactory service, their consequences may be aggravated by surprise. Therefore, the means at our disposal for preventing failures of these types are more important than all the other features of the design of the dams, and no "calculated risk" or even remote possibility of such failures can be tolerated.

Influence of Geological Factors on Mechanics of Piping

The failure of a dam by piping ranks among the most serious accidents in civil engineering. In addition to causing disastrous floods downstream, it is likely to result in collapse of the dam itself as well as in extensive damage to the subsoil for a considerable depth. Therefore, the conditions that lead to failure by piping and the means for avoiding the danger deserve special attention.

A typical failure caused by piping is illustrated in Fig. 63.1. The dam, of the slab-and buttress type, rested on a reinforced-concrete base slab provided with an upstream cutoff wall 9 ft deep and a downstream cutoff 7 ft deep. Failure occurred suddenly by a rush of water beneath the dam. A 52-ft gap was left in the subsoil and was bridged over by the structure.

If a dam rests on a perfectly homogeneous, cohesionless subsoil, the factor of safety with respect to piping can be computed as shown in Article 24. The theory presented in that article, in agreement with laboratory tests, leads to the following conclusions: (1) the head h_c at which piping occurs is independent of the grain size of the subsoil, and (2) failure occurs almost instantaneously as soon as the hydraulic head becomes equal to the critical head at which the seepage pressures lift the ground adjacent to the downstream edge of the structure. Piping failures of this kind are referred to as piping due to heave.

In reality, most piping failures occur at hydraulic heads h'_c much smaller than the head h_c computed on the basis of theory. They occur from a few to many years after the reservoir is filled for the first time. Moreover, the ratio h'_c/h_c decreases rapidly with decreasing grain size. The conspicuous and almost universal time-lag between application of head and failure indicates that most piping failures are caused by a process that reduces the factor of safety with respect to piping gradually and inconspicuously until the point of failure is reached. The only process that can produce such results is subsurface erosion, progressing along a narrow belt toward the reservoir. As will be pointed out under the next subheading, such a process cannot take place in a homogeneous body of cohesionless sand. In nonhomogeneous material, the locations of lines of least resistance against subsurface erosion, and the hydraulic gradient required to produce a continuous channel along these lines, depend on geological details which cannot be ascertained by any practicable means. The factors of safety with respect to piping by heave and by subsurface erosion can be compared to those against failure by bending of a wooden beam in an intact state and after it has been weakened to an unknown extent by termites. The value of the factor of safety under the latter condition cannot be determined by rational procedures.

Mechanics of Subsurface Erosion

The destruction of dams by piping is usually so complete that the sequence of events can seldom be reconstructed. However, subsurface erosion can also be induced by careless pumping from open sumps or by natural events such as the tapping of bodies of ground water by the erosion of river banks. These processes commonly leave evidence that remains open to inspection. Therefore, they constitute the principal sources of our knowledge of the characteristics of subsurface erosion. The following paragraphs contain abstracts of the records of pertinent observations.

Figure 63.2 represents a cross section through a gently inclined blanket of gravel that rests on a deep bed of very fine uniform loose sand. At A a pit was dug for the foundation of a new machine. Although the pit was surrounded by sheet piles that extended to a considerable depth below final grade, the pump discharged a mixture of sand and water. The quantity of sand removed was far in excess

Fig. 63.2. Diagram illustrating underground erosion produced by pumping mixture of sand and water from sump *A*. Sinkhole at *B* 300 ft distant from *A*.

of the volume of the pit. Before final grade was reached, the building collapsed. At the same time a sink hole, 3 ft deep and 20 ft in diameter, appeared at *B*, at a distance 300 ft from the pit. Between *A* and *B* the ground surface was intact. Hence, the loss of ground can be accounted for only by soil transportation in a relatively narrow subterranean conduit. It is most likely that the conduit was located immediately below the gravel blanket, because the slightly cemented gravel was capable of forming an unsupported roof.

In the Rhineland pumping was kept up for 13 years in a sand pit. The bottom of the pit was located between 16 and 20 ft below the original water table. During this period three of the springs that discharged into the sump cut backward and eroded tunnels in the slightly cohesive sand. Each tunnel terminated in a sink hole on the ground surface. The largest tunnel was 3 to 6 ft wide and in its length of 170 ft had an average grade of only 6%. The sink hole above the end of this tunnel was 8 ft deep and 35 ft in diameter.

In another instance an open cut was excavated for the construction of a sewer. The excavation passed through fairly stiff clay into fine sand that was drained by pumping from an open sump. While pumping proceeded, a narrow strip of the ground surface subsided about 1 ft. The formation of the trough started at the sump and gradually proceeded to a distance of about 600 ft. The width of the trough increased from a few feet at the sump to more than 10 ft at the farther end.

Examples of underground erosion due to natural causes are also not uncommon. On the east bank of the Mississippi River near Memphis a large-scale subsidence occurred after the high water of 1927. At this location the river bank rises in a bluff about 100 ft high. Without any warning a strip at the top of the bluff about 700 ft long and 100 ft wide started to subside at the rate of 1 ft per hr. The pavement that covered the ground surface remained horizontal and fairly intact for a period of about 30 hr. During the following two months the subsidence increased to as much as 60 ft and the subsided surface broke up, as shown in Fig. 63.3. The trough-like depression was caused by the failure of the roof above the intake section of an underground sand flow (Terzaghi 1931).

Although the piping phenomena described in the preceding paragraphs took place in very different soil formations, they all had two important features in common. First, the material overlying the eroded soil always possessed at least a trace of cohesion, sufficient to form a roof over the erosion tunnel. Since an unsupported roof cannot be maintained in homogeneous cohesionless sands, such materials are not subject to subsurface erosion unless, of course, they lie beneath an artificial roof such as the base of a concrete dam.

The second feature common to all the examples is that the subsidence of the roof always occurred at a great distance from the discharge end of the tunnel. This fact indicates that the erosive capacity of a spring increases as the length of the tunnel increases. The reason is illustrated by the flow nets in Fig. 63.4. The thin dash curves indicate equipotential lines, or contour lines of the water table, whereas the solid curves represent the flow lines. The dash-dotted lines indicate the boundary of the intake area. With increasing length of the tunnel, the number of diverted flow lines increases. Thus, the discharge from the spring becomes greater, and the rate of erosion increases.

Progressive subsurface erosion starting at springs near the toe of a dam also proceeds as shown in Fig. 63.4 along lines leading toward the reservoir. The frequent occurrence of springs at the downstream edge is known to everyone who has had experience with dams. If

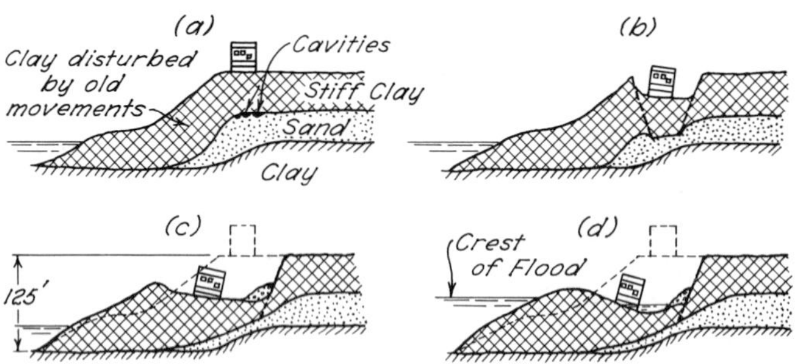

Fig. 63.3. Diagrams illustrating large-scale subsidence due to underground erosion. (*a*) Incipient state. (*b*), (*c*), and (*d*) Subsidence after 24 hr, two months, and one year, respectively.

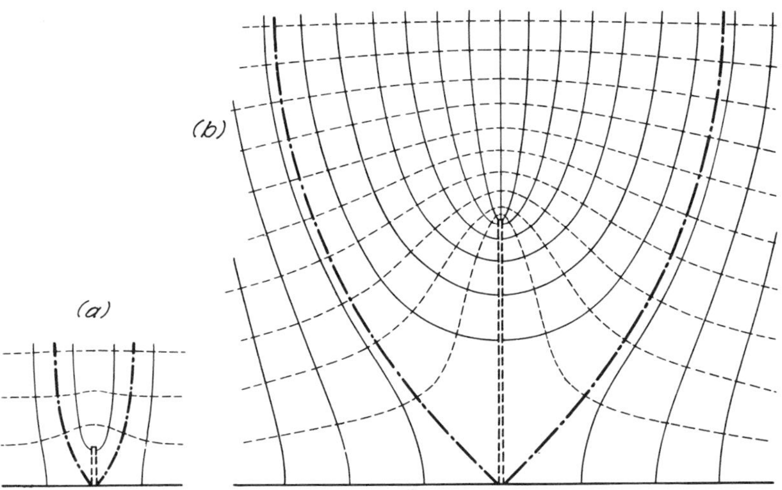

Fig. 63.4. Flow nets illustrating increase of intake area of spring as length of eroded channel increases. (*a*) Incipient state. (*b*) After erosion has proceeded to considerable distance from spring.

a spring is powerful enough to start erosion in the first place, the erosion will almost certainly become more serious as time goes on, because the flow from a given spring increases with the length of the eroded tunnel (Fig. 63.4). Finally, the dam will fail by piping.

Empirical Rules for Estimating Factor of Safety

On account of the frequency and the serious consequences of failures of dams by piping, empirical rules for evaluating the factor of safety against piping were established long before the mechanics of the process were clearly understood. The first rules of this kind (Bligh 1910) were set up after the catastrophic failure in 1898 of the Narora Dam on the Ganges River in India.They were derived from a compilation of case records, and were based on the assumption that the sole cause of piping was erosion along the surface of contact between the soil and the base of the dam. The path that a water particle followed along this surface was called the *line of creep.* If the length L of the line of creep was such that the average hydraulic gradient $i = h/L$ was less than a certain critical value for the foundation material, the dam was believed to be safe. The quantity

$$C_c = \frac{L}{h_{cr}} \tag{63.1}$$

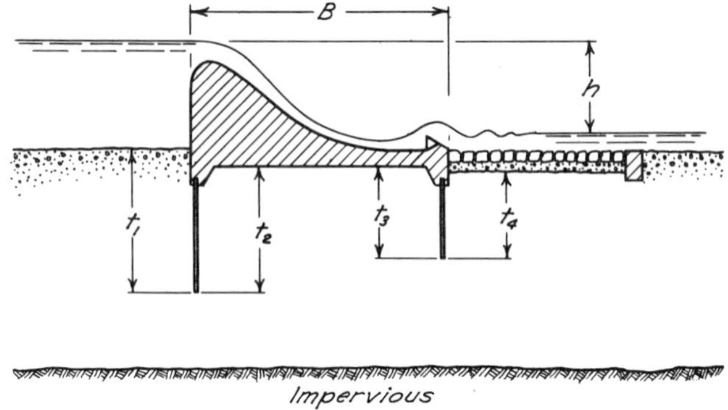

Fig. 63.5. Diagram indicating dimensions used for computation of length of line of creep.

was called the *creep ratio.* The value h_{cr} represented the greatest height to which the water level in the reservoir could rise with reference to tailwater level without producing failure by piping. The available failure records indicated that the ratio C_c increased with increasing fineness of soil from about 4 for gravel to about 18 for fine sand and silt.

The first step in designing a dam on the basis of Eq. 63.1 was to estimate the creep ratio C_c of the subsoil. This was done by means of a table containing the values of C_c for the principal types of soil. The required length L of the creep line was then obtained by multiplying the creep ratio C_c by the hydraulic head h_{cr} created by the dam. The foundation was laid out in such a manner that the length of the creep line was at least equal to L. For example, the length of the line of creep for the dam shown in Fig. 63.5 is

$$L = t_1 + t_2 + B + t_3 + t_4 = B + \Sigma t$$

and this distance must be at least as great as $C_c h_{cr}$.

During the next 30 years it was gradually recognized that vertical sections of the line of creep contribute more toward reducing the danger of piping than horizontal sections of equal length. The difference is due to the fact that the subsoil of dams is commonly of sedimentary origin, and sedimentary deposits are always much less permeable in the vertical direction than in the horizontal directions (Article 11). If k_h and k_v are, respectively, the coefficients of permeability in the horizontal and vertical directions, the loss in head per unit of length of vertical sections of the line of creep is roughly equal to the ratio k_h/k_v times that of horizontal sections. The value of the ratio ranges between 2 or 3 and almost infinity, depending on the details of stratification and the importance of the variations of the permeability in the vertical direction.

To take account of the greater efficiency of vertical sections of the line of creep, the original procedure was modified by the assumption that every horizontal section of the line of creep was only one third as effective as a vertical section of the same length. On this assumption, the equation

$$C_w = \frac{\frac{1}{3}B + \Sigma t}{h_{cr}} \tag{63.2}$$

was obtained. The value C_w is known as the *weighted creep ratio*. Since Eq. 63.2 corresponds approximately to the ratio $k_h/k_v = 3$, it is obvious that it does not take into account the wide range of values that this ratio can have in the field.

Table 63.1

Weighted Creep Values C_w (Eq. 63.2)	
Very fine sand or silt	8.5
Fine sand	7.0
Medium sand	6.0
Coarse sand	5.0
Fine gravel	4.0
Medium gravel	3.5
Coarse gravel including cobbles	3.0
Boulders with some cobbles and gravel	2.5

From E. W. Lane (1935).

Table 63.1 is an abstract of a list of safe values for C_w, based on a digest of about 280 dam foundations of which 24 had failed (Lane 1935).

The line-of-creep approach to the problem is purely empirical. Like every other procedure based solely on statistical data, it leads to design with an unknown factor of safety. Experience and experiments have shown that the values of C_w (Eq. 63.2) are widely scattered from the statistical average for a given soil. The values of C_w contained in Table 63.1 represent maximum rather than average values, and the values of h_{cr} obtained by means of Eq. 63.2 and Table 63.1 represent the smallest heads at which piping under field conditions has ever occurred. For a single row of sheet piles driven into homogeneous sand the weighted creep ratio C_w at failure by piping through heave would be approximately 0.67 irrespective of the grain size. Yet, under field conditions, dams have failed at creep ratios ranging from 2.5, for boulders with cobbles and gravel, to 8.5 for fine sand and silt. This discrepancy indicates that the real factor of safety against piping by heave of dams designed on the basis of Eq. 63.2 and Table 63.1 is very high. On the other hand, the wide scattering of the values of C_w from the statistical average indicates that, against failure by subsurface erosion, the factor of safety of some existing dams designed by the empirical rules must be excessive; that of others may be barely tolerable; and an unprecedented combination of several unfavorable circumstances may even lead to failure.

Means for Avoiding Piping

Theory and experience lead to the following conclusions. Practically all piping failures on record have been caused by subsurface erosion involving the progressive removal of materials through springs; this condition invalidates the theory of piping by heave (Article 24). The factor of safety with respect to piping by subsurface erosion cannot be evaluated by any practicable means. However, if the removal of subsurface material is reliably prevented, the conditions for the validity of the theory are satisfied and the critical head can be computed. It is, moreover, very much greater than the critical head for subsurface erosion. The means for preventing subsurface erosion depend on the importance of the project and the pattern of stratification of the subsoil.

The foremost requirement is to avoid, by appropriate design of the foundation, local concentrations of flow lines. Such concentrations were, for example, responsible for the failure of Hauser Lake Dam in Montana (Fig. 63.6*a*). The subsoil consisted of 66 ft of gravel. The water was retained by a skin of steel plates supported by a steel framework that rested on large footings. The presence of the footings produced a local concentration of flow lines, as shown in the figure. The dam failed in 1908, one year after the first filling (Sizer 1908). Since it did not fail immediately, the cause was undoubtedly spring erosion. A second example is shown in Fig. 63.6*b*, which represents a section through a dam across the Elwha River in Washington. The structure rested on gravel and coarse sand underlain by bedrock. While the reservoir was being filled, large springs developed at the downstream toe. In order to reduce the flow, a row of sheet piles was driven to a depth between 30 and 40 ft, at a distance

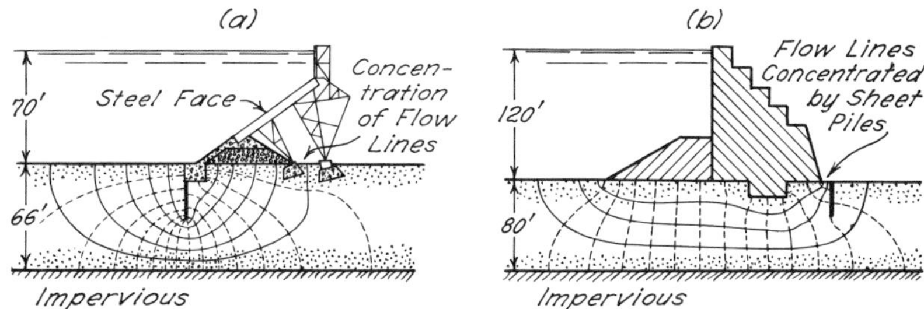

Fig. 63.6. Flow nets showing concentration of flow lines responsible for failure by piping of two dams. (*a*) Hauser Lake Dam, Mont. (*b*) Elwha River Dam, Wash.

of 8 ft from the toe. This obstruction caused a concentration of flow lines, as shown in the figure, and subsurface erosion occurred. The dam failed before the sheet-pile wall was completed.

If local concentrations of flow lines are avoided, design on the basis of Eq. 63.2 will be acceptable from the point of view of safety but the factor of safety may range from a high value to one close to unity; the real value depends on factors that are unknown. Hence, on important projects, provisions should be made to eliminate the possibility of subsurface erosion by one or more of the following procedures: (1) lowering the piezometric levels beneath the downstream edge of the foundation by means of relief wells; (2) establishing the downstream portion of the foundation on an inverted filter; and (3) observing the terrain downstream from the foundation during the first filling of the reservoir and covering with inverted filters the areas where springs begin to come out of the ground. This procedure was followed, for example, on Vermilion Dam (Terzaghi and Leps 1960). On minor projects such elaborate provisions may not be economically justified and design on the basis of Eq. 63.2 may be appropriate.

The greatest difficulties associated with preventing subsurface erosion are encountered in sedimentary deposits in which layers of inorganic silt are in direct contact with layers of clean coarse sand or gravel. Erosion occurs in the silt and the silt is carried into suspension into the relief wells or toward the springs. If the exit of the sand layer is covered with a filter fine enough to prevent the escape of silt, the filter obstructs the flow of the water out of the sand layer. Such conditions of stratification often exclude the possibility of reliably preventing subsurface erosion by means of filters, and the foundation of even an important structure must be designed on the basis of a conservative interpretation of Eq. 63.2. If relief wells are installed, special precautions must be taken (Art. 64).

Safety with Respect to Sliding

The potential surface of sliding in the subsoil of a concrete dam may be located in very permeable material such as clean sand, in a soil of intermediate permeability such as silt, or else in clay which is practically impermeable. In the following discussion, only the two extreme possibilities are considered.

If the surface of sliding is located in sand, the total sliding resistance S (pounds) per lineal foot of the dam is

$$S = (P - U) \tan \phi$$

in which

P = total vertical pressure on base of dam due to weight of dam and vertical component of water pressure on the inclined faces of the dam (lb/lin ft)
U = total neutral pressure on base of dam (lb/lin ft)
ϕ = angle of friction between concrete and sand

Since the value of $\tan \phi$ is always at least 0.6, and since the neutral pressure U can usually be reduced to a very small value by means of suitable drainage provisions, it is seldom difficult to eliminate the danger of sliding.

On the other hand, if the substrata contain horizontal layers of soft clay, or if the dam rests on a thick clay stratum, it may be very difficult to establish adequate resistance against sliding. After the clay beneath a dam becomes consolidated, sliding is resisted both by cohesion and by friction. However, because of the low permeability of the clay, consolidation proceeds very slowly and, furthermore, the rate of consolidation can seldom be reliably forecast. Therefore, it is commonly advisable to assume that $\phi = 0$ conditions still prevail at the end of the period of construction (Article 18).

In order to make the dam shown in Fig. 63.8 safe against sliding before the underlying clay consolidated, its base width was increased from 110 to 250 ft by a reinforced concrete apron on the upstream side. Since the apron was an integral part of the dam, sliding was resisted by cohesion over the full length of 250 ft. The factor of safety increased steadily because of consolidation of the clay under the weight of the dam itself as well as the weight of the water above the apron. To make the weight of the water effective the under side of the apron was drained.

Settlement Considerations

If a dam is of a rigid type or contains rigid elements, a settlement forecast is needed in advance of construction to determine whether joints are required between various parts of the structure and, if so, how much movement should be anticipated. The methods for making the forecasts are no different from those that have been described for estimating the settlement of buildings (Article 41). To supplement the available information concerning the future settlement, the observational method can often be used to advantage (see page 294). The following history of a dam across the Svir River in Russia will serve as an example (Graftio 1936).

The dam includes a reinforced-concrete powerhouse section (Fig. 63.7) and a plain concrete overflow section (Fig. 63.8). It rests on a deposit of stratified heavily precompressed clay at least 300 ft thick. According to Article 13, forecasts of settlement due to consolidation of precompressed clay are always very unreliable. Furthermore, the schedule of operations precluded the possibility of making elaborate soil investigations before construction. Therefore, it was decided to base the preliminary settlement computations on the results of tests of a few representative samples obtained from a test shaft. The forecast showed that a joint was necessary between the powerhouse and the adjacent overflow section and that no rigid connection could be tolerated between the body of the dam and the adjoining aprons.

The preliminary settlement computations also indicated that filling the reservoir would cause the powerhouse to tilt upstream about 1°. Since the turbines had to be installed before the reservoir was filled,

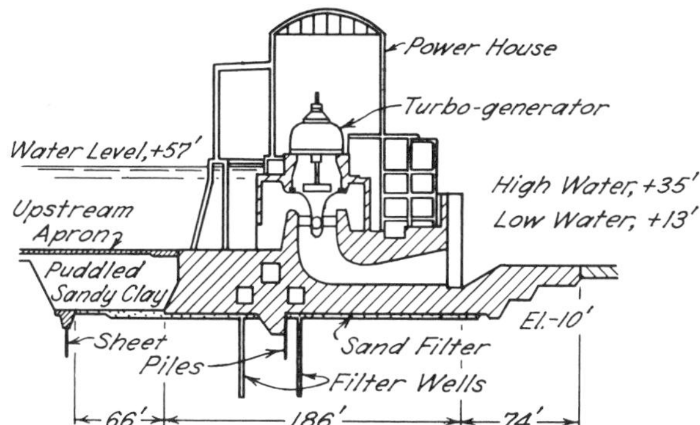

Fig. 63.7. Section through powerhouse portion of dam resting on thick deposit of stratified clay. Svir III, USSR (after Graftio 1936).

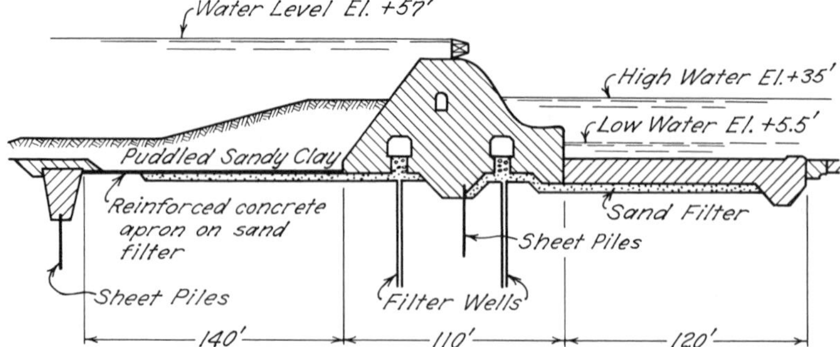

Fig. 63.8. Section through dam resting on thick deposit of clay. Stability against sliding increased by weighted concrete apron on upstream side. Svir III, USSR (after Graftio 1936).

and the computed tilt was far in excess of the value considered admissible by the turbine designers, it was decided to install the turbine shafts out of plumb in such a manner that they would become vertical when the reservoir was filled. In order to secure a more accurate value for the tilt, the results of the preliminary soil tests were used as a basis for computing the displacement of many points on and beneath the ground surface at different stages of construction. As construction proceeded, the displacements were measured. It was found that the real displacements were consistently equal to 0.35 times the computed displacements. Therefore, the turbine shafts were mounted with a downstream tilt of 0.35° and, when the reservoir was filled, the shafts were practically vertical.

Most dams of the rigid type are located on alluvial deposits of great depth, with a fairly regular pattern of stratification. Therefore, the prerequisites for an accurate settlement forecast are commonly satisfied.

Loss of Water due to Seepage

Compliance with the requirement that the loss of water should not exceed a specified amount calls first of all for a knowledge of the coefficients of permeability k_h and k_v of the subsoil. The information is needed even when the feasibility of the project is being studied. Below the water table k_h should be determined by pumping tests whereas k_v is estimated on the basis of the boring records. The sediments located above the water table should be assigned the highest values compatible with the results of the pumping tests supplemented by the boring records.

If bedrock is located at a moderate depth, massive concrete cutoffs carried to the rock are quite commonly used. If the depth to bedrock is too great to permit installation of a cutoff to rock, the principal alternatives are grout curtains or upstream blankets (Article **61**).

Selected Reading

Terzaghi, K. (1929*a*). "Effect of minor geologic details on the safety of dams," *Am. Inst. Min. and Met. Eng.*, Tech. Publ. 215, pp. 31–44.

Lane, E. W. (1935). "Security from under-seepage—masonry dams on earth foundations," *Trans. ASCE,* **100,** pp. 1235–1351.

ART. 64 SUPERVISION OF DAMS DURING CONSTRUCTION

Aim and Scope of Supervision

The performance of the foundation for a building or a bridge (Chapter 9) depends primarily on the statistical average of the physical properties of the principal strata underlying the site. If, in the design of the foundations, the uncertainties involved in the results of the subsurface exploration have been properly taken into consideration, all the essential details of the design can be covered by plans and specifications prepared in advance of construction. Therefore, in most instances it is not necessary to make any decisions on the job that would affect the safety of the foundation.

By contrast, the successful performance of a dam depends on many details of design and construction that cannot be adequately covered by the plans and specifications before the contract is let and the dam site stripped. These include, among others, the layout of the temporary installations for draining the site before the start of filling operations, the procedure for compacting impervious construction materials in the proximity of very uneven surfaces of contact, the layout and depth of grout holes in defective rock underlying such surfaces, and the details of the means for eliminating the possibility of piping. Closing the corresponding gaps in the plans and specifications requires that vital decisions be made on the job, sometimes on short notice, by the resident engineer in cooperation with the contractor. Yet, neither of these individuals is necessarily aware of the consequences of his decisions. Therefore, the performance of the completed structure can be very different from what the designer anticipated unless he is given the opportunity to keep in close contact with the job until the reservoir is filled for the first time. The designer should also make sure that a continuous record is kept of all those significant details of design and construction that were not covered by the original plans and specifications.

Sumping Operations

The construction of a dam is commonly preceded by excavation and unwatering of the site. If the site is located on rock, nothing is known in advance of excavation about the microtopography of the rock surface or about the locations of the points where springs will come out of the rock. Hence, the layout of the sumps and drains must be prepared on the job. If the drains are placed where they are most convenient and if they are inadequately grouted after they have served their purpose, failure by piping may occur, many years after completion, by subsurface erosion along the drains. Therefore, the layout of the drains and the technique for subsequently grouting them should be subject to approval or modification by the designing engineer after he has inspected the stripped site. The drains as installed should be shown in the field records.

Contact Areas

After the site has been stripped, the bottom of the excavation may turn out to be very uneven. Nevertheless, the contractor is required to place and compact the fill in thin layers. Since the contractor cannot do this on an uneven surface with standard equipment, he will be tempted to fill the depressions with inadequately compacted material unless he receives special instructions from the designer on how to proceed, and unless he is supervised closely and without intermission until a working area is established large enough for operation of the standard compaction equipment. Special instructions must also be issued by the designer for the compaction of the fill material adjacent to very uneven slopes, after the slopes have been stripped and the designer has examined them.

Construction Materials

The design of an earth dam presupposes that each portion of the dam will be reasonably homogeneous. The most detrimental deviations from homogeneity are the presence of sandy layers across the impervious portions or silty layers across the pervious portions. The avoidance of such deviations requires continuous sampling and testing operations while filling proceeds. Adequate specifications for these operations can be prepared by the designer only after he has had an opportunity to examine the pattern of stratification of the borrow material in the field. Unsuitable material must be rejected.

Before construction, the physical properties of the materials in the borrow pits may be known only along vertical lines that may be hundreds of feet apart. Therefore, during construction it may turn out that the available quantities of materials with the specified properties are much smaller than the designer anticipated. If such a condition is encountered, the designer must be notified without delay, whereupon he will alter the boundaries between the zones. No decision of this kind should be left to the field personnel.

In order to make sure that the compressibility of dumped rockfill is reasonably uniform, the specifications commonly state the percentage of "fines" that should be permitted in the fill. However, if the quarry is located in rock containing shear zones, or if the spacing between joints changes over short distances, it is rather difficult to determine whether or not the contractor is complying with the specifications. Reasonably adequate assurance can be obtained only if conscientious inspectors supervise the loading operations in the quarry. They classify and accept or reject the individual batches during loading. Even under these conditions, strict adherence to the specifications cannot be expected.

Means for the Prevention of Piping

Piping can be prevented by properly designed relief wells and inverted filters. If the subsoil is fairly homogeneous and well graded, the installation of these features is a routine operation that can be carried out in accordance with plans and specifications prepared before the contract is let.

If the subsoil contains continuous strata of very fine, uniform sand or rock flour overlying thick strata of clean sand or gravel, the lines of potential subsurface erosion are located along the boundary between the two types of material. If a filter stretching across the boundary has everywhere the same composition, it is either too coarse to prevent the washing out of the fine particles or too fine to permit free discharge of the water out of the coarse stratum. If such a boundary or boundaries are encountered in filter wells, the lengths of the pervious sections of the walls of the wells should be limited to the central portion of the outcrops of the coarse-grained strata. The thicknesses and elevations of the strata may change from well to well. Therefore, a detailed record should be kept of the sequence of the strata that were encountered in each well while it was being drilled, and the specific procedures for the installation of each well must be decided on the job.

The most difficult and unfavorable conditions are encountered in ice- or water-laid sediments having an erratic pattern of stratification and containing layers and lenses of very fine sand or rock flour in direct contact with coarse-grained and very pervious materials. In formations of this kind it may be impossible to prevent continuous and excessive discharge of silt into at least some of the filter wells. Furthermore, while the reservoir is being filled for the first time, large springs discharging silt-laden water may emerge downstream from the row of filter wells. Prevention of the silt discharge from such springs may require patient experimentation in the field because the pattern of seepage toward the springs is and will remain unknown regardless of the number of observation wells, inasmuch as the well records always leave a wide margin for interpretation.

In some instances, the observations during the first filling of the reservoir may show that minor additions to the original drainage provisions will satisfy all essential safety requirements (Terzaghi and Leps 1960). At other sites, on geologically similar deposits, exasperating difficulties may be encountered (Terzaghi 1961*b*).

Whatever the subsoil conditions may be, the means for prevention of piping must be fully and permanently adequate. Otherwise, sooner or later a catastrophic failure may ensue. Therefore, the efforts to stop subsurface erosion must be continued until they are successful. If, during the first filling of the reservoir, a powerful spring breaks out of the ground discharging sediment-laden water, it may even be necessary to stop further filling of the reservoir until the spring is brought under control.

Selected Reading

Bjerrum, L. (1960). "Some notes on Terzaghi's method of working," *From theory to practice in soil mechanics,* New York, John Wiley and Sons, pp. 22–25.

Terzaghi, K. (1960*c*). "Report on the proposed storage dam south of Lower Stillwater lake on the Cheakamus river, B.C.," *From theory to practice in soil mechanics,* New York, John Wiley and Sons, pp. 395–408. Reproduced job report exemplifying close supervision of job and attention to detail.

Chapter 12

PERFORMANCE OBSERVATIONS

ART. 65 AIM AND SCOPE OF PERFORMANCE OBSERVATIONS

Introduction

Observations in the field serve two general purposes. In the first place, they permit the elimination, during the construction period, of defects in design due to inevitable gaps in knowledge of subsoil conditions at the time the plans were prepared. Secondly, they provide information during and after construction concerning the effects of the construction operations on the subsoil, and the corresponding effects of the changes in the subsoil upon the structure. According to their specific functions, field observations can be classified and described in six categories.

Observations Which Serve to Detect Signs of Impending Danger

It has often been stated that an accident in connection with earthwork construction took place without warning. Actually, it should have been said that the symptoms of the pending accident escaped the attention of the observers, because they did not anticipate the possibility of an accident and failed to watch for its symptoms by sufficiently sensitive means of observation. In many instances, even conspicuous signs of impending failure have escaped the attention not only of laymen but also of engineers. Two days before the occurrence of a catastrophic landslide in Switzerland, the bees left their hives, and the cattle became restless and fled for safety. Yet, the human inhabitants of the village located in the path of the slide were taken by surprise. While a road fill was being constructed above a gentle clay slope in Germany, the laborers claimed that the clay was becoming "alive" and that a slide might occur. The slide did occur, on the day after a commission of engineers had examined the site to investigate the condition reported by the laborers, and had arrived at the conclusion that the slope was safe.

Most accidents which occur as a surprise are due to hydrostatic uplift and seepage forces. All forecasts regarding these forces are more or less uncertain. Furthermore, experience has shown that permeability conditions are likely to change for months or years after construction. Hence, whenever the possibility exists that water pressure or seepage forces might cause an accident, it is the duty of engineers to observe and control the hydraulic conditions until they become stationary.

The failures of slopes by sliding, and complete foundation failures, are preceded by displacements which increase at a higher rate than the rate of increase of the stresses, or which increase at a practically constant rate under constant stresses. These symptoms can be detected by observations on reliable reference points in the zone of potential movement. Such reference points, for example, have been installed by railroad engineers to stop traffic before the occurrence of slides. Automatic stop signals are adjusted to function as soon as the horizontal displacement of any of the reference points exceeds a certain value.

Field Observations to Furnish Vital Information during Construction

In many instances, the safety of a dam requires the drainage of adjoining natural ground by means of wells, shafts, or tunnels. The information that can be obtained from borings is seldom reliable enough to serve as the sole basis for the layout of the drainage measures. Therefore, only the more urgent drainage provisions are usually established at once, and the balance constructed after seepage conditions become known as a result of water pressure observations during the first filling of the reservoir.

Another example of the use of field observations to secure vital design information was described in Article 63. It was known, by means of preliminary computations based on soil tests, that the filling of a reservoir would cause an appreciable tilt of the turbine shafts in the power house. In order to ascertain a reliable value for the tilt, accurate values of the soil properties were computed in accordance with the results of settlement and heave observations on a number of reference points during construction. The tilt was then recomputed with the aid of the corrected values and the turbine shafts were constructed with a corresponding tilt in the opposite direction. While the reservoir was being filled, the shafts became vertical.

Field Observations Prior to Underpinning

At the present state of our knowledge, excessive settlement of a structure should not occur, because at least the order of magnitude of the settlement can be predicted before the beginning of construction. If such settlement should occur without being anticipated because of inadequate soil reconnaissance, the first step toward remedial measures is to explore the subsoil by means of borings and soil tests. If the results leave any doubt concerning the seat of settlement, subsurface reference points must be established and observed until the seat of settlement becomes known. Otherwise the money invested in underpinning may be wasted.

In connection with one structure, it was found after underpinning that the shear cracks in the walls were due to unequal compression of excessively thick mortar joints. In another instance, it was found, again after underpinning, that the seat of settlement was many feet below the base of the underpinning piers.

Field Observations as a Means for Improving Methods of Construction

Some construction procedures have been strongly influenced by erroneous theoretical conceptions. Many others still leave a wide margin for improvement. Improvement of the procedures can be accomplished only by digesting and profiting by the results of pertinent field observations. This approach is no more than the intelligent use of the method of trial and error practiced with adequate facilities. For example, the bracing for subway cuts in Berlin was designed for several decades on the assumption, based on an erroneous theoretical conception, that the distribution of pressure was hydrostatic. In 1936, it was found by field measurements that the real distribution was roughly parabolic. On the basis of this finding, a more suitable arrangement of the bracing was adopted.

In connection with the construction of open cuts and tunnels in cities, the ground surface should be permitted to settle no more than the least amount compatible with the general construction procedure. By observing, for instance, the effects of modifications in procedure, it was possible to reduce the magnitude of the settlements associated with the construction of the initial system of subways in Chicago to a small part of that which occurred at the outset of the project.

Field Observations for Accumulating a Stock of Local Experience

In cities, the local experiences pertaining to foundations are summarized in building codes which contain tables of allowable soil pressures and rules for determining the allowable loads on piles. In order to avoid the misapplication of these tables and rules, and to increase their usefulness, building authorities should accumulate case histories. These histories, for example, should contain the records of settlements, associated with the allowable soil pressure, of footings of different types and sizes, located at different depths below the ground surface. Without such supplementary information, unsatisfactory foundation behavior will continue to occur with undiminished frequency in spite of building codes, because settlement depends on many factors other than the load per unit of area or the load per pile (Articles 40 and 41).

Field Observations for Providing Evidence in Lawsuits

Lawsuits frequently arise from conflicts between the owner and contractor because of defects in the finished structure, or between the owner and a neighbor because of damage to the latter's property during construction. In either case, a fair decision is to be expected only if the causes and the nature of the mishap are known. If the contractor or owner can prove that he has anticipated the undesirable condition, has observed its progress during construction, and has done everything possible to avoid it, he is in a much more favorable position than if the condition has taken him by surprise. The element of surprise not only injures his professional reputation, but it may also injure his financial standing. In several instances, excessive settlement has been ascribed to defective pile foundations on the grounds that the number and quality of piles were inadequate. By observations on subsurface reference points, it was possible to prove that the seat of settlement was far below the points of the piles. This evidence changed the entire legal aspect of the situations.

Field Observations for Checking Theories

It has been emphasized repeatedly that no new theory in soil mechanics should be accepted for practical use without ample demonstration by field observations that it is at least reasonably accurate under a variety of conditions. One or two sets of observations cannot be considered conclusive evidence. Field observations for checking theories should be attempted only on jobs where the soil conditions are unusually clear and simple and are thoroughly known. However, when this condition is satisfied, even very elaborate investigations such as the measurement of soil pressure on the base and sides of the structures are fully justified.

Large permanent construction organizations such as the U.S. Army Engineers, the U.S. Bureau of Reclamation, the Bureau of Public Roads, and certain public utility corporations have a considerable interest in reducing the cost of construction by improving design. They are among the principal beneficiaries of progress in theoretical knowledge, and are justified in the expenditure of considerable sums in carrying out extensive field observations. Even small projects, however, occasionally offer exceptional opportunities for significant additions to our knowledge.

Conclusion

In recent years the practice of making field observations has increased rapidly, among both public corporations and individual contractors, with very beneficial effects on design and construction. In the present state of technique, adequately planned field observations can be expected to reduce the risk of accident by surprise to a small fraction of the risk in former days. This condition will not fail to have a decisive influence on legal procedure in cases which deal with tunnels, cuts, dams, and foundation accidents.

From the point of view of technique, field observations can be divided into four principal groups: measurement of displacements, of porewater pressure, of the load carried by struts and other bracing, and of earth pressure by means of pressure cells.

In order to prepare a satisfactory layout for any type of observation, the designer must have a clear conception of the purpose of the observations, and he must be able, in a general way, to anticipate the results. Otherwise, he is likely to require observations at points where they are not needed, and to fail to specify observations at points where essential information could be obtained. Thus, the records would contain unnecessary duplication, as well as gaps in the information.

The installation of reference points and observation wells can be made by any competent engineer or contractor on the basis of detailed specifications. The readings are a matter of routine.

The measurement of the loads carried by struts requires the capacity to adapt the general procedures to local conditions. Therefore, such measurements should be made by an engineer well trained in full-scale testing.

The installation of devices for measuring porewater pressure in clay, and pressure cells for measuring earth pressure, requires an intimate knowledge of all of the factors which are likely to influence the functioning of the measuring devices. A single oversight, or a slight defect in the installation, may ruin the job. Therefore, the installation of such devices cannot be handled as a routine matter. It requires continuous and painstaking supervision by a competent engineer who has a thorough knowledge of the physical processes involved, and of all the peculiarities of the instruments.

ART. 66 MEASUREMENT OF DISPLACEMENTS

Purpose and Type of Observations

Field observations may serve to detect the displacement or deformation of a structure supported by a foundation, to locate the seat of settlement in the ground beneath a structure, to detect symptoms of the impending failure of a slope, and to reveal the deformation of a flexible structure, such as a tunnel tube, which is entirely buried in the ground.

Vertical displacements are usually associated with the settlement or heave of structures. Observations to determine the amount of the movement may or may not be combined with measurements that serve to locate the seat of settlement or heave. Horizontal displacements occur if a structure is acted upon by horizontal forces such as water or earth pressure. In addition, under the influence of both vertical and horizontal forces, the structure may tilt. The tilt may be measured by observing the horizontal movement of two rows of reference points at different elevations, or the settlement of two rows of reference points separated horizontally. The failure of a slope, unless it is caused by sudden loss of shearing resistance due to excess hydrostatic pressure, is preceded by warping of the ground surface. The warping may be detected by observation of reference points located along lines which are likely to experience the greatest deviation from their original position. Since the location of these lines can only be estimated, the layout of the system of reference points requires sound judgment regarding the deformations that will occur. Location of the surface of sliding and observation of the rate of slip along that surface require installation of vertical flexible tubing of which the shape can be surveyed by special instruments from time to time. The measurement of the deformation of buried structures requires only periodic determination of the inside dimensions in the approximate directions of maximum lengthening and shortening. The dimensions can be ascertained by means of very simple apparatus.

Settlement Observations

The purpose of settlement observations is to provide information concerning the amount, rate, and distribution of settlement. The settlement of the base of the structure, and of selected points at some depth below the base, may be observed. Commonly, only observations of the first category are made. Furthermore, the observations may be needed for only a comparatively short time to determine, for example, whether a structure settles excessively during the excavation for an adjacent basement, or they may be required to extend with great accuracy for many years to permit comparing the settlement of a building under its own weight with the settlement predicted on the basis of theory and soil tests.

The first requirement for adequate settlement observations is a suitable bench mark. Preferably, even for a series of observations of short duration, the bench mark should be founded on rock or on some stratum which does not settle. In many instances, however, it is not possible to construct such a bench mark. It is then customary to establish the bench mark on an existing building at some distance from the structure under observation. If no reliable settlement record is available for the building, at least two and preferably three bench marks should be established on buildings of different ages, located on different sides of the structure under observation. Because of the inevitable subsidence of the ground surface in the vicinity of the loaded area, the shortest distance from any of the bench marks to the building under observation should never be less than twice the width of the building. In such cities as Boston or New Orleans, a bench mark cannot be relied upon unless it is established on the bottom of a deep drill hole which penetrates a hard stratum. In Cambridge, Massachusetts, for example, a large area along Massachusetts Avenue settled through distances ranging from 0 to 2 ft in a period of 40 yrs. It is likely that similar widespread and unequal subsidence occurs in every city built upon beds of clay or silt. Settlement observations referred to bench marks on nearby structures therefore can do no more than provide information about the difference between the settlement of the given structure and the general subsidence of the surrounding area. Under these circumstances, it would be desirable for municipal authorities to construct a few permanent bench marks on the bottom of drill holes, and to check the elevation of the auxiliary bench marks from time to time.

A bench mark suitable for long-time precision levels on a structure located above deep compressible deposits is shown in Fig. 66.1 (Bjerrum et al. 1965.) Since the outer casing may be compressed by the forces of negative skin friction due to the settlement of the surrounding soil, the bench mark is established at the top of an inner rod that is not influenced by the strain in the casing.

The reference points on the structure should be readily accessible to the observer and should be well protected against injury. If the

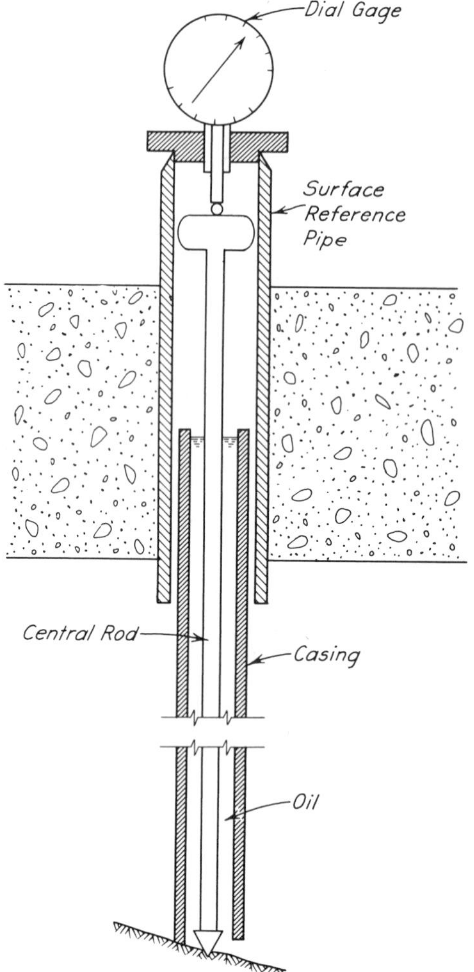

Fig. 66.1. Deep bench mark and surface reference point for long-time precision levels (after Bjerrum et al. 1965).

period of observation is short and the purpose of the surveys is to ascertain the movements associated with adjacent construction, marks or scratches on the walls and columns of the structure under observation may suffice as reference points. On the other hand, if the observations are to be continued for some time, protection against corrosion and wear must also be provided. The number of points should be sufficient to permit tracing in a fairly accurate manner the curves of equal settlement, as shown in Figs. 56.7, 56.8, and 69.2. In order to satisfy this requirement, at least two thirds of the reference points should be located in the interior of the structure. In structures of ordinary sensitivity, one reference point should be established for every 200 ft^2 of area.

Observations can be made by means of a surveyor's level or by means of a water level. Use of the surveyor's level has been widespread in the past and is generally satisfactory for observing reference points on the outside of a structure. It is inconvenient, however, in the interior where there are many obstructions such as columns and partitions. The accuracy of settlement readings obtained by the use of the surveyor's level cannot be considered much greater than one-eighth inch.

Greater accuracy can be obtained, and greater convenience in cramped quarters, by the use of the water level shown in Fig. 66.2*a* (Terzaghi 1938*b*). It consists of two glass tubes connected by a rubber hose. The entire system is filled with water. In each tube, the position of the water level is measured by means of a micrometer screw. The vertical distances z_0 are constant and the same for both tubes. The distances z_1 and z_2 are measured by means of the micrometer screws.

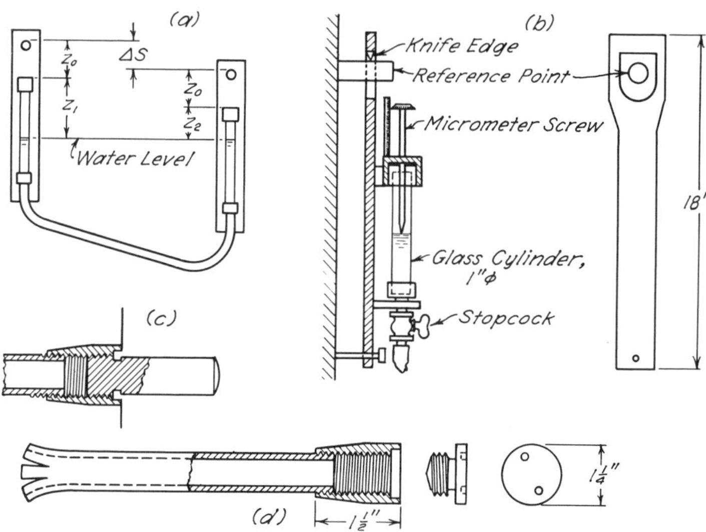

Fig. 66.2. (*a*) Schematic diagram of water level. (*b*) Detail of gage glass and micrometer screw. (*c*) Reference point. (*d*) Detail of wall pipe and protective cap for use when reference point is not in place (after Terzaghi 1938*b*).

Hence, the difference between the elevation of two reference points is equal to $z_1 - z_2$. The error inherent in the use of this device is approximately 0.002 in. In order to eliminate systematic errors and to provide a check on individual readings, it is desirable to determine the difference in elevation between two reference points with the glass tubes in one position, and then to repeat the observations after exchanging the glass tubes on the two reference points. Care should be taken that the entire hose is either in the sun or in shadow, because the difference in density of the water caused by the accompanying temperature difference can introduce a considerable error. Differences in atmospheric pressure at the tubes may also lead to significant errors. By eliminating the micrometer screw and reading the water level against a graduated scale on the glass tube, a very simple water level can be constructed which can be read with an accuracy of about 0.05 in.

The reference point illustrated in Fig. 66.2*d* satisfies the requirements of accessibility and permanence. It consists of a short piece of pipe entirely embedded in the wall. Normally, the end of the pipe is covered by a brass plug which is flush with the surface of the wall. To make a settlement observation, the plug is removed and temporarily replaced by a cylindrical extension (Fig. 66.2*c*).

Regardless of the type of instrument, it is desirable to make a complete circuit starting at the bench mark, including all the reference points, and returning to the bench mark. In this manner, the ultimate error is determined. The value of the ultimate error should always be included in the settlement record.

In order to measure the settlement of the crest of a dam or of the bottom of an inspection gallery, it may be convenient to establish a permanent pipe line equipped at intervals with vertical transparent tubes. The water level in the tubes serves as a reference level from which the vertical distance to reference points on the structure or on the pipe line can be measured.

Regardless of the character of the settlement survey, observations should be made on selected points at sufficiently short intervals of time to permit the construction of reliable time-settlement curves for the points. A general survey of all the reference points should be made immediately after they are established before any construction operations are carried out. If the observations are intended to disclose settlements caused by nearby excavation or construction, additional surveys should be made at frequent intervals, in some instances as often as daily, while the construction is actively going on. Thereafter, the intervals may be lengthened until it is evident that movements have ceased. It is not uncommon, however, for the period of readjustment to extend over several months or even a year; hence, observations should not be discontinued prematurely. If the observations are for the purpose of determining the settlement of a structure under its own weight, the first set of observations should be made before the foundation carries any load. General surveys should also be made once or twice during construction, once after construction when all the dead loads have been applied, once as soon as the live loads have been added, and thereafter about once a year until the settlement stops.

If it is desired to determine the seat of settlement, or the distribution of the compression of the soil along vertical lines, it is necessary to locate subsurface reference points at various depths. In order to establish a subsurface reference point, a hole may be drilled to the required depth and cased with $2\frac{1}{2}$-in. pipe. The lower 2 or 3 ft are filled with concrete and a 1-in. pipe is lowered into the hole and forced into the concrete. The portion of the pipe above the concrete should be well greased. The casing is then withdrawn until the bottom is about 2 ft above the top of the concrete. The upper end of the 1-in. pipe serves as the reference point. A suitable cap must be provided on the top of the casing to protect the upper end of the reference pipe. A convenient form of subsurface reference point is the *Borros point* (Bjerrum et al. 1965). In this device the inner rod terminates in a point containing three prongs that can be forced outward into the soil to form a support preventing movement between the bottom of the reference point and the surrounding soil (Fig. 66.3).

Before subsurface reference points are installed, exploratory borings should be made to permit an estimate of the locations of the seats of settlement. At least one reference point should be established at the upper and lower boundary of each stratum which may represent a seat of settlement.

In many instances it is necessary to determine the settlement of the base of a fill due to the compression of the underlying soil, whereas the compression of the fill itself is insignificant. Under these circumstances *settlement plates* (Fig. 66.4) are commonly installed on the ground surface before the fill is placed. The size of the plate depends on the compressibility and uniformity of the surficial materials underlying the fill. The plate supports a flange to which a section of pipe, usually about 5 ft long, is attached. As the fill is built up, additional sections are coupled to the pipe. The elevation of the top of each section must be determined as soon as it is installed, and also just before the next section is added; in this manner the settlements produced by the increments of fill can be evaluated. Inasmuch as the riser pipe interferes with filling and compacting operations, the fill for a few feet around the pipe must be placed and compacted by hand. The accuracy of the results is on the order of 0.5 in.

In connection with the construction of earth dams and other high fills, the settlements originating within the fill itself may be as significant as those of the foundation. A series of settlement plates at differ-

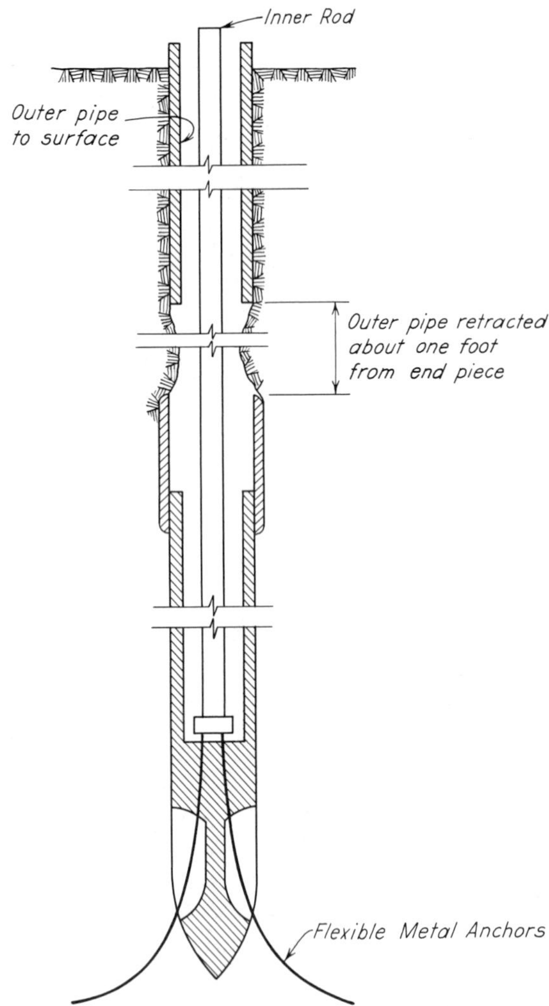

Fig. 66.3. Borros subsurface reference point (after Bjerrum et al. 1965).

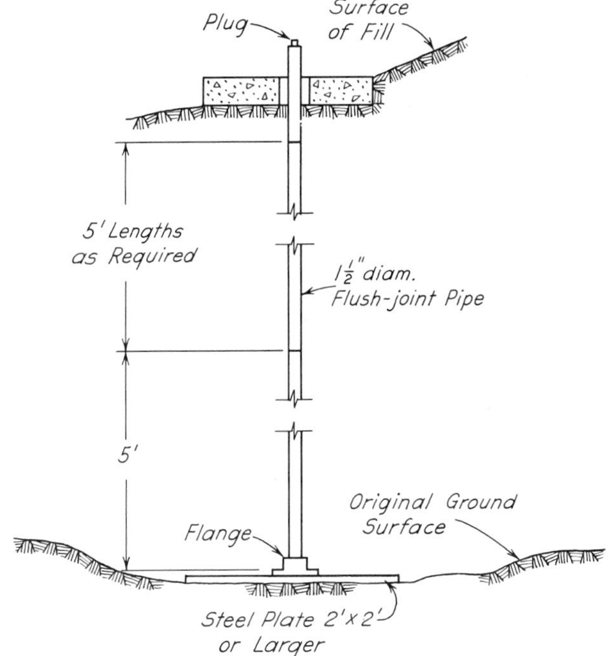

Fig. 66.4. Settlement plate for determining settlement of base of fill.

ent depths would provide the required information but would seriously interfere with the placement of the fill. One of several devices for permitting observations at several levels by means of a single installation has been developed by the U.S. Bureau of Reclamation. The installation (Fig. 66.5*a*) consists of a series of pipes in which, at intervals of 5 or 10 ft, are inserted sleeves from which protrude crossarms that serve the function of settlement plates. The locations of the sleeves, and hence the levels of the crossarms, are detected and measured by a settlement torpedo (Fig. 66.5*b*) containing a set of pawls that engage the bottom of each sleeve (USBR **1963**). The device has been used successfully on many large dams.

Settlements within the interior of a fill can also be measured by means of a water-level gage (Mallet and Pacquant **1951**). Such a gauge (Fig. 66.6) eliminates the need for the riser pipes that interfere with filling operations. Although the principle of such a device is simple, meticulous attention to details is required for satisfactory results. The presence of air bubbles in the lines leads to intolerable errors and must be avoided by circulating water through the pipes before an observation is made. The overflow chamber at the location of the observation cell must be vented to atmospheric pressure. Since most high fills experience horizontal stretching as well as settlement, the conduits between the cell and the measuring point must be able

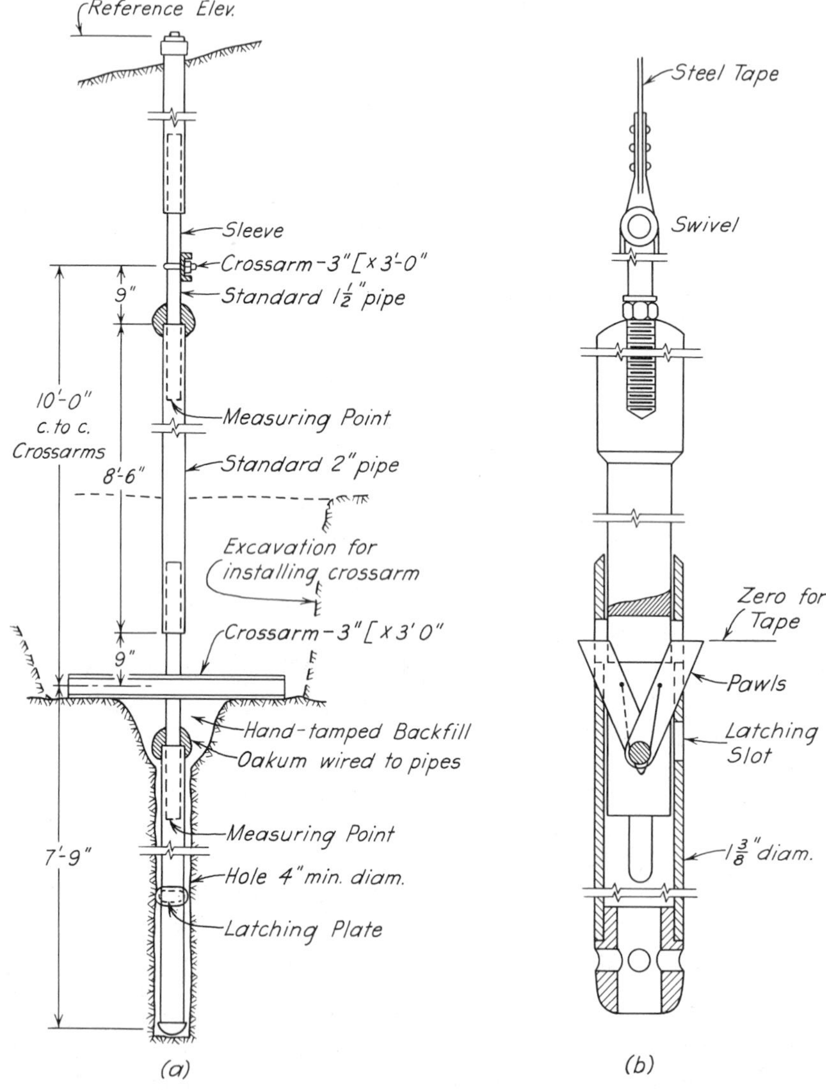

Fig. 66.5. Equipment developed by U.S. Bureau of Reclamation for measuring settlements within a dam. (*a*) Arrangement of crossarm reference points. (*b*) Measuring torpedo (after USBR 1963).

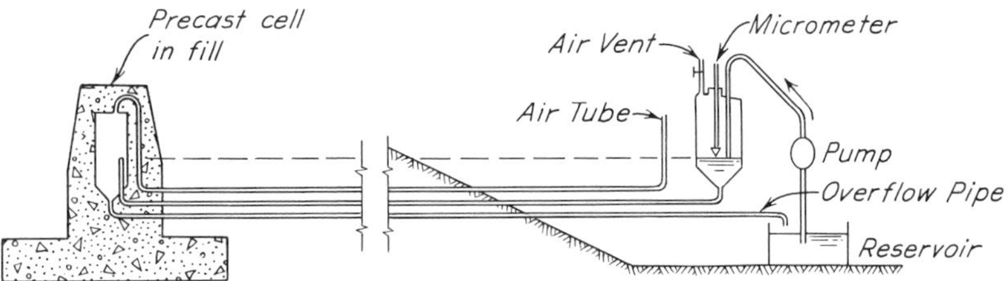

Fig. 66.6. Water-level gage for measuring settlement at point inside a fill (after Mallet and Pacquant 1951).

to accommodate the movements of the fill without breaking. The accuracy of the results is usually not greater than about 0.5 in.

Observations of the Horizontal Displacement of Structures

This group of measurements includes the observation of deflections of the crest of retaining walls or cofferdams, and of the tilt of such structures.

Measurement of the horizontal deflections of a line such as the crest of a cofferdam requires a good transit, a solid foundation to support the transit while the observations are being made, and at least two reliable fixed points to which the survey can be tied. Preferably, the transit should be set up in such a manner that readings can be made by holding a rule horizontally against the reference points. By the use of this procedure, readings can be made with an accuracy of about one sixteenth of an inch, or two millimeters. The probable magnitude of the anticipated displacements must be considered when selecting the location for the solid foundation to support the transit.

The horizontal displacements of retaining walls can be measured by means of a transit as described above. However, more accurate measurements can be obtained with less effort with the aid of horizontal rods arranged as shown in Fig. 66.7. One end of each rod is anchored in the stationary part of the backfill at a considerable distance behind the wall. The middle portion of the rod is housed in a pipe *c*, and the outer end in a pipe *d* which is embedded in the concrete wall. Measurements are made of the distance from the front end of this pipe, which serves as the reference point, to the free end of the rod. If considerable accuracy is desired, the rule used to measure the distance may be equipped with a vernier.

If displacements are due entirely to a tilting movement, they may

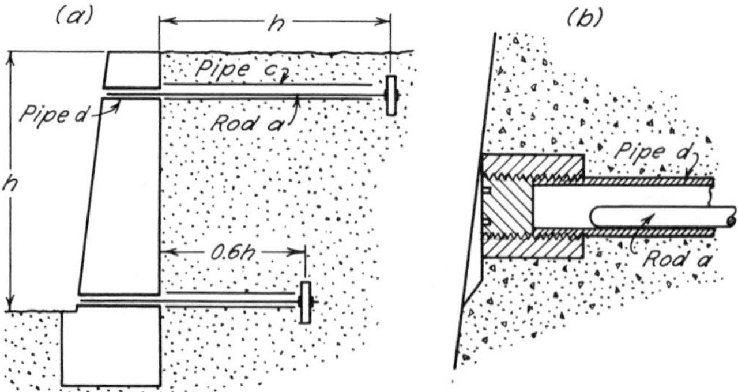

Fig. 66.7. **(*a*) Diagram of installation for observing movement of retaining wall. (*b*) Detail of reference rod and protective plug at face of wall.**

be observed by means of a plumb bob. The angular displacements cause a change of the distance between a reference point on the wall and the position of the freely suspended weight. Tiltmeters of different design have been used extensively in connection with concrete dams. By the use of some of these devices, the tilt can be measured with great accuracy.

Observations of the Distortion of Slopes

These observations serve to detect the danger of sliding. The measurement of the progressive deformation of slopes is made difficult because the upper layer of soil has a tendency to creep, even if the slope as a whole possess adequate stability. In moderate climates the creep may extend to a depth of as much as three feet. Therefore, it is necessary to prevent contact between a reference rod buried in the slope and the soil for a distance of three feet from the upper surface of the slope. One of the several methods for detecting movement of the slope is to drill a hole four or six inches in diameter to a depth of five feet, and to line it with casing. A two-inch reference pipe is then inserted in the hole and driven for an additional distance of three feet below the bottom. The upper end of the reference pipe should project slightly above the casing, but should not extend so far that a removable cover cannot be provided for the casing. Horizontal displacement of the pipe is measured with a transit as described previously. To facilitate the measurements, the reference points should be located on straight lines. The transit observations should always be supplemented by level readings.

The most suitable position for the reference pipes, with respect to the slope, depends upon whether a slope failure or base failure is anticipated. If a slope failure is likely to occur, the pipes may be located as shown in Fig. 66.8*a*. In connection with a base failure, the arrangement shown in Fig. 66.8*b* is satisfactory. In soft or plastic clay, only base failures need be considered.

If the possibility of a base failure exists during the excavation of an open cut, underground reference points may be established a few feet below the final grade. A base failure is always preceded by a marked heave of the bottom of the cut. If such a heave is observed, remedial measures can be taken before an accident occurs. For these observations, only level readings are required.

One type of underground reference point is illustrated in Fig. 66.9. It consists of a plate fixed to four blades that are forced into the soil at the bottom of a bored hole (Bjerrum et al. 1965). The walls of the hole are kept open by a mixture of bentonite and water colored with a pigment that permits finding the hole as excavation proceeds. The heave is measured by lowering a rod of known length through the soft filling until it makes contact with the plate, and by determining the elevation of the top of the rod. Under the conditions prevailing on a construction job it is sometimes difficult to locate the colored filling, but the accuracy of the observation is on the order of 0.1 in.

The horizontal movements preceding or associated with the instability of slopes may also be investigated by successive surveys of the shape and position of flexible vertical casings installed in the ground. The surveys are made by lowering a device containing a pendulum that indicates the deviation of the device from the vertical. An initial survey is made of the deviations from verticality at closely spaced vertical intervals, and successive surveys are carried out to determine

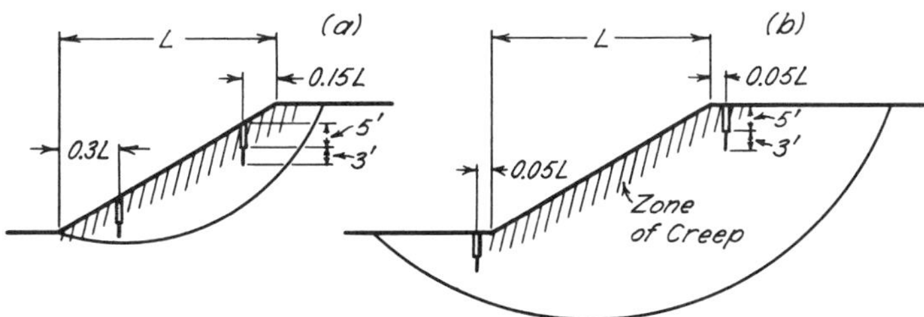

Fig. 66.8. Position of reference points to detect movement of slope. (*a*) If slide along toe circle is likely. (*b*) If base failure may occur.

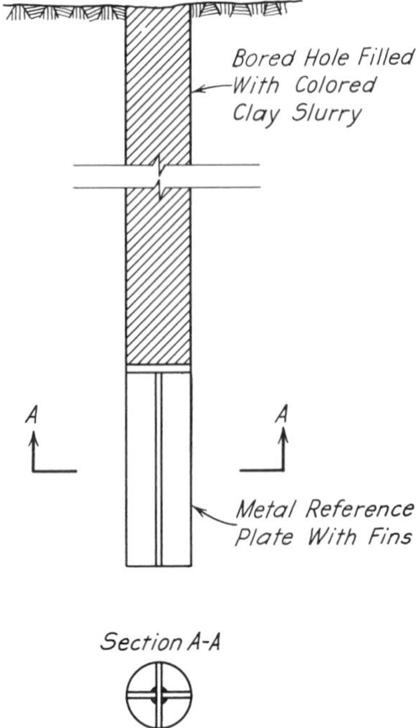

Fig. 66.9. Underground reference point for observation of heave.

the changes in inclination at the same levels. The changes in inclination can be integrated to determine the deflection of the casing in the time between the two sets of readings. Several types of sensitive inclinometers have been developed (Koch et al. 1952, Wiegmann 1954, Wilson and Hancock 1960). The accuracy of the observations is not limited by the sensitivity of the inclinometers, however, but by the requirement that successive readings be made with the same orientation of the instrument and at the same point in the casing. The arrangement used most widely to satisfy this requirement consists of a grooved casing in which observations are made by means of the *Wilson slope indicator.*

The slope indicator consists of a pendulum of which the tip makes contact with a resistance coil, subdividing it into two resistances that form one half of a Wheatstone bridge. The other half, contained in a portable control box, includes a precision potentiometer of which the readings are proportional to the inclination of the instrument in the plane of the pendulum. The instrument (Fig. 66.10*a*) supports four spring-mounted wheels in the plane of the pendulum. These

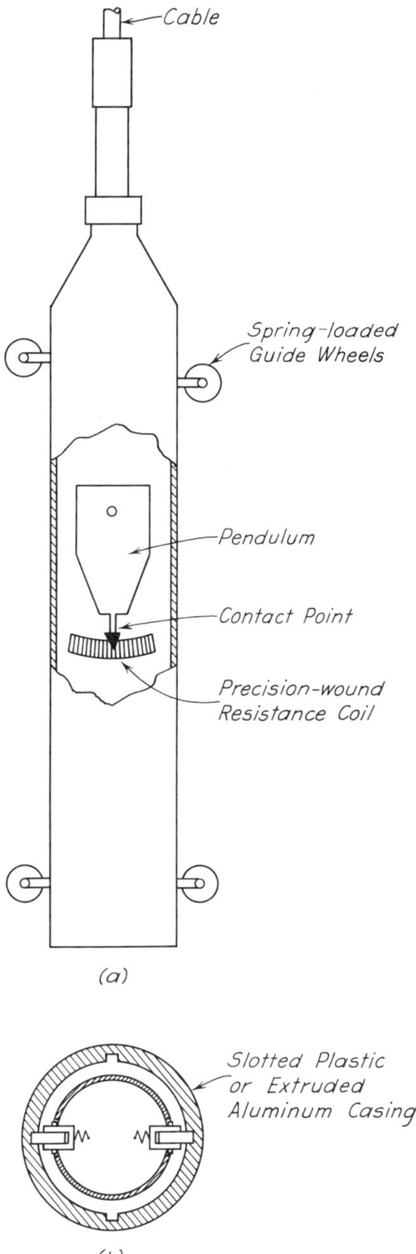

Fig. 66.10. Diagrammatic sketch of Wilson Slope Indicator. (*a*) View of instrument. (*b*) Cross section showing instrument in slotted casing (after Wilson and Hancock 1960).

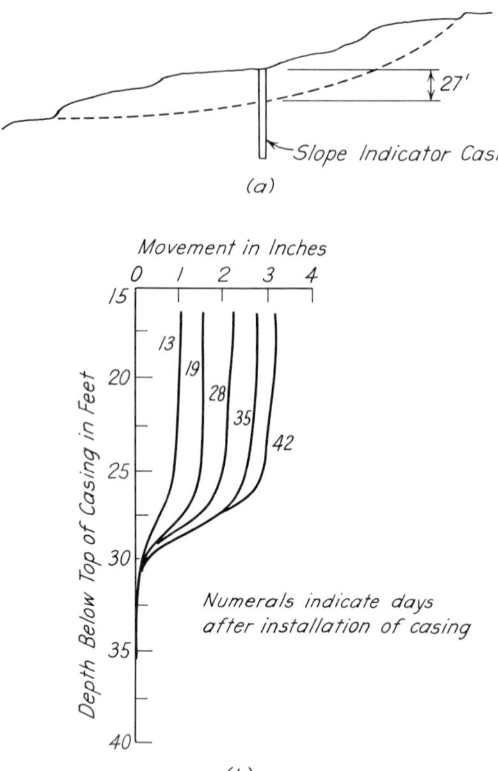

Fig. 66.11. **(*a*) Cross section through slide in Hawaii showing location of casing for slope indicator. (*b*) Results of series of observations disclosing zone of movement concentrated near 27 ft.**

wheels travel in the grooves of the casing. The casing itself consists either of plastic or anodized aluminum with an inside diameter of $2\frac{7}{8}$ in. It contains two sets of slots, in planes at right angles to each other (Fig. 66.10*b*). The slots permit reliable orientation of the instrument in successive surveys. By carrying out each set of observations in the two planes at right angles to each other, the resultant inclination can be calculated.

The results of a series of observations at the location of one landslide are shown in Fig. 66.11. The location of the surface of sliding and the rate of displacement are clearly shown. The accuracy of the deflections depends to some extent on the shape of the deflection curve. If the displacements occur within one or two feet, the movement of the top of a casing 100 ft deep, with respect to the bottom, can be determined with an accuracy of about $\frac{3}{8}$ in. On the other hand, if the distortions are spread over a zone with a thickness of 10 ft or more, the accuracy may be on the order of 1 in.

Observations of the Deformation of Flexible Tunnels and Culverts

If the temporary lining of an approximately circular tunnel or culvert consists of closed rings or ribs, valuable information regarding earth-pressure conditions can be obtained by measuring several diameters at a given section immediately after installation of the lining, and at regular time intervals thereafter. It is generally most convenient to measure the horizontal and vertical diameters, and the two diameters inclined to them at 45°. The measurements can be made with a steel tape.

The measurement of the gradual deformation of permanent tunnel linings, such as concrete tubes, requires a more refined method because the deformations are smaller. One procedure which has been used with success is to utilize an invar wire by means of which the change in distance between two permanent reference blocks can be observed. On one end of the wire is fastened a metal ring which bears against the pin of one reference block, as shown in Fig. 66.12. On the other end is fastened a light-weight metal cylinder with accurately machined ends. The cylinder is provided with a hook to which is attached a small spring balance. In order to make a measurement, a string is lashed to the spring balance and to the second reference block in such a manner that tension in the spring balance is a predetermined amount, such as 15 lb. When the required tension is obtained, the distance is measured by means of a pointed steel scale from a fixed point on the second block to the end of the metal cylinder. In order to provide a check on the length of the wire or to permit its replacement if it should become kinked or damaged, a standard set of blocks

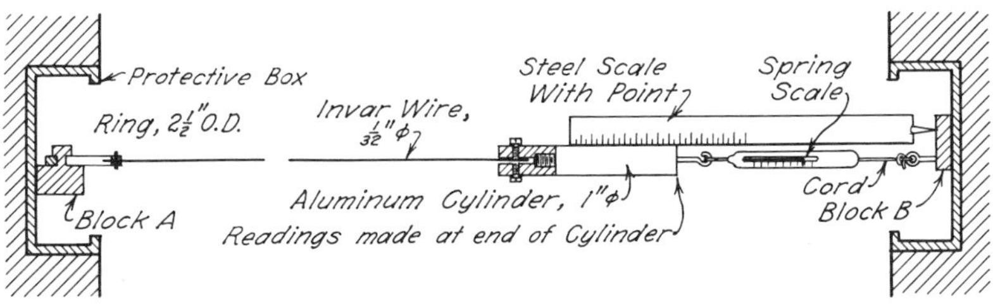

Fig. 66.12. Equipment for observing change in diameter of permanent lining of tunnel tubes.

should be installed permanently at some accessible place where the distance between the standard blocks is not subject to change. Readings of the standard length should be made before and after each set of field observations. For distances up to about thirty feet, measurements can be made with this device with an accuracy of about 0.02 in.

Selected Reading

Kjellman, W., T. Kallstenius, and Y. Liljedahl (1955). "Accurate measurement of settlements," *Proc. Royal Swedish Geot. Inst.*, No. 10.

Shannon, W. L., S. D. Wilson, and R. H. Meese (1962). "Field problems: field measurements," Chapter 13 in *Foundation engineering,* G. A. Leonards, ed., New York, McGraw-Hill, pp. 1025–1080.

Bjerrum, L., T. C. Kenney, and B. Kjaernsli (1965). "Measuring instruments for strutted excavations," *ASCE J. Soil Mech.*, **91,** No. SM1, pp. 111–141.

Wilson, S. D. and C. W. Hancock, Jr. (1965). "Instrumentation for movements within rockfill dams," *ASTM Special Tech. Publ. 392,* pp. 115–130.

ART. 67 MEASUREMENT OF EARTH PRESSURES

Purpose and Means of Observations

Observations of earth pressures are made to determine the magnitude and distribution of the contact pressure between soils and structures, in order to verify or improve the basis for design; to determine the magnitude and distribution of stresses in earth masses such as the subgrades of highway and airport pavements; and to provide information about the loads carried by individual members of temporary and permanent supports for the soil during or after construction.

Attempts to measure contact pressures against the plain face of concrete masonry have usually been made by means of *pressure cells* embedded in the concrete in such a manner that the contact face between the soil and the cell is flush with the face of the masonry. The results may be misleading because of errors due to imperfections in the pressure cells themselves, and because of the small size of the contact area between soil and cell. Errors associated with imperfections in the cells can be avoided by proper design and installation. Those due to the small contact area may be so great as to necessitate the use of a totally different system of measurement such as isolating a large section of a buried structure and measuring the total load acting upon it. The use of pressure cells for measuring the stresses in the interior of an earth mass also leads to errors unless the cells are so designed and installed that their presence does not itself alter the state of stress in the mass. Whenever the earth pressure acting against a support is carried by a simple structural system, such as the struts in an open cut (Article 48), the magnitude and distribution of the pressure can be investigated most conveniently and reliably by measuring the loads in the struts.

Pressure Cells for Measurement of Contact Pressures

Since the contact face between the soil and the cell is flush with the face of the masonry and the cell is entirely embedded in concrete, the shape of the cell is irrelevant. However, any displacement of the contact face between the cell and the soil changes the pressure on the contact face. The error becomes excessive if the ratio of displacement to diameter exceeds about $\frac{1}{1000}$ (Taylor 1947). The earliest cells consisted of flat circular boxes filled with a liquid. The contact face was a flexible membrane. The liquid pressure was measured with an ordinary Bourdon gage. Although the absolute deflection of the membrane was small, it was nevertheless large enough to cause an important change of pressure. Furthermore, the cells were extremely sensitive to changes in temperature.

The next stage (Goldbeck and Smith 1916) in the development of pressure cells is represented by the *Goldbeck cell,* which consists of a circular box 5.4 in. in diameter and 1.5 in. thick. The area of contact with the soil is 10 in^2. In this device, illustrated in Fig. 67.1*a*, the contact face is supported at its center by a metal contact button. Behind the remainder of the contact face is a chamber into which air under pressure can be admitted. When the air pressure within the inner chamber is just sufficient to balance the external pressure on the contact face, the contact face moves away from the contact button and an electrical circuit is broken. The break is indicated by the extinction of an electric light or by the drop in reading of an ammeter.

Although the Goldbeck cell represents a great improvement over the previous Bourdon type cells, it possesses a number of disadvantages. Chief among these is the outward movement of the contact face required to break the electric circuit. Therefore, the indicated pressure is too large. On one field installation, the pressure observed by means of Goldbeck cells was 80% in excess of the load determined by more accurate methods. In many instances, the break in the electrical circuit is not sharply defined but consists of a gradual diminution of current corresponding to a considerable range in pressure. Condensation of water from the compressed air or other sources commonly fouls the contact points and renders the cells inoperative. In order to avoid these difficulties, various modifications have been at-

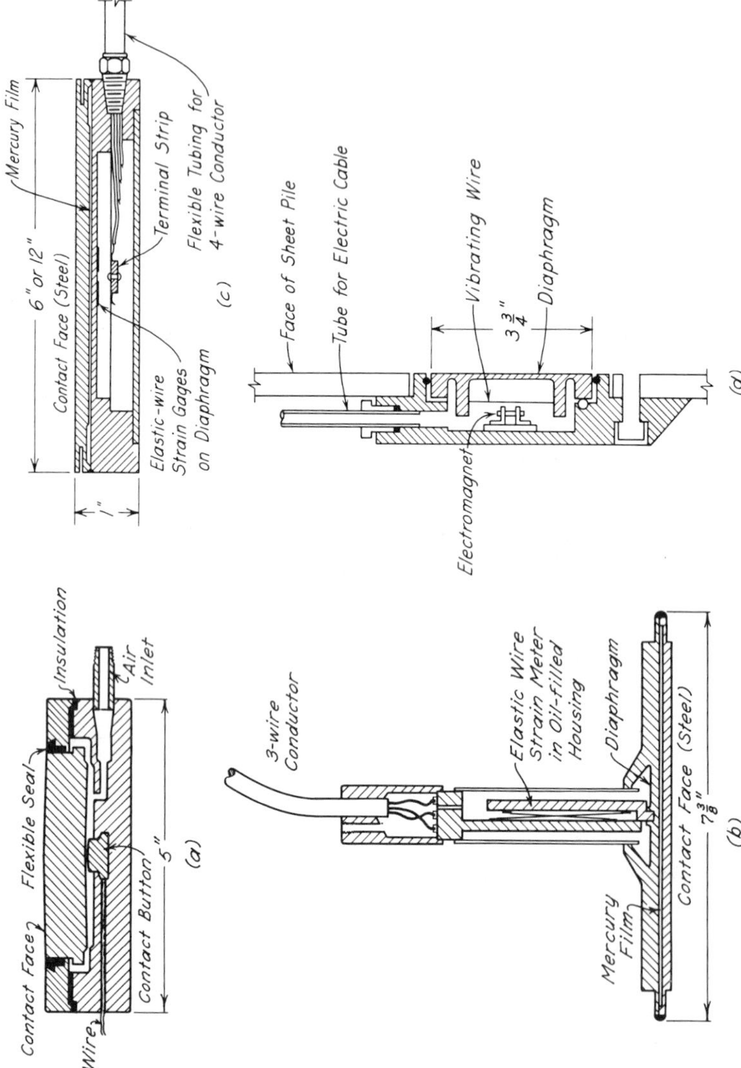

Fig. 67.1. Earth-pressure cells. (*a*) Goldbeck cell. (*b*) Carlson cell. (*c*) Waterways Experiment Station cell. (*d*) Adaptation of vibrating-wire cell for measuring pressure against face of steel sheet pile.

tempted, but the results have not been encouraging and new approaches to the problem have been made.

In the improved cells, in contrast to the Goldbeck type, the readings can be made without previously producing a displacement by artificial means. The deflections of the contact face are very small. The record of long-time reliability of most of the cells under field conditions has, however, been rather unsatisfactory. Among those with a high degree of demonstrated dependability are the *Carlson cell*, the *Waterways Experiment Station cell*, and the *vibrating wire cell.*

The Carlson cell (Fig. 67.1*b*) consists essentially of two flat steel plates 7 in. in diameter separated by a film of mercury approximately .02 in. thick (Carlson and Pirtz 1952). A load exerted against the steel plates produces a corresponding pressure in the mercury. The central part of the upper steel plate has a reduced thickness so that it acts as a relatively flexible diaphragm which deflects upward due to the increase in mercury pressure and actuates a Carlson strain meter. The strain meter consists of two coils of steel wire mounted on porcelain spools attached to a steel frame. The deflection of the diaphragm increases the tension in the wires of one coil and reduces the tension by the same amount in the other. The changes in tension cause a change in the ratio of the electric resistances of the two coils, which can be measured by means of a Wheatstone bridge. The change in resistance ratio is a measure of the deflection of the diaphragm, and hence of the contact pressure against the cell. On the other hand, a change in temperature either increases or decreases the tension in the wires of both coils by the same amount, and consequently has no effect upon the resistance ratio. The entire strain meter unit is housed in a stem and may be read by electrical means from a remote point. The accuracy of the readings may be influenced by changes in resistance of the connecting cables.

The effective modulus of elasticity of the Carlson cell is almost equal to that of concrete. Consequently, errors due to deflection of the contact face are not important. The capacity of Carlson cells ranges from about 600 to 100,000 lb. For a particular installation, a cell should be selected which has the smallest capacity not likely to be overloaded, in order to obtain the greatest possible sensitivity. The sensitivity is approximately 1% of the capacity.

The Waterways Experiment Station (WES) cell (Fig. 67.1*c*) is similar in principle to the Carlson cell, except that the deflection of the diaphragm that constitutes the contact face is measured by means of elastic-wire strain gages bonded to the inside of the diaphragm (Woodman 1955). The electrical circuitry eliminates the likelihood of error due to changes in resistance of the connecting cables, but the tendency to long-time creep of the cement bonding the strain gages to the diaphragm may lead to instability.

The vibrating wire cell (Fig. 67.1*d*) operates on the principle that the deflection of the diaphragm changes the tension in an elastic wire stretched between two posts affixed to the diaphragm, and thereby causes a change of the natural frequency of vibration of the wire. A combined permanent and electromagnet is mounted near the wire. To make an observation, an electrical impulse is sent through the electromagnet which causes the wire to vibrate. The vibration of the wire in the field of the permanent magnet sets up an electromotive force in the coils of the electromagnet with a frequency equal to that of the vibrating wire. The electromotive force is amplified and its frequency determined by means of a portable frequency-measuring instrument (Bjerrum et al. 1965). The deflection of the diaphragm is proportional to the change in the square of the frequency. Temperature produces the same strain in the diaphragm as in the vibrating wire and, consequently, requires no compensation. The long-time performance of such cells has been very satisfactory (Cooling 1962). The cell shown in Fig. 67.1*d* was constructed to measure the contact pressure against the face of a steel sheet pile; the calibration of the cell was not impaired by the vibrations due to pile driving.

Measurement of Contact Pressure Against Large Areas

A pressure cell of any of the types in common use provides an area of contact which is very small compared to the total area of contact between masonry and soil, although cells of the WES type have been installed with diameters as great as 30 in. (Thayer 1966). Under field conditions the contact pressure is never uniformly distributed because the soil is never homogeneous. Furthermore, construction operations induce local stresses which may deviate widely from the average contact pressure. Therefore, a second stage in the development of the technique of measuring contact pressure has consisted in developing methods for measuring the pressures on large areas.

One method for measuring the pressure over large areas is illustrated in Fig. 67.2*a*, which represents an installation for the measurement of contact pressure on the base of a tunnel section of the Chicago subway. In this installation, the contact faces are heavily reinforced concrete slabs within a framework of steel channels. The slabs are isolated from the remainder of the concrete invert by means of corkboard, which is sufficiently compressible to carry not more than an insignificant part of the ultimate pressure. Between each slab and

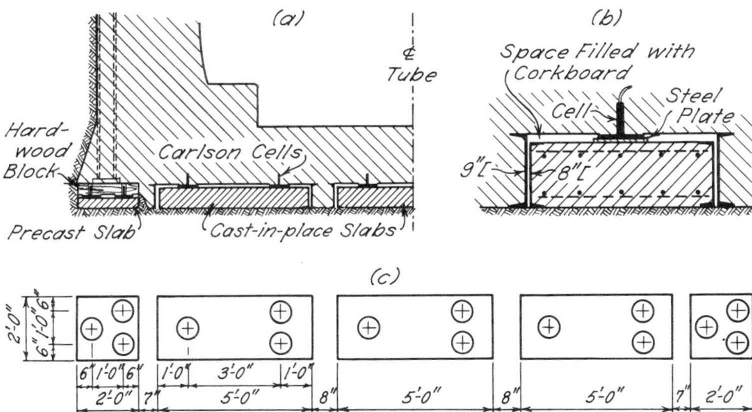

Fig. 67.2. Installation for measuring contact pressure on base of Chicago Subway tunnel. (*a*) Half-section through invert showing measuring slabs. (*b*) Section through slab. (*c*) Plan showing arrangement of slab (after Terzaghi 1943*a*).

the invert are located three Carlson cells by means of which the load carried by the contact face may be determined.

In general, this installation operated in a successful manner and modifications of the principles involved can be expected to produce reliable results under other conditions. Preferably, measuring slabs such as those shown in Fig. 67.2*a* should be precast, or else cast as long as possible before the slabs are subjected to earth pressure. Otherwise, the deflection of the green concrete slabs is apt to cause a decrease in the indicated pressure similar to that associated with the deflection of the contact faces of the early fluid cells. After hardening, the slabs should be practically rigid. The slabs in the Chicago installation were not quite rigid enough. Therefore the single cells did not carry as much of the load as the double cells combined. It seems probable that four cells per slab instead of three would have been more satisfactory.

Carlson cells, when used to measure reactions in this manner, constitute one of several varieties of *load cells*. Another variety, which makes use of vibrating-wire strain gages, is described subsequently in connection with the measurement of strut loads in open cuts.

Figure 67.3 suggests a method for installing similar slabs to measure the pressure exerted by earth against a vertical wall. When the wall is constructed, a vertical recess is provided for the slab, and the load cells are embedded in the concrete of the wall. After the concrete has set and the forms are removed, the face of each cell is covered

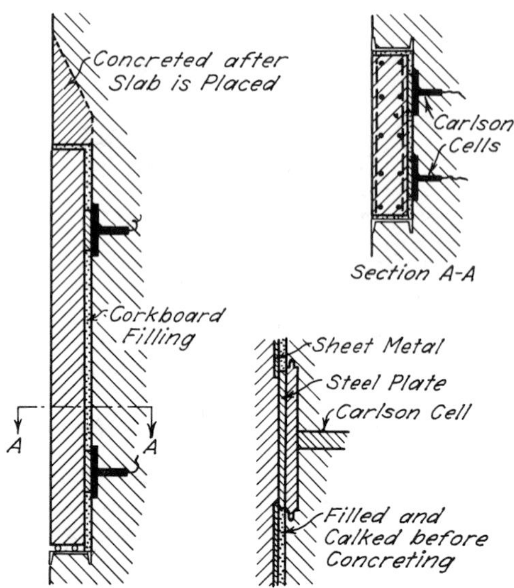

Fig. 67.3. Suggested method for measuring earth pressure against back of retaining wall.

by a circular metal disk which serves to transmit the load from the slab to the cells. The face of the concrete between the disks is covered with corkboard. The precast slab is placed in the recess. At the bottom, it is supported by rollers which offer a very small resistance to a slight lateral movement of the wall. The clearance between the walls of the recess and the slab is filled with corkboard.

If it should be more convenient to cast the slab in place, a space must be left above the recess, as shown in Fig. 67.3, to provide room for placing the concrete.

Measurement of Pressure in Interior of Earth Masses

If pressure cells are used for measuring the pressure in the interior of a body of earth, they must satisfy the additional condition that their presence should have no marked influence on the state of stress in the earth. A cell represents a rigid core in a compressible medium. Both theory and experience have shown that the ratio between diameter and thickness must be greater than about five if the cell is installed for the purpose of measuring the vertical pressure on a horizontal plane within a fill such as a dam (Taylor 1947). The same cell would lead to erroneous results if used to measure the horizontal pressure against a vertical plane, because the long vertical dimension of the cell would resist the vertical strain in the adjacent soil and radically change the state of stress.

Since there is no possibility for installing cells of any type in undisturbed soil without producing a radical change in the state of stress in the soil to a considerable distance from the cell, pressure cells can be used only in artificial fills.

Measurement of Load in Struts and Other Temporary Supports

The most reliable measurements for providing information about the load carried by individual members of a system of temporary supports for a mass of soil are those on members subject to pure compression, such as the horizontal struts in open cuts. The load in compression members can be determined either by computation from data obtained by the use of a strain gage, or else by transferring the load from the members onto a suitable measuring device. In a few instances, attempts have been made to estimate loads by observing the deflection of beams subject to bending, but the results of the estimates are not reliable because the deflection of the beams induces arching in the soil adjoining the deflecting support and relieves the pressure to an unknown extent.

Whatever methods of measurement are used, the observations should be made on several independent profiles, in order to get a conception of the deviation of the loads on individual profiles from the average (Article 48).

Until recently the reliability of strain gages involving electrical circuitry has not been satisfactory for field use, largely because of the sensitivity of the gages to moisture and the practical impossibility of providing dependable protection against damage and moisture under the adverse conditions prevailing on construction jobs. For these reasons, portable 10-in. mechanical strain gages were used for many of the early measurements of strut loads. The operation was time-consuming and required great skill on the part of the operator (Peck 1941). More recently, the vibrating-wire strain gage has been developed into a rugged, simple, but dependable instrument that considerably improves the ease and accuracy of the work. An adaptation suitable for use in strut-load measurements is illustrated in Fig. 67.4 (Bjerrum et al. 1965). If the struts are steel wide-flange or H sections, two gages are used, one mounted on each side of the web at the neutral axis of the strut. To avoid the influence of nonuniform distribution of stress, the gages are located not closer to the end of

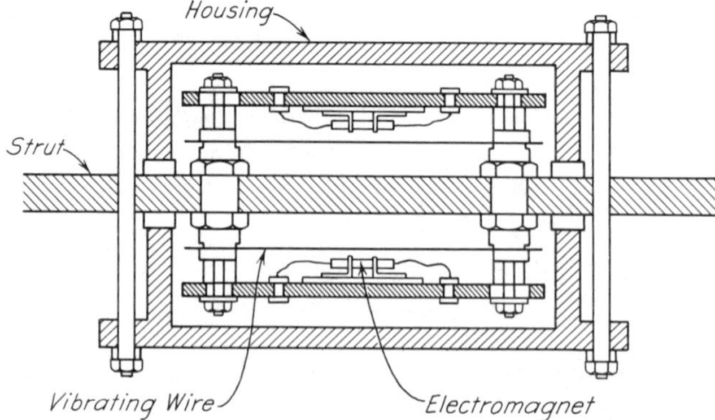

Fig. 67.4. Adaptation of vibrating-wire strain gage for measurement of load in steel strut (after Bjerrum et al. 1965).

a strut than about six times the depth of the strut between flanges. The gages are installed by drilling two holes through the web of the strut and setting a post in each hole by means of threaded nuts. To the posts are fastened, one on each side of the web, a tensioned wire and an electromagnet. Each gage is then covered by a protective housing and connected by a cable to an electric outlet plug fastened to the web of the strut at a convenient location near the end, where the operator may plug in the cable from the frequency-measuring instrument when an observation is to be made. Since the tensioned wires are located symmetrically with respect to the neutral axis, the average compressive strain in the strut is obtained merely by averaging the results of the two gages. If the struts are unsymmetrical, several gages may be needed to establish the distribution of compressive stress across the section. The error in the strut loads determined by means of vibrating-wire strain gages does not usually exceed 10%. If possible, a final set of readings should be made on the unstressed struts after their removal from the excavation, to detect and permit correction for any drift in the zero-reading of the gages.

Strut-mounted strain gages are relatively inexpensive and may be considered expendable. They cannot be used, however, if the elastic properties of the strut are unknown or not constant. Hence, they cannot be used to determine reliable values of loads in timber struts. Fairly reliable measurements can be made by a hydraulic jacking procedure. The equipment is shown in Fig. 67.5. The force between the strut and the wale is transferred to a pair of interconnected hydraulic jacks. The load supported by the jacks is indicated by a pressure gage. The hydraulic pressure is increased by increments until the jacks exert sufficient force to produce a narrow gap between the end of the strut and the wale. The width of the gap is measured by means of dial gages. After the separation between strut and wale reaches about 0.1 in., the hydraulic pressure in the jacks is reduced by increments.

In order to compute the load in the strut, a curve (Fig. 67.6*a*) is plotted which shows the relation between the load corresponding to the pressure in the jacks and the separation between strut and wale. Because of friction in the jacks, the curve encloses a hysteresis loop. At a given value of the separation, the load actually carried by the jacks corresponds roughly to a point midway between the two sides of the hysteresis loop. The locus of points representing the load carried by the jacks at different values for the separation is a straight line which intersects the axis of zero separation at a point corresponding to the load carried by the jacks before the strut was compressed due to jacking. This load is approximately equal to the load in the strut before jacking was started.

The strut load determined by this procedure is too great by the amount of force necessary to relieve from stress the portion of the

Fig. 67.5. Equipment for measuring strut load by means of hydraulic jacks.

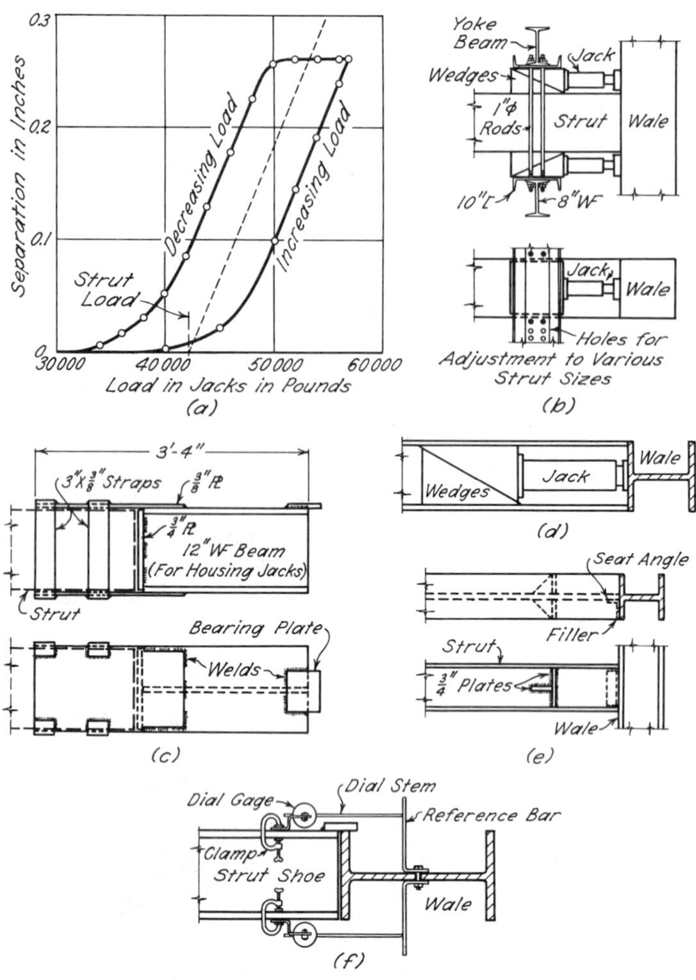

Fig. 67.6. (*a*) **Relation between load carried by jacks and separation between strut and wale.** (*b*) **and** (*d*) **Methods for providing reaction for jacks if strut load is small.** (*c*) **Shoe for wood strut.** (*e*) **Bracket for steel strut.** (*f*) **Method for measuring separation (after Peck 1941).**

strut adjacent to the jacks. The error, however, is negligible unless the strut is very short and very rigid. In practice, the strut is likely to begin to separate from the wale at one point of contact before the others, and it is necessary to average the values of strut load obtained by means of measurements at each of the four corners of the strut during separation.

In order to use the jacking procedure, a sufficiently strong reaction must be provided for the jacks. The methods shown in Fig. 67.6*b* and *d* have been successfully used for measuring strut loads which did not exceed about 15 tons. They have the advantage that the struts need not be prepared in advance. If the strut loads are greater, preparations for measurements must be made before the strut is inserted in the cut. For wood struts, shoes (Fig. 67.6*c*) can be fitted over the end of the strut to house the jacks. Brackets (Fig. 67.6*e*) may be welded to steel struts. Provision should be made to prevent the end of the strut from falling if, for some reason, the jacks should suddenly release their load. The separation between strut and wale can be measured by means of dial gages reading to 0.001 in., supported as shown in Fig. 67.6*f*.

Experience has indicated that the error associated with the jacking method may be on the order of 20 to 30% of the strut load. More reliable and much more convenient measurements of strut loads, under circumstances where strain gages are not suitable, can be obtained by inserting a load cell between the end of the strut and the wale, as shown in Fig. 67.7*a*. One variety of load cell consists of a short steel cylinder inside of which are mounted three vibrating-wire gages. The ends of the cylinder are closed by water-tight plates (Fig. 67.7*b*). The electrical cable is brought out of the cell through a water-tight port. Each cell is calibrated in a testing machine. Such a load cell has the advantages that it is rugged and reliable, and can be used even under water (Bjerrum et al. 1965). On the other hand, it is relatively expensive, requires advance preparation by the contractor for its insertion, and is somewhat sensitive to eccentricity of loading.

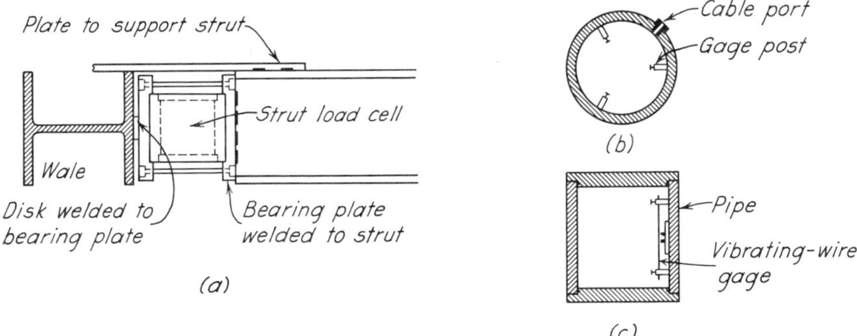

Fig. 67.7. (*a*) Load cell of vibrating-wire type adapted for measuring load in strut. (*b*) and (*c*) Sections through load cell (after Bjerrum et al. 1965).

The error in measured strut loads is considered to be on the order of 20%.

Selected Reading

Ward, W. H. (1955). "Techniques for field measurement of deformation and earth pressure," *Proc. Conf. on Correlation between Calculated and Observed Stresses and Displacements in Structures,* Inst. Civil Engrs. London, Paper No. 3, Group 1, pp. 28–40.

Burke, H. H. (1960). "Garrison dam test tunnel: investigation and construction," *Trans. ASCE,* **125,** pp. 230–267. Extensive use of mechanical strain gages.

Cooling, L. F. (1962). "Field measurements in soil mechanics," *Géot.,* **12,** No. 2, pp. 77–103.

NGI (1962). "Vibrating-wire measuring devices used at strutted excavations," *Norwegian Geot. Inst. Tech. Rept. No. 9.* Detailed discussion of vibrating-wire instruments.

Sikso, H. A. and C. V. Johnson (1964). "Pressure cell observations—Garrison dam project," *ASCE J. Soil Mech.,* **90,** No. SM5, pp. 157–179.

Bjerrum, L., T. C. Kenney, and B. Kjaernsli (1965). "Measuring instruments for strutted excavations," *ASCE J. Soil Mech.,* **91,** No. SM1, pp. 111–141.

ART. 68 MEASUREMENT OF POREWATER PRESSURE

Fundamental Requirements

When the subsoil is fairly permeable, the porewater pressure can be determined readily by observing the piezometric level in an open standpipe or observation well, because every change in hydrostatic pressure produces an almost simultaneous change of the water level in the well. If, for example, the porewater pressure increases in the soil surrounding the open lower end of an observation well, a hydraulic gradient into the well is created. Consequently, water flows rapidly into the well until equilibrium is reached; the water level in the well then corresponds to the porewater pressure that would exist in the soil if there were no observation well. The presence of the well has virtually no influence on the porewater pressures near the point of measurement.

On the other hand, when the subsoil is quite impermeable, the presence of a device for measuring the porewater pressure may so radically alter the pressure near the point of measurement that the results of the observations are utterly misleading. If, in order to indicate a change in pressure, even a fairly small flow of water must occur into or out of the measuring device, the time required to reach equilibrium may be intolerably long. While water is flowing into the instrument, the pore pressure at the measuring point is smaller than it would be if the instrument were not present. If the porewater pressure in the mass of soil should decrease because of natural events or construction activities before equilibrium is reached, the indicated pore pressure would still continue to rise until the general pore pressure dropped below the locally depressed value near the instrument, whereupon water would tend to flow out of the instrument into the soil. The local pore pressure at the instrument would then become greater than the value that would exist if no measuring device were present.

Hence, in order to avoid erroneous and meaningless results, an instrument for measuring porewater pressures in an impervious soil must react almost instantaneously, without requiring a significant movement of the surrounding porewater. Piezometers utilizing closed hydraulic systems or electrical measuring devices have been devised to satisfy these requirements.

The type of installation best suited for a given site and purpose is determined largely by the *hydrostatic time lag* of the installation (Hvorslev 1951). This quantity is defined as the time required for the installation to adjust itself almost completely to a change in pore pressure. Since theoretically the time for complete adjustment is infinitely long, practical requirements are based on the time required for 90% of the equalization to take place. The time lag depends on the quantity of flow required to produce a response in the apparatus, on the permeability of the soil, and on the dimensions of any filters surrounding the pervious tip.

The suitability of a particular type of equipment also depends on the physical requirements of the site. Piezometers consisting of open standpipes, for instance, can be observed only if the tops of the standpipes are accessible. Hence, they may be impractical for measuring the porewater pressures in the core of an earth dam while construction of the dam is going on. A closed hydraulic system that may be read remotely may be more appropriate. On the other hand, in a closed hydraulic system it is not feasible to measure large negative pore pressures because of the tendency of air and water vapor to come out of solution and form bubbles in the measuring system. The formation of such bubbles, known as *cavitation,* greatly increases the time lag of the system. Even if the porewater pressure is positive at the point where the measurement is to be made, negative pressures in the hydraulic system cannot be avoided unless all the pipe lines and the gage house are located no higher than the piezometric level at the point under investigation.

If the soil is not saturated, the pores are filled partly with water and partly with air, and the pressures in the liquid and gaseous phases are different. The difference is small if the degree of saturation is close to 100%, but may be extremely large for low degrees of saturation. The significance of the measurements then depends to a considerable extent on the *air entry value* of the porous tip or filter of the apparatus in contact with the soil (Bishop et al. 1964). The air entry value is the amount by which the air pressure on one side of a saturated filter must exceed the water pressure on the other side in order to force air through the filter. If the difference $u_g - u_w$ between the air pressure and the water pressure in the soil exceeds the air entry value, air enters the saturated filter, water is drawn from it into the soil, and the measured pressure corresponds to the pore-air pressure rather than to the porewater pressure. Filters with a high air entry value are, therefore, required for measuring porewater pressures in partly saturated soils. This condition has rarely been satisfied in installations made thus far in the cores of earth dams compacted on the dry side of optimum and, consequently, most such measurements are invalid. If the negative porewater pressure is very large, even the use of a filter with high air entry value does not prevent cavitation and the accumulation of air and water vapor between the filter and the measuring device, whereupon reliable measurements are no longer possible.

If the subsoil consists of permeable layers separated by less permeable ones, the piezometric levels in the permeable layers may differ. A piezometer intended to measure the porewater pressure in any one layer must be carefully sealed in that layer alone. Otherwise flow alongside the installation will take place from one permeable layer to another and will invalidate the results of the observations.

The foregoing discussion leads to the conclusion that, unlike the observations of displacements or even of earth pressures, the selection and installation of instruments for measuring porewater pressures cannot be considered a routine matter but, on the contrary, require a thorough knowledge of soil mechanics, experience, and meticulous attention to detail. Otherwise, the observations may be of no value or may even be misleading.

The following sections contain descriptions of some of the more common types of measuring devices and of the precautions required during installation. This information is followed by a summary of the hydraulic time lags for a variety of conditions, to permit judging the types appropriate for a particular purpose.

Open Piezometers or Observation Wells

If the coefficient of permeability k is greater than about 10^{-4} cm/sec, corresponding to that of a clean or slightly silty fine sand, the measurement of porewater pressure can be made in observation wells. If the waterbearing stratum is fairly homogeneous, ordinary well points (Article 47) about 2 in. in diameter may be driven or jetted to a depth of several feet below the lowest estimated position of the water table. The well points and their riser pipes serve as piezometric tubes in which the water rises to the level of the free water surface. The elevation of the upper end of each riser pipe must be established because these ends serve as the reference points. If a high degree of accuracy is not required, the depth to the water level may be determined by inserting a thin wooden stick after its surface has been coated with white chalk; the surface of the immersed portion turns gray. Another method is to lower a pair of wires exposed at their lower ends. An electric circuit, detected by means of a galvanometer, is closed when the two ends are immersed in the water.

If the waterbearing stratum contains one or more layers having a relatively low permeability, it is necessary to install separate piezometric tubes extending to each permeable horizon. The well points should not be pushed or driven through the feebly permeable layers because the openings in the perforated lower section are likely to become clogged. Moreover, they should not be inserted in an uncased drill hole because the outside diameter of the pipe is likely to be less than the diameter of the hole. It is advisable to drill and case a hole about 6 in. in diameter from the ground surface to a point near the bottom of the pervious horizon under investigation. The lowest 3 ft should be filled with a clean sand having an effective grain size two or three times that of the sand in the ground, and a piezometric tube consisting of a 1-in. pipe should be pushed about 1 ft into the sand. Alternatively a well point may be lowered to the bottom of the hole and the sand dropped into the hole until a filter about 3 ft high has been established beside and above the perforated section. During this operation a downward flow of water in the casing should be maintained. The casing is then withdrawn to a short distance above the top of the pervious layer while additional sand is placed. From this level to a point a short distance above the top of the feebly pervious stratum, the hole is backfilled with an impermeable material as the casing is withdrawn. The backfill may consist of a tight packing of clay at a water content intermediate between the plastic and liquid limits, deposited a few chunks at a time and carefully tamped. After a few inches of clay have been placed, it may be more expedient to form the seal of a thick slurry of portland cement grout. In any event, the closest supervision is required to avoid careless workmanship which can ruin the installation.

In less permeable materials the hydrostatic time lag of an ordinary open standpipe may become excessive. It can be reduced by making the area of contact between the filter and the surrounding soil as large as possible and by reducing the diameter of the riser pipe to the least dimension within which accurate soundings can be made. Moreover, in feebly permeable soil, the presence of bubbles of gas in the filter, the porous tip, and elsewhere in the apparatus increases the reaction time; since the formation of gas is enhanced by electrolytic action of metals, it is advantageous to eliminate metallic components. These refinements have been incorporated in the porous-tube piezometer proposed by A. Casagrande and widely used.

The Casagrande piezometer (Fig. 68.1) consists essentially of a pervious tubular tip located at the depth where the porewater pressure is to be measured, connected to a standpipe in which the position of the water level is observed. The tip is a hollow cylinder of porous fine grade Norton ceramic tubing 1 to 2 ft long, with outside and inside diameters of 1.5 and 1.0 in., respectively. The standpipe is of plastic Saran tubing with an outside diameter of $\frac{1}{2}$ in. No metallic materials are used in the assembly. To provide a large area of contact between the instrument and the soil, the porous tube is embedded in a column of standard Ottawa sand.

The piezometer is installed in a drill hole of which the lower portion is made by driving a 10-ft length of 2-in. casing in a single section without any coupling or drive shoe, and by cleaning out the casing without washing below the bottom. In this manner, a tight contact is assured between the casing and the soil. As the Ottawa sand is installed and the piezometer set into position, the casing is retracted about 5 ft. The Saran tubing is sealed against the inside of the casing by two layers of bentonite introduced into the casing in the form of plastic balls about $\frac{3}{8}$ in. in diameter and tamped with a specially designed cylindrical weight provided with a central vertical hole that permits the weight to slide up and down around the standpipe.

The details of the installation have been worked out with great care (A. Casagrande 1949, 1958) and should be followed strictly unless local conditions require or justify deviations. Under some conditions the exacting and tedious procedures for forming the bentonite seals may be replaced by the use of a relatively quick-setting mixture of kaolinite and AM-9 (Lambe 1959).

The position of the water table may be observed by means of a sounding cable consisting of two insulated wires bared at the lower end. The cable is lowered into the standpipe until the bare ends make contact with the water, whereupon an electrical circuit is completed.

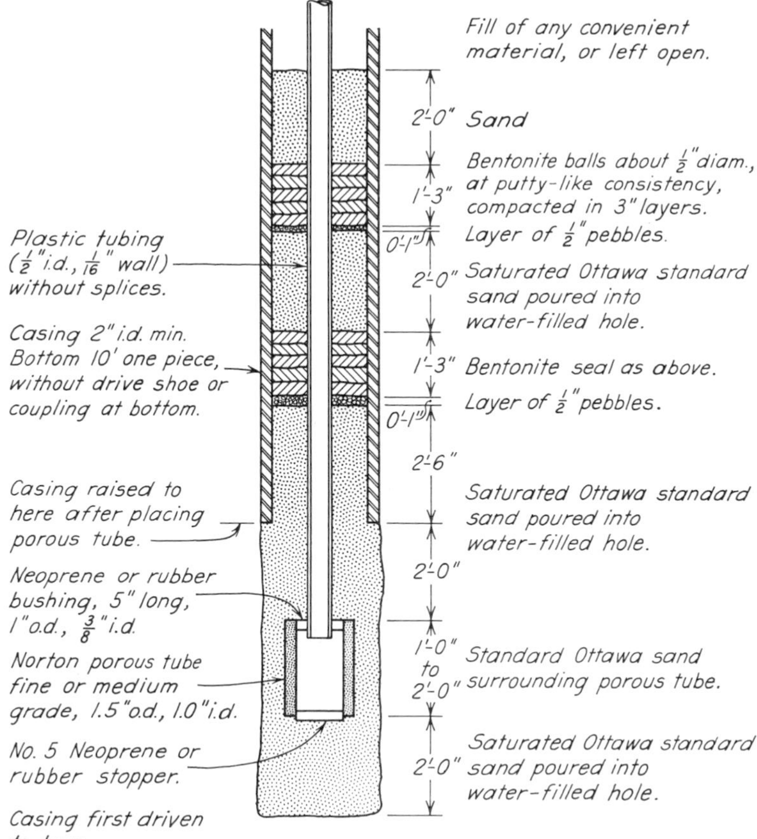

Fig. 68.1. Casagrande open-standpipe porous tube piezometer (after A. Casagrande 1949).

The contact is indicated by a galvanometer. To keep the cable taut a lead wire is wound spirally around the cable near the bottom. Care must be taken that a false indication is not given by drops of water adhering to the walls of the riser pipe. A more refined sounding device can be constructed of shielded microphone cable (A. Casagrande 1958).

The rather elaborate procedures necessary for sealing the Casagrande piezometer may be avoided under favorable subsoil conditions by the use of devices that can be pushed into the ground. The Geonor piezometer (Fig. 68.2) is representative of equipment of this type. A conical metal point facilitates advancing the piezometer. The porous

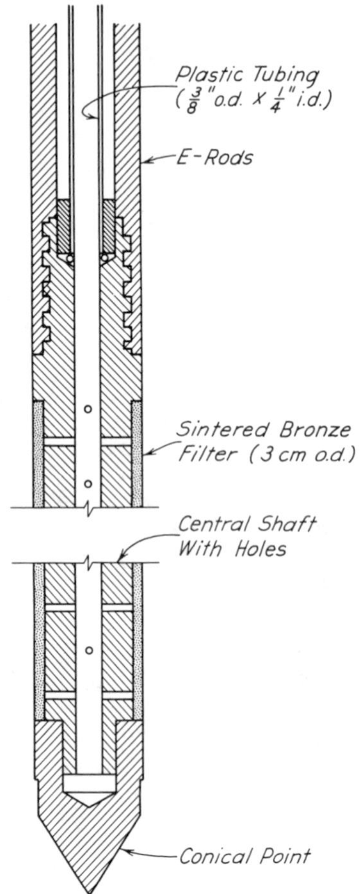

Fig. 68.2. Open piezometer of Geonor type (after Bjerrum et al. 1965).

element is a sintered bronze cylinder having the same external diameter as the upper part of the tip, and having the same diameter as the standard E-rods by means of which the assembly is pushed or jacked into the ground. In order to keep the hydrostatic time lag as small as practicable, the porous element is relatively long, and the plastic tubing has the smallest diameter in which reliable soundings can be made. In soft soils the piezometer can be pushed for its entire length. In more resistant soils, a hole may be drilled for part of the depth and the piezometer pushed below the hole only as far as necessary to assure an adequate seal against the E-rods.

Closed Hydraulic Systems

An open standpipe can be converted to a closed hydraulic system by attaching a pressure gage to the top of the standpipe, provided the equilibrium position of the water level in the standpipe is above the level of the pressure gage. However, except in permeable soils, the hydrostatic time lag is not likely to be appreciably reduced unless all parts of the system are completely filled with water, because the compressibility of inclusions or bubbles of air or water vapor is so great that substantial flows of water are required to produce a response. Furthermore, although a water-filled system is theoretically capable of measuring at least small negative pressures, the tendency of air to come out of solution or of gas to accumulate is likely to render an initially saturated system unresponsive. Hence, closed systems intended to measure small or negative pressures are customarily provided with means for flushing out the accumulated air.

Equipment developed by the U.S. Bureau of Reclamation for measurement of porewater pressures in dams and other embankments, and used widely by many organizations, is shown in Fig. 68.3*a* (USBR 1963). It consists of a plastic tip in which are embedded two porous disks through which the pore pressure communicates with the measuring system. The disks are connected to a chamber into which two plastic tubes are led. The tubes permit circulation of de-aired water through the tip in order to dislodge and remove air bubbles. The tip is placed in a pocket excavated below the surface of the fill during construction of the embankment, and the tubing is laid in a nearly horizontal trench to a gage house, usually on the downstream slope of the dam. At the gage house are located the pressure gage and a pump for circulating the de-aired water. In order to make an observation, water is circulated through the system until all free air is removed, whereupon the circulation lines are closed and connected to air-free Bourdon gages. If the de-airing has been successful, the gages quickly come to equilibrium.

Experience with such systems has been generally satisfactory except for the shortcoming that, until recently, the measurements of negative pressures have probably indicated pore-air rather than porewater pres-

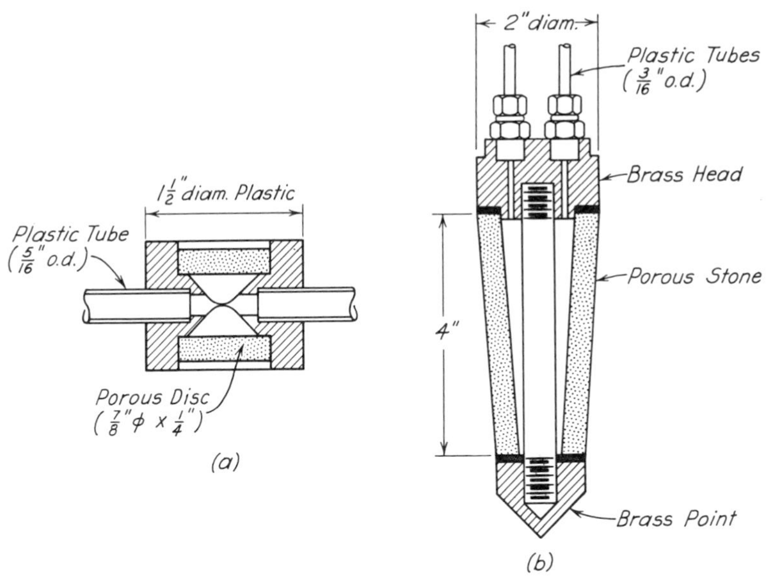

Fig. 68.3. Closed-system hydraulic piezometers. (*a*) USBR plastic tip for embankments (after USBR 1963). (*b*) Bishop type (after Bishop et al. 1960).

sures because of the use of porous stones with too low an air entry value. In recent installations this shortcoming has been corrected. A serious practical difficulty is the vulnerability of the tubing to damage during construction. Unless proper allowances are made, the tubing may also be damaged by the deformations of the embankment, especially those that cause the tubing to stretch. If the length of tubing between the tip and the gage house is very long, the volume change of the tubing may increase the hydrostatic time lag.

Various modifications of the USBR type have been developed. That shown in Fig. 68.3*b* (Bishop et al. 1960) is intended to reduce to a minimum the necessity for flushing to remove air. It is provided with a ceramic tube having an air entry value of about 30 $lb/in.^2$, tapered to improve the initial contact with the soil when inserted in a hole previously formed in the fill by means of a steel mandrel. Since some varieties of plastic tubing are slightly pervious to air and others to water, and since some possess deficient toughness or tend to deteriorate with time, the selection of the most suitable tubing deserves attention (Bishop et al. 1964). Installations of these devices in the cores of several earth dams were still operating successfully in 1964 after 4 years, and required de-airing only about once per year.

The necessity for an adequate seal around the tubing, to avoid pervious paths that would alter the distribution of porewater pressure near the point of measurement, becomes even more vital with decreasing permeability of the soil. The trenches containing the tubing for gages of the USBR type should be backfilled and compacted by hand through the impervious part of the dam or embankment. The backfill should consist of plastic clay somewhat wetter than the Proctor optimum moisture content. If the piezometers are installed in drill holes, procedures similar to those described in connection with the Casagrande piezometer should be followed. Particular care is required to make sure that the two lines of tubing are separated far enough to permit the sealing material to surround each tube.

Electrically Indicating Piezometers

In principle, a piezometer with extremely small hydraulic time lag can be constructed by providing a waterproof chamber separated from the porous tip by a diaphragm of which the deflection can be measured by a strain gage read by means of an electric circuit. The long-term performance of most devices of this type has not been satisfactory, principally because of instability or creep of the gages or of the materials used to bond them to the diaphragms, and the eventual leakage of water into the chamber. These and other difficulties are similar to those discussed in Article 67 in connection with pressure cells.

Thus far, the best record of performance has been achieved by piezometers containing strain gages of the vibrating-wire type (Article 67). The Maihak gage, embodying this feature, is shown in Fig. 68.4 (Brooker and Lindberg 1965). The porous tip for most installations has consisted of sintered metal saturated with neatsfoot oil. Inasmuch as the air-entry value of these tips is low, on the order of 1.5 lb/in.^{2}, pore-air rather than porewater pressures have probably been measured in such installations whenever the soil was unsaturated and the difference between pore air and porewater pressures was more

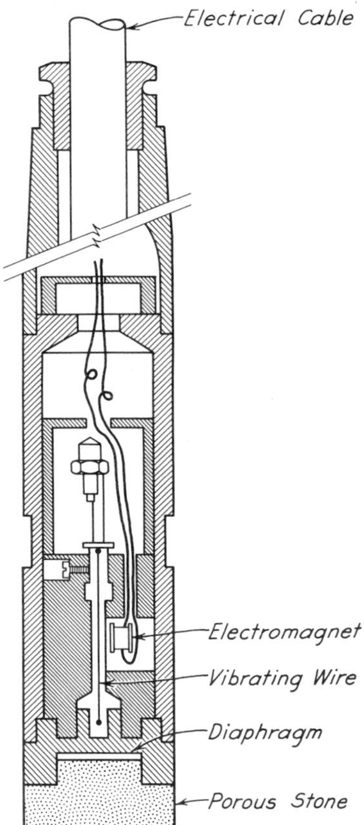

Fig. 68.4. Electrically operated pore-pressure gage of vibrating-wire type (after Brooker and Lindberg 1965).

than a few pounds per square inch. The measurement of positive pore pressures has been satisfactory.

Replacement of the sintered metal tip by a water-saturated porous tip with a high air-entry value should permit correct measurement of negative porewater pressures for short-term installations. However, inasmuch as air can be expected to come out of solution and gradually accumulate in the tip, and since no means of flushing out the air are available, the device may eventually begin to read the pore-air pressure (Bishop et al. 1964).

The same care is required to seal electrically observed piezometers in trenches or drill holes as must be exercised in the installation of closed-system hydraulic devices.

Air-Actuated Piezometers

In the electrically indicating piezometers the porewater pressure acts against and deflects the face of a diaphragm; the deflection of the diaphragm serves as the measure of the porewater pressure. In air-actuated piezometers the diaphragm is subjected on its back side to air pressure that balances and measures the water pressure. The sensitivity of the instrument depends on the magnitude of the deflection imposed on the diaphragm in order to balance the air pressure. The instruments are of two types.

In both types, two air tubes lead to a chamber behind a flexible diaphragm. In piezometers of the *bubbler type,* air is introduced under slowly increasing pressure into the inlet tube. As long as the air pressure is less than the water pressure, the diaphragm keeps the entrance to the outlet tube closed. When the air pressure becomes equal to the water pressure, the diaphragm is displaced slightly and air escapes through the outlet tube from which it is discharged just below the surface of a vessel filled with water. The appearance of the air bubbles indicates that the porewater pressure is equal to the air pressure shown by a gage on the inlet tube (Warlam and Thomas 1965).

In a second type, the outlet and inlet tubes normally communicate with each other behind the diaphragm. When an observation is to be made, air pressure is slowly increased in the inlet tube while the end of the outlet tube is closed. When the pressure in the tubing reaches the porewater pressure, the diaphragm actuates a valve that blocks the flow of air between inlet and outlet lines; the reading of a pressure gage on the outlet line is then equal to the porewater pressure (Wilson 1966).

Air-actuated piezometers avoid the necessity of de-airing water-filled tubes and of protecting the tubing and gages from freezing.

They also avoid electrical circuitry. Their time lag is relatively small, but has not yet been fully evaluated.

Selection of Type of Piezometer

As a rule, the piezometer selected for a particular purpose should be the simplest that will satisfy the requirements. With increasing sophistication of the piezometer or of the measuring system, the probability of malfunction and of eventual failure increases, as does the cost.

The influence of the physical restrictions at the site on the type of piezometer that should be selected has already been mentioned. Not only is the type of instrument of concern, but so also are the relative ease or difficulty of providing an effective seal, and the extent to which the installation may interfere with the construction operations.

In all instances the hydrostatic time lag for the installation deserves careful consideration and may eliminate certain types of piezometers. The order of magnitude of the time required for 90% response of piezometers of several types, located in homogeneous soils, can be obtained from Fig. 68.5. The significance of the hydraulic time lag depends to a considerable extent on the nature of the anticipated fluctuations of the porewater pressure. For example, according to Fig. 68.5, the time lag for 90% response of a Geonor open standpipe in a soil having a coefficient of permeability of 10^{-7} cm/sec is about 5 days. If the purpose of the installation is to determine the porewater pressure in a natural deposit in which fluctuations of the pressure are not likely to be significant, and if the instrument can be left in position for several days, use of the Geonor piezometer would be appropriate. On the other hand, if the intention is to make a detailed survey of piezometric conditions over a wide area by inserting the instrument at a given location, waiting for equilibrium, and then withdrawing the instrument and moving to a new location, a time lag of more than a few minutes would not be tolerable and the instrument would not be suitable. Furthermore, if the water pressure at the point of measurement should be subject to daily fluctuations, as might occur in the reservoir above the dam for a power house, a hydrostatic time lag of three days would completely obscure the real variations of porewater pressure and the observations would have no value whatsoever. To obtain satisfactory results under these conditions, an installation with a time lag of no more than 30 to 60 minutes would be required. According to Fig. 68.5, a closed hydraulic piezometer would be needed.

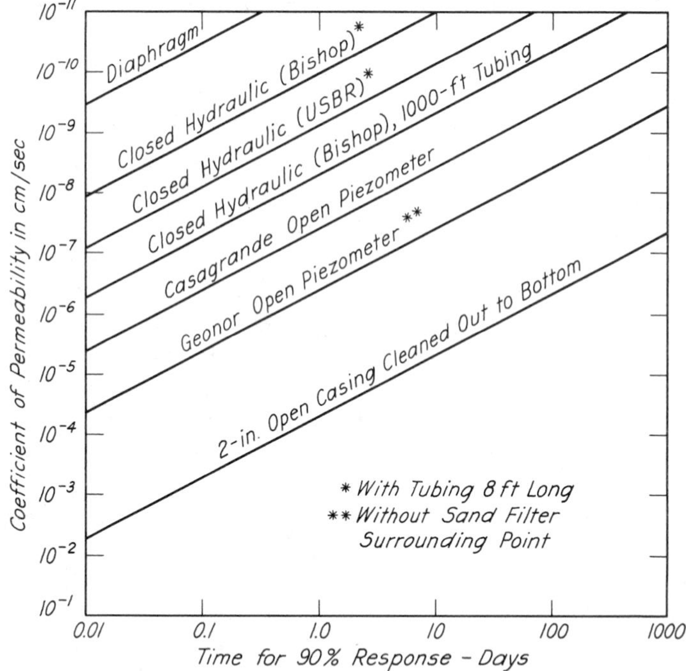

Fig. 68.5. Approximate response times for various types of piezometers (after Hvorslev 1951, Penman 1961, Brooker and Lindberg 1965, and others).

It is apparent that the requirements of each installation must be given careful consideration. A thorough knowledge of the subsurface conditions and of the fundamentals of groundwater flow is necessary for a proper choice of instrument. Moreover, unless the installation is made with the greatest care, and with intelligent consideration of field conditions rather than blind adherence to a set of rules, even the most refined instruments may lead to completely erroneous results or may cease to function altogether. Hence, the installation of piezometers, except in pervious, fairly homogeneous soils, cannot be delegated to the construction forces but must be done or supervised at every step by an experienced man who appreciates the significance of all the requirements for success.

Selected Reading

Detailed instructions concerning the installation and observation of piezometer tips of the USBR type are found in the *Earth Manual* (1963), 1st edition, revised reprint, Denver, pp. 620–672. The reference also describes and provides instructions for the installation of open piezometers of the Casagrande type, as modified for use by the U.S. Bureau of Reclamation, and gives details of an electrical water-level sounding device suitable for use in the Casagrande piezometer.

Details of the Casagrande piezometer and step-by-step procedures for its installation are reported in the Appendix to the paper by A. Casagrande (1949): "Soil mechanics in the design and construction of the Logan airport." *J. Boston Soc. Civil Engrs.,* **36,** No. 2, pp. 192–221, and reprinted in *Contributions to soil mechanics, 1941–1953.* Boston Soc. Civil Engrs., pp. 198–205.

Several useful papers concerning pore pressures and their measurement are contained in *Pore pressure and suction in soils,* London, Butterworths, 1961. This volume represents the proceedings of a conference organized under the same title by the British National Society of the International Society of Soil Mechanics and Foundation Engineering in 1960.

ART. 69 RECORDS OF FIELD OBSERVATIONS

Introduction

The preceding chapters have demonstrated the vital role of field observations in earthwork and foundation engineering. The full benefit of the observations cannot be gained, however, unless the records containing the information are kept in an intelligent and conscientious manner. Quite often, measurements warning of approaching disaster have been ignored because they were recorded in field books or in complicated tables, or were not brought to the attention of an engineer in a position to appreciate their significance. In many instances potentially valuable information has had to be discarded as worthless because of a few omissions that escaped attention at the time the records were filed. Many records are useless because the data are so poorly presented that it is not worthwhile to spend the time required to organize and digest them.

To be useful, records must be kept in such a manner that the data can be understood by any engineer without further inquiry and without the chance for misinterpretation. The following paragraphs summarize the minimum requirements that field records should satisfy.

Key Plan and Boring Information

Every field record should contain a key plan showing the location and elevation of each observation point and its relation to the principal features of the project. Once a letter or a numeral has been assigned to a point, the designation should never be changed, because the record of the change is likely to become lost. The key plan should also contain a full description of the principal bench marks and of the datum to which elevations are referred. In addition, it should show the location of all borings.

A digest of the soil conditions should be shown in a single sheet containing simplified cross sections with verbal descriptions of the principal formations supplemented by representative numerical values of the pertinent soil properties.

Dimensions and Numerical Data

The results of all observations should be assembled in one document in tabular form. Above every column of numerical data should be inscribed the exact meaning of the quantity represented by the numerals. In this connection, nothing should be taken for granted. It should be remembered that the dimensions in which the results of measurements are expressed vary from place to place, and at the same place from time to time.

The numerical data should be complete. In one instance, a record of the observations of water pressure gages was received. It contained the results of the readings, but not the elevation of the gages.

Frequency of Observations

If observations are made too often, money is wasted and the records become too cumbersome. If they are made too seldom, the records contain gaps that are not discovered until too late.

It is generally advantageous to make frequent observations on a given job until the character of the phenomenon under observation becomes known. Thereafter, the frequency can be decreased without reducing the value of the results. Fig. 69.1*a* represents a section through a shipway, and Fig. 69.1*b* represents the effect of filling and emptying an adjacent shipway upon the piezometric level in the permeable stratum located beneath both shipways. When the adjacent shipway was filled for the first time, readings of piezometric level were made every six hours. The results showed that the lag between the change of free water level and that of the piezometric level was unimportant. Therefore, as often as the shipway was again filled and emptied, readings were made only immediately before and after the shipway was filled, immediately before and after it was emptied, and once a week while it remained empty.

The diagram (Fig. 69.1*b*) also disclosed that the changes in water level in an observation well showed an intolerable lag behind the corresponding changes in free water level. Therefore, the observation-

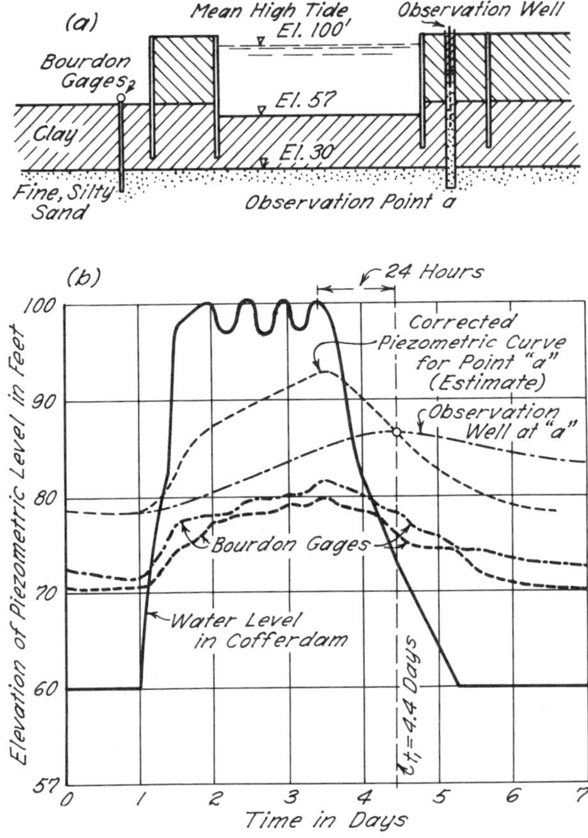

Fig. 69.1. (*a*) **Cross-section through cofferdams in clay overlying sand.** (*b*) **Piezometric levels corresponding to water levels in sand (after FitzHugh et al. 1947).**

well readings were discontinued. This experience illustrates one advantage of a graphical presentation of the data. If the data had merely been tabulated, the defects of the observation-well readings might have escaped attention.

In one instance, the tabulated results of measurements of the deflection of a large wall indicated that, even after three months, the deflection was still increasing by a considerable amount each day. The trend of the increase was obscured by the inevitable errors of observation. Yet, after the readings had been plotted as a function of time in a small-scale diagram, it became evident that the average rate of deflection was decreasing rapidly, and that the wall was approaching a state of equilibrium. It was also evident that the time-deflection curve could have been constructed with sufficient accuracy if readings had been made only once in five days, rather than daily.

In order to obtain the maximum amount of information from a given set of observations, the observer should always be given detailed instructions concerning the type of information expected. At least part of the decision regarding the frequency of the observations should then be left to his judgment. If the general trend of readings can be anticipated, it is advisable to prepare a tentative graph showing the expected results, and to indicate on the graph all the points that should be verified by observation.

Digest of Field Data

The term *digest* indicates the presentation of the contents of a field record in such a form that an engineer who is not familiar with the job can learn with a minimum of effort all the essential findings. A satisfactory digest presents the essential data graphically, to a small scale. The graphs should be accompanied by a key plan and a brief, descriptive text. A large sheet covered with numerical data only discourages and confuses the reader.

To avoid waste of potentially valuable information, the following procedure is advisable. As soon as a set of observations is made, the essential results should be introduced into graphs plotted to such a scale that they cover the entire period of the proposed observations. If the man in charge of the observations is not competent to select the type and scale of these graphs, he should be properly instructed by his superior. If periodic reports are to be sent to headquarters, they should contain not only the complete data but the small-scale graphs.

The benefits to be derived from the graphical procedure for presenting the results of observations are illustrated by Figs. 69.2 to 69.5. Figure 69.2 represents the settlement of a raft foundation corresponding to three different stages of loading. The top of the raft is located at a depth of about 10 feet below the original ground surface. To a depth of about 20 feet below the bottom of the raft, the soil consists of silt and fine silty sand which rests on a thick bed of fairly stiff clay. Settlement observations were made during construction once every few weeks on 34 reference points, uniformly distributed over the raft. If the results of the observations had been assembled in tables, few engineers would have had the patience to analyze the data. Therefore, it was decided to represent them by plotting curves of equal settlement. In Fig. 69.2, the left-hand diagrams represent the curves of equal settlement for three typical stages of loading and the right-hand diagrams the corresponding conditions of loading. During the first stage (*a*), while the load was still very small, the distribution of the settlement seemed to have no relation whatsoever to the distribution of the loads. During the second stage (*b*), at an intermediate state of loading, the settlement of the middle part became more pronounced than that of the two ends. Under full load (*c*), the settlement assumed the character of a gentle, trough-like subsidence.

In order to keep track of the rate of settlement, the settlements of several points were plotted as a function of time. One of these diagrams is shown in Fig. 69.3*a*. Because of the very nonuniform rate of loading, these diagrams merely showed that the settlement increased. However, when the settlement was plotted against the unit load on the raft (Fig. 69.3*b*), it was found that after a small initial settlement, the settlement increased almost linearly with the unit load. Figures 69.2 and 69.3*b* led to the following interpretation. During excavation, the very top of the uppermost stratum became soft, but a small surcharge sufficed to reconsolidate this top layer. Therefore, the settlement under the low, initial load was relatively small and reflected only the local variations in this stratum. This accounts for the erratic character of the left-hand diagram in Fig. 69.2*a*. As the load increased, the additional settlement was due to a slight compression of the soil located beneath the seat of the initial settlement. The regular shape of the settlement trough (Fig. 69.2*c*) indicates that the compressible stratum is statistically fairly homogeneous. The shape of the load-settlement curve (Fig. 69.3*b*) demonstrates that

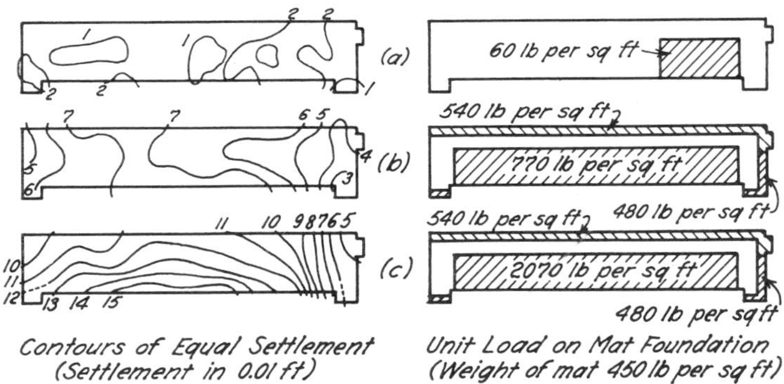

Fig. 69.2. Settlement of raft foundation at three stages of loading.

the ultimate bearing capacity of the subsoil substantially exceeds the maximum load applied; otherwise the load-settlement curve would be concave downward.

Curves corresponding to those in Fig. 69.3 are shown in Fig. 69.4 for a reference point on the base of a large grain elevator during the period when it was filled for the first time. Again, the time-load and time-settlement curves (*a*) indicate only that the settlement increases. The plot of unit load against settlement (Fig. 69.4*b*), in contrast, clearly discloses the imminence of a foundation failure. As a matter of fact, the structure overturned and was completely destroyed, but the accident came as a surprise because the curve shown in the figure was not plotted until after the catastrophe. Had the plot been made, the approaching disaster would have been obvious and the load could have been maintained at a value smaller than the capacity of the bins until the strength of the subsoil became adequate because of consolidation of the underlying clay.

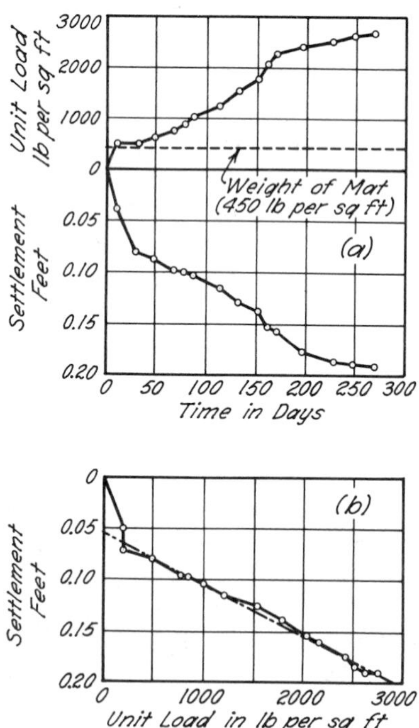

Fig. 69.3. (*a*) Relation between settlement, load and time for point on raft foundation. (*b*) Relation between settlement and load for same point.

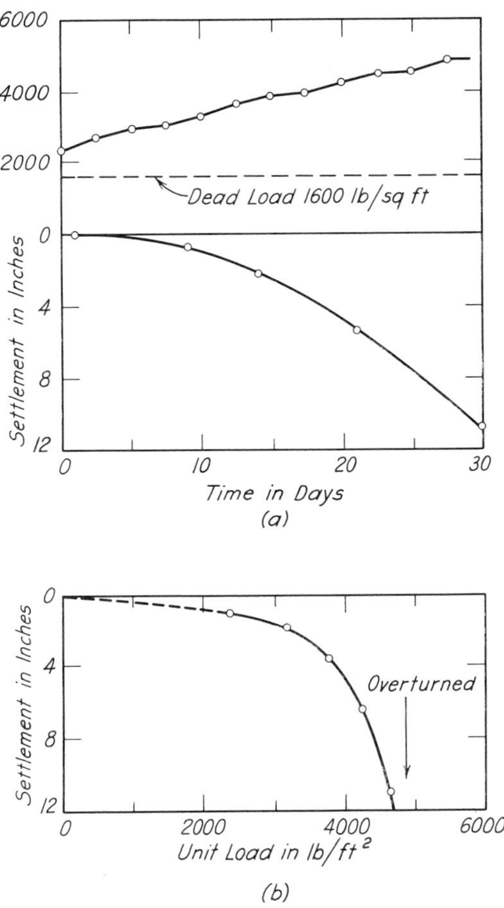

Fig. 69.4. (*a*) Relation between settlement, load and time for point on base of grain elevator above deposit of soft clay. (*b*) Relation between average unit load and settlement of structure at the center of gravity of the loaded area.

Figure 69.5 is a graphical representation of the results of hydrostatic pressure measurements at the base of the clay stratum located below the floors of the shipways shown in Fig. 69.1*a*. The piezometric elevations were determined by means of observation wells capped by Bourdon gages. In Fig. 69.5, the observation points are indicated by small circles. When presented in tables, the results of the observations merely show that the piezometric elevations vary considerably from point to point. However, when the data are used to plot curves of equal piezometric elevation as shown in Fig. 69.5, one sees at a glance

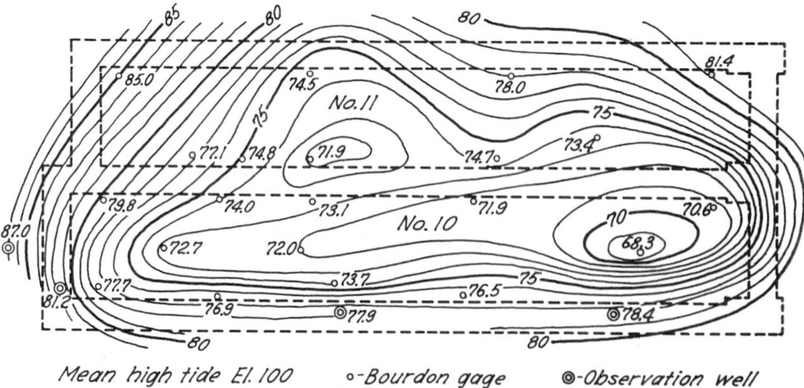

Fig. 69.5. Curves of equal piezometric level for sand stratum below floor of shipways.

the intensity and the distribution of the forces that tend to lift the clay stratum together with the shipway floors.

After the job is finished, no further digest is required because all the essentials are already contained in small-scale graphs such as those shown in Figs. 69.2 to 69.5. To these graphs should be added the key plan, the condensed boring records, a condensed account of all the field tests such as loading or pile driving tests, and a few pages of text containing a summary of all the observations in the field together with a statement of the findings. In their totality, these pieces of information constitute the digest of the field records.

If the data in a condensed record have been obtained from the results of field measurements by a method of computation which involves some assumption, as, for example, the computation of the distribution of earth pressure on the sides of a cut on the basis of measured strut loads, the information upon which the computation was based should appear on the same sheet that shows the results in graphical form.

After the job is finished, the condensed records should be prepared in duplicate. One set should be kept in a file devoted to such records. The other should be filed with the original field data. The second copy should contain all the references required for locating the original data without excessive loss of time.

Every satisfactory professional paper dealing with an individual construction job is essentially an abstract of the digested field records. Therefore, the technique of digesting field records may be learned by studying such professional papers. Several examples are listed at the end of this article.

Selected Reading

The following professional papers are essentially refinements of digests of field records originally prepared for the control of a project during construction. They serve as examples of the technique and value of preparing such digests.

Terzaghi, K. (1942). "Shield tunnels of the Chicago subway," *J. Boston Soc. Civil Engrs.*, **29**, pp. 163–210.

Peck, R. B. (1943). "Earth pressure measurements in open cuts, Chicago subway," *Trans. ASCE,* **108,** pp. 1008–1036.

Terzaghi, K. (1943*a*). "Liner-plate tunnels on the Chicago (Ill.) subway," *Trans. ASCE,* **108,** pp. 970–1007.

FitzHugh, M. M., J. S. Miller, and K. Terzaghi (1947). "Shipways with cellular walls on a marl foundation," *Trans. ASCE,* **112,** pp. 298–324.

Casagrande, A. (1949). "Soil mechanics in the design and construction of the Logan airport," *J. Boston Soc. Civil Engrs.*, **36,** No. 2, pp. 192–221.

Terzaghi, K. and R. B. Peck (1957). "Stabilization of an ore pile by drainage," *Proc. ASCE,* **83,** No. SM1, paper 1144.

Zeevaert, L. (1957). "Foundation design and behavior of Tower Latino Americana in Mexico City," *Géot.*, **7,** No. 3, pp. 115–133.

Terzaghi, K. (1958*b*). "Design and performance of the Sasumua dam," *Proc. Inst. Civil Engrs.*, London, **9,** Apr., pp. 369–394; **11,** Nov., pp. 360–363.

Mansur, C. I. and R. I. Kaufman (1960). "Dewatering the Port Allen lock excavation," *ASCE J. Soil Mech.*, **86,** No. SM6, pp. 35–55.

Terzaghi, K. and T. M. Leps (1960). "Design and performance of Vermilion dam," *Trans. ASCE,* **125,** pp. 63–100.

Klohn, E. J. (1961). "Pile heave and redriving," *ASCE J. Soil Mech.*, **87,** No. SM4, pp. 125–145.

Terzaghi, K. and Y. Lacroix (1964). "Mission Dam. An earth and rockfill dam on a highly compressible foundation," *Géot.*, **14,** pp. 14–50.

REFERENCES

Abbott, M. B. (1960). "One-dimensional consolidation of multi-layered soils," *Géot.*, **10**, pp. 151–165.

Agerschou, H. A. (1962). "Analysis of the Engineering News pile formula," *ASCE J. Soil Mech.*, **88**, No. SM5, pp. 1–11.

Akagi, T. (1960). *Effect of desiccation and ring friction on the apparent preconsolidation load of clay.* M.S. thesis, Univ. of Illinois, Urbana, 114 pp.

Aldrich, H. P. (1965). "Precompression for support of shallow foundations," *ASCE J. Soil Mech.*, **91**, No. SM2, pp. 5–20.

Ambraseys, N. N. (1960). "On the seismic behavior of earth dams," *Proc. 2nd Int. Conf. Earthquake Eng.*, Tokyo, **1**, pp. 331–358.

Andresen, A. and N. E. Simons (1960). "Norwegian triaxial equipment and technique," *Proc. ASCE Research Conf. on Shear Strength of Cohesive Soils,* pp. 695–709.

AREA (1933). "Use of portable cribbing in place of rigid retaining walls and the utility of the different kinds of cribbing," Committee Report, *Proc. Am. Rwy. Eng. Assn.*, **34**, pp. 139–148.

AREA (1955). "Soil engineering in railroad construction," *Proc. Am. Rwy. Eng. Assn.*, **56**, pp. 694–702.

ASCE (1961). "Sonic pile-driver shows great promise," *Civ. Eng.*, **31**, No. 12, Dec., p. 52.

Atterberg, A. (1908). "Studien auf dem Gebiet der Bodenkunde" (Studies in the field of soil science), *Landw. Versuchsanstalt,* **69.**

Atterberg, A. (1911). "Über die physikalishe Bodenuntersuchung und über die Plastizität der Tone" (On the investigation of the physical properties of soils and on the plasticity of clays), *Int. Mitt. für Bodenkunde,* **1**, pp. 10–43.

Atterberg, A. (1916). "Die Klassifikation der humusfrien und der humusarmen Mineralböden Schwedens nach den Konsistenzverhältnessen derselben." (The classification of Swedish mineral soils with little or no humus content, according to their consistency limits), *Int. Mitt. für Bodenkunde,* **6**, pp. 27–37.

Babbitt, H. E. and D. H. Caldwell (1948). "The free surface around, and interference between, gravity wells," *Univ. of Illinois Eng. Exp. Sta. Bull. 374*, 60 pp.

Baker, B. (1881). "The actual lateral pressure of earthwork," *Min. Proc. Inst. Civil Engrs.,* London, **65**, pp. 140–186; Discussions pp. 187–241.

Barberis, M. C. (1935). "Recent examples of foundations of quay walls resting on poor subsoil, studies, results obtained," *16th Int. Congr. Navigation,* Brussels, 2nd section, 3rd communication.

Barentsen, P. (1936). "Short description of a field-testing method with cone-shaped sounding apparatus." *Proc. 1st Int. Conf. Soil Mech.,* Cambridge, Mass., **1**, pp. 7–10.

Barkan, D. D. (1962). *Dynamics of bases and foundations.* New York, McGraw-Hill, 434 pp.

Barron, R. A. (1948). "Consolidation of fine-grained soils by drain wells," *Trans. ASCE,* **113**, pp. 718–742.

Baumann, P. (1942). "Design and construction of San Gabriel dam No. 1," *Trans. ASCE,* **107**, pp. 1595–1634.

Baumann, P. (1960). "Cogswell and San Gabriel dams," *Trans. ASCE,* **125**, Part II, pp. 29–57.

Belcher, D. J. (1945). "The engineering significance of soil patterns," *Photogrammetric Engineering,* **11**, No. 2, pp. 115–148.

Berbower, R. F. (1959). "Subsidence problem in the Long Beach Harbor District," *ASCE J. Waterways and Harbors Div.,* **85**, No. WW2, pp. 81–97.

Bertram, G. E. (1963). "Rockfill compaction by vibratory rollers," *Proc. 2nd Panamerican Conf. on Soil Mech. and Found. Eng.,* Brazil, **1**, pp. 441–455.

Beskow, G. (1935). "Tjälbildningen och Tjällyftningen med Särskild Hänsyn till Vägar och Järnvägar" (Soil freezing and frost heaving with special application to roads and railroads), *Sveriges Geologiska Undersokning,* Stockholm, Series Cv, No. 375, 242 pp.

Biot, M. A. (1941). "General theory of three-dimensional consolidation," *J. Appl. Phys.,* **12**, pp. 155–164.

Bishop, A. W. (1948). "A new sampling tool for use in cohesionless sands below ground water level," *Géot.,* **1**, No. 2, pp. 125–131.

Bishop, A. W., (1954). "The use of pore-pressure coefficients in practice," *Géot.,* **4**, pp. 148–152.

Bishop, A. W. (1955). "The use of the slip circle in the stability analysis of slopes," *Géot.,* **5**, pp. 7–17.

Bishop, A. W. (1960). "The principle of effective stress," *Norwegian Geot. Inst. Publ. 32,* Oslo, pp. 1–5.

Bishop, A. W. (1966). "The strength of soils as engineering materials," *Géot.,* **16**, pp. 91–128.

Bishop, A. W., I. Alpan, G. E. Blight, and I. B. Donald (1960). "Factors controlling the strength of partly saturated cohesive soils," *Proc. ASCE Research Conf. on Shear Strength of Cohesive Soils,* pp. 503–532.

Bishop, A. W. and L. Bjerrum (1960). "The relevance of the triaxial test to the solution of stability problems," *Proc. ASCE Research Conf. on Shear Strength of Cohesive Soils,* pp. 437–501.

Bishop, A. W. and D. J. Henkel (1962). *The measurement of soil properties in the triaxial test.* 2nd ed., London, Edward Arnold, 228 pp.

Bishop, A. W., M. F. Kennard, and A. D. M. Penman (1960). "Pore-pressure observations at Selset dam," *Proc. Conf. on Pore Pressure and Suction in Soils.* London, Butterworths, pp. 91–102.

Bishop, A. W., M. F. Kennard, and P. R. Vaughan (1964). "Developments in the measurement and interpretation of pore pressure in earth dams," *Trans. 8th Congr. Large Dams,* Edinburgh, **2**, pp. 47–72.

Bishop, A. W. and N. R. Morgenstern (1960). "Stability coefficients for earth slopes," *Géot.,* **10**, pp. 129–150.

Bjerrum, L. (1954). "Geotechnical properties of Norwegian marine clays," *Géot.,* **4**, pp. 49–69.

Bjerrum, L. (1955). "Stability of natural slopes in quick clay," *Géot.,* **5**, No. 1, pp. 101–119.

Bjerrum, L. (1960). "Some notes on Terzaghi's method of working," *From theory to practice in soil mechanics,* New York, John Wiley and Sons, pp. 22–25.

Bjerrum, L. (1966). *Mechanism of progressive failure in slopes of overconsolidated plastic clays and clay shales.* Third Terzaghi Lecture presented before the ASCE, Miami, Feb. 1, 1966.

Bjerrum, L. and A. Eggestad (1963). "Interpretation of loading tests on sand," *Proc. European Conf. on Soil Mech. and Found. Eng.,* Wiesbaden, **1,** pp. 199–203.

Bjerrum. L. and O. Eide (1956). "Stability of strutted excavations in clay." *Géot.,* **6,** pp. 32–47.

Bjerrum, L., T. C. Kenney, and B. Kjaernsli (1965). "Measuring instruments for strutted excavations," *ASCE J. Soil Mech.,* **91,** No. SM1, pp. 111–141.

Bjerrum, L., S. Kringstad, and O. Kummeneje (1961). "The shear strength of a fine sand," *Proc. 5th Int. Conf. Soil Mech.,* Paris, **1,** pp. 29–37.

Bjerrum, L. and N. E. Simons (1960). "Comparison of shear strength characteristics of normally consolidated clays," *Proc. ASCE Research Conf. on Shear Strength of Cohesive Soils,* pp. 711–726.

Bleifuss, D. J. and J. P. Hawke (1960). "Design and construction problems," *Trans. ASCE,* **125,** Part II, pp. 275–294.

Bligh, W. G. (1910). "Dams, barrages, and wiers on porous foundations," *Eng. News,* **64,** pp. 708–710.

Bogdanović, L., D. Milović, and Z. Certić (1963). "Comparison of the calculated and measured settlements of buildings in New Belgrade," *Proc. European Conf. on Soil Mech. and Found. Eng.,* **1,** pp. 205–213.

Bolt, G. H. (1956). "Physico-chemical analysis of the compressibility of pure clays," *Géot.,* **6,** pp. 86–93.

Boreli, M. (1955). "Free-surface flow toward partially penetrating wells," *Trans. American Geophysical Union,* **36,** No. 4, pp. 664–672.

Bozozuk, M. (1962). "Soil shrinkage damages shallow foundations at Ottawa, Canada," *Eng. J. Canada,* **45,** pp. 33–37.

Brinkhorst, W. H. (1936). "Settlement of soil surface around foundation pit," *Proc. 1st Int. Conf. Soil Mech.,* Cambridge, Mass., **1,** pp. 115–119.

Brooker, E. W. and D. A. Lindberg (1965). "Field measurement of pore pressure in high plasticity soils," *Proc. Int. Research and Engineering Conf. on Expansive Clay Soils,* **2,** Texas A. and M. Univ.

Brown, F. S. (1941). "Foundation investigation for the Franklin Falls dam," *J. Boston Soc. Civil Engrs.,* **28,** pp. 126–143.

Bruggen, J. P. v. (1936). "Sampling and testing undisturbed sands from boreholes," *Proc. 1st Int. Conf. Soil Mech.,* Cambridge, Mass., **1,** pp. 144–160.

Bruns, T. C. (1941). "Don't hit timber piles too hard," *Civ. Eng.,* **11,** pp. 726–728.

Buchanan, S. J. (1938). "Levees in the lower Mississippi valley," *Trans. ASCE,* **103,** pp. 1378–1395; Discussions pp. 1449–1502.

Buisman, A. K. (1943). *Grondmechanica* (Soil mechanics). 2nd ed., Delft, 281 pp.

Burke, H. H. (1960). "Garrison dam test tunnel: investigation and construction," *Trans. ASCE,* **125,** pp. 230–267.

Burmister, D. M. (1956). "Stress and displacement characteristics of a two-layered rigid base soil system: influence diagrams and practical applications," *Proc. Hwy. Res. Board,* **35,** pp. 773–814.

Cadling, L. and S. Odenstad (1950). "The vane borer," *Proc. Swedish Geot. Inst.* No. 2, 88 pp.

Cambefort, H. (1955). *Forages et sondages* (Borings and soundings). Paris, Eyrolles, 396 pp.

Caquot, A. and J. Kerisel (1948). *Tables for the calculation of passive pressure, active pressure and bearing capacity of foundations.* Transl. from the French by Maurice A. Bec., Paris, Gauthier-Villars, 120 pp.

Carlson, L. (1948). "Determination in situ of the shear strength of undisturbed clay by means of a rotating auger," *Proc. 2nd Int. Conf. Soil Mech.,* Rotterdam, **1**, pp. 265–270.

Carlson, R. W. and D. Pirtz (1952). "Development of a device for the direct measurement of compressive stress," *J. American Concrete Inst.,* **49**, pp. 201–215.

Carpenter, J. C. and E. S. Barber (1953). "Vertical sand drains for stabilization of muck-peat soils," *Proc. ASCE,* **79**, Separate No. 351, **17** pp.

Carslaw, H. S. and J. C. Jaeger (1959). *Conduction of heat in solids.* Oxford, Clarendon Press, 2nd ed., 510 pp.

Cary, A. S., B. H. Walter, and H. T. Harstad (1943). "Permeability of Mud Mountain core material," *Trans. ASCE,* **108**, pp. 719–728; Discussions pp. 729–737.

Casagrande, A. (1931). "Discussion: A new theory of frost heaving," *Proc. Hwy. Res. Board,* **11**, pp. 168–172.

Casagrande, A. (1932*a*). "Research on the Atterberg limits of soils," *Public Roads,* **13**, pp. 121–136.

Casagrande, A. (1932*b*). "The structure of clay and its importance in foundation engineering," *J. Boston Soc. Civil Engrs.,* **19**, No. 4, p. 168.

Casagrande, A. (1934). "Discussion: The shearing resistance of soils," *J. Boston Soc. Civil Engrs.,* **21**, No. 3, pp. 276–283.

Casagrande, A. (1935*a*). "Discussion: Security from under-seepage masonry dams on earth foundations," *Trans. ASCE,* **100**, pp. 1289–1294.

Casagrande, A. (1935*b*). "Seepage through dams," *J. New England Water Works Assn.,* **51**, No. 2, pp. 131–172.

Casagrande, A. (1936*a*). "Characteristics of cohesionless soils affecting the stability of slopes and earth fills," *J. Boston Soc. Civil Engrs.,* **23**, No. 1, pp. 13–32.

Casagrande, A. (1936*b*). "The determination of the pre-consolidation load and its practical significance," *Proc. 1st Int. Conf. Soil Mech.,* Cambridge, Mass., **3**, pp. 60–64.

Casagrande, A. (1947). "The pile foundation for the new John Hancock building in Boston," *J. Boston Soc. Civil Engrs.,* **34**, pp. 297–315.

Casagrande, A. (1948). "Classification and identification of soils," *Trans. ASCE,* **113**, pp. 901–992.

Casagrande, A. (1949). "Soil mechanics in the design and construction of the Logan airport," *J. Boston Soc. Civil Engrs.,* **36**, No. 2, pp. 192–221.

Casagrande, A. (1958). *Piezometers for pore pressure measurements in clay.* Mimeographed. Harvard Univ., Div. of Engineering and Applied Physics, Pierce Hall, Cambridge, Mass., 9 pp.

Casagrande, A. (1960). "An unsolved problem of embankment stability on soft ground," *Proc. 1st Panamerican Conf. Soil Mech. and Found. Eng.,* Mexico, **2**, pp. 721–746.

Casagrande, A. (1965). "Role of the 'calculated risk' in earthwork and foundation engineering," *ASCE J. Soil Mech.,* **91**, No. SM4, July, pp. 1–40.

Casagrande, A. and R. E. Fadum (1940). "Notes on soil testing for engineering purposes," *Harvard Univ. Grad. School of Engineering Publ. 268,* 74 pp.

Casagrande, A. and R. C. Hirschfeld (1960). "Stress-deformation and strength characteristics of a clay compacted to a constant dry unit weight," *Proc. ASCE Research Conf. on Shear Strength of Cohesive Soils,* pp. 359–417.

Casagrande, A. and S. D. Wilson (1951). "Effect of rate of loading on strength of clays and shales at constant water content," *Géot.,* **2,** pp. 251–263.

Casagrande, L. (1936). "Settlement observations on structures of the 'Reichsautobahnen'," *Proc. 1st Int. Conf. Soil Mech.,* Cambridge, Mass., **3,** pp. 104–106.

Casagrande, L. (1949). "Electro-osmosis in soils," *Géot.,* **1,** No. 3, pp. 159–177.

Casagrande, L. (1962). "Electro-osmosis and related phenomena," *Revista Ingenieria,* Mexico, **32,** No. 2, pp. 1–62 (Spanish and English text).

Cedergren, H. R. (1967). *Seepage, drainage, and flow nets.* New York, John Wiley and Sons, 489 pp.

Chellis, R. D. (1961). *Pile foundations,* 2nd ed., New York, McGraw-Hill, 704 pp.

Chen, L. S. (1948). "An investigation of stress-strain and strength characteristics of cohesionless soils by triaxial compression tests," *Proc. 2nd Int. Conf. Soil Mech.,* Rotterdam, **5,** pp. 35–43.

Clarke, N. W. B. and J. B. Watson (1936). "Settlement records and loading data for various buildings erected by the Public Works Department, Municipal Council, Shanghai," *Proc. 1st Int. Conf. Soil Mech.,* Cambridge, Mass., **2,** pp. 174–185.

Clevenger, W. A. (1958). "Experiences with loess as a foundation material," *Trans. ASCE,* **123,** pp. 151–169.

Close, U. and E. McCormick (1922). "Where the mountains walked," *Nat. Geog. Mag.,* **41,** pp. 445–464.

Collins, J. J. (1935). "New type sea wall built for subsiding lake shore in Venezuela," *Eng. News-Record,* **114,** No. 12, pp. 405–408.

Cooling, L. F. (1962). "Field measurements in soil mechanics," *Géot.,* **12,** No. 2, pp. 77–103.

Cooling, L. F. and H. Q. Golder (1942). "The analysis of the failure of an earth dam during construction," *J. Inst. Civil Engrs.,* London, **20,** No. 1, pp. 38–55; Discussions, Supplement to No. 8, pp. 289–304.

Corps of Engineers (1960). "Stability of earth and rockfill dams," *Manual EM 1110-2-1902,* 27 Dec., 67 pp.

Costes, N. C. (1956). "Factors affecting vertical loads on underground ducts due to arching," *Hwy. Res. Board Bull. 125,* pp. 12–57.

Coulomb, C. A. (1776). "Essai sur une Application des Règles des Maximis et Minimis à quelques Problèmes de Statique Relatifs à l'Architecture" (An attempt to apply the rules of maxima and minima to several problems of stability related to architecture). *Mém. Acad. Roy. des Sciences,* Paris, **3,** p. 38.

Crandell, F. J. (1949). "Ground vibration due to blasting and its effect upon structures," *J. Boston Soc. Civil Engrs.,* **36,** pp. 245–268.

Culmann, C. (1875). *Die graphische Statik* (Graphic statics). Zurich, Meyer and Zeller, 644 pp.

Cummings, A. E. (1940). "Dynamic pile driving formulas," *J. Boston Soc. Civil Engrs.,* **27,** pp. 6–27.

Cummings, A. E., G. O. Kerkhoff, and R. B. Peck (1950). "Effect of driving piles into soft clay," *Trans. ASCE,* **115,** pp. 275–285.

D'Appolonia, E. (1953). "Loose sands—their compaction by vibroflotation," *ASTM Special Tech. Publ. 156,* pp. 138–154.

D'Appolonia, E. and J. A. Hribar (1963). "Load transfer in a step-taper pile," *ASCE J. Soil Mech.,* **89,** No. SM6, pp. 57–77.

D'Appolonia, E. and J. P. Romualdi (1963). "Load transfer in end-bearing steel H-piles," *ASCE J. Soil Mech.,* **89,** No. SM2, pp. 1–25.

Darcy, H. (1856). *Les fontaines publiques de la ville de Dijon* (The water supply of the city of Dijon). Dalmont, Paris, 674 pp.

Dawson, R. F. (1959). "Modern practices used in the design of foundations for structures on expansive soils," *Colo. School of Mines Quarterly,* **54,** No. 4, pp. 67–87.

De Beer, E. and A. Martens (1957). "A method of computation of an upper limit for the influence of heterogeneity of sand layers in the settlement of bridges," *Proc. 4th Int. Conf. Soil Mech.,* London, **1,** pp. 275–282.

Deere, D. U. (1957). "Seepage and stability problems in deep cuts in residual soils, Charlotte, N.C.," *Proc. Am. Rwy. Eng. Assn.,* **58,** pp. 738–745.

DiBiagio, E. and L. Bjerrum (1957). "Earth pressure measurements in a trench excavated in stiff marine clay," *Proc. 4th Int. Conf. Soil Mech.,* London, **2,** pp. 196–202.

Dupuit, J. (1863). *Études théoriques et pratiques sur le mouvement des eaux dans les canaux découverts et à travers les terrains perméables* (Theoretical and experimental studies of the flow of water in open channels and through permeable ground). 2nd ed., Paris, Dunod, 304 pp.

Eggestad, A. (1963). "Deformation measurements below a model footing on the surface of dry sand," *Proc. European Conf. on Soil Mech. and Found. Eng.,* Wiesbaden, **1,** pp. 233–239.

Endo, M. (1963). "Earth pressure in the excavation work of alluvial clay stratum," *Proc. Int. Conf. Soil Mech. and Found. Eng.,* Budapest, pp. 21–46.

ENR (1929). "Reconstruction of Lafayette dam advised," *Eng. News-Record,* **102,** pp. 190–192.

ENR (1937). "Foundation of earth dam fails," *Eng. News-Record,* **119,** p. 532.

ENR (1941). "Foundation failure causes slump in big dike at Hartford, Conn.," *Eng. News-Record,* **127,** p. 142.

Fadum, R. E. (1941). *Observations and analysis of building settlements in Boston.* Sc.D. thesis, Harvard Univ.

Fadum, R. E. (1948). "Concerning the physical properties of clays," *Proc. 2nd Int. Conf. Soil Mech.,* Rotterdam, **1,** pp. 253–254.

Fahlquist, F. E. (1941). "New methods and technique in subsurface explorations," *J. Boston Soc. Civil Engrs.,* **28,** No. 2, pp. 144–160.

Feld, J. (1943). "Discussion: Timber friction pile foundations," *Trans. ASCE,* **108,** pp. 143–144.

Fellenius, W. (1927). *Erdstatische Berechnungen* (Calculation of stability of slopes). Berlin, (Revised ed. 1939, 48 pp.)

Fellenius, W., F. Blidberg, L.v. Post, and J. Olsson (1922). *Statens Järnvägars Geotekniska Kommission 1914–22, Slutbetänkande* (State Railways Geotechnical Committee 1914–22, Final Report). Stockholm, 180 pp.

FitzHugh, M. M., J. S. Miller, and K. Terzaghi (1947). "Shipways with cellular walls on a marl foundation," *Trans. ASCE,* **112,** pp. 298–324.

Flaate, K. S. (1964). "An investigation of the validity of three pile-driving formulae in cohesionless material," *Norwegian Geot. Inst. Publ. 56,* pp. 1–12.

Flaate, K. S. (1966). *Stresses and movements in connection with braced cuts in sand and clay.* Ph.D. thesis, Univ. of Illinois, Urbana, 264 pp.

Forchheimer, P. (1917). "Zur Grundwasserbewegung nach isothermischen Kurvenscharen" (Concerning groundwater movement in accordance with isothermal families of curves), *Sitzber. kais. Akad. d. Wiss., Wein, Abt. IIa,* **126,** pp. 409–440.

Fucik, E. M. and R. F. Edbrooke (1960). "Ambuklao rockfill dam, design and construction," *Trans. ASCE,* **125,** Part 1, pp. 1207–1227.

Fülscher, J. (1897–1899). "Der Bau des Kaiser Wilhelm-Kanals" (Construction of the Kaiser Wilhelm canals), *Zeitschrift für Bauwesen,* **47,** 1897, column 117–142, 275–304, 405–454, 525–586; **48,** 1898, column 41–82, 205–282, 441–490, 693–752; **49,** 1899, column 99–126, 269–304, 425–464, 621–675.

Gaunt, G. C. (1962). "Marina City—foundations," *Civ. Eng.,* **32,** Dec., pp. 61–63.

Geuze, E. C. W. A. (1948). "Critical density of some Dutch sands," *Proc. 2nd Int. Conf. Soil Mech.,* Rotterdam, **3,** pp. 125–130.

Gibson, R. E. and P. Lumb (1953). "Numerical solution of some problems in the consolidation of clay," *Proc. Inst. Civil Engrs.,* London, **2,** Part 1, pp. 182–198.

Gibson, R. E. and J. McNamee (1963). "A three-dimensional problem of the consolidation of a semi-infinite clay stratum," *Quart. J. Mech. and Appl. Math.,* **16,** Part 1, pp. 115–127.

Gilboy, G. (1928). "The compressibility of sand-mica mixtures," *Proc. ASCE,* **54,** pp. 555–568.

Glanville, W. H., G. Grime, E. Fox, and W. W. Davies (1938). "An investigation of the stresses in reinforced concrete piles during driving," *Dept. Sci. Ind. Research, Bldg. Research Sta., England,* Tech. Paper 20, 111 pp.

Glossop, R. (1960). "The invention and development of injection processes," Part 1: 1802–1850, *Géot.,* **10,** No. 3, pp. 91–100; Part 2: 1850–1960, *Géot.,* **11,** No. 4, pp. 255–279.

Glossop, R. and A. W. Skemptom (1945). "Particle-size in silts and sands," *J. Inst. Civil Engrs.,* London, Paper 5492, Dec. 1945, pp. 81–105.

Gnaedinger, J. P. (1961). "Grouting to prevent vibration of machinery foundations," *ASCE J. Soil Mech.,* **87,** No. SM2, pp. 43–54.

Godskesen, O. (1936). "Investigation of the bearing-power of the subsoil (especially moraine) with 25 × 25-mm pointed drill without samples," *Proc. 1st Int. Conf. Soil Mech.,* Cambridge, Mass., **1,** pp. 311–314.

Goldbeck, A. T. and E. B. Smith (1916). "An apparatus for determining soil pressures," *Proc. ASTM,* **16,** Part 2, pp. 309–319.

Golder, H. Q. (1948). "Measurement of pressure in timbering of a trench in clay," *Proc. 2nd Int. Conf. Soil Mech.,* Rotterdam, **2,** pp. 76–81.

Golder, H. Q. (1965). "State-of-the-art of floating foundations," *ASCE J. Soil Mech.,* **91,** No. SM2, pp. 81–88.

Golder, H. Q. and G. C. Willeumier (1964). "Design of the main foundations of the Port Mann bridge," *Eng. J. Canada,* **47,** No. 8, pp. 22–29.

Gottstein, E. v. (1936). "Two examples concerning underground sliding caused by construction of embankments and static investigations on the effectiveness of measures provided to assure their stability," *Proc. 1st Int. Conf. Soil Mech.,* Cambridge, Mass., **3,** pp. 122–128.

Gould, J. P. (1960). "A study of shear failure in certain Tertiary marine sediments," *Proc. ASCE Research Conf. on Shear Strength of Cohesive Soils*, pp. 615–641.

Graftio, H. (1936). "Some features in connection with the foundation of Svir 3 hydro-electric power development," *Proc. 1st Int. Conf. Soil Mech.*, Cambridge, Mass., **1**, pp. 284–290.

Gray, H. (1945). "Simultaneous consolidation of contiguous layers of unlike compressible soils," *Trans. ASCE*, **110**, pp. 1327–1344.

Grim, R. E. (1953). *Clay mineralogy.* New York, McGraw-Hill, 384 pp.

Guelton, M., P. Baldy, and C. Magne (1961). "La Barrage de Serre-Ponçon, Conception d'Ensemble" (Serre-Ponçon dam, concept of the project), *Travaux*, **45**, pp. 298–315.

Haefeli, R. (1950). "Investigation and measurements of the shear strengths of saturated cohesive soils," *Géot.*, **2**, No. 3, pp. 186–208.

Hall, C. E. (1962). "Compacting a dam foundation by blasting," *ASCE J. Soil Mech.*, **88**, No. SM3, pp. 33–51.

Hansen, B. (1965). *A theory of plasticity for ideal frictionless materials.* Copenhagen, Teknisk Forlag, 471 pp.

Hansen, J. Brinch (1961). "A general formula for bearing capacity," *Ingeniøren*, **5**, pp. 38–46; also Bull. 11, Danish Geotechnical Inst.

Harr, M. E. (1962). *Groundwater and seepage.* New York, McGraw-Hill. 315 pp.

Hazen, A. (1892). "Physical properties of sands and gravels with reference to their use in filtration," *Rept. Mass. State Board of Health*, p. 539.

Hendron, A. J. (1963). *The behavior of sand in one-dimensional compression.* Ph.D. thesis, Univ. of Illinois, Urbana, 283 pp.

Henkel, D. J. (1960). "The shear strength of saturated remolded clays," *Proc. ASCE Research Conf. on Shear Strength of Cohesive Soils*, pp. 533–554.

Hertwig, A., G. Früh and H. Lorenz (1933). "Die Ermittlung der für das Bauwesen wichtigsten Eigenschaften des Bodens durch erzwungene Schwingungen" (The determination by means of forced vibrations of soil properties of special importance for construction work). *Degebo. Veröffentlichung*, **1**, 45 pp.

Hetényi, M. (1946). *Beams on elastic foundation.* Ann Arbor, Univ. of Michigan Press, 255 pp.

Hilf, J. W. (1948). "Estimating construction pore pressures in rolled earth dams," *Proc. 2nd Int. Conf. Soil Mech.*, Rotterdam, **3**, pp. 234–240.

Hirashima, K. B. (1948). "Highway experience with thixotropic volcanic clay," *Proc. Hwy. Res. Board*, **28**, pp. 481–494.

Holmsen, P. (1953). "Landslips in Norwegian quick-clays," *Géot.*, **3**, pp. 187–200.

Holtz, W. G. and H. J. Gibbs (1956*a*). "Engineering properties of expansive clays," *Trans. ASCE*, **121**, pp. 641–677.

Holtz, W. G. and H. J. Gibbs (1956*b*). "Triaxial shear tests on pervious gravelly soils," *ASCE J. Soil Mech.*, **82**, No. SM1, Paper No. 867, 9 pp.

Horn, H. M. and D. U. Deere (1963). "Frictional characteristics of minerals," *Géot.*, **12**, pp. 319–335.

Hough, B. K., Jr. (1938). "Stability of embankment foundations," *Trans. ASCE*, **103**, pp. 1414–1431.

HRB (1958). "Landslides and engineering practice," Committee on landslide investigations. *Hwy. Res. Board Special Rept. 29*, 232 pp.

Hubbard, P. G. (1955). "Field measurement of bridge-pier scour," *Proc. Hwy. Res. Board*, **34**, pp. 184–188.

Huber, W. G. (1960). "Kenney and Cheakamus dams," *Trans. ASCE,* **125,** Part II, pp. 255–265.

Humphreys, J. D. (1962). "The measurement of loads on timber supports in a deep trench," *Géot.,* **12,** pp. 44–54.

Huntington, W. C. (1957). *Earth pressures and retaining walls.* New York, John Wiley and Sons, 534 pp.

Hvorslev, M. J. (1937). "Über die Festigkeitseigenschaften gestörter bindiger Böden" (On the strength properties of remolded cohesive soils), *Danmarks Naturvidenskabelige Samfund, Ingeniørvidenskabelige Skrifter,* Series A, No. 45, Copenhagen, 159 pp.

Hvorslev, M. J. (1948). *Subsurface exploration and sampling of soils for civil engineering purposes.* Waterways Exp. Sta., Vicksburg, Miss., 465 pp.

Hvorslev, M. J. (1951). "Time lag and soil permeability in ground water measurements," *Corps of Engrs. Waterways Exp. Sta., Vicksburg, Miss., Bull.* 36, 50 pp.

Hvorslev, M. J. (1960). "Physical components of the shear strength of saturated clays," *Proc. ASCE Research Conf. on Shear Strength of Cohesive Soils,* pp. 169–273.

IISEE (1965). "The Niigata earthquake 16 June, 1964, and resulting damage to reinforced concrete buildings," *IISEE Earthquake Rept. No. 1, Int. Inst. Seismology and Earthquake Eng.,* Tokyo, 62 pp.

Ireland, H. O. (1955). *Settlements due to foundation construction in Chicago, 1900–1950.* Ph.D. thesis, Univ. of Illinois, 128 pp.

Ishihara, K. and Y. Yuasa (1963). "Earth pressure measurements in subway construction," *Proc. 2nd Asian Regional Conf. on Soil Mech. and Found. Eng.,* Tokyo, pp. 337–343.

Jacobus, W. W., Jr. (1963). "Hydro-Quebec's big, beautiful Manicouagan 5 hides in the bush," *Eng. News-Record,* **171,** Oct. 24, pp. 38–45.

Jacoby, H. S. and R. P. Davis (1941). *Foundations of Bridges and Buildings.* New York, McGraw-Hill, 3rd ed., 523 pp.

Janbu, N. (1953). "Une analyse énergétique du battage des pieux à l'aide de paramètres sans dimension" (An energy analysis of pile driving with the use of dimensionless parameters), *Ann. Inst. Tech. du Bâtiment et des Travaux Publics,* Nos. 63–64. *Norwegian Geot. Inst. Publ. 3.*

Janbu, N. (1954*a*). "Application of composite slip surfaces for stability analysis," *Proc. European Conf. on Stability of Earth Slopes,* Sweden, **3,** pp. 43–49.

Janbu, N. (1954b). "Stability analysis of slopes with dimensionless parameters," *Harvard Soil Mech. Series No. 46,* 81 pp.

Janbu, N., L. Bjerrum, and B. Kjaernsli (1956). "Veiledning ved løsning av fundamenteringsoppgaver" (Soil mechanics applied to some engineering problems), in Norwegian with English summary, *Norwegian Geot. Inst. Publ. 16,* 93 pp.

Jennings, J. E. (1953). "The heaving of buildings on desiccated clay," *Proc. 3rd Int. Conf. Soil Mech.,* Zurich, **1,** pp. 390–396.

Jimenez-Quiñones, P. (1963). *Compaction characteristics of tropically weathered soils.* Ph.D. thesis, Univ. of Illinois, 135 pp.

Johnson, A. W. and J. R. Sallberg (1962). "Factors influencing compaction test results," *Hwy. Res. Board Bull. 319,* 148 pp.

Johnson, H. L. (1940). "Improved sampler and sampling technique for cohesionless materials," *Civ. Eng.,* **10,** pp. 346–348.

Jurgensen, L. (1934). "The application of elasticity and plasticity to foundation problems," *J. Boston Soc. Civil Engrs.*, **21**, pp. 206–241.

Kaufman, R. I. and W. C. Sherman, Jr. (1964). "Engineering measurements on Port Allen Lock," *ASCE J. Soil Mech.*, **90**, No. SM5, pp. 221–247.

Kerisel, J. (1964). "Deep foundations basic experimental facts," *Proc. Deep Foundations Conference*, Mexico, **1**, pp. 5–44.

Kezdi, A. (1965). "General report on deep foundations," *Proc. 6th Int. Conf. Soil Mech.*, Montreal, **3**, pp. 256–264.

King, F. H. (1899). "Principles and conditions of the movements of ground water," *U.S. Geol. Soc. 19th Ann. Rept.*, Part 2, pp. 59–294.

Kjaernsli, B. and N. Simons (1962). "Stability investigations of the north bank of the Drammen river," *Géot.*, **12**, No. 2, pp. 147–167.

Kjellman, W., T. Kallstenius, and Y. Liljedahl (1955). "Accurate measurement of settlements," *Proc. Royal Swedish Geot. Inst.*, No. 10, 33 pp.

Kjellman, W., T. Kallstenius, and O. Wager (1950). "Soil sampler with metal foils," *Proc. Swedish Geot. Inst.*, No. 1, 76 pp.

Klenner, C. (1941). "Versuche über die Verteilung des Erddruckes über die Wände ausgesteifter Baugruben" (Tests on the distribution of the earth pressure over the walls of braced excavations), *Bautechnik*, **19**, pp. 316–319.

Klohn, E. J. (1961). "Pile heave and redriving," *ASCE J. Soil Mech.*, **87**, No. SM4, pp. 125–145.

Koch, J. J., R. G. Boiten, A. L. Biermasz, G. P. Roszback, and G. W. v. Santen (1952). *Strain gauges: theory and application.* Philips Industries, Eindhoven, Holland, 95 pp.

Kögler, F. (1933). "Discussion: Soil mechanics research," *Trans. ASCE*, **98**, pp. 299–301.

Kolb, C. R. and W. G. Shockley (1959). "Engineering geology of the Mississippi valley," *Trans. ASCE*, **124**, pp. 633–645.

Kyrieleis, W. and W. Sichardt (1930). *Grundwasserabsenkung bei Fundierungsarbeiten* (Groundwater lowering for foundation construction). 2nd ed., Berlin, J. Springer, 286 pp.

Lacroix, Y. (1956). *Measurements of earth pressure against bracing of Inland Steel Building excavation, Chicago, Ill.* Paper presented before the ASCE, Pittsburgh, Oct. 19, 1956.

Ladd, G. E. (1935). "Landslides, subsidences and rock-falls," *Proc. Am. Rwy. Eng. Assn.*, **36**, pp. 1091–1162.

Lambe, T. W. (1951). *Soil testing for engineers.* New York, John Wiley and Sons, 165 pp.

Lambe, T. W. (1959). "Sealing the Casagrande piezometer," *Civ. Eng.*, **29**, No. 4, p. 256.

Lambe, T. W. (1960). "Structure of compacted clay," *Trans. ASCE*, **125**, pp. 682–705.

Lambe, T. W. and H. M. Horn (1965). "The influence on an adjacent building of pile driving for the M.I.T. Materials Center," *Proc. 6th Int. Conf. Soil Mech.*, Montreal, **2**, pp. 280–284.

Lane, E. W. (1935). "Security from under-seepage—masonry dams on earth foundations," *Trans. ASCE*, **100**, pp. 1235–1351.

Lane, K. S. and P. E. Wohlt (1961). "Performance of sheet piling and blankets for sealing Missouri River reservoirs," *Proc. 7th Congr. on Large Dams*, Rome, **4**, pp. 255–279.

Larsen, E. S. and H. Berman (1934). "The microscopic determination of the nonopaque minerals," 2nd ed. *U.S. Dept. of Interior Bull. 848,* 266 pp.

Laursen, E. M. (1955). "Model-prototype comparison of bridge pier scour," *Proc. Hwy. Res. Board,* **34,** pp. 188–193.

Lebedeff, A. F. (1928). "Methods of determining the maximum molecular moisture holding capacity of soils," *Proc. 1st Int. Conf. Soil Science,* Washington, **1,** pp. 551–560.

Lee, C. H. (1953). "Building foundations in San Francisco," *Proc. ASCE,* **79,** Separate **325,** 32 pp.

Legget, R. (1950). "Discussion: Effect of driving piles into soft clay," *Trans. ASCE,* **115,** pp. 319–322.

Lo, K. Y. (1962). "Shear strength properties of a sample of volcanic material of the valley of Mexico," *Géot.,* **12,** pp. 303–316.

Lockwood, M. G. (1954). "Ground subsides in Houston area," *Civ. Eng.,* **24,** No. **6,** pp. 48–50.

Loos, W. C. (1936). "Comparative studies of the effectiveness of different methods for compacting cohesionless soils," *Proc. 1st Int. Conf. Soil Mech.,* Cambridge, Mass., **3,** pp. 174–179.

Lorenz, H. (1934). "Neue Ergebnisse der dynamischen Baugrunduntersuchung" (New results of dynamic investigations of foundation soils), *Zeitschrift des Vereins deutscher Ingenieure,* **78,** pp. 379–385.

Lowe, J. (1960). "Current practice in soil sampling in the United States," *Hwy. Res. Board Special Rept. 60,* pp. 142–154.

Lumb, P. (1965). "The residual soils of Hong Kong," *Géot.,* **15,** No. **2,** pp. 180–194.

Lundgren, H. and K. Mortensen (1953). "Determination by the theory of plasticity of the bearing capacity of continuous footings on sand," *Proc. 3nd Int. Conf. Soil Mech.,* Zurich, **1,** pp. 409–412.

Lyman, A. K. B. (1942). "Compaction of cohesionless foundation soils by explosives," *Trans. ASCE,* **107,** pp. 1330–1348.

Lysmer, J. and F. E. Richart, Jr. (1966). Dynamic response of footings to vertical loadings," *ASCE J. Soil Mech.,* **92,** No. SM1, pp. 65–91.

Mallet, C. and J. Pacquant (1951). *Les barrages en terre* (Earth dams). Paris, Eyrolles, 346 pp.

Mansur, C. I. and R. I. Kaufman (1960). "Dewatering the Port Allen lock excavation," *ASCE J. Soil Mech.,* **86,** No. SM6, pp. 35–55.

Mansur, C. I. and R. I. Kaufman (1962). "Dewatering," Chapter 3 in *Foundation engineering,* G. A. Leonards, ed., New York, McGraw-Hill, pp. 241–350.

Marsal, R. J. (1960). "Earth dams in Mexico," *Proc. 1st Panamerican Congress on Soil Mech. and Found. Eng.,* Mexico, **3,** pp. 1294–1308.

Marsal, R. J. and M. Mazari (1962). *El Subsuelo de la Ciudad de México* (The subsoil of Mexico City). Universidad National Autónoma de México, Facultad de Ingenieria, 2nd ed. 614 pp.

Mason, A. C. (1952). "Open-caisson method used to erect Tokyo office building," *Civ. Eng.,* **22,** pp. 944–947.

Masters, F. M. (1943). "Timber friction pile foundations," *Trans. ASCE,* **108,** pp. 115–140.

McNary, J. V. (1925). "Earth pressure against abutment walls measured with soil pressure cells," *Public Roads,* **6,** pp. 102–106.

Means, R. E. (1959). "Buildings on expansive clay," *Colo. School of Mines Quarterly,* **54,** No. 4, pp. 1–31.

Meem, J. C. (1908). "The bracing of trenches and tunnels, with practical formulas for earth pressures," *Trans. ASCE,* **60,** pp. 1–23. Discussions pp. 24–100.

Megaw, T. M. (1951). "Foundations at Poole power station," *Géot.,* **2,** pp. 280–292.

Mehta, M. R. (1959). *Stresses and displacements in layered systems.* Ph.D. thesis, Univ. of Illinois, 33 pp.

Meigh, A. C. and I. K. Nixon (1961). "Comparison of in-situ tests of granular soils," *Proc. 5th Int. Conf. Soil Mech.,* Paris, **1,** p. 499.

Meyerhof, G. G. (1951). "The ultimate bearing capacity of foundations," *Géot.,* **2,** pp. 301–332.

Meyerhof, G. G. (1955). "Influence of roughness of base and ground-water conditions on the ultimate bearing capacity of foundations," *Géot.,* **5,** pp. 227–242.

Meyerhof, G. G. (1956). "Penetration tests and bearing capacity of cohesionless soils," *ASCE J. Soil Mech.,* **82,** No. SM1, Paper 866, pp. 1–19; Discussion, **83** (1957), No. SM1, Paper 1155, pp. 11–17.

Meyerhof, G. G. (1965). "Shallow foundations," *ASCE J. Soil Mech.,* **91,** No. SM2, pp. 21–31.

Mitchell, J. K. (1961). "Fundamental aspects of thixotropy in soils," *Trans. ASCE,* **126,** Part 1, pp. 1586–1620.

Mohr, H. A. (1943). "Exploration of soil conditions and sampling operations," *Harvard Univ. Grad. School of Engineering, Soil Mechanics Series 21,* 3rd revised ed., 65 pp.

Mohr, H. A. (1964). "The Gow caisson," *J. Boston Soc. Civil Engrs.,* **51,** No. 1, pp. 75–94.

Monahan, C. J. (1962). "John Day lock and dam: foundation investigations," *Proc. ASCE,* **88,** No. PO4, pp. 29–45.

Moore, R. W. (1961). "Observations on subsurface exploration using direct procedures and geophysical techniques," *Proc. 12th Annual Symp. on Geology as Applied to Highway Engineering, U. of Tenn. Eng. Exp. Sta. Bull. 24,* pp. 63–87.

Moran, Proctor, Mueser and Rutledge (1958). *Study of deep soil stabilization by vertical sand drains.* U.S. Dept. of Commerce, Office Tech. Serv., Wash., D.C. 192 pp.

Morgenstern, N. R. and V. E. Price (1965). "The analysis of the stability of general slip surfaces," *Géot.,* **15,** pp. 79–93.

Moum, J. and I. Th. Rosenqvist (1957). "On the weathering of young marine clay," *Proc. 4th Int. Conf. Soil Mech.,* London, **1,** pp. 77–79.

Müller, F. (1898). *Das Wasserwesen der niederländischen Provinz Zeeland* (The water system of the Dutch province of Zeeland). Berlin, Ernst and Son, 612 pp.

Müller, P. (1939). "Erddruckmessungen bei mechanisch verdichteter Hinterfüllung von Stützkörpern" (Measurements of the earth pressure exerted by mechanically consolidated backfills of abutments), *Bautechnik,* **17,** pp. 195–203.

Murphy, E. C. (1908). "Changes in bed and discharge capacity of the Colorado River at Yuma, Ariz.," *Eng. News,* **60,** p. 344.

Muskat, M. (1937). *The flow of homogeneous fluids through porous media.* New York, McGraw-Hill, 763 pp. Reprinted by J. W. Edwards, Ann Arbor, 1946.

Neill, C. R. (1964). "A review for bridge engineers," *Canadian Good Roads Assn., Ottawa, Tech. Publ. No. 23.*

Newland, D. H. (1916). "Landslides in unconsolidated sediments," *N.Y. State Museum Bull. 187,* Albany, pp. 79–105.

Newmark, N. M. (1942). "Influence charts for computation of stresses in elastic foundations," *Univ. of Illinois Eng. Exp. Sta. Bull. 338,* 28 pp.

Newmark, N. M. (1960). "Failure hypotheses for soils," *Proc. ASCE Research Conf. on Shear Strength of Cohesive Soils,* pp. 17–32.

Newmark, N. M. (1965). "Effects of earthquakes on dams and embankments," *Géot.* **15**, No. 2, pp. 139–160.

NGI (1962*a*). "Measurements at a strutted excavation, Oslo subway, Enerhaugen South," *Norwegian Geot. Inst. Tech. Rept. No. 3,* 71 pp.

NGI (1962*b*). "Measurements at a strutted excavation, Oslo subway, Grønland 1, *Norwegian Geot. Inst. Tech. Rept. No. 1,* 67 pp.

NGI (1962*c*). "Measurements at a strutted excavation, Oslo subway, Vaterland 1," *Norwegian Geot. Inst. Tech. Rept. No. 6,* 76 pp.

NGI (1962*d*). "Measurements at a strutted excavation, Oslo subway, Vaterland 2," *Norwegian Geot. Inst. Tech. Rept. No. 7,* 56 pp.

NGI (1962*e*). "Measurements at a strutted excavation, Oslo subway, Vaterland 3," *Norwegian Geot. Inst. Tech. Rept. No. 8,* 56 pp.

NGI (1962*f*). "Vibrating-wire measuring devices used at strutted excavations," *Norwegian Geot. Inst. Tech. Rept. No. 9,* 151 pp.

NGI (1965). "Measurements at a strutted excavation, Oslo subway, Grønland 2," *Norwegian Geot. Inst. Tech. Rept. No. 5,* 124 pp.

Nonveiller, E. (1965). "The stability analysis of slopes with a slip surface of general shape," *Proc. 6th Int. Conf. Soil Mech.,* Montreal, **2**, pp. 522–525.

Nordlund, R. L. (1963). "Bearing capacity of piles in cohesionless soil," *ASCE J. Soil Mech.,* **89**, No. SM3, pp. 1–35.

Ohde, J. (1938). "Zur Theorie des Erddruckes unter besonderer Berücksichtigung der Erddruck Verteilung" (On earth-pressure theory with special consideration to earth-pressure distribution), *Bautechnik,* **16**, pp. 150–159, 176–180, 241–245, 331–335, 480–487, 570–571, 753, 761.

Osterberg, J. O. (1940). "A survey of the frost-heaving problem," *Civ. Eng.,* **19**, pp. 100–102.

Osterberg, J. O. (1952). "New piston type soil sampler," *Eng. News-Record,* **148**, Apr. 24, pp. 77–78.

Osterberg, J. O. (1957). "Influence values for vertical stresses in a semi-infinite mass due to an embankment loading," *Proc. 4th Int. Conf. Soil Mech.,* London, **1**, pp. 393–394.

Parsons, J. D. (1959). "Foundation installation requiring recharging of ground water," *ASCE J. Constr. Div.,* **85**, No. CO2, pp. 1–21.

Parsons, J. D. (1966). "Piling difficulties in the New York area," *ASCE J. Soil Mech.,* **92**, No. SM1, pp. 43–64.

Peck, R. B. (1940). "Sampling methods and laboratory tests for Chicago subway soils," *Proc. Purdue Conf. on Soil Mech.,* pp. 140–150.

Peck, R. B. (1941). "The measurement of earth pressures on the Chicago subway," *ASTM Bull. 111,* pp. 25–30.

Peck, R. B. (1943). "Earth-pressure measurements in open cuts, Chicago subway," *Trans. ASCE,* **108**, pp. 1008–1036.

Peck, R. B. (1948). "History of building foundations in Chicago," *Univ. of Ill. Eng. Exp. Sta. Bull. 373,* 64 pp.

Peck, R. B. (1953). "Foundation exploration—Denver Coliseum," *Proc. ASCE,* Separate **326**, 14 pp.

Peck, R. B. (1954). "Foundation conditions in the Cuyahoga River valley," *Proc. ASCE,* **80,** Separate 513, 20 pp.

Peck, R. B. (1958). "A study of the comparative behavior of friction piles," *Hwy. Res. Board Special Rept. 36,* 72 pp.

Peck, R. B. (1961) "Records of load tests on friction piles," *Hwy. Res. Board Special Rept. 67,* 418 pp.

Peck, R. B. (1965). "Pile and pier foundations," *ASCE J. Soil Mech.,* **91,** No. SM2, pp. 31–38.

Peck, R. B. and S. Berman (1961). "Recent practice for foundations of high buildings in Chicago," *Symp. on The Design of High Buildings,* Univ. of Hong Kong, pp. 85–98.

Peck, R. B. and F. G. Bryant (1953). "The bearing-capacity failure of the Transcona elevator," *Géot.,* 3, pp. 201–208.

Peck, R. B. and H. O. Ireland (1958). "Discussion: Experiences with loess as a foundation material," *Trans. ASCE,* **123,** pp. 171–179.

Peck, R. B., H. O. Ireland, and C. Y. Teng (1948). "A study of retaining wall failures," *Proc. 2nd Int. Conf. Soil Mech.,* Rotterdam, 3, pp. 296–299.

Peck, R. B. and W. C. Reed (1954). "Engineering properties of Chicago subsoils," *Univ. of Illinois Eng. Exp. Sta. Bull. 423,* 62 pp.

Penman, A. D. M. (1953). "Shear characteristics of a saturated silt, measured in triaxial compression," *Géot.,* **3,** pp. 312–328.

Penman, A. D. M. (1961). "A study of the response time of various types of piezometer," *Proc. Conf. on Pore Pressure and Suction in Soils,* London, Butterworths, pp. 53–58.

Petersen, J. S., C. Rohwer, and M. L. Albertson (1955). "Effect of well screens on flow into wells," *Trans. ASCE,* **120,** pp. 563–585.

Peterson, R., J. L. Jaspar,, P. J. Rivard, and N. L. Iverson (1960). "Limitations of laboratory shear strength in evaluating stability of highly plastic clays," *Proc. ASCE Research Conf. on Shear Strength of Cohesive Soils,* pp. 765–791.

Plantema, G. and C. A. Nolet (1957). "Influence of pile driving on the sounding resistances in a deep sand layer," *Proc. 4th Int. Conf. Soil Mech.,* London, **2,** pp. 52–55.

Poland, J. F. (1958). "Land subsidence due to ground-water development," *ASCE J. Irr. and Drainage Div.,* **84,** Paper 1774, 11 pp.

Pollack, V. (1917). "Über Rutschungen im Glazialen und die Natwendigkeiteiner Klassifikation loser Massen" (Concerning slides in glacial soils and the need for a classification of soils), *Jahrb. geol. Reichsanstalt,* Wien, **67,** p. 456.

Polubarinova-Kochina, P. Ya (1962). *Theory of ground water movement.* Translated from the Russian by J. M. R. de Wiest, Princeton Univ. Press, 613 pp.

Porter, O. J. (1936). "Studies of fill construction over mud flats including a description of experimental construction using vertical sand drains to hasten stabilization," *Proc. 1st Int. Conf. Soil Mech.,* Cambridge, Mass., **1,** pp. 229–235.

Prandtl, L. (1921). "Über die Eindringungsfestigkeit (Härte) plastischer Baustoffe und die Festigkeit von Schneiden" (On the penetrating strengths (hardness) of plastic construction materials and the strength of cutting edges), *Zeit. angew. Math. Mech.,* **1,** No. 1, pp. 15–20.

Prentis, E. A. and L. White (1950). *Underpinning,* 2nd ed., New York, Columbia Univ. Press, 374 pp.

Press, H. (1933). "Die Tragfähigkeit Pfahlgruppen in Beziehung zu der des Einzelpfahles" (The bearing capacity of pile groups in relation to that of the single pile), *Bautechnik,* **11**, pp. 625–627.

Proctor, R. R. (1933). "Four articles on the design and construction of rolled-earth dams," *Eng. News-Record,* **111**, pp. 245–248, 286–289, 348–351, 372–376.

Rankine, W. J. M. (1857). "On the stability of loose earth," *Phil. Trans. Roy. Soc.,* London, **147,** Part 1, pp. 9–27.

Rehbock, Th. (1931). *Wasserbauliche Modellversuche zur Klärung der Zuiderzee* (Hydraulic model tests for the clarification of the Zuyder Zee). Netherlands, The Hague, 282 pp.

Reissner, H. (1924). "Zum Erddruckproblem" (Concerning the earth-pressure problem), *Proc. 1st Int. Congress of Applied Mechanics,* Delft, pp. 295–311.

Richart, F. E. Jr. (1960). "Foundation vibrations," *ASCE J. Soil Mech.,* **86,** No. SM4, pp. 1–34.

Rose, A. C. (1924). "Practical field tests for subgrade soils," *Public Roads,* **5,** No. 6, pp. 10–15.

Rosenqvist, I. Th. (1946). "Om leires kvikkakfighet" (On the sensitivity of clays). *Meddelelser fra Vegdirektören,* No. 3, p. 24.

Rosenqvist, I. Th. (1953). "Considerations on the sensitivity of Norwegian quick clays," *Géot.,* **3,** pp. 195–200.

Rutledge, P. C. (1939). *Compression characteristics of clays and application to settlement analysis.* Sc.D. thesis, Harvard Univ.

Rutledge, P. C. (1947). "Review of the cooperative triaxial shear research program of the Corps of Engineers," *Soil mechanics fact finding survey,* Progress Report, Waterways Exp. Sta., Vicksburg, Miss., pp. 1–178.

Samsioe, A. F. (1931). "Einfluss von Rohrbrunnen auf die Bewegung des Grund-Wassers" (Influence of anisotropy on the flow of ground water), *Zeitschrift fuer angewandte Mathematik und Mechanik,* **11,** pp. 124–135.

Sanglerat, G. (1965). *Le pénétromètre et la reconnaissance des sols* (The penetrometer and soil exploration). Paris, Dunod, 230 pp.

Scheidig, A. (1931). *Versuche über die Formänderung von Sand und ihre Anwendung auf die Setzungsanalyse von Bauwerken* (Tests on the deformation of sand and their application to the settlement analysis of buildings), M. S. thesis, Vienna.

Scheidig, A. (1934). *Der Loss* (Loess). Dresden, 233 pp.

Schmertmann, J. H. (1953). "Estimating the true consolidation behavior of clay from laboratory test results," *Proc. ASCE,* **79,** Separate 311, 26 pp.

Schmertmann, J. H. and J. O. Osterberg (1960). "An experimental study of the development of cohesion and friction with axial strain in saturated cohesive soils," *Proc. ASCE Research Conf. on Shear Strength of Cohesive Soils,* pp. 643–694.

Schneible, D. E. (1924). "Some field examples of scour at bridge piers and abutments," *Better Roads,* **24,** Aug., p. 21.

Schultze, E. and K.-J. Melzer (1965). "The determination of the density and the modulus of compressibility of non-cohesive soils by soundings," *Proc. 6th Int. Conf. Soil Mech.,* Montreal, **1,** pp. 354–358.

Seed, H. B., R. J. Woodward, Jr., and R. Lundgren (1962). "Prediction of swelling potential for compacted clays," *ASCE J. Soil Mech.,* **88,** No. SM3, pp. 53–87.

Seed, H. B. and R. W. Clough (1963). "Earthquake resistance of sloping core dams," *ASCE J. Soil Mech.*, **89,** No. SM1, pp. 209–242.

Seed, H. B. and K. L. Lee (1966). "Liquefaction of saturated sands during cyclic loading," *ASCE J. Soil Mech.*, **92,** No. SM6, pp. 105–134.

Seed, H. B., J. K. Mitchell, and C. K. Chan (1960). "The strength of compacted cohesive soils," *Proc. ASCE Research Conf. on Shear Strength of Cohesive Soils,* pp. 877–964.

Seiler, J. F. and W. D. Keeney (1944). "The efficiency of piles in groups," *Wood Preserving News,* **22,** No. 11, pp. 109–118.

Shannon, W. L. and S. D. Wilson (1964). *Report on Anchorage area soil studies, Alaska.* U.S. Corps of Engrs., Alaska Dist., 70 pp. plus appendices.

Shannon, W. L., S. D. Wilson, and R. H. Meese (1962). "Field problems: field measurements," Chapter 13 in *Foundation engineering,* G. A. Leonards, ed., New York, McGraw-Hill, pp. 1025–1080.

Sharpe, C. F. S. (1938). *Landslides and related phenomena.* New York, Columbia Univ. Press, 136 pp.

Sherard, J. L. (1953). "Influence of soil properties and construction methods on the performance of homogeneous earth dams," *U.S. Bureau Reclamation, Tech. Memo. 645,* 244 pp.

Sherard, J. L., R. J. Woodward, S. F. Gizienski, and W. A. Clevenger (1963). *Earth and earth-rock dams,* New York, John Wiley and Sons, 725 pp.

Sibley, E. A. and G. Yamane (1965). "A simple shear test for saturated cohesive soil," *Proc. 5th Pacific Area National Meeting, ASTM, Seattle.*

Sikso, H. A. and C. V. Johnson (1964). "Pressure cell observations—Garrison dam project," *ASCE J. Soil Mech.*, **90,** No. SM5, pp. 157–179.

Simons, N. E. (1960*a*). "Comprehensive investigations of the shear strength of an undisturbed Drammen clay," *Proc. ASCE Research Conf. on Shear Strength of Cohesive Soils,* pp. 727–745.

Simons, N. E. (1960*b*). "The effect of overconsolidation on the shear strength characteristics of an undisturbed Oslo clay," *Proc. ASCE Research Conf. on Shear Strength of Cohesive Soils,* pp. 747–763.

Simons, N. E. (1963). "Settlement studies on a nine storey apartment building at Økernbråten, Oslo," *Proc. European Conf. on Soil Mech. and Found. Eng.,* Wiesbaden, **1,** pp. 179–191.

Simpson, W. E. (1934). "Foundation experiences with clay in Texas," *Civ. Eng.,* **4,** pp. 581–584.

Sizer, F. L. (1908). "Break in the Hauser Lake dam, Montana," *Eng. News,* **59,** p. 491.

Skempton, A. W. (1942). "An investigation of the bearing capacity of a soft clay soil," *J. Inst. Civil Engrs.,* London, **18,** pp. 307–321; Discussions pp. 567–576.

Skempton, A. W. (1944). "Notes on the compressibility of clays," *Quart. J. Geol. Soc.,* London, **C,** pp. 119–135.

Skempton, A. W. (1948). "The $\phi = 0$ analysis of stability and its theoretical basis." *Proc. 2nd Int. Conf. Soil Mech.,* Rotterdam, **1,** pp. 72–78.

Skempton, A. W. (1951). "The bearing capacity of clays," *Proc. British Bldg. Research Congress,* **1,** pp. 180–189.

Skempton, A. W. (1953). "Discussion on piles and pile foundations," *Proc. 3rd Int. Conf. Soil Mech.,* Zurich, **3,** p. 172.

Skempton, A. W. (1954). "The pore-pressure coefficients *A* and *B*," *Géot.*, **4**, pp. 143–147.

Skempton, A. W. (1957). "Discussion: The planning and design of the new Hong Kong airport," *Proc. Inst. Civil Engrs.*, London, **7**, pp. 305–307.

Skempton, A. W. (1959). "Cast in-situ bored piles in London clay," *Géot.*, **9**, pp. 153–173.

Skempton, A. W. (1960). "Terzaghi's discovery of effective stress," *From theory to practice in soil mechanics,* New York, John Wiley and Sons, pp. 42–53.

Skempton, A. W. (1961*a*). "Effective stress in soils, concrete and rocks," *Pore Pressure and Suction in Soils,* London, Butterworths, pp. 4–16.

Skempton, A. W. (1961*b*). "Horizontal stresses in an over-consolidated Eocene clay," *Proc. 5th Int. Conf. Soil Mech.*, Paris, **1**, pp. 351–357.

Skempton, A. W. (1964). "Long-term stability of clay slopes," *Géot.*, **14**, No. 2, pp. 77–101.

Skempton, A. W. and L. Bjerrum (1957). "A contribution to the settlement analysis of foundations on clay," *Géot.*, **7**, pp. 168–178.

Skempton, A. W. and P. Cattin (1963). "A full-scale alluvial grouting test at the site of Mangla dam," *Proc. Symp. on Grouts and Drilling Muds in Engineering Practice,* London, Butterworths, pp. 131–135.

Skempton, A. W. and D. J. Henkel (1955). "A landslide at Jackfield, Shropshire, in a heavily overconsolidated clay," *Géot.*, **5**, No. 2, pp. 131–137.

Skempton, A. W. and R. D. Northey (1952). "The sensitivity of clays," *Géot.*, **3**, No. 1, pp. 30–53.

Skempton, A. W. and W. H. Ward (1952). "Investigations concerning a deep cofferdam in the Thames Estuary clay at Shellhaven," *Géot.*, **3**, pp. 119–139.

Smith, E. A. L. (1960). "Pile driving analysis by the wave equation," *ASCE J. Soil Mech.*, **86**, No. SM4, pp. 35–61.

Smith, R. and R. B. Peck (1955). "Stabilization by grouting on American railroads," *Géot.*, **5**, pp. 243–252.

Smith, T. W. and G. V. Stafford (1957). "Horizontal drains on California highways," *ASCE J. Soil Mech.*, **83**, No. SM3, 26 pp.

Sokolovski, V. V. (1960). *Statics of soil media.* Translated from Russian by D. H. Jones and A. N. Schofield. London, Butterworths, 237 pp.

Sörensen, T. and B. Hansen (1957). "Pile driving formulae, an investigation based on dimensional considerations and a statistical analysis," *Proc. 4th Int. Conf. Soil Mech.*, London, **2**, pp. 61–65.

Sowers, G. F. (1953). "Soil and foundation problems in the southern Piedmont region," *Proc. ASCE,* **80**, Separate 416, 18 pp.

Sowers, G. F. (1962). "Shallow foundations," Chapter 6 in *Foundation engineering,* G. A. Leonards, ed., New York, McGraw-Hill, pp. 525–632.

Spilker, A. (1937). "Mitteilung über die Messung der Kräfte in einer Baugrubenaussteifung" (Note on the measurement of the forces in a braced excavation), *Bautechnik,* **15**, pp. 16–18.

Stamatopoulos, A. C. and P. C. Kotzias (1965). "Construction and performance of an embankment in the sea on soft clay," *Proc. 6th Int. Conf. Soil Mech.*, Montreal, **2**, pp. 566–570.

Staniford, C. W. (1915). "Load tests of lagged piles," *Eng. News,* **74**, pp. 76–77.

Steele, I. C. and J. B. Cooke (1960). "Salt Springs and Lower Bear River concrete face dams," *Trans. ASCE,* **125**, Part II, pp. 74–116.

Steuermann, S. (1939). "A new soil compacting device," *Eng. News-Record,* **123**, pp. 87–88.

Stevens, W. C., N. C. Yang, M. S. Kapp, A. Lier, and E. Fasullo (1965). "The LaGuardia Airport runway extensions," *Trans. N.Y. Acad. Sci. Ser. II,* **27,** No. 3, pp. 324–336.

Swiger, W. F. (1941). "Foundation tests for Los Angeles steam plant," *Civil Eng.,* **11,** pp. 711–714.

Syffert, O. (1929). *Erddrucktafeln* (Earth-pressure tables). Berlin, J. Springer, 12 pp.

Taber, S. (1930). "Freezing and thawing of soils as factors in the destruction of road pavements," *Public Roads,* **11,** pp. 113–132.

Taylor, D. W. (1937). "Stability of earth slopes," *J. Boston Soc. Civil Engrs.,* **24,** pp. 197–246.

Taylor, D. W. (1947). "Review of pressure distribution theories, earth pressure cell investigations and pressure distribution data," *Soil mechanics fact finding survey, Progress Report,* Waterways Exp. Sta., Vicksburg, Miss., pp. 179–332.

Taylor, D. W. (1948). *Fundamentals of soil mechanics.* New York, John Wiley and Sons, 700 pp.

Teixeira, A. H. (1960). "Typical subsoil conditions and settlement problems in Santos, Brasil," *Proc. 1st Panamerican Conf. on Soil Mech.* Mexico, **1,** pp. 149–177.

Teng, W. C. (1962). *Foundation Design.* New Jersey, Prentice-Hall, 466 pp.

Terzaghi, K. (1922). "Der Grundbruch an Stauwerken und seine Verhütung" (The failure of dams by piping and its prevention), *Die Wasserkraft,* **17,** pp. 445–449. Reprinted in *From theory to practice in soil mechanics,* New York, John Wiley and Sons, 1960, pp. 114–118.

Terzaghi, K. (1925). "Structure and volume of voids of soils," Pages 10, 11, 12, and part of 13 of *Erdbaumechanik auf Bodenphysikalisher Grundlage,* translated by A. Casagrande in *From theory to practice in soil mechanics,* New York, John Wiley and Sons, 1960, pp. 146–148.

Terzaghi, K. (1927). "Wellpoint method for handling excavation of foundation pit at new sewage pumping station, Lynn, Massachusetts," *J. Boston Soc. Civil Engrs.,* **14,** No. 7, pp. 389–397.

Terzaghi, K. (1929*a*). "Effect of minor geologic details on the safety of dams," *Amer. Inst. Min. and Met. Engrs. Tech. Publ. 215,* pp. 31–44.

Terzaghi, K. (1929*b*). "The mechanics of shear failures on clay slopes and the creep of retaining walls," *Public Roads,* **10,** No. 10, pp. 177–192.

Terzaghi, K. (1929*c*). "Soil studies for the Granville dam at Westfield, Mass." *J. New Engl. Water Works Assn.,* **43,** pp. 191–223.

Terzaghi, K. (1930). "Die Trägfahigkeit von Phahlgründungen" (The bearing capacity of pile foundations), *Bautechnik,* **8,** Nos. 31, 34, pp. 475–478, 517–521.

Terzaghi, K. (1931). "Underground erosion and the Corpus Christi dam failure," *Eng. News-Record,* **107,** pp. 90–92.

Terzaghi, K. (1934*a*). "Large retaining-wall tests," *Eng. News-Record,* **112,** pp. 136–140, 259–262, 316–318, 403–406, 503–508.

Terzaghi, K. (1934*b*). "Retaining-wall design for Fifteen-Mile Falls Dam," *Eng. News-Record,* **112,** pp. 632–636.

Terzaghi, K. (1935). "The actual factor of safety of foundations," *Structural Eng.,* **13,** pp. 126–160.

Terzaghi, K. (1936*a*). "Effect of the type of drainage of retaining walls on the earth pressure," *Proc. 1st Int. Conf. Soil Mech.,* Cambridge, Mass., **1,** pp. 215–218.

Terzaghi, K. (1936*b*). "The shearing resistance of saturated soils," *Proc. 1st Int. Conf. Soil Mech.*, Cambridge, Mass., **1**, pp. 54–56.

Terzaghi, K. (1938*a*). "Die Coulombsche Gleichung für den Scherwiderstand bindiger Böden" (The Coulomb equation for the shear strength of cohesive soils), *Bautechnik,* **16**, pp. 509–512. Translated by L. Bjerrum in *From theory to practice in soil mechanics,* New York, John Wiley and Sons, 1960, pp. 174–180.

Terzaghi, K. (1938*b*). "Settlement of structures in Europe and methods of observations," *Trans. ASCE,* **103,** pp. 1432–1448.

Terzaghi, K. (1941*a*). "Undisturbed clay samples and undisturbed clays," *J. Boston Soc. Civil Engrs.*, **28,** No. 3, pp. 211–231.

Terzaghi, K. (1941*b*). "General wedge theory of earth pressure," *Trans. ASCE,* **106,** pp. 68–97.

Terzaghi, K. (1942). "Shield tunnels of the Chicago subway," *J. Boston Soc. Civil Engrs.* **29,** pp. 163–210.

Terzaghi, K. (1943*a*). "Liner-plate tunnels on the Chicago (Ill.) subway," *Trans. ASCE,* **108,** pp. 970–1007.

Terzaghi, K. (1943b). *Theoretical Soil Mechanics.* New York, John Wiley and Sons, 510 pp.

Terzaghi, K. (1950). "Mechanism of landslides," *Application of Geology to engineering practice, Berkey Vol.,* Geological Society of America, pp. 83–123. Reprinted in *From theory to practice in soil mechanics,* New York, John Wiley and Sons, 1960, pp. 202–245.

Terzaghi, K. (1955*a*). "Influence of geological factors on the engineering properties of sediments," *Economic Geology.* Fiftieth Anniversary Volume, pp. 557–618.

Terzaghi, K. (1955*b*). "Evaluation of coefficients of subgrade reaction," *Géot.,* **5,** No. 4, pp. 297–326.

Terzaghi, K. (1958*a*). "Consultants, clients and contractors," *J. Boston Soc. Civil Engrs.,* **45,** No. 1, pp. 1–15.

Terzaghi, K. (1958*b*). "Design and performance of the Sasumua dam," *Proc. Inst. Civil Engrs.* London, **9,** pp. 369–394.

Terzaghi, K. (1960*a*). "Discussion: Salt Springs and Lower Bear River concrete face dams," *Trans. ASCE,* **125,** Part II, pp. 139–148.

Terzaghi, K. (1960b). "Memorandum concerning landslide on slope adjacent to power plant, South America," *From theory to practice in soil mechanics,* New York, John Wiley and Sons, pp. 410–415.

Terzaghi, K. (1960*c*). "Report on the proposed storage dam south of Lower Stillwater lake on the Cheakamus river, B.C.," *From theory to practice in soil mechanics,* New York, John Wiley and Sons, pp. 395–408.

Terzaghi, K. (1960*d*). "Stabilization of landslides," Series of memoranda contained in *From theory to practice in soil mechanics.* New York, John Wiley and Sons, pp. 409–415.

Terzaghi, K. (1961*a*). "Discussion: Horizontal stresses in an over-consolidated Eocene clay," *Proc. 5th Int. Conf. Soil Mech.,* Paris, **3,** pp. 144–145.

Terzaghi, K. (1961*b*). "Past and future of applied soil mechanics," *J. Boston Soc. Civil Engrs.,* **68,** pp. 110–139.

Terzaghi, K. (1962). "Dam foundation on sheeted granite," *Géot.,* **12,** No. 3, pp. 199–208.

Terzaghi, K. (1963). "Discussion: Atomic power, a failure in engineering responsibility," *Trans. ASCE,* **128,** Part 5, pp. 56–57.

Terzaghi, K. and O. K. Fröhlich (1936). *Theorie der Setzung von Tonschichten* (Theory of settlement of clay layers). Leipzig, 166 pp.

Terzaghi, K. and Y. Lacroix (1964). "Mission dam—An earth and rockfill dam on a highly compressible foundation," *Géot.,* **14,** pp. 14–50.

Terzaghi, K. and T. M. Leps (1960). "Design and performance of Vermilion dam," *Trans. ASCE,* **125,** pp. 63–86.

Terzaghi, K. and R. B. Peck (1957). "Stabilization of an ore pile by drainage," *Proc. ASCE,* **83,** No. SM1, paper 1144, 13 pp.

Thayer, D. P. (1966). "Dynamic stress cells for Oroville dam," *U.S. Comm. on Large Dams, Newsletter,* Issue 20, May, pp. 7–8.

Thorfinnson, S. T. (1960). "Underseepage control at Ft. Randall dam," *Trans. ASCE,* **125,** pp. 792–806.

Thornley, J. H. (1951). *Foundation design and practice.* New York, Columbia Univ. Press, 298 pp.

Todd, D. K. (1959). *Ground water hydrology.* New York, John Wiley and Sons, 336 pp.

Tomlinson, M. J. (1957). "The adhesion of piles driven in clay soils," *Proc. 4th Int. Conf. Soil Mech.,* London, **2,** pp. 66–71.

Tomlinson, M. J. (1963). *Foundation design and construction.* New York, John Wiley and Sons, 749 pp.

Trautwine, J. (1937). *Civil Engineer's Reference Book.* 21st ed., Ithaca, N.Y., 1514 pp.

Turnbull W. J. (1948). "Utility of loess as a construction material," *Proc. 2nd Int. Conf. Soil Mech.,* Rotterdam, **5,** pp. 97–103.

Turnbull, W. J. and W. G. Shockley (1958). "Compaction of earth dams in the Corps of Engineers, U.S. Army," *Trans. 6th Int. Congr. Large Dams,* New York, **3,** pp. 317–331.

Turneaure, F. E. and E. R. Maurer (1913). *Principles of reinforced concrete construction.* 2nd ed., New York, John Wiley and Sons, 429 pp.

U.S. Bureau of Reclamation (1947). "Laboratory tests on protective filters for hydraulic and static structures," *Earth Materials Laboratory Rept. EM-132,* Denver, 28 pp.

U.S. Bureau of Reclamation (1960). *Design of small dams.* Wash., D.C., 725 pp.

U.S. Bureau of Reclamation (1963). *Earth manual,* 1st ed. revised, Wash., D.C., 783 pp.

Vesić, A. B. (1963). "Bearing capacity of deep foundations in sand," *Hwy. Res. Record No. 39,* pp. 112–153.

Vey, E. (1957). "Frictional resistance of steel H-piling in clay," *ASCE J. Soil Mech.,* **83,** No. SM1, Paper 1160, pp. 1–31.

Wafa, T. A. (1961). "Field tests for grouting Nile alluvials under the Aswan High dam," *Proc. 7th Congr. on Large Dams,* Rome, **2,** pp. 191–214.

Walker, F. C. and W. W. Daehn (1948). "Ten years of pore pressure measurements," *Proc. 2nd Int. Conf. Soil Mech.,* Rotterdam, **3,** pp. 245–250.

Ward, W. H. (1955). "Techniques for field measurement of deformation and earth pressure," *Proc. Conf. on Correlation between Calculated and Observed Stresses and Displacements in Structures,* Inst. Civil Engrs. London, Paper No. 3, Group 1, pp. 28–40.

Warlam, A. A. and E. W. Thomas (1965). "Measurement of hydrostatic uplift pressure on spillway weir with air piezometers," *ASTM Special Tech. Publ. 392,* pp. 143–151.

Wellington, A. M. (1888). "Formulae for safe loads of bearing piles," *Eng. News,* **20,** pp. 509–512.

WES (1963). "Development and evaluation of soil bearing capacity foundations of structures, field vibratory tests data," *Corps of Engrs., Waterways Exp. Sta. Tech. Rept. No. 3-632,* Rept. **1,** 18 pp.

Whitaker, G. M. J. and R. B. Colman (1965). "The design of piles and cylinder foundations in stiff, fissured clay," *Proc. 6th Int. Conf. Soil Mech.,* Montreal, **2,** pp. 347–351.

White, E. E. (1958). "Deep foundations in soft Chicago clay," *Civ. Eng.,* **28,** pp. 816–818.

White, L. and E. A. Prentis (1940). *Cofferdams.* New York, Columbia Univ. Press, 273 pp.

White, L. S. (1953) "Transcona elevator failure: eye-witness account," *Géot,* **3,** pp. 209–214.

Wiegmann, D. (1954). "Der Erddruck auf verankerte Stahlspundwände, ermittelt auf Grund von Verformungsmessungen am Bauwerk" (Earth pressure on anchored steel sheet-pile walls, determined from deformation measurements on the structure), *Mitt. Hannoverschen Versuchsanstalt für Grundbau und Wasserbau,* **5,** p. 79.

Willis, E. A. (1946). "Discussion: A study of lateritic soils," *Proc. Hwy. Res. Board,* **26,** pp. 589–591.

Wilson, G. and H. Grace (1942). "The settlement of London due to underdrainage of the London clay," *J. Inst. Civil Engrs.,* London, **19,** pp. 100–127.

Wilson, S. D. (1966). Personal communication.

Wilson, S. D. and C. W. Hancock, Jr. (1960). "Horizontal displacements of clay foundations," *Proc. 1st Panamerican Conf. Soil Mech.,* Mexico, **1,** pp. 41–64.

Wilson, S. D. and C. W. Hancock, Jr. (1965). "Instrumentation for movements within rockfill dams," *ASTM Special Tech. Publ. 392,* pp. 115–130.

Woodman, E. H. (1955). "Pressure cells for field use," *Waterways Exp. Sta. Bull. 40,* 33 pp.

Woods, K. B., R. D. Miles, and C. W. Lovell, Jr. (1962). "Origin, formation, and distribution of soils in North America," Chapter 1 in *Foundation engineering,* G. A. Leonards, ed., New York, McGraw-Hill, pp. 1–65.

Woodward, R. J., R. Lundgren, and J. D. Boitano (1961). "Pile loading tests in stiff clays," *Proc. 5th Int. Conf. Soil Mech.,* Paris, **2,** pp. 177–184.

Wu, T. H. (1957). "Relative density and shear strength of sands," *ASCE J. Soil Mech.,* **83,** No. SM1, Paper No. 1161, 23 pp.

Wu, T. H. and S. Berman (1953). "Earth pressure measurements in open cut: contract D8, Chicago subway," *Géot.* **3,** pp. 248–258.

Yong, R. N. and B. P. Warkentin (1966). "Soil freezing and permafrost," Chapter 12 in *Introduction to soil behavior,* New York, MacMillan, pp. 391–428.

Zangar, C. N. (1953). "Theory and problems of water percolation," *U.S. Bureau of Reclamation, Eng. Monograph No. 8,* 76 pp.

Zeevaert, L. (1953). "Pore pressure measurements to investigate the main source of surface subsidence in Mexico City," *Proc. 3rd Int. Conf. Soil Mech.,* Zurich, **2,** pp. 299–304.

Zeevaert, L. (1957). "Foundation design and behavior of Tower Latino Americana in Mexico City," *Géot.,* **7,** No. 3, pp. 115–133.

Zunker, F. (1930). "Das Verhalten des Bodens zum Wasser" (The behavior of soils in relation to water), *Handbuch der Bodenlehre,* Berlin, **6,** pp. 66–220.

AUTHOR INDEX

SUBJECT INDEX

LADY MARGARET HALL LIBRARY
OXFORD

LADY MARGARET HALL LIBRARY
OXFORD